PROCEEDINGS OF THE 1ST ASIAN ROCK MECHANICS SYMPOSIUM: ARMS '97
A REGIONAL CONFERENCE OF ISRM/SEOUL/KOREA/13-15 OCTOBER 1997

Environmental and Safety Concerns in Underground Construction

Edited by
HI-KEUN LEE
Seoul National University, Korea

HYUNG-SIK YANG
Chonnam National University, Kwangju, Korea

SO-KEUL CHUNG
Korea Institute of Geology, Mining and Materials, Taejon, Korea

VOLUME 1

A.A.BALKEMA / ROTTERDAM / BROOKFIELD / 1997

The texts of the various papers in this volume were set individually by typists under the supervision of each of the authors concerned.

Authorization to photocopy items for internal or personal use, or the internal or personal use of specific clients, is granted by A.A.Balkema, Rotterdam, provided that the base fee of US$1.50 per copy, plus US$0.10 per page is paid directly to Copyright Clearance Center, 222 Rosewood Drive, Danvers, MA 01923, USA. For those organizations that have been granted a photocopy license by CCC, a separate system of payment has been arranged. The fee code for users of the Transactional Reporting Service is: 90 5410 910 6/97 US$1.50 + US$0.10.

Published by
A.A.Balkema, P.O.Box 1675, 3000 BR Rotterdam, Netherlands (Fax: +31.10.413.5947)
A.A.Balkema Publishers, Old Post Road, Brookfield, VT 05036-9704, USA (Fax: 802.276.3837)

For the complete set of two volumes, ISBN 90 5410 910 6
For Volume 1, ISBN 90 5410 911 4
For Volume 2, ISBN 90 5410 912 2

© 1997 A.A.Balkema, Rotterdam
Printed in the Netherlands

ENVIRONMENTAL AND SAFETY CONCERNS IN UNDERGROUND CONSTRUCTION
VOLUME 1

Table of contents

Preface — XI

Acknowledgement — XIII

Organization — XV

Heat, gas, water flow and contaminant transport problems in waste disposal

Contaminant transport from waste depository in stochastic rock media — 3
Y.Ohnishi, M.Tanaka, H.Tajika, M.A.Soliman, Z.Ismail & K.Ando

The influence of industrial objects being placed in underground space on rock massifs — 9
J.M.Kazikaev

Preliminary analysis of a conceptual radwaste repository — 13
H.K.Moon, M.K.Song & S.I.Choi

Numerical modeling of groundwater contamination with volatile material with multi-component flow formulation — 19
K.Itoh, H.Tosaka, S.Mori & Y.Otsuka

Slurry fracture injection for the disposal of large volumes of low-toxicity wastes — 25
M.B.Dusseault & R.A.Bilak

Solution cavern disposal for low-level radioactive and toxic civil waste — 33
M.B.Dusseault, B.Davidson & M.S.Bruno

Disposal of smelter slag into a mined-out pit: A rock stability study — 41
R.Ciccu & P.P.Manca

Heat, gas, water flow and contaminant transport problems in caverns and tunnels

3D flow simulation for optimizing environment of tunneling faces — 49
B.Y.Kim, Y.D.Jo & S.Nakayama

R&D policy for underground space use and energy storage in Sweden — 55
B.T.Sellberg

Effect of thermal hysteresis on rock mass around openings for storage of low temperature materials *Y. Inada, N. Kinoshita, T. Shimogochi, Y. Kohmura & T. Matsuo*	61
Development of a simulation model for the vehicle tunnel ventilation using network theories *C. Lee, S. Lee, D. Baek & S. Moon*	67
Field measurements and numerical analysis on the efficiency of water curtain boreholes in underground oil storages *K. J. Lee & H. K. Lee*	73
A study on the variation of surface and groundwater flow systems related to tunnel excavation in the DONGHAE mine area *D. H. Lee, D. W. Ryu, T. K. Kim & H. K. Lee*	79
How to evaluate jointed rock masses with respect to water-sealing *A. Okamoto, Y. Nakazawa, K. Kojima & M. Hasegawa*	85
Fracture detection/characterization techniques for the groundwater control around the oil storage caverns *O. Choi, H. Tosaka, K. Kojima & A. Okamoto*	93

Blasting and machine excavation

Study on estimate of damage zone caused by blasting *M. Tezuka, Y. Kudo, H. Matsuda, A. Hasui & K. Nakagawa*	101
Mechanical excavation of hard rock tunnels without environmental disturbance: Effectiveness of TBM-driven pilot tunnels in non-blasting rock breaking method *M. Murata & K. Yokozawa*	107
Geological input as applied to TBMs: A case study of the Kelinchi Transfer Tunnel, Malaysia *M. Sundaram & I. Komoo*	113
New mechanical hard rock tunneling by FON drilling method and FASE method *T. Noma, T. Tsuchiya & M. Hada*	119
Qualified underground technologies for excavation of the Super-KAMIOKANDE cavern *T. Nakagawa, J. Yamatomi, G. Mogi, T. Takemura & K. Tsurumi*	125
Fragmentation of rock by cutting tools *M. C. Seifabad*	131
Choice of optimal operating conditions of the mining machine during tunnel drivage in geodynamically hazardous zones *A. N. Shabarov, N. V. Krotov & A. P. Zapryagaev*	137
Numerical analysis on the effect of pre-notches of borehole on fracture propagation *W. K. Song*	141
Rock fragmentation with plasma blasting method *K. W. Lee, C. H. Ryu, J. H. Synn & C. Park*	147
Blast vibration mechanism for underground blasting *T. N. Singh & C. Sawmliana*	153

Subsidence and ground control

Some incorrect problems of biharmonic equations arising in land subsidence mechanics V.I.Dimova	161
The prediction of tunnelling induced settlements in weak rock H.Asche	165
Predicting underground stability using a hangingwall stability rating E.Villaescusa, C.Scott & D.Tyler	171
Reclamation and construction in abandoned mining districts in Japan: Review of practical experiences and countermeasure instruction N.Kameda, N.Mori, T.Esaki, Y.Jiang & G.Zhou	177
Underground injection technology for protection of the Wieliczka Salt Mine in Poland A.Gonet, J.Stopa, S.Stryczek & S.Rychlicki	183
A study on the mechanism of chimney subsidence H.K.Moon, D.H.Huh & B.C.Kim	187
Underground constructions: Estimation of their influence on the earth surface deformations and stability of buildings in Odessa, Ukraine K.K.Pronin, E.A.Cherkez & V.I.Shmouratko	193
Problems of the surface subsidence above old workings V.N.Zemisev	197
An inverse problem approach in mining subsidence A.Constantinescu & D.Nguyen Minh	199
Calculation of surface subsidence associated with oil and gas production from multilayered deposits with abnormal pore pressure I.A.Garagash	205

Questions and measures for safety in the design and construction of tunnels

Design of undersea and under-river tunnel linings N.N.Fotieva & N.S.Bulychev	211
Study of stability for tunnel group in alternative stratified rock M.Xie, S.Yan, Y.Li & J.Yu	217
Effect of mud slurry of surrounding soil in using pipe jacking H.Shimada & K.Matsui	221
Technical analysis on the cost-saving in Norwegian rock excavation M.K.Kim, E.Broch & B.Nilsen	227
Spatial discreteness of geological environment and of underground drainage constructions in Odessa, Ukraine E.A.Cherkez, T.V.Kozlova & V.I.Shmouratko	233
Deformations of the shallow tunnels in flysch rock mass K.Thiel & L.Zabuski	239

Effective face distance in an Indian tunnel through jointed and weak rock masses 245
R.K.Goel, A.Swarup & A.K.Dube

The numerical simulation to the construction with double shields driving in the opposite directions 251
X.Zeng & Q.Zhang

3D FEA on effects of shield tunnelling on the adjacent deep piles 255
L.Ruan & Y.Li

Measurement control method and expert system for tunneling by fuzzy set theory 261
H.Chikahisa, K.Matsumoto, H.Nakahara, M.Tsutsui & S.Sakurai

Reinforcement of railway tunnel in fault zone 267
Y.D.Kwon, N.S.Park, E.R.Ha & D.Y.Oh

River diversion tunnelling works of Bakun Hydroelectric Project in Sarawak, Malaysia 275
W.R.Jee & J.H.Koo

Questions and measures for safety in the design and construction of caverns and openings

Cusp catastrophe of the rockbursts induced by ore pillars and its forewarning 287
H.Li, Z.Xu, X.Xu & Y.Wang

Numerical stability analysis of a large cavern in weak rock 293
R.J.Fowell & S.J.Ma

Numerical analysis of gas-liquid 2-phase fluid behavior around underground rock caverns storing pressurized gas 299
H.Suenaga, H.Tosaka & K.Kojima

Utility of fly ash as mine fill – A geotechnical study 305
J.M.Kate

Excavation analysis of a large-scale power station cavern by micromechanics-based continuum model for jointed rock mass 311
H.Yoshida, H.Horii & K.Kudo

Development of design criteria for low temperature gas storage 317
U.E.Lindblom & R.Glamheden

Pre-feasibility study on compressed air energy storage in Korea 323
D.S.Lee, J.Y.Kim, I.Y.Han & S.J.Hong

Stability evaluation of an underground opening used for special experiment 327
Y.Mizuta & Y.Kato

Feasibility study on ACC compressed air energy storage system by water-sealing method 333
S.Hibino, E.Koda, K.Nakagawa & Y.Uchiyama

Stope design in highly fractured area 339
S.J.Jung & H.Bogert

A study on the design of the shallow large rock cavern in the Gonjiam underground storage terminal *E.S.Park, H.Y.Kim & H.K.Lee*	345
Estimation of floor bearing capacity underneath full size pillars in longwall panels *Y.P.Chugh & D.Dutta*	353

Modelling techniques for safety evaluation: Rock characterization

Comparison of Hoek cell and a laboratory made cell in performing triaxial tests on intact and jointed specimens *M.Gharouni-Nik*	361
Influence of porosity on the absorption of moisture of sandstone and siltstone *H.Z.Anwar, H.Shimada, M.Ichinose & K.Matsui*	367
Relationship between microcrack density and mechanical properties in granite *S.E.Lee & H.M.Park*	371
Geological engineering research carried out on the Romania's territory concerning the geomechanical properties of the volcano-sedimentary rocks *E.Marchidanu*	377
The dynamic deformation moduli of some metamorphic rocks from Sri Lanka *U.de S.Jayawardena*	383
Mechanical properties of shales for estimation of damage zone dimensions around waste repositories *P.A.Nawrocki, M.B.Dusseault, B.Davidson & M.Kim*	389
Thermal conductivity of saturated quartz-illitic and smectitic shales as a function of stress and temperature *D.A.MacGillivray, B.Davidson & M.B.Dusseault*	395
Triaxial extension tests of cemented soils and soft rocks *K.Tani*	401
The acoustic emission properties of faulted rock under compression *D.Liu & K.Zhu*	407
A comparison of the Barton-Bandis joint constitutive model with the Mohr-Coulomb model using UDEC *R.Bhasin & N.Barton*	413
Shear behaviour of soft joints using large-scale shear apparatus *B.Indraratna, A.Haque & N.Aziz*	421
Studies on contact mechanism and closure behaviour of rock joints *C.Xia & Z.Sun*	427
Effect of structural anisotropy on deformation properties of granite under cyclic loading *H.Kusumi, Y.Mine & K.Nishida*	433
Postfailure lateral deformation of rock specimen under the triaxial compression test *T.Saito, S.Murata & H.Takehara*	439

Anisotropic behaviour of schistose rocks and effect of confining pressure on them 445
M.H.Nasseri, K.S.Rao & T.Ramamurthy

Development of a micromechanical crack model based on crack information 451
S.Jeon

Development of a numerical tool for the treatment of the data supplied by compression tests 459
J.P.Tshibangu K.

The effect of a polyaxial confining state on the behavior of two limestones 465
J.P.Tshibangu K.

A laboratory evaluation of grout jointed specimens composed of different rocks 471
R.K.Srivastava, K.K.Sharma & D.S.Soni

Analysis of nonlinear stress-strain behavior of intact rock 477
M.K.Kim & P.K.Lee

The influence of stress ratio and confining pressure on the weakly cemented sandstone 483
H.R.Nikraz & M.Press

Strength and geophysical behaviour of metagraywacke rock 489
C.S.Gokhale, J.M.Kate & A.M.Deshmukh

Direct observation of progressive microcrack development in relaxation tests on granites and their main component minerals 495
Y.S.Seo, N.Fujii & Y.Ichikawa

Fatigue behavior of cyclically compressed rock under the confining pressure 501
C.I.Lee, J.O.Im & J.J.Song

Detection of mode of failure of sandstone through image processing 507
D.Chakravarty & S.K.Pal

Sub-critical damage in brittle rock around underground storage caverns 513
J.F.Shao, G.Duveau, M.Sibai, M.Bart & N.Hoteit

A study on the measurement of the shear strength of the Seoul granite by the multiple direct shear test 519
D.Y.Kim, J.S.Yoon, H.S.Lee, H.I.Yoon

Fundamental fracture modes of granite using new testing devices 525
M.P.Luong & N.Hoteit

A new empirical failure criterion using data from triaxial tests for intact rocks 531
C.Park, C.Park & Y.Park

Strength properties and their relations with abrasiveness of some Indian rocks 537
A.K.Giri, C.Sawmliana, T.N.Singh & D.P.Singh

Author index 543

Preface

It is my great honor to host this 1st ARMS (Asian Rock Mechanics Symposium) from the 13th to the 15th of October 1997 in Seoul, Korea. I greatly appreciate the sponsorship by the International Society for Rock Mechanics. Although this is a regional symposium, 216 abstracts have been submitted since the first call for papers and of these 160 papers were finally selected. The authors numbered 364 from more than 27 countries, showing that this symposium is as international as it is regional. I would like to express my deepest gratitude to all the authors who have made the 1st ARMS successful. The topic of this symposium is 'Environmental and Safety Concerns in Underground Construction' with 12 specific themes under two main topics: Environmental concerns and Safety concerns.

The need for the use of underground space is affected by: 1. The need to conserve nature; 2. The rise in land value (normally in small countries); 3. Utilization of advantages for underground space, and 4. The swift progress of high technology in underground development area. As concerns the advantages for utilizing underground space, the physical characteristic of a rock mass, for example, can be summarized by its resistance to heat, its shield against heat, its consistent temperature, its air-tightness, and its protection against fire, acoustic noise, vibration and radioactive waves. The maximization of underground space utilization depends on how to use such advantages effectively.

The topic of this symposium, Environmental concerns and Safety concerns, is the most important issue regarding the construction of GEOTOPIA (or underground city) which is the primary concern of underground construction. I sincerely hope that a wide variety of technical and cultural exchanges will be participated in by the more than 500 professionals gathered here during this 1st ARMS.

Hi-Keun Lee
Chairman of 1st ARMS

Acknowledgement

The Organizing Committee of the 1st Asian Rock Mechanics Symposium gratefully acknowledge the financial contributions of:

Hyundai Engineering Co., Ltd
Sunkyung Engineering & Construction Co., Ltd
Ssangyong Engineering & Construction Co., Ltd
Jinro Engineering Co., Ltd
Suh Jun Trading Co., Ltd
Dong Ah Construction Co., Ltd
Byuck San Engineering Co., Ltd
Daelim Industrial Co., Ltd
Daeduk Consulting & Construction Co.
Kumho Construction & Engineering Co., Ltd
Kolon Construction Co., Ltd
Dong Seo Engineering Co., Ltd
Daewoo Co., Ltd
Samsung Heavy Industries Co., Ltd
Batu Engineering Co., Ltd
Seong Ha Geology Engineering Co., Ltd
The Korean Federation of Sciences and Technology Societies
Korea Institute of Geology, Mining and Materials
The Korean Institute of Mineral and Energy Resources Engineers

Organization

ADVISORY COMMITTEE (National Coordinate Committee)
Prof. Shunsuke Sakurai, ISRM President
Dr Chin-Der Ou, ISRM Vice President
Prof. Jun Sun, CSRME President (China)
Dr Shri N. B. Desai, ICISRM President (India)
Dr Koichi Sassa, JCISRM President (Japan)
Prof. Hi-Keun Lee, KSRM President (Korea)
Prof. Ooi Teik Aun, SEAGS President (SE Asia)

ORGANIZING COMMITTEE
Chairman
Prof. Hi-Keun Lee, Seoul National University
Co-chairman
Prof. Chung-In Lee, Seoul National University
Vice-chairman
Prof. Han-Uk Lim, Kangwon National University
Secretary General
Dr So-Keul Chung, Korea Institute of Geology, Mining and Materials

Committee Members
Prof. Hyo-Taek Chon, Seoul National University
Dr Kong-Chang Han, Korea Institute of Geology, Mining and Materials
Dr Sung-Wan Hong, Korean Institute of Construction Technology
Dr Seung-Hwan Jung, Korea Institute of Geology, Mining and Materials
Dr In-Ki Kim, Korea Institute of Geology, Mining and Materials
Dr Kwang-Soo Kwon, Korea Institute of Geology, Mining and Materials
Dr Kyung-Won Lee, Korea Institute of Geology, Mining and Materials
Prof. Hyun-Koo Moon, Hanyang University
Dr Chang-Ha Ryu, Korea Institute of Geology, Mining and Materials
Dr Hee-Soon Shin, Korea Institute of Geology, Mining and Materials
Dr Choon Sunwoo, Korea Institute of Geology, Mining and Materials
Prof. Hyung-Sik Yang, Chonnam National University
Prof. Ji-Sun Yoon, Inha University
Prof. Yoshiaki Mizuta, Yamaguchi University
Prof. Michio Kuriyagawa, Tsukuba University

Prof. Katsuhiko Sugawara, Kumamoto University
Dr Satoshi Hibino, Central Research Institute of Electric Power Industry
Prof. Toshiaki Saito, Kyoto University
Dr Ryuichi Iida, Japan Engineering Center
Prof. Sijing Wang, Academica Sinica, Inst. Geology
Prof. Bingjun Fu, Secretary General, Chinese Soc. Rock Mech.
Prof. Xiurun Ge, Academician Chinese Academy of Engineering
Prof. T. Ramamurthy, Indian Institute of Technology, Delhi

STEERING COMMITTEE
Secretary General: Dr So-Keul Chung
Technical Program: Prof. Hyung-Sik Yang
Council Meeting: Prof. Hyung-Dong Park
Conference Hall Management: Dr Chang-Ha Ryu
Treasurer: Dr Hee-Soon Shin
Administration: Dr Choon Sunwoo
Accommodations: Prof. Yeonjun Park
Tour and Excursion: Dr Chulwhan Park
Technical Exhibition: Dr Ho-Young Kim
Documentation: Dr Jung-Ho Synn
Advertisement: Dr Sung-Oong Choi

Heat, gas, water flow and contaminant transport problems in waste disposal

Contaminant transport from waste depository in stochastic rock media

Y. Ohnishi, M. Tanaka & H. Tajika
School of Civil Engineering, Kyoto University, Japan

M. A. Soliman
National Center for Water Research, Delta Barrages, Egypt

Z. Ismail
Department of Civil Engineering, University Malaya, Kuala Lumpur, Malaysia

K. Ando
Design Department, Civil Engineering Technical Division, Obayashi Co., Japan

ABSTRACT : Stochastic continuum modeling of groundwater flow and transport was conducted for soil media nearby environment to predict the contamination movement under steady state condition. From artificially established "perfectly known model", we picked up the measured hydraulic conductivity by setting the detection limit and performed the Monte Carlo type stochastic simulations. Two dimensional models were used in conjunction with hydraulic conductivity. Fifty realizations of hydraulic conductivity were generated conditionally by kriged simulation and sequential indicator simulation, and flow and transport were computed numerically for each realization. We compared the two stochastic methods with and without extrapolation under detection limit, classical deterministic approach (permeability is set to be homogeneous) and "perfectly known model". We conclude that the continuum stochastic approach is well suited for the interpretation and simulation of flow and transport.

1 INTRODUCTION

For the past few decades, groundwater flow and transport have a direct impact to many geo-environmental engineering problems. Therefore, transport analysis is very important in relation to environmental protection and resource utilization. One particular concern is the potential movement of contaminant due to nuclear waste disposal in the fractured rock media.

The modeling of flow and transport in fractured rocks can be done deterministically or stochastically, using either discrete or continuum representations. Deterministic methods are widely popular, because they concern unique geologic and/or hydrologic setting. However, stochastic methods are more realistic in that they account for uncertainty and gaps in the data. This is done by viewing the data as a sample from a population with statistical properties that can be inferred from the data.

Classical continuum models treat the fractured medium as a porous equivalent. In its crudest form, the continuum approach calls for the assignment of average (effective, equivalent) properties (permeability, porosity, dispersivity) so relatively large blocks of rock. In this way, details of heterogeneity (spatial variability) on scales smaller than the block size are lost. Furthermore, the block must be large enough to exceed a so-called representative elementary volume (REV). A REV is a hypothetical volume whose properties are not affected by small changes in volume. The associated macroscopic properties vary smoothly enough over the domain of interest to allow treating them by differential calculus. Unfortunately, measurements are often strongly scale dependent and thus non unique (Neuman, 1994). If a REV exists on some scale, it may be too large to be of practical interest (Neuman, 1987).

2 OBJECTIVES

This paper deals with the application of a stochastic continuum modeling approach to flow and transport using the artificially established "perfectly known model". One of the objectives is how conditional simulation keep the heterogeneity observed at the measured points, and how well it was represented using the statistical properties, i. e. mean, variance and the correlation structure. Another main objective is how to extrapolate the data under the detection limit which are often observed at the field measurement.

3 PERFECTLY KNOWN MODEL

As mentioned above, the so called perfectly known model is generated for reference of simulation as shown in Fig. 1. This model is of course artificially established by random generation with mean of log equal to -5.0 and variance of log is set to 1.0. During conditioning which is similar to measure the values at field, we set the detection limit for conditioning points. If the measured value from true model is smaller than the detection limit, the data of this point is set to exactly detection limit. Fig. 2 introduce the histogram of conditioning data with detection limit. 1.0E-6 is set to be detection limit. In field measurement, we face the detection limit because of limit of time and/or equipment limitation in many cases. For example, at Fanay-Augeres (Cacas et al., 1990), almost one third of the 2.5m packer tests yielded flow rates below the detection limit of the measuring instruments.

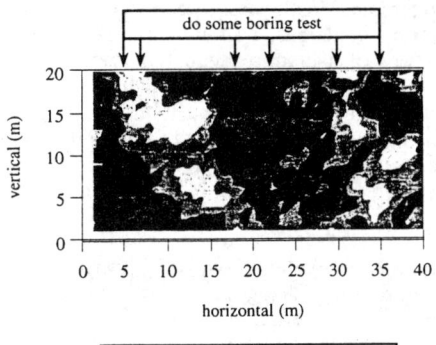

Fig. 1   "Perfectly known model" hydraulic conductivity map and location of "boring test"

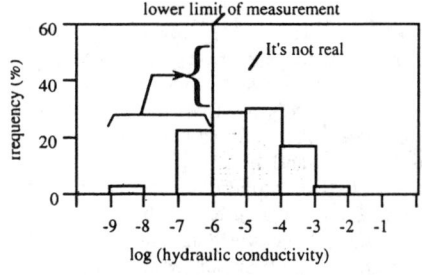

Fig. 2   Histogram of measured hydraulic conductivity

4 STOCHASTIC APPROACH FOR HYDRAULIC CONDUCTIVITY

In order to apply geostatistical concepts to the data, we approach the following three different ways.

- Classical homogeneous approach using the mean hydraulic conductivity from measured value.
- Kriged simulation without extrapolation under detection limit.
- Sequential indicator simulation with extrapolation under detection limit.

Some geostatistical concepts relevant to the present study are briefly outlined. At the simulations, the number of realizations which are created for numerical flow simulation is somehow arbitrary. Cacas et al. (1990) generated twenty realizations, and Dverstorp et al. (1992) created thirty realizations. In this paper, fifty realizations are generated and analyzed with respect to their spatial and statistical properties. The simulations are performed on a rectangular grid of 20 columns and 40 rows, with cells of size 1.0m x 1.0m as in Fig. 3.

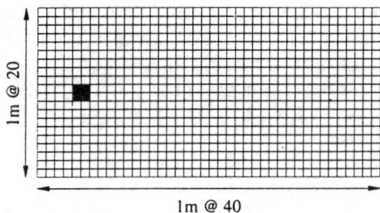

Fig. 3   Finite element mesh

4.1 CASE 1 Homogeneous approach

This is simply average over the measured values. In this case, 1.0E-4.93 m/sec is calculated from the measured data. The mean value is not equal to 1.0E-5 which is mean of the perfectly known model, because of the detection limit which is influenced by the values lower than detection limit.

4.2 CASE 2 Kriged simulation

The data is a collection of randomly varying hydraulic conductivities at various points in space. Thus, only one realization of the random field, the "perfectly known model" in this paper, is available. In adopting the notion of ergodicity it is assumed that the spatial average and second moments of this single realization converge to the corresponding ensemble moments as sample size increases. Several valid theoretical variogram models, i.e. models of a positive definite type, can be fitted to the experimental variogram, such as linear, spherical, exponential, or pure-nugget. The variogram used in this study is the spherical model, written as

$$\gamma_{(s)} = \begin{cases} C_1\{\frac{3}{2}(\frac{s}{a}) - \frac{1}{2}(\frac{s}{a})^3\} & , s \leq a \\ C_1 & , s > a \end{cases}$$

where C_1 is sill or variance of the data and a is range or separation distance beyond which there is no more correlation between the data. In the presence of a nugget effect, a nugget term is added to the equation, and C_1 is set equal to sill minus nugget. A nugget effect may be due to a variability of data values at a separation distance smaller than the minimum distance between data locations, and/or the presence of measurement errors.

The experimental variogram may reveal some degree of anisotropy in the spatial correlation among data, i.e. the variogram (or covariance) depends not only on the length of the separation vector, but also on its direction. Thus the variogram parameters for different directions are different from each other. The spatial correlation along the direction of sedimentation of a high permeability channel, for example, will be greater than the spatial correlation in the transverse direction. The type of anisotropy most commonly considered is geometric, where the sill is constant but the range varies with direction. The anisotropy is statistical, and does not refer to the local nature of the random variable, e.g. the local hydraulic conductivity. The random variable may still be a scalar, and this is how we treat it in this study.

Fig. 4 indicates the measured variogram and fitted theoretical one using spherical model with 8m range and 1.0 sill.

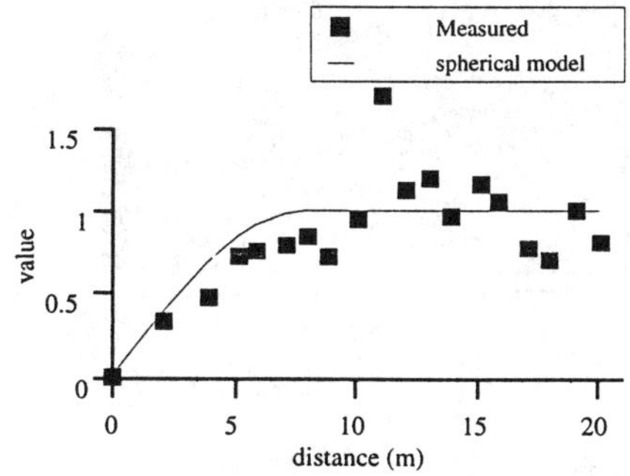

Fig. 4 Semivariogram of hydraulic conductivity

4.3 CASE 3 Sequential indicator simulation

There are several ways to treat values below the detection limit. However, the spatial distribution of very low hydraulic conductivities has considerable impact on the movement of tracers, which trend to migrate around such areas. It is preferable to extract and use all possible information from the data. We apply indicator simulation by using all data including the value under detection limit. In the indicator approach the data at each location x are transformed into a binary variable (x,z_c) consisting of only 1's and 0's, depending on whether the data value is below or above the indicator (cutoff) value z_c. This method has the important property that local hard indicator data originating from both local hard data and ancillary information that provides hard inequality constraints can be analyzed jointly (Journel and Alabert, 1990)

The first step of indicator simulation is the calculation of indicator variograms for a choice of cutoff values. The choice of the number of cutoffs is a balance between computational demand and a proper definition of the cumulative distribution function.

From the cumulative probability of the original data, three different cutoff values were set as 1.0E-6.0, E-4.0 and E-2.8. The first cutoff value is the detection limit of hydraulic conductivity measurements and the second is the median of the data. A spherical variogram using the second cutoff with 5m range and 0.29 sill, fitted to the omni-directional sample variogram, is shown in Fig. 5.

Extrapolation below the first cutoff value is done with a power model of the form,

$$\phi^*_{\omega,z_{sl},z_l}(z) = \left(\frac{z - z_{sl}}{z_l - z_{sl}}\right)^\omega$$

where z_{sl} is a specified lower limit, z_l is the first cutoff, and is a power parameter. We set the latter equal to 1, which yields a uniform distribution between the lower limit, arbitrarily set equal to 1.0E-8 m/s, and the first cutoff. Extrapolation beyond the last cutoff is done with a hyperbolic model of the form

$$\phi^*_{\omega,\lambda}(z) = 1 - \frac{\lambda}{z^\omega}$$

where

$$\lambda = z_p^\omega (1 - p)$$

and is a scaling parameter, z_p is a pre-calculated quantile value, and is a power. Here we set the power equal to 1.02, which is obtained from the measured data above the last cutoff. Using this power value, the mean Z value above the p-quantile value is equal to the mean of the measured data above the last cutoff. The estimated cumulative distribution function increases slowly above the last cutoff, thereby preserving the statistical properties and the positive skewness apparent from the original data. This method of extrapolation leads to estimates with a finite probability of having hydraulic conductivity values larger than the highest values measured. One realization of Case 2 and Case 3 is shown in Fig. 6. With this method, the full variability of the measured hydraulic conductivities is preserved, and even amplified. Whereas the sampled data do not fall below the detection limit, some of the simulated values do. This is due to the extrapolation of the estimated cumulative probabilities below the first cutoff value (detection limit) as discussed before. The only significant difference between the measured and simulated histograms is the wider spread (tail) of the latter below the detection limit.

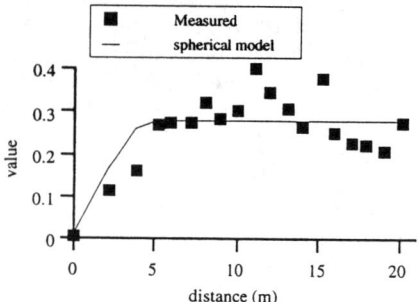

Fig. 5   Indicator semivariogram

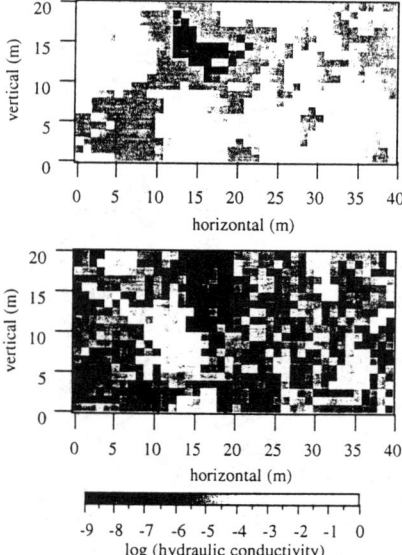

Fig. 6   Gray scale of realizations of hydraulic conductivity field
above : case 2
below : case 3

5 FLOW AND TRANSPORT SIMULATIONS

Steady state numerical flow and transport simulations were performed with fifty hydraulic conductivity realizations each. A finite element numerical model was used for this purpose, together with the grid of square isoparametric elements same as simulated grid. In Fig. 7, the top of the grid represents the water table. And Dirichlet boundary conditions are prescribed on both side. The dark mesh on the left side is assumed to be the contaminant source in the model.

5.1   Flow modeling

Each element is assigned a scalar hydraulic conductivity that varies from one realization to another. The modeled area is thus modeled as a continuum at a high degree of spatial resolution, with small-scale heterogeneity represented in some detail.

5.2   Transport modeling

To simulate transport, Eulerian Lagrangian finite element method was adapted using the same grid that was used previously for flow. Eulerian Lagrangian method is proposed by many researchers, e.g. Neuman (1981, 1984a) is based on interpolating the flow velocities at the element nodes which are estimated from the flow model.

Neglecting the third dimension in our model means that we do not allow the contaminant to advect nor disperse in direction transverse to the two-dimensional flow domain. We set the effective porosity, longitudinal and transverse dispersivity constant because we want to study the spatial variability of hydraulic conductivity in this paper.

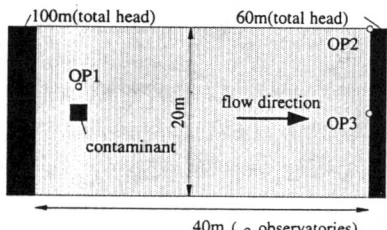

Fig. 7   Analysis model

6 RESULTS

The flow and transport simulations are evaluated by the outflow from the right hand side boundary, and concentration measured after 100 hour releasing at three points i. e. 3m above of the releasing point (OP1 as an observation point 1), surface of right hand side boundary (OP2), and same vertical level as sources at right hand side boundary (OP3).

6.1   Flow

The statistics of the flow result are summarized in

Table 1. The result of outflow over the right hand side boundary is more or less a good fit with the "perfectly known model" with some variation (Fig.8). This is because the result is average over the boundary and strict boundary conditions. The variation of outflow from simulation indicates the uncertainty with statistically homogeneous field.

Table 1 Statistics of the quantity of flow
(unit log(quantitiy of flow))

case	quantity of flow	
	mean	std.dev.
1	-0.143	-
2	-0.722	0.468
3	-0.498	0.673
reference	-0.435	-

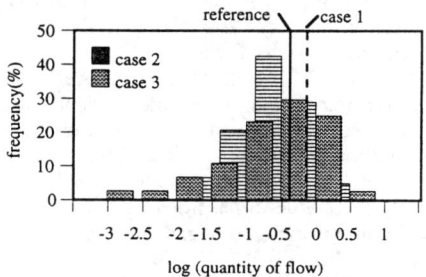

Fig. 8 histogram of quantity of flow
by Monte Carlo simulation

Table 2 Statistics of the transport
(unit log(concentration))

case	observatory 1	
	mean	std.dev.
1	-0.513	-
2	-0.690	0.126
3	-0.584	0.153
reference	-0.531	-

case	observatory 2	
	mean	std.dev.
1	-11.9	-
2	-14.1	0.997
3	-13.8	1.423
reference	-13.7	-

case	observatory 3	
	mean	std.dev.
1	-11.5	-
2	-13.7	0.957
3	-12.9	1.124
reference	-12.5	-

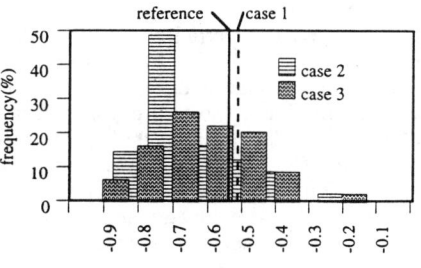

Fig. 9 Histogram of concentration
by Monte Carlo simulation (OP1)

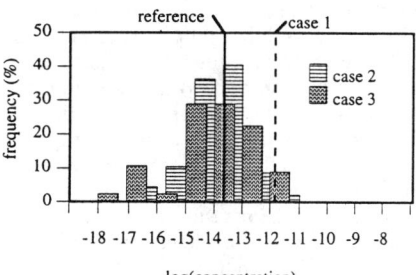

Fig. 10 Histogram of concentration
by Monte Carlo simulation (OP2)

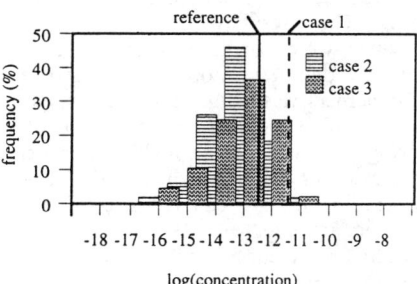

Fig. 11 Histogram of concentration
by Monte Carlo simulation (OP3)

6.2 Transport

The summary of the result is shown in Table 2, and histograms of concentration at each OP are also shown in Fig. 9, Fig. 10 and Fig. 11.

The histograms of concentration at OP1 with two simulations indicates that the concentration obtained from the perfectly known model is in the middle of range. At this point, the homogeneous model shows similar concentration, because it is perpendicular to the head gradient and thus concentration to governed by transverse dispersion.

The simulated concentration at OP2 and OP3 indicates the same trend. Fig. 10 and Fig. 11 show the results of the homogeneous, simulation with and without detection limit, and true models. Best fit is simulations extrapolating under the detection limit. Also the range of simulated concentration by Monte Carlo simulation indicates the uncertainty of the field. It means that the contaminant movement is possibly evaluated by the range of concentration.

7 CONCLUSIONS AND RECOMMENDATION

This paper presents an application of stochastic continuum modeling to the analysis of two dimensional flow and transport in extremely variable media. The stochastic continuum model successfully simulates the flow and transport. From the measured hydraulic conductivity, the spatial distribution of hydraulic conductivities has been extracted.

The measured hydraulic conductivities are viewed as random scalars representing a stochastic process. As almost one third of the measurements are assumed to yield values below the detection limit of the instrumental device, indicator analysis has been employed to include them.

To preserve the full variability of the hydraulic conductivity and to account for its uncertainty at unsampled locations, multiple and equally likely stochastic realizations of its spatial distribution were generated. The resulting sets of hydraulic conductivities were conditioned on the measured data and were thus less smooth than kriged results. Each grid element was assigned a value of hydraulic conductivity which might be different from values in neighboring elements. Thus a highly heterogeneous medium was generated, in which extensive flow pass may be represented by connected cells of large hydraulic conductivities.

Fifty different stochastic realizations of hydraulic conductivity were generated for input into a numerical flow and transport model. The results of the flow model such as outflow to the boundary and transport analysis were analyzed statistically. We observed that sequential indicator kriging produce the best fits in both outflow to the boundary and concentration at three different OPs with reference to perfectly known model.

We are planning to couple other information such as head to evaluate more close results to the measured value, and apply to the real field data. We will also try quantitative evaluation of additional measurements which will decrease uncertainty or indicate where the key area is to contribute to field planning.

REFERENCES

Cacas, M. C., E. Ledoux, G. de Marsily, A. Barbreau, P. Calmels, B. Gaillard, & R. Margritta, 1990a. Modelling fracture flow with a stochastic discrete fracture network: Calibration and validation, 1, The flow model, *Water Resour. Res.*, 26(3), : 479-489.

Cacas, M. C., E. Ledoux, G. de Marsily, A. Barbreau, P. Calmels, B. Gaillard, & R. Margritta, 1990b. Modelling fracture flow with a stochastic discrete fracture network: Calibration and validation, 2, The transport model, *Water Resour. Res.*, 26(3), : 491-500.

Delhomme, J. P., 1979. Spatial variability and uncertainty in groundwater flow parameters: A geostatistical approach, *Water Resour. Res.*, 15(2), : 269-280.

Deutsch, C. V., & A. G. Journel, 1992. *GSLIB: Geostatistical software library and user's guide*, Oxford University Press.

Dverstorp, B., & J. Andersson, 1989. Application of the discrete fracture network concept with field data: Possibilities of model calibration and validation, *Water Resour. Res.*, 25(3), : 540-550.

Freeze, R. A., 1975. A stochastic-conceptual analysis of one-dimensional groundwater flow in nonuniform homogeneous media, *Water Resour. Res.*, 11(5), : 725-741.

Journel, A. G., & F. G. Alabert, 1990. New method for reservoir mapping, *J. of Pet. Technology*, 16(2), : 212-218.

Neuman, S. P., 1981. A Eulerian-Lagrangian finite numerical scheme for dispersion-convection equation using conjugate space-time grid. *J.Comp. Phys.* 41 : 270-294.

Neuman, S. P., 1984a. Adaptive Eulerian-Lagrangian finite element method for advection-dispersion. *Int. J. Num. Meth. In Engng.*, ASCE, 114 : 59-75.

Neuman, S. P., 1984b. Role of geostatistics in subsurface hydrology, In "Geostatistics for Nature Resources Characterization." *Proc. NATO-ASI, Part 1*, Reidel, Dordrecht, The Netherlands,: 787-816.

Neuman, S. P., 1987. Stochastic continuum representation of fractured rock permeability as an alternative to the REV and fracture network concepts, *Proc. 28th US Symposium on Rock Mechanics*, Tucson, : 533-561.

Neuman, S. P., & A. Rouleau, 1994. *Fractured Rocks: Characterization, Flow and Transport*, E3 short course, Tucson.

Ohnishi, Y., & M. A. Soliman, 1996. Transport from waste-depository in stochastic soil media, *Proc. 2nd Int'l Congress on Environmental Geotechnics, Kamon (ed.)*, : 281-286.

The influence of industrial objects being placed in underground space on rock massifs

J. M. Kazikaev
Moscow Mining State University, Russia

ABSTRACT: Preliminary results of study of interaction between underground production plants and surrounding rock massif are given in the report. Characteristic features of underground environment being of importance for problems' solution are described here.

A problem of disposing ecologically harmful and hazardous production plants and storages arouses the growing interest due to the factors as follows: rock massif presents the more favourable conditions for construction; use of rock massif in the capacity of constructional elements meets requirements of safety, durability and economy; high ecological standards are also satisfied.

In connection with solution of the said problem problems of improvement of confidence level of data on underground space and efficiency of engineering geology investigations; elaboration of concepts of regularities of development of mechanical, physico-chemical and thermal processes in rock massifs due to their interaction with underground structures as well as migration of toxic and radioactive pollutants in underground water acquire special actuality.

Complex investigations carried out by the author during several year period are aimed at achieving the following goals:

1. search of methods of natural environment conservation, elimination or reduction of negative impact on it when disposing harmful production plants and facilities underground;

2. increase of reliability and efficiency of production plant construction and operation in underground space.

Some statements obtained in the course of the fulfilled complex investigations are set forth in the given report.

Utilization of underground space not only for the purpose of burial of toxic and radioactive wastes or construction of depositories for them or some other similar materials is not our only concern. I consider a problem of construction of industrial objects in underground environment to be more actual and complicated. The latters are characterized by the greater dynamics of forms and intensity of their affecting upon environment. And what is more they interact with rock massif being, in the present event, the natural environment. Analysis of the problem showed that reliable measures on environmental protection could not be elaborated at the level of consideration of only rock massif properties. So a statement about necessity to take up studying properties and responses of a system of the higher level rather than the rock massif one has been put forward. Regional geological structure where a production facility is being disposed was suggested to be considered such a system.

At the initial stage of the problem solution an assessment of principal possibility of underground disposing of harmful production plants and facilities acquires great importance. A classification of underground space in accordance with the said condition and requirements has been developed. Not presenting it here in all the details we can note that this classification is a double-level one: at the first level indexing of hydrogeological structure is done and that of the rock massif is carried out at the second one. The classification gives a possibility of solving problems of two groups, namely:

- to give preliminary assessment of feasibility and efficiency of underground environment utilization for construction of harmful production plants and facilities there;

to determine requirements for methods of underground construction of the said purpose the meeting of which can eliminate negative impact on natural environment or reduce it to a minimum;

$1°$. Using the classification characteristics of sites being the most favourable for disposing harmful production plants can be formulated. For example, these are sites within hydrogeological massifs and lower structural levels of volcanogenic basins characterized by hydrogeological closeness, inward underground water flow. Here stable monolithic rocks such as magmatic and metamorphic ones occurring below regional fracture zone look promising for disposing production facilities.

It is known that rock massif screening effect is assessed by means of seepage coefficient. The regularities of this factor variation are studied very thoroughly lately especially in connection with mineral deposit exploitation in complicated engineer-geological, hydrogeological conditions and due to construction of underground storages of gas, oil, toxic and radioactive wastes.

Studies carried out by a number of specialists including our investigations have found out regularities describing influence of stress-strain state of rock massifs on their permeability. However, only rocks of porous or fractured types were considered. Double porosity rocks refer to the more general and complex types.

In the result of carried-out studies a property of additive logarithms of seepage coefficient of double porosity rocks has been determined which is described by the following equality:

$$\ln k = \ln k_T + \ln k_\Pi \qquad (1)$$

where
k = rock massif seepage coefficient; k_T = fracture rock seepage coefficient; k_Π -porous rocks seepage coefficient.

Using this property our associate Yu. S. Osipenko has obtained general regularity of variation of seepage coefficient of double porosity rocks. Here not only influence of rock stress-strain state but also change of their fracture ratio could be taken into account:

$$k(M_T, p) = cM_T e^{\xi_T \beta_K (p-p_0)} \qquad (2)$$

where
$k(M_T, p)$ -seepage coefficient with current value of M_T and p, m/day; c -coefficient depending on fracture width, their orientation in space and relative to seepage flow direction, m²/day; M_T -current value of fracture modulus, g/m; ξ_T -dimensionless coefficient depending on fracture space structure; β_K -coefficient of fracture compressibility, 1/Pa; p -current effective pressure, Pa; p_o -standard effective pressure, Pa.

The experiments have shown that application of the stated relationship allows to build a mathematical model which indicates adequate real distribution of rock seepage properties in natural conditions.

2°. Underground space classification is based on the assumption of multilevel structural heterogeneity of geological sites (and rock massifs as their component). The author's long-term studies in this field have allowed him to reveal the most characteristic features and peculiarities of this system. Some of them are given below:

1. Structural heterogeneity of the site is an effect of its geological history and conditions of formation. It should be noted that lithological structure of the site does not add any substantial changes into general picture of structural heterogeneity giving only a slight correction of it or its effects. Intrusive rocks make an exception where factors of structural heterogeneity may differ from those of the enclosing rock massif.

2. The structural heterogeneity gives evidence of several systems of fractures that divide rock massif into separate structural blocks which are not interconnected or are weakly intercemented. Structural block dimensions at macrolevel and within geological space that presents some interest for us can vary from fractions of centimeters to tens and hundreds of meters and are present in any scale simultaneously when blocks of minor dimensions compose the greater ones. Due to this rock massif appears as structurally heterogenic multilevel structure. A number of levels of heterogeneity in space that presents interest for us may reach 5 to 6 and more.

3. The structural heterogeneity varies spasmodically. It means that if dimensions of blocks of adjacent levels are ℓ_i and ℓ_j, respectively (for example $\ell_i < \ell_j$) then structural heterogeneities satisfying unequality $\ell_i < \ell_k < \ell_j$ can not exist.

4. At various levels of structural heterogeneity the known factors of rock properties and state manifest differently both quantitatively and qualitatively. It is general knowledge that rock strength of small rock specimens is sometimes much higher than that of the massif. This also can be said about elastic properties, heat conductivity, volume mass etc. However, there are some factors which manifest themselves qualitatively in different ways. For example, hydrogeological characteristic feature at higher structural levels (hundreds of meters, kilometers) manifests at all its aspects. At the same time at the lower levels (tens of centimeters, meters) it can be manifested partially. And at such levels as centimeters and less it can not be discovered at all (for example, in dense rocks). Manifestation of tectonic situation can also be interpreted ambiguously at different structural levels.

Such factor as rock stress state also presents interest. While giving quantitative assessment of stress characteristics such units of measurement as kg/cm², t/m², Pascal, i.e. definite forces should be related to area of 1 square centimeter or 1 square meter. With old system of measurement some reasonings about stresses expressed, for instance, in megatons per square kilometers could be hypothetically assumed. Though in scientific and technical literature it can hardly be met. However, in SI system of units even such possibility is excluded as everything must be referred only to area of 1 square meter, pascal (Pa), for instance. If now it is correlated with a concept of multilevel structural heterogeneity of rock massifs then it can result in the following. At structural levels of the high orders ($\ell>1$m) area of force application should meet a requirement S>>1m. In the opposite case features defining this level of heterogeneity are lost. Consequently, stresses measured in pascals loose their physical sense.

In the above-said example imperfection of two used concepts are demonstrated. Firstly, the shown contradictions of the adopted systems of measurements and nature of some states of medium and processes occurring in it. Secondly, insolvency of the notion "stress" which has no physical confirmation in nature. It is not by chance that some concepts have been suggested (by Academician Ye. I. Shemyakin, for example) to give up notions "stresses" and "deformations" and turn to application of really existing notions "forces" and "displacements" while studying geomechanic processes.

At the same time numerous studies show that values of stresses (natural stresses, for instance) at various structural levels of rock massif differ considerably similar to the event demonstrated by experimental data plots at Figs.1 and 2.

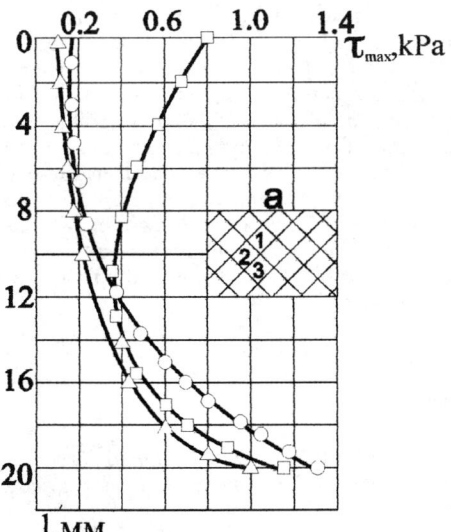

Fig. 1. Variation of value of maximum tangential stresses along structural element length:
a - model scheme; 1,2,3 - structural elements

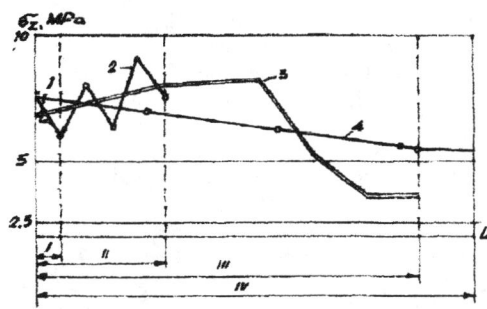

Fig.2. Variation of stresses in horizontal plane at various structural levels

It can be explained not only by influence of scale factor and some other peculiarities of structural heterogeneity but also by specific features of nature of the said environment. Therefore the observed feature should be taken into consideration when solving various geomechanical problems if one operates with such notion as "rock stress state".

5. With spasmodical character of transition from one structural level to another which is common for all the factors the transition itself can take different forms, namely: natural, spontaneous and mixed ones.. Physical and mechanical characteristics of rocks etc can represent the first form of this transition. Characteristic feature of this form of bonds of rock massif properties at two or more levels of structural heterogeneity is such that it can be expressed by functional, statistical or empirical relationship.

The second form is the form of spontaneous transition of massif properties from one structural level to another which can be realized only in the event of occurrence of qualitatively new element at the higher adjacent level. It can be a zone of tectonic failure, a zone of rock breaking, an aquifer etc. It is clear that such elements impart qualitatively new characteristics to rock massif which the latter lacked at the previous structural level.

As a rule both forms of transition of factors from one level of structural heterogeneity to another can be met at selected for study sites of great extension (hundreds and thousands of meters). Consequently, they should be considered in combination with their interference taken into account.

6. Up to now we referred to absolute characteristics of rock massif structural heterogeneity and their manifestations. For solution of geomechanical problems relative characteristics of rock massif structural state being known among specialists as "scale effect" exhibiting itself at various scale levels, are important and taken into account. In this

case scale effect is assumed to be meant as ratio of dimension of an object, being studied in rock massif, to structural block dimension. Physical, mechanical and other characteristics of rock massif are governed substantially by the scale level. This was observed by many researchers and was considered in our paper (Kazikaev 1987). At Fig. 3 a plot of relationship of rock strength against scale level is shown.

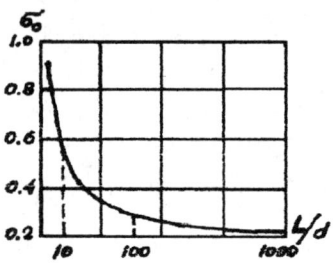

Fig.3. Massif strength variation at various scale levels

The curve is plotted for structural blocks of 0,25-0,3m size which are the most widely spread in hard rocks. Rock strength values are laid along the vertical axis and scale effect is laid along the horizontal one, i.e. relationship of linear dimension of the object under study (L) against dimension of the structural block (D). For convenience of representation at this axis not absolute values of relationships L/d are indicated but ln (L/d). Three characteristic zones are conventionally selected and the first of them denotes a zone of action of a single working (gallery, tunnel) where 1<L/d<10; the zone is limited by value of ln(L/d) = 1; the second one is a zone of influence of a single chamber and here 10<L/d<100, respectively, zone boundary is on ln(L/d) = 2; the third zone is the zone of influence of an production facility which is disposed in underground space, here 100<L/d<1000 and its boundary is on ln (L/d) = 3. The obtained curve indicates regularity of the massif strength factor variation in one and the same point during transition from one scale level to another. Together with this a meaning of a point is also changed as it is transforming gradually into a system of concentric circles as transition to the higher scale levels takes place.

These are brief fragmentary results of the fulfilled systematization of notions and the more exact definitions of concepts of geological space - that environment where production plants of the enhanced ecological danger are placed. One can start studying interference of industrial object and underground environment only after the said concepts are assumed.

3°. Analysis of the problem shows that technogenic processes being in progress at hazardous underground production plant may produce negative impact upon all principal geological fields such as hydrogeological, geomechanical, geochemical, and thermal ones. These statements were assumed as primary ones while developing procedures of assessment of technogenic process effects.

In connection with underground disposition of hazardous production plants a problem of determination of density of heat flow to rock massif from underground plant surface and of temperature field parameters of the enclosing massif was under solution. As was to be expected while superimposing the known decisions, obtained for non-stationary process of heat spreading, over geological environment and rock massif in particular, the above-said features of this environment acquire decisive importance. Moreover, rock massif features allow to control a reverse process to the greater extent by enclosing environment effect on underground plant temperature field using both passive and active engineering methods.

Similar to the previous problem some problems of estimation of prediction of changes of hydrogeological conditions, natural geochemical fields, rock massif geomechanical state have been solved.

In the result of investigations a statement of necessity of the problem complex study has been confirmed. It means that together with study of rock massif, its changes caused by production plant disposition it is necessary to determine the character and degree of environmental influence on technological processes of a disposed production plant, structures ands building materials, equipment and mechanisms. Moreover, formulation of problems of development of principally new technologies for obtaining known products or kinds of energy optimized over ecological, energy-saving, materials-consumption, economic factors is rightful with taking into account underground space specific conditions and interference of underground environment and technological processes.

REFERENCE

Kazikaev, J.M. 1987. *Geomechanical processes during ore mining by complex method.* Moscow: Nedra

Preliminary analysis of a conceptual radwaste repository

H.K. Moon & M.K. Song
School of Geosystem, Environmental and Civil Engineering, Hanyang University, Korea

Soo-Il Choi
Kolon Engineering and Construction Co., Ltd, Korea

ABSTRACT: Geological disposal of radioactive wastes deep in underground vaults is not the choice but the inevitable. In the case of high-level radioactive waste disposal the high thermal output of radioactive waste can cause a significant problem on the stability of disposal tunnels. To evaluate the effects of thermal loading on the disposal vaults, we have simulated the heat transfer in jointed rock mass with various boundary conditions and analyzed the thermo-mechanical stresses. The results have shown that the stability of the tunnels is highly dependent on the orientation of joints. It is shown that after disposal the rock mass is expanded initially, but is contracted as temperature goes down, and a disturbed zone is developed widely around the vault. In the case of environmental problem, presented in this paper is the long-term impact of the radioactive waste on the natural environment. The transport analysis based on the advection-adsorption mechanism has shown a strong influence of joint connectivity on the solute concentration, transport time and migration. The results have shown an increasing arrival time of nuclides with increased depth of disposal vaults and justified the concept of a deep repository.

1 INTRODUCTION

The rapidly increasing demand on electricity and the growing share of nuclear power in the total electricity supply in Korea have brought on increasing accumulation of radioactive wastes. In order to carry out geological disposal of radioactive wastes, the authorities concerned must, first of all, develop a reliable and comprehensive waste management program. Site characterization comes first of course. The potential hazards to people and natural environment must be evaluated in the program for variable options of disposal schemes. Shallow ground disposal can be a method for low radioactive waste, while deep underground disposal in stable geological formations may be a choice for highly hazardous wastes to guarantee the highest degree of safe isolation.

This paper presents heat transfer from the waste canister to buffer and rock masses, and the groundwater flow and solute transport in the rock mass having a large number of joints. Numerical experiments using joint network method are performed. The the residence time and concentration of nuclide particles in the jointed rock mass, and the particle arrival time to the ground surface are discussed.

2 THERMAL PROBLEM

The heat transfer and thermal stress analysis are important, because the swelling capacity and cation exchange capacity of buffer and backfill materials (e.g. bentonite) decrease as the temperature goes up. At a high temperature vitrified glasses are crystallized and become brittle (Cohen, 1977). The stability of rock mass depends on the thermal stress which in turn depends on the temperature change in the mass. In order to evaulate the effects of the layout and depth of disposal tunnels on temperature distribution, a two dimensional finite element program is used for the transient analysis of heat transfer and thermo-elastoplastic stress analysis.

2.1 Heat transfer analysis

In this section, the effects of the depth and spacing of disposal tunnel on the temperature in the neighborhood of the tunnel are investigated.

For the problem of high-level nuclear waste, the heat condition will be of a transient nature, as heat is transferred from the canister to a large rock mass at time-varying rate. The power output q(t) is a decreasing function of time t and approximated as:

$$q(t) = q_0 \exp(-\beta t) \qquad (1)$$

where β is 0.024 per year, q_0 is initial thermal output and is assumed to be 235 W/m^3 in this study.

In order to investigate the effects of thermal output from the canister, the Swedish concept KBS3 is used for numerical experiment. Figure 1 shows the dimension of disposal vaults and canister. The height of the domain is 60 m and the width varies from 25 m to 50 m. The top and bottom boundaries have a constant temperature, and the left and right boundaries are insulated.

It is highly recommended for the site of nuclear waste repository that the rock doesn't have discontinuities such as cracks, joints and faults. But in most cases, it is difficult to find such a rock mass. In the case of nuclear waste repository the area for disposal site is very large, so that it is even more difficult to find intact rock mass.

In this study, it is assumed that the base rock is granite and has 4 joint sets. A total of 2700 joints are contained in the domain. The mechanical and thermal properties of jointed rock mass are estimated theoretically (Moon, 1987; Jue, 1991).

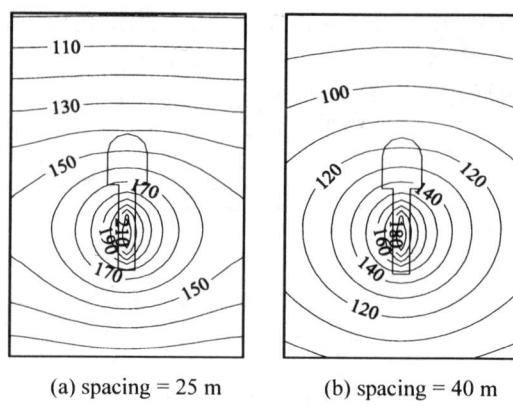

(a) spacing = 25 m   (b) spacing = 40 m

Figure 2. Temperature distribution at 20 years after disposal of radioactive waste (unit : degree in Celcius).

Figure 3 shows that the variation of temperature around the canister. The temperature rises to its maximum value at about 20 years after disposal and converges to its original temperature at 200 years.

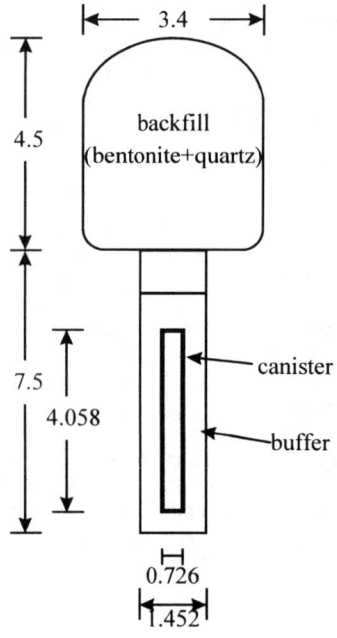

Figure 1. Disposal vault and canister (unit : m).

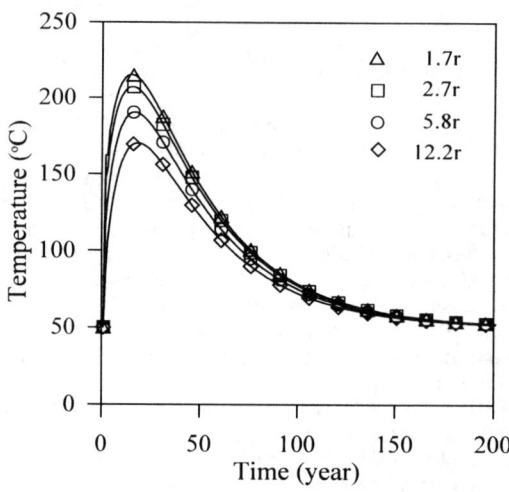

Figure 3. Variation of temperature with time at 4 different locations around the canister.

Figure 2 shows the temperature distribution in the vicinity of the vault at 20 years after disposal. The effect of tunnel spacing on the temperature distribution is such that if the spacing is decreased, the maximum temperature increases. The maximum temperature in the case (a) is above 220 °C, but is about 190 °C in the case (b).

In order to design the layout of disposal tunnels, one of the design criteria is that the maximum temperature must not exceed the allowable temperature to maintain the mechanical stability of rock and the chemical stability of buffer and backfill material.

There are many parameters to be tested in the

design, e.g., the depth, size, spacing of the vaults. In this study, the depth and spacing are chosen for the parameter study. Figure 4 shows the effects of the depth and spacing on the maximum temperature of rock. In the case of the vaults at depth 1000 m and 500 m in granite rock, the maximum temperature is decreased by 12 % as the spacing increases from 25 m to 50 m. For comparison, the same analysis is performed for rock salt. The maximum temperature in rock salt is decreased by 30 %. It seems reasonable to conclude that the spacing should be minimized to it's optimal value, but the constraint is that the maximum temperature should not exceed the allowable temperature taking into account the mutual interaction between heat sources (i.e. the containers).

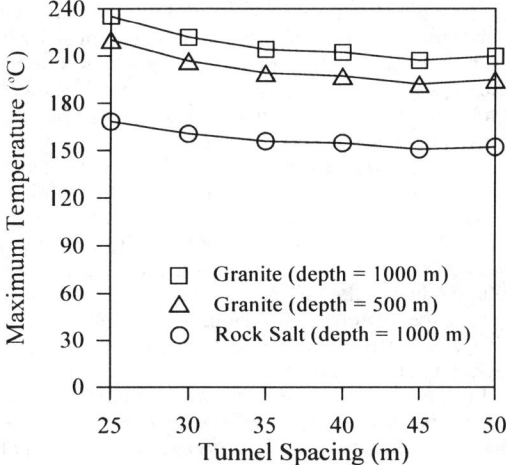

Figure 4. Variation of maximum temperature with tunnel spacing.

2.2 Thermal stress analysis

In this study, the following equations are used to estimate the in-situ stress in the repository area.

$$\sigma_v = 0.0233Z + 1.36,$$
$$\sigma_h = 0.0138Z + 2.78 \tag{2}$$

where σ_v and σ_h are the vertical and horizontal stresses in MPa, and Z is depth in meter.

The same joint pattern as in the previous section is used to this thermo-elastoplastic analysis. The elastic and strength properties of the jointed rock mass having 4 sets of joints are estimated using an analytical scheme. The essential feature of the scheme is integrating in sequential manner the individual effects of joints into the equivalent properties.

After the waste canisters are disposed in the canister boreholes, they are filled and compacted with buffer material for tight embedding of waste container and for sealing boreholes, shafts and tunnels. Stress redistribution due to excavation of the tunnel and borehole is taken into account first, and then the thermal stress due to thermal output from the canister is added.

Figure 5 shows the change of the state of principal stresses before and after disposal. Figure 6 shows the variation of vertical stress around the canister. The increase of the stress in the vicinity of the canister is dramatic in early stage of disposal, but such high stress concentration is decreased in the late stage. In this parameter study it is found that the tunnel spacing has less impact on the final state of stresses compare to the influence of the excavation induced stress and its thermal disturbance.

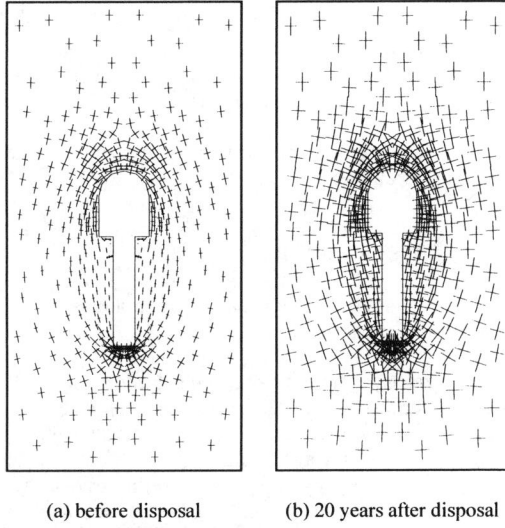

(a) before disposal (b) 20 years after disposal

Figure 5. Principal stress plot around the disposal vault.

We expect that the most dangerous period may be the time of highest temperature. On the contrary, the result shown that the later stage of disposal can be more dangerous, because the contraction following the initial expansion of rock mass has occurred. The heating-cooling process in jointed rock mass has developed the disturbed zone.

The disturbed zone affects hydraulic safety of the disposal vaults. The aperture of joints can be changed. Such phenomenon is also discovered in the physical experiment of hydraulic conductivity at the Stripa project (Pusch, 1993). The hydraulic conductivity is increased by the factor of two, when the rock sample are heated and cooled. It is important to note that the heating-cooling process deteriorate the flow resistance of the rock mass.

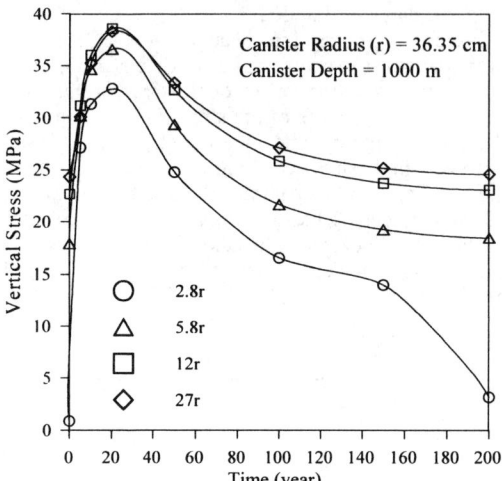

Figure 6. Variation of vertical stress with time around the canister.

3 SOLUTE TRANSPORT

3.1 *Groundwater flow and solute transport in jointed rock masses*

In rock mass hydraulics, the contribution of intact rock material may be minimal compared to the joints which usually constitute preferential paths of flow. The flow through many joints, therefore, is strongly affected by the orientation, frequency, length, connectivity, and other geometrical properties which are highly variable. Hence it is reasonable to take a statistical approach rather than the deterministic for describing the joint geometry.

In this study, a stochastic joint generating program is developed for simulating and visualizing the joints in a window of any desired view direction. The base angle, i.e. the relative orientation of the joints projected in the window, is employed to construct the two-dimensional view of a jointed rock mass (Lee, 1992). The joint trace length is assumed to follow the log-normal distribution, the orientation follows the Arnold's hemispherical normal distribution. The joints are located randomly according to the Poisson distribution. The full scale rock mass tested in this study is 200 m x 500 m in size and contains three sets of joints.

Once the joint map and the boundary conditions are established, the connectivity of the joints must be examined in order to form a network, eliminating the isolated joints, closed-loop joints and the dead-end segments of joint which do not contribute to the flow. The method by Priest & Samaniego (1983) is followed to identify the connected joints and their intersection points (nodes) using a matrix $[m_{ij}]$, where the subscripts i and j denote the node and joint numbers. The component m_{ij} takes either the value "0" (unconnected) or "1" (connected). The flow of water through the network is analyzed first by solving the following simultaneous equations for hydraulic head at all nodes.

$$h_i = \sum_j C_{ij} h_j \qquad (3)$$

where subscripts i and j denote the node number, h is hydraulic head, and C is hydraulic conductance between nodes. For the problem having a very large number of nodes, the relaxation method of solving the system of equations has caused memory problem. The sparse matrix solving method is proved quite successful in this case.

Using the head difference Δh and the conductance C between nodes, volumetric flow rate $Q = C\Delta h$ is computed for every individual joints which form the network. The hydraulic boundary conditions applied to the 200 m x 500 m model are as follows: (i) Water table on the top (ground surface), (ii) no flow across the bottom (impervious datum plane), and (iii) a constant head commonly on the left- and right-hand boundaries.

In the course of this network analysis, it is conceived that the water inflow may depend on the size of the domain being analyzed. If the external boundary is far from the opening and if the medium is homogeneous, the inflow, of course, should be independent of the domain size. But, a close inspection of the five network models indicates that the distribution of the flow paths is neither homogeneous nor isotropic near the opening.

Therefore, four square domains with side length 100 m, 200 m, 300 m and 400 m are taken from the individual network model and tested for the boundary effects. The results are shown in Figure 7, where the volumetric flow rate is normalized to the rate obtained from the 500 m full scale model.

In the network analysis, the inflow is always overestimated when the size of the test domain is small. In the case of 600 joints in 100 m x 100 m domain, the inflow into the opening can be as high as 5.6 times that of the 500 m x 500 m domain. Since the network analysis is sensitive to the choice of the domain size, it may lead to erroneous results for the inflow, unless the inhomogeneity of the network configuration is properly taken into account.

The flow rate is a controlling factor in advective transport of non-reactive (non-adsorbing) solutes.

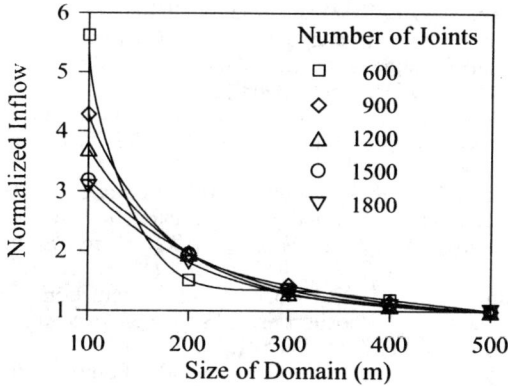

Figure 7. Variation of normalized inflow with increased test domain (opening dia. = 30 m).

But, in the case of adsorptive chemical species or radionuclides, diffusion in the rock matrix and the adsorption on the joint surface will have significant influence on the transit time. The relative rate of water flow to the transport rate of a reactive solute can be expressed by the retardation factor R_d:

$$R_d = v_s / v_R \qquad (4)$$

where v_s and v_R are the seepage velocity of water and the transport rate of the reactive solute, respectively. For non-reactive solutes, the retardation factor is unity and the solute is transported at a seepage velocity. If the joints constitute the main channels for solute transport, the retardation factor depends on the channel geometry can be expressed as follows (Rasmuson and Neretnieks, 1986; Krishnamoorthy et al., 1992):

$$R_d = 1 + (2K_a / b) \text{ for planar channel,}$$
$$R_d = 1 + (2K_a / a) \text{ for cylindrical channel} \qquad (5)$$

where K_a is the surface sorption coefficient, b is the aperture of the planar channel, and a is the radius of the cylindrical channel.

3.2 *The effects of depth of disposal vaults*

A simulated rock mass model is used to study the migration of radionuclides escaping from a hypothetical repository. The repository consists of rows of disposal vaults extending 150 m horizontally. The repository horizon is located at somewhat shallow depth, such as 50 m, 100 m and 150 m for comparison purpose. The aperture assumed is 0.1 mm for all joints in the model. The average joint spacing is around 14 m. The radionuclides taken for this study are ^{90}Sr and ^{129}I. The surface sorption coefficient is 3.7 x 10^{-4} m for ^{90}Sr, and 4.1 x 10^{-7} m for ^{129}I. A total of 10000 particles are released to the joints crossing the repository horizon in proportion to the flow rate. Complete mixing is imposed on the intersection of two joints.

In order to find out the migration paths of the nuclides dissolved in the water, an upward flow is artificially initiated by placing a sink on the ground surface. This sink behaves like a pumping well and is subjected to a pumping pressure of 0.01 atm, then the pumping pressure is doubled to 0.02 atm for comparison. Figure 8 shows the variation of the first arrival time of the nuclides with increasing depth of the disposal vaults. It is noteworthy that the relation between the first arrival time and the depth is linear for both sorbing and non-sorbing cases.

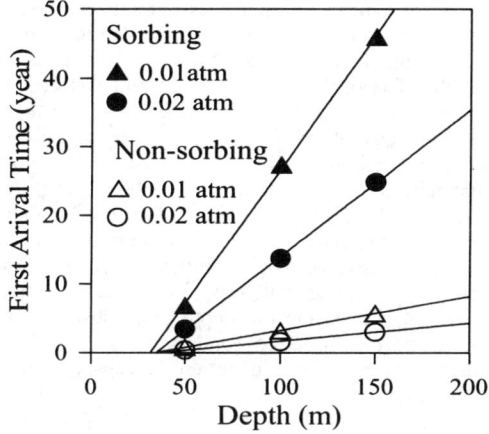

Figure 8. Variation of the first arrival time with increasing repository depth.

The difference in transit time between the sorbing and non-sorbing solutes is quite significant. This transport analysis based on advection and adsorption shows that more than 45 years are required for the first nuclide to reach the ground surface 150 m above the disposal vaults. Vapor and water flow due to high thermal loading at the vault horizon may also induce the upward movement of solutes. The analysis must, of course, be extended to take into account other transport mechanisms, such as diffusion into the rock matrix, sorption onto the micro fissures and chemical reactions between solutes and rock.

4 CONCLUSION

Geological disposal of radioactive wastes deep in stable rock formations is inevitable in Korea, where 11 nuclear power plants are currently in operation with a plan to increase the number up to 27 by the year 2006. Crystalline rock mass is the choice for permanent repository site in the future. The long-term impact of the heat generated from canisters on the stability of disposal vaults and the influence of the highly hazardous waste on the environment should be evaluated at the early stage of conceptualization and preliminary design. Solute transport with groundwater in jointed rock masses is the most important mechanism through which the radioactive nuclides escaped from disposal vaults can reach the biosphere.

Presented and tested in this paper are transient two-dimensional heat transfer analysis, thermo-mechanical analysis, analysis for water flow and solute transport in a jointed rock mass. In the case of thermal problem, the disturbed zone due to thermal loading alters the state of stresses and affects the overall safety of the disposal vaults and canisters. In the cases of groundwater flow and solute transport, the connectivity of the joints has a significant influence on both water flow and solute transport. Channellized solute transport is observed in the tested model. The arrival time of reactive solutes such as ^{90}Sr and ^{129}I to the ground surface is predicted based on the advection, diffusion and adsorption. The arrival time is increased almost linearly by increasing the depth of disposal vaults. The concept of deep underground repository utilizing the rock mass as natural barrier is justified quantitatively.

ACKNOWLEDGEMENTS

The financial support from Kolon Eng. & Const. Co. Ltd. is gratefully acknowledged. This study is also supported by the Korea Science and Engineering Foundation (KOSEF) under grant number 961-0407-022-2

REFERENCES

Cohen, B. L. 1977. The Disposal of Radioactive Wastes from Fission Reactors, *Scientific American*, 236:21-31.

Jue, K. S. 1991. Thermal conductivities of discontinuous rock masses and computer modelling study on thermal stress analysis, MS. Thesis. Hanyang University, Seoul, Korea, p.94.

Krishnamoorthy, T. M., R. N. Nair, & T. P. Sarma 1992. Migration of radionuclides from a granite repository. *Water Resource Research*, 28:1927-1934.

Lee, K. 1992. Stochastic modelling of joint geometry and its application to the assessment of underground cavern stability. MS. Thesis. Hanyang University, Seoul, Korea.

Moon, H. K. 1987. Elastic moduli of well-jointed rock masses, Ph.D. Dissertation, Univ. Utah, p. 284.

Priest, S. D. & A. Samaniego 1983. A model for the analysis of discontinuity characteristics in two dimensions. *Proceedings 5th ISRM Congress*. V. 2:F199-F207.

Pusch, R. 1993. Rock-backfill interaction in radwaste repositorries, *Comprehensive rock engineering*, V. 5:565-581.

Rasmuson, A. & I. Neretnieks 1986. Radionuclide transport in fast channels in crystalline rock. *Water Resource Research*, 22:1247-1256.

Samaniego, A. & S. D. Priest 1984. The prediction of water flows through discontinuity networks into underground excavations. *Proceedings of ISRM Symposium-Design and Performance of Underground Excavations*:157-164.

Song, M.K., Jue, K. S. & Moon, H. K. (1994). A theoretical and numerical study on channel flow in rock joints and fracture networks. *Journal of Korean Rock Mechanics Society*, 4:1-16.

Numerical modeling of groundwater contamination with volatile material with multi-component flow formulation

Kazumasa Itoh, Sei-ichiro Mori & Yasunori Otsuka
OYO Corporation, Omiya, Japan

Hiroyuki Tosaka
The University of Tokyo, Department of Geosystem Engineering, Japan

ABSTRACT: In numerical prediction of groundwater contamination by volatile organic material that has small dissolution rate into water, it is necessary to calculate dissolution and volatilization in addition to contaminant flow. In this paper, numerical model of contaminant flow has been introduced based on multi component convection-diffusion formulation. And, some numerical studies have been carried out about dissolution and volatilization. Through these numerical experiments, Validity of dissolution and volatilization model has been shown.

1 INTRODUCTION

Recently, groundwater contamination becomes one of serious environmental problems. Especially, a lot of contamination cases with volatile organic material discharged from large industrial plants or small laundry factories are reported world widely.

In order to reconstruct and predict flow and transport of contaminant in groundwater and evaluate effect of remediation, numerical simulation with solute convection-diffusion has been utilized in many cases. However, in case of groundwater contamination with materials, which are slightly soluble to water and volatile in air, conventional numerical simulation can not be applied directly.

Contamination with soluble and volatile material would widely spread because of volatilization in unsaturated zone and diffusion in air and resolution into groundwater in unsaturated and saturated zone according to rainfall.

In order to calculate the movement of contaminant material, the authors have been developing numerical modeling method based upon multi-component flow formulation. In this paper, numerical formulation of contaminant flow has been introduced and some numerical experiments are presented on dissolution and volatilization problems.

2 NUMERICAL FORMULATION

2.1 Basic Formulation

For numerical simulation of contaminant transport, the authors has been developing numerical expression with multi-component flow formulation. In this method, the flow of water-gas-contaminant three phase, and five components, i.e. pure contaminant, water, air, contaminant solute, and evaporated contaminant, has been introduced as the numerical model. Volatilization and resolution from pure material to water, resolution from vapor to water has been modeled as mass transfer between phases.

The basic equations of multi-component system are as follows;

(1) Material balance of water

$$\nabla \frac{kk_{rcw}}{\mu_{cw}B_{cw}} \nabla \psi_{cw} + \nabla \frac{kk_{rcc}}{\mu_{cc}B_{cc}} \nabla \psi_{cc} - q_{ws}^{cw} - q_{ws}^{cc}$$

$$= \frac{\partial}{\partial t}\left(\frac{\phi S_{cw}}{B_{cw}} + \frac{\phi S_{cc}}{B_{cc}}\right) \quad (1)$$

(2) Material balance of air

$$\nabla \frac{kk_{rca}}{\mu_{ca}B_{ca}} \nabla \psi_{ca} - q_{as}^{ca} = \frac{\partial}{\partial t}\left(\frac{\phi S_{ca}}{B_{ca}}\right) \quad (2)$$

(3) Material balance of pure contaminant

$$\nabla \frac{kk_{rcc}R_{cc}}{\mu_{cc}B_{cc}} \nabla \psi_{cc} - f_{cs}^{cc-cw} - f_{cs}^{cc-ca} - f_{cs}^{cc-r} - q_{cs}^{cc}$$

$$= \frac{\partial}{\partial t}\left(\frac{\phi S_{cc}R_{cc}}{B_{cc}}\right) \quad (3)$$

(4) Material balance of dissolved contaminant in water phase

$$\nabla \frac{kk_{rcw}R_{cw}}{\mu_{cw}B_{cw}}\nabla\psi_{cw} + \nabla D_{cw}\nabla\left(\frac{R_{cw}}{\alpha_{cw}}\right) + f_{cs}^{cc-cw} + f_{cs}^{cw-ca}$$

$$-f_{cs}^{cc-r} - q_{cs}^{cw} = \frac{\partial}{\partial t}\left(\frac{\phi S_{cw}R_{cw}}{B_{cw}}\right) \quad (4)$$

(5) Material balance of evaporated contaminant in air phase

$$\nabla \frac{kk_{rca}R_{ca}}{\mu_{ca}B_{ca}}\nabla\psi_{ca} + \nabla D_{ca}\nabla\left(\frac{R_{ca}}{\alpha_{ca}}\right) + f_{cs}^{ca-cc} - f_{cs}^{ca-cw}$$

$$-f_{cs}^{ca-r} - q_{cs}^{ca} = \frac{\partial}{\partial t}\left(\frac{\phi S_{ca}R_{ca}}{B_{ca}}\right) \quad (5)$$

where, k : permeability of ground [L^2]
k_r : relative permeability of each phase[-]
μ : viscosity of each phase [$ML^{-1}T^{-1}$]
B : formation volume factor [-]
R : volumetric resolution/volatilization ratio
D : diffusion coefficient [LT^{-2}]
q : sink/source term [T^{-1}]
p : porosity [-]
S : saturation of each phase [-]
ψ : potential of each phase [MLT^{-2}]

Subscriptions cw, cc, and ca denote contaminated water phase, chemical contaminant phase and contaminated air phase, respectively.

In these equations, unknown parameters are ψ_{cw}, y_{cc}, y_{ca}, S_{cw}, S_{cc}, S_{ca}, R_{cw}, R_{ca}, and R_{cc}. Supplementary relationships are used in order to reduce unknown parameters.

$$\psi_{cw} = P_{cw} - \rho_{cwR}gZ = P_{cc} - P_{c,cw} - \rho_{cwR}gZ \quad (6)$$

$$\psi_{cc} = P_{cc} - \rho_{ccR}gZ \quad (7)$$

$$\psi_{ca} = P_{ca} - \rho_{caR}gZ = P_{cc} + P_{c,ca} - \rho_{caR}gZ \quad (8)$$

$$S_{cw} + S_{cc} + S_{ca} = 1 \quad (9)$$

where, $P_{c,cw}$: capillary pressure of water
$P_{c,ca}$: capillary pressure of air
ρ_{cwR} : density of water under pressure
ρ_{ccR} : density of chemical under pressure
ρ_{caR} : density of air under pressure

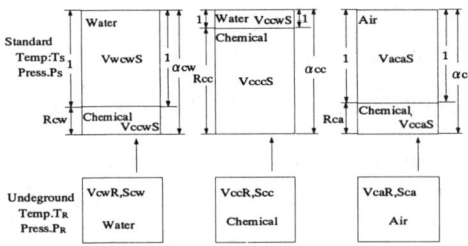

Figure.1 Definition of concentration

Using equations (6) - (9) and, considering R_{cc} as constant, five unknown parameters are selected as P_{cc}, S_{cw}, S_{ca}, R_{cw}, and R_{ca}.

In this formulation, volumetric resolution or volatilization ratio R and volumetric concentration α has been defined as shown in Figure.1.

According to the definition, R and α are written as follows.

$$R_{cw} = \frac{V_{ccws}}{V_{wcws}} \quad (10)$$

$$R_{cc} = \frac{V_{cccs}}{V_{wccs}} \quad (11)$$

$$R_{ca} = \frac{V_{ccas}}{V_{acas}} \quad (12)$$

$$\alpha_{cw} = \frac{(V_{wcws} + V_{ccws})}{V_{wcws}} \quad (13)$$

$$\alpha_{cc} = \frac{(V_{wccs} + V_{cccs})}{V_{wccs}} \quad (14)$$

$$\alpha_{ca} = \frac{(V_{acas} + V_{ccas})}{V_{acas}} \quad (15)$$

Here, relationship between R and α is calculated by the next equation.

$$\alpha = 1 + R \quad (16)$$

In this method, formation volume factors are defined from apparent formation volume factor B', defined as follows,

$$B'_{cw} = \frac{V_{cwR}}{(V_{wcwS} + V_{ccwS})} \quad (17)$$

$$B'_{cc} = \frac{V_{ccR}}{(V_{wccS} + V_{cccS})} \quad (18)$$

$$B'_{ca} = \frac{V_{cca}}{(V_{acaS} + V_{ccaS})} = \frac{P_{ref}}{P_{ca}} \quad (19)$$

$$B_{cw} = \frac{V_{cwR}}{V_{wcwS}} = \frac{(V_{wcwS} + V_{ccwS})}{V_{wcwS}} \cdot \frac{V_{cwR}}{(V_{wcwS} + V_{ccwS})}$$

$$= \alpha_{cw}B'_{cw} = (1 + R_{cw})B'_{cw} \quad (20)$$

$$B_{cc} = \alpha_{cc}B'_{cc} = (1 + R_{cc})B'_{cc} \quad (21)$$

$$B_{ca} = \alpha_{ca}B'_{ca} = (1 + R_{ca})\frac{P_{ref}}{P_{ca}} \quad (22)$$

where, P_{ref} : reference pressure
P_{ca} : total pressure of air phase

Phenomena of dissolution and volatilization can be considered as material exchange between two phases shown as f in basic equations.

From this formula, pressure of contaminant phase, saturation of water phase and gas phase, volumetric content of solute and gaseous contaminant can be calculated with Finite Difference Method.

2.2 Modeling of dissolution and volatilization

In order to construct mathematical model of dissolution and volatilization, the authors has introduced three kind of material exchange between two phases.
1) Dissolution of chemical from chemical phase to water phase
2) Volatilization of chemical from chemical phase to air phase
3) Dissolution of chemical from air phase to water phase

And, for the modeling of material exchange, analytical solution of transient diffusion is introduced.

1) Numerical modeling of dissolution from chemical to water

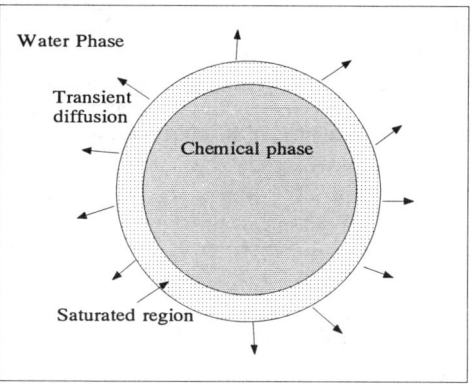

Figure 2 Schematic diagram of dissolution

Schematic diagram of dissolution is shown in Figure 2.

Considering dissolution from chemical to water as one dimensional diffusion, concentration of solute chemical in water is calculated from Fick's second law.

$$\frac{\partial C}{\partial t} = D_{cw}\frac{\partial^2 C}{\partial Z^2} \quad (23)$$

From analytical solution of (23) with initial condition $C = C_{sw}$, and boundary condition $C = C_{sat}$ (at $Z=0$), $C = C_{sw}$ (at $Z=\infty$), distribution of concentration in water phase is calculated as (24).

$$\frac{C - C_{sw}}{C_{sat} - C_{sw}} = \text{erfc}\left(\frac{Z}{2\sqrt{D_{cw}t}}\right) \quad (24)$$

From this equation, dissolution rate from saturated region through unit area is shown as equation (25).

$$N_\ell = -D_{cw}\frac{\partial C}{\partial Z}\bigg|_{Z=0} = \sqrt{\frac{D_{cw}}{\pi t}}(C_{sat} - C_{sw}) \quad (25)$$

And, average dissolution rate between time $t - t + \Delta t$ through unit area is shown by equation (26).

$$u_{cs} = \frac{1}{t}\int_t^{t+\Delta t} N_\ell dt = 2\sqrt{\frac{D_{cw}}{\pi \Delta t}}(C_{sat} - C_{sw}) \quad (26)$$

As a result, average dissolution rate can be calculated by multiplying contact area (A).

$$f_{cs}^{cc-cw} = 2A\sqrt{\frac{D_{cw}}{\pi \Delta t}}\left(\frac{R_{cc}}{1+R_{cc}} - \frac{R_{cw}^{(n+1)}}{1+R_{cw}^{(n+1)}}\right) \quad (27)$$

And, if contact area is assumed to be proportional to m_1th power of the volume of chemical phase, and volumetric ratio of dissolution to water and chemical phase, (27) has been transformed to next equation.

$$f_{cs}^{cc-cw} = \lambda(V(1 - S_{cw}^{(n+1)} - S_{ca}^{(n+1)}))^{m_1} \frac{S_{cw}^{(n+1)}}{S_{cw}^{(n+1)} + S_{ca}^{(n+1)}}$$

$$\cdot \sqrt{\frac{D_{cw}}{\pi \Delta t}}\left(\frac{R_{cc}}{1+R_{cc}} - \frac{R_{cw}^{(n+1)}}{1+R_{cw}^{(n+1)}}\right) \quad (28)$$

2) Numerical modeling of volatilization

For numerical modeling of volatilization of chemical component to air phase, transient diffusion formulation is also applied. Schematic diagram of volatilization is shown in Figure 3.

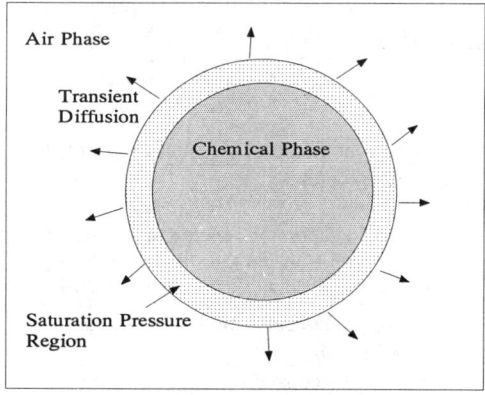

Figure 3 Schematic diagram of volatilization

After the same process as dissolution, volumetric ratio of volatilization from chemical phase is expressed as next equation.

$$f_{cc}^{ca-cc} = \lambda(V(1-S_{cw}^{(n+1)}-S_{ca}^{(n+1)}))^{m_1} \frac{S_{ca}^{(n+1)}}{S_{cw}^{(n+1)}+S_{ca}^{(n+1)}}$$
$$\cdot \sqrt{\frac{D_{ca}}{\pi \Delta t}} \left(\frac{R_{ca(sat)}}{1+R_{ca(sat)}} - \frac{R_{ca}^{(n+1)}}{1+R_{ca}^{(n+1)}} \right) \quad (29)$$

Volumetric concentration in air phase in equation (29) has been expressed with total pressure and saturation pressure.

$$f_{cc}^{ca-cc} = \lambda(V(1-S_{cw}^{(n+1)}-S_{ca}^{(n+1)}))^{m_1} \frac{S_{ca}^{(n+1)}}{S_{cw}^{(n+1)}+S_{ca}^{(n+1)}}$$
$$\cdot \sqrt{\frac{D_{ca}}{\pi \Delta t}} \left(\frac{P_{cs}}{P_{cc}^{(n+1)}+P_{c,ca}} - \frac{R_{ca}^{(n+1)}}{1+R_{ca}^{(n+1)}} \right) \quad (30)$$

On the other hand, material loss in chemical phase has been calculated with conditional equation of gas. In this condition, pressure of evaporated chemical has been considered as saturation pressure in saturated region. Mol number of evaporated chemical has been calculated as next equation.

$$n = \frac{P_{cs} f_{cc}^{ca-cc}}{RT} \quad (31)$$

where, R : gas constant
T : temperature

And, volumetric volatilization rate in liquid chemical phase is calculated as,

$$f_{cc}^{cc-ca} = \frac{xP_{cs} f_{cc}^{ca-cc}}{\rho_c RT}$$
$$= \lambda(V(1-S_{cw}^{(n+1)}-S_{ca}^{(n+1)}))^{m_1} \frac{S_{ca}^{(n+1)}}{S_{cw}^{(n+1)}+S_{ca}^{(n+1)}}$$
$$\cdot \frac{xP_{cs}}{\rho_c RT} \sqrt{\frac{D_{ca}}{\pi \Delta t}} \left(\frac{P_{cs}}{P_{cc}+P_{c,ca}} - \frac{R_{ca}^{(n+1)}}{1+R_{ca}^{(n+1)}} \right) \quad (32)$$

3) Numerical modeling of dissolution from gas phase to water phase

Dissolution of evaporated chemical to water has been modeled with Henry's law.
$$c_w = HP_c \quad (33)$$
where, c_w : weight concentration of chemical in water phase
H : Henry's constant

In this model, material exchange between gas phase and water phase is caused only by dissolution into water, and concentration of chemical in water calculated by Henry's law is the upper limit of dissolution.

With Henry's law and transient diffusion model, material exchange rate between gas and water phases is calculated by the next equation.

$$f_{cs}^{cw-ca} = \lambda(VS_{ca}^{(n+1)})^{m_1} \frac{S_{cw}}{1-S_{ca}}$$
$$\cdot \sqrt{\frac{D_{cw}}{\pi \Delta t}} \left(\frac{HR_{ca}^{(n+1)}(P_{cc}+P_{c,ca})}{\rho_c(1+R_{ca}^{(n+1)})} - \frac{R_{cw}^{(n+1)}}{1+R_{cw}^{(n+1)}} \right) \quad (34)$$

And, volumetric exchange rate in water phase is calculated by similar method to volatilization modeling.

$$f_{cs}^{ca-cw} = \lambda(VS_{ca}^{(n+1)})^{m_1} \frac{\rho_c RTS_{cw}^{(n+1)}}{x(1-S_{ca}^{(n+1)})} \frac{(1+R_{ca}^{(n+1)})}{R_{ca}^{(n+1)}(P_{cc}+P_{c,ca})}$$
$$\cdot \sqrt{\frac{D}{\pi \Delta t}} \left(\frac{HR_{ca}^{(n+1)}(P_{cc}+P_{c,ca})}{\rho_c(1+R_{ca}^{(n+1)})} - \frac{R_{cw}^{(n+1)}}{1+R_{cw}^{(n+1)}} \right) \quad (35)$$

3. NUMERICAL EXPERIMENT OF MATERIAL EXCHANGE

In order to validate numerical modeling of material exchange phenomena, the authors carried out some numerical experiment of material exchange, in which convection term by flow was not considered.

At first, simulation of dissolution was carried out. In this calculation, initial volumetric ratio of water phase and chemical phase was set to 1:1, and diffusion coefficient for dissolution was changed. Result of simulation is shown in Figure 4.

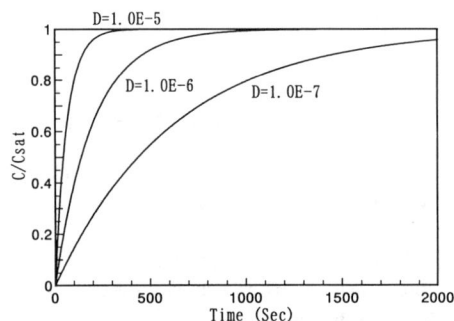

Figure 4 Result of simulation about volatilization

From this result, it is shown that numerical modeling of dissolution with transient diffusion model would be valid, and diffusion coefficient shows significant influence on dissolution rate.

And, for the next step, simulation of volatilization was carried out. In this case, initial volumetric ratio of chemical phase and gas phase was set to 1:1, and initial pressure in gas phase was set to 1.033 kgf/cm^2. Chemical material was assumed to be PCE, whose saturation pressure is about 20 mmHg. In this

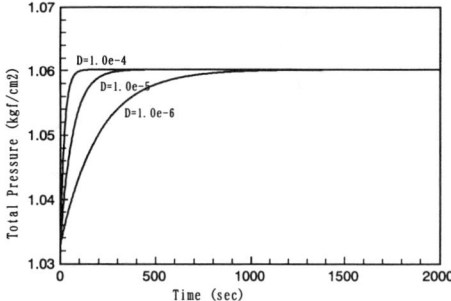

Figure 5 Result of simulation about dissolution

simulation, transient total pressure of gas phase was calculated.

Result of volatilization simulation is shown in Figure 5.

In this case, partial pressure of air in gas phase was not changed, although, partial pressure of evaporated chemical in gas phase changed according to volatilization. As a result, transient change of pressure is similar to concentration change in dissolution simulation.

For the next case, initial saturation was changed in dissolution simulation. Result of calculation is shown in Figure 6. In this figure, calculated concentration is normalized by saturation concentration.

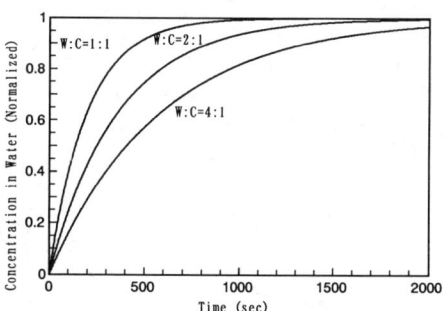

Figure 6 Effect of initial saturation to dissolution

From this result, if saturation ratio between water and chemical phases is 4:1, chemical concentration in water phase does not come to saturated condition in 2,000 seconds. So, saturation of each phase has great influence on transient dissolution phenomena.

Next, the authors carried out three phase simulation of dissolution and volatilization. In this case, initial saturation of water, chemical and air phase was changed. Results of this simulation are shown in Figure 7 and Figure 8.

Figure 7 shows transient change of air phase total pressure, and Figure 8 shows transient change of water phase chemical concentration. In this calculation, initial saturation of each phase is different in each case. From this result, initial saturation variation has significant effect on dissolution and volatilization

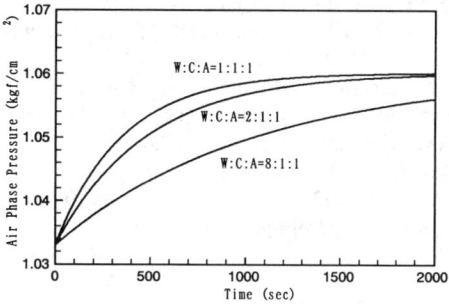

Figure 7 Result of three phase simulation
(Air Phase Pressure)

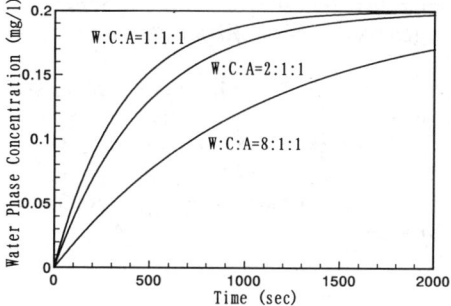

Figure 8 Result of three phase simulation
(Water Phase Concentration)

From this result, saturation ratio has also great influence on dissolution and volatilization process. And, comparing these results with results of two phase case, dissolution and volatilization speeds were reduced in three phase case.

These material exchange phenomena has been able to modeled and calculated, however, in practical application, the assumed parameters, such as diffusion coefficient in transient diffusion model must be identified by laboratory experiment.

4 CONCLUSION

Followings are the result of this study,
1) Numerical model of contaminant flow with transient dissolution and volatilization with multi component flow formulation was constructed.
2) In the model, dissolution and volatilization processes were modeled as transient diffusion. And, some numerical studies were carried out for verification and sensitivity analysis.

This work will be continued in the following points.
1) Determination of diffusion coefficient in dissolution and volatilization with laboratory experiment and numerical studies.
2) Development of three dimensional flow simulation with capability of expressing natural geological hydrological conditions.

REFERNCES

Aziz K. and Settari A. 1979, Petroleum Reservoir Simulation. Elsevier Applied Science Publishers

Dagan G, Hornung U, and Knabner P, 1991 Mathematical Modeling for flow and transport through porous media   Kluwer Academic Publications.

Tosaka H, Itoh K, Ebihara M, Inaba K, Itoh A, and Kojima K. 1996,   Comprehensive Treatment of Groundwater Pollution by Multi-component, Multi phase Convection/Diffusion Modeling. Journal of Groundwater Hydrology   Vol.38 No.3 pp.167-180

Slurry fracture injection for the disposal of large volumes of low-toxicity wastes

Maurice B. Dusseault & Roman A. Bilak
Terralog Technologies Incorporated, Waterloo, Ont. & Calgary, Alb., Canada

ABSTRACT: Slurry fracture injection (SFI) disposal of large volumes of oil field wastes was first attempted in Canada in 1988-1990. A mature technology has evolved, and SFI has become the preferred approach for environmentally secure disposal of non-hazardous wastes in Alberta and Saskatchewan. The rock mechanics aspects of SFI include the initiation and maintenance of a large volume fracturng process at depth, the evolution of stresses, and the changes in transport properties that take place in the reservoir. Environmental security is demonstrated in SFI operations by monitoring and analyzing the process in various ways so that waste containment can be demonstrated and so that the SFI process can be optimized in a systematic manner.

1 INTRODUCTION

Initial attempts in the 1960's to modify oil-fiel hydraulic fracturing for radioactive waste disposal using cement slurries met with poor results: it was thought that optimum security required impermeable strata (e.g. slates) and cement slurries with agents to reduce water bleed-off and increase slurry viscosity. The consequences were thin fractures propagating great distances in unpredictable ways with large vertical growth components. At about this time, regulatory agencies stipulated waste water injection well pressures less than the fracture gradient (typically 0.7 to $1.05\sigma_v$ in sedimentary basins). This limited injection to clear liquids under Darcy flow conditions, obviating large-scale solids injection.

Massive fracturing for thermal enhanced oil recovery improved the understanding of subsurface fracture response in the 1970's and 80's, and led us to recommend massive slurry fracture injection (SFI) of oily sand wastes without additives into high porosity, high permeability sandstones bounded by low permeability shales. This approach leads to short, contained fractures with minimal vertical growth and restricted lateral propagation. With regulatory permission, Mobil Oil Canada disposed $\approx 10,000$ m^3 of oil-contaminated fine-grained sand in Saskatchewan in 1988-1990, using a 35 m thick quartzose sandstone of 30% porosity. A slurry of density 1.08-1.12 g/cm3 was episodically injected at BHP (bottom-hole pressures) above σ_v at a depth of 690 m in a vertical well. Pressure responses and occasional leveling surveys at the surface were used for monitoring.

Regulatory agencies are justifiably loath to change existing regulations to allow for special cases unless convinced that the process is viable and can meet claims. However, a dynamic heavy oil industry in the 1990's in Alberta began producing ever larger annual volumes of oily sand through a production practice known as "Cold Production" (Geilikman et al. 1997, Geilikman and Dusseault 1996; Dusseault et al. 1995). This was expensive to dispose under increasingly rigorous environmental standards. Experience in steam fracturing and detailed geological knowledge suggested that SFI could be rationally implemented, and verification of solids containment in the target reservoir using surface tiltmeter installations was developed in 1993-1994 (Dusseault et al. 1993). Afterward, the Alberta Energy Utilities Board permitted regulatory variances for properly conducted, monitored and analyzed SFI projects.

Since then, SFI has become the preferred approach for disposal of massive quantities of sand. Regulatory permission was aided by the fact that natural solid wastes were being returned to where they came from: deep porous sandstone reservoirs. The State of California has recently permitted SFI on a demonstration basis, and other jurisdictions are studying case histories and regulatory practices for successful and environmentally secure SFI.

2 SFI PROCEDURE

SFI follows the sequence shown in Fig 1. Injection pressures (p_{inj}) are measured at hole bottom (BHP), and the horizontal axis is time. High rate injection of clear water is used (A) to initiate hydraulic fracture (B), after which p_{inj} declines slightly to conditions approaching stable injection (C) into a short fracture because of high natural fluid bleed-off rates. Solids slurry is made using a screening and blending procedure and introduced into the injection stream gradually until the desired density is

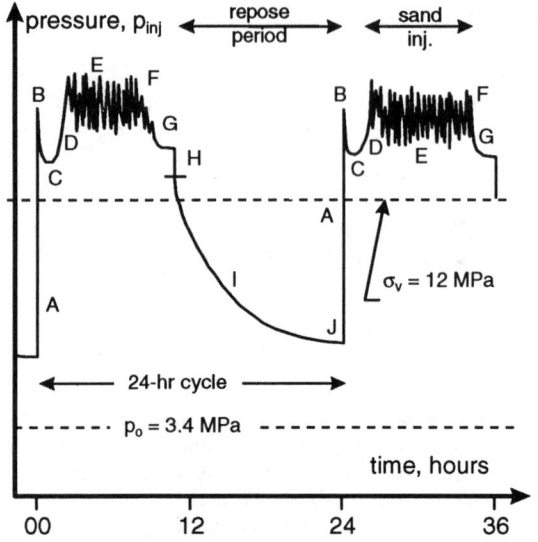

Fig. 1: Typical SFI data for fine-grained sand and waste water slurry at 540 m depth.

reached. This generates an increase in p_{inj} to values of 1.05-1.35σ_v (D), depending on the site and slurry properties. Stable SFI continues for a period of 4-12 hours (E), during which time p_{inj} typically exhibits fluctuations of ±5-6%. The slurry is gradually replaced with clear water and from 5-40 m^3 is flushed through the system (F). The well is shut-in (G), and a sharp drop to fracture closure pressure (H) takes place. Porous medium flow then takes over from fracture flow, the pressure decays with time (I), and the SFI cycle is repeated the next day (J).

SFI success depends on fracture re-initiation and propagation in a formation of high permeability so that wastes stay near the injection point. Various factors are optimized: slurry density, rates, injection periods, and so on. Recently, SFI has been successfully modified for "slops" (oily liquids with a small amount of fine-grained solids, Dusseault and Bilak 1997). Also, slops can be introduced under careful control into massive sand SFI, provided that formation transmissivity is not impaired, and providing that permeability blockage is not so severe that fractures tend to propagate out-of-zone.

3 THE MANNVILLE GROUP

To date, all SFI projects in Canada use target formations in the Mannville Group of clastic sediments, an Early Cretaceous (115 MYBP) 70-200 m package of poorly consolidated sandstone, siltstones and shales. SFI is viable in many other geological situations, but an introduction to the geology and lithostratigraphy of this region is warranted, as this may be taken as an ideal case.

The Western Canadian Sedimentary Basin in Alberta and Saskatchewan has virtually ideal conditions for SFI (Fig 2). Strata are flat-lying and largely unfaulted, fluid flow is horizontal and slow (cm per year), thick ductile horizontal shales protect shallow potable waters, oil-free permeable sandstones are widespread, and the sandstones are cohesionless albeit dense (n ≈ 30%).

The Mannville Group is underlain by older clastics and carbonates which dip gently to the southwest. Before the shallow Cretaceous sea invaded the Interior Plains, a mature landscape existed with shallow, broad valleys forming a drainage pattern flowing toward the northwest (≈Az 330°). Uplands were topped by residual soils formed by weathering, and the valley traces contained typical sand and gravel valley fill. Some of these valley fills are ideal SFI targets.

A major feature was a wide (200-300 km) shallow (200 m deep maximum) trough which led to accumulation of the sediments which now form the vast oil sands and heavy oil deposits which generate the large amounts of produced sand (>150,000 m^3 in 1996). This trough, bounded to the west and southwest by a resistant ridge of rock and to the east and northeast by the Canadian Shield, was formed by slow dissolution of the shallow edge of a thick wedge of salt (Prairie Evaporites).

Marine transgression generated channel and blanket sands, prograding deltaic deposits, estuarine accretion plain deposits, and shallow marine shales being laid over the erosion surface, filling in the topography and creating a series of almost horizontal beds. Horizontal sands and shales may be traced for hundreds of km in all directions, with beds thinning to the NE. The sand structures are ideal SFI targets where oil free, but oil saturated reservoirs may become SFI sites after depletion.

The Mannville Group underlies at least 5×10^5 km^2 in Alberta and Saskatchewan and is overlain by the Early to Late Cretaceous Colorado Group, composed almost entirely of shales and clayey siltstones. This marine sediment package possesses great lateral continuity, and the thickness and ductile nature of these smectitic shales give environmental security to waste disposal projects in the Mannville Group. Slurry waters are constrained to flow laterally, and vertical flow is essentially impossible. Also, the effective sealing characteristic of the shales means that the pore fluids in the underlying Mannville Group are ancient (>30 MYBP), with distant provenance.

Depending on the site, there may be other Cretaceous and Tertiary strata above the Colorado Group. Throughout the region, there are 0-200 m of glacial sediments containing potable water aquifers.

4 INJECTION PARAMETERS, MONITORING

Initial geomechanics information is collected from geophysical logs, seismic data, and cores. Permeabilities, pressures, compressibilities, transmissivities, fracture opening pressure, and fracture closure pressure are determined using step-rate injection tests, pressure fall-off and recovery tests, and reservoir engineering data. The rock mechanics properties of strata are estimated from geophysical data (sonic wave velocities), experience, previously published data, and calculated compressibilities. Parameters depend on interval thickness, on whether the sand is clayey or not, and if the SFI stratum is oil-free, oil-saturated, or partially depleted.

Laterally continuous zones of 10 m or thicker with porosity of 28-30% and mean permeability >500 mD are the best SFI targets. Such a zone can accommodate large volumes of sand (>25,000 m^3), and will easily dissipate pressures. Any permeable zone above the target depth will also blunt upward fracture propagation through rapid bleed-off of pressures.

Fracture behavior is determined from step-rate tests (Dusseault and Bilak, 1997), and both step-rate and pressure fall-off tests are used to assess formation flow capacity (injectivity and transmissivity). Figure 3 is a step rate test for the case discussed below.

During active SFI, which may take place daily for months and be repeated annually, injection rates and WHP (wellhead pressure) for the SFI well, BHP in the SFI well and any observation wells, slurry density, sand-fluid concentrations, and surface tilt from an array of tiltmeters are recorded continuously. Cumulative volumes of water,

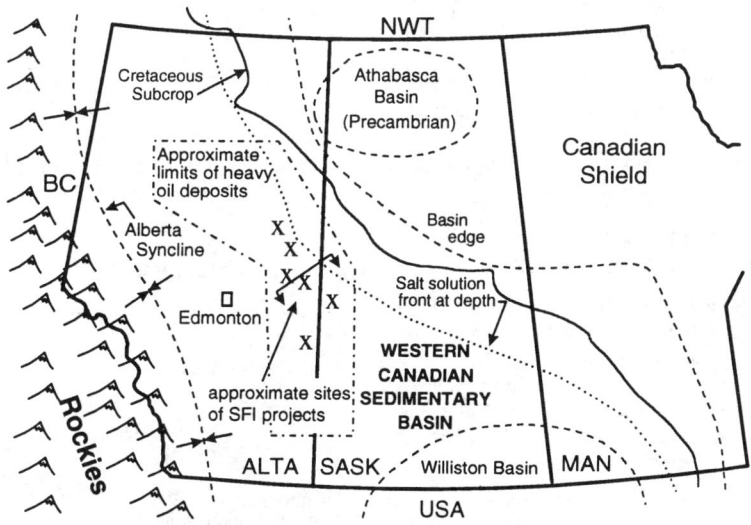

The Western Canadian Sedimentary Basin

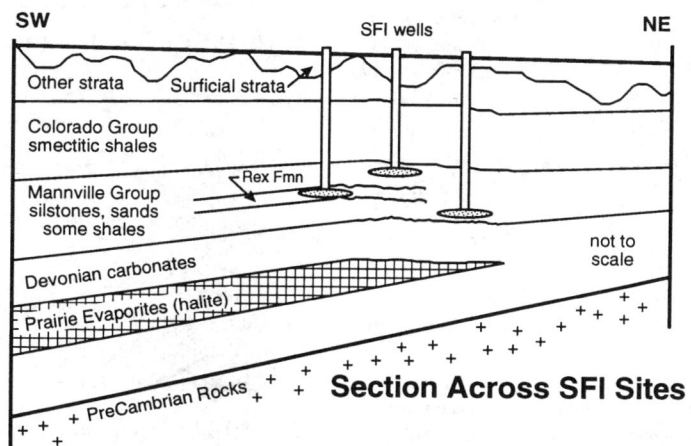

Fig 2: SFI Sites and General Lithostratigraphy

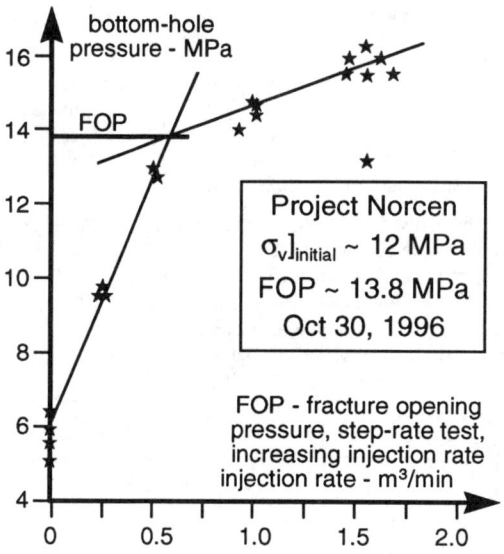

Figure 3: Fracture Pressure Step-Rate Test

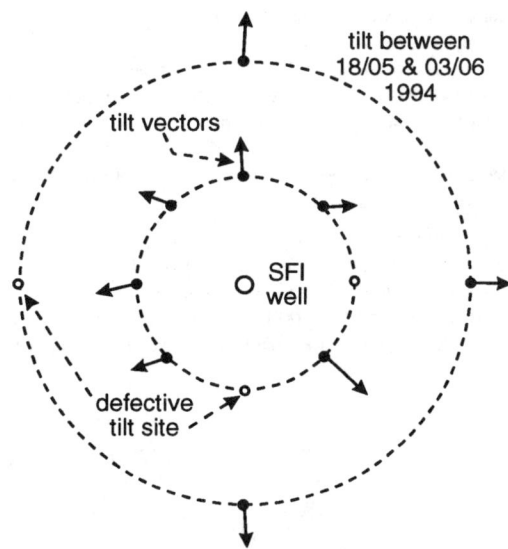

Fig. 4: Tilt data for a two-week SFI period showing mainly horizontal fracturing

sand, and slops (oily fluids) are determined from injection rates, slurry density, and direct volume measurements.

Tilt data are analyzed to estimate fracture orientation (azimuth), attitude (vertical, horizontal, inclined), extent (length, width), and volume (aperture times area). An example of tilt vectors over a time interval is presented in Fig 4; the outward pointing vectors indicate that fracture is dominated by horizontal attitude, and no evidence for vertical fracturing is present in this example.

After each SFI episode, BHP and surface tilt are collected until the next SFI episode. Pressure fall-off analysis (Fig 5) permits reassessment of transmissivity and flow regime, and tilt data allow a direct estimate of the gradual reservoir volumetric diminution. Determining reservoir behavior after each episode tracks reservoir parameter evolution, and this is used to guide the injection-relaxation strategy. Step-rate injection tests may be used occasionally to verify opening and closure pressure.

Given the number of projects to date, only one project is described in detail. The project described in detail is located about 200 km east of Edmonton, Alberta.

5 PROJECT NORCEN

The fine- to medium-grained Rex (Clearwater) Formation sand at a depth of 536-554 m was chosen as the best SFI stratum for Project Norcen. Downhole transducers (35 MPa capacity) were installed in two observation wells and in the SFI well. Surface displacement was monitored by a 12-tiltmeter array in concentric circles of 100, 200, and 400 m radius from the SFI well. From July 25 to August 31, 1996, 14,925 m^3 of slurry containing 3,288 m^3 of oily sand and some slops were injected.

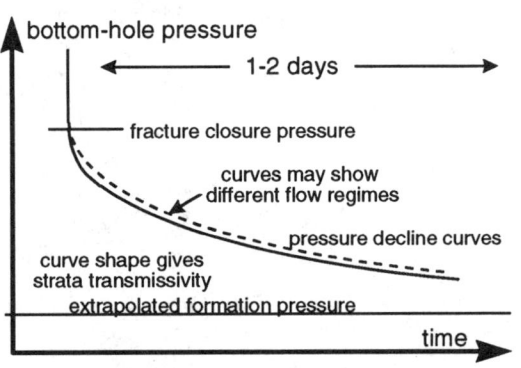

Fig 5: Pressure Fall-off Curve Analysis

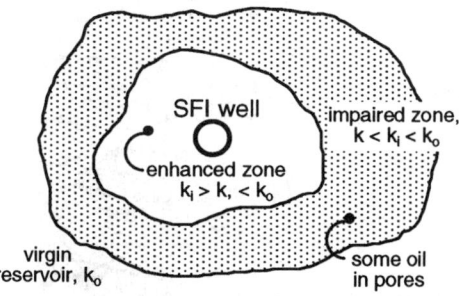

Fig. 6: Differing hydraulic conductivity zones generated around a SFI well

Prior to SFI operations, the Rex had the following characteristics: 990 mD permeability, 3.4 MPa formation pressure (unusually low), and a flow capacity of 22.7 D-m. The well showed an exceptionally large skin coefficient (27), and this masked the step-rate data interpretation of fracture extension and closure pressures (Fig 3). At this depth in this region, the vertical total stress at 536 m is about 11.5-12 MPa and the lateral total stress for an undepleted reservoir is about 10-11 MPa.

Injection was carried out episodically from July 25 to August 31, 1996, with total injected volumes of 3,288 m^3 of sand and some slops, giving 14,925 m^3 of total SFI slurry (Table 2). Injection rates varied between 1.5 and 1.8 m^3/min and slurry densities varied from 1.17 to 1.25 g/cm^3 initially, but values of 1.5-1.65 m^3/min and 1.17 g/cm^3 proved to be optimum. The back-calculated skin coefficient initially was in the range 5 to 28, but optimization of SFI operations (Dusseault et al., 1997) resulted in negative skin factors (-2.5 to -5) for the latter two thirds of the injection period, indicating that the near-wellbore environment had become "enhanced", probably through dilation and emplacement of a region of higher permeability waste sand. An average BHP of 13-16 MPa indicates that formation uplift dominated SFI activity, a condition of horizontal to sub-horizontal (low-angle inclined) fracture planes, confirmed by tiltmeter analyses.

Fall-off test analysis showed that reservoir permeability of about 1 Darcy dropped to about 100 mD around the well (representing a composite response of the wellbore to the pressure fall-off tests) and then stayed more or less in the same range. The wastes being injected were oily produced sands and slops; these materials partially blocked pore throats in a zone around the wellbore beyond the zone of enhancement. In fact, back-calculated permeabilities change with time even if no active injection is taking place because of pressure decline and rearrangement of the oily phase within the pore structure.

Extrapolated equilibrium formation pressure was also determined by fitting a decline model to the data analysis from fall-off curves, shown in Fig 5. These data are reflective of a combination of far-field and intermediate-field responses, combined with evolution of the pressure response into different flow regimes with time and with continued SFI episodes. It is important to remember that the SFI process with solids and oily fluids is generating a dynamic spatiotemporal evolution of reservoir behavior; tracking that evolution in a fully quantitative manner remains a challenging task, although better models are becoming available (Zhang, 1997),. The far-field formation pressure is best determined through adjacent monitoring wells, or through a prolonged fall-off test (several days).

Our conceptual permeability model of the zones generated during this phase of the project is shown in Fig 6. An enhanced zone of higher permeability is surrounded by a region of hydraulic conductivity impairment because of pore throat blocking; beyond this zone, the reservoir is at its original unimpaired state. These zones do not have sharp boundaries, but grade into one another.

Shut-in pressure fall-off analysis after a SFI episode permits an estimate of fracture closure pressure. The fracture gradient was increasing during this time, as expected, considering the large volumes of solids being injected, but values also indicated a vertical fracture tendency, with a gradual evolution toward a tendency for sub-horizontal fractures.

Tiltmeter analyses for a number of selected periods are presented in Table 1. First, note that the values for surface deformation and tilt are extremely small, but the overall tilt response was consistent with the variations in formation flow behavior (this will be the subject of a more detailed article).

6 ROCK MECHANICS ISSUES IN SFI

6.1 *Plasticity in the Target Stratum*

Although a fracture is clearly created during high rate SFI, true fracture length is very short, no more than 10-20 m, because the SFI zone is a high porosity sandstone with extremely large permeability. Seepage forces are high during injection, providing confining force to the target stratum and the previously emplaced sand. However, pore pressures are high and reduce effective stress. We believe

TABLE 1: Approximate Fracture Parameters from Tiltmeter Analysis

Calendar Period, 1996	Shape and azimuth	Type (attitude)	Maximum Distance (m)	Calculated Depth (m)	Δz (max) (mm)	$\Delta \theta$ (max) (μ-rads)
Jul 27-30	radial, no azimuth	sub-horizontal to inclined	< 150	525-587	0.025	0.12
Aug 14-17	elongated, NE-SW	sub-horizontal to inclined	< 200	549	0.037	0.21
Aug 19-21	elongated, NE-SW	inclined	< 200	555	0.009	0.11
Aug 27-30	elongated, NE-SW	inclined	50-100	540-557	0.050	0.33

plasticity processes dominate sand emplacement, and some continued deformation after SFI ceases suggests that plastic extrusion of waste sand from the near-wellbore environment takes place until equilibrium is reached.

6.2 Strains and Bending in Overlying Strata

At present, a consistent solution to stresses and strains in the overburden rocks is not available, but is needed to simulate the bending stresses and displacements in overlying low-permeability rocks to assess alterations in stress and the development of slip along bedding planes.

Induced bending strains in the overburden beam will lead to a focusing of σ_v and a reduction in σ_h above the waste pod (Fig 7). This creates a tendency for vertical fractures to rise, but this tendency is counter-acted by the high bleed-off of the target stratum, which tends to blunt vertical rise. Ideally, another high-permeability stratum exists above the injection stratum, which guarantees fracture blunting unless agents are added to the slurry to reduce bleed-off, something which is not advised.

Overburden bending leads to bedding plane slip as strains are accommodated (Fig 8). This is known from cases of compaction, where slip of weak planes can lead to casing shear (Schwall and Denney, 1994). Given the symmetry of the waste pod, confirmed by tilt meter monitoring, shear slip magnitude is the least just above the SFI well. In operations to date, no casing impairment has been observed in the SFI well, nor in adjacent monitoring wells, which are generally at least 100 m distant.

Bending strains attenuate with distance above the waste pod because of deformation spreading. At the surface above a 500 m deep SFI waste pod, the deformation is spread out over a kilometer, and curvature attenuates to values well below any possibility of bedding plane slip. Furthermore, slip along horizontal planes does not lead to permeability impairment of overlying seals.

6.3 Stress Changes and Rotations

SFI alters the virgin stresses in and around the target zone because massive quantities of solids are forced into fractures. If vertical fractures were initially predicated by the virgin stress regime ($\sigma_{hmin} < \sigma_v$), the permanent lateral strains imposed by SFI lead to the condition $\sigma_{hmin} \to \sigma_2$, $\sigma_v \to \sigma_3$, and dominantly horizontal growth develops. Typical SFI p_{inj} values of 1.05-$1.35\sigma_v$ show that fractures lift the overburden, and also show that there is a stress concentration effect associated with the waste emplacement (Fig 9). Horizontal fracture dominance, restricted vertical growth, and rapid bleed-off serves to contain the solids in the near well-bore environment.

Stress rotation is normally observed in SFI during the first few cycles. However, the waste zone grows radially with time, and strains and curvature generate stress changes which may lead to components of vertical fractures being generated, although the SFI process remains dominated by horizontal emplacement. Cycling between vertical and

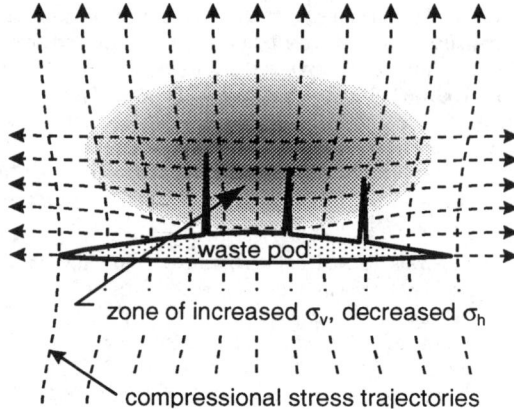

Fig. 7: Stress trajectory focussing around an emplaced solids waste body

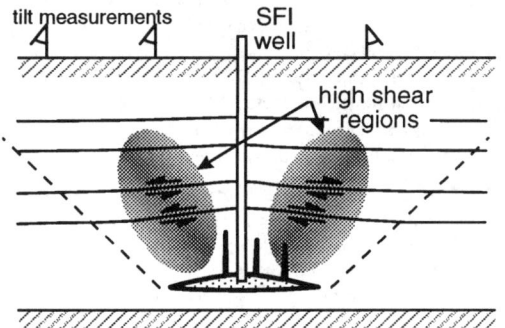

Fig. 8: Deformation patterns and shear slip zones above a solids waste pod

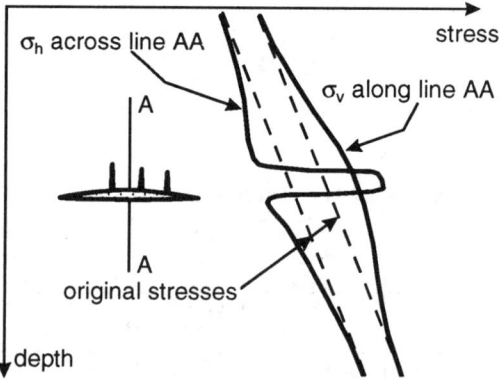

Fig. 9: Stress profiles along centre-line of the emplaced solids waste pod

horizontal fracturing has been observed and analyzed for fluid fracturing (Dusseault and Simmons, 1982), but this cycling is not observed for SFI fracturing because the solids injection leads to permanent straining and does not allow stresses to relax by fluid drainage during a shift from vertical to horizontal fracturing.

The size of the zone affected by the SFI waste emplacement is limited, but stress equilibrium dictates that any stress changes in an interval are equilibrated by opposite stress changes in adjacent intervals, leading to induced stress distributions similar to those sketched in Fig 9.

7 CLOSURE

Slurry fracture injection has proven to be an environmentally secure means for disposal of large volumes of solid waste. The major physical aspects of the process are apparent through monitoring, and solid-fluid coupling behavior is necessary to understand stress changes, pressure response, and the large-scale changes in these factors which accompany the process through time.

SFI is implemented in an excellent sedimentary environment in Canada, but similar strata are available in many areas around the world. Even sites which depart from ideal conditions can accommodate substantial volumes of wastes in a single well. Offshore sedimentary basins may be used, and the ocean provides another security barrier in the unlikely even of fracture breaching (which should never be an issue with proper monitoring).

We believe that SFI can be used for large-volume, low-level, solid radioactive wastes. These wastes could be slurried in aqueous solution with one part waste to 5-10 parts shales fragments, and injected at depths in excess of 1500 m at appropriate sites. Given the right geological conditions, million-year security is easily attained, giving ample time for decay of radioactive species under conditions of minimal flow rates.

8 REFERENCES

Bruno MS, Bilak RA, Dusseault MB & Rothenburg L 1995. Economic disposal of solid oil field wastes through slurry fracture injection, Soc.. of Pet. Eng., SPE #29646, pp 313-320.

Dusseault MB, Bilak R & Rothenburg L 1993. Inversion of Surface Displacement to Monitor In-Situ Processes. Int. J. of Rock Mechanics, Mineral Sciences and Geomechanics Abstracts.

Dusseault MB, Bilak RA, Bruno MS & Rothenburg L, 1994. Disposal of Granular Solid Wastes in the Western Canadian Sedimentary Basin. Deep Well Injection Symp. Berkeley, CA.

Dusseault MB & Bilak RA 1997. Disposal of dirty liquids using slurry fracture injection, Soc. Of Petroleum Engineers, SPE #37907, Proc. SPE/EPA E & P Environmental Conf, Dallas.

Dusseault MB & Geilikman MB, and Roggensack WD 1995. Practical requirements for sand production implementation in heavy oil applications, Int. Heavy Oil Symp., Calgary, pp. 103-112.

Dusseault, MB & Simmons, JV 1982. Injection-induced stress and fracture orientation changes, Canadian Geotechnical J., 19, 4, pp. 483-493.

Geilikman MB & Dusseault MB 1996. Fluid-rate enhancement from massive sand production in heavy oil reservoirs, in J. of Petr. Science & Engineering, Special Issue: Near Wellbore Formation Damage and Remediation, 568.

Geilikman MB, Dullien FAL & Dusseault MB 1997. Erosional creep of fluid-saturated granular media, J. of Eng. Mechanics of ASCE, in press.

Schwall GH & Denney CA, 1994. Subsidence induced casing deformation mechanisms in the Ekofisk Field. Proc. EUROCK '94, Balkema, Rotterdam, 507-516.

Zhang L 1997. Analysis of Fluid Flow to Horizontal and Slant Wells. PhD Thesis, University of Waterloo, Department of Earth Sciences.

Solution cavern disposal for low-level radioactive and toxic civil waste

Maurice B. Dusseault
Porous Media Research Institute, Waterloo, Ont., Canada

Brett Davidson
Brett Davidson Consulting Inc., Cambridge, Ont., Canada

Michael S. Bruno
Terralog Technologies USA Inc., Arcadia, Calif., USA

ABSTRACT: Waste placement in a designed accessible repository probably represents the technology of preference for high-level radioactive materials. For other hazardous wastes, slurry fracture injection and salt cavern disposal are preferred for reasons of cost. Salt cavern disposal has exceptionally high environmental security: there are many suitable sites, salt has low permeability, a dense waste slurry can be designed and placed, salt will close viscously around placed solid wastes, and the rock mechanics properties of salt and designed slurries are well known.

A salt cavern is created by solutioning. Solid wastes in granular form are slurried in a saturated brine and placed through injection procedures into the cavern. When full, the cavern is sealed and decommissioned. This article discusses rock mechanics and environmental aspects of rational waste disposal practice using salt caverns.

1 INTRODUCTION

Vast quantities of money and technical interest have been focused on geological repositories for high-level solid radioactive waste. Such programs have greatly advanced rock mechanics and porous media flow analysis. Currently, the favored solution is placement as canisters in mined repositories in igneous rock (Canada and Sweden), shale (Belgium and France), unsaturated tuff (United States), or salt (Germany, United States). Including all research efforts, high level waste placement will cost at least $US30,000/tonne, maybe much more.

However, high-level wastes comprise a small volume of the total amount of radioactive wastes, and economic disposal means are also required for large volumes of civil and military non-radioactive wastes. If chemical or thermal detoxification cannot be economically used, a repository is required for solid wastes. Realistically, there exist six options: permanent warehousing, surface landfills or quarries, ocean dumping, old or specially built mines, slurry fracture injection, and salt solution cavern placement. We do not consider the first three as genuine solutions. Mine placement, even without physical and chemical pretreatments, appears to be expensive but secure. Slurry fracture injection is the subject of a companion paper in the proceedings of this conference (Dusseault and Bilak, 1997). This article will address rock mechanics and environmental issues in salt cavern waste placement, which we will refer to as **SCD** (Solution Cavern Disposal).

2 THE PROPOSED APPROACH

We have shown that an engineered slurry of granular solid waste, salt, brine, and other additives will compact under stress to a low porosity, low permeability pod (e.g. Dusseault and Davidson, 1997). Solution caverns in salt will close with time through slow creep processes; salt does not fracture during slow creep, and a low hydraulic conductivity ($<10^{-16}$ m/s to brine) is maintained. Dense slurries will flow and fill all cavern voids, yet not show turbulent mixing with clear brine in the cavern. Intact salt will close around the wastes, expelling brine, and a pod of suitably engineered waste will approach 2-3% porosity, with hydraulic conductivity below 10^{-12} m/s.

Based on laboratory research, modeling, and the experience base afforded by existing storage and brine caverns, an engineering approach is proposed. A cavern is dissolved in a salt stratum using a multiply-cased hole (Fig 1), providing a brine-filled cavity. Solid wastes are slurried in brine with granular salt and other additives, proportioned based on waste toxicity level and creep-compaction behavior of the mix. The slurry is placed through a central tailpipe, and displaced brine is disposed in a deep well. The cavern is fully filled, and close-out techniques used to enhance the sealing capabilities of the intact salt. The site is decommissioned, and slow cavern closure eventually results in a waste pod of low permeability entombed in intact salt of even lower permeability. Details of site investigation needs, placement and

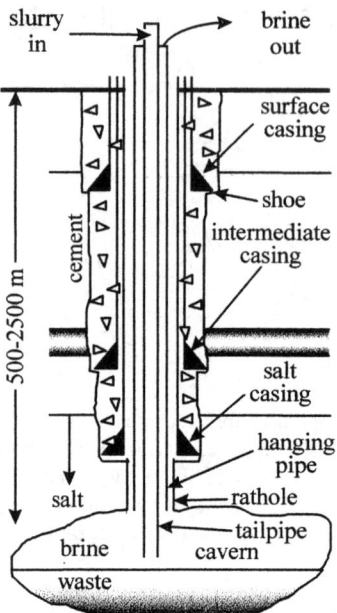

Figure 1: Cased Hole and a Salt Cavern

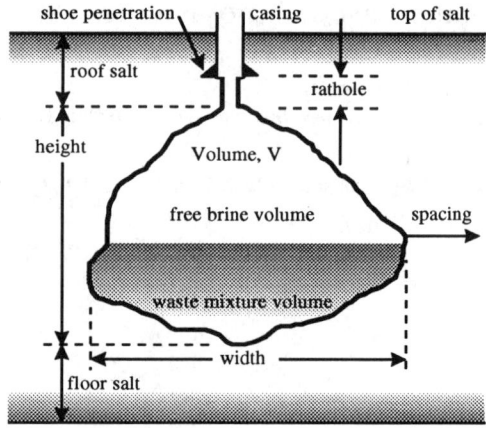

Figure 2: Geometric Design Parameters

cavern close-out approaches have been published elsewhere (Davidson and Dusseault, 1997).

Slurry placement has economic and rock mechanics advantages, compared to dry placement in a "dry" cavern. Slurry placement is economical, wastes do not have to be dried, and a suitable mix is more easily formulated. Furthermore, slurry placement guarantees better cavern filling, caverns are more stable with brine pressure support, and back-pressure can be used to increase support. In a "dry" cavern, natural seepage from high hydraulic gradients along clay and limestone seams means that the waste pod will become brine-saturated in any case.

3 ROCK MECHANICS ISSUES IN SCD

Cavern design, closure simulation and monitoring, stress-compaction behavior of the waste mixture, and overlying strata integrity are the rock mechanics issues.

Cavern Design: The parameters are shape, volume and spacing. These are dictated by roof beam stability, strata geometry (thickness), and the design SCD volume. The latter may be controlled by regulatory agencies, based on waste volumes, toxicity and facilities life. Available volume is equal to the volume of salt dissolved minus the closure volume during cavern development and filling. Figure 2 shows dimensions which can be controlled. Technology exists to create caverns of various shapes, but ellipsoidal caverns are, in general, most stable.

Cavern Closure: Rate controlling parameters are stress, temperature, cavern shape and size, internal pressure, visco-mechanical salt properties, and mechanical properties of thick non-salt interbeds and overburden strata. Closure is more rapid at high stresses and temperatures, in a strongly non-linear manner. For example, a cavern at 2000 m will close ~20 to 50 times more rapidly than one at 1000 m. The temperature at 2000 m is approximately 30°C higher than at 1000 m, and the stress is nominally twice as high.

Closure Monitoring: While the cavern is open (brine-filled), sonar techniques, brine efflux, deformation monitoring and pressure transient methods (Bérest et al., 1996) allow volumetric closure rate to be estimated. After the cavern is filled with waste but still accessible from the surface, brine efflux and deformation monitoring may be used. After decommissioning, remote geophysical techniques (3-D seismic, gravimetry, and deformation) may be used to monitor closure.

Waste Compaction: The important parameters are the stress-density-permeability-time relationships for the compacting waste mixture, determined through laboratory tests and verified by field monitoring.

Overburden Stability: Cavern geometry, overburden proximity (Fig 2), and overburden mechanical properties are dominant parameters. The roof beam must redistribute vertical stresses without caving or collapsing, which could block the placement well. Proper design will eliminate the possibility of caving-to-surface, thus this will not be discussed. Of greater concern is that excessive flexure of the overlying rocks will open joints and generate flow paths from the immediate overburden to higher horizons.

4 ENVIRONMENTAL ASPECTS OF SCD

Once placed, the only waste vector to the biosphere is through fluid flow. Time-dependent hydromechanical properties and geochemical adsorptivity are of interest.

Hydromechanical Properties: Salt, non-salt interbeds, and overburden permeabilities are functions of porosity, grain size, fracture network properties, and interconnectivity. The waste pod permeability is governed by its formulation and state of porosity; fully compacted waste mixtures with 50% salt will have permeabilities several orders of magnitude lower than surrounding porous strata. Large formation fluid volumes in sediments assure that any fugitive brines enter into a vast aqueous reservoir where other processes occur to minimize risk.

Geochemical Adsorptivity: Waste mixtures are formulated to have a high inherent adsorptivity through use of clay minerals (shales) or zeolites. These materials retain polyvalent metal cations or organic species through surface adsorption. Also, overlying strata contain large quantities of clay minerals with adsorptive capacities.

Flow Behavior: The interstratification of beds and the location and sources of hydraulic head govern macroscopic flow behavior. In flat-lying strata, alternating low and high permeability beds direct flow laterally (Fig 3). In the hydrogeological column, shallow waters are fresh and waters near the salt cavern are saturated brine ($\rho \approx 1.20$ g/cm^3), giving density-stabilized flow which is resistant to perturbations. Long horizontal flow paths result, and if fugitive toxins are released, diffusive and dispersive dilution reduce concentrations. As slowly migrating fluids approach near-surface exit points, massive fresh water recharge provides further dilution capacity.

5 CAVERN DESIGN AND CLOSURE BEHAVIOR

Geological stratification may limit the size and affect cavern shape design. The thickness of upper and lower salt barriers and cavern spacing can be chosen directly by geomechanical modeling, but may be predicated by the degree of environmental security required. In general, small caverns are best for highly toxic wastes, larger caverns or existing brine solution fields for large volumes of non-hazardous wastes.

For example, in Ontario, Canada, a chemical company is planning to use an old brine field at 460 m depth for the disposal of non-toxic, saturated brine mineral sludges currently stored in surface ponds (Davidson et al., 1994). Their abandoned brine field has a storage potential of ~7-10×10^6 m^3. Interconnections exist between most wells in the brine field, and 3-D seismic has shown the existence of caving to 120 m above the top of salt. SCD is cheaper than new surface pond development, and will stabilize the old brine field against further subsidence and eliminate any cave-to-surface possibility.

Because of creep behavior and viscous stress redistribution, natural salt has favorable properties for waste disposal. Viscoplastic behavior means that conventional rock mechanics concerns may be inappropriate, or must be modified.

Under high deviatoric and confining stress, salt creeps and does not fracture unless strain rates are in excess of 10^{-5}- 10^{-6} s^{-1}. Mine and cavern closure measurements indicate strain rates slower than 10^{-8}- 10^{-9} s^{-1}, and most salt mines have creep rates in the range 10^{-10}- 10^{-12} s^{-1}. Thus, even a deep cavern will not have an extensive fractured zone beyond a low-confining-stress skin around the opening.

Fractures in salt anneal under stress (Allemandou and Dusseault, 1993). Core specimens with obvious fracturing were subjected to confinement at $\sigma_3 \sim$ 20-28 MPa for 72-96 hours at 23-25°C. Rapid strain tests ($>10^{-5}$ s^{-1}) showed that the "stiffness" of these annealed specimens is over twice that of the core, and the yield stress was 15-20% higher (Fig 4). Volumetric strain measurement during annealing and post-test X-ray tomography showed that porosity is reduced and microfractures close. Also, careful examination of specimens subjected to 5-10% axial creep strain at rates of 10^{-8}- 10^{-9} s^{-1} and stresses on the order of σ_3 = 5-10 MPa, σ_1 = 20-30 MPa, showed no substantial microfissure development. For confining stresses of $\sigma_3 >$ 3-4 MPa and strain rates $<10^{-7}$ s^{-1}, we do not believe that a strain limit for natural salt exists beyond which fracturing occurs; creep fracturing reported

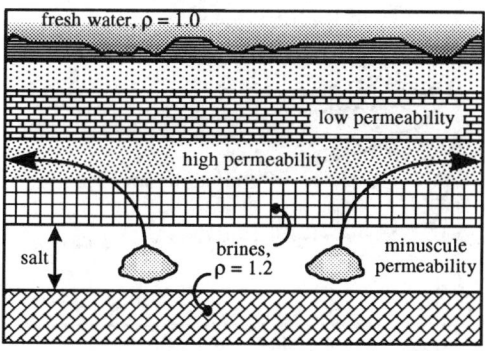

Figure 3: Stratification and Lateral Flow

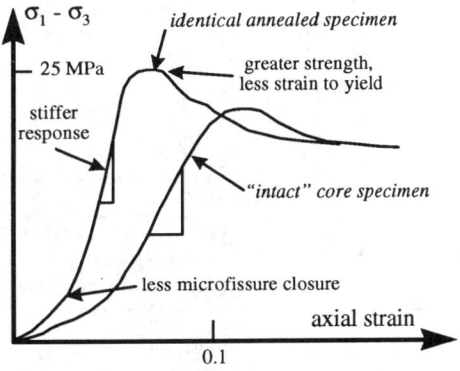

Figure 4: Microfissure Annealing in Salt

for laboratory strains exceeding 15% is the result of specimen barreling and end-cap punching.

In situ, natural stress fields are redistributed so that the regional stresses "flow" around the salt cavern (Fig 5). A thin elastoplastic skin may exist, but the cavern is surrounded by a viscoplastic zone where stress differences are much less than predicted by elastic theory. In a brine-filled or waste-filled cavern, a support pressure is applied to the salt. Intact salt is of exceedingly low permeability, hence pressure permeation into salt micropores is so slow that pressure support acts in an effective stress manner, reducing the deviatoric stress and slowing creep rates. Thus, stable caverns with pressure support ($p_i = \rho g z = 11.8 \cdot z$ for brine) may be created at much greater depths ($z > 2000$ m) than the deepest salt or potash mines ($z = 1000$-1200 m).

These facts give rise to several novel possibilities. First, through back-pressure maintenance it is feasible to create caverns at great depth (>2500 m) yet keep them from closing excessively rapidly. Second, once a cavern is filled, it is possible to accelerate closure rate through pumping to maintain low pressure. Because of the strongly non-linear creep response of salt to deviatoric stresses, reducing brine support from hydrostatic to zero may increase the closure rate by a factor of 10-30 times. Both possibilities may be useful for waste management.

As the salt cavern closes, overlying strata are flexed downward. The induced curvature may open joints to become flow paths (Fig 6). To mitigate this, a suitable lithostratigraphy with competent roof rocks is desired to spread outward the induced strain and reduce curvature. Optimum cavern shape is a vertical ellipsoid to allow strains to spread laterally through horizontally-dominated creep. Roof curvature predictions are amenable to numerical modeling. Cavern filling with a properly formulated waste mixture will reduce ΔV_{max} to ~20% or less of its original volume.

Based on previous experience and knowledge of salt be-

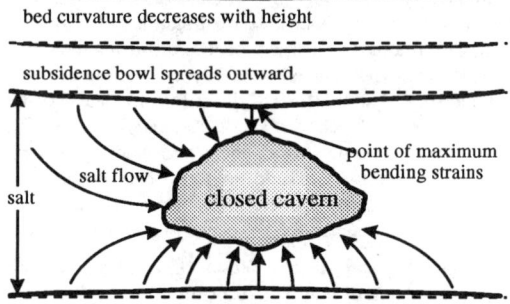

Figure 6: Bed Curvature Above Cavern

havior, salt, we conclude that a secure cavern can be designed and used as a SCD repository with full mechanical competence and environmental security.

6 WASTE MIXTURE BEHAVIOR

Brine-saturated granular salt subjected to stress undergoes creep-compaction; even under modest stresses (<10 MPa) it will eventually approach a condition of low porosity (<3%) and correspondingly low permeability. Figure 7 shows laboratory oedometric results at room temperature for granular halite from the Sifto Goderich Mine in Ontario. Clearly, salt continues to compact slowly even under low stresses. Granular salt creep is dominated by microscopic-scale mass transfer of halite from stressed contacts to free pore walls, a process referred to as pressure solution or fluid-assisted diffusional transport. All natural salts (NaCl) behave in a similar manner.

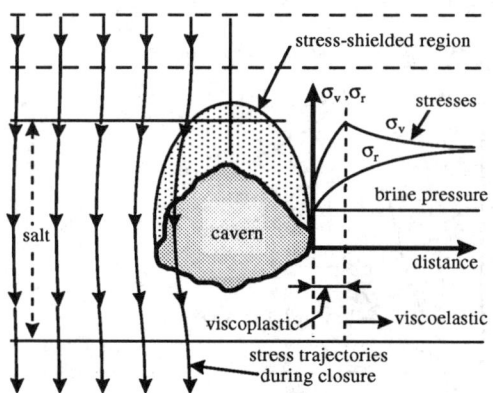

Figure 5: Stresses Around a Salt Cavern

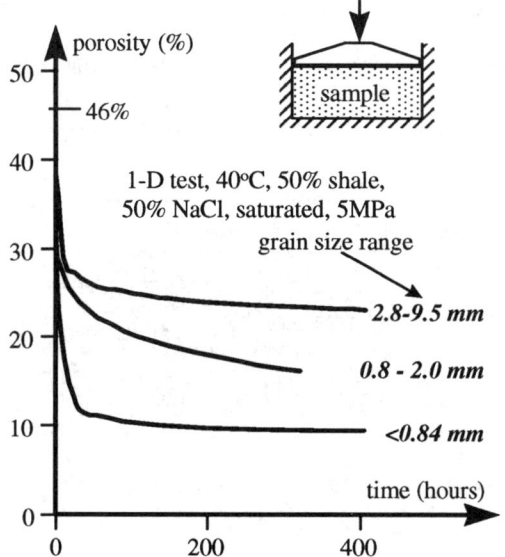

Figure 7: Salt & Shale Compaction

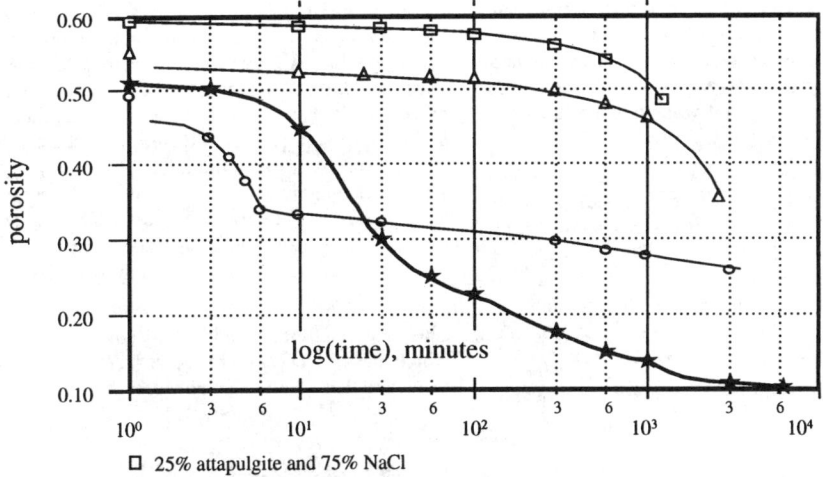

□ 25% attapulgite and 75% NaCl
▲ 50% kaolinite and 50% NaCl
○ 50% ground limestone and 50% NaCl
★ 50% Queenston Shale and 50% NaCl

Figure 8: Creep-Compaction of Granular Salt With Additives

Figure 8 shows the effect of various additives on creep compaction. Admixing shale fragments actually accelerates the creep rate, and these effects occur more rapidly at elevated temperatures. It is apparent that a mixture of ~50% salt and ~50% shale fragments (with some amount of waste material) will eventually approach a condition where the intergranular porosity is 2-3%. (Note that this porosity value does not include the porosity within the shale chips themselves, which becomes isolated as salt creeps and fills the interstices around grains.)

A SCD waste mixture is initially at a porosity of 25-40%, depending on the granulometry. A cavern will typically have a height of 20-60 m and the filling time may be several years, therefore the mixture undergoes self-weight creep-compaction well before cavern closure begins to apply lithostatic stress. The self-weight creep-compaction rate is under investigation, but it appears that an average porosity of 10-20% will be achieved before lithostatic stressing. During this time, expelled brine is withdrawn through the well, and pressure reduction may be used to accelerate closure.

As the cavern closes on the waste mixture, stress transfer begins; this accelerates waste compaction and slows cavern closure. Until porosity is low, the waste pod "stiffness" is minimal, and compaction continues. Although currently under investigation, we believe that a correctly formulated compacting mixture will only begin to substantially retard cavern closure when low porosities (< 5-8%) are reached.

It is feasible to mathematically analyze salt cavern closure behavior with an internally compacting waste mixture. To do this, three-dimensional finite element methods are used, and the constitutive behavior of salt, wastes, and overlying strata are required. The closure predictions may be verified in the field through monitoring. Figure 9 is a conceptual prediction of waste densification in an SCD operation: specific values for the time axis depend on details of depth, temperature and cavern geometry.

7 ENVIRONMENTAL CONSIDERATIONS

During cavern filling and pre-sealing closure, expelled brines are used to formulate more slurry, and excess brine is injected into permeable strata below the salt horizon. These brines may have some amounts of toxic materials, but we note that the solubility of materials in a

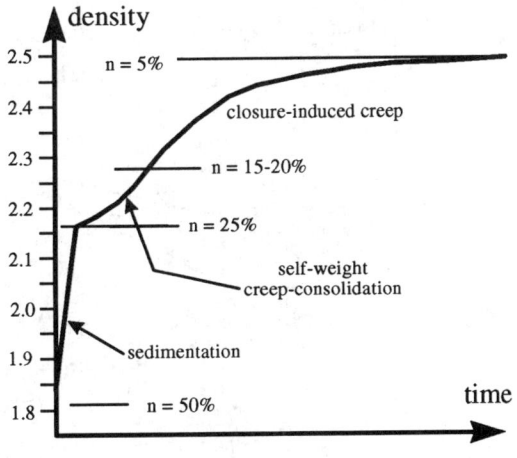

Figure 9: Conceptual Waste Densification

saturated anoxic brine is less than in surface water, particularly in oxygenated and slightly acidic rainwater. The volume of brine to be disposed is somewhat less than the cavern volume; compared to other liquid disposal operations, this is a trivial volume and is inconsequential considering the porosity in situ.

Between sealing and final cessation of closure, a small amount of brine (<10% of the cavern volume) will continue to be slowly expelled into surrounding strata through slightly permeable interbeds. There may also be pressure increases (Bérest et al., 1996) in the cavern, but this attenuates with time, and appreciable amounts of formation fluids cannot flow back into the cavern, nor through it. The solid wastes are of course immobile.

Once the SCD facility is fully closed, the low permeability waste mixture is encased in a salt stratum of even lower permeability. The waste is held in place as a solid under overburden stresses, and cannot be remobilized under any realistic scenarios. The salt strata are bounded by beds of higher permeability with saturated pore fluids and low pressure gradients, therefore leachate generation and transport will be minuscule. SCD in a geologically stable site not subject to active solutioning means security for geological times (>10^6 years). This figure sounds high, but fully stable salt strata older than 200×10^6 years are found worldwide in horizontally stratified sedimentary basins.

There is a number of additional environmental factors which make SCD yet more secure from an environmental view, and these factors are now presented.

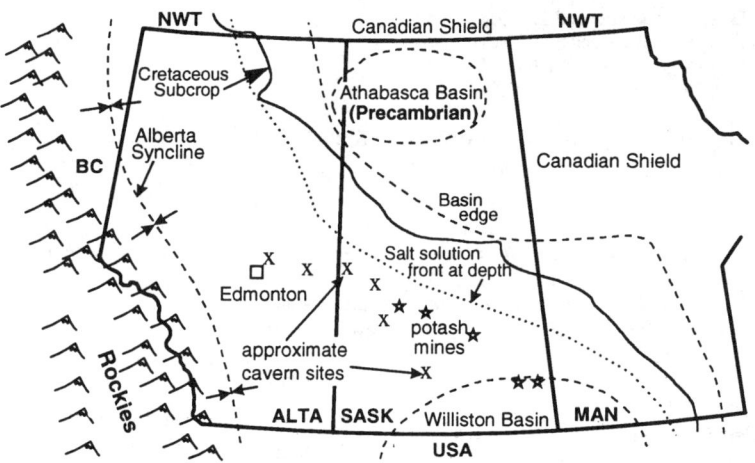

The Western Canadian Sedimentary Basin

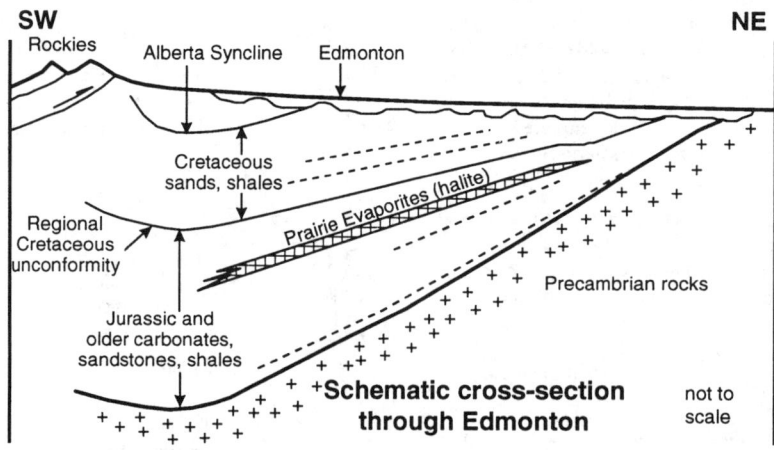

Figure 10: Salt Strata in Alberta and Saskatchewan

Suppose that some fugitive leachate does exit the salt stratum. In a horizontally stratified sedimentary environment, flow is constrained to follow flat-lying sandstone and limestone beds. For example, in Canada (Alberta and Saskatchewan, Fig 10), deep, thick Devonian salts are widespread and for the most part geologically stable in non-faulted regimes. (The up-dip northeastern edge of the salt strata is undergoing slow but active solutioning.) In the overlying strata, flow is horizontal and at rates of millimetres to centimetres per year. Formation brine age in many cases exceeds 30×10^6 years, and the distance to exit points is hundreds of kilometres.

Clayey shale fragments may be added to the waste mixture to provide adsorptive capacities for metallic cations or organic molecules entering aqueous solution in the waste pod. Specific proportions of waste and shale depend on waste type and shale properties. Also, during flow in formation waters, any fugitive toxic species will contact clay minerals interstitially in sandstones, as beds of shale at depth, and as stiff clays nearer the surface. Given flow paths of tens to hundreds of kilometres for optimum sites, the retentive and retardant capacities of the sediments, combined with dispersion, diffusion and dilution, obviate significant biosphere interaction.

We believe that the environmental security of SCD is so large that it is irresponsible to incur the risks associated with shallow placement, given such an alternative. The only realistic environmental danger is a surface spill, and this is a risk attendant to any waste disposal approach.

8 MONITORING AND REGULATORY CONTROL

Leaving aside transportation and site management issues, the regulatory issues related to rock mechanics involve verification that wastes are properly formulated and placed and that the cavern closure onto the wastes is proceeding as designed. Conventional methods such as rate meters, slurry densimeters, waste mixture analysis, pressure sensors in the tubing and return annulus, monitor wells, and other methods must be chosen and implemented properly. Cavern closure rate and the response of surrounding strata are major aspects.

Direct cavern volume tracking using volume balance of placed wastes and displaced brines is necessary, and precision volume rate meters are installed on injection lines and outflow lines. On the surface, a net of survey monuments and tiltmeters is installed so that surface ground movements can be traced with a high precision. During the entire life of the SCD facility, these data are collected and analyzed to determine closure rates (Dusseault et al., 1993). While the cavern remains substantially brine-filled, the waste-brine interface location is measured using open-hole sounding methods and cavern volume is estimated using sonar methods and transient pressurization techniques. Thus, four independent methods can be used to track closure; surface displacement, volume balance, sonar, and transient pressure pulse testing (Fig 11).

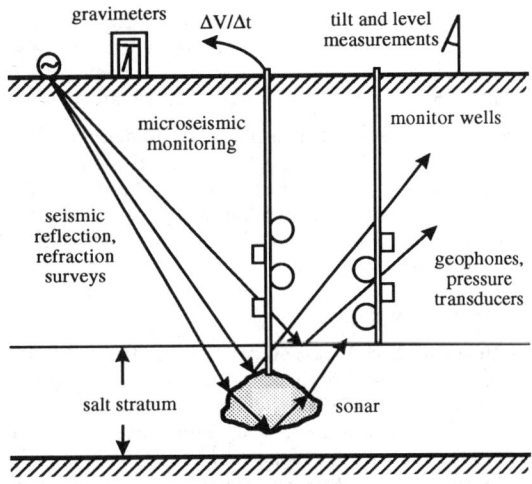

Figure 11: Cavern Closure Monitoring

During final filling stages, outflow lines are monitored to insure that no solids are returning to the surface. If the cavern is depressed to accelerate closure, pumped volumes are accurately measured. At this time, sonar or pressure techniques are no longer possible, but surface deformation inversion can be used.

Once sealed and decommissioned, long-term slow final closure (perhaps 10% of the original volume) is monitored using surface deformations, and three-dimensional seismic surveys can be used to examine velocity changes in the waste pod. Wave velocity will continue to rise slowly as the porosity declines.

During cavern creation, filling and closure, if there is any concern as to the behavior of the overburden, this is best monitored through the deformation field and by using microseismic monitoring. Excessive flexure of overburden will result in accelerated microseismic activity, and the amount of flexure is back-calculated from the surface deformation data.

9 FINAL REMARKS

Solution cavern disposal is a relatively economic and environmentally secure method of disposing of toxic wastes. The formulation of the waste mixture is a key factor to assure a low final porosity, waste dilution, and entrapment. Cavern design and operation in salt is an area where there are now excellent material constitutive behavior data, suitable numerical models, and a large amount of operating experience. The rock mechanics of SCD can be fully modeled and verified in practice through appropriate monitoring.

The major rock mechanics issues are the closure rate of the cavern and the structural and flow integrity of the overlying rocks. Stability of the cavern itself is not an issue if a specially designed cavern in a suitable geological environment is envisioned. A major feature of the use of salt caverns is that salt is of extremely low permeability and experiences self-annealing under stress.

We suggest that the high degree of environmental security afforded by SCD warrants consideration as a technology of preference for toxic wastes. Other methods, such as slurry fracture injection, are more suitable for non-toxic or mildly toxic wastes, and are substantially more economic than SCD.

10 REFERENCES

Allemandou X & Dusseault MB_ 1993. Healing Processes and Transient Creep of Salt Rock. *Int. Symp. on Hard Soils and Soft Rocks*, Athens, Greece.

Bérest P, Bergues J, Brouard B, Durup G & Guerber B 1996. A tentative evaluation of the MIT. *Proc. Spring Meeting Solution Mining Res. Inst.*, Houston TX, 23p.

Bérest P, Brouard B & Durup G 1996. Behavior of sealed solution-mined caverns. *Proc. Eurock '96*, Turin Italy. Balkema, Rotterdam, XXXX.

Davidson BC, Dusseault MB & Demers RJ 1994. Solution Cavern Disposal of Solvay Process Waste. *Proc. 1st Int. Conf. on Waste Geotechniques*, Edmonton, Alberta.

Davidson BC & Dusseault MB 1997. Salt solution caverns for Petroleum Industry wastes. Proc. *SPE/EPA Expl. And Prod. Environmental Conf.*, Dallas TX, SPE #37889, 51-61.

Dusseault MB, Bilak R & Rothenburg L 1993. Inversion of Surface Displacement to Monitor In-Situ Processes. *Int. J. of Rock Mech., Min. Sci. and Geomech. Abst.*,

Dusseault MB & Bilak RA 1997. Slurry fracture injection for the disposal of large volumes of low-toxicity wastes. *Proc. ARMS '97*, in press.

Dusseault MB & Davidson BC 1997. Salt solution cavern disposal for toxic solid mining wastes. *Proc. Tailings and Mine Waste '97*,

Veil, J., Elcock, D., Raivel, M., Caudle, D., Ayers Jr., R.C., and Grunewald, B.: "Preliminary Technical and Legal Evaluation of Disposing of Nonhazardous Oil Field Waste into Salt Caverns," report, *U.S. Department of Energy Office of Fossil Energy, Contract W-31-109-ENG-38*, Argonne National Laboratory, Washington, D.C., U.S.A., (1996).

Disposal of smelter slag into a mined-out pit: A rock stability study

R.Ciccu & P.P.Manca
Department of Geoengineering and Environmental Technologies, University of Cagliari, Italy

ABSTRACT: In view of the next decommissioning of the present slag disposal landfill by a Sardinian smelter, a new site has been singled out in an old mining area nearby, where a large pit exists having suitable storage capacity and favourable geometric features for the placement of industrial waste. However the site is characterised by the presence of a major fault and of some underground works up to few metres below the surface. Therefore, particular attention had to be focused on the possible collapse of the foundation rock under the increasing weight of the future overlying material, eventually causing the rupture of the impermeable layer required for the complete sealing of the waste bulk. As a consequence, pollution from residual heavy metals is feared to contaminate the surrounding groundwater. In order to assess the stability conditions during the years and to evaluate the risk of failure, a geotechnical study has been carried out using the FLAC code which enabled to predict the stress and strain behaviour of the rocks involved.
The paper illustrates the results of the investigation and discusses the various aspects of the problem.

1 STATEMENT OF THE PROBLEM

About 130,000 tons per year of slag are currently produced by a lead-and zinc smelter located in the South West of Sardinia. At present they are disposed of in a dump site near the plant, which however has almost reached its full storage capacity.

A new site has been found in an area about 15 km away where both underground and open pit excavations have been carried out in the past for the production of a complex ore, beneficiated in a centralised processing plant.

After the progressive closure of the mining activity, a relatively large extension of land has been left abandoned for years, causing a severe damage to the environment consisting in landscape deterioration and in soil and water pollution.

Now a general reclamation plan has been set up for the recovery of the land over the area surrounding the city of Iglesias, in which the decontamination of soil, the protection of ground water resources and the re-vegetation of pit slopes, tailings embankments and waste rock dumps are envisaged.

This involves important earth moving operations for reshaping the surface and placing suitable soil layers before plantation. In particular, some large open pits should be filled in order to avoid the accumulation of rain water that may become contaminated upon contact with the ore-bearing strata, thus further aggravating the risk of water pollution by heavy metals.

This situation suggests that a combined solution of the two problems can be found by using the available space for placing the smelter residues in such a way as to rebuild the original surface and to create an artificial hill, achieving the advantage of burying the exposed metal-bearing area and of diverting the stream water away.

Of course the landfill body must be completely isolated since the slag material is potentially harmful and, in agreement with the Italian legislation, an impermeable mantle must be placed over the landfill area before and after dumping, spreading and compacting of the waste bulk by layers.

The stability over time of the landfill's foundation strata is of a great importance for the feasibility of the project since the duration and reliability of the impermeable cover can be adversely affected due to shear or tensile forces generated by differential displacement, to which the materials used are poorly resistant.

In the site selected for the new landfill, the rock movement under the load resulting from the own weight of the foundation strata and of the future overlying deposit is likely to be affected by the presence of a major fault and of some openings underground, the collapse of which would produce a certain subsidence at surface.

Therefore a geotechnical study had been carried out in order to define the problem and to predict the pattern of expected displacement.

2 FEATURES OF THE PRESENT PIT

The dump site consists of a pit having a roughly elliptical shape, about 450 m long and 35 m wide at the base, with steep walls, one of them corresponding to the fault's plane and the other to a series of benches left by the mining activity. The average depth is about 50 m from the contour line at the border.

The bottom is rather irregular, reflecting the state of the excavation at the time of closure some twenty years ago. Slope is generally stable except for some points where landslide phenomena have occurred in the form of block fall. Subsidence due to caving of some sections of the development drifts underneath have also been recorded. No water sources are present in the inside area of the pit and only some seepage occasionally appears on the walls.

The prevailing rock exposed is a Cambrian sandstone on the walls whereas at the bottom the zinc-bearing ore rich in pyrite inside a carbonate gangue is found.

3 LANDFILL DESIGN

The waste material will be disposed of by horizontal layers until reaching the final outline after about ten years of activity. Each layer is delimited by a perimeter embankment, 5 m high, built with waste material from previous mining available in the surroundings.

Preliminary work will consist in the modelling of the entire area by removing the loose blocks and fragments and by smoothing the surface inside the pit by earth moving and levelling according to a suitable profile allowing the flow of water to the drain points. Whenever required, a cave-in of the near-surface sections of the upper underground galleries will be induced by blasting.

Then the entire area of the landfill will be prepared following the conventional method suggested by the consolidated experience in the field. A bed of clay 0.5 m thick will be placed over a sand layer and then the impermeable blanket is applied, consisting of two HDPE sheets sandwiching a sand seam into which a set of pierced drain pipes is buried for the collection of the percolate leaks in case of rupture of the upper sheet. Above the sealing blanket, the main percolate drainage system is laid, again consisting of a network of pierced pipes hosted inside a sand bed (Acquater, 1997).

4 GEOLOGIC FEATURES

The original ore body consisted of a series of lens-shaped masses containing sphalerite with abundant pyrite, inside a carbonate seam hosted between the Cambrian sandstone formation. The deposit is genetically connected to a main fault and it is crossed by some secondary faults and tectonic disturbances.

Based on the results of reconnaissance drilling, it was first decided to mine the deposit underground by sub-level caving from development drifts vertically spaced about 40 m and connected by raises, further divided into 6-8 m sub-levels for the sake of the mining method. However, when the production activity was started by enlarging the upper sub-levels over the entire width of the mineralised area, it was soon evident that the roof could not remain stable and tended to collapse. Moreover the working environment was very bad due to high temperature and to the presence of sulphur dioxide from the combustion of the pyrite, in spite of the attempts to improve the quality of the air by forced ventilation.

It was then decided to mine the ore at surface by opening a pit that was progressively sunk until the cessation of the activity due to the impoverishment of the ore. Regarding the hydrogeology of the area, the reservoir rocks are represented by the carbonate formations into which karst phenomena are well developed, whereas the other geologic formations (sandstones and schists) are basically impervious. Therefore it is believed that no adverse interference exists with the major water sources present in the neighbours of the mine, as confirmed by stream sampling and water table survey (Miniere Iglesiente, 1997).

5 GEOMECHANICAL STUDY

In order to assess the present stability conditions and

to predict the behaviour of the foundation rock during the formation of the landfill and after the completion of the waste deposit, a geomechanical study has been carried out examining the situation at three different steps:
- at the end of the past activity underground;
- on cessation of the subsequent mining operations at surface;
- after the future completion of the landfill.
Results are illustrated here below.

5.1 Material characterisation

After a careful and detailed survey, the rocks occurring in the area have been classified according to the methods of analysis currently applied for the prediction of stability conditions over long time in view to the development of excavation works.

The most interesting data are those obtained by following the RMR rating criterion proposed by Bieniawski, according to which rocks can be classified into five groups delimited by the corresponding range of variation of the RMR index, each of them showing a distinct geomechanical behaviour.

The picture provided by the analysis for the rocks into consideration is somewhat variable from place to place although both the barren rock and the ore-bearing formation can be included into class IV (moderately unstable) and only occasionally into classes III (stable) or V (unstable).

Class IV is that including weathered and intensely fractured rocks showing progressive instability over time.

Accordingly, suitable actions have been devised in order to eliminate the risk of instability on the pit slopes, consisting in the accurate scaling of the dangerous zones taking into account the orientation of the three main families of fractures recognised in the area, on which the kind of instability (either toppling or planar or wedge sliding) chiefly depends (Manca et al. 1993).

The problem is rather more complicated at the bottom of the pit due to the presence of the openings underground and thus a more detailed stress-strain analysis had to be carried out using the 2D-FLAC code.

Concerning the slag to be disposed of, the relevant properties are the following:
- Weight of the unit volume (loose): 18.5 kN/m^3
- Angle of shear resistance: 38°

The stability analysis for circular slide carried out using the conventional Spencer method for the designed slopes of 35° showed that the stockpiled waste will remain stable with a minimum safety factor of 1.373

5.2 Computer simulation

The 2-D FLAC code for the assessment of the stress-strain pattern in the landfill's foundation rocks was applied taking into consideration a study volume 210 x 200 m^2 in cross section, inside which a grid of 15,000 elementary cells was arranged as shown in figure 1.

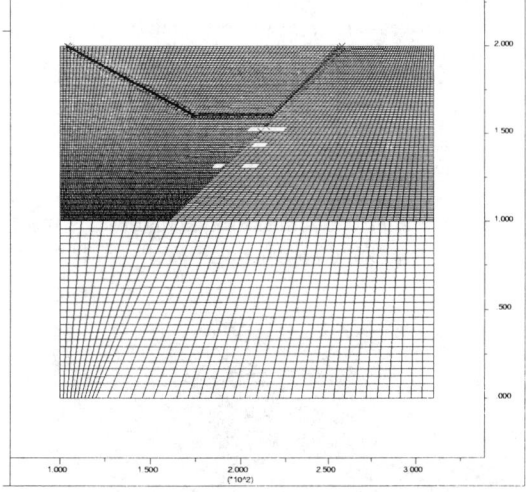

Figure 1. Gridpoints for the application of the FLAC code

The model adopted was that of a Mohr-Coulomb elasto-plastic behaviour assuming the parameters given in the following table 1.

The assumed fault properties are reported in table 2.

Table 1. Physical and geomechanical characteristics of the ore body and the country rock.

Property	Value	
	Ore body	Country rock
Volumic mass [kg/m^3]	4,000	2,500
Tensile strength [kPa]	100	100
Compressive strength [kPa]	770	770
Elastic modulus [GPa]	0.5	0.5
Poisson coefficient	0.3	0.3
Bulk modulus [MPa]	195	195
Shear modulus [MPa]	240	240
Angle of internal friction [°]	35	35
Cohesion [kPa]	200	200

Table 2. Geomechanical characteristics of the fault.

Property	Value
Shear stiffness [GPa/m]	0.2
Normal stiffness [GPa/m]	0.2
Angle of internal friction [°]	20
Cohesion [kPa]	0.2

The analysis has been carried out on different cross sections of the ore body. In particular two most critical situations where excavation underground was considerably extended sideways leaving an exposed roof exceeding 15 m have been studied in details. In the first case (section A) the upper sub-level was later destroyed by the open cast mining, whereas in the second case (Section B) three sub-levels still exist below the bottom of the present pit.

5.3 Stress-strain assessment

Computer results are summarised in figures 2 to 7 for different stages of development.

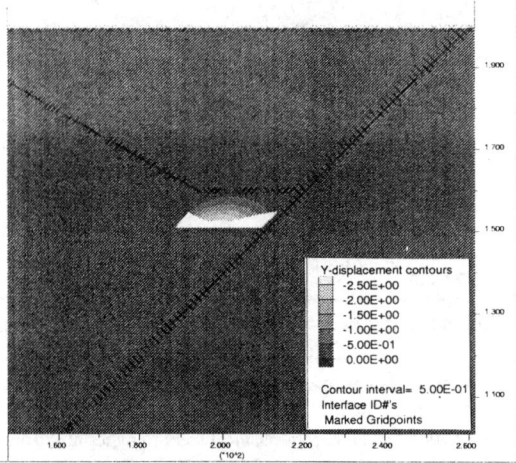

Figure 2. Section A. Pattern of displacement after the underground excavation. Contour interval: 0.05 m

For section A no further analysis was necessary, since the two enlarged galleries have been found to collapse and therefore the landfill's foundation can be considered stable in that area, provided that the precarious openings are completely filled in order to prevent further caving-in. This can be done by inspecting the real state of things underground by means of drill holes or by geo-radar survey, since a direct access could be too dangerous. If necessary,

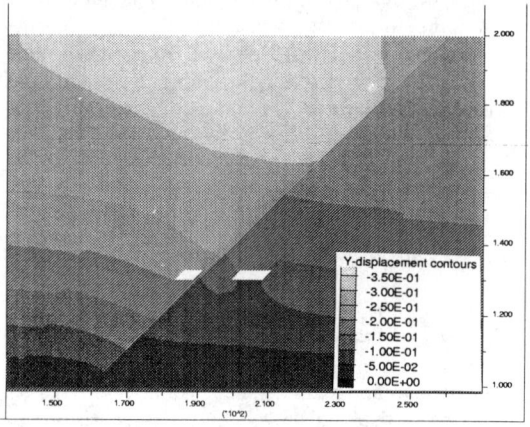

Figure 3. Section B. Pattern of displacement after the excavation of the lower sub-level. Contour interval: 0.05 m

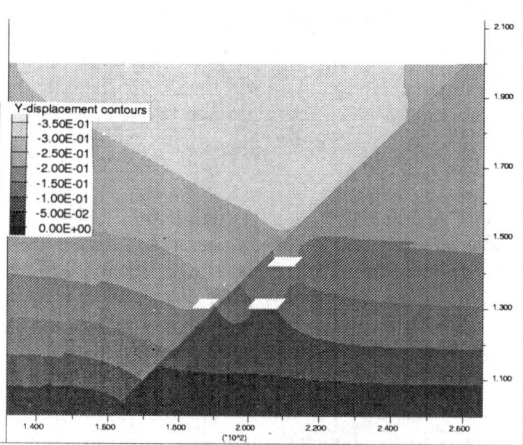

Figure 4. Section B. Pattern of displacement after the excavation of the intermediate sub-level. Contour int.: 0.05 m

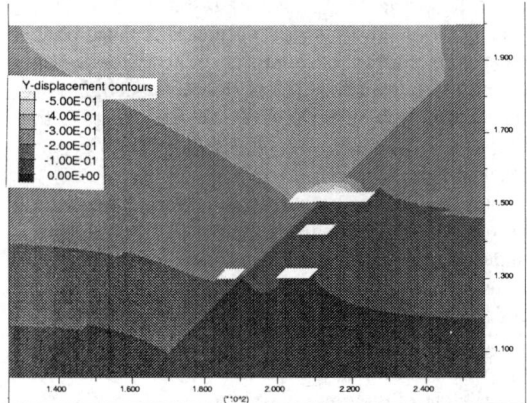

Figure 5. Section B. Pattern of displacement after the excavation of the upper sub-level. Contour interval: 0.10 m

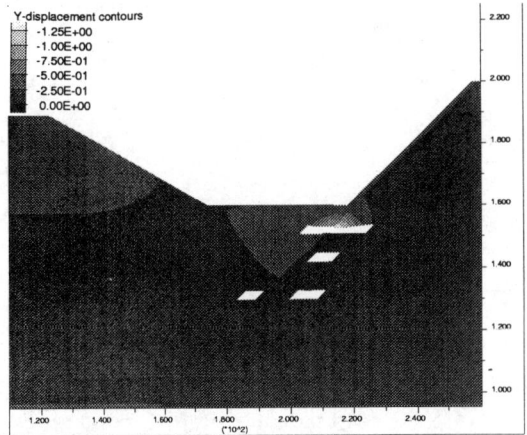

Figure 6. Section B. Pattern of displacement at the ultimate pit stage. Contour interval: 0.25 m

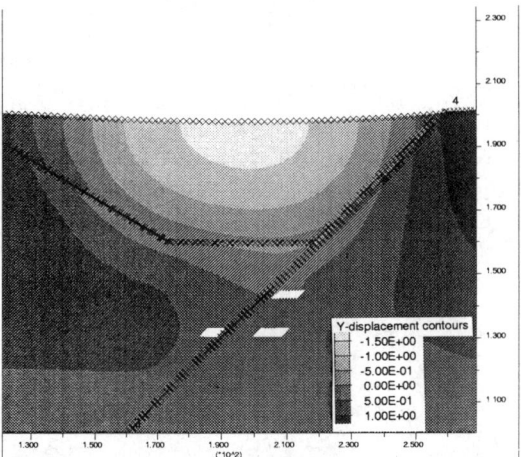

Figure 7. Section B. Pattern of displacement after total back-filling of the upper sub-level and the future completion of the landfill. Contour interval: 0.50 m.

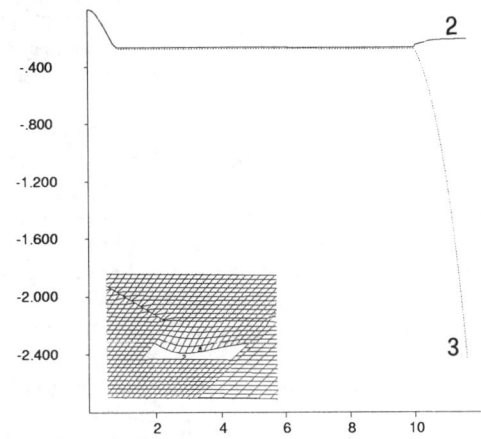

Figure 8. History of roof (3) and floor (2) dispacements (in meters) vs. number of steps (10^3) for the galleries of section A.

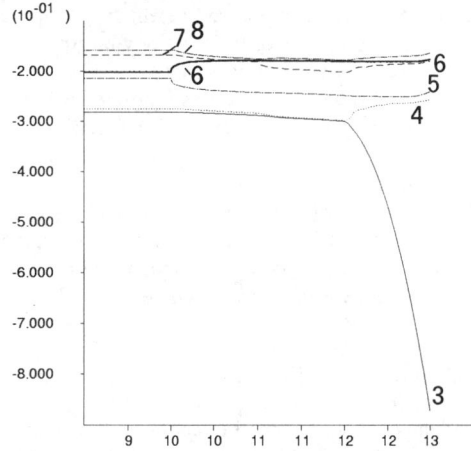

Figure 9. History of roof and floor dispacements (meters) vs. number of steps (10^3) for the galleries of section B: (7,8) lower sub-level, (5,6) intermediate sub-level, (3,4) upper sub-level.

autogenous back-filling can be induced by means of taylored blasts.

5.4. *Convergence prediction*

Displacement of roof and floor midpoints in each of the sub-level drifts has also been assessed: For the two sections examined, FLAC silulation results are reported in figures 8 and 9.

Graphs show that, while a stable condition is reached for the lower sub-level, the roof of the upper gallery collapses. The effect of each additional opening in upwards sequence is also clearly evident.

Steady-state convergence is of the order of few mm for the lower gallery, slightly smaller for the intermediate one and increases rapidly for the upper gallery.

6. DISCUSSION AND CONCLUSIONS

The most important aspects put into light by the computer analysis can be summarised as follows:

- The underground stopes corresponding to the upper sub-level drifts close to the present pit bottom are to be considered as unstable over long time and they are likely to have caved-in meanwhile, at least in part;

- the galleries at a depth exceeding 10 m from the pit bottom show stability conditions and appear to have been little influenced by possible collapse events above them;

- in-pit filling with the waste slag according to the landfill design does not further affect the situation below the presently stable levels. Actually, stresses generated by the landfill load on the foundation rocks are considerably lower than those originally present before undertaking the excavation at surface, since the density of the ore-bearing rock removed is about two times higher than that of loose slag to be disposed of.

Local instability phenomena so far occurred can be attributed to the concomitant influence of:

- too small thickness of rock between the pit bottom and the gallery's roof beneath;
- roof exposure exceeding 10 m in width;
- presence of the fault.

All these adverse factors are present in the upper sub-levels along the axis of the ore body but they are reduced in importance deeper below.

Although all suitable actions ensuring the complete filling of unstable openings is recommended, it can reasonably be assumed that the process of ground settling initiated twenty years ago as the consequence of the mining activity has already reached a steady state. This is confirmed by computer simulation results and it is witnessed by the rapidity with which the first instability events have occurred.

ACKNOWLEDGEMENTS

The preliminary study and the landfill's design have been carried out by Aquater on behalf of ENI-Risorse, to whom the Authors are grateful for the permission to publish a summary of the geo-technical study.

REFERENCES

Aquater SpA: Relazione tecnica generale al progetto esecutivo Discarica tipo 2B in località Genna Luas, Confidential, 1997.

Manca P.P., Marcello A., Massacci G. 1993. Fenomeni di subsidenza da coltivazioni minerarie: una panoramica della situazione in Sardegna. *Atti del II Congresso Italo-Brasiliano d'Ingegneria Mineraria*, S. Paolo (Brasil), pp 499-525.

Miniere Iglesiente SpA: Studio geologico e idrogeologico dell'area Genna Luas e Funtanaperda. Internal report, 1997.

Heat, gas, water flow and contaminant transport problems in caverns and tunnels

3D flow simulation for optimizing environment of tunneling faces

B.Y. Kim & Y.D. Jo
Korea Institute of Geology, Mining and Materials (KIGAM), Taejon, Korea

S. Nakayama
Kyoritsu University, Fukuoka, Japan

ABSTRACT : In general, tunneling workings are located in deep underground isolated from the main ventilation networks. Workers in tunneling faces are exposed to the air contaminated mainly by the contaminants from diesel exhaust or after-gas of explosives. To tackle with this matter, it is essential to figure out the exact phenomenon of gas movement in the concerned spaces. This paper presents the trend of gas movement in the dead-end workings using diesel equipments to provide informations for designing adequate ventilation system. For analyzing of gas movement, a newly developed software(3D-Flow) for 3-dimensional flow movement was used which is based on computerized fluid dynamics.

1. INTRODUCTION

Ambient air of tunneling faces, which are usually isolated from the main ventilation system, are severely contaminated mainly by diesel exhaust or after-gas of explosives etc. Especially, toxic contaminants such as NOx and diesel particulate matter(DPM) known as a carcinogens are definitely hazardous material for the workers' health. Heavy duty diesel equipments are widely employed in recent tunneling workings for taking advantage of their merits in view of mobility and efficiency.

According to the survey of local mines, the average production of CO and NOx of diesel equipments are ranging 155-267 ppm and 210-315 ppm respectively. It was known that the gas concentration increases along with engine revolution and the governing contaminant is NOx instead of CO. But, probably in the near future, DPM will be the more critical material to be tackled with.

To dilute the contaminated air, mechanical ventilation is required, but it is not easy to design a reasonable ventilation method and flow rate. As of now, the required ventilation flow rate has been calculated based on average volume concept but, it is not believed as a scientific approach.

This paper presents how to figure out the trend of gas movements in tunneling faces and calculate the appropriate ventilation flow rate in scientific manner. For predicting gas movement, a 3-dimensional analyzing software was used.

The remarkable outcome of a series of simulations is that the gas concentration in the dead end faces ranges 5 - 15 % of the immediate concentration out of the engine exhaust pipes.

2. CHARACTERISTICS OF DEAD-END AIR FLOW

In case of blowing type ventilation system, the jet flow from the ventilation duct is not an ideal free jet because of complicated surroundings such as confined wall and equipments employed. Practically, the jet from the duct mouth along the side wall appears as a wall flow, and the flow at the face is scattering and then forms return flow to the beneath of the duct.

To simulate such a complicated flow by the computerized fluid dynamics, a special algorithm is required which can count on both of free jet and wall jet concurrently.

3. SIMULATION PROGRAM

3-dimensional simulation program(3D-Flow) for air movement was coded by FORTRAN and C-language based on k-e high Reynolds number turbulent flow model which is commonly accepted world wide. This program can treat laminar flow, turbulent flow, heat transfer and gas movements. Geometrical conditions such as fluid, free end and symmetrical section can be counted on.[1]

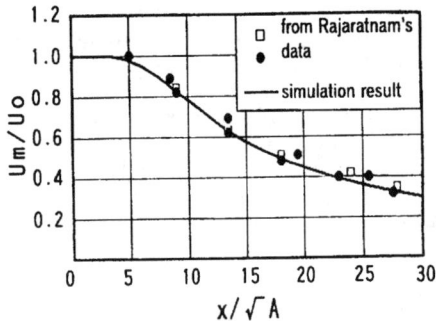

Fig. 1. Distribution of Maximum Velocity of Wall Jet

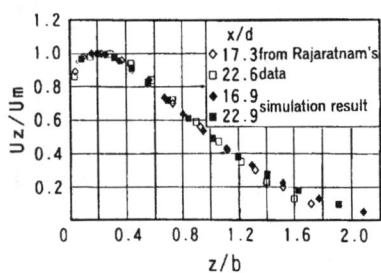

Fig. 2. Distribution of Horizontal Velocity perpendicular to Wall

The reliability and accuracy of this program had been verified by a model test. Figure 1. shows that simulated attenuation of maximum velocity on the axis of wall jet is almost coincident with the test result presented by Rajratnam. Um and Uo represents maximum flow velocity at the distance x from duct mouth on the axis of jet and flow velocity at the duct mouth respectively. x and A represents distance from the duct mouth and sectional area of duct respectively.

The velocity perpendicular to the wall near the duct appeared to the same result as in figure 2, so that it was confirmed that this program is practically applicable.

4. APPLICATION AND DISCUSSIONS

For the practical application of this program(3D-Flow), simulation has been done on a model case as presented on figure 3. Firstly, gas movement in case of no ventilation has been simulated. Gas distribution and flow vectors on any section can be displayed and printed out in color.

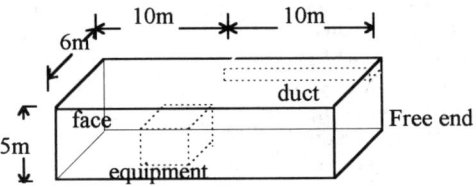

Fig. 3. Calculation model of heading

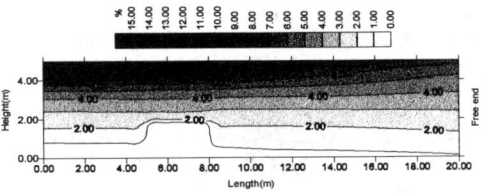

Fig.4. Gas movement in longitudinal section 3.9m from left wall in 20m long tunnel without auxiliary ventilation

Figure 4 shows the simulated result by 3D-Flow program. Originally, a color display of flow vector and gas concentration is available, but in this paper, only a contour line of gas concentration is presented. Gas percentage on the figure 4 represents the percentage against the original gas concentration at the mouth of exhaust pipes. Figure 4 illustrates that even there is no ventilation, air is circulated by the convection due to the temperature difference between ambient air and diesel exhaust.

The maximum concentration of NOx inside the heading face appeared to only 15 % of immediate gas concentration out of exhaust pipes in steady state. And the concentration at the height of workers' respiring range appeared to less than one third of the maximum concentration at roof zone. Considering the average NOx concentration from exhaust pipes ranges more or less 300 ppm in Korean mines, NOx concentration of this face can be estimated about 15 ppm which exceeds the permissible level of 5 ppm.

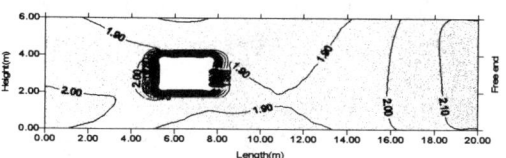

Fig.5. Gas movement in horizontal section 1.5m height from floor in 20m long tunnel without auxiliary ventilation

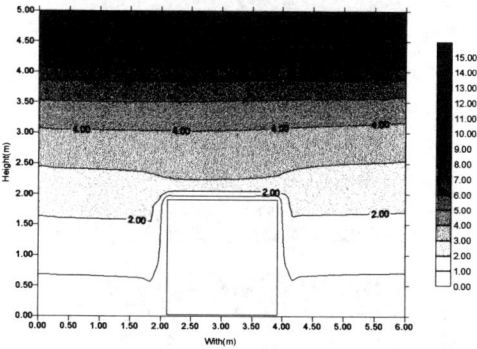

Fig.6. Gas movement in cross section 7m behind of face in 20m long tunnel without auxiliary ventilation

Figure 5 shows the gas distribution at the height of 1.5m from the floor. The gas concentration of human height appeared to more or less 2 % of concentration from exhaust pipes, but it is recognizable that the concentration around the machine is lower than that of face or rear side of the machine. It is probably because of the convection is more active at around the machine. Actual NOx concentration in this case is expected to 6 ppm which is slightly higher than permissible level, 5 ppm. Figure 6 shows the gas distribution on the cross section at 7.25 meters behind of face. It is noticeable that gentle convection is occurring and gas concentration contours are lying horizontally.

From the above simulation results, it is recognized that, even in short distance headings, the operation of diesel equipments should not be allowed when there is no appropriate ventilation.

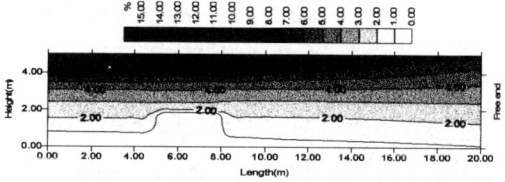

(a) 20m long heading

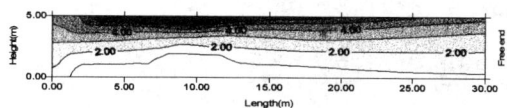

(b) 30m long heading

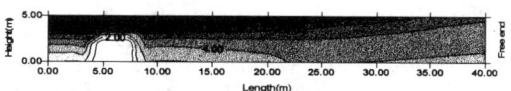

(c) 40m long heading

Fig. 7. Gas movement in longitudinal section at 3.9m from left wall in case of no ventilation

In figure 7, it is noticeable that the maximum concentration of NOx does not exceed 15% of the initial concentration from exhaust pipes regardless of the change of tunnel length. And the concentration at the space where workers are moving around is slightly increasing according to the tunnel length but, it is ranging from 2 to 5% only.

What we can remark on this study is that the way of calculation of required ventilation flow rate has to be reconsidered. As of now, the ventilation flow rate has been calculated on the basis of average concentration concept without considering the convection. It was obviously over calculation.

Table 1 shows the big difference between two way of calculation in case of 300ppm NOx is produced from a diesel equipment of which exhaust air volume is 7.5 m^3/min. The permissible level of NOx is 5 ppm.

As shown in table 1, the ventilation flow rate in conventional way was over calculated about 20 times. This fact was verified at the actual headings as well. Gas concentrations of all the headings we surveyed were not exceeded the permissible level even though the ventilation flow rate was much lower than the flow rate calculated based on average concentration concept.

Table 1. Comparison of Ventilation Air Flow

	based on average conc.	based on simulation
Dilute factor (d)	$d = g/g_p$ = 300/5 = 60	$d = (g \times 5\%)/g_p$ = (300×0.05)/5 = 3
Ventilation flow rate (Q)	Q= 7.5×60 = 450m3/min	Q=7.5×3 = 22.5 m3/min

g : gas conc. of exhaust,
g_p: permissible level of gas

Accordingly, it is suggested that the ventilation flow rate in dead-end headings should be calculated based on the informations provided by 3D-Flow simulations instead of conventional average concentration concept.

To figure out the general pattern of air movement and gas distribution in dead-end faces where auxiliary ventilation is practicing, another series of simulation were carried out on the same model as shown in figure 3. The auxiliary fan blows 65 m^3/min at the velocity of 3m/sec. The diesel equipment is exhausting 7.5 m^3/min of 150°C hot gases.

Fig. 8 shows a cross section 3m behind of face. The right side half shows lower gas concentration than the left side half. But right side roof shows the most dense gas concentration.

Figure 9 presents gas movement in a longitudinal section 1m from right wall, a space between right side wall and diesel equipment. In this section, it is noticeable that large part of air is recirculating in-between face and duct and the highest concentration appears at the mid. part of roof between face and duct.

Figure 10 presents gas movement of a longitudinal section 1 m from left side wall, a space between left wall and diesel equipment. In this section, it is noticeable that air came down from the face is moving

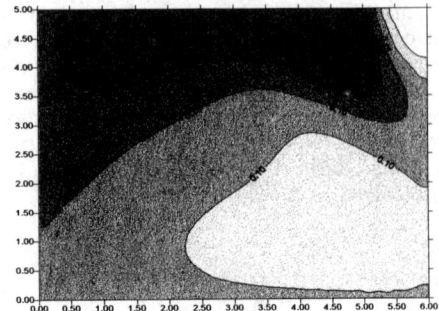

Fig. 8. Gas movement in a cross section 3m behind of face

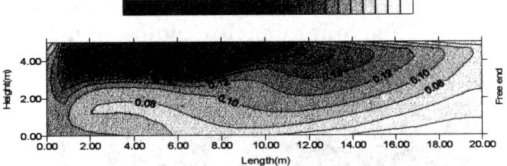

Fig. 9. Gas movement in a longitudinal section 5m from left side wall

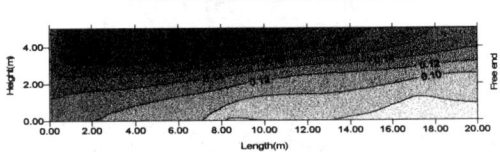

Fig. 10. Gas movement in a longitudinal section 1m from left side wall

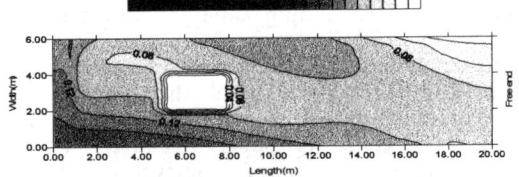

Fig. 11. Gas movement in a horizontal section 1.7m height from floor

up to roof and free end direction very rapidly.

Figure 11 presents gas movement of a horizontal section 1.7m height from floor which is considered as a height of human respiration. In this section, it is noticeable that air is flowing anti-clock wise from right side of face and moving up to the rear side. The highest gas concentration appears at the corner of face and left side wall and around the duct mouth. Considering that average NOx concentration from the exhaust in local mines is 300 ppm approximately, the gas concentration of ambient air in this model is expected to be acceptable level from 0 to 0.45% of original gas concentration of exhaust which can be converted from 0 to 1.5 ppm. The highest concentration appears at the roof in-between face and duct mouth.

NOx concentration of the height where workers are moving around is expected to be less than 0.45ppm(0.15% of 300ppm).

5. CONCLUSION

For optimizing the working environment of tunneling faces, it is essential to figure out 3-dimensional movement of air and gas. 3D-Flow program coded based on computerized fluid dynamics was proven as an useful tool to solve these problems. According to simulations using this software, following results are derived.

1) In dead-end headings, even there is no ventilation, the maximum concentration of gas reaches less than 15% of the immediate concentration of engine exhaust in steady state.

2) The minimum required ventilation of headings should be reasonably calculated based on informations derived from computer simulation instead of conventional way based on average concentration concept. It was recognized that, as of now, the ventilation flow rate has been 20 times over calculated by the conventional method

3) The general pattern of air movement and gas distribution of dead-end faces under the auxiliary ventilation was appeared as follows;

a) Gas concentration at the height where workers are repiring reaches only 30% of the maximum concentration appeared at the roof zone.

b) Space between diesel equipment and right side wall where ventilation duct is installed shows better environment than the other side.

c) In longitudinal sections, the maximum gas concentration is appeared at the roof zone in-between face and duct mouth.

d) In horizontal sections, the maximum gas concentration distributes at the corner of face and left side wall that is the opposite side of duct.

4) It is known that 5% of ventilation flow rate calculated by conventional method is enough to dilute the diesel exhaust in the dead-end headings as far as the tunnel length is not too long.

5) To maintain the clean environment of the faces, it is important not to put obstacles at the corner of face and left side wall(opposite side of duct).

REFERENCES

1. BY Kim, 1996, "Research on 3 Dimensional Air Flow Simulation in Underground Workings", Tunnel & Underground Space, Vol., No. P250-259, Journal of Korean Society for Rock Mechanics
2. BY Kim, 1995, "A Study on Environmental Measures for Application of Mobile Diesel Equipments in Underground", Korea Institute of Mineral and Energy Resources Engineers,. No.4
3. B.Y. Kim, 1995, "Diesel Equipments in Underground Excavation Workings using Diesel Equipments", Tunnel and Underground Space, Vol.5, No.1, Journal of Korean Society for Rock Mechanics
4. B.Y.Kim,1996, "Study on Characteristics of Diesel Particulate Matter and it's Measurement and Evaluation" Tunnel and Underground Space, Vol.6, No.1, Journal of Korean Society for Rock Mechanics
5. Suhas V. Patankar, 1980, "Numerical Heat Transfer and Fluid Flow", Hemisphere Publishing Corp.
6. Hans M. Mathisen, 1994, "Verification of Tool for Energy Calculation in Rock Caverns", Proceedings of Int.Symp.on Underground Openings for Public Use, Gjovic, Norway, p316
7. Even Thorbergsen, 1994, "PC Program for Design of Public Halls in Rock Caverns Modeling and User Interface", Proceedings of Int.Symp.on Underground Openings for Public Use, Gjovic, Norway, p327
8. J. W. Oberholzer C. F. Meyer, 1995,

"The Evaluation of Heading Ventilation Systems Through the Use of Computer Simulations", Proceedings of 7th US Mine Ventilation Symp.
9. D.J. Bruner, 1995, "Example of the Application of Computational Fluid Dynamics Simulation to Mine and Tunnel Ventilation", Proceedings of 7th US Mine Ventilation Symp.

R&D policy for underground space use and energy storage in Sweden

B.T. Sellberg
Swedish Council for Building Research, Stockholm, Sweden

ABSTRACT: Efficient underground energy storage facilities were developed in Sweden during the last two decades. The stores are based in eskers and deep sedimentary rock aquifers. Duct storage, in archean rocks is another promising technology. Heating and cooling applications are the most important.
    Environmental issues of underground utilization are extremely important. Topics like LCA and maintenance are of growing interest lately. Pooling underground energy storage technology with underground utilization has been favourable, especially in the exchange of experience through international collaboration.

1 INTRODUCTION

The development of energy storage technology has concentrated on implementation and market introduction of aquifer storage and duct storage in archean rocks during the 1990's. Successful facilities have been set into operation. The applications are mainly for heat and cooling purposes. The good experience from these facilities shows the importance of Research, Development and evaluation of Demonstration Plants as well as international collaboration.

R&D on Underground utilization within the framework of infrastructural development has been supported by the Swedish Council for Building Research (abbreviated as BFR) since the 1990's. The social and psychological aspects of Man being underground are important; as well as the technical development and safety aspects. Large road and railway tunnels and rock caverns in urban areas are at the moment in the design and construction phase.

2 R&D PROGRAM ON THERMAL ENERGY STORAGE IN SWEDEN

Underground Thermal Energy Storage is a technology with great future potential. One of the main problems with the utilization of solar energy systems is the storage. There are enormous thermal energy resources on earth, but in many cases, at the wrong place and time, (Sellberg, 1997 b).
    The development of underground thermal energy has been successful in Sweden during the 1990's, especially for heating and cooling applications in combination with heat pumps and cooling machines. Systems for storing heat and cold are developed also without heat pumps.
    The technology requires a throrough knowledge of applied rock and soil mechanics. The most promising and developed systems are: Aquifers, rock caverns and duct stores in soft soils and archean rocks. Simulation models have been developed for these storage types (*aquifers, duct stores in clays and archean rocks*) for theoretical and general design support. A research group at the University of Lund, Sweden, has worked with the development of these models for many years, (Hellström, 1991, Probert, et. al, 1994). The results of these developments has been very successful and the models are well known and used now worldwide.

2.1 Aquifers.
    The BFR has supported R&D on aquifer thermal energy storage (ATES) since the mid 1970's. The R&D has been successful. Thus, more than 20 ATES plants are now in operation. Majority part of

these plants are located in confined aquifers of sedimentary rocks in the southern part of Sweden and in unconfined aquifers in eskers. An evaluation study performed by the BFR (1994) showed that this technology is one of the most promising and shows good economy, especially for the combined use of heat and cold storage. Thus, the projects realized after 1991 are commercial. (Andersson, 1994,1997).

The main environmental impacts of ATES are normally limited to local temperature changes in the aquifer and its close surroundings. A total Environmental Impact Assessment (EIA) study has not yet been conducted for the aquifer storage plants. However, many of the ATES projects are to some extent motivated by positive environmental effects. For example, the direct systems save electricity and decrease the use of refrigerants and cooling machines.

2.2 Rock Caverns

The storage of hot water in rock caverns has been found to be a promising technology, favouring both *short term and seasonal energy storage* and for reducing the *power demand*. The caverns have to be large enough - at least some 100 000 m³. However, blasting of new rock caverns is expensive and the systems have to be large. Thus, large solar systems are for instance suitable. A combination of a small cavern and borehole storage was recently suggested. (Nordell, et. al. 1994).

The use of abandoned oil storage caverns have been tested in some cases in Sweden lately to improve the economy. However, the operational and environmental problems with the residual oil are not totally resolved as yet. The residual oil causes some problems with the heat exchangers.

2.3 Duct stores

Research and Development activities on duct stores in clays have been on the BFR agenda for nearly ten years now. The main activities are performed by the Swedish Geotechnical Institute (SGI).Thus, a test field for high temperature (35 - 90° C) storage in soft clay has been in operation for five years with extensive monitoring. Overall operating results show that seasonal heat storage in soft clay has good prospects of functioning satisfactorily. The heat store itself, as well as the integration of the heat store into a heating supply system, may be cost-effective. The choice of installation method is largely influenced by the local geological conditions at the site.(Gabrielsson, et. al., 1997).

Several duct storage projects in archean rocks are in operation in Sweden since the 1980's. The BFR has supported this technology since the very beginning. The principle of the borehole thermal energy store is that the thermal energy is charged and stored in a rock volume, which is penetrated by boreholes. The geometry of the store is important and computer programs have been developed to optimize the geometry and economy of the store. This newly developed tool is very important in increasing the cost-effectiveness of the borehole stores. Thus, the thermal properties of the rock enable storage of large quantities of thermal energy in large rock volumes within a fairly small area on the ground (Nordell, 1994)

3 POLICY FOR R&D ON THE URBAN UNDERGROUND

3.1 General Aspects

The BFR has - for many years- taken an active part in R&D support on several projects, planning activities and working groups concerning the field of Underground Technology. The Technology Development and Planning activities on Underground Space Use are important for urban sustainable development. During the last few years we have studied the psychological and medical aspects of Man being underground, for example.

An extensive program for construction of large road and railway tunnels and rock caverns in urban areas is now underway and the architectural design and safety aspects are being addressed. The general objectives of the R&D program are to achieve a better and more effective use of the excellent building material - the archean rock - that underlies nearly the whole country.

Since the 1960's Sweden had performed a large construction program for the storage of fossil fuels in rock caverns. Large caverns were built for commercial and strategic use. The current international policy has caused some change in the use of these caverns. The change of operation to thermal energy storage has been a topic of discussion for the last few years.

3.2 Urban Underground Space Use

In the city of Stockholm there is at the moment an extensive construction program on underground use; the most apparent ones are for transportation purposes. Thus, we are discussing a partly underground ring road around the entire city. "Intelligent Underground Utilization" is a method to use the ground more efficiently in densely populated areas.

The information, planning and design procedures have caused extensive discussions amongst the population. From this experience we learnd the importance of better planning, information, dissemination and legal instruments. Consequently, several R&D projects on these issues have been initiated lately. There are few permits for underground construction in Sweden up to now, and the handling of applications has generated some problems.

The wide spectra of the underground space use in the Stockholm area are illustrated by the following examples:
- The Saltsjö Tunnel, a sewage TBM tunnel at a depth of 65 m under the central part of Stockholm.
- Super tube pilot installation in clayey soils sealed with concrete, the tube contains pipes for district heating, water supply and sewage.
- Subsurface pipes for electricity transmission lines in densely populated areas. The energy losses could be utilized for heating purposes.
- Storm water tunnel under the central parts of the City of Stockholm.

The utilization of the subsurface space of Stockholm is common, and the need for improved planning process is urgent. (Sellberg, 1996, a).

Using the underground for the right purpose and leaving the above ground for its proper purposes are among the delicate problems and concerns facing urban planners in the near future. In many countries the planners/ developers must submit an Environmental Impact Assessment study before starting large underground construction works. One common experience from these studies is that the use of underground space has a very low impact on the environment in comparison with the corresponding project above ground. (Sellberg, 1997,b).

3.3 Maintenance

Maintenance needs for subsurface facilities are extremely low in comparison to aboveground construction. However, it could be hazardeous to neglect them. The most common maintenance problems concern groundwater leakage and rock fall in storage caverns. This could cause decreased reinforcement, and therefore must be observed and repaired continuously.

3.4 Life Cycle Assessment (LCA)

LCA´is a very interesting topic to discuss in conjunction with underground space use. One way to address this issue related to restoration of underground space would be to create a list of new ways to use subsurface space. This matter was discussed at a seminar in Stockholm in November 1996. (SGI, 1996). More than 50 different uses were tabled; from stores to various activities. Oil stores converted to coal and heat storage have been performed in full-scale projects. Restoration of underground construction in a broad application is a very challenging issue for R&D and creative thinking.

The storage of fossil fuels in rock caverns is well-known for various oil products. Efforts have also been made in Sweden and other countries to store natural gas in rock caverns, even as liquid natural gas, (LNG). It is a delicate problem,which has been studied thoroughly.With unlined rock walls, the opening of joints is be detrimental, since these would allow the cold liquid to migrate into the warmer environment. Therefore a suggestion is to use a tight membrane and thermal insulation and a concrete wall as a protection.(Margen, Lindblom, 1988)

4 INTERNATIONAL ENERGY AGENCY (IEA)

4.1 General Information on the IEA

The IEA was established in 1974 as an autonomous body within the framework of the Organization for Economic Co-operation and Development (OECD) to implement an international energy programme The future promises fundamental changes in the global energy balance. Higher economic growth rates will result in non-industrialized nations accounting for more than half of the world energy demand by the end of the century. The growing environmental concern will also have a strong influence on the nature of national and global energy policy. The global environmental viewpoint motivates the IEA to broaden the scope of interest to non-member countries, because the energy use of these countries plays an increased role in many ways. Thus, Energy Efficiency, Environmental Concern, Economy - Sustainable Development - are important goals for the IEA work.

The IEA organization has different "levels" of collaborative work. All levels, from governmental representatives to research workers are involved in the collaboration. There are different groups; Working Parties and Committees. (Sellberg, 1997a).

4.2 The Implementing Agreement (IA) on Energy Conservation through Energy Storage (ECES)

An IEA IA is a framework which facilitates the initiation, implementation, monitoring and review of international collaborative efforts. The IAs can encompass any phase of the technology cycle research, development and demonstration; validation of technical, environmental, and economic performance. The IEA has no central funds for financing the collaboration, thus, the activities within the collaboration are financed by the participating countries. The IA works with technical collaborative projects in annexes which are financed by the participating countries. The main advantage is that the participating countries will receive all the information from every country but each country is only responsible for its own cost. Now the IA deals with matters on Underground Thermal Energy Storage Systems (UTESS), Electrical Storage and Storage in Phase Change Materials (PCM) and Thermochemical Reactions. The main R&D efforts have been on UTESS. Experience and knowledge of Geology, Energy Technology and related techniques are necessary in this field of research.

5 CONCLUSIONS

The specific environmental benefits of an intelligent use of the urban underground should be implemented on a broad basis. One way to facilitate this is to emphasize the environmental benefits of subsurface space use more clearly than today. A "total comparison" with the above ground construction is recommended.

Extensive and organised international collaboration within the field of underground space use is therefore of greatest importance to resolve complicated questions better than is possible today. There are a wide scope of R&D issues from various disciplines that have to be solved. Interdisciplinary collaboration on a national and international basis is therefore suggested. The important interdisciplinary pooling of geotechnology people with energy people is already established in the UTESS technology.

The technology and way of thinking in the underground construction business has easily shown to be transferable to underground energy storage. In the energy field there is an extensive and well-experienced international collaboration under the IEA umbrella. This organised collaboration has been in operation for more than 20 years by now. It has its bureaucracy, but also successes. I would like to suggest the use of the IEA as a model for organising international collaboration within the multi-technology issue of urban underground utilization.

6 ACKNOWLEDGEMENTS

The author wants to thank the Drs Cam Mirza, Strata Engineering Corp., Canada., and Bo Nordell, Luleå Technical University, Sweden for their valuable comments and advice on the manuscript.

7 REFERENCES

Andersson, O. (1994): Aquifer Thermal Energy Storages in Sweden - Experiences so far and market potential. Proc. 6th Int'l Conf. On Thermal Energy Storage. Calorstock '94. Espoo, Finland

Andersson, O.: (1997): ATES Utilization in Sweden - an overview. Proc. MEGASTOCK '97. Sapporo, Japan.

Gabrielsson, A., Moritz, L. and Lehtmets, M. (1997): Heat Storage in soft clay at 35-90°C - Long Term Experience. Proc. MEGASTOCK '97, Sapporo, Japan.

BFR (1994): Thermal Energy Storage. Basis for Research Program 1993-1996. Swedish Council for Building Research, G:5 1994, Stockholm, Sweden (in Swedish)

Hellström, G.(1991):Ground Heat Storage. Thermal Analysis of Duct Storage Systems.University of Lund.

Margen, P., Lindblom, U.(1989): Storage of LNG in insulated rock caverns: Technology and Economics. Gecon, Report R:3, 89. Gothenburg, Sweden.

Nordell, B.(1994): Borehole Heat Store Design Optimization. Thesis. Luleå University of Technology,1994:137 D. Luleå, Sweden.

Nordell, B, Ritola, J. Sipilä,K. and Sellberg, B. (1994): The Combi Heat Store - a Combined Rock Cavern/Borehole Heat Store. Tunnelling and Underground Space Technology, Vol. 9, No.2 pp.243-249. Elsevier. UK.

Probert, T., Claesson, J. And Hellström, G. (1994): Thermohydraulic modelling of aquifer storage systems. Proc. 6th Int'l Conf. On Thermal Energy Storage. Calorstock'94. Espoo. Finland.

Sellberg, B. (1996 a): Swedish R&D plans, policy and results on the urban underground. Proc. North American Tunneling '96, Ozdemir (ed.). Balkema, Rotterdam, The Netherlands.

Sellberg, B. (1996 b): Environmental Benefits: a Key to Increased Underground Space Use in Urban Planning. Editorial.Tunneling and Underground Space Technology, Vol. 11, No. 4. Pergamon

Sellberg, B. (1997a): International Energy Agency Implementing Agreement on Energy Conservation through Energy Storage. Proc. MEGASTOCK '97. Sapporo, Japan

Sellberg, B.(1997b): Challenge for Energy Storage in the 21st Century. Proc. MEGASTOCK '97. Sapporo, Japan.

Swedish Geotechnical Institute (1996): Bergrum till salu. (Rock Caverns for sale) . Documentation. Linköping. (In Swedish).

Effect of thermal hysteresis on rock mass around openings for storage of low temperature materials

Yoshinori Inada & Naoki Kinoshita
Department of Civil and Environmental Engineering, Ehime University, Japan

Takahumi Shimogochi & Yiuchi Kohmura
Research and Development Institute, Takenaka Co., Japan

Takashi Matsuo
Research and Development Institute, Nissan Construction Co., Ltd, Japan

ABSTRACT: When low temperature materials such as LNG, LPG and frozen food, actually are stored in openings excavated in rock mountain, as the quantity of LNG and LPG change continually, the rock mass around openings will receive the effects of thermal hysteresis of low temperatures. Therefore, obtaining the strength and deformation characteristics of rocks after undergoing thermal hysteresis of low temperatures becomes important for discussing the stability of the openings. In this study, strength and deformation characteristics of rocks which had undergone thermal hysteresis of low temperatures were investigated. Using these values, stress distribution around openings were analyzed theoretically, and the affect of thermal hysteresis of low temperatures on cracks and the cracked zone around openings are discussed. Then, countermeasures for preventing leakage of gas and liquid from openings and reduction of thermal stress around openings are also discussed.

1 INTRODUCTION

The demand for LNG and LPG as substitutes for oil has tended to increase, and frozen food also shows a similar tendency with the increase of two income families and the great variety and rapid increase of frozen food. With these points as background, it is expected there will be an increase in the need for storage establishments in the near future.

The authors have proposed and discussed temporary storage of low temperature materials such as a LNG, LPG, frozen food etc. in openings excavated in rock mountain from the view points of efficient utilization of land and preservation of the environment. In this case, the authors have shown by theoretical analysis that thermal stress occurs around openings due to the thermal shrinkage of rock, and that cracks develop in a radial direction (Inada et al. 1991, 1995).

When low temperature materials actually are stored in openings, as the quantity of LNG and LPG change continually, the rock mass around openings will receive the effects of thermal hysteresis of low temperatures. Therefore, obtaining the strength and deformation characteristics of rock after undergoing thermal hysteresis of low temperatures becomes important for discussing the stability of the openings. In this study, strength and deformation character-

istics of rocks which has low porosity and high porosity were investigated after undergoing thermal hysteresis of low temperatures using a thermal cycle apparatus. Using these values, stress distribution around openings were analyzed theoretically, and the affect of thermal hysteresis of low temperatures on cracks and the cracked zone around openings are compared and discussed.

Then, countermeasures for preventing leakage of gas and liquid from openings and reduction of thermal stress around openings which become important problems when low temperature materials are stored in openings are also discussed.

2 STRENGTH AND DEFORMATION CHARACTERISTICS OF ROCKS AFTER UNDERGOING THERMAL HYSTERESIS

2.1 *Rocks used for experiments*

The rocks used for the experiments were granite (obtained in Miyakubo, Ehime, Japan) and tuff (obtained in Utsunomiya, Tochigi, Japan). In the case of granite, "rift plane", "grain plane" and "hardway plane" were determined by measuring elastic wave propagation velocity of the block of rock. Then a specimen was taken by core drill in the direction of intersecting perpendicularly to the hardway plane. In the case of tuff, the depositional surface was also determined by measuring the elastic wave propagation velocity of the block of rock, and a specimen was taken by core drill in the direction of intersecting perpendicularly to the depositional surface. Specimens were formed φ30×60mm for the uniaxial compression test, and φ30×30mm for the radial compression test. Specimens were prepared as "Dry" by drying for one week in a desiccator after air drying for a week, and as "Wet" by being kept in a desiccator filled with destilled water for 5 hours using a vacuum pump. The physical properties of these specimens are shown in Table 1.

Table 1 Physical properties of rocks.

Rocks	Porosity (%)	Moisture content ratio (%)	Degree of saturation (%)	Bulk specific gravity	True specific gravity
granite(Dry)	1.99	0.16	20.67	2.625	2.673
granite(Wet)	1.99	0.16	70.86	2.653	2.673
tuff(Dry)	37.09	1.39	5.76	1.560	2.446
tuff(Wet)	37.09	24.01	99.62	1.908	2.446

2.2 Experimental method

The specimens received thermal hysteresis of low temperatures by using a thermal cycle apparatus by the following method. The Wet specimens were covered with wet cloths and polyethylene film for protection from drying and set in the thermal cycle box. The Dry specimens were set in as is. Thermometers were set up at one point each in the center of specimens Wet and Dry and at 27 points in the thermal cycle box. Then the specimens were cooled from 15°C to -160°C as shown in Fig.1 assuming LNG storage. Cooling and heating sources were liquid nitrogen and a heater, as the temperature drops and rises linearly. The cooling rate was set at 1°C/min for avoiding the effects of thermal impact (Yamaguchi et al. 1970). The specimens were kept at -160°C for 60min making certain that the temperature at the center of the specimen was -160°C and heated to 15°C at a rate of 1°C/min and kept at 15°C for 60min. We call the process mentioned above as "1 cycle". Specimens received thermal hysteresis of low temperatures up to a maximum of 10 cycles.

The uniaxial compression test and the radial compression test were carried out. The tests were carried out at 15°C and -160°C. When the specimens were cooled to -160°C, the cooling conditions were the same as those of thermal hysteresis of low temperatures.

2.3 *Results and considerations*

Fig.2(a),(b) show the compressive strength after undergoing thermal hysteresis of low temperatures. In all cases, it was found that compressive strength

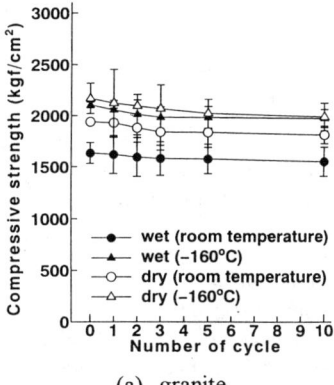

(a) granite

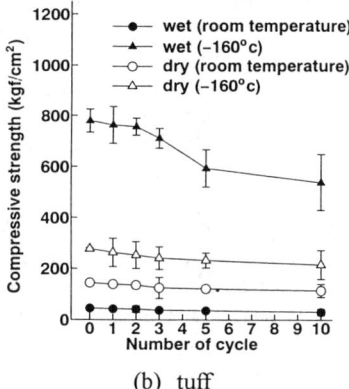

(b) tuff

Fig.2 Compressive strength of rocks.

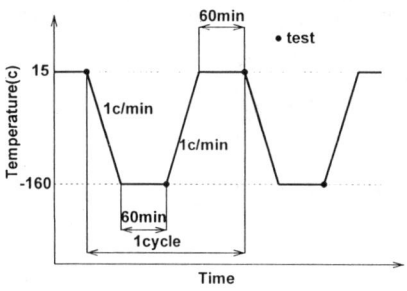

Fig.1 Schematic diagram of thermal hysteresis.

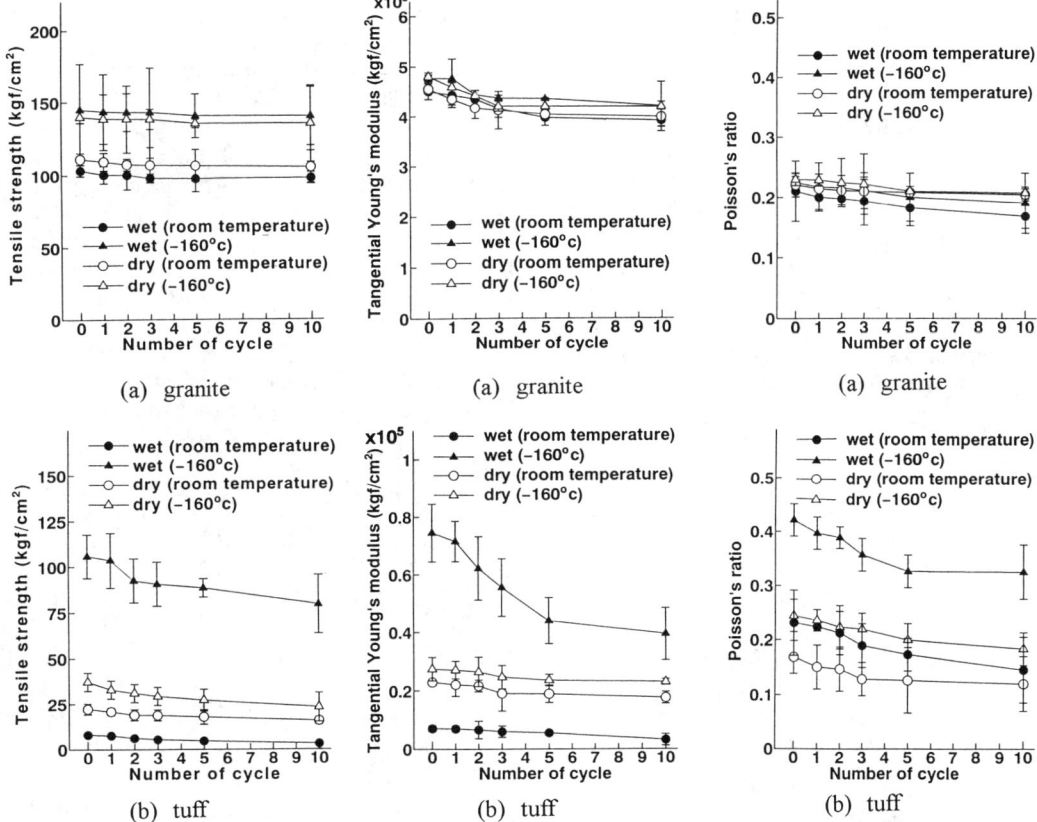

Fig.3 Tensile strength of rocks.

Fig.4 Tangential Young's modulus of rocks.

Fig.5 Poisson's ratio of rocks.

decreases with the increasing number of thermal hysteresis. One of the reasons for this is the enlargement of microcracks due to differences of thermal shrinkage among rock-forming mineral grains when the rock receives thermal hysteresis. However, it is found that the ratio of decreasing compressive strength decreases with the increasing number of thermal hysteresis. From this fact, it is supposed that the value will converge to a constant value. The compressive strength for Wet of both rocks at 15℃ is lower than that of the Dry. At -160℃, the value of the strength of both rocks is larger than that at 15℃ and especially in the case of tuff, the strength for Wet is larger than that of the strength for Dry. This is thought to be because rock-forming mineral grains shrink and harden with falling temperatures and tuff which has high porosity was affected by freezing pore water.

Fig.3(a),(b) show tensile strength. In all cases, the tensile strength decreases with the increasing number of thermal hysteresis. The value at -160℃ is larger than that at 15℃. At 15℃, the strength for Wet of both rocks is lower than that of the strength for Dry. However, at -160℃, the strength for Wet of both rocks is larger than the strength for Dry as in the case of compressive strength.

Fig.4(a),(b) show the tangential Young's modulus at 30% of fracture stress which was obtained from the stress-strain curve of the compression test. In all cases, the tangential Young's modulus decreases with the increasing number of thermal hysteresis. This seems to be due to microcrack expansion in rocks. A large difference in the value of tangential Young's modulus between the Dry and the Wet is not seen in granite. However, in tuff, the value of Wet is larger than that of Dry at -160℃.

The values of Poisson's ratio at 30% fracture stress are shown in Fig.5(a),(b). The value becomes lower with the increasing number of thermal hysteresis. The value at -160℃ is larger than that at 15℃.

3 THERMAL BEHAVIOR OF ROCK MASS AROUND OPENINGS AFTER UNDERGOING THERMAL HYSTERESIS

3.1 Temperature distribution

It is supposed that an opening is excavated in granite rock mountain to a depth of 100m beneath the ground surface with a diameter of 10m, and LNG of -162°C was stored within. Temperature distribution around an opening was analyzed by using FDEM (Inada et al. 1983). Thermal properties of granite at low temperatures used for analysis obtained by examination were shown in Table 2.

Table 2 Thermal properties of granite.

	Heat capacity (cal/(cm^3·°C))	Thermal diffusivity ($\times 10^{-3}$ cm^2/s)	Thermal conductivity ($\times 10^{-3}$ cal/s·cm·°C)
granite	0.541	11.0	7.64

Fig. 6 shows the change of temperature distribution around the opening with time. It is found that the temperature gradient is extremely sharp at the beginning of storage, but becomes gentler with time. The temperature enters a semi-steady state after 1 year.

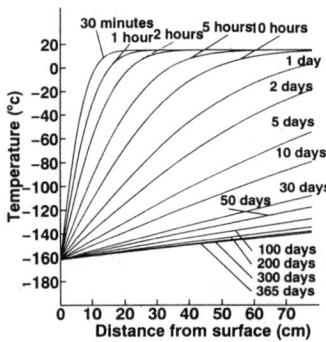

Fig.6 Temperature distribution around openings in the case of LNG storage.

3.2 Stress analysis

Cracks around openings caused by thermal stress were analyzed using temperature distribution by CAM (Inada et al. 1987) which is the applied method of the FEM. In this study, it is supposed that the rock mass around openings has undergone the affects of thermal hysteresis of low temperatures as the quantity of LNG changes continually. Actually, temperature distribution around the opening is a concentric circle condition, and temperature of the rock mass is not uniform. But, in this analysis, it is supposed that the whole rock mass undergoes thermal hysteresis from 15°C to -160°C for the sake of convenience. The physical properties of granite used for analysis are those after undergoing thermal hysteresis of 10 cycles from experiments mentioned above. For comparison, the case in which rock mass has not undergone thermal hysteresis was analyzed. Physical properties of granite used for analysis are shown in Table 3.

Table 3 Physical properties of granite used for analysis after undergoing thermal hysteresis.

Temperature (°C)	Expansion coefficient (1/°C)$\times 10^{-4}$	Young's modulus (kgf/cm^2)$\times 10^6$	Poisson's ratio	Compressive strength (kgf/cm^2)	Tensile strength (kgf/cm^2)
20~ 10	0.0000	0.309	0.162	-1550.0	95.9
10~ 0	0.1510	0.392	0.163	-1573.7	98.1
0~ -10	0.1490	0.393	0.165	-1597.3	100.2
-10~ -20	0.1450	0.395	0.167	-1621.1	102.4
-20~ -30	0.1400	0.396	0.168	-1644.6	104.5
-30~ -40	0.1360	0.398	0.170	-1668.3	106.7
-40~ -50	0.1320	0.400	0.171	-1691.9	108.8
-50~ -60	0.1270	0.402	0.173	-1715.6	111.0
-60~ -70	0.1230	0.403	0.175	-1739.2	113.1
-70~ -80	0.1190	0.405	0.176	-1762.9	115.3
-80~ -90	0.1140	0.407	0.178	-1786.5	117.4
-90~ -100	0.1100	0.408	0.179	-1810.2	119.6
-100~ -110	0.1060	0.410	0.181	-1833.8	121.7
-110~ -120	0.1010	0.412	0.182	-1857.5	123.9
-120~ -130	0.0970	0.413	0.184	-1881.1	126.0
-130~ -140	0.0930	0.415	0.185	-1904.8	128.2
-140~ -150	0.0880	0.417	0.187	-1928.4	130.3
-150~ -160	0.0840	0.418	0.189	-1952.1	132.5
-160~ -170	0.0800	0.420	0.190	-1975.7	134.5

The results of stress analysis are shown in Fig.7(a),(b). It was found that cracks occurred due to shrinkage of rock mass around openings and developed with time. In the case of undergoing thermal hysteresis, crack length after 1 year is about 12.5m. In the case of not undergoing thermal hysteresis, it is about 9.5m. The reasons for this are

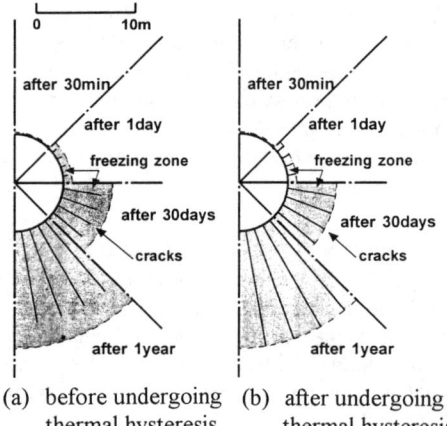

(a) before undergoing thermal hysteresis (b) after undergoing thermal hysteresis

Fig.7 Cracks and cracked zone around openings in the case of LNG storage.

considered to be that the thermal stress occurring in the rock mass decreased because tangential Young's modulus decreased after undergoing thermal hysteresis, but, the strength of granite also decreased.

4 PREVENTION OF LEAKAGE OF LIQUID AND GAS AND REDUCTION OF THERMAL STRESS

4.1 Analysis method

As mentioned above, when low temperature materials are stored in openings, cracks occur due to shrinkage of the rock mass. Therefore, countermeasures for preventing leakage of liquid and gas from the opening becomes important. Also, countermeasures for reduction of thermal stress becomes important to control the occurring cracks. In this study, a "resin lining system" was proposed to prevent the leakage. Using the results of the tests of strength, deformation characteristics and thermal properties of resin at low temperatures, the temperature distribution and cracks around openings caused by thermal stress were analyzed. From the results of the analysis, the effects of resin lining for preventing leakage are discussed. Then, a "combination lining system of resin and adiabatical material" is also discussed as a countermeasure for preventing leakage and reduction of thermal stress around openings at the same time.

Temperature distribution around an opening was analyzed by using FDEM which was developed for adoption of composite materials (Inada et al. 1996). Cracks around openings caused by thermal stress were analyzed by CAM. Thermal and physical properties of resin (that is, urethane resin) and adiabatical material (that is, rigid urethane foam) used for analysis are shown in Table 4 and Table 5.

In this analysis, it is assumed that 3cm of resin was set up on the surface of the opening as a "resin lining system" when frozen food of -60°C was stored in the opening. The case of in which an adiabatical material layer of 10cm on the surface of the opening and a resin layer of 3cm on the surface of the adiabatical material as a "combination lining system of resin and adiabatical material" were set up is also assumed.

4.2 Results and considerations

Results of stress analysis are shown in Fig.8-10. Fig.8 shows the results for the case of granite rock mass only. From this figure, it is found that cracks occurred due to shrinkage of rock mass around openings and developed with time. Fig.9(a) shows the results for the case of setting up an urethane resin layer. It is found that cracks occurred slowly compared to the case of rock mass only. But, cracks occur in the lining after 30 days caused by tensile stress. Fig.9(b) shows the results of analysis of the case of setting up a waterproof sheet between the rock mass and the resin, so that the thermal behavior

Table 4 Thermal properties of resin and adiabatical material.

	Heat capacity (cal/(cm$^3 \cdot$°C))	Thermal diffusivity ($\times 10^{-3}$cm^2/s)	Thermal conductivity ($\times 10^{-3}$cal/s·cm·°C)
urethane resin	0.173	0.44	0.08
rigid urethane foam	0.009	5.19	0.05

Table 5 Physical properties of urethane resin used for analysis.

Temperature (°C)	Expansion coefficient (1/°C)$\times 10^{-4}$	Young's modulus (kgf/cm^2)	Poisson's ratio	Compressive strength (kgf/cm^2)	Tensile strength (kgf/cm^2)
20~ 10	0.0000	175.90	0.433	-13.8	114.4
10~ 0	0.2815	332.55	0.439	-28.2	152.0
0~ -10	0.5630	489.20	0.445	-41.9	188.5
-10~ -20	0.7468	1051.46	0.442	-55.7	255.0
-20~ -30	0.9306	1613.72	0.437	-69.4	325.0
-30~ -40	1.1144	2175.95	0.435	-83.1	395.0
-40~ -50	1.2982	2738.24	0.432	-96.9	460.0
-50~ -60	1.4820	3300.50	0.429	-110.6	503.0
-60~ -70	1.4554	3491.20	0.442	-157.6	523.0
-70~ -80	1.4288	4875.20	0.419	-204.5	537.5
-80~ -90	1.4022	6259.20	0.414	-251.5	554.0
-90~ -100	1.3756	7643.20	0.409	-298.4	570.0
-100~ -110	1.3490	9317.76	0.406	-345.4	583.0
-110~ -120	1.3175	10992.20	0.406	-387.8	590.0
-120~ -130	1.2860	12666.70	0.407	-430.1	600.0
-130~ -140	1.2545	14341.20	0.408	-472.5	610.0
-140~ -150	1.2230	16015.70	0.409	-514.8	612.5
-150~ -160	1.1915	17690.20	0.409	-557.2	618.0
-160~ -170	1.1600	19364.60	0.410	-599.5	629.0

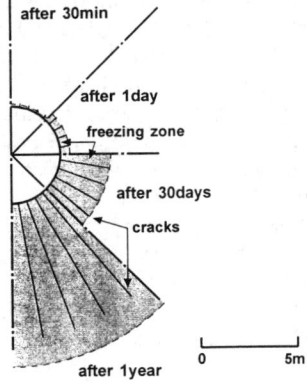

Fig.8 Cracks and cracked zone around opening in case of rock mass only.

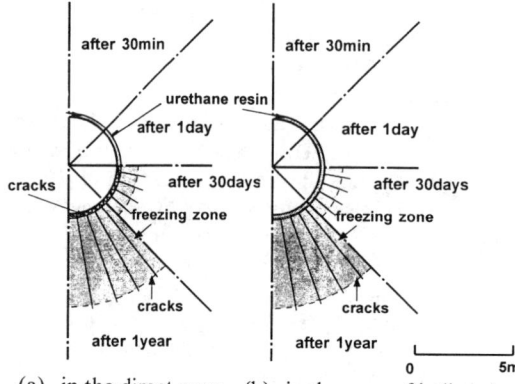

(a) in the direct case (b) in the case of indirectly

Fig.9 Cracks and cracked zone around opening in case of resin lining system.

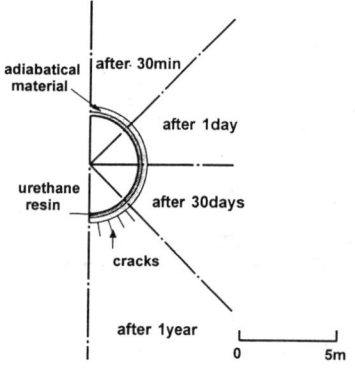

Fig.10 Cracks and cracked zone around opening in case of combination lining system.

of rock mass and the resin are independent of each other. It was found that cracks occurred due to the shrinkage of rock mass around openings, however, the layer of resin is stable after 1 year. Fig.10 shows the results for the case of a "combination lining system of resin and adiabatical material". In this case, it is found that the cracks were developing around the opening radially. But, development of cracks was controlled by setting up adiabatical materials. On the other hand, in the layer of resin, the theoretical fracture stress does not occur and the layer is stable after 1 year. From, these facts, the "resin lining system" and the "combination lining system of resin and adiabatical material" will become effective ways to prevent leakage and reduce thermal stress.

5 CONCLUSION

The main results obtained in this study are as follows:

1. Compressive and tensile strength of rocks decrease with the increasing number of thermal hysteresis. However, the ratio of decreasing decreases. From this fact, it is supposed that the value will converge to a constant value.
2. The values of tangential Young's modulus and Poisson's ratio of rocks after undergoing thermal hysteresis have the same tendency as those of compressive and tensile strength.
3. To make the thermal behavior of rock mass and a lining of resin independent, the "resin lining system" will be an effective way to prevent leakage of liquid and gas from an opening. The "combination lining system of resin and adiabatical material" will be also an effective way to reduce thermal stress and prevent leakage of liquid and gas at the same time.

REFERENCES

Inada, Y. & Kohmura, Y, ; Low temperature storage of gasses in rock caverns with a water curtain system, *Proc. of 7th Int. Cong. on Rock. Mech.(ISRM)*, 111-114, 1991.

Inada, Y., Kinoshita, N. & Seki, S. ; Thermal behavior of rock mass around openings affected by low temperature, *Proc. 8th Int. Cong. on Rock Mech.(ISRM)*, 721-724,1995.

Yamaguchi, U. & Miyazaki, M. ; A study of the strength or failure of rocks heated to high temperature, *J of Mining and Metallurgical Inst. of Japan*, **86**, **986**, 346-351, 1970.

Inada, Y. & Shigenobu, J. ; Temperature distribution around underground openings excavated in rock mass due to storage of liquid natural gas, *J of Mining and Metallurgical Inst. of Japan*, **99**, **1141**, 179-185, 1983.

Inada, Y. & Taniguchi, K. ; Plastic zone around underground openings excavated in rock mass due to storage of liquid natural gas, *J of Mining and Metallurgical Inst. of Japan*, **103**, **1192**, 365-372, 1987.

Inada, Y. & Seki, S. ; Application of resin lining system for countermeasures for preventing leakage from openings in low temperature materials storage, *J. of Japan Soc. of Civil Eng.*, **554**, III-37, 259-268, 1996.

Development of a simulation model for the vehicle tunnel ventilation using network theories

C. Lee & S. Lee
Department of Mineral Engineering, Dong-A University, Korea

D. Baek & S. Moon
Kolon Engineering and Construction Co., Ltd, Korea

ABSTRACT: Recent demand for longer vehicle tunnels and more complex ventilation schemes necessitate an optimization tool for the ventilation system. This paper aims at developing a ventilation system simulation model adopting distinctive local conditions. The model, NETVEN, is designed to calculate pressure, air speed and pollutant concentration profiles regardless of the tunnel type and the ventilation method. Applications are also made to semi-transverse and complex ventilation systems and the results are presented.

1. INTRODUCTION

Over recent years vehicle tunnels have gotten longer and their ventilation systems tend to have more complex schemes. This trend is expected to continue as more private companies show interests in the major SOC projects. To meet the safety as well as the environmental requirements in tunnels, ventilation system should be optimized and the elements in the system must be in compliance with all the standards and regulations. Analyzing the variables affecting tunnel ventilation and selecting an optimal system are not straightforward and a simple rule of thumb does not exist. Among the various tools for this purpose, simulation is recognized to be the best measure due to the capability of the simultaneous sensitivity analysis of multiple variables.

This paper aims at developing a simulation model of the vehicle tunnel ventilation which takes into account the local conditions. Therefore, the ultimate goal is to develop a tool to optimize the vehicle tunnel ventilation system.

2. EXISTING SIMULATION MODELS

The currently available models can be grouped into two categories; one for the normal state of ventilation and the other for the emergent states such as the accident and fire. Among the existing models, the only model which has ever been applied to local tunnel design is TUNVEN developed by the US Federal Highway Association, which calculates quasi-steady-state longitudinal air velocities and pollutant concentrations. It seeks for numerical solutions of the one-dimensional momentum equation and the advective diffusion equation. However, the model has several limitations on its applicability as well as the model itself. It is applicable only to the I-type tunnel and the source has to be modified to treat most of the ventilation equipment. Furthermore, it calculates THC concentrations and the conversion to smoke concentrations is not straightforward. Without mentioning the different local vehicle characteristics, these limitations prompt the efforts for localizing the ventilation system design technology.

3. MODEL DEVELOPMENT

The first requisite for the model development is that users should be able to use regardless of the tunnel shape and ventilation system. To accomplish this goal, the ventilation system is transformed to a network without tree branches and then network theories are utilized to obtain solutions for air velocities, pressures and pollutant concentrations in steady-state as well as quasi-steady-state conditions. Other requisites are as follows:

1. It can handle the variable speed axial-flow fan.
2. Emission rates are modeled on the basis of the local vehicles and regulations.
3. Each segment of the tunnel can have different characteristics.
4. Smoke concentrations are estimated as the visibility instead of the gravimetric concentration.
5. Modules containing the data on the physical characteristics and the emission rates of vehicles are programmed to be easily accessed and modified by users.

As shown in Figure 1., the simulation model, NETVEN, is composed of four major modules; modules for branch data input, ventilation rate calculation, pollutant concentration calculation, and output manipulation, respectively. The model is written in Visual Basic and every step for running the model is menu driven.

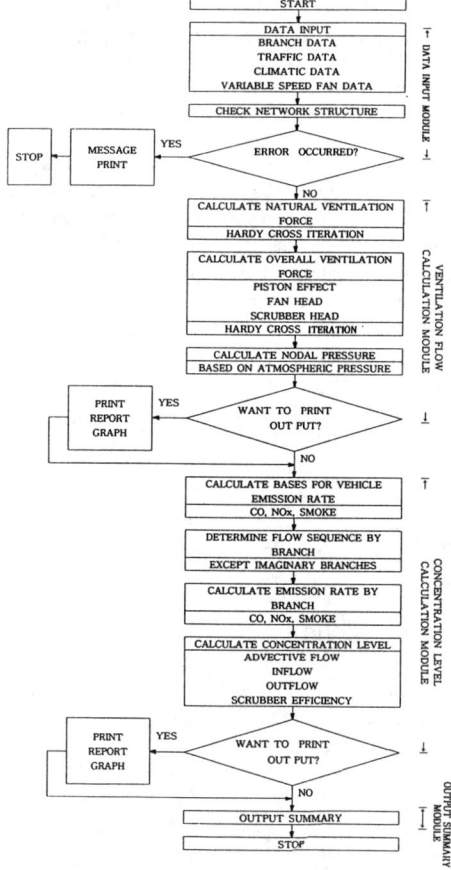

Figure 1. NETVEN flowchart

3.1 Module for the branch data input

There are four windows for data input of branch, traffic, weather and axial-flow fan. Each window asks users to type in data interactively or load external data files. Tunnel is modeled as a network with branches and nodes. In constructing a network, user must be cautious that tree branch should not be included; tree branch is one without either incoming or outgoing branches connected. To avoid the occurrence of tree branches, imaginary branches are inserted between the branches extended to the atmosphere. Air velocities are calculated in branches while pressures are estimated at nodes.

1. branch data:
   branch and node numbers,
   physical characteristics,
   friction and shock loss factors,
   jet fan and precipitator characteristics,
   traffic rate and vehicle speed.
2. traffic data:
   percentage of vehicle types,
   vehicle frontal area,
   number of lanes.
3. weather data:
   atmospheric pressure.
   wet and dry temperatures,
4. variable speed fan data:
   characteristic curve.

3.2 Module for the air velocity calculation

Assuming incompressibility and one-dimensional flow, the balance equation of air flow in tunnel can be expressed as follows:

$$P_{natural} + P_{piston}(U) + P_{fan}(U) + P_{port}(U) + P_{loss}(U) = 0$$

$P_{natural}$: natural force P_{piston} : traffic-induced force
P_{fan} : force by fans P_{port} : force by ports
P_{loss} : pressure loss U : air velocity

Hardy Cross algorithm is used to solve the equations on pressure and air quantity based upon the Kirchhoff's 1st and 2nd laws, and fundamental meshes are determined by minimum spanning tree algorithm. As the first step for solution, the air velocity is initialized by the natural ventilation and then the traffic-induced and mechanical ventilation forces are added to calculate the air velocity profiles in tunnel. After completing the calculation in all branches, pressure at each node is calculated sequentially in the order of the air flow.

3.3 Module for the pollutant concentration calculation

Gaseous and particulate pollutants discharged by vehicles are transported and mixed by the air flow in tunnel and mixing is accelerated by eddies and instantaneous fluctuation of the air velocity. Since the airflow in tunnel can be described as one-dimensional, the pollutant transfer is expressed by the following Fick's diffusion equation.

$$\frac{\partial c}{\partial t} = -\frac{\partial uc}{\partial x} + E_x \frac{\partial^2 c}{\partial x^2}$$

c : concentration u : air velocity.
E_x : turbulent diffusion coeff x : distance
t : time

Sato and Koso (1985) documented well the relative importance of the convective and turbulent transfer in the equation.

"... where longitudinal air flow exists to some extent, the convective transfer is predominantly larger than the diffusive transfer. The diffusive transfer ... plays a substantial role only when convective transfer diminishes due to stagnant air flow or the diffusive transfer is enhanced by a large local concentration gradient."

The turbulent diffusion coefficients in tunnel vary between 0.18 and 0.50 m^2/s (Sato, et al, 1985) and this is very similar to the range of 0.10~0.48 m^2/s by Lee and Yang(1994) who studied diffusion in coal mine airways with frequent equipment movement. If more frequent vehicle movement is expected, the coefficient is likely to be even smaller. Therefore, in NETVEN, only the convective diffusion of smoke, CO and NOx is modelled as follows:

$$\frac{d(uc)}{dx} = q_{in}c_{in} - q_{ex}c + I$$

q_{in} : inflow rate q_{ex} : outflow rate
c_{in} : conc. in inflow I : vehicle emission rate

While the air velocity is estimated for the branch and the pressure calculation is at the node, concentration prediction is based on the zone which is defined as a segment of branch approximately 10m long. In NETVEN a FDM solution of the above equation is modelled and at first this is applied to a branch of which intake portal is exposed to the atmosphere. If two or more branches are tied, selection may be made arbitrarily.

3.4 Module for the output manipulation

Simulation output is summarized in tables and graphs as follows:

1. Summary of the network structure
2. Summary of the network analysis
3. Summary of the air velocity and pressure calculations
4. Summary of the overall vehicle emission rate calculations
5. Summary of the pollutant concentrations (smoke, CO and NOx)

4. APPLICATIONS

NETVEN has been applied to various ventilation systems of the tunnels either in operation or under construction. Two tunnels with semi-transverse and complex ventilation systems are simulated and the results are presented.

4.1 Semi-transverse ventilation system

Hwangryong tunnel in Pusan is 1860m long and employs the inflow semi-transverse ventilation system. The air driven by large axial-flow fans enters the tunnel through the numerous ports in the ventilation duct and flows with vehicle movement. The ventilation duct is separated by a wall at the midpoint and a fan located at each portal blows the air into each half of the duct. Figure 2. show the tunnel network.

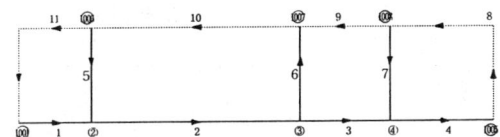

Figure 2. Network of Hwangryong tunnel

1. Input

The static pressure differences between the duct and the roadway determine the port flow rate. The velocity profile tends to decrease from the portals to the bulkhead. However due to the significant turbulences developed by frequent vehicle passages, in most of the cases uniform port flow is observed. Ports are described as Branches 22 through 41 in Figure 2. and all the flow rates are assumed to be identical. This assumption is modeled by treating the ports as fixed-quantity branches. The total inflow from the two main fans is measured to be 320m³/s and 16m³/s flows through each port. Traffic rate, 794.4veh/hr and speed, 64.6km/hr, are kept constant throughout the tunnel. Atmospheric pressure and temperatures are 100KPa, 25.8℃(dry) and 23℃(wet). The input data collected on July 21, 1995 are arranged in Table 1.

Table 1. Input data for Hwangryong tunnel

(a) branch data

Branch	length (m)	area (m²)	darcy friction coeff	shock loss factor	wind speed (m/s)	fixed Q (m³/s)	CO (ppm)
1	88.6	73.5	0.03	0.5	-2.3		2.3
2~20	88.6	73.5	0.03				
21	88.6	73.5	0.03	1.0	5.5		
22~30	1					16	
31	1						
32	1						
33~41	1					16	
42~59	88.6	40	0.03				
60~61	1	40	0.03			160	2.85

(b) traffic data

Vehicle type	(%)	Vehicle frontal area	
passenger car	71	small vehicle	2.31 m²
small bus	3		
heavy bus	11	heavy vehicle	7.11 m²
small truck	0		
medium truck	14	number of lanes	2
heavy truck	1		

2. Output

As shown in Table 2., the natural ventilation head which is constant throughout the tunnel induces 130m³/s and the total ventilation rate increases from the entrance to the exit; from 100.92m³/s to 420.92m³/s. Since the low traffic rate of 794.4veh/hr creates low piston effect, without the inflow from the ports, all the other ventilation forces pulls in only 100m³/s of the air. The piston effect is proportional to the squares of the vehicle speed relative to the air speed. Therefore, near the entrance where the air velocity is relatively low, the piston effect is large, 3.70Pa and decreases down to 2.01Pa at the exit showing the highest velocity profile.

CO concentrations in the air coming through the ports and the portals are measured to be 2.85ppm and 2.30ppm. As shown in Figure 3., simulated concentrations of CO tend to be higher than the observed data. Compared with the relatively small differences varying from -2.4 to 3.8ppm, differences become fairly larger near the exit. This can be explained as follows:

(1) The instantaneous fluctuation may contribute to this.
(2) As stated earlier, only advective transfer is modeled in NETVEN, even though large concentration gradient at the exit causes diffusive transfer more pronounced. However, this phenomenon is usually localized and affects only a small portion of the tunnel.

Table 2. Simulated ventilation rates for Hwangryong tunnel

Branch	natural vent rate (m³/s)	total vent rate (m³/s)	pressure at the end (Pa)	pressure loss (Pa)	piston effect (Pa)
1	-130.10	100.92	99998.6	1.99	3.70
2	-130.10	116.92	10001.7	0.45	3.60
4	-130.10	148.92	10007.3	0.74	3.41
6	-130.10	180.92	10011.9	1.09	3.23
8	-130.10	212.92	10015.3	1.50	3.05
12	-130.10	276.92	10018.1	2.54	2.70
14	-130.10	308.92	10017.2	3.17	2.54
16	-130.10	340.92	10014.7	3.86	2.38
18	-130.10	372.92	10010.4	4.61	2.23
21	-130.10	420.92	10018.0	5.88	2.01
22~41		16			

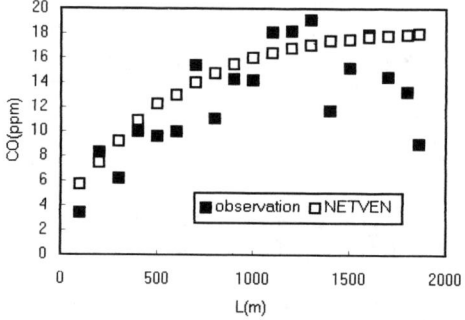

Figure 3. Comparisons of CO for NETVEN and observations

4.2 Complex tunnel

Tunnels other than the I-type are expected to appear soon particularly in the urban underground expressway currently being studied. The tunnel in Figure 4. with slip roads is one of the candidates and is yet to be found locally. The tunnel is quoted from Velde(1988) and its ventilation is simulated by NETVEN.

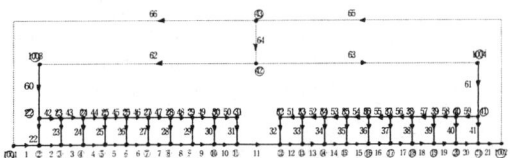

Figure 4. Network of the complex network tunnel

1. Input

Table 3. show the physical characteristics and fan locations. The equivalent resistance area of vehicle is defined as the product of the drag coefficient and the vehicle frontal area is of particular importance in determining the piston effect. It is assumed to be $4m^2$ for the heavy-duty vehicle and $1.0m^2$ for the small vehicle. The heavy-duty vehicle traffic fraction is 11% and the air density is $1.2kg/m^3$. Wind blows at the speed of 6.45m/s along the exits of Branches of 4 and 6.

Table 3. Physical characteristics and ventilation system of the complex network tunnel

physical characteristics	
length 　main tunnel 　slip road x-sectional area hydraulic diameter	1765 m 75m, 310m, 215m 52.8~106.3 m^2 6.4~8.8 m
ventilation system (complex longitudinal)	
jet fan installation location	1, 2, 4, 6, 7 branch

Table 4. Input data for the complex network tunnel

Brch	length (m)	area (m^2)	darcy friction coeff	vehicle speed (km/hr)	traffic rate (veh/hr)	shock loss factor		
						ent	exit	junct
1	785	85.4	0.019	23	4190	0.6		0.05
2	500	106.3	0.018	23	5566			-0.02
3	275	85.4	0.019	18	3862			0.02
4	205	85.4	0.017	60	5118		0.8	
5	75	42.1	0.021	23	1376	0.6		
6	310	52.8	0.020	60	1704		0.8	
7	215	52.8	0.020	18	1256	0.6		0.1

2. Output

As shown in Table 5., wind pressure at Branches 4 and 6 is 25Pa (same as the wind velocity of 6.45m/s) and it causes the airflow rate ranging from 113.83 to 352.79m^3/s. The simulated total airflow in the main tunnel consisting of Branches 1 through 4 is 401.79m^3/s at the entrance and 656.25m^3/s at the exit. The piston effects vary significantly among the branches between 0.88 and 17.63Pa. This is due to the varying relative vehicle speeds.

The ambient air is assumed to contain 10ppm of CO at Branches 1 and 5ppm at Branches 5 and 7. At the exit of the main tunnel, the concentrations of smoke, CO and NOx are $1.64 \times 10^{-3} m^{-1}$, 31.72ppm and 2.94ppm, respectively, while $2.62 \times 10^{-3} m^{-1}$, 49.75ppm and 5.22ppm are found to exist at the exit of Branch 6 corresponding to an outgoing slip road. The concentration profiles in the main tunnel in Figure 5. show drastic drops at the junctions with incoming slip roads. This indicates that the slip road plays an important role as a major ventilation head source and its effects on the entire network have to be analyzed carefully.

Compared with Velde (1988), NETVEN results in higher airflow rates in all branches. In Velde, data on the external wind direction and the shock loss factors at the junctions are not clearly stated. Therefore, in running NETVEN, wind is assumed to blow in the same direction as the traffic and this results in a relatively large ventilation head of 25Pa compared with the piston effect and the jet fan thrust. Lower CO concentrations from NETVEN are due to the different bases for the emission rates. Velde uses data for the European vehicles of the 1980's, while the emission data for the vehicles manufactured since February 2, 1991 in compliance with the local regulations are modeled in NETVEN.

Table 5. Simulated ventilation rate for the complex network tunnel

Branch	natural vent rate (m^3/s)	total vent rate (m^3/s)	pressure loss (Pa)	piston effect (Pa)	jet fan thrust (Pa)
1	183.47	401.79	49.63	3.78	15.10
2	297.30	556.42	17.37	1.01	11.86
3	135.33	288.20	23.88	1.45	
4	352.79	656.25	7.78	12.79	26.14
5	113.83	164.62	100.59	0.58	
6	161.97	278.21	59.26	17.63	14.84
7	217.46	368.04	71.61	0.87	14.04

Table 6. Comparisons of CO concentrations for NETVEN and Velde

Branch	NETVEN (ppm)		Velde (ppm)	
	IN	OUT	IN	OUT
1	10.0	34.91	10	64
2	25.74	37.35	51	83
3	37.35	46.35	83	108
4	27.17	31.72	70	77
5	5.0	9.18	5	10
6	37.35	49.75	83	94
7	5.0	11.92	5	14

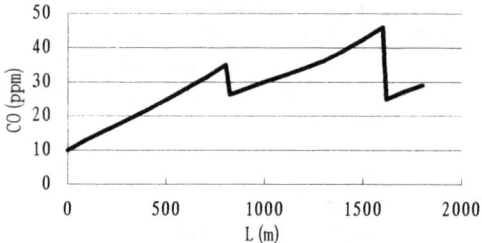

Figure 5. Concentration profiles of CO

5. CONCLUSIONS

This paper aims for development of a simulation model of the vehicle tunnel ventilation system which is one of the key elements in constructing and managing long tunnels. The model, NETVEN, solves many problems found in the existing models. NETVEN can simulate the pressure, airflow rate and concentration profiles of smoke, CO and NOx in the ventilation system regardless of the tunnel shape and the system type. Network theories such as minimum spanning tree algorithm and Hardy Cross iteration are adopted in the model.

NETVEN is applied to two tunnels employing semi-transverse and complex ventilation systems. In both cases, calculations of the airflow rate and the pollutant concentrations converge fast. Comparisons with the observed data and the other model show that differences are not significant and most of the discrepancies appear to be due to the different bases for the emission rates, temporal fluctuations of the air velocity and the pollutant profiles at a fixed point, and the ignorance of the diffusive transfer near the exit.

REFERENCES

Lee, C. & W. Yang, 1994, A study on the diffusion coefficients of diesel exhaust in coal mine airways using a tracer gas, *J. of the Korean Institute of Mineral and Energy Resources Engineers.* 31/5:483-490.

Lee, C., Yang, W. & S. Lee, 1996, Simulation modelling of the pollutant concentration in vehicle tunnels, *J. of Korean Society for Rock Mechanics, Tunnel and Underground Space.* 6/1:57-63.

Sato, N., Ohta, Y. & K. Komatsu, 1985, Discharge of exhaust pollutant from portal of one-way traffic automobile tunnel with exhaust shaft, *Proceedings of the 5th International Symposium on the Aerodynamics and Ventilation of Vehicle Tunnels*:H2-445-H2-460. Lillie, France.

Velde, K. te, 1988, A computer simulation for longitudinal ventilation of a road tunnel with incoming and outgoing slip roads, *Proceedings of the 6th International Symposium on the Aerodynamics and Ventilation of Vehicle Tunnels:* C3-179-C3-201. Durham, UK.

Jacques, E. J., 1991, Numerical simulation of complex road tunnels, *Proceedings of the 7th International Symposium on the Aerodynamics and Ventilation of Vehicle Tunnels:*467-487. Brighton, UK.

Ferro, V., Borchiellini, R. & V. Giaretto, 1991, Description and application of a tunnel simulation model, *Proceedings of the 7th Inter national Symposium on the Aerodynamics and Ventilation of Vehicle Tunnels:* 487-512. Brighton, UK.

ns Underground Construction, Lee, Yang & Chung (eds)
Field measurements and numerical analysis on the efficiency of water curtain boreholes in underground oil storages

K.J.Lee
Korea Petroleum Development Corporation, Seoul, Korea

H.K.Lee
Department of Mineral and Petroleum Engineering, Seoul National University, Korea

ABSTRACT: This study was undertaken to suggest the suitable design conditions of water curtain system through the analysis on the pressure down of boreholes by hydraulic tests carried out in construction fields of underground oil storages. The influence by hydraulic conductivities of rock mass around boreholes on the pressure down of water curtain boreholes was analysed. The relation between the arrays of boreholes and its pressure down was also analysed. Groundwater flow analysis on crude oil and LPG storages by finite difference method was carried out to simulate results of field tests and to investigate the distribution of hydraulic gradient around cavern. As the results, hydraulic tests showed that the pressu‧ down of boreholes was inverse proportional to the hydraulic conductivity of the surrounding rock mass. The rate of pressure down of boreholes was not influenced by water curtain system more than 20m over cavern and was proportional to the interval of boreholes. The hydraulic gradient around cavern was proportional to the distance and interval of boreholes and its value was not satisfactory to oil tightness condition in case of no water curtain system.

1. INTRODUCTION

The basic principle to reserve oil in underground unlined cavern is that the pore pressure of rock mass around cavern should be higher than inner pressure of cavern in order to flow groundwater into cavern from the surrounding rock mass. In case constructing cavern in crystalline rock mass, preexisted open cracks or new cracks are developed into rock mass around cavern by blasting. To prevent gas leakage through these cracks, the rock mass should be improved by methods lowering of hydraulic conductivity with grouting or installing of water curtain boreholes. Water curtain system which maintains natural groundwater table and raises the pore pressure of rock is a popular method used to ensure hydraulic stability of cavern. The design of water curtain system is very important economically because water curtain increase the cost in construction and operation of a facilities. For the suitable design of water curtain, various conditions including geology of the site, engineering characteristics of rock mass, natural water table, hydraulic conductivity etc. should be analysed. This study was undertaken to suggest the suitable design conditions of water curtain system through the analysis on hydraulic test carried out in construction fields of underground oil storages and the verification of the test results by numerical analysis.

2. GEOLOGY OF THE SITES

The hydraulic tests were carried out in construction fields of five sites. The geological conditions of sites are all different and types of rocks in five sites are as follows. A site consists of the daebo granite intruded at the jurassic in mesozoic age. B site consists of the gneiss of kyonggi metamorphic complex in precambrian age. C-1 site consists of the granite intruded at the late cretaceous in mesozoic age. C-2 site consists of the tuff formed by volcanic activities at the late cretaceous in mesozoic age. D site consists of the graniticdiorite intruded at the cretaceous in mesozoic age.

The engineering characteristics and physical properties of the representative rocks in each site are as below table 1~3.

Table 1. Engineering characteristics of in situ rocks

Content	R.Q.D(%)	R.M.R	Q-system	Remarks
A site	93	61	6.9	
B site	90	63	7.0	
C site	86	68	16	
	96	78	44	
D site	94	82	11	

Table 2. Physical properties of rock specimens

Content	Specific gravity (g/cm3)	porosity (%)	seismic velocity P-wave(m/s)	seismic velocity S-wave(m/s)
A site	2.763	0.79	3,900	2,400
B site	2.678	1.10	4,300	2,100
C site	2.584	0.71	4,600	2,500
	2.713	0.11	5,800	2,600
D site	2.747	0.65	5,700	3,200

3. THE METHODS OF FIELD MEASUREMENTS.

The hydraulic conductivity of rock mass around boreholes was measured with injection fall off test immediately after installation of boreholes.

The draw down in a borehole was checked for about forty five minutes after water injection for fifteen minutes into the borehole. The hydraulic conductivity of rock mass around a borehole was calculated with measured data by horner plot method as below equation (1).

$$K = \frac{2.303Q}{4\pi Lm} \quad (1)$$

In the equation (1), K is hydraulic conductivity, Q is quantity of injection water, L is test interval of a borehole, m is the slope of tangent line of draw down plots according to time on semi-log sheet.

In the mean time, the draw down and pulse tests were carried out for measurement of efficiency of boreholes after cavern excavation.

The draw down tests were carried out to read the pressure down of boreholes for 3~4 days with no injection of water into boreholes. And the pulse tests were carried out to be divided into two groups of every other injection and observation boreholes. The pressure down of boreholes was checked in observation boreholes with injection of water into injection boreholes. Fig.1 shows section of boreholes and cavern.

4. MEASUREMENT RESULTS AND CONSIDERATION

4.1 The analysis on pressure down of boreholes according to hydraulic conductivity of rock mass

Fig.2 shows results of the draw down tests according to the hydraulic conductivity of rock mass in underground storages for crude oil and products. The pressure down of boreholes increases as hydraulic conductivity of rock mass decreases and its values are in the range of 0~1.3kg/cm^2.

Fig.3 shows results of the pulse tests according to hydraulic conductivity of rock mass in underground storages for crude oil and products. The pressure down of boreholes increases as hydraulic conductivity of rock mass decreases and its values are in the range of 0~1.1kg/cm^2. As the above results, the pressure down of boreholes was inversely proportional to the hydraulic conductivity of rock mass. The reason is considered as the rate of underground water flow into boreholes from the surrounding rock mass that has high hydraulic conductivity is balanced to the flow rate from boreholes to cavern in contrast with that the rate of underground water flow into boreholes from the surrounding rock mass that has low hydraulic conductivity is less than the flow rate from boreholes to cavern.

Table 3. Mechanical properties of rock specimens

Content	Compressive strength(kg/cm2)	Tensile stren-gth(kg/cm2)	Shear stren-gth(kg/cm2)	Cohesion (kg/cm2)	Internal fric-tion angle	Young's modulus (kg/cm2)	Poisson ratio
A site	1,501	88	196	204	54○	5.20x105	0.22
B site	1,490	120	210	196	42○	4.65x105	0.24
C site	2,060	110	238	195	47○	4.69x105	0.38
	3,040	200	390	371	57○	8.39x105	0.41
D site	1,746	105	241	244	58○	7.28x105	0.26

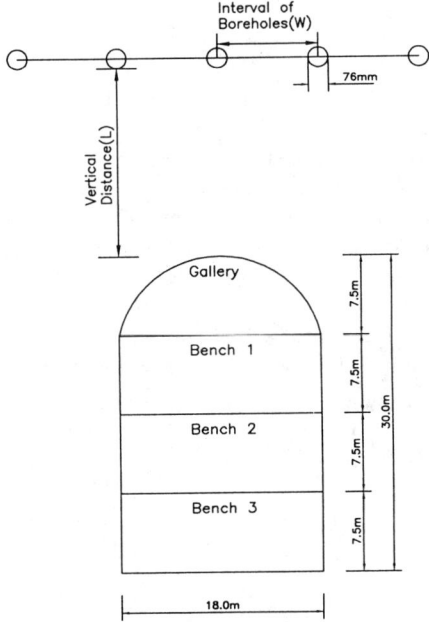

Fig. 1 Section of boreholes and cavern

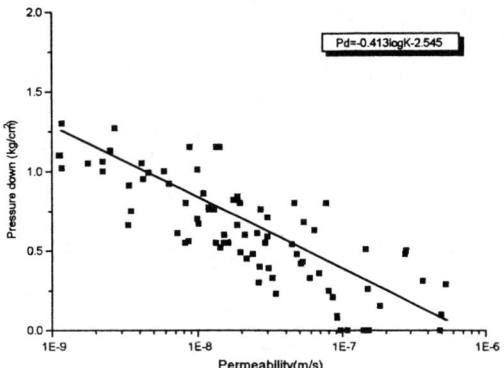

Fig. 2 Pressure down of boreholes by drawdown test

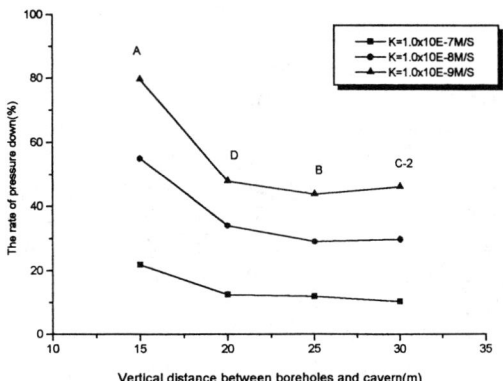

Fig. 4 The rate of pressure down of boreholes according to the vertical distance

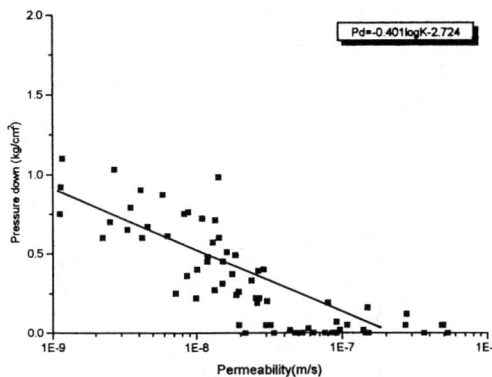

Fig. 3 Pressure down of boreholes by pulse test

4.2 The analysis on pressure down of boreholes according to vertical distance between boreholes and cavern.

Fig.4 shows the analysis results on pressure down of boreholes which has initial pressure of about two kg/cm^2 and hydraulic conductivity of about each 1.0 x 10E-7m/s, 1.0 x 10E-8m/s, 1.0 x 10E-9m/s in the surrounding rock mass. In Fig.4, the rate of pressure down of boreholes was constant with no relation to the vertical distance that is in the range of 20~30m, but it was rapidly increased in case that the vertical distance is 15m. This is considered as the reason that groundwater flow is easier as the vertical distance between boreholes and cavern is closer. As the results in cases that the vertical distance is 20~30m, the rate of pressure down of boreholes was depended on hydraulic characteristics of rock mass

Table 4. The rate of pressure down of boreholes according to the distance of water curtain boreholes

Permeability(m/s)	A storage(15m)	B storage(20m)	C storage(25m)	D storage(30m)
1.0x10E-7	21.9%	12.5%	11.93%	10.4%
1.0x10E-8	55%	34.1%	29.6%	29.8%
1.0X10E-9	79.6%	47.7%	43.69%	46.4%

rather than on the vertical distance of boreholes. Table 4. shows the rate of pressure down of boreholes in the above three values of hydraulic conductivity according to the vertical distance between boreholes and cavern as the result of the draw down tests.

4.3 The analysis on pressure down of boreholes according to interval of boreholes

The installed interval of boreholes influences not only tightness conditions of cavern but economical construction and operation of a facilities. In this section, the rate of pressure down of boreholes by pulse tests with condition that intervals of boreholes of sites are different one another were analysed according to the hydraulic conductivity of the surrounding rock mass. The analysis was carried out in three sites of B, C-2, D storages. Intervals of boreholes in these sites are each 10, 14, 21m. As the analysis results, the rate of pressure down of boreholes decreases as the interval of boreholes decreases. Table 5 shows the rate of pressure down of boreholes according to the interval of boreholes in the pulse tests.

Table 5. The rate of pressure down of boreholes according to the interval of water curtain boreholes

Permeability (m/s)	B storage (10m)	D storage (14m)	C-2 storage (21m)
1.0x10E-7	0.0%	3.14%	6.4%
1.0x10E-8	6.31%	15.93%	17.65%
1.0x10E-9	20.78%	35.7%	39.3%

5. UNDERGROUND WATER FLOW ANALYSIS

5.1 Summary

The validaty of the results of hydraulic tests was reviewed through numerical analysis by finite difference method. The mathematical model of program consists of partial differential equation on three dimensional flow of groundwater in porous media as, below.

$$\frac{\partial}{\partial x}(K_{xx}\frac{\partial h}{\partial x}) + \frac{\partial}{\partial y}(K_{yy}\frac{\partial h}{\partial y}) + \frac{\partial}{\partial z}(K_{zz}\frac{\partial h}{\partial z}) - W = S_s\frac{\partial h}{\partial t} \quad (2)$$

In the above equation (2), K_{xx}, K_{yy}, K_{zz} are hydraulic conductivity in each x, y, z direction, h is head, S_s is storage coefficient, t is time.

The solution of equation (2) is calculated to h(x,y,z,t). As the analysis results, the difference of values between field tests and numerical analysis showed less than only 10% as below table 6.

Table 6. The comparision of pressure down by numerical analysis with field measurements

Permeability (m/s)	Pressure down(kg/cm^2)	
	field measurements	numerical method
1.0x10E-7	0.07	0.075
1.0x10E-8	0.36	0.393
1.0x10E-9	0.75	0.786

5.2 The analysis on the pressure down of boreholes according to the arrays of boreholes by numerical method(clude oil carven)

Because field tests were carried out for only 3~4 days, it is extremely short term compared with construction period of a facilities. therefore, the results of field tests don't show hydraulic behaviour of boreholes in long term. In this section, the pressure down of boreholes in long term was analysed in total 28 cases with design conditions of boreholes (distance : 15, 20, 25, 30m, interval : 10, 14, 21m) including draw down tests and pulse tests by numerical method. Fig.5 plots values of the pressure down in the rock mass in mid-position of boreholes in 12 cases of design conditions of boreholes. The pressure down of boreholes in numerical analysis increases as the distance of boreholes decreases and the values in case of 15m distance were analysed to higher than other cases.

But, in the numerical analysis, the pressure down in cases that the distance of boreholes is higher than 20m were showed a trend related to the distance in contrast with the result that was not related to the distance in field tests. The reason is considered as the difference between continuous media supposed in the model and inhomogeneous media in the natural rock mass. The pressure down was calculated in the range of 0.395~1.03kg/cm^2 according to the intervals of boreholes. Fig.6 shows the results of numerical analysis on the draw down and the pulse tests in

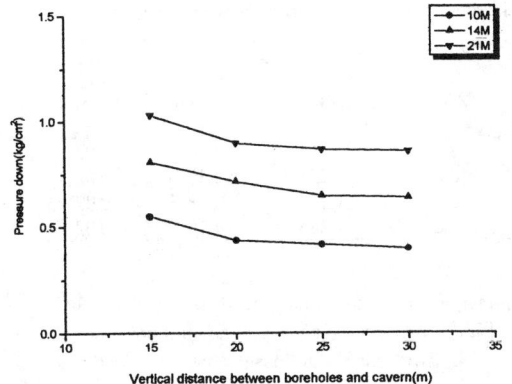

Fig. 5 Pressure down of boreholes according to the vertical distance in numerical analysis

case that the interval of boreholes is each 10, 14, 21m. The values were 0.795~1.567 kg/cm² in the pulse tests and 1.989~2.133kg/cm² in the draw down tests. The values calculated with steady state condition were about three times as large as the results measured in field tests. that is, the pressure down of boreholes measured after the passage of four days in field was about 30% of the value calc-

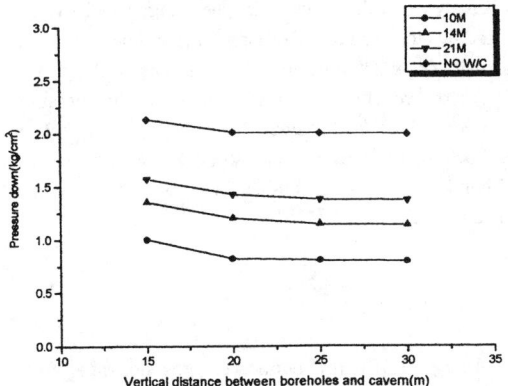

Fig. 6 Pressure down of boreholes according to the vertical distance in numerical analysis (draw down and pulse test)

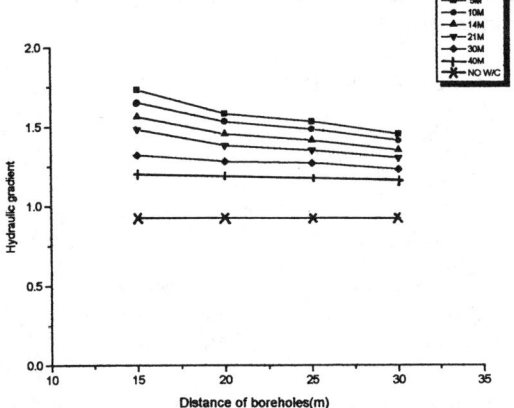

Fig. 7 Distribution of hydraulic gradient in the rock mass over cavern according to distance

ulated with steady state condition by numerical analysis, and the value after sixty days was about 85%. Table 7 shows the results calculated by numerical analysis.

5.3 The hydraulic gradient analysis in the surrounding rock mass over cavern according to the arrays of boreholes

The hydraulic gradients in the rock mass over cavern in total 28 cases according to the arrays of boreholes mentioned in the above section were calculated with an operation condition with inner pressure of two kg/cm² in cavern. Fig. 7 shows that the hydraulic gradient has a trend increasing as the distance and the interval of boreholes decreases.

This means that the hydraulic gradient in the rock mass is increased because groundwater flows more easily as the distance and the interval of boreholes approaches. The calculated values were in the range of 1.23~1.73 according to the arrays of boreholes, but its value in no water curtain boreholes was 0.92.

Table 8. shows the analysis results of hydraulic gradient in the rock mass over cavern.

Table 7. The pressure down of boreholes according to arrays of water curtain boreholes

(unit : kg/cm²)

Interval Distannce	Injection			Pulse test			Draw down test
	10m	14m	21m	10m	14m	21m	no w/c
15m	0.548	0.806	1.03	1.003	1.353	1.567	2.133
20m	0.435	0.715	0.897	0.824	1.201	1.422	2.008
25m	0.415	0.645	0.865	0.805	1.145	1.375	1.998
30m	0.395	0.635	0.855	0.795	1.135	1.365	1.989

Table 8. The distribution of hydraulic gradient in the rock mass over cavern according to arrays of water curtain boreholes

Distance \ Interval	5m	10m	14m	21m	30m	40m	no w/c
15m	1.73	1.65	1.56	1.48	1.31	1.20	0.92
20m	1.58	1.53	1.45	1.38	1.28	1.18	0.92
25m	1.53	1.48	1.41	1.35	1.27	1.18	0.92
30m	1.45	1.41	1.35	1.30	1.23	1.16	0.92

5.4 The analysis on pressure down and hydraulic gradient according to inner pressure of cavern and the interval of boreholes(LPG cavern)

The pressure down and the hydraulic gradient in the rock mass over cavern was analysed with design conditions of B storages. The analysis conditions are as follows. Distance of borehole is 25m, interval of borehole is 10m, section of cavern is 18x27m, level of boreholes is -90msl, water pressure of borehole is $10kg/cm^2$, inner pressure of cavern is $1kg/cm^2$ in construction term, and $7.9kg/cm^2$ in operation term, hydraulic coefficient is $2.0 \times 10E-9 m/s$, others are isometric, steady state conditions.

As the analysis results, the pressure down of boreholes were each $1.62kg/cm^2$ and $0.8kg/cm^2$ in construction and operation condition in case that the interval of boreholes is 10m, and each $3.32kg/cm^2$ and $1.7kg/cm^2$ in case that it is 20m. But, the pressure down in case of no water curtain boreholes were each $9.98kg/cm^2$ and $4.66kg/cm^2$ in construction and operation condition.

In the mean time, the hydraulic gradient in the rock mass over cavern was analysed to each 4.95, 4.33, 1.22 in construction term and each 2.25, 1.97, 0.5 in operation term in three cases that the interval of boreholes are 10, 20m, no water curtain boreholes. From these results, we can know that the boreholes in B LPG storages should be installed necessarily.

6. CONCLUSIONS

The main results analysed in this study are summarized as follows.

1) As the results of the draw down and the pulse tests carried out in five fields, The pressure down of boreholes increases as the hydraulic conductivity of the surrounding rock mass decreases.

2) The pressure down of boreholes showed the highest value in case that the distance of boreholes is 15m, but the value was not influenced to the distance any more in case that the distance is 20~30m.

3) The trend on pressure down of boreholes analysed by numerical method was well corresponded with the results measured by hydraulic tests in fields. The pressure down of boreholes measured after the passage of four days in field was about 30% of the value calculated with steady state condition by numerical analysis, and the value after sixty days was about 85%.

4) As the results of numerical analysis on operation condition of crude oil storages, the hydraulic gradient in rock mass over cavern increases as the distance of boreholes decreases, and the analysed values were in the range of 1.23~1.73 according to design conditions of boreholes, but its value in no water curtain boreholes was 0.92.

5) As the results of numerical analysis on operation condition of LPG storages, the hydraulic gradient in the rock mass over cavern was 2.25 in the interval of boreholes of 10m, but it was 0.5 in no water curtain boreholes.

REFERENCES

B. Aberg, 1977, Prevention of gas leakage from unlined reservoirs in rock, the first Int. sympo. on storage in excavated rock caverns, Rockstore 77, Stockholm, Sweden, September 5~8

B. Aberg, 1989, Pressure distribution around stationary gas bubbles in water saturated rock fractures, proc. Int Conf. Storage of gas in rock caverns, Trondheim, Norway, june 26~28

C. O. Soder, 1994, Water curtain in gas storage an experimental study, Goteborg, Sweden

H. Kjorholt, 1991, Gas tightness of unlined hard rock caverns, Norges tekniske hogskole universitetet I trondheim, Norway, May

A study on the variation of surface and groundwater flow systems related to tunnel excavation in the DONGHAE mine area

Dae-Hyuk Lee, Dong-Woo Ryu, Taek-Kon Kim & Hi-Keun Lee
Department of Mineral and Petroleum Engineering, Seoul National University, Korea

ABSTRACT: The purpose of this study was to effectively manage the excavation process of the transport tunnel in the DONGHAE mine area by investigating the variation of surface and groundwater flow systems around the tunnel and neighbouring villages. Thus, the effects of excavation and water-proofing on the water system was studied by following methods: a naked eye survey of the tunnel and surface outcrop, a joint survey, core drilling, measurements of the surface water quantity, evapotranspiration and precipitation, a rock hydraulic approach, and numerical analysis. From the above approaches, we concluded that the exhaustion of the surface water was not caused by the tunnel excavation but by a continuous drought that last 1 year, and that the effect of the excavation on the groundwater system was minimized by an effective water proofing process.

1. INTRODUCTION

One of the problems related to underground excavation is the variation of the surface and subsurface flow systems which causes the drying up of surface water, the inflow-into-tunnels, and the variation of groundwater head distribution, and flow direction. The inflow of groundwater makes the operation of excavation equipment very difficult and slows the excavation process. The variation of in-situ stresses induced by the discharge of groundwater may trigger the subsidence of the surface.

This paper presents many approaches to the investigation of the variation of the surface and subsurface flow related to the excavation process of a transport tunnel in the DONGHAE mine area.

The groundwater flow through the major joints of the tunnel found in the range of 1680m to 1900m starting at the adit delayed the operation of TBM and the excavation works. The epoxy grouting was performed to reduce the discharge rate of the discharge area and was followed by additional cement grouting and shotcrete for complete water proofing.

A discharge area of about 1000m was confirmed inside the tunnel by a naked eye survey, a joint survey and an outcrop survey which characterized the discontinuities and the surface geology, respectively. We used the trajectory method for the approximate evaluation of the surface flow rate in the Choneun Valley. We divided the neighboring area of the Choneun Valley into the farmland and the forest, and measured the surface hydraulic conductivity of each subarea. The analysis of the evapotranspiration and water budget was accomplished based on the monthly weather report data from the Samchuk Province. After installing a water levelmeter into the four boreholes, we measured the variation of the water level over time. The hydraulic conductivity of the basin rock was determined through the in-situ borehole pressure test. The MODFLOW package was used to create a numerical simulation of the flow system in the drainage area including the tunnel.

2. TOPOLOGY, GEOLOGY AND TUNNEL SPECIFICATION

The studied drainage area is located at the Choneun Valley in Miro-meun Samchuk Kangwon-do. Fig. 1 shows the contour and 3-D geological profile of the tunnel. The transport tunnel was excavated by TBM with a 13.6 km length and 4.5 m diameter. Instead of soil compaction, the concrete was coated 1.2 m thick for the roadbed work to prevent the inflow. In addition, epoxy grouting was completed against the tunnel wall for the water-proofing process along approximately a 200m interval of the 2km tunnel line that passes across the drainage area. The thickness of the cap rock was about 115m above the tunnel section crossing the Choneun Valley in the drainage area. The rock mass between the surface and the tunnel was composed of Samhwa granite and limestone. Samhwa granite, of which the tunnel was excavated, had intruded into the limestone geology. Fig. 2 shows the detailed surface geology in the drainage area.

3. THE RESULTS OF GEOLOGICAL SURVEY

3.1 Joint spacing and RQD

Assuming that the discontinuities occur at random the mean spacing was estimated with a core depth of 1~14m at borehole no. 3. In the interval of 1 4~51m, the joint spacing was over 1 m. The core logging of 14 meters long found that 27 separate pieces had a length of more than 0.1m.

Let $\overline{x_t}$ be the mean total discontinuity spacing and then the total discontinuity frequency λ_t is $\lambda_t = 1/\overline{x_t}$. If L is the total length of the borehole core, the total number, n_{int}, of the separate intact pieces is $\lambda_t = 1/\overline{x_t}$ given by the total number of discontinuities $n_{int} = \lambda_t L = \dfrac{14}{\overline{x_t}}$.

The probability, $P(X_t \rangle x)$, that a random value of the total discontinuity spacing X_t is greater than x is, for negative exponential spacings, given by

$$P(X_t \rangle x) = e^{-\lambda_1 x} \quad (1)$$

Noting that in this case $x = 0.1 m$ and equating the numerical proportion $14/n_{int}$ with $P(X_t \rangle x)$ gives $\dfrac{27\,\overline{X_t}}{14} = e^{-0.1/\overline{x_1}}$ or $\overline{x_t} = 0.5185 e^{-0.1/\overline{x_1}}$.

Solving this equation gives $\overline{x_t} = 0.405\,m$, $\lambda_t = 2.47\,m^{-1}$. The theoretical RQD for a general threshold value t, was first derived by Priest and Hudson (1976) in the following equation:

$$TRQD_t = 100\,e^{-\lambda t}(1 + \lambda t) \quad (2)$$

Equation (2) for $t = 0.1\,m$ gives $TDQD_t = 97\,\%$. It was found that $\overline{x_t}$ was $1.02\,m$, λ_t was $0.98\,m^{-1}$, TRQD was 99% up to the core depth of $28\,m$, and $\overline{x_t}$ was $0.985\,m$, λ_t was $1.02\,m^{-1}$, TRQD was 99% up to the core depth $47\,m$ of borehole no. 4.

The results of the joint survey inside the tunnel wall were approximately consistent with those in borehole no. 3 and no. 4, which made it possible to assume the joint pattern to be isotropy.

3.2 The results of the joint survey in the outcrop and the tunnel wall

The discharge interval in the tunnel was from 1680m to 1900m. The interval covered the major fractured joints of which the apertures were 1~3 cm minus the infilling material. The joint wall was unweathered, and the spacing is about $1\,m$. The joints exposed to the tunnel wall had the average dip and strike of $70°\,/N60°\,E$. Although the tunnel line was described to pass through the Ep area (Pungchon granite) as shown in Fig. 2, it must be noted that it passed through Samhwa granite to the depth of $110\,m$. The discontinuity orientation obtained from the surface outcrop survey was plotted on a lower hemisphere projection to group the three joint sets and determine those orientation properties. We defined a major joint set and two minor joint sets to have the orientation properties as follows :

Joint set 1 : dip and dip direction - 86° / 149°
Joint set 2 : dip and dip direction - 81° / 119°
Joint set 3 : dip and dip direction - 81° / 096°

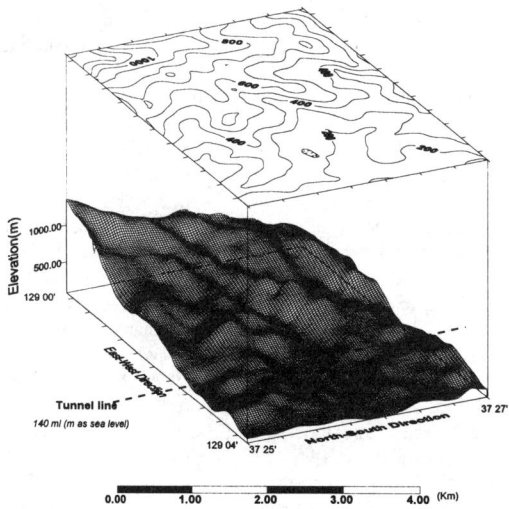

Fig. 1. The topographical map and 3-D view of the studied area.

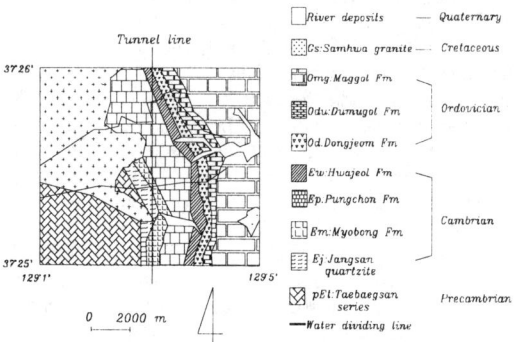

Fig. 2. The surface geological map of Samchuk area including the experimental basin.

Table 1. The length of daylight time, and adjustment coefficient according to the latitude.

Month / Latitude	1	2	3	4	5	6	7	8	9	10	11	12
30	10.4	11.0	11.9	12.8	13.6	14.1	13.9	13.1	12.3	11.4	10.6	10.2
37.5	9.8	10.6	11.8	13.1	14.1	14.8	14.4	13.5	12.4	11.2	10.1	9.5
40	9.6	10.5	11.8	13.2	14.3	15.0	14.6	13.6	12.4	11.1	9.9	9.3
Adjustment coefficient	0.82	0.88	0.98	1.09	1.18	1.23	1.20	1.13	1.03	0.93	0.84	0.79

The above results approximately equal to the mean discontinuity orientation $70°/N60°E$ obtained by the above joint survey on the fractured joints inside the tunnel. The mean spacing of the discontinuities was estimated 1.255 m by the scanline survey. Supposing that the distribution of discontinuities is a negative exponential distribution, the mean spacing can be obtained from the equation below (Priest and Hudson, 1976).

$$\mu_{XL} = \frac{1 - e^{-\lambda L} - \lambda L e^{-\lambda L}}{\lambda(1 - e^{-\lambda L})} \quad (3)$$

where μ_{XL} is the mean spacing, λ is the estimated discontinuity frequency (m^{-1}), and L is the length of scanline (m). Using the equation (3), the estimated discontinuity spacing is 1.255 m, which is equivalent to the spacing 1.02 and 0.98 in borehole no. 4.

4. HYDROLOGICAL OBSERVATION

4.1 Analysis of water budget

To evaluate the potential evapotranspiration, we applied Thornthwaite's method which represents the function of the weather factors. Thornthwaite's empirical equation is as follows:

$$PE = cT_m^a \quad (4)$$

where PE is potential evapotranspiration (mm), a and c are constants depending on the location, and T_m is average monthly temperature (℃).

The value of a can be determined from annual heat index, I, as follows:

$$a = 67.5 \times 10^{-8} I^3 - 77.1 \times 10^{-6} I^2 + 0.0179 I + 0.492$$

$$I = \sum_{m=1}^{12}\left[\frac{T_m}{5}\right]^{1.51} \quad (5)$$

Provided that daytime is 12 hours and a month is 30 days, Equation (4) can be expressed simply as follows :

$$PE = 1.62b\left[\frac{10T_m}{I}\right]^a \quad (6)$$

where b is adjustment coefficient depending on the longitude.

The average daytime data per month is divided by the standard 12 hours to calculate the adjustment coefficient based on the reported data of Samchuk. In equation (5) and (6), the months when the average temperature is less than 0℃ was excluded.

We used Thornthwaite's empirical equation above to estimate and compare differences over time of the potential evapotranspiration based on the monthly and annual rainfall from 1985 to December 1994. The total rainfall from September, 1993 to July, 1994 was only 47.4 mm. It was far behind the average monthly rainfall, 110.3 mm. This period had continuous dry weather causing a drought and, notably, the precipitation in July of 1994 was only 23% of the annual average value. The monthly mean of the potential evapotranspiration was 62 mm which is close to the yearly mean and the difference of minimum and maximum is about 10%. The soil-moisture storage capacity, SS is supposed to be 42.8 mm to evaluate the portion of the actual evapotranspiration to the potential. The remainder after subtracting the soil-moisture storage capacity and the potential evapotranspiration from the monthly mean precipitation was added to the soil-moisture storage capacity of the proceeding month to give the surplus of the precipitation. If the estimated SS exceeds the monthly precipitation, the remainder after subtracting the potential evapotranspiration from the precipitation is assumed to be the actual SS, and the precipitation exceeds the evapotranspiration, the part of SS compensates for the shortage. It was concluded that soil moisture provided water contents sufficient enough for the evapotranspiration, because the value after subtracting the precipitation from the potential evapotranspiration exceeds SS except for June. In the Samchuk region, the actual evapotranspiration was 55%, the ratio of PE to P. The lost quantity was defined as the value after subtracting the surface water outflow from the precipitation. As the seepage and infiltration of the surface outflow was usually ignored in the water budget analysis, the lost quantity may be equal to the actual evapotranspiration. But, if these quantities can not be ignored, the difference is the subsurface recharge. The analysis of the weather report said that 42% of the total precipitation was lost quantity of which 43% (18% of the total precipitation) was the subsurface inflow, the other 57% was evapotranspiration. But, groundwater recharge depends on field conditions: we considered the recharge after the evaluation of the lost quantity.

Table 2. The evaluation of lost quantities using Coutagne and Turc equations, and comparison with the evapotranspiration

	Input data	Lost quantity (mm)		Ratio to rainfall (%)		Ratio of potential transpiration to rainfall (%)
		Coutagne	Turc	Coutagne	Turc	
Average during the last 10 years	T:12.5℃ P:1350mm	638	633	47	48	55
During 1994 year	T:13.6℃ P:1083mm	649	635	60	59	71

Table 2 shows that the difference between the actual evapotranspiration and the potential was 7~8% during the past 10 years and 11~12% in 1994. So, we considered the difference of 7~8% as the groundwater recharge, which was consistent with the results from the stream flow measurement.

4.2 Stream flow measurement

The trajectory method was used to evaluate the surface flow in the Choneun Temple Valley and the results are summarized in Table 3. The five measurements were performed from Aug., 1994 to Feb., 1995 and the average of the four results from each measurement was determined. The flow of the upper stream was measured to be 240 m^3/day, which was measured during the dry weather of Aug. 1994, and the flow of Nov., 1994 was measured to be 340 m^3/day after rainfall. The third measurement was performed when 25 days elapsed after the second and the amount of precipitation was about 100 mm. The effect of precipitation recovered the surface flow to be 400 m^3/day. If the flow is 300~400 m^3/day in the drainage area, it can be concluded that it is considered as a normal stream.

5. MEASUREMENT OF HYDRAULIC CONDUCTIVITY

5.1 Hydraulic conductivity of surface soil

A part of the quantity after subtracting the loss quantity due to evapotranspiration from the precipitation is surface flow, and the other part is the liverbed water or seepage into the rock. So, it is very important to measure the hydraulic conductivity in surface soil. A transparency acril pipe with a diameter of 8~10 cm was inserted into the soil as deep as possible and the hydraulic head is kept at 1 m. We applied a falling head test to measure the head drops over time. Hydraulic conductivity can be calculated by Darcy's law as follows:

$$K = 2.3 \frac{L}{t_2 - t_1} \log_{10} \frac{H_1}{H_2} \quad (7)$$

where K is the hydraulic conductivity (cm/min), L is the thickness of formation (cm), and H_1 and H_2 are the heads at a time t_1 and t_2, respectively.

Fig. 3 shows the result of the measurement in the farmland of the drainage area. The relationship between head drop and time is bilinear as shown in Fig. 3. In the initial stage, the head drop needs to saturate the soil layer and in later stage the head drop is kept constant after complete saturation. So, the abscissa of an intersection point of two regressed lines is the saturation time. The line after the saturation is used to obtain the hydraulic conductivity as follows:

$K = 1.701 \times 10^{-5}$ cm/sec in the farmland
$K = 8.14 \times 10^{-4}$ cm/sec in the colluvium

From the hydraulic conductivities in two kinds of soil layers, the soil layer of the drainage area is thought to have the intermediate hydraulic conductivity and be inadequate for serving as an aquifer. From this condition, this surface soil layer has the ability of water storage and although, in rainfall, surface flow runs fast to downward streams the liver-bed water flows out with delay.

5.2 In-situ pressure test

The in-situ pressure test (injection fall-off test) was performed to measure hydraulic conductivities of rock masses.

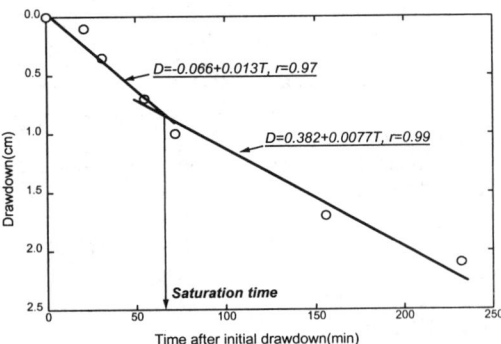

Fig. 3. The result of hydraulic conductivity measurements in the surface soil layer of the farmland of the drainage area.

Table 3. The result of measurements for surface water flow during seven months.

Date	#	Dia(cm)	y	%	x(cm)	x(in)	Flow(gpm)	Flm(gpm)	Flow(lpm)	Discharge m³/day	Specific Discharge m³/day/km²
13/8 1994	1	10.2	7	0.31	23.8	9.37	121	38.0	167.2		
	2	10.2	6.5	0.36	22.0	8.66	121	43.9	193.3		
	3	10.2	7.5	0.26	20.0	7.87	108	28.6	125.9		
	4	10.2	7	0.31	24.5	9.65	135	42.4	186.5		
							Average	38.2	168.2	242.2	78.3
04/11 1994	1	10.2	7	0.31	33.0	12.99	176	55.2	243.2		
	2	10.2	6.5	0.36	27.0	10.63	142	51.5	226.8		
	3	10.2	7	0.31	34.0	13.39	182.5	57.3	252.2		
	4	10.2	6.5	0.36	27.0	10.63	142	51.5	226.9		
							Average	80.3	237.3	341.6	110.4
29/11 1994	1	10.2	7	0.31	35	13.78	189	59.3	261.1		
	2	10.2	4.5	0.56	25	9.84	135	75.4	332.2		
	3	10.2	6.5	0.36	29	11.42	149	54.0	238.0		
	4	10.2	7	0.31	30	11.81	162	50.8	223.8	322.3*	
							Average	59.9	277.1	399.1	128.98
12/01 1995	1	10.2	9.2	0.098	5	1.97	27	2.65	11.66	16.79	5.43
07/02 1995	1	10.2	9	0.118	5	1.969	27	3.18	13.99	20.14	6.51

Drainage area : 3.09 km², Gradient : 466/1000

* This measuring point was located downstream about 100 meters away from the Choneun Temple

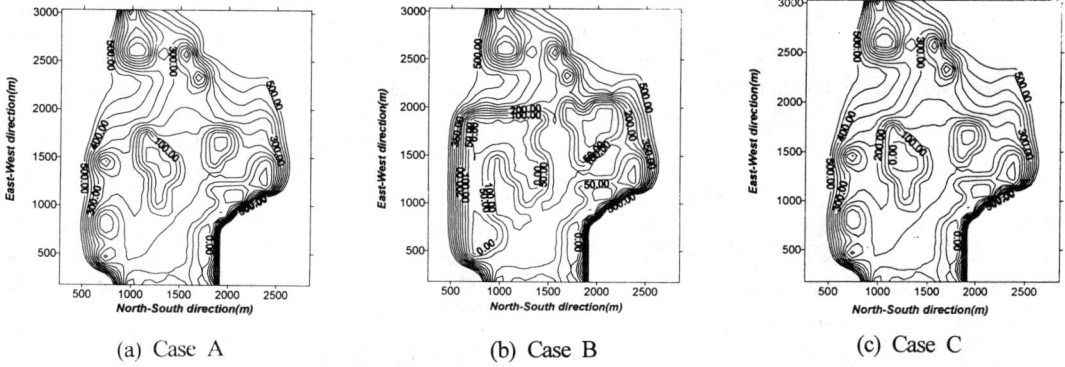

(a) Case A   (b) Case B   (c) Case C

Fig. 4. The head distribution at stress period 2 (winter, 1993) in layer 1 of the 3-D model.

As the variation of pressure is measured over time after the increase of the borehole water pressure to a given value, this method has the advantage in the rockmass of low hydraulic conductivity.
Hydraulic conductivity is defined as follows:

$$K = \frac{2.3Q}{4\pi L m} \quad (8)$$

where L is the interval of test (m), and m is the slope of the line on the Horner plot, which is equal to $\triangle P$ at the point that $\log(t_P + \triangle t)/\triangle t$ is 1 cycle. t_P is total injection time and $\triangle t$ is the time through the pressure drop.

The geometric mean of measured hydraulic conductivities was $6.52 \times 10^{-7} cm/\sec$. It is 25 times smaller than the hydraulic conductivity of the colluvium and 1240 times smaller than that of the farmland. It is thought that the hydraulic conductivity of the basin rock is usually 100 times smaller than that of the surface soil. In comparison with the surface soil, the recharge of the infiltration water through the surface soil layer takes much more time and it is expected that only excavation in the basin rock does not affect the surface water flow.

6. NUMERICAL MODELLING

6.1 Input data

Numerical modelling was performed to obtain the information of overall head distribution and flow patterns based upon measured geological and hydrological properties of the drainage area. First, 2D plane model was used to simulate the head distribution and the flow direction. Then 3D models were classified and applied to each cases as follows:
1) A 3D model without any tunnel to simulate the surface and subsurface flow conditions - Case A
2) A 3D model with an ungrouted tunnel to simulate the effects of tunnel excavation - Case B
3) A 3D model with a grouted tunnel to simulate the effects of grouting for water-proofing - Case C
All measured and calculated data was inputed into the numerical analysis. The simulated period was 1 year from May, 1993 to June, 1994, which included dry weather. The results of the modelling were presented in terms of the head distribution, the head drop and the flow direction and quantity.

6.2 The results of modelling

Fig. 4 shows the head contours in Case A, B and C during the last time step of the stress period 2, early winter of 1993, respectively. Layer 1 is the shallow subsurface layer of which the upper stream is covered with granite but most of site is covered with limestone. In Fig. 4 (b), the variation of the head along the tunnel line was found to be more distinct than in Case A. It was concluded that the groundwater flow into the tunnel caused the head drop. The difference between the Case A and C is, however, very small. The water proofing was very effective in restricting the flow into the tunnel.

Fig. 5 shows head distribution contour in the vertical section (from west to east) at the last time step of stress period 2. In Fig. 5 (a) and (c) the black solid circle represents the tunnel. The right boundary is the upper stream of the Choneun Temple and the left is the lower part of the stream. The head distributions of the lower stream are the same, but the head drop around the tunnel in Case C is more dramatic than in Case A. It was due to the prevailing water flow into the tunnel. But its effected zone did not reach the shallow surface layer. In this section along the valley, the subsurface flow to the downstream is more dominant than into the tunnel. So, it may be concluded that water proofing with grouting has been performed effectively.

7. CONCLUSIONS

The effects of the excavation for the transport tunnel on the surface and subsurface flow are thought to be negligible from the results of this study. From the analysis of the water budget, surface flow, and the variation of hydraulic head, it is found that the water proofing with grouting was effective. The numerical simulation using MODFLOW confirmed these conclusions.

8. REFERENCES

Ssangyong Eng. Company 1991. *A report on geological survey of adit and brigde area for B/C structure basic design in Donghae mine area.* 15-16
C.W.Fetter 1988. *Applied hydrology.* London: Merrill Publishing Company. 285-295
Stephen D. Priest 1993. *Discontinuity analysis for rock engineering.* Chapman & Hall. 121-141
J.H.Sunwoo 1994. *Hydrology.* Dongmyungsa. 77-93
Korea Water Resource Corp. (KOWACO) *Annual report on water resource*
KIGAM 1993. *A report on the groundwater flow in 2nd tunnel for Masan industrial water*

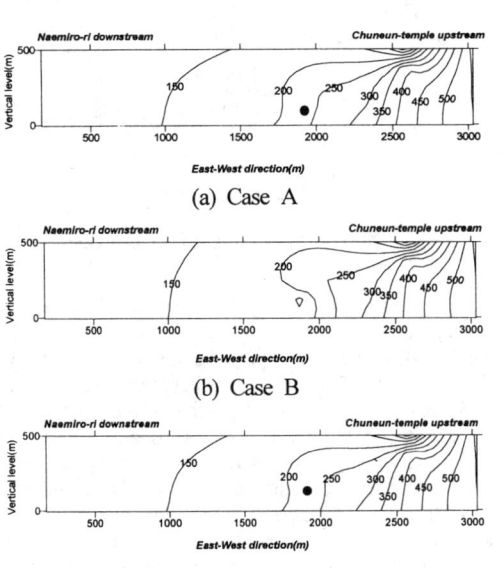

Fig. 5. The head distribution at stress period 2 (winter, 1993) in the crosssection along the valley.

How to evaluate jointed rock masses with respect to water-sealing

A. Okamoto
Japan Underground Oil Storage Co., Ltd, Tokyo, Japan

K. Kojima
Faculty of Engineering, University of Tokyo, Japan

Y. Nakazawa
Mitsui Kushikino Mining Co., Ltd, Japan

M. Hasegawa
Shimizu Corporation, Japan

ABSTRACT : The first rock caverns for oil storage in Japan have been constructed in Kuji, Kikuma and Kushikino. Although many phenomena could not maintain the necessary groundwater level through some cracks and fractures which occurred during the process of actually constructing the rock caverns, we were able to complete the storage facilities into service by the various measures to render water-sealing effective. As a result, it was confirmed that the observation of the flow trends through water passage is most important to secure the water-sealing system. For this reason, we made the classification and evaluation system to be composed mainly of "Evaluating Water passages" and "Geological Conditions such as Joint Spacing etc.", how to evaluate jointed rock masses with respect to water-sealing. This paper describes the outline of such classification for aiming and helping the development of water-sealing technology.

1 INTRODUCTION

The first rock caverns for oil storage in Japan have been constructed in Kuji, Kikuma and Kushikino[1] (Fig.1). Although many phenomena could not maintain the necessary groundwater level though some cracks and fractures which occurred during the process of actually constructing the rock caverns, we were able to complete the storage facilities and put them into service by the various measures to render water-sealing effective. Additionally, with the objective of reproducing the phenomena, a three-dimensional time-depended groundwater analysis was carried out.
The result was that calculated values conformed well with date measured from the initial investigation up to the entry into service of the rock caverns.

This paper describes the outline of various which were taken to secure the water-sealing system, during their construction and operations for aiming and helping the development of water-sealing technology.

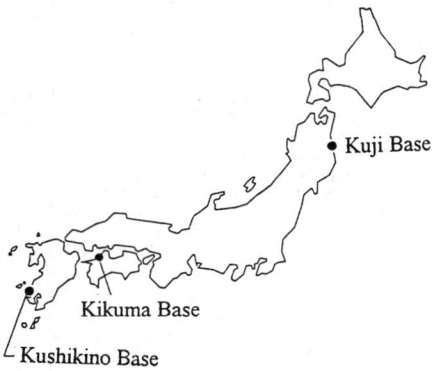

Figure 1. Location of Underground Oil Storage Bases

Table 1. Result of water-Sealing Management

	Kuji Base state	Kuji Base measures	Kikuma Base state	Kikuma Base measures	Kushikino Base state	Kushikino Base measures
geology and hydraulic ・rock and category 2)	H M L Kuki-granite 95%(2% 31% 66%)		H M L Shinki-Ryoke }97%(63% 32% 5%) granite		H M L andesite(LB) 76%(63% 32% 5%)	
・joint spacing	I II III IV V — — 5% 82% 13%		I II III IV V — 2% 65% 31% 2%		I II III IV V 13% 34% 22% 26% 4%	
・crack,fault,fracture	eleven faults and intrusive rock		twenty two fractures		fourteen faults and intrusive rock	
・seepage water	・main water passage:faults, intrusive rock and surround ・water level fall rapidly and widely ②③ ・rock surface wet	・design base change ⑤ cement grouting ↓ add clay grouting (wide area)	・water passage: limited fractures ・water level fall a little ・some part of rock surface dry		・water passage:limited fractures and intrusive rock ・water level of some part of No.9 fault fall already ⑩ ・many part of rock surface dry	・design base change ⑪ nature water-sealing system ↓ artificial water sealing system (partially)
・coefficient of permeability average high permeability zone	7.6×10⁻⁶cm/sec >10⁻⁴cm/sec		8.1×10⁻⁷cm/sec <10⁻⁵cm/sec		1.5×10⁻⁷cm/sec >10⁻⁴cm/sec	
water balance ('84~'94average) ・rainfall (A) ・voparization (B) ・river flows (C) ・D=A-B-C ・D' (m³/day)	some rivers dry ② 1,324 mm/year 321 " 382 " 621 " (2,331 m³/day)	soil moisture (non damage)	1,234 mm/year 636 " 324 " 274 " (466 m³/day)		2,154 mm/year 596 " 681 " 877 " (1,938 m³/day)	
seepage water (E) ・after excavation ・before oil in ・after oil in	design 3,760t/day estimate 10,000t/day 3,400t/day (max.4,700t/day) 6,200 " [max.7,400t/day when flood water tunnel ①] 4,500 " (decreasing)	target <6,000t/day ・from inside cavern cement grouting ・from outside cavern clay grouting ⑤⑥	design 290t/day a little seawater 210t/day 340t/day 250t/day	・special grouting(decrease) ・cement grouting test (a little effect)	design 240t/day increase when change artificial water-sealing system 1,320t/day (max.2,300t/day) 1,370t/day 1,000t/day (decreacing)	・for high permeability zone length30-40m invert cement grouting ⑫ (effect) ・for joint spacing category III cement grouting (a little effect)
injection water (F) ・after excavation ・befor oil in ・after oil in	design 2,880t/day store water in water tunnel when excavated 700t/day 3,200 " [increase when flood water tunnel] 1,800 " (decreasing)		design 150t/day store water in water tunnel when excavate 80t/day 140 " [increase when flood water tunnel] 110 "	・improvement of water quality (a little effect)	・high permeability zone >3,000t/day ・water leakage outside 250t/day 300 " 220 " (decreasing)	・curtain grouting for water passage ⑫ ・adopt artificial water-sealing system(partially)⑫
groundwater level	・some observation wells not be able to keep limited water level ④ ・the area where ground water level fall rapidly ③	add water injection borehole (keepable)	・almost recovery when flood water tunnel (except one well) ⑲ ・demonstration plant water level fall	・cement grouting (a little effect) ・clay grouting (effect)	・necessary to confirm recovery ground water level	・add observation well to fault (effect) ・piezometer (confirm) ⑫
airtigh test	・boosting cavern pressure by compressed air (0.3kg/cm²) ・unsaturated some part of rock where ground water level fall rapidly ⑦⑧	・add water injection boreholl and cement grouting (effect)	・boosting cavern pressure by compressed air (0.3kg/cm²) ・saturated all of rock		・boosting cavern pressure by compressed air (0.3kg/cm²) ・unsaturated some part of rock where ground water level fell rapidly and slowry ⑭	・water-sealing level set up (effect) ⑮ ・add water injection borehole and cement grouting (effect) ⑯
observe unsaturation of rock after oil in	drying rock surface ⑨	・add water injection boreholl (effect)	・no drying		・drying rock surface ⑬	・add water injection borehole and cement grouting (effect)

 phenomenon

2 OUTLINE OF THE WATER-SEALING MANAGEMENT SYSTEM

2.1 Adoption of water-sealing system and necessity of water-sealing management system

Unlike Scandinavian countries where rock caverns are developed in early years, Japan, perhaps, has one of the most complex geological setting in the world. For this reason, in order not to let oil or gas leak out the cavern, an artificial water-sealing system (Fig.2) is adopted.

The mechanism of the water-sealing effect is as follows : in the rock mass below groundwater level, joints, cracks, and other crevices are filled with natural or artificial groundwater. If the groundwater pressure within the surrounding rock mass is greater than the pressure within the storage cavern, a constant groundwater flow can be set set up towards the cavern by discharging the seepage water from the cavern. With wrapping up the crude oil and gas in this groundwater flow, safe storage can be achieved without leakage from the cavern. However, some part of the rock easily allows water to flow out, while others do not, through cracks and fractures in natural rock, and so the groundwater level cannot be constantly maintained in natural conditions. Therefore, the water-sealing management system in particular, is most important and necessary until the completion of the rock cavern.

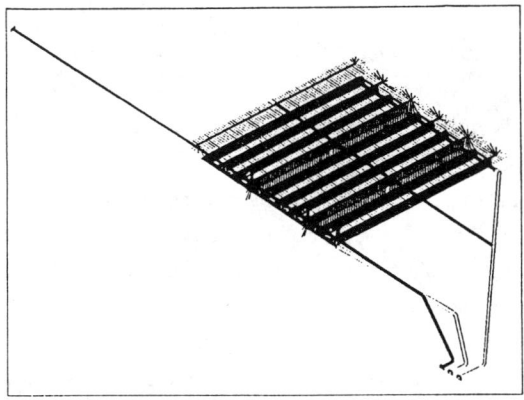

Figure 2. Artificial Water-Sealing System (Kuji Base)

2.2 Framework of the water-sealing management system

The most important objective of the water-searing management is to secure the necessary groundwater level without risking a leakage. For this reason, the level of undergroundwater (which is most important the rock cavern) and its changes caused by the construction work were continuously monitored through observation wells.

The water level was maintained by cement grouting or increasing the number of the water injection boreholes, when necessary, until the completion of the construction. The water balance surrounding the rock cavern is also important. The seepage water in the rock cavern is supplied by rain and injected water. The rainfall around the cavern area, the river flows and vaporization were observed to survey how much of these natural flows come into the cavern. The flow trend was continuously observed and monitored.

The effect of the excavation to plants on the ground surface and other ground conditions has also been observed.

3 PHENOMENA

3.1 Examples of phenomena

Various phenomena not envisaged at the design stage occurred during the process of actually constructing the caverns, causing serious problems to the water-sealing system. The state of the management of water-sealing system is shown in Table 1.

3.2 Characteristics of phenomena

The phenomena, classified from the standpoint of evaluating water passage, are as follows :

1. water passages, which proved that the rock masses were not homogeneously porous, existed in rock masses surrounding the caverns at Kuji and

Kushikino Bases. These water passage were found to be the dominant obstacle in water-sealing.

2. At Kuji and Kikuma Bases, where artificial water curtains by means of boreholes were employed the amount of water injected was small. The diameter (76mm ϕ) of each borehole was not so large or the number of boreholes not so sufficient that the probability of continuously encountering connected joints which form water passages was found to be low. Hence unsaturated zones are considered to remain in the rock masses surrounding the caverns.

3. During the process of grouting in order to stop seepage, joints from which water had subsided or through which water scarcely flowed tended to be overlooked, thus becoming latent water passages and forming unsaturated pockets.

4. Once the water table fell in highly permeable rock masses, it is difficult to restore the previous water-level by means of water-injection through boreholes. Consequently, unsaturated zones are formed.

5. The water level in an observation well can be considered to be a result of hydrological balance in the borehole as a water container. Changes in the water level can be classified into four groups depending on the situation of the observation well :

(1) The rock mass surrounding the observation well comprises a zone of uniform permeability.

(2) Continuous joints derived form the weathered zone near the surface of the mountain intersect the observation well.

(3) Continuous joints connecting to the water-sealing tunnels or rock caverns intersect the observation well.

(4) Not only the features of both (2) and (3) are given, but also joints the top and the bottom of the observation well are often blocked or opened up.

6. At the Kuji storage facility, an earthquake resulted in changes in the amount of seepage, the amount of water supplied and the water table.

7. Water passages have a tendency to become narrower.

4 HOW TO EVALUATE JOINTED ROCK MASSES WITH RESPECT TO WATER-SEALING

4.1 Background

A detailed analytical survey on the rock mass in which the phenomena occurred revealed the following :

Of the continuous joints in the rock mass which became water passages, there are :

1. Joints which lose all seepage water rapidly or within a short period of time. These were found in the faults and the high permeability zone of Kuji Base. Also, those joints in the high permeability zone which caused a partial drop in the water table below the floor of the rock cavern at Kushikino Base, fall into this category.

2. Joints where the flow of groundwater decreases with time elapsing and completely dry up unless recharged during the construction process.

3. Joints which are constantly filled with groundwater, and to which Darcy's law is applicable. These were found at Kikuma and Kushikino Base. They are the three types of joints which act as water passages ; the phenomena in question occurred in type 1 and 2. Formation of water passages is complicated, and should be considered to be network of pipes. In this paper, type 1, 2 and 3 are referred to as categories L, M and S, respectively.

Fig. 3 physically shows signification of L, M and S.

4.2 Evaluation of joints from the standpoint of effective water-sealing gives a relationship between the phenomena and the categories L, M and S

A large number of surveys and measurements were carried out in order to ensure the effectiveness of water-sealing. Geological conditions such as joint spacing, existence of faults and fractured zones and the degree of cracking, as well as the hydrological condition of seepage, were investigated in classifying the rock mass at the three Bases.

These criteria are shown in Table 2.

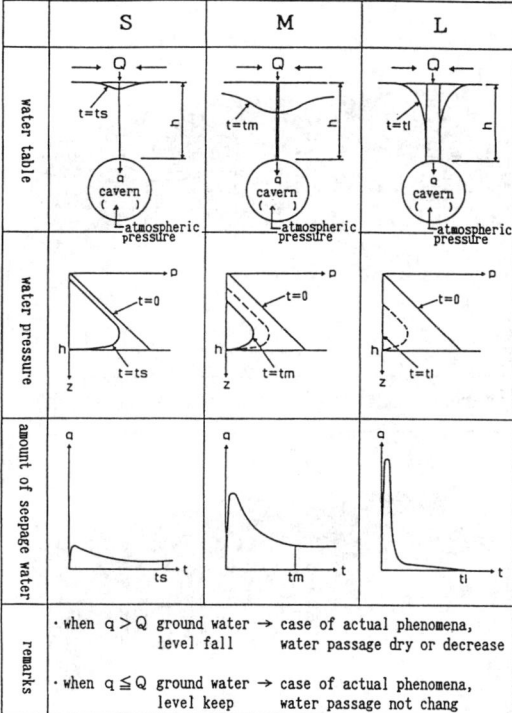

Figure 3. Physical Signification of L, M and S
 ＊Case of single joint

Table 2. Criterion of Classification

「Water passage」

rank	judgment factor		
	continuity of seepage water	water level fall	amount of seepage water
S	no change	small	little
M	decrease	middle	middle
L	drying	large	much

- do not judge only situation of water passage, but it is important to grasp a water vein
- set up S S and L L according degree

「Joint spacing」

rank	criterion
I	>100cm
II	40～100cm
III	20～40cm
IV	5～20cm
V	<5cm

- rock classification for deep underground cavern

「Crack」・「Fault・Fracture」

rank	judgment factor		
	water level fall	amount of seepage water	scale
joint I	small	little	small
joint II	middle	middle	middle
fault・fracture	large	much	large

- when fault, fracture and boundary are small scale, clssify category joint I or II

4.3 Classification and evaluation system

Table 3 summarizes the investigation. An analysis gave the relationship between the rock classification and the phenomena that occurred, and the aforementioned ctegories L, M and S, leading to a method of evaluating rock mass and joints in terms of the effectiveness of water-sealing. In this paper, this method is termed a water passage model for evaluating rock mass and joints. According to Table 3, jointed rock mass corresponding to L and M requires measures to render water sealing effective. Fig. 4 shows examples of crssified water passages.

Table 3. Combination of factor
(water passage model for evaluating jointed rock mass)

		coefficient of permeability of joint small ← → large		
	joint spacing	joint I	joint II	fault・fracture
coefficient of permeability of rock mass small ↑ ↓ large	I	S S	S S	S S
	II	S	M	L
	III	S	M	L
	IV	M	L	L
	V	L L	L L	L L
remarks		appear often inside thick line		

5 REPRODUCING THE PHENOMENA WITH NUMERICAL ANALYSES[3)]

With the objective of reproducing the phenomena, a

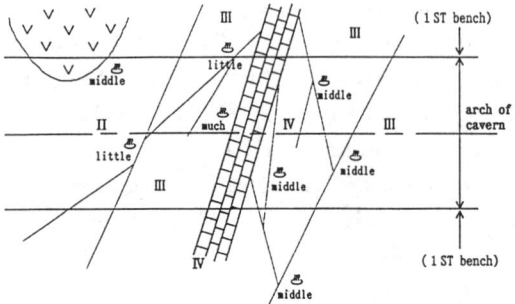

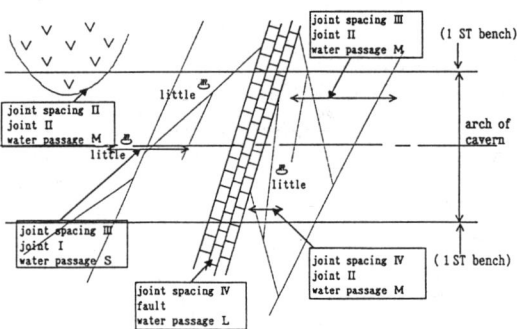

Figure 4. Examples of Classified Water Passage

three-dimensional time-dependent groundwater analysis with two-component dual-phase was carried out, according to the conditions supplied by the water passage model. The result was that calculated values conformed well with date measured from the initial investigation up to the entry into service of the caverns. Hence, the water passage model used in this paper is appropriate for examining the effectiveness of water-sealing. Additionally, a prediction analysis was carried out for each of the cases of : a) a year of heavy rainfall, b) a year of drought, c) an earthquake, and d) when the capacity to suply water for water-sealing drops. It was found that the model served as a valuable criterion in determining the effectiveness of water-sealing.
Fig. 5～7 examples show.

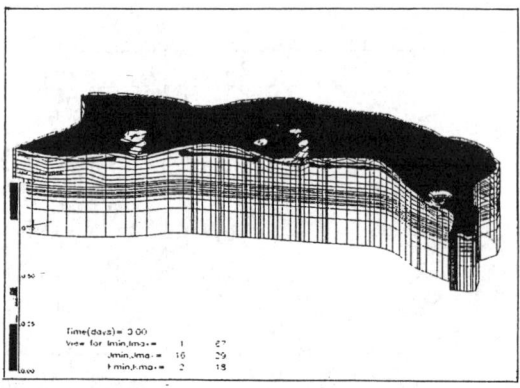

Figure 5. Saturation (before excavation)

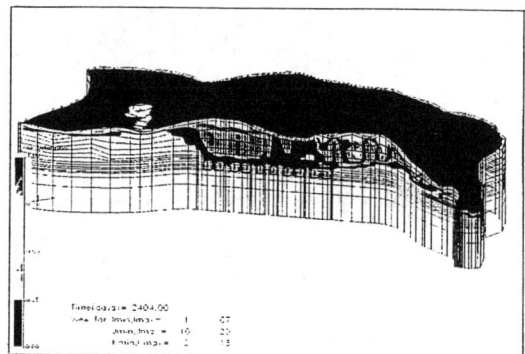

Figure 6. Saturation (when the excavation of cavern completed)

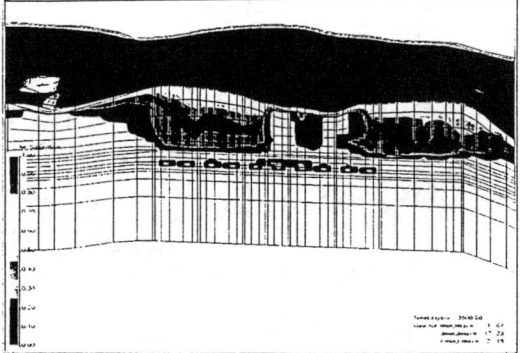

Figure 7. Saturation (after oil in···Kuji Base)

6 CONCLUSION

A detailed analytical survey on the jointed rock masses in which the phenomena occurred during the process of actually constructing the three water-sealing rock caverns revealed the followwing :

1. The classification and evaluation system (Table 3) was made to be composed mainly of "Evaluating water passages" and "Geological Conditions such as joint Spacing etc".

In this paper, this method is termed a water passage model (Table 3) for evaluating jointed rock masses. It is found that the model served as a valuable criterion in determining the effectiveness of water-sealing.

2. The discovery at an early as possible stage of joints of categories L and M as shown in the Table 3 is most important to secure the water-sealing system.

3. The excavation of the rock cavern without lowering the water table and the planning countermeasures to be taken in the event that the water table dose fall are also important in managing water-sealing.

It is expected that this model would be improved through accumulating advanced data etc.

7 ACKNOWLEDGMENT

The author would like to thank Dr. Hiroyuki TOSAKA and Mr. Tuyoshi YAMAISHI for reproducing the phenomena with numerical analyses.

REFERENCES

1) T. MAKITA : Outline of Underground Oil Storage Plant, *Journal of the Mining and Materials Processing Institute of Japan* 107 (1991) No.13

2) T. MAKITA, Y. FUKUTAKE, N. HOSHINO, K. IGUCHI and K. NIINO : Rock Classification for Deep Underground Caverns, *Journal of the Japan Society of Engineering Geology* 32-5, (1991)

3) A. OKAMOTO, H. TOSAKA, T. TANI, T. YAMAISHI and H. KOBAYASHI : History Matching Simulation of Surface and Underground Flow around Underground Oil Plant during the Construction, *Symposium of Journal of the Japan Society of Engineering Geology*, (1995)

Fracture detection/characterization techniques for the groundwater control around the oil storage caverns

O. Choi, H. Tosaka & K. Kojima
Tokyo University, Japan

A. Okamoto
Japan Underground Oil Storage, Co., Ltd, Tokyo, Japan

ABSTRACT: Fracture system is studied statistically using the data obtained in the site of underground rock caverns for oil storage. Over 1500 points of observation on the cavern walls, occurrences of fracture widths and spacings show fractal and exponential aspects, while negative binomial function may be applicable to evaluate the degree of clustering of them. Areal variations of statistical properties are also found which can be used for reconstructing 3-D structure of fractures.

1 INTRODUCTION

From engineering point of view, occurrences of fractures/faults and associated groundwater flows are the main concerns to be prospected in order to estimate costs and reduce risks for underground utilization. Many researchers have been studying the nature of fracture system using statistics, geometry, or numerical methods. However, most of our observations for fractures are limited to 1-D or 2-D, so that real nature of 3-D structure of fracture system is still remained within imagination or theory.

In Japan, oil storage plants which utilize underground rock caverns were constructed in three sites by 1994 and they are in operation today. From the initial stage of site investigation through completion of construction, continuous and thorough compilation of descriptions of geology, measurements of petrophysical and hydraulic properties, geometries of fracture systems had been done in those sites. Such data are very rare and useful to validate our historical knowledges be generally applicable or not, and to extend it to 3-D considerations.

In this paper, the authors discuss the statistical properties of fracture system by using a lot of spatial data obtained in an underground oil storage plant. Three dimensional aspects will be discussed separately in other papers.

2 STUDY AREA AND DATA ACQUISITION

One of the underground oil storage sites was selected for studying fracture structure. In the region, rock caverns are constructed in massive Kuki granite of the early cretaceous. As shown in Figure 1, ten oil storage caverns are lined up along north-south direction. For convenience of measuring fracture geometry, X, Y, and Z axes are set with the origin placed at the southern tip of the westernmost cavern. Fractures appeared on the cavern walls at excavation are described and recorded with other information as development figures for each cavern in this site. Figure 2 is the example of the developments.

Utilizing it, we made the data files by measuring the positions (X, Y, Z) and widths of a fracture at 4 points, i.e., top of the ceiling, two points on side walls, and the basement. Figure 3 shows relationship between the positions of fractures in averaged Y and the averaged widths of them. Strike and dip were calculated using the geometrical data above with least square approximation.

The data we used is basically one dimensional along caverns, but it can be extended to 2-D or to 3-D data if proper correlation is made. In Figure 4, areal distribution of fractures at the level of the mid depth of the cavern are shown.

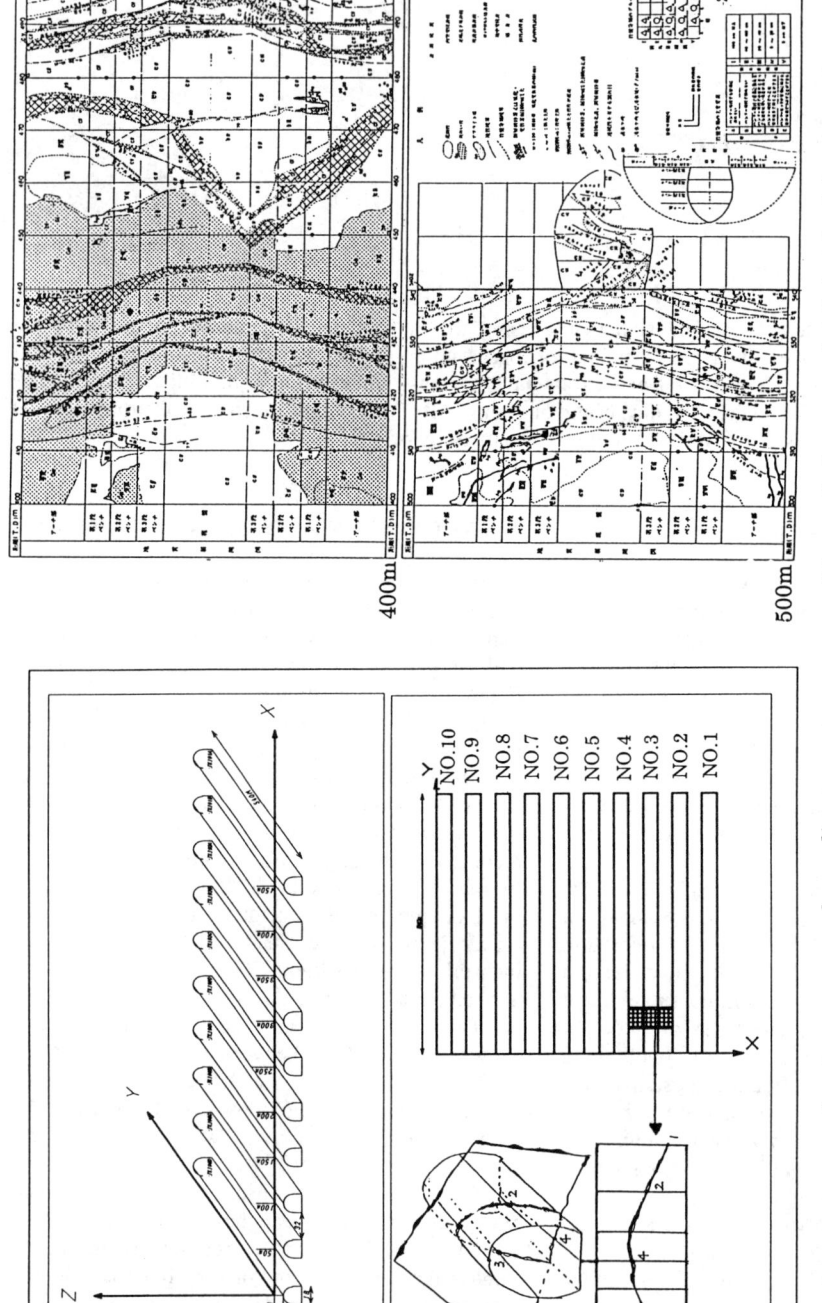

Figure 2. An example of sketched fractures in a cavern development.

Figure 1. Arrangement of rock caverns and coordinate system for data acquisition.

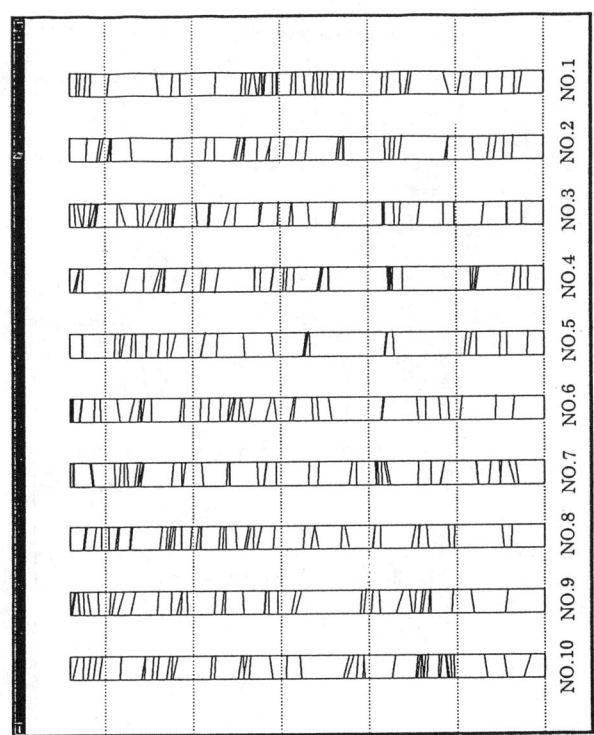

Figure 4. Areal distribution of fractures.

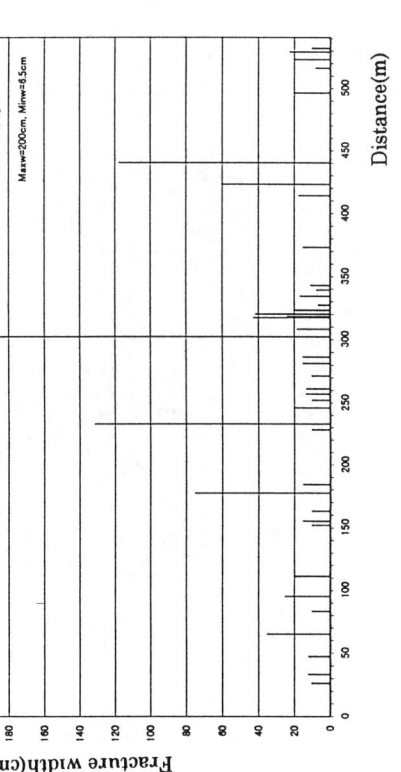

Figure 3. Location vs width of observed fracture.

3 STATISTICAL PROPERTIES OF FRACTURES

3.1 Basic statistical properties

In Table 1, basic statistical properties are shown for each cavern and for all the caverns. The width, i.e., thickness or aperture, is the transverse length of fractured zone for large faults, while for small ones it is the single aperture. The strikes of fractures are the angles measured from the north clockwisely.

The characteristics of the fracture system of this site can be summarized as follows.
(1) The average number of fractures found in a caverns along Y is approximately 39, and the standard deviation is about 4.
(2) The averaged value of width is about 30cm and the difference among caverns is small.
(3) The average spacing between two adjacent fractures is about 14m, with minimum and maximum values of 0.5m and 77.5m, respectively.
(4) Dominant direction commonly exists in E-W direction for all the caverns.
(5) The dips of fractures/faults are almost constant with about 70°N.

3.2 Distribution of fracture width

In Figure 5, cumulative numbers of fractures $N(w)$ having width greater than w are plotted against width w. The distribution over 1500 sample points matches to the power function much better than the logarithmic or exponential ones. This indicates that the fracture widths in 3-D region of this site distribute as fractal with the dimension of about 1.5. For each cavern, the frequency distribution, not cumulative one, can also be approximated by power function, but the fractal dimension varies as shown later. Since the advocation of the concept of 'fractal' by Mandelbrot, a lot of statistical studies found that the distribution of 2-D trace lengths of fractures might be close to fractal. For this case, we validated that it may be true for wide 3-D points.

3.3 Fracture spacing

In Figure 6, cumulative number of fractures $N(l)$ having the spacing greater than l are plotted against the spacing l, in which the exponential function fits much better than the power or logarithmic one. This indicates us that the positions of fractures might be distributed spatially according to Poisson's.

Table 1. Basic statistical properties of fractures.

Caverns No.	No. Of Frac	Frac. Width Mean (m)	S.D (S)	Frac. spacing Mean (m)	S.D (S)	Dip of Frac. Mean (degree)	S.D (S)	Strike of Frac. Mean (degree)	S.D (S)
NO.1	40	0.29	0.38	13.0	12.8	72.1	7.3	93.0	6.4
NO.2	34	0.43	0.44	14.7	14.5	74.2	7.6	97.2	5.6
NO.3	42	0.32	0.35	12.4	11.4	73.7	8.6	94.6	9.7
NO.4	39	0.32	0.30	13.5	18.7	74.2	6.6	97.1	7.5
NO.5	29	0.30	0.54	18.1	21.5	72.6	7.0	95.7	5.6
NO.6	38	0.32	0.47	13.6	12.3	74.1	7.9	94.1	8.4
NO.7	41	0.31	0.47	12.6	11.9	73.0	8.1	91.3	9.0
NO.8	43	0.25	0.35	11.6	11.9	73.1	9.7	94.9	7.2
NO.9	38	0.31	0.39	13.3	13.4	70.6	8.3	95.3	8.6
NO.10	43	0.16	0.13	12.2	12.4	69.8	8.8	95.2	9.5
T. Mean	38.7	0.30	0.38	13.5	14.1	72.8	8.0	94.8	7.7
S.D (S)	4.37	0.07	0.11	1.83	3.35	1.52	0.93	1.77	1.52

[T.Mean:Total Mean, S.D(S):Standard Deviation]

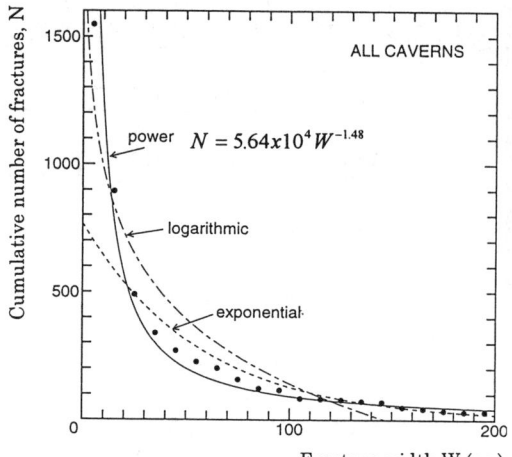

Figure 5. The cumulative distribution of fracture widths for all caverns.

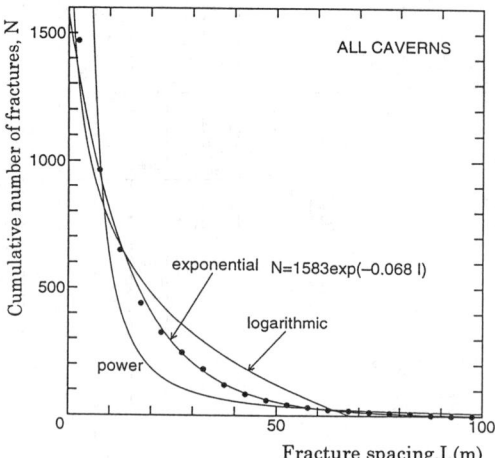

Figure 6. The cumulative distribution of fracture spacings for all caverns.

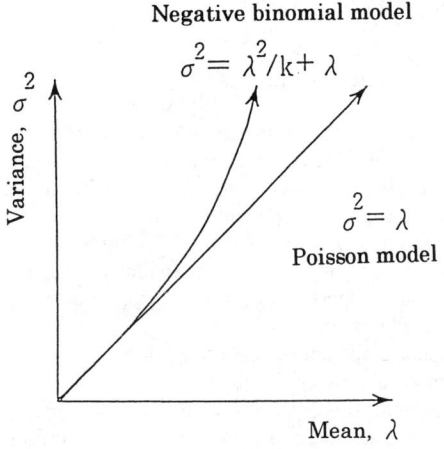

Figure 7. The relation between mean and variance and the patterns of distribution of the fractures.

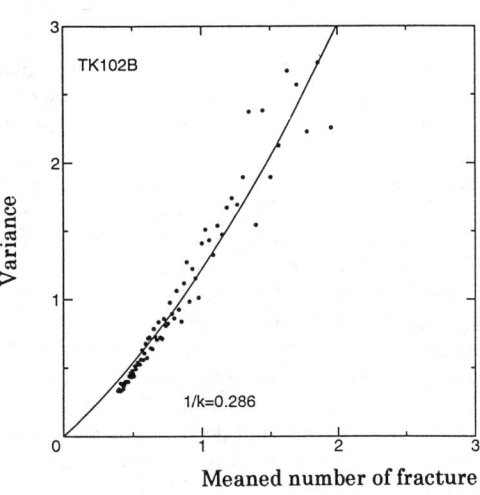

Figure 8. The typical clustered patterns of the fractures observed at a cavern.

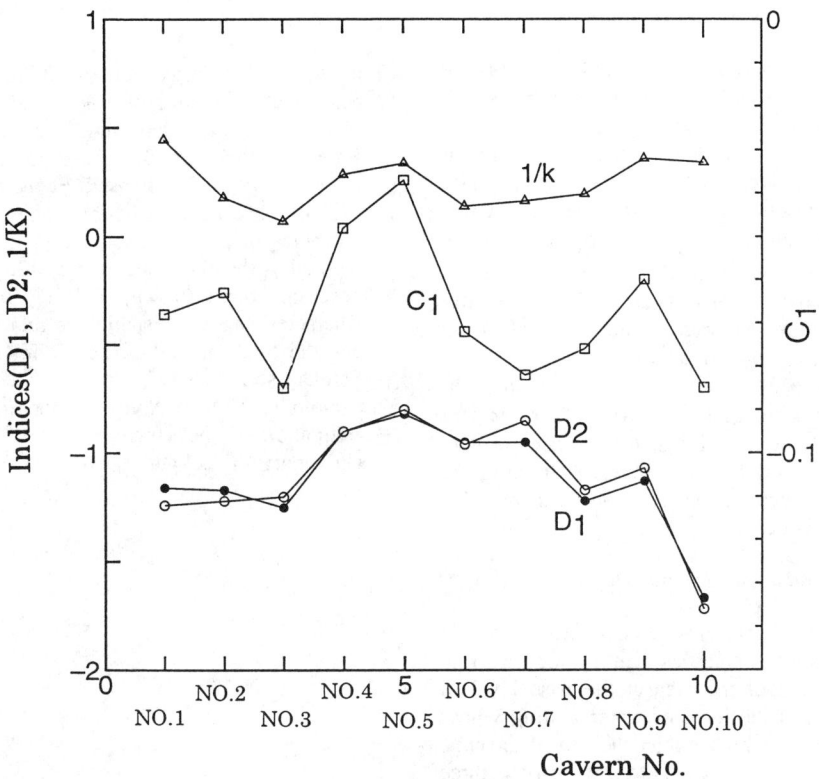

Figure 9. Comparison of statistical indices for all caverns.

3.4 Fracture position

As shown in Figure 3, the occurrences of fractures in this area are generally not so random, but in some part, are concentrated or clustered. Yamamoto(1995) made examinations on the quantification of clustering by using the negative binomial distribution. He divided a scan line L into N intervals with length l, and measured the number of fractures $n(l)$ and calculated the mean $\lambda(l)$ and variance $\sigma^2(l)$. By changing the interval length, he obtained many sets of mean and variance. If cross-plotted as shown in Figure 7, the samples from Poisson's distribution will make a straight line, while for clustered distribution the plots deviate to the upper side of the straight line, which can be refered to as the negative binomial distribution.

The result of plots for this site in Figure 8 indicates us that deviation is mostly to the upper side of the straight line, i.e., clustered to some degree, although variation is observed areally as mentioned in the following section.

3.5 Areal variation of statistical properties

In the study, we derived several indicators for the statistical properties of the fracture system of this area. In Figure 9, the fluctuation of them is shown cavern by cavern where $1/k$ is the clustering index calculated from the negative binomial distribution, C_1 is the order of exponential function for spacing, D_1 is the order of power function, i.e., the absolute value is equivalent to the fractal dimension, for cumulative width distribution, while D_2 is the order of power function for the interval width distribution.

Looking in the Figure 9, we see that the tendencies of variation of the four indices resemble showing the fluctuation of geometry of fracture system in this site. In reconstructing the 3-D fracture structure, this type of areal variation should be considered.

3.6 3-D structure and groundwater flow

There are several fractures having large width in this area. Their continuities can be traced from the plane section of the area at the base level of caverns, top, and the level of water sealing whose level is about 20m higher than the top of caverns.

In the future, we want to reconstruct three dimensional fracture structure from geometric and geostatistical aspects of fracture systems. We will consider the hydrological/hydraulical observation records of groundwater seepage and groundwater level, and numerical simulation.

4 CONCLUSIONS

In this paper, we presented the results of statistical analysis of spatial distribution of the fracture systems in an underground rock cavern site. And the following results were obtained.

We summarize our results as follows.
(1) Fracture widths distribute spatially showing fractal property, but the fractal dimension varies areally.
(2) Fracture spacing distribution along scan lines shows exponential shape.
(3) Negative binomial function may be applied to evaluate the order of spatial concentration or clustering of fractures.

REFERENCES

Baecher, G. B(1983) : Statistical analysis of rock mass fracturing, Math. Geo. Vol.15, No.2, 329-347

Choi Okgon, H. Tosaka and K. Kojima(1996) : Statistical properties of fracture system observed in an underground rock cavern site, Proc. of 1996 Jap. Soc. Eng. Geol., 169-172.

Koike K., Noguchi Y., Iwasaki H. & Kaneko K. (1996) : Spatial distribution analysis of attributes of rock joints using geomathematical methods, Shigen-to-Sozai, vol.112, 907-914

Villaescusa, E and Brown, E. T(1990) : Characterizing joint spatial correlation using geostatistical methods, Rock joints, Barton & Stephansson, 115-122

Yamamoto H.(1995) : Study on the interscale estimation of rock fractures, Ph.D thesis, University of Tokyo.

Blasting and machine excavation

Study on estimate of damage zone caused by blasting

M.Tezuka – *The Kansai Electric Power Co., Inc., Japan*
Y.Kudo & H.Matsuda – *Dept. of Civil Engineering and Architecture, Tokuyama College of Technology, Japan*
A.Hasui – *Technical Research Institution, Hazama Corporation, Japan*
K.Nakagawa – *Dept. of Civil Engineering, Yamaguchi University, Japan*

ABSTRACT: Core specimens were obtained from a retained wall to evaluate the damage region caused by blasting in a rock cavern. After reconstructing the separated cores in their original positions, P waves propagated normal to core axes were measured. P wave's recovery was found at about 0.6 m from the blast hole. After the measurement some core specimens were cut into two pieces parallel to the core axis on which thin section analysis was performed. Thin section analysis was used to observe damaged region in detail. The observation of thin section analysis supports the results that primary wave velocity is an appropriate measure for evaluating damage zones. Critical dynamic stress which might have caused damage in rock mass was evaluated from the results based on the measurement.

1. INTRODUCTION

In tunnel blasting, serious damage is induced around the blast hole. The more effective the fragmentation inside the cavern is, the stronger the outgoing stress waves are. Those stress waves cause serious damage to the rock mass which is supposed not to be damaged. Although many papers have been written about blast-induced damage, there still is little work done about the damage near field rock (Yang et al. 1993). Blast-induced damage hole results in reduction of elastic constants in rock cavern design and additional supports might be required. Therefore it is important to know how far the effects of blasting is reached and how serious the blast-induced damage is.

This study focuses on the evaluation of damage zone caused by blasting. Damaged zone induced in hard tuff is evaluated through P wave velocity propagated in drilled cores and optical microscopic observation of thin sections.

2. EXPERIMENTAL METHOD

The experiments performed in this study consists of two parts, i.e., in-situ experiment and laboratory experiment. Core specimens were prepared by blasting and drilling, and blast-induced cracks were investigated by P wave measurement and optical microscopic observation.

2.1 *In-situ experiment*

The in-situ experiment was performed at the extension site of Oku-tataragi pumped storage power station, located 90 km north-west of Osaka (Fig. 1). The test site located in a rock cavern which has 200m overburden. The rock mass at the test site is hard tuff though the main part of the rock mass consists of rhyolite. The uni-axial compressive strength of intact rock ranges from 233 MPa to 273 MPa. Fig. 2 shows the process of the test. Before blasting, two boreholes were drilled. Both holes were used for borehole camera monitoring before and after the test. The downward hole was filled with cement mortar after pre-blasting observation by borehole camera. Those arrangements were made in order to distinguish the pre-existing cracks from blast-induced cracks. Cracks found in cement mortar after blasting must be the blast-induced cracks. Two accelerometers are set along the stress wave travelling line in order to evaluate the peak particle velocity. Fig. 2a shows the arrangement of blasting holes. Rock mass between the blasting holes and free surface were removed by blasting (Fig. 2b). After removing the fragments, core specimens were sequentially drilled in the direction away from the blast hole (Fig. 2c). The drilled cores were numbered in order from upward to downward, #1 to #19. Fig. 3 shows the vertical profile of the experimental layout. Each

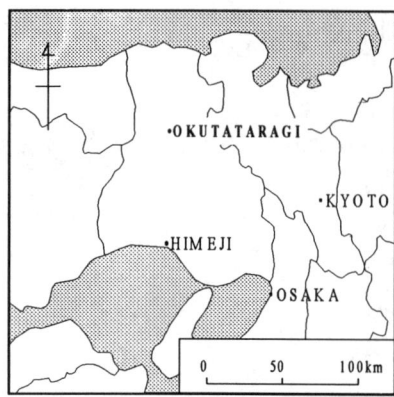

Fig. 1. Location of the test site

bottom of the blast hole from #1 to #4 was loaded with 0.9 kg explosive, while #5 and #6 blast holes were loaded with 1.1 kg explosive, respectively.

Fig. 4 shows the sketch of the drilled cores. All cores were broken into many pieces, especially near the blast holes, because of blast-induced cracks and/or natural joints. Separated pieces were collected and reconstructed into their original shapes using cyano-acrylate paste to suit the convenience of subsequent test in laboratory.

2.2 *Measurement in laboratory*

Fig. 5 shows experimental arrangement for P wave measurement. Since P waves are sensitive to the existence of open cracks, two kinds of P waves propagated normal to the core axis were used, one of which propagate within the horizontal plane and the other propagate within the vertical plane.

After the measurement, two core specimens, #10 and #19, which kept the original structures as they were before, were cut into two pieces along the vertical plane including the core axes. Photographs of both cut surfaces were taken so as to cover the whole area of the core. Each photograph was converted into a digital image using photo CD and reconstructed into its original structure from the fragments. Each image of one plane was converted into a mirror image that could be compared with the image of another plane easily. Pieces of core specimens were immersed in dye and then were carefully ground into thin sections for optical microscopic observation. Large thin sections (100 mm x 70 mm) of base planes consistent with cut plane, were made to check the microstructures made from blasting.

3. EXPERIMENTAL RESULTS

Fig. 6 shows the cracks found on the monitoring plane after the blasting. Fig. 7 shows the results of observation from borehole camera and cement mortar core after the test. In the borehole camera monitoring, cracks found both before and after the test might be pre-existing cracks, and cracks found in drilled cement mortar core might be blast-induced cracks. Crack identification should be done carefully because new cracks might be induced in the process of blasting and boring might enhance crack opening and make the obscure pre-existing crack visible. In that sense, it must be concluded that, from the cement mortar core observation, there was no blast-induced cracks beyond 1.4 m from the blast hole.

Fig. 8 shows the sketch drawn based on the observation by mosaic image of cut surface using photo CD. Crack I, II, III, IV shown in Fig. 8 are the cracks with orientation parallel to the lineament of the rock mass, cracks found in crashed zone, horizontally oriented crack, and vertically oriented crack, respectively.

Fig. 9 shows the P waves' profile obtained from the measurement of the reconstructed core

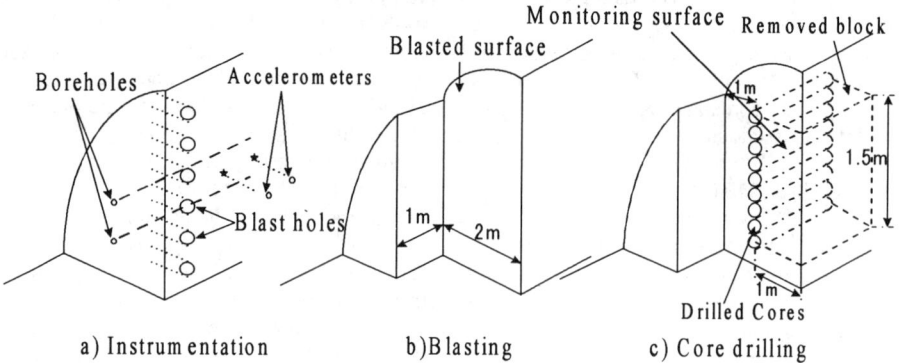

Fig. 2. Process of the in-situ test.

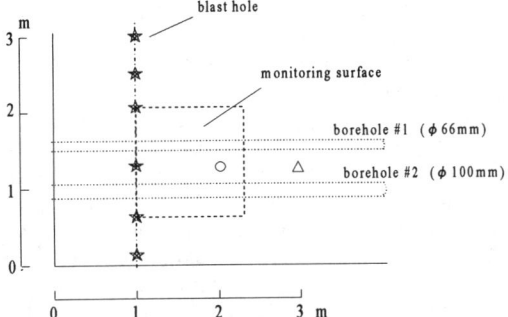

Fig. 3. Vertical profile of experimental layout.

specimen #10. Longitudinal axis is the distance from the blast hole. P waves within about 30 cm were not obtained because the rock could not be reconstructed from many small fragments. P wave velocity and wave amplitude give available information on inner structure of the core. P wave propagated in intact rock has higher amplitude and velocity than the wave propagated in damaged rock. In this study, wave velocity gives more reliable information than amplitude does since amplitude information is a bit ambiguous due to contact of surface's condition between the rock specimen and the transducer. As expected, the closer the measured positions to the blast hole are, the slower the P wave velocities are. This means that the density of the blast-induced cracks decrease with

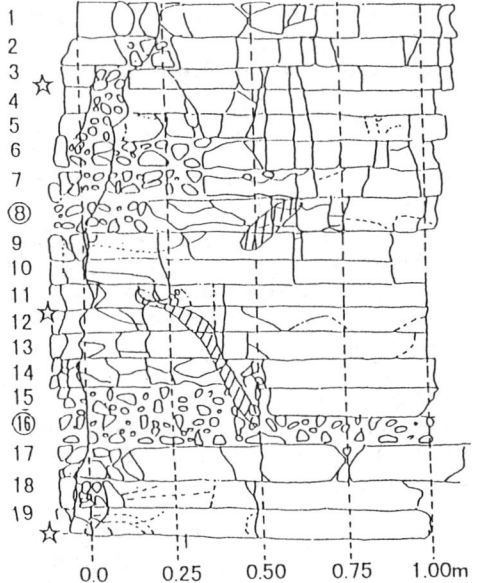

Fig. 4. A sketch of the drilled cores. The cores numbered with circle were used as monitoring boreholes for pre-blasting observation.

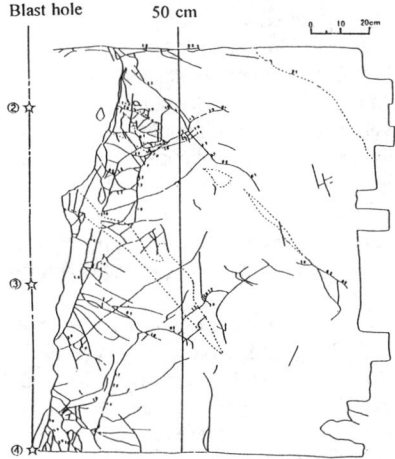

Fig. 6. Cracks found on the monitoring plane after blasting.

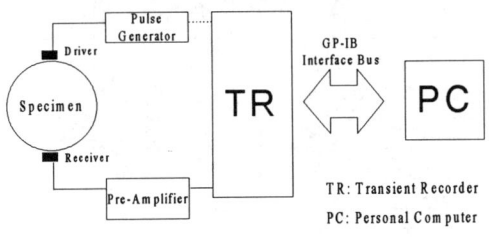

Fig. 5. Experimental arrangement for P wave measurement.

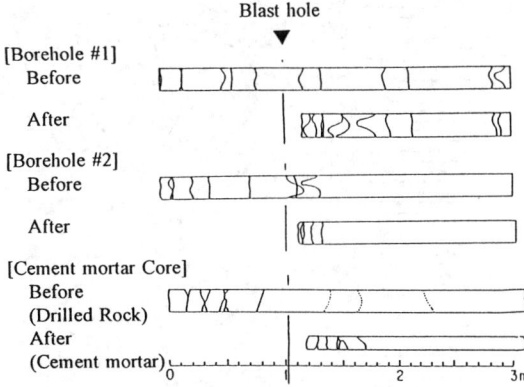

Fig. 7. Results of borehole camera observation and cement mortar core observation.

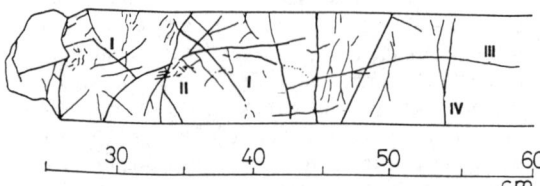

Fig. 8. Cracks found on cut plane

distance. The complicated wave shape near the wave front of vertically propagated wave is that of a chipped shape due to continuous drilling in vertical direction. It is obvious that there is a damage zone around the blast hole even if it is not visible to the naked eyes.

Fig. 10 and Fig 11 show the relationship between P wave propagated vertically and horizontally and distance from blast hole for the boring core #10 and #19. Since P wave is sensitive to cracks oriented normal to wave orientation, P waves propagated vertically and horizontally give information mainly on the cracks oriented in horizontal plane and vertical plane, respectively. P wave recovery to its original value for the core #10 is found at the point 54 cm for the waves propagated vertically and 64 cm for the wave propagated horizontally.

Fig 12 shows the P wave velocities for all boring cores tested. Roughly speaking, P wave velocities beyond 60 cm recover to their original velocity, 5200 m/s which was measured at the test site before blasting. Of course, there are some lower values than the original value, but those values were found for the paths near natural joints or partially damage zone. Therefore, as long as we could judge from the P wave's point of view, blast-induced crack zone is within about 60 cm from the blast hole.

The result of the optical microscopic observation for boring core #10 is shown in Fig. 13. Because of immersed dye, the cracks which have enough aperture were all dyed. Those dyed cracks appeared in thick solid lines, while the small cracks which cannot be penetrated by dye are shown in thin solid lines. Comparison of Fig. 13 with Fig. 8 indicates that more cracks can be detected by closer observation. The fabric found by core observation is similar to the fabric observed by optical microscope. The only exception identified between the two is that Crack III found in Fig. 8 is not a blast-induced crack but a calcite vein. Besides this point, it is indicated that P wave velocity propagated in boring core is an appropriate measure for blast-induced cracks.

4. DISCUSSION

P wave velocity measurement and the optical microscopic observation indicate that damage zone which is characterized by stress-induced cracks is at most within 60 cm from the blast hole. Of course, the size of damage zone depends on many factors, e.g. rock types, spacing of blast holes, charge weight, and so on.

Peak particle velocity has been used to generalize the effect of the blast damage, although this factor does not include a very important factor like strain rate. Peak particle velocity was calculated by integrating the acceleration measured at 1 m

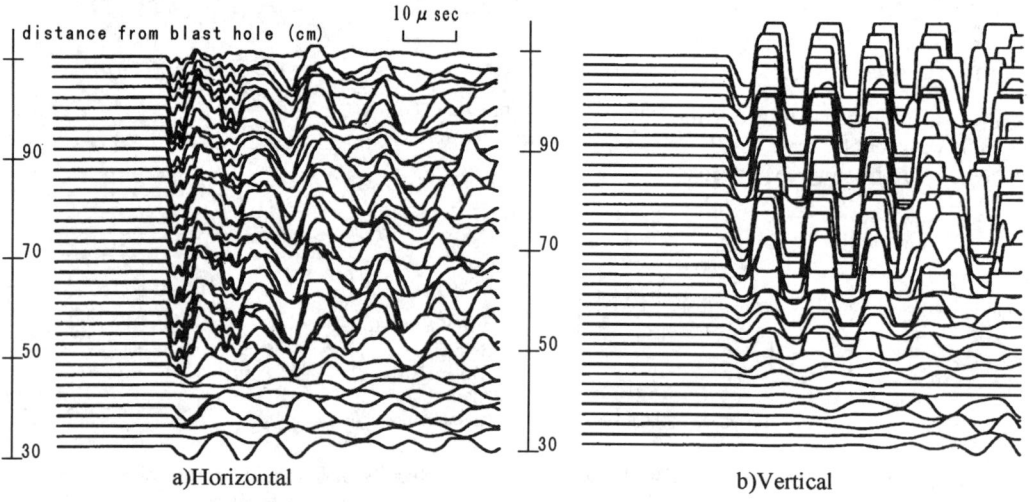

Fig. 9. Temporal changes in the seismogram of P waves.

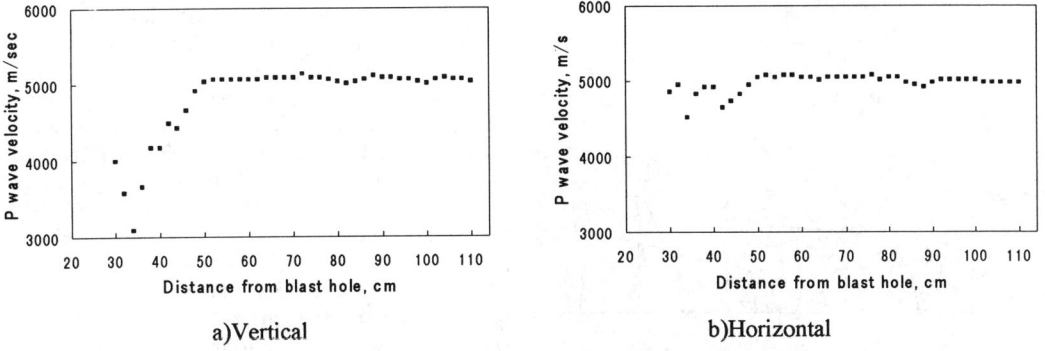

Fig. 10. P wave velocities transmitted in the boring core #10.

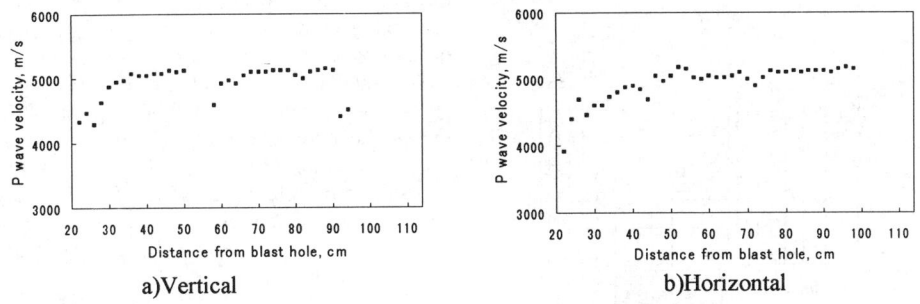

Fig. 11. P wave velocities transmitted in the boring core #19.

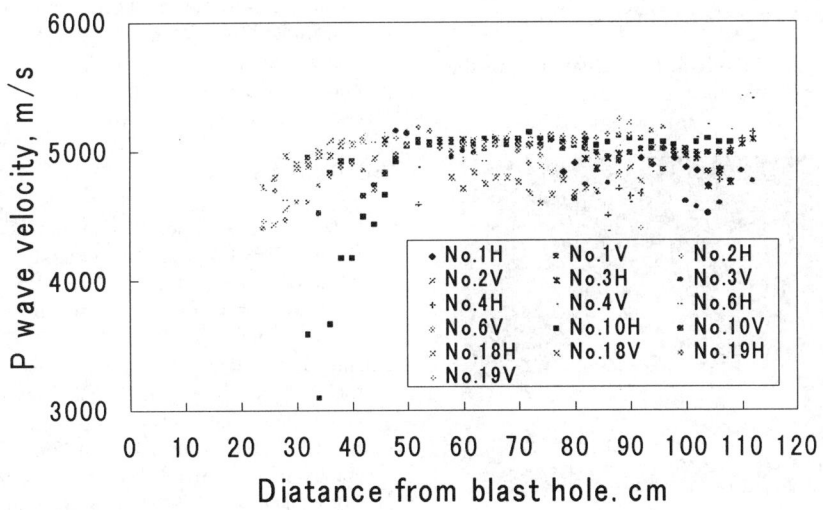

Fig. 12. P wave velocities measured for all cores.

and 2 m from the blast hole. Calculated peak particle velocity including extrapolated line for near field rock mass is shown in Fig. 14. Evaluated peak particle velocity at 50 cm from the blast hole is about 2.50 m/s. Dynamic stress σ is given by

$$\sigma = \rho \, cV$$

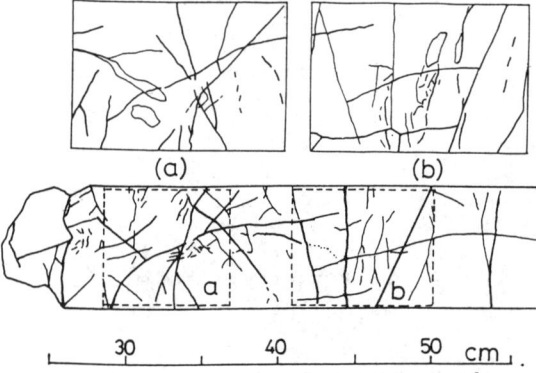

Fig. 13. The results of optical microscopic observation for core specimen #10.

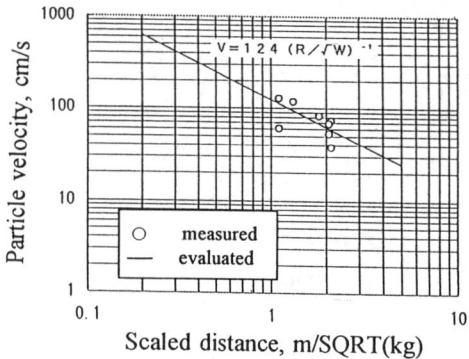

Fig. 14. Peak particle velocities evaluated from the acceleration measurement.

Fig. 15. Typical results which shows the dominating role of joints. The test was performed in the different site from the site of the present study.

where ρ, c, V are rock mass density, P wave velocity, and peak particle velocity, respectively. Substituting measured values for the in-situ rock mass, $\rho = 2650$ kg/m^3, c = 5300 m/s and the evaluated peak particle velocity at 50 cm for above equation, the dynamic stress which might have induced at the edge of damage zone is 340 kgf/cm^2 (21 MPa). This is a pretty high value compared with the static tensile stress which is presumed from its uniaxial compressive strength, 2400 kgf/cm^2. On the other hand, it is well known that the dynamic strength of rock exceeds the static strength by as much as ten times. Although further research on dynamic strength of the rock might be required, the method used in this study and the threshold value obtained appear to give some acceptable predictions for damage zone evaluation of hard rock.

It should be noted that the rock studied in this report was a hard rock with few joints. Fig. 15 shows the core specimens obtained from another site in the same district. In this case, joints play a dominating role in blasting, for the rocks are with well developed joints.

CONCLUSION

Damaged zone caused by blasting was evaluated by P wave velocity measurement of drilled core and optical microscopic observation. The damage zone evaluated was at most 60 cm for hard tuff rock mass. P wave velocity propagated in core specimen gives an appropriate measure for blast-induced cracks

REFERENCES

Yang R.L., P. Roscque, P. Katsabanis and W. F. Bawden 1993, Blast damage study of blast vibration and damage in the area adjacent to blast hole, Proc. 4th Int. Symp. Rock Fragmentation by blasting, 137-144.

Rinehart J. S. 1965, Dynamic fracture strength of rocks. Proc. 7th Symp. Rock Mech., Pennsylvania State University.

Grady D. E. and Kipp M. E. 1979, The micromechanics of impact fracture of rock, Int. J. Rock Mech. Min. Sci. & Geomech. Abstr., 16, 293-302.

Mechanical excavation of hard rock tunnels without environmental disturbance: Effectiveness of TBM-driven pilot tunnels in non-blasting rock breaking method

Masanobu Murata
Honshu-Shikoku Bridge Authority, Tokyo, Japan

Keiichirou Yokozawa
Japan Construction Method and Machinery Research Institute, Shizuoka, Japan

ABSTRACT: The Maiko Tunnel is located at the Kobe end of the Akashi Kaikyo Bridge. It is composed of a pair of three-lane tunnels with a cross-sectional area of excavation of 140 m². The length of the tunnel is 3 km including 0.6 km of hard rock section of granite, whose uniaxial compressive strength is around 2,000 kgf/cm². The tunnel in this hard rock section passes directly under a densely populated area, and covering depth is fairly small. The tunneling work therefore must be done very carefully to avoid environmental problems such as noise, vibration and ground deformation.
In order to satisfy these requirements, twin pilot tunnels 5 meters in diameter were driven with TBMs in the first stage. Then, these pilot tunnels were expanded to the design section by controlled blasting and mechanical excavation.
These tunneling works were completed successfully without disturbing the surrounding area.
This report explains and discusses the mechanical excavation through the hard rock with pilot tunnels.

1. INTRODUCTION

The Maiko Tunnel is a 3.3-km-long pair of three-lane highway tunnels located at the Honshu end of the Akashi Kaikyo Bridge on the Kobe-Naruto route of the Honshu-Shikoku Bridge Highway.

The Rokko granite section where hard rock tunneling was carried out is about 600 m long, one-third the total tunnel length. As shown in Figure 1, this tunneling section is divided into the mid-south work section (about 390 m long), which runs from an inclined shaft bottom to Yatani, and the south work section (about 210 m long), which runs from Yatani toward an inclined shaft bottom. Since earth cover in the granite section was only 10 to 20 m and the tunneling section was directly below a residential area, tunnel excavation had to be carried out under a number of constraints because of the need to minimize adverse effects on the surroundings. It was therefore essential to employ mechanical excavation instead of conventional blasting excavation.

Another characteristic of the tunnel excavation in the granite section was that two 5-m-diameter pilot tunnels were driven mechanically in the tunnel cross section.

In this study, of the excavation methods applied to this granite section, the non-blasting rock breaking method, which generates relatively low noise and vibration, is considered, and the rate of excavation attained by the method is calculated from records of excavation. At the same time, superiority of free faces made by TBM-driven pilot tunnels in mechanical excavation through hard rock to those made by conventional excavation methods is verified.

2. TOPOGRAPHY AND GEOLOGY

The Rokko granite formation is made up, from ground surface downward, of increasingly hard materials, namely, soil, weathered granite, and granite. The surface layer contains decomposed granite and is discolored at some places. The weathered rock stratum is thin and is generally highly jointed, relatively hard rock of classes C_M to C_H. The unconfined compressive strength and seismic velocity of the rock are 500 to 2,000 kgf/cm² and 2 to 5 km/sec, respectively.

It can be said that on the whole the geological conditions are favorable because there is no major fault and relatively hard rock is distributed widely. Since, however, the tunnel section with earth cover depths of 10 to 20 m was to pass directly below or near residential districts, care had to be taken to minimize adverse effects, particularly vibration, on the surroundings during tunnel excavation (see Figure 3).

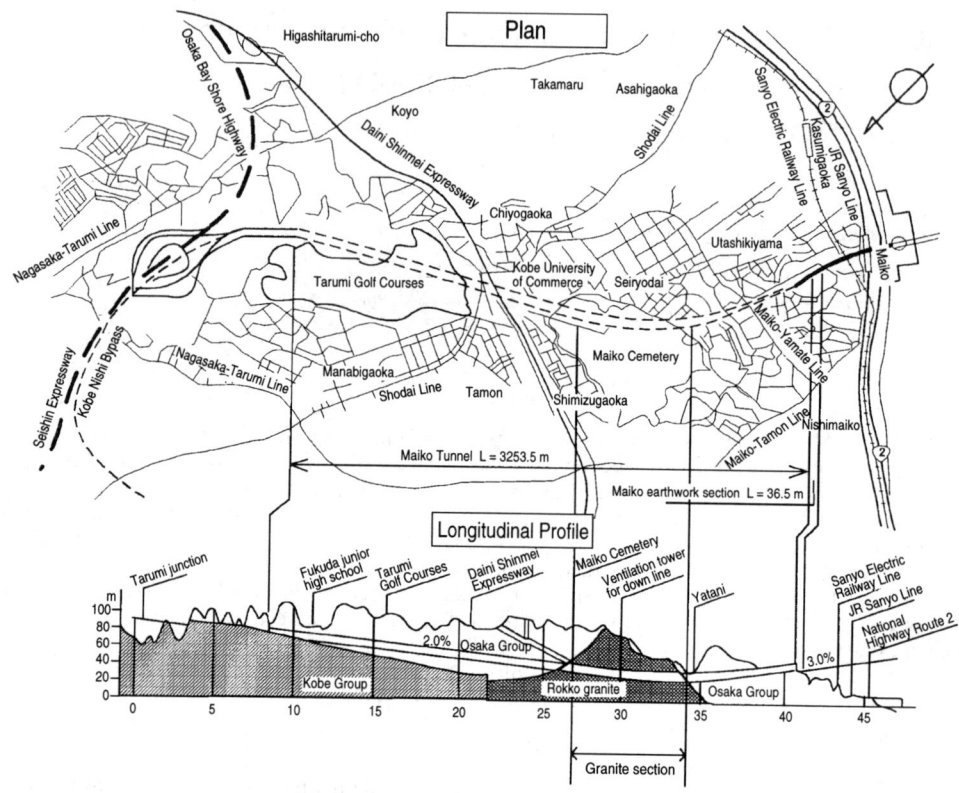

Figure 1. Granite section of the Maiko Tunnel.

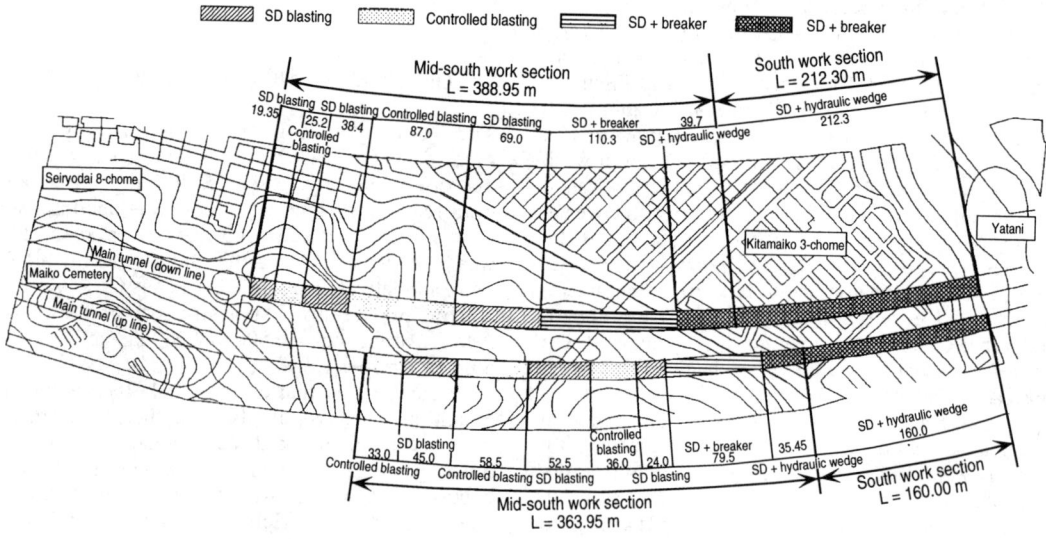

Figure 2. Excavation patterns in the granite section.

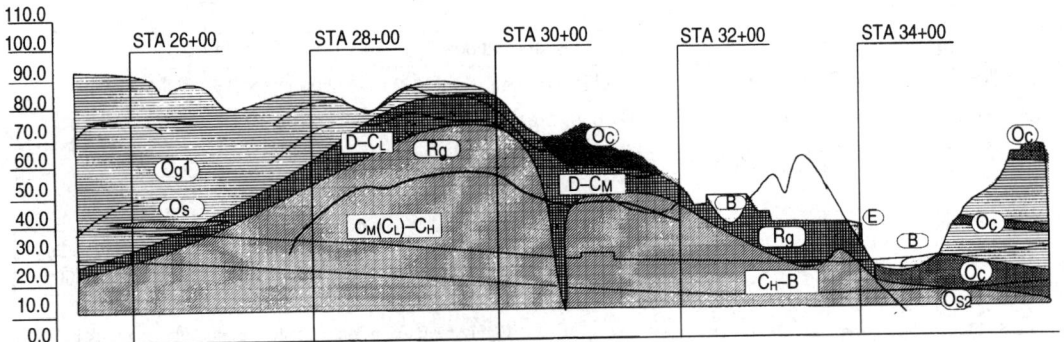

Figure 3. Geological profile of the granite section.

Table 1. Restrictions on vibration during Maiko Tunnel excavation.

	Daytime	Nighttime
Blasting	7:00 a.m. to 8:00 p.m.	Prohibited
Breaker	7:00 a.m. to 9:00 p.m.	Prohibited
Target level of vibration	55 dB (private house)	

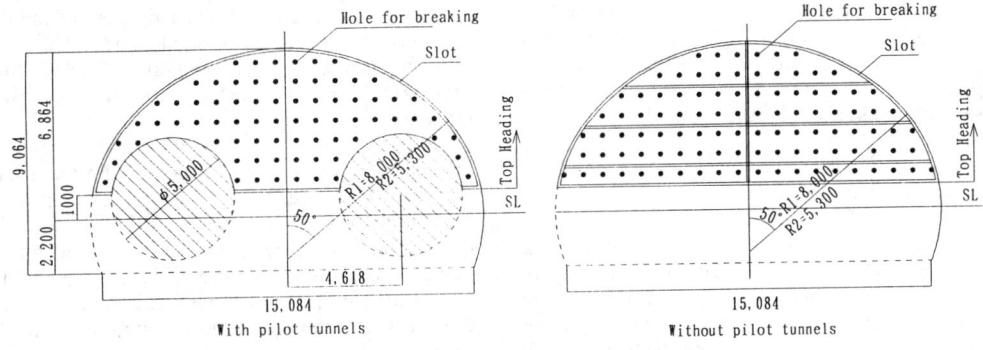

Figure 4. Cross sections of tunnels excavated by the non-blasting rock breaking method.

3. CONSTRAINTS ON EXCAVATION

The environmental conditions imposed a number of constraints on the tunnel excavation in the granite section, as shown in Table 1.

Because of these constraints, all excavation work in the south section directly below a residential area had to be carried out mechanically by the non-blasting rock breaking method.

4. EXCAVATION RATE ATTAINED BY NON-BREAKING ROCK BREAKING METHOD

In this section, the rate of excavation attained by the non-blasting rock breaking method for the top heading is calculated. At the same time, the rate of excavation attainable in the case where there is no pilot tunnel is calculated, and the rates of excavation in the two cases are compared. Through this comparison of excavation rates, the superiority of the pilot-tunnel approach in non-blasting rock breaking excavation is considered.

4.1 Cross-sectional area of excavation

Figure 4 shows the cross-sectional area of excavation carried out by the non-blasting rock breaking method in the granite section and that used

Table 2. Equipment used.

Use	Name	Specifications	Quantity
Slot drilling	SD drill jumbo	Model II, $\phi 60 \times 5$	2 booms × 2 = 4 units
Rock breaking	Hydraulic wedge	MTS800, breaking force 800t	2 units
Secondary breaking	Hydraulic breaker	3t class	1 unit

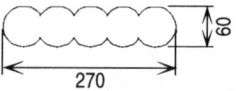

Figure 5. Slot configuration.

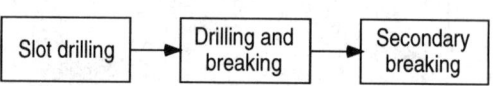

Figure 6. Excavation procedure in non-blasting rock breaking method.

to calculate the rate of excavation attainable in the case where no pilot tunnel is driven.

The cross-sectional area of top heading excavation through the Class C rock excluding that of the pilot tunnels is 52 m². The total length of drilled slots is 26.5 m.

4.2 Equipment used

For non-blasting rock breaking excavation, slot drilling machines for slot drilling, hydraulic wedges for rock breaking, and a hydraulic breaker for secondary breaking were used. Table 2 lists the equipment used. Figure 5 shows the configuration of a slot formed in a single drilling cycle.

4.3 Excavation procedure

The excavation procedure by the non-blasting rock breaking method is shown in Figure 6. As the figure indicates, in the non-blasting rock breaking method slot drilling, rock breaking, and secondary breaking were performed in that order. Subsequent steps such as mucking are the same as the NATM procedure for mountain tunnels.

4.4 Calculation of Rate of Excavation

Rates of excavation were calculated on the basis of the results of a fact-finding survey conducted by researchers of Japan Construction Method and Machinery Research Institute at the site of the Maiko Tunnel, records kept by the contractors, technical data on the slot drilling method, and cost estimation data. Calculated rates of excavation that can be achieved with or without pilot tunneling are shown in Table 3.

4.5 Discussion

The calculated rates of excavation attainable by the non-blasting rock breaking method were 5.3 m³/hr for the pilot-tunneling case and 3.2 m³/hr for the non-pilot-tunneling case, the former being about 1.7 times as high as the latter. Thus, the calculated rates of excavation based on field data indicate that pilot tunneling increases the rate of excavation.

For comparison between the pilot-tunneling case and the non-pilot-tunneling case, the time required for excavating one cubic meter by each of the three methods of excavation was calculated. The results thus obtained are shown in Figure 7. Comparison of the calculated values reveals the following tendencies:

a) In the case where the pilot tunnels are driven, the total length of slot required is shorter because slot drilling is needed only along the periphery of the tunnel cross section. Slotting work can be reduced because free faces can be formed by pilot tunnels. As shown in Figure 7, the time required for slot drilling in the pilot-tunneling case is about one-third of the time required in the non-pilot-tunneling case.

b) The rate of secondary breaking is increased by use of pilot tunneling. Secondary breaking rates were calculated from data obtained from free faces formed by the main tunnel with both pilot tunnels and slots and by the ventilation tunnel with slots only. The differences in secondary breaking rates probably reflect the difference between the two 5-m-diameter pilot tunnels at the bottom right and left of the top heading and the 6-cm-wide slots. As shown in Figure 7, the time required for secondary breaking in the pilot-tunneling case is about one-third the time required in the non-pilot-tunneling case.

The above results are thought to indicate that pilot tunneling helped to reduce the required slot length and improve the efficiency of secondary breaking.

Table 3. Excavation rates attained by non-blasting rock breaking method.

	Item		Pilot Tunnel Used	Pilot Tunnel Not Used	Description
	A	Cross-sectional area of excavation (m^2)	52		Class C rock in the mid-south section; top heading
				70	excluding the faces of pilot tunnels (18 m^2).
	B	Advance per cycle (m)	1.2	1.2	
	C	Total slot length (m)	26.5		Past data for Class C rock.
Workload				81.2	SD Method: Technical Data, p. 28.
	D	No. of drilled holes (hole)	82	110	Average of past data: 0.8 m×0.8 m
	E	Volume of excavation (m^3)	62.4	84	A×B
	F	Depth of hole (m)	1.4	1.4	From past data.
	G	No. of breaking cycles (cycle)	1	1	One cycle of rock breaking per cycle length (1.2 m).
	H	No. of slot drilling machines (unit)	4	4	Three-lane cross section is large enough to accommodate two jumbos; see Table 2.
	I	No. of breakers (unit)	2	2	Two units can operate simultaneously.
Capability	J	Slot configuration (m)	0.27	0.27	Model SD II is a 5-drill machine; 0.27 m wide, 60 mm in diameter; see Table 2 and Fig. 5.
	K	Slot drilling (m/min)	0.1	0.1	The rate of drilling is a measured value.
	L	Breaking time per hole (min/hole)	7	7	Drilling and breaking can be done by a single machine; actual time including the time required for moving into position.
	M	Breaking capability (m^3/hr)	50	20	Calculated from past data submitted by the contractor.
	N	Preparation (min)	30	30	Marking, moving into position, installation, etc.
	O	Slot drilling (min)	294	902	C÷K÷H×B÷ J Time required in the case of 4-boom simultaneous operations.
Time Required	P	Preparation for breaking (min)	10	10	Removal of slot drilling machines and replacement of machines are repeated.
	Q	Breaking (min)	287	385	D÷I÷G×L
	R	Secondary breaking (min)	10	10	Removal of breakers and replacement of machines are repeated.
	S	Fracturing (min)	75	244	E÷M×60
	T	Total (min)	706	1581	
	U	Cycle time per m (m/min)	588	1318	
	V	Rate of Excavation (m^3/hr)	5.3	3.2	

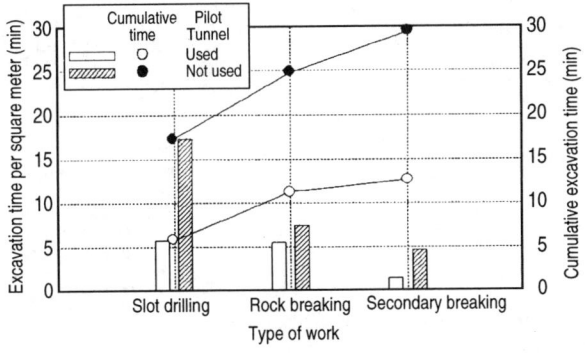

Figure 7. Excavation time per square meter by type of work.

5. CONCLUDING REMARKS

Mechanical excavation of the Maiko Tunnel, which had to be driven through hard granite without disturbing the surroundings, was carried out by making utmost use of the latest tunneling technology in Japan. The excavation was successfully completed, while achieving higher rates of excavation than in conventional mechanical excavation and without disturbing the surrounding area.

Pilot tunneling was employed primarily for the convenience of mucking. It is believed, however, that by paying attention to the rate of mechanical excavation, superiority of the pilot tunneling approach has been confirmed. Although this paper dealt only with evaluation based on the rate of excavation, in practice it is important to evaluate not only the rate of excavation but also economy.

The Maiko Tunnel was excavated by use of tunnel boring machines and mechanical excavation through hard rock primarily because of the environmental constraints and construction schedule, though under normal conditions the blasting method would be most efficient and economical. As the number of urban tunnels increases, however, it is likely that there will be a growing number of cases where tunnel excavation has to be carried out under various constraints, as in the case of the Maiko Tunnel. It is the authors' sincere hope that the data on the Maiko Tunnel reported here will help tackle such difficult tunneling problems.

REFERENCES

First Construction Bureau, Honshu-Shikoku Bridge Authority March 1995. *1994 Maiko Tunnel Study Report, Part 9* (in Japanese).
First Construction Bureau, Honshu-Shikoku Bridge Authority March 1996. *1995 Maiko Tunnel Study Report, Part 10* (in Japanese).
Japanese Association for SD Method May 1995. *SD Method: Technical Data* (in Japanese).

Geological input as applied to TBMs: A case study of the Kelinchi Transfer Tunnel, Malaysia

Mogana Sundaram
Peabody Resources Corp. (M) Sdn. Bhd., Malaysia

Ibrahim Komoo
Department of Geology, National University of Malaysia, Bangi Selangor, Malaysia

ABSTRACT: The Kelinchi Transfer Tunnel is the first tunnel in Malaysia bored by a Tunnel Boring Machine (TBM). Since the mechanical excavation technique is highly sensitive to geology, a comprehensive surface geological mapping was carried out to predict the probable ground conditions that would be encountered during the tunnel excavation. Unfavourable geological conditions that will be encountered, such as fault zones, shear zones, high water ingress, fractured and massive ground, were predicted with the help of satellite imagery, aerial photographs and field mapping, together with a limited number of exploratory borehole logs. Almost 75% of the ground conditions predicted from the above geological exploration methods were actually encountered during tunnel excavation. In addition, machine downtime, penetration rate and power consumption associated with the ground conditions are also analysed and presented.

1 INTRODUCTION

Tunnels in Malaysia are mostly related to hydroelectric dams and highways. These tunnels have been constructed with the conventional method of excavation, drill and blast. The Kelinchi Transfer Tunnel is the first of its kind in Malaysia to be constructed with a fullface TBM.

The Kelinchi Transfer Tunnel forms a part of a water supply scheme to meet the domestic and industrial water demand in the state of Negeri Sembilan in West Malaysia. This tunnel (6.2 km long, 3.5 m diameter and 1 : 330 gradient) will deliver untreated raw water from Kelinchi reservoir to the Sg. Terip treatment plant by gravity flow. The treated water will then be delivered to the fast growing western region of the Peninsular Malaysia.

The influence of geological conditions on TBM performance has long been recognised, not only in terms of ground support but also in the overall efficiency of the machine (McFeat-Smith & Tarkoy 1979; Tarkoy 1987; Nelson 1993). Hence a poor understanding of the ground conditions in TBM bored tunnels can lead to disastrous consequences. The performance of TBM is affected by rock characteristics and machine parameters which will ultimately influence downtime, support, muck disposal, ancillaries, labour and organisation (Fowell 1993). The effects of rock mass and material properties are well recognised (Bamford 1986; Movinkel & Johannessen 1986).

This paper focuses on the influence of ground conditions on TBM performance and the applicability of a combination of methods to predict ground conditions.

2 TUNNEL BORING MACHINE (TBM)

A reconditioned TBM which had worked in various ground conditions in Australia was utilised to bore Kelinchi Transfer Tunnel. Major modifications were made to several components including cutterhead diameter, cutter configuration, thrust capacity and drive motors. The specifications of the TBM are stated in Table 1.

Table 1. TBM Specifications

TBM model	Robbins 116-181
Diameter	3.5 m
Max. Power	671 kW
No. of cutters	4(394 mm) & 25 (413 mm)
Max. thrust per cutter	213 kN
Stroke length	1.2 m
RPM	10.08

In Kelinchi, the work was carried out 6 days a week. Each day there were two ten-hour production shifts and a four-hour maintenance period. This gave a total of 120 production and 24 maintenance hours per week. The tunnel was bored from an outfall portal. Standard and hard steel discs with constant section ends were used while tungsten carbide button discs were used from km 3.5 to km 0.

3 GEOLOGY AND GEOTECHNICAL DATA

The tunnel is located in the main range which forms the backbone of the Malay Peninsula. It comprises predominantly medium to coarse grained to porphyritic Mesozoic granite. The maximum overburden above the tunnel is 500 m. The location and geology surrounding the tunnel site are shown in Figure 1.

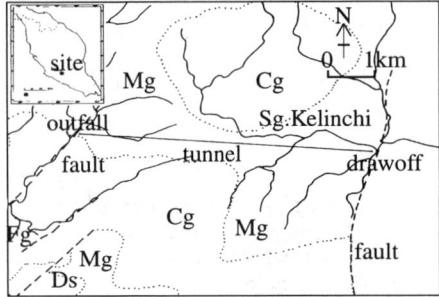

Figure 1. Site location and geology.
Cg - coarse grained Granite, Mg - medium grained Granite, Fg - fine grained Granite, Ds - Dinding Schist

In the Kelinchi Transfer Tunnel, the available information to the tunnelling contractor during bidding was only from 6 boreholes (2 at each portal and 2 in the middle), together with geotechnical data (Table 2). Dense tropical jungle, depth of cover and rugged topography had limited the number of holes drilled. A detailed surface geological mapping was carried out to obtain rock mass properties. Due to time constraint, no geophysical survey and additional boreholes were carried out.

Table 2. Geotechnical data of the site

Tests	No. of tests	Range	Mean
UCS, MPa	13	130-245	182
Poisson ratio	9	0.17-0.28	0.22
E, kN/mm²	13	42-74.4	60.58
ό, MPa	2	11-13.9	12.4
Is50, MPa	210	2.8 - 21.81	8.97
Porosity, %	2	0.2-0.3	0.25
Swelling, %	3	0.006-0.023	0.012
MC, %	3	0.1-0.2	0.13
Id4, %	4	97.5-99.6	98.93
Rock type - Granite	3	Q 20-30, KF 10-15, P 20-30, B 10-15	Q 25, KF 12.5, P 25, B 12.5

UCS - unconfined compressive strength; E - Young's modulus; ό - Brazilian tensile Strength; Is50 - Point load strength; Id4 - Slaking Durability Index 4th cycle; MC- moisture content; Q-Quartz, KF-Alkali Feldspar, P-Plagioclase, B-Biotite.

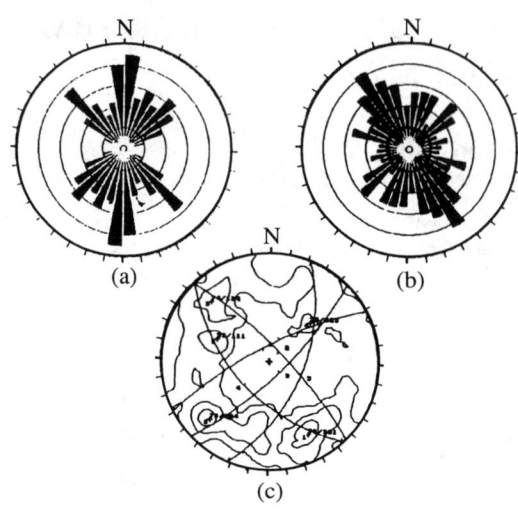

Figure 2. Rosettes and streoplot.
(a) Rosette plot of lineaments from aerial photographs and Landsat imagery (160 lineaments)
(b) Rosette plot of field joint data (531 measurements)
(c) Streoplot of field joint data with major planes (577 measurements)

4 GEOLOGICAL MAPPING

Geological mapping was commenced with the interpretation of aerial photographs and Landsat imagery. Filtered images of Landsat were used in the identification of linear geological features. The combination of aerial photographs and Landsat imagery complemented each other in identifying regional structures. Figure 2 shows the rosette of linear features obtained from both the remote sensing methods.

Streams within 1 km on either side of the proposed tunnel were mapped. All pertinent features of the discontinuities and other geological features were recorded. A rosette diagram and stereoplot of the discontinuities measured are shown in Figure 2.

Not all lineaments mapped in aerial photographs and Landsat imagery were able to be identified in the field due to the scarcity of outcrops, thick weathering, boulder cover on streams and dense vegetation. Similarities existed between the lineaments from remote sensing and field mapping data hence it was interpreted that the joints and shears observed in the field were an indication of faults and shears (Tjia 1972; Hancock 1986).

The mean orientation of the joints was matched with the average trend of linear features obtained from the Landsat imagery and aerial photographs. In this manner the dip of the lineament was identified and projected to tunnel level. The interpreted and observed faults are marked on Figure 3. Some of the

Table 3. Predicted geological features and associated geotechnical problems

Predicted geological features	No of locations /total length	Anticipated problems with machine tunnelling
Faults and shear zones blocks (very poor rock)	38 nos.	Gripping problem, support, dislocation of at intersection of faults
Massive (Good to excellent rock)	3400m (55%)	Low penetration, high cutter costs
Fractured (inclusive of fault zones - fair to very poor rock)	2800 m (45%)	High vibration, fatigue failures of cutters in strong fractured ground
Water ingress	km 6.2 - 5.3, 3.7 - 3.2, 2.4 - 0.6	Water ingress along these zones causing delay in support installation; electrical problems

Table 4. Geotechnical problems associated with rock mass conditions encountered

Geological features encountered in tunnel	No of locations / total length	Problems encountered	Action/Solution
Fault/fault zone more than 1m wide	53 nos. (428 m or 7%)	Bearing pressure failures at springlines, steering and gripping problems, crown and side wall failures	Ground support, shotcrete to increase wall strength, short strokes to position gripper pads on better ground
Sudden entry into massive rock from fault/fractured zone	3 nos.	Steering difficulty, very slow progress, bearing pressure failure	Cut with low thrust, timber lagging between gripper pad and wall
Fault/joint intersections at crown/springline	7 nos.	Failure at crown, springlines, gripping difficulty	Roof support behind cutterhead, shotcrete wall strength to improve gripping
Water ingress	km 6.2-5.0, 2.2-1.8, 1.2-0.7	Washing of muck through cutter head, cutter motor tripping, increased electrical locomotive breakdown	Water relief holes, hand mucking, ground support, water proofing electrical motors and switches, use of diesel locomotives
Massive rock*	2910m (47%)	Low penetration rate, high cutter costs, cuttermotor bolt breakage, grippers slipping	Keep cutting at maximum propel pressure, button cutters
Massive rock*, with open joints and water	km 1.8 - 0.6	Falling of rock blocks on cutterhead, bucket and disc damage, blocked chute, cutterhead stalling, damage of TBM conveyor, high water ingress up to 50 l/sec from single location.	Hand breaking to remove fallen blocks, support at crown with mesh and steel sets as exposed behind head, unscheduled maintenance of buckets and cutterhead.
Fractured rock/shear zone*	2840m (46%)	Good penetration rate	Roof support concurrently with cutting

*zones less than 10 m are not included

geotechnical characteristics of the faults and their surroundings were obtained directly in the field.

From the mapping and analysis, it was observed that: i) the tunnel is located in a single rock type i.e. medium to coarse grained granite with minor aplite and pegmatite intrusion less than 1.5 m in thickness; ii) all crevices in the topography had a lineament running through it; iii) the highest concentration of lineaments on the surface generally produced the lowest topographic relief and iv) the faults were brittle in nature.

An examination of rock core samples showed no sign of weathering below 50 m, alteration (mostly chloritization) was limited to fault and shear zones only, and high water intake (up to 30 lugeons) was found only in fractured zones.

5 PREDICTION OF GEOTECHNICAL PROBLEMS

The approximate location of faults, shears and the likely problems to occur as a result of encountering these features can only be predicted in general terms and these are summarised in Table 3. The width of the fractured and massive zones were interpolated from surface to tunnel level.

6 ACTUAL CONDITIONS AND PROBLEMS ENCOUNTERED

The TBM bored section of the tunnel was back-mapped as the work progressed where all pertinent engineering geological properties of the discontinuities were recorded. The rock mass was classified using NGI's Q-System. The major geological features encountered are presented in Figure 3.

From this study it was found that almost 75 % of the structures predicted and associated geotechnical problems were encountered in the tunnel although the locations sometimes deviated by up to 250 m from the predicted locations. This is not surprising as the measurements were made only on small geological planes during surface mapping and most of the structures were deduced from remotely sensed data. Table 4 summarises the actual geological conditions encountered and their associated geotechnical problems in the tunnel.

Initially a probe drill rig was attached to the TBM gripper carriage. Difficulties in collaring, unreliable penetration rates and the frequency of encountering ground conditions in the cutting face that differed from those predicted by probe hole logs ultimately resulted in the abandonment of probe drilling. Further it was evident from surface and tunnel mapping results that changing ground conditions ahead of cutting face could be predicted by examining the trend of changes in ground already exposed by excavation. Guidelines to trigger the need to carry out forward probing were developed and included.

7 ANALYSIS OF TBM PERFORMANCE

The above mentioned geological factors contributed to a significant amount of downtime. This was computed based on Tarkoy (1987). The degree of fracturing along the tunnel is represented by RQD (Figure 3). Figure 3 also shows water ingression in tunnel, weekly net penetration, downtime, utilisation, production and relative power consumption. Other rock mass properties such as joint sets, joint roughness, joint alteration, strength and orientation, also influence the TBM utilisation and performance (Mogana 1997).

Weekly TBM performance associated with ground conditions (A to E) along selected sections in tunnel are summarised in Table 5.

A-*Massive rock* (good to excellent): In massive rock, though high utilisation was able to be achieved (41.8-85.0%), the rate of penetration was low (1.0 to 1.6 m/h). Here the cuttability is controlled by the hardness and strength of the rock. The average weekly production was 75m. Cutter usage was higher in massive ground compared to fractured and faulted ground and the downtime was between 13.3 to 29.4 %.

B-*Fractured rock* (poor to fair rock): Good net penetration was achieved in this condition (2.5-2.9 m/h) with the major downtime due to cutter change (18.5-25.4%).Tight to partly opened joints with calcite infill facilitated the breakage of rock. TBM conveyor maxi-

Table 5. TBM performance in relation to ground conditions

Ground condition	Typical Section (km)	Downtime (range in %)				Er	Pnet	U	Adv.
		1	2	3	4				
A	4.1-3.9, 2.8-2.3	0.0-0.8	13.3-29.4	0.0-0.8	0.3-2.2	1.6-2.0	1.0-1.6	41.8-85.0	51-119
B	5.4-5.0	0.4-0.6	18.5-25.4	1.3-4.3	0.0-0.6	0.6-0.7	2.5-2.9	34.0-44.2	117-135
C	2.2-1.0, 0.9-0.6	0.0-13.9	1.6-25.6	0.0-23.6	0.0-12.3	0.8-1.9	0.7-2.1	0.0*-62.1	46-71
D	3.5-3.3, 1.0-0.9	45.8-76.0	0.0-11.3	0.0	0.0-1.0	0.7-1.5	0.9#-2.3	18.4-46.2	34-76
E	4.7-4.2	0.2-0.6	12.3-15.1	0.0-0.2	0.5-0.6	0.8	2.5-2.7	39.4-44.8	119-166

1 = Tunnel Support; 2 = Cutter change; 3 = TBM Conveyor; 4 = Locomotives; Er - Relative energy consumption (MJ/m^3); Pnet (m/h)- Net penetration rate; U (%) - Machine Utilisation; Adv.(m) - weekly advance. * - downtime due to unscheduled maintenance work; # - entering massive ground from fault/fractured zone

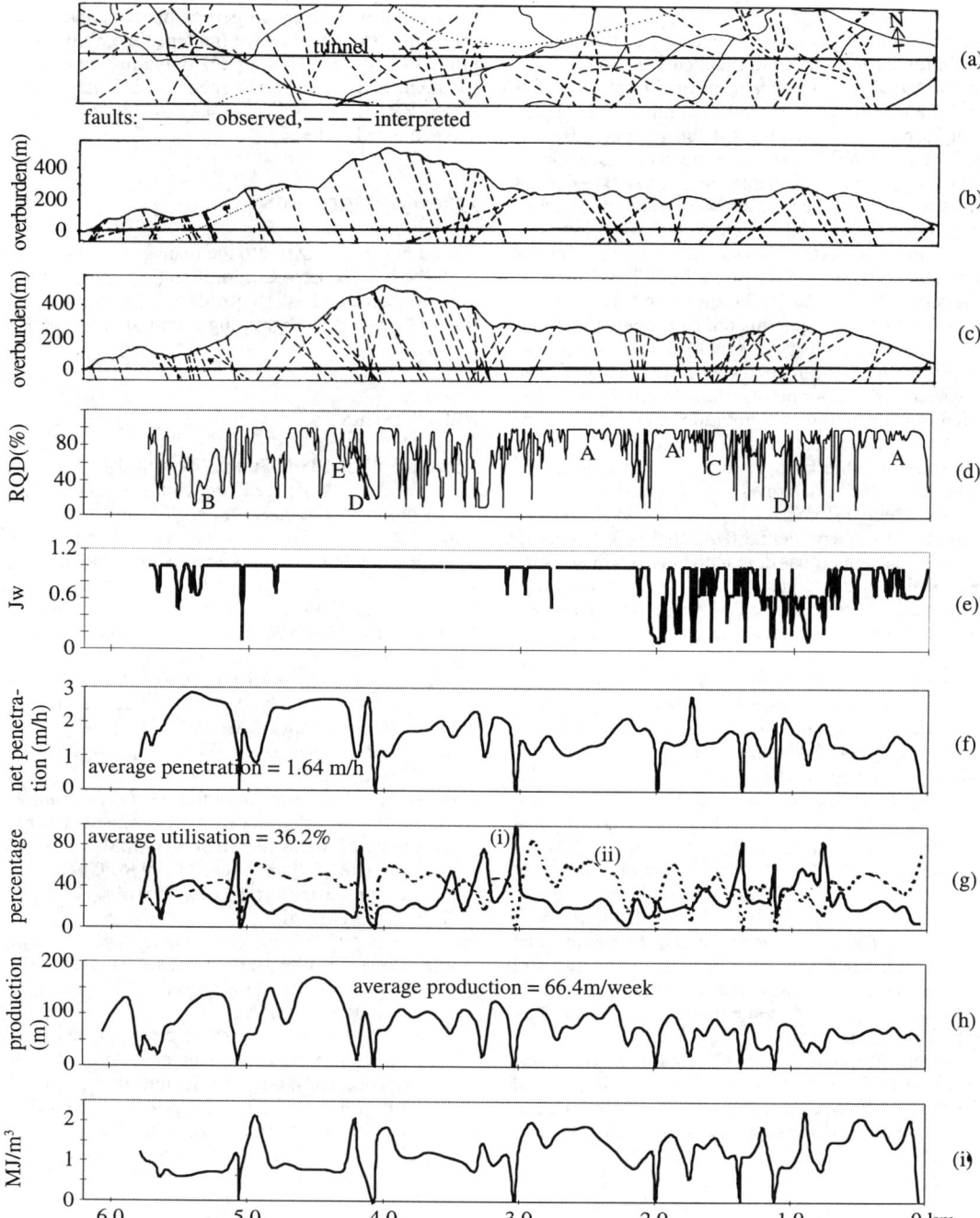

Figure 3. Geology and TBM performance along tunnel alignment
(a) Surface geology; (b) Geology as predicted; (c) Geology as encountered; (d) RQD {A-massive, B-fractured, C-massive with opened joints + water, D-fault zone, E-shear zone};
(e) Water ingress (Jw-joint water reduction factor, Q-System); (f) Weekly net penetration (m/h);
(g) Weekly downtime (i-related to ground conditions) and utilisation -ii; (h) Weekly production;
(i) Relative energy consumption (MJ/m^3)

mum downtime was only 4.3 %. Tunnel support was minimal and mostly was concentrated on the roof.

C-*Massive rock* (good to excellent) with open joints and water: The occurrence of opened joints up to 0.4 m without infill resulted in the downtime in tunnel support, locomotives and the TBM conveyor belt. Bucket and cutter damage was more severe in this zone. Gushing of water from these opened joints was experienced initially. However the flow declined with time to a constant rate of about 10 l/s. Here, the downtime was higher in all aspects: tunnel support 0.0-43.5%, cutter change 1.6-25.6%, TBM conveyor 0.0-23.6 % and locomotives 0.0 -12.3%. Penetration rate between 0.7-2.1m/h was achieved in this mixed ground condition.

D-*Fault Zones* (very poor rock): The width of fault zones varied from 2 m up to 100 m. These fault zones, however, had a water ingress less than 10 l/sec. It differed from the initial interpretation. The sealing off of the fault zone is believed to be due to presence of clay minerals, illite, kaolinite and montmorillonite. The materials in the fault zones were crushed and altered and the average strength was less than 25 MPa. The stand-up time at a particular fault zone was less than 5 minutes and sometimes demanded instantaneous support behind the cutterhead. In fault zones, tunnel support made up most of the downtime between 45.8 - 76.0 %. Utilisation between 18.4 to 46.2 % was achieved.

E-*Shear zone* (fair to good rock): Good progress was achieved in shear zones. Closely spaced tight joints and shear planes with calcite and chlorite infill facilitated breakage of rock along pre-existing weak planes. Weekly net penetration rate was between 2.5 to 2.7 m/h. Highest production was also achieved in this zone with weekly production up to 166 m and monthly 524 m. The major downtime in this area was due to cutter change (12.3-15.1%). This area warranted only roof support which was done concurrently with cutting.

The average disc consumption for this tunnel was 0.2 disc/m. The disc wear was higher in massive and strong fractured ground compared to sheared and faulted ground. This shows that cutter wear is not only dependent on mineral contents but also on the rock hardness and strength.

Power consumption in massive ground (RQD> 75) was 3 times higher compared to poor very poor ground (RQD < 25). The power consumption doubled in massive ground compared to good to fair (RQD 25-75%) ground.

8 CONCLUSIONS

The study revealed that the predictions from surface geological mapping is in close agreement (almost 75 %) with the actual ground conditions encountered. Hence this can be used as a tool to predict ground conditions reliably in tropical terrains. The study confirmed that rock mass properties have a great influence on TBM performance (cutter usage, utilisation, progress and penetration rate) thus due importance should be given to the assessment of rock mass properties apart from routine drilling, sampling, logging and geotechnical tests.

ACKNOWLEDGEMENT

The authors are grateful to the management and staff of Peabody Resources Corporation (M) Sdn. Bhd. for their guidance and assistance during the entire duration of sitework and granting permission to publish this article.

REFERENCES

Bamford, W.E. 1986. Cuttability and drillability of rock. Civil College Technical Report. *The Institution of Engineers Australia:* July.

Fowell, R.J. 1993. The mechanics of rock cutting. In. J.A Hudson (ed.), *Comprehensive Rock Engineeing,* Vol. 4, 155-176. Pergamon Press, Oxford.

Hancock, P.L. 1985. Brittle microtectonics: principles and practice. *Journal of Structural Geology*, Vol. 7, Nos. 3/4.437-457.

McFeat-Smith, I. & Tarkoy, P.J. 1979. Assessment of Tunnel Boring Machine performance. *Tunnels & Tunnelling,* December. 33-37.

Mogana, S.N. 1997. *The influence of engineering geological properties of rock on TBM performance. A case study from Kelinchi Transfer Tunnel, Seremban*. Ph.D. Thesis, National University of Malaysia. (in preparation)

Movinkel, T. & Johannessen, O. 1986. Geological parameters for hard rock boring. *Tunnels & Tunnelling,* April.

Nelson, P.P. 1993. TBM performance analysis with reference to rock properties. In. J.A Hudson (ed.), *Comprehensive Rock Engineering*, Vol. 4, 261-291. Pergamon Press, Oxford.

Tarkoy, P.J. 1987. Practical geotechnical and engineering properties for Tunnel Boring Machine performance analysis and prediction. *Transportation Research Record 1087, Transportation Research Board*, National Research Council. 62-78.

Tjia, H.D. 1972. Fault movement, reoriented stress field and subsidiary structures. *Pacific Geology.* Vol. 5. 49-70.

New mechanical hard rock tunneling by FON drilling method and FASE method

Tatsuya Noma, Toshiro Tsuchiya & Mitsutaka Hada
Fujita Corporation Technical Research Institute, Yokohama, Japan

ABSTRACT: A new rock-fracturing excavation method for hard rock tunneling is developed using a slot made by continuous hole drilling and fracturing toward the slot using a rubber-tube-type fracturing machine. This paper gives an outline of the method of making a slot by continuous hole drilling, the fracturing machine and its system, and the actual tunnel excavating procedure.

1 Introduction

It is well known that blasting is the most effective and least costly method of fracturing and excavating rock mass. However, this method involves tremendous shock waves and noise, and is not suitable for building tunnels near residential areas. A great deal of the construction now under way in Japan is near residential areas.

The Kaminiko-tunnel, now being excavated at Kure City in Japan, is just such a case.

This tunnel has these restrictions:
1) There is a cluster of houses in the neighborhood of the working site.
2) There are many rocks over the excavation route.
   Even if protective measures are taken, such rocks may fall because of the vibrations due to blasting.

So, the blasting method cannot be used for all parts of the tunnel excavation. And, the mountainous area of the tunnel route consists of granite rock with compressive strength greater than 200MPa, making simple conventional mechanical excavation such as with a partial face machine impossible. Consequently, after forming a slot (free face) by drilling continuous holes at the tunnel face, excavation of the tunnel is carried out using a rock-fracturing method that fractures rock toward the slot.

While a variety of slot formation methods have been developed [1], problems relating to the need for specialized equipment, formation efficiency, and the continuity of the slot remain. This tunnel uses general purpose equipment in the slot method, and a continuous hole drilling method was developed which was superior to conventional methods in terms of efficiency and accuracy of continuity.

Furthermore, we are also currently working on a new rock-fracturing method. Many static (non-blasting) fracturing methods such as the expansive agent method, hydraulic wedge method, and pressurizing method using gas or water have been developed [2]. However, all of these methods meet with problems of safety and require fracturing machines of too large a scale. In this paper, static fracturing is defined as producing cracks in the rock bed to reduce its strength (primary fracturing) and then completely fracturing it with a breaker or ripper (secondary fracturing) [3].

The authors have been investigating hydraulic-based fracturing methods using high-pressure rubber tubes as a means of primarily fracturing rock bed with ease and efficiency [4]. We tried to expand this method for tunnel excavation, and applied it to this tunnel.

This paper summarizes the methods of the continuous hole fast drilling system and new rock fracturing system.

2 Continuous Holes Fast Drilling System

In this slot formation method, single holes are drilled continuously in order to maximize the capacity of the general purpose drill. When drilling the continuous holes, there is a tendency for the hole curves to lean toward the existing hole next to the rod bit. To prevent this, a drilling system was developed whereby a SAB(Spinning Anti-Bend) rod is inserted into the existing hole next to where the hole is to be drilled, and the bit is brought into contact with and knocks this rod.

This prevents gaps from forming between the bit

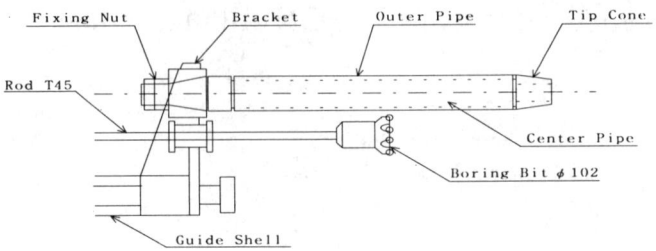

Fig.1 Concept of SAB rod

and the SAB rod and the continuity of the slot is maintained. The construction of this SAB-rod allows it to be rotated, and the reduction in friction during drilling brought about by the contact and knocking makes high-speed drilling possible. Figure 1. shows the concept of the SAB rod system, and Figure 2. shows the continuous hole construction procedure. The procedure is as follows:
① Insert the SAB rod into the existing hole.
② Commence drilling. Drill while rotating the bit and knocking it against the SAB rod. The rotation power of the bit will rotate the SAB rod, thereby making high-speed drilling possible.
③ Drilling is carried out to the designated depth.
④ Two continuous holes are formed, the SAB rod is inserted into the second hole, and the procedure is repeated, forming a slot (free face).

Furthermore, as the SAB rod can be rotated, wear on the rod and bit are reduced, and the SAB rod itself wears more evenly, leading to a longer service life. This method involves attaching the SAB rod to a bracket on the tip of the feed. Insertion and withdrawal are carried out using the slide on the feed. The continuous hole drilling capacity in hard granite (such as was encountered in this tunnel) using a 102 mm diameter bit, was approximately 3.5 ~4.0 m^2/h with a hole depth of 1.1 m. Figure 3. shows the relationship between continuous hole drilling ability and rock strength.

The features of this method are not limited to high operation capacity. Ease of attachment/ removal of the SAB rod and reduced congestion of the workspace can be expected because drilling of rock-fracturing holes and rock bolting can be carried out by the same machine.

Furthermore, the only consumable part of the SAB rod is the thick pipe, making it very economical in terms of parts replacement.

This method was named "FON (Fast, Onside and Nonpareil) Drill Method" because its high operation capacity and accuracy of continuity.

Figure 4. shows the state of continuous hole drilling.

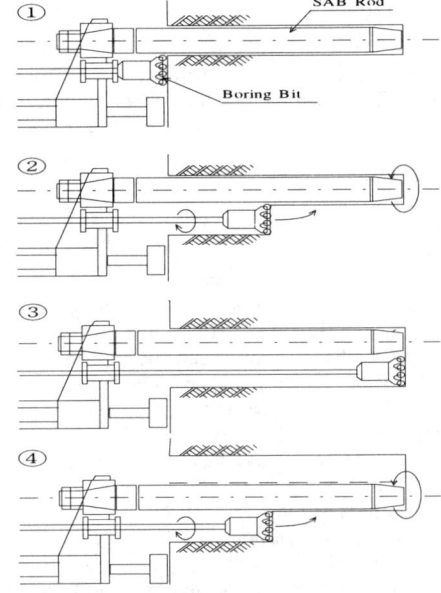

Fig.2 Procedure of continuous hole drilling

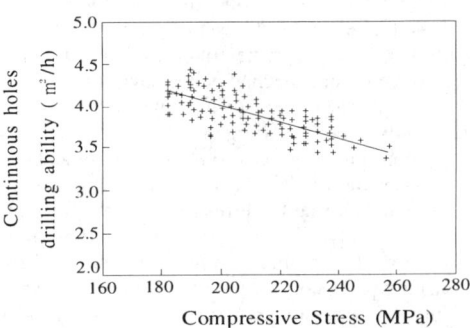

Fig.3 Relationship between continuous hole drilling ability and rock strength

Fig.4 Continuous hole drilling

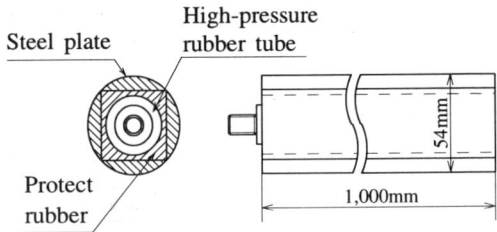

Fig.5 Concept of Aqua-Splitter

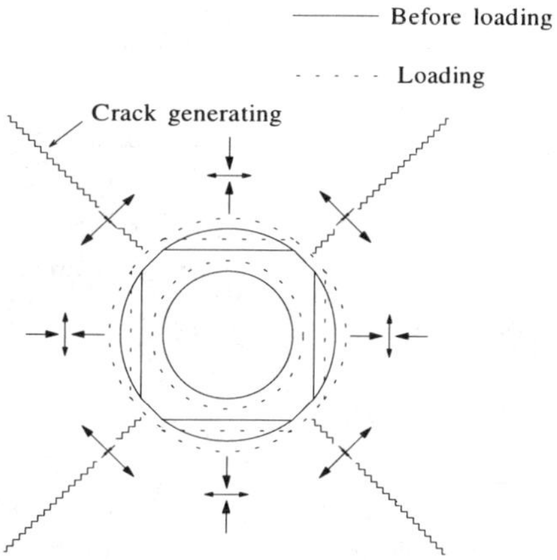

3 Fracturing Machine and System

We sought to develop a fracturing system that could break down masses of rock quietly, easily, efficiently, safely and economically. In this paper, we give an outline of a rubber-tube-type fracturing machine (hereinafter called the Aqua-Splitter) and its system.

3.1 *Rubber-tube-type Fracturing Machine*

Figure 5. illustrates the concept of the rubber-tube-type fracturing machine. The Aqua-Splitter consists of a high-pressure rubber tube, a rubber protector, and a steel loading plate. Rubber is the principal material of the Aqua-Splitter because it is lightweight (80N), yet has high crushing force (maximum working pressure: 50 MPa). Due to the device's configuration, it can be used repeatedly and the direction of cracking can be controlled. Also, many Aqua-Splitters can be used at the same time (10 Aqua-Splitters can be used for 1 unit), thereby improving the efficiency of fracturing work.

The mechanism of the Aqua-Splitter is as follows: First, the Aqua-Splitter is inserted into a bore hole. Next, by providing high liquid pressure to the high pressure tube, the high pressure tube and the rubber protector are expanded and transmit the pressure to the loading plate. Figure 6. shows a typical model of the motion of the high-pressure rubber tube, the rubber protector, and the loading plates and also shows the generated principle stresses in the rock mass caused by high liquid pressure as analyzed by the finite element method. According to this figure, the compressing force at the top of the loading plates which are arranged at right angles to each other results in tension at the point between them. This tension fractures the rock mass. The 90-degree angle between the loading plates allows tension to be exerted in four directions,

Fig.6 Mechanism of fracturing

enabling the direction of fracturing to be controlled, so rock mass fracturing to be effectively performed.

In addition, the rubber protector protecting the high pressure tube prevents the tube from being damaged upon abrupt pressure release during rock mass fracturing so that it can be used repeatedly.

3.2 *Fracturing System*

Figure 7. outlines the fracturing system. The system consists of a hydraulic unit, a control microcomputer, and the Aqua-Splitter mentioned previously. The hydraulic unit converts oil pressure, which is generated by a hydraulic pump, into water pressure, and at the same time, amplifies the water pressure using an oily water exchange booster. The purpose of exchanging oily water is to protect the rubber against oil-induced deterioration.

A control microcomputer is introduced to control the pressure and water-supply rate. Figure 8. shows the control flow of pressure. The system applies pressure to the Aqua-Splitter through the oily water exchange booster, which consists of a piston and primary and secondary cylinders. By installing a

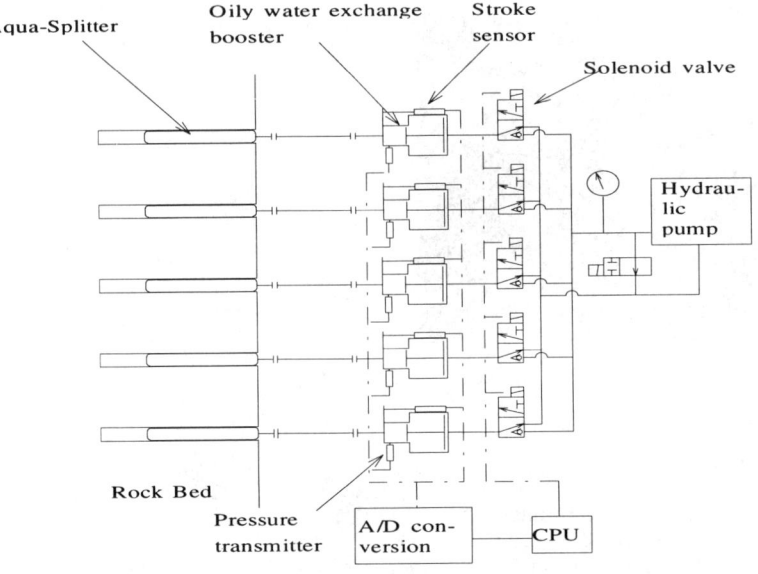

Fig.7 Fracturing system

pressure transmitter in the cylinder of the oily water exchange booster, it is possible to monitor and control the amount of pressurized water to be supplied to the Aquq-Splitter, based on pressure readings and the piston ejection volume, which is

To be more specific, loading stops when a large crack is produced in the rock bed, at which time the pressure drops sharply. Loading also stops automatically when the stroke reaches a preset level even if the pressure does not drop. This control method prevents the fracturing machine from expanding excessively, thereby markedly reducing the likelihood of damage to the machine.

This method was named FASE (Fujita Aqua-Splitting Excavation) Method.

4 Tunnel Excavating Procedure

In this tunnel, the excavating procedure includes the following tasks: marking the tunnel face, drilling fracturing holes, drilling continuous holes, primary fracturing by Aqua-Splitters, secondary fracturing by braker, mucking, shotcreting, and rock bolting.

Figure 9. shows the pattern of fracturing holes and continuous holes of this tunnel. Where compressive strength was greater than 150 MPa along the tunnel route, slots were drilled along the periphery and at the center of the tunnel, one line was in a vertical direction, two lines were in a horizontal direction, to form 6 divided blocks, then each block was fractured by the Aqua-Splitter.

The total drilling line was 62m, and a 102 mm diameter bit was used for drilling. The total number of drilling holes was about 700 because of the lapping zone for continuous holes. Although many drilling holes are needed for slot formation, the continuous hole drilling capacity of the new method

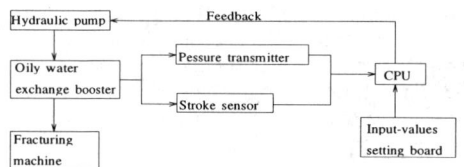

Fig.8 Control flow of pressure

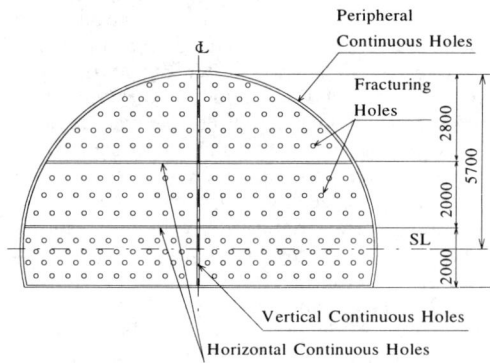

Fig.9 Continuous and fracturing hole pattern

is as high as 3.5 to 4.0 m² /h as mentioned previously, and a 3-boom jumbo was used for this tunnel, so the rate of continuous hole drilling is 27% for the cycle time as shown in Figure 10. Because a 102 mm diameter bit was used, the width of the free face was broader than in the conventional method (which uses a 65 mm diameter) and the slot continuity was very accurate. The total cycle time could therefore be reduced.

After continuous hole drilling, the tunnel face was fractured by the Aqua-Splitter as shown in Figure 11. The Aqua-Splitter's rock fracturing ability is 5MN. This is less than that of the conventional method (the hydraulic wedge has a rock fracturing ability of 10MN). For this reason, 17 to 20 fracturing holes were drilled horizontally in a straight line for each block. Then 17 to 20 Aqua-Splitters were inserted into the fracturing holes, and 2 hydraulic units were used to create pressure, thereby generating a long and massive crack in one operation.

Primary fracturing was therefore effective, since a long and massive crack could be generated in one operation.

The horizontal interval along the line of fracturing holes was about 40 to 50 cm. This interval allowed easy secondary fracturing.

The pressure applied by the Aqua-Splitter for fracturing the rock was about 20 to 30 Mpa and pressure was applied for about 2 minutes. The process of applying pressure using the Aqua-Splitter was repeated more than 30 times. Figure 12. shows crack generating.

Comparing this method to the hydraulic wedge, a conventional engineering method, it has the following advantages: a long and massive crack is opened in one operation; the rock-fracturing holes are small in diameter; and there is no need for excess length of rock-fracturing holes.

However, there was danger of the tunnel face collapsing because of generating a long and massive crack, so it was necessary to keep away from the tunnel face. Consequently, the work of inserting the Aqua-Splitter into the fracturing hole and removing it from the hole was carried out not by manpower but by machinery. For this reason, the inserting and removing machine shown in Figure 13., was mounted on the guide shell of the jumbo. By using this machine, the task of inserting and removing the Aqua-Splitter could be performed from the driver's seat of the jumbo. The mechanism of this machine is as follows: Grip the adapter, which is connected to the bottom of the Aqua-Splitter by the clamp, then slide the guide shell to insert or remove the Aqua-Splitter. By adopting this

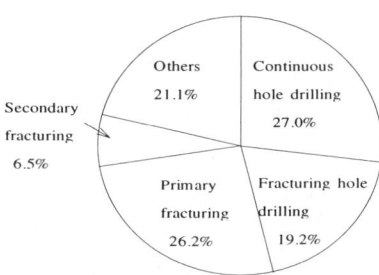

Fig.10 Cycle time

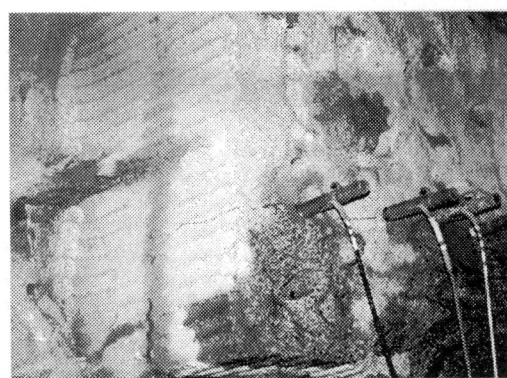

Fig.12 Crack generating

Fig.11 Primary fracturing by Aqua-Splitter

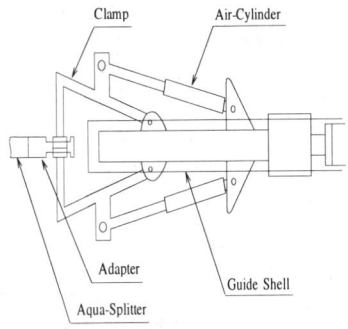

Fig.13 Inserting and removing machine

Fig.14 Secondary fracturing

machine, primary fracturing could be carried out safety.

After these processes have been repeated, secondary fragmentation is carried out using a breaker, which completes the excavation procedure. Figure 14. shows secondary fracturing.

5 Conclusion

We have outlined the new rock-fracturing excavation method for hard rock tunneling. This method has two key points. One point is the method of continuous hole drilling with a SAB rod. The advantages of this method are not only efficiency and accuracy of continuity, but the option of using a general purpose jumbo. Another point is the primary fracturing using the Aqua-Splitter. The advantages of this method are that it generated a long and massive crack in one operation, and rock fracturing holes are small in diameter and there is no need for excess length.

Because this method hardly damages the rock around the tunnel being excavated, it seems
that this method is suitable for tunnel excavating
for various waste.

Acknowledgments: The Aqua-Splitter is developed by the co-operation with BRIDGESTONE CORPORATION. The authors are grateful to Mr. Nakayama, Products Development Department, and other staffs.

REFERENCES

1) Hagimori, K., Furukawa, K., Nakagawa, K., Yokozeki, Y 1991: Study of non-blasting tunneling by slot drilling method, Proc. 7th ISRM Int. Congr. on Rock Mechanics, 1991, Aachen, pp. 1001-1004.

2) Nakagawa, K. 1987.: Recent method of excavating of rock mass by non-blasting, "Kyoryo", No. 2, pp. 42-46. (in Japanese)

3) Hagimori, K. 1990.: Study on tunneling into hard rock using a slot drill, Doctoral Thesis of Nagoya University. (in Japanese)

4.) Noma, T., Hada, M., Kadota, S., Murayama, H., Ueda, S., 1995. : Development and practical application of the static rock-mass fracturing method using hydraulic pressure, Proc. 8th ISRM Int. Congr. on Rock Mechanics, Tokyo, pp. 653-656

Qualified underground technologies for excavation of the Super-KAMIOKANDE cavern

T. Nakagawa
Australia Representative Office, Mitsui Mining and Smelting Co., Ltd, Melbourne, Vic., Australia

J. Yamatomi & G. Mogi
Department of Geosystem Engineering, University of Tokyo, Japan

T. Takemura & K. Tsurumi
Mitsui Mining and Smelting Co., Ltd, Tokyo, Japan

ABSTRACT: This paper presents the qualified and informed underground technologies applied for excavation of the Super-KAMIOKANDE cavern. The dome-shaped cavern has dimensions of 40 m in diameter and 57.6 m in height and has been excavated at a depth of about 1,000 m below the overlying mountains at Kamioka Mine, Gifu, Japan. Before and during the construction of the large cavern, various properties and behavior of the in-situ rock mass were measured and monitored, and also used for a reasonable systematic rock reinforcement in order to realize a safe operation. A newly developed elastic FEM analysis is also presented, which can obtain a three-dimensional solution of stress and displacement under a general loading condition by using a two dimensional/axisymmetric finite element model. Because of its simplicity, it has been utilized satisfactorily for predicting the rock behavior during excavation as well as for estimating the in-situ moduli of rock mass deformation.

1 INTRODUCTION

The Super-KAMIOKANDE (KAMIOKA Nucleon Decay Experiment) is belonged to the second generation of a scientific program organized by the Institute for Cosmic Ray Research of University of Tokyo to detect astronomical neutrinos and investigate the validity of "the Grand Unified Theories" in a subsurface observatory excavated at Kamioka Mine, a leading underground lead and zinc mine located in the central mountainous region of Japan.

In 1982, University of Tokyo and Mitsui Mining & Smelting Co., Ltd. constructed its first underground observatory called KAMIOKANDE, a large dome with a height of 22.5 m and a diameter of 19 m, in which a 4,500 m^3 pure water tank with 1,000 detectors called photo-multipliers was installed (Saito & Namekawa 1983). The deep and competent underground environment of Kamioka Mine was selected as the best site in Japan to locate the observatory by reason of isolation to avoid any contamination and disturbance caused by usual cosmic ray events. It has been operated since 1983 and established a worldwide reputation in the fields of elemental particle physics and astrophysics by its successful results in detecting the cosmic neutrinos caused by the burst of the supernova in the Large Magellanic Cloud.

For further improvement in accuracy and sensitivity of observation, in November 1991, University of Tokyo ordered Mitsui Mining & Smelting Co., Ltd. to construct a ten times larger underground dome, Super-KAMIOKANDE, installing a 50,000 tonne water tank with 11,200 detectors. The super dome has dimensions of 40 m in diameter and 57.6 m in height and has been excavated at a distance of some 150 m from the former one at +350 mEL, about 1,000 m below overlying mountain called Mt. Ikenoyama with +1,368 mEL peak in which major part of the Mozumi orebody of Kamioka Mine located (Tsurumi et al. 1995). The excavation was completed in August 1994, followed by the water tank set up in November 1995, and then the observation started from March 1996.

Geological surveys and various in-situ rock measurements have been conducted since 1986, for deciding a preferable site for the new observatory at Kamioka Mine. The site is situated closer to the synclinal axis in the central mass of complicated folding structure of the Hida gneisses, being composed of amphibolite, hornblende gneiss and so-called the "Inishi migmatite" partly with a little limestone and skarn, which are fresh and competent enough with very little mineralization. The properties of rock at the construction site are given in Table 1 (Takemura et al. 1995).

During the construction, various kinds of rock

measurements have been conducted and monitored in order to guarantee a safe operation and realize a reasonable rock supports, along with computer-aided numerical analysis of the stability.

This paper presents qualified and informed underground technologies applied for the Super-KAMIOKANDE construction, and also presents a newly developed elastic FEM analysis for axisymmetric cavern under an skew-axisymmetric state of stress, which has been fully utilized to predict surrounding rock behavior and evaluate stability during excavation of the cavern.

Table 1. Properties of rock at the construction site

Property	Value
Name of rock	amphibolite gneiss
Density	0.026 MN/m^3
Compressive strength	149 MPa
Tensile strength	9.7 MPa
Young's modulus	49,900 MPa
Elastic wave velocity	4,930 m/s
Initial state σ_1	28.8 MPa (N309E68E)
of stress σ_2	18.9 MPa (N 60E 8W)
σ_3	6.1 MPa (N153E20E)
RQD index	84 %
RMR	84 - 94

2 EXCAVATION OF SUPER-KAMIOKANDE

2.1 Outlines of Super-KAMIOKANDE excavation

Excavation of Super-KAMIOKANDE consists of the huge main dome cavern with a volume of 69,000 m^3 for the detector tank, access tunnels and a decline ramp to the dome and several small caverns for related facilities such as water purification and degassing system (Figure 1). A total volume of 78,000 m^3 was successfully excavated within net 32 months as scheduled.

ANFO and emulsion explosives with NONEL detonators were, though, fully used for the cavern excavation, trackless mining systems and NATM concepts were also applied from the beginning of the construction.

Since the huge volume of rock should be excavated in addition to high stress condition and brittle nature of the rock mass prone to cause any rockburst hazards, the main cavern was excavated step by step as shown in Figure 2. Block 1, the rooftop portion of the cavern, was first excavated followed by Block 2, 3, and 4, from the top to the bottom. The most critical stress concentrations and related troubles were anticipated during excavation of Block 1, consequently, it was subdivided into the sub-sections from ① to ⑧, as depicted in Figure 2. Namely, in order to relieve any local stresses and allocate them to the surrounding rock mass to form a stabilized ground arch as early as possible, the peripheral zone of Block 1 was initially removed and subsequently the internal core was mined out.

Modern semi-automated trackless mobile mining machinery, almost the same types as operating at Kamioka Mine, were utilized for the construction.

Blasted materials, summed up to about 120,000 m^3, was all mucked and hauled up by 12 tonne LHD's through the access tunnels and the ramp way to the temporary stockpiling space located closer to the dome at +350 mEL, where water purification & degassing facilities will be installed after completing the excavation. And thereafter, the stockpiled material was hauled horizontally through "Atotsu" main adit by 20 tonne underground trucks and finally dumped in a waste disposal.

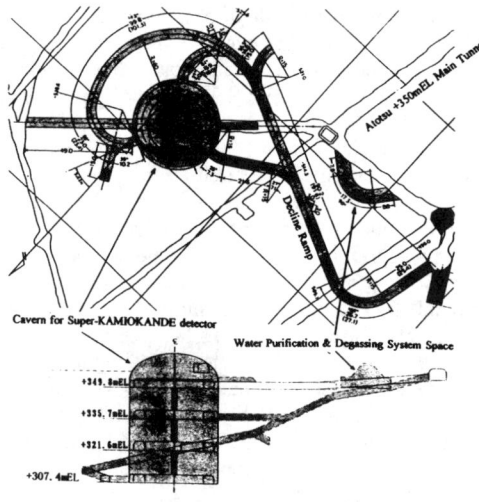

Figure 1. Overview of Super-KAMIOKANDE with related access tunnels and observatory facilities.

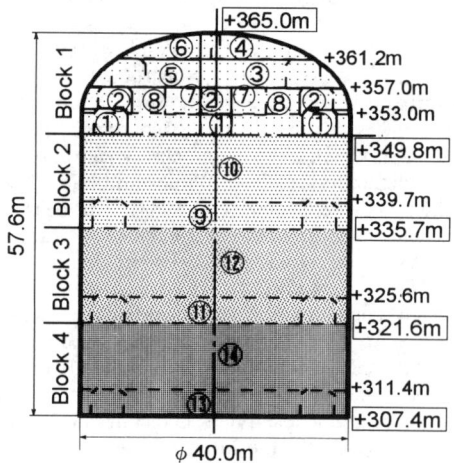

Figure 2. Excavation sequence of Super-KAMIOKANDE cavern for the detector tank.

2.2 Excavation and support of the roof part

For excavating Block 1, the roof part of the cavern, overhand cut & fill stoping method was applied. Because it was required to carve the roof in a hemispheroidal shape with a height of 12 m, the blasted rock was piled and used as a working floor for more precisely finishing up the curved roof, in addition to supporting the ground stress during the excavation.

Double layered 16 cm thick wet-shotcrete mixed with 80 kg/m^3 steel-fiber were sprayed on the peripheral surface of the dome. Cement grouted cablebolts (ϕ 15.2 mm $\times$ L 8 m, 7 strand PC steel, 2 wires/hole) were installed at every 4 m^2 space, and cement grouted rockbolts (ϕ 22 mm $\times$ L 2 m, steel rebar) were installed at every 1 m^2 space except for the location of cablebolts. Those supports were performed soon after the peripheral surface of the dome was exposed. During excavation, temporary shotcrete and rockbolts were installed properly on the immediate roofs and walls to maintain working condition as safe as possible.

Excavation procedure was as follows; firstly approaching the dome at +349.8 mEL to make a circular tunnel, ①, at 2 m inside of the periphery of the dome followed by smooth blasting to finish it, and then the immediate roof of the circular tunnel, ②, was stoped at two stages upward to stand on the piles at +357 mEL. On that level, internal rock pillar, ③, was horizontally stoped to expose a half of the flat immediate roof at +361.2 mEL, from which pre-cablebolts were installed before blasting upward, ④, with parallel drillholes, whose length were precisely determined, to make curved surface of the rooftop. After installing shotcrete and rockbolts on the half of the curved rooftop, the other half, ⑤ and ⑥, was excavated in a same manner using pre-cablebolts (Figure 3).

Finally, the internal core of Block 1, ⑦ and ⑧, was excavated by bench blasting along with removing the piled rock, once used as the working floor.

In order to stabilize the roof and minimize the overbreak, in total, 211 pre-cablebolts were installed. Special caution was required to make the position and direction as well as the anchor length of cablebolting as precise as possible, since they are expected to work not only temporarily but also as permanent supports after completing the cavern and during observatory stage. For the examination of the anchorage of the installed cablebolts, pulling test was carried out and up to 30 tonne capacity was confirmed.

2.3 Excavation and support of the cylindrical part

The cylindrical part below +349.8 mEL were divided into three blocks (Block 2,3,4) and excavated by the modified sublevel stoping method from upper sublevel to lower one.

Prior to excavation of Blocks 2, 3, and 4, from +349.8 mEL until the lower +307.4 mEL, the bottom level of the main dome, a spiral decline was tunneled and branched at sublevels, +335.7 mEL and +321.6 mEL. Procedure of excavating Block 2 was as follows; firstly a circular tunnel was developed 2 m inward of the cavern's periphery at +335.7 mEL, followed by smooth blasting to finish it and the designed support were completed. Then the sublevel was horizontally opened up leaving the minimum size of two or three pillars, and 10.1 m long vertical blasting holes were drilled parallel from +339.7 mEL to +349.8 mEL, and they were drilled with closer spacing at its perimeter. After raising a 4 m^2 slot by stage blasting, rock mass was stoped by longhole blasting and the peripheral wall was carved by longhole smooth blasting. Rock support on the peripheral wall was conducted along with lowering the slope of blasted rock pile. Block 3 and 4 were subsequently excavated in the same manner as Block 2.

Rock supports for cylindrical part of the cavern were almost the same as those performed in the roof part, except that cablebolts were installed at every 6 m^2 space, and 2.35 m long rockbolts were installed to stick 19 cm out of the shotcreted wall of the cavern. Those bolt heads were to be directly connected to the tie-rods welded onto the side wall of the detector tank and filled with lining concrete later. Cablebolts were also installed at every 4 m^2 space on the bottom floor and its perimeter at +307.4 mEL. Temporary rock supports were also properly conducted to keep operational safety.

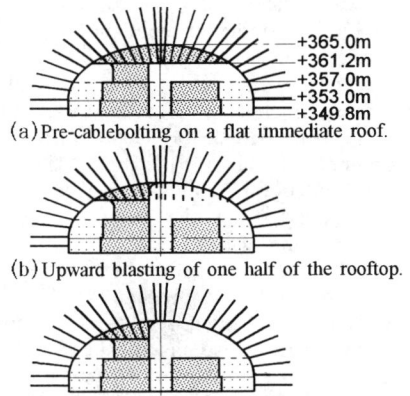

Figure 3. Excavation procedure for the rooftop of the cavern with pre-cablebolt installation.

2.4 Grouting

Before the construction, some 80 m³ of cement grouting against a typical fault located 12 m N.W. from the site was conducted for minimizing underground water seepage into the main cavern. During the excavation below +335.7 mEL, a systematic pre-grouting at the faces of the decline ramp and sublevels inside of the cavern was performed by drilling 9 m - 12 m long, four to six check holes after every 2 or 3 blasts. The outer zone of the cavern below +335.7 mEL, were also curtain-grouted downward from each sublevel before its longhole blasting. Cement/sodium silicate mixture with a little amount of potassium phosphate was used as grouting material for quicker hardening. Totally, a length of 2,885 m grouting holes were drilled and a volume of 373 m³ grouting mixture was injected to minimize a flow volume of ground water leaked into the site as low as 0.8 m³/min.

2.5 Quality control of the construction

Quality of the construction to satisfy various requirements has been checked in many ways. Only the certified materials could be used for the construction. In-situ pulling strength of rockbolts and cablebolts and specimen strength of shotcrete and cement paste have been tested periodically to assure the long term stability of the cavern.

Among those quality requirements, keeping the size and shape of the cylindrical part of the cavern within the tight limitations has been most severely controlled, because the detector tank had direct contact with the cavern via lining concrete.

Distances from the central vertical axis to the finished surface of the cavern below +349.8 mEL were measured at 2,615 points in a regular pattern using an optical range finder. The average radius resulted in 20.447 m (0.447 m overbreakage) with 0.266 m standard deviation.

3 INFORMED CONSTRUCTION WITH ROCK MONITORING

3.1 Rock mass measurements and monitoring

In order to monitor the rock mass behavior during the construction, we have measured rock mass deformations, convergences, stress changes, axial loads induced in rockbolts and so on. These data were found also useful for modifying the support strategy. Geological aspects of freshly exposed rock surface were observed and sketched after every blasting (Figure 4). A very competent gneiss with fewer cracks was found in S.E. half domain of the cavern, while cracks and joints with N45E strike and 45 - 90W dip were predominant and a small amount of limestone was mixed in gneiss in N.W. half domain.

The convergence of the cavern was measured in 18 lines every week using a precise optical range finder. 10 extensometers (E1, E3 - E10; 20 m long, anchors at 1, 3, 6, 10, 15, 20m deep, E2; 30 m long, anchors at 5, 10, 15, 20, 25, 30m deep), 13 instrumented rockbolts (R1 - R13; 8 m long, gages at 0.5, 1, 2, 3, 5, 7.5m deep), and 2 stress sensors with 8 component gages were installed in-situ rock surrounding the cavern as shown in Figures 5 and 6. Those data were automatically picked up every 6 hours by the data-logger at the underground monitoring room close to the site, and finally transferred through a tele-communication line connected to a modem of a computer located on surface (Tsurumi et al. 1996).

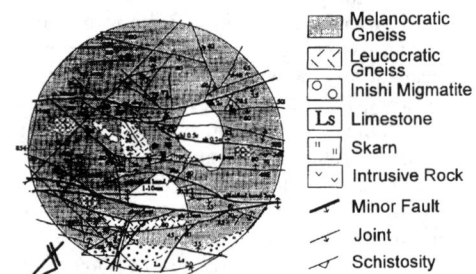

Figure 4. Geological sketch of the immediate roof of the cavern at +339.7 mEL

3.2 Rock noise and rockburst phenomena

During excavating upward in eastern domain of the roof part of the cavern between the levels of +353 mEL and +357 mEL, about 10 m³ of pyramid-shaped rock was spalled off by rockburst from post-smooth blasted surface of the dome.

A rock noise was heard just after blasting the temporary slender rock pillar which had been functioning as a final support in the eastern half domain of the dome between +357 mEL floor and a flat immediate roof at +361.2 mEL. The rock noises were emitted for about an hour, gradually spreading co-axially from the ex-pillar position further and deeper into the rock mass.

A rock noise was also heard when the face of the access tunnel reached to the side wall position of the cylindrical part of the cavern at +321.6 mEL.

In either case, rock noise and rockburst did not cause disastrous rock collapse, so the ordinary pattern rock support was enough to cope with those local phenomena.

3.3 Redesigning of rock support reflecting the monitoring data

Most violent concentration and relocation in the state of stress were expected to occur while the roof part of the cavern was excavated and some instability of the surrounding rock was anticipated.

After completing the excavation of the roof part of the cavern (Block 1), the extensometer E1, E3, and E5 showed 7 mm, 9 mm, and 16 mm in displacement of the wall respectively. Then during the excavation of the cylindrical part of the cavern between +335.7 mEL and +349.8 mEL (Block 2), although E1 showed only a slight change, E3 increased its displacement by 24 mm to 33 mm and E5 increased by 8 mm to 24 mm (Figures 7 and 8). These values are about three times as large as those predicted by FEM analysis calculated beforehand.

The extensometers E3, E4, E5, and E6, at +352 mEL indicated that in S.E. part of the cavern, only shallow zone at most 3 m deep showed relatively large deformation with local spalling breakage, while in N.W. part, the zone of large deformation expanded deeper into 6 - 10 m from the peripheral surface.

Values of rockbolt tensions in R12, R3, and R7, increased to 10 - 18 tonne within 3 m deep in excavating Block 1, and R10 and R11 indicated 18 tonne within 3 m deep in excavating Block 2.

Therefore, it was concluded, in order to improve the stability and secure a larger safety margin, the support pattern should be reexamined and consequently redesigned to increase the number and density of cablebolts in the S.E. portion and the N.W. portion of the dome. In Block 1, 33 of 15 m cablebolts/16 m^2, and 65 of 5 m cablebolts/20 m^2 were installed additionally on both areas of N70E - N160E and N110W - N10E. In Block 2, 15 of 15 m cablebolts/ 12 m^2, and 50 of 5 m cablebolts/12 m^2 were installed in the same areas as Block 1 (Figure 9).

Although concerning the upper half of the main cavern, Blocks 1 and 2, rock support was redesigned, we have encountered none of serious changes in rock mass deformations measured by extensometers during excavation of the lower half, Blocks 3 and 4. At last, in August 1994, we have successfully finished excavation and all the monitored data almost leveled off and stabilized after excavation.

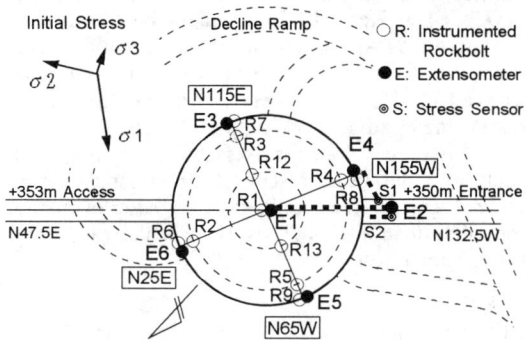

Figure 5. Schematic diagram of the arrangement of rock monitoring apparatuses above +349.8 mEL.

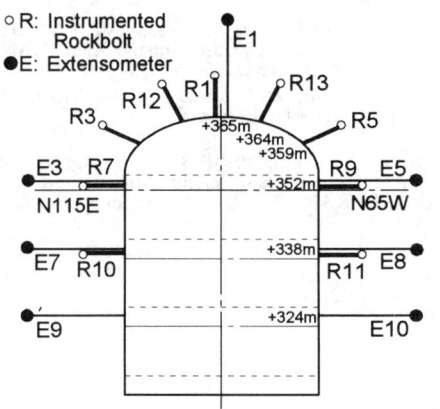

Figure 6. Schematic diagram of the arrangement of rock monitoring apparatuses in the N115E - N65W vertical section

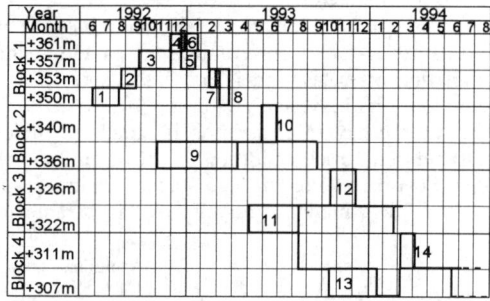

Figure 7. Excavation schedule of the cavern.

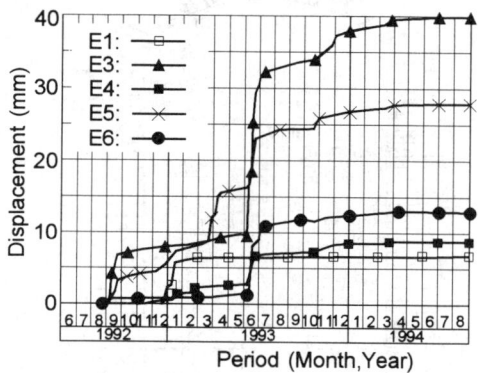

Figure 8. Displacement of the peripheral wall of the cavern monitored by the extensometers.

4 A NEWLY DEVELOPED FEM ANALYSIS FOR AXISYMMETRIC CAVERN

An elastic FEM analysis is newly developed which can obtain a three-dimensional solution of stress and displacement under a general loading condition by using a two dimensional/axisymmetric finite element model. The proposed FEM technique takes advantage of geometrical axisymmetry of the cavern and the elastically competent features of surrounding rocks. These make it possible, by using the Fourier expansion technique, to decompose the external loads into symmetric and skew-symmetric components and then apply the principle of superposition to synthesize the three-dimensional solution from the separately obtained answers under the corresponding simplified loads.

The proposed FEM analysis has advantages of its simplicity and ease in maneuvering on a personal computer with a minimum time of calculations, so it has been utilized repeatedly at construction site office for various purposes such as estimating stability, predicting behavior of the surrounding rock mass, and reversely inferring the deterioration of rock deformation character by inputting updated data obtained from in-situ monitoring apparatuses.

An attempt to estimate deformation moduli of the in-situ rock mass surrounding the cavern at the completion of excavation is presented here as one example of the FEM analysis usage.

Firstly the cavern excavation is simulated under the same initial rock condition as shown in Table 1 using Poisson's ratio of 0.25 except inputting an assumed deformation moduli as a parameter.

Secondly calculated displacement by the analysis at corresponding locations to the four extensometers installed at +352 mEL of the cavern at an early stage of its excavation were selected, and then compared with measured displacement data obtained from those extensometers. Repeating these procedures by changing parameter value and adapting the concept of least square method will lead to find the answer. With the above procedures the in-situ rock mass deformation moduli was estimated as approximately 9,000 MPa, about 18 % as small as that of an intact rock specimen for laboratory experiment.

5 CONCLUSION

The Super-KAMIOKANDE cavern was successfully completed as scheduled, even though its stress condition was anticipated severe due to some 1,000 m deep overburden. Moreover, it also satisfied the strict requirement of quality and safety such as no loss time of injuries. We suppose, the qualified mininig technologies historically developed at Kamioka Mine might be the key of the success as well as the informed and modernized excavation might be useful for detecting any unpredictable problems and realizing a reasonable/optimum rock support design.

A newly developed axisymmetric elastic FEM analysis under general state of stress satisfactorily utilized and reversely estimated the value of in-situ rock deformation moduli as 18 % small as that of the specimen test.

REFERENCES

Saito, S. & M. Namekawa 1983. Excavation of the underground observatory at Kamioka Mine. *J. of MMIJ.* 99 (1142): 268-272.

Takemura, T., S. Fujii & T. Nakagawa 1995. Rock behavior observed in excavating Super-KAMIOKANDE cavern. *Proc. of Autumnal Conf. of MMIJ.* T-7 : 7-8.

Tsurumi, K., S. Fujii & T. Nakagawa 1995. Excavation of Super-KAMIOKANDE cavern. *J. of MMIJ.* 111 (6): 381-386.

Tsurumi, K., Y. Hirabayashi & T. Nakagawa 1996. Rock measurements and monitoring in excavating Super-KAMIOKANDE cavern at Kamioka Mine. *Proc. of Autumnal Conf. of MMIJ.* A-①-8 : 29-32.

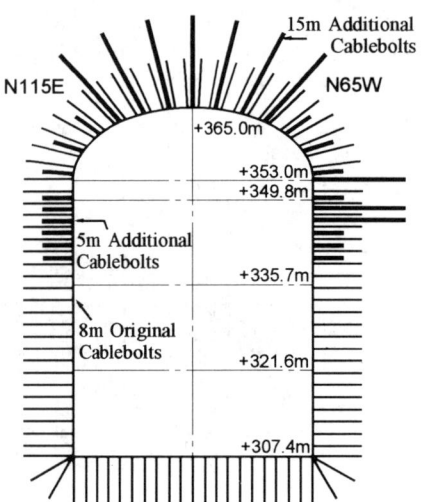

Figure 9. Schematic diagram of the redesigned cablebolts pattern in the N115E - N65W vertical section.

Fragmentation of rock by cutting tools

M.Cheraghi Seifabad
Department of Civil Engineering, K.N.Toosi University, Tehran, Iran

ABSTRACT: The fragmentation of rock by cutting tools forms an important aspect of rock mechanics, rock excavation and other energy excavating processes. The complex phenomenon of the rock cutting process is influenced by various parameters, such as cutting tool parameters (type, rake angle, sharpness, etc.), motion parameters, and rock parameters (rock type, strength, and fracture parameters, etc.). There is a growing concern that rock fracture toughness is an inherent index-like property directly related to rock cuttability. Attempts have been made to find relationship between rock fracture toughness and the cuttability parameters.

1 INTRODUCTION

Rock cutting machinery requires capital investment, especially for large scale machines such as a roadheader. Production is highly depended on performance and stability of the system. It is therefore very important to determine the most suitable cutting machine, and to assess the performance of the machine for particular rocks. An excavating process is influenced by many factors such as rock properties, machine operating parameter, i.e. thrust, RPM etc. As a result, prediction or assessment of cutting performance of a rock cutting machine should based on these factors.

In the search for a suitable measure of rock properties to assess rock cutting performance of rock cutting, many methods have been proposed. The UCS has been the most widely used parameter.

The SE (Specific Energy) is proposed by Teale (1965) as a parameter for measuring the relative efficiency of various cutting tools, machines and cutting processes in a given rock. It has been used by many investigators to assess the relative resistance of various rocks to a given tool or machine. A number indentation probes have also been proposed, which normally involve an indenter with some kind of configuration, being forced into the surface of the rock. A typical example of such a device is the NCB cone indenter (the NCB cone indenter, 1977).

Rock cutting with drag bits, is characterized by the penetration of the cutting tool into the rock, which results in crushing the rock around the tool. As the penetration continues cracks are developing and propagation of such cracks result in rock fragments, which are removed by further movement of the cutting tool. This fracture mechanism suggests that rock cutting performance should be related to rock fracture properties, i.e. rock fracture toughness.

2 FRACTURE MECHANICS RELATED TO ROCK CUTTING

Nelson et al (1985) have shown a good relationship between critical energy release rate and field penetration index (field penetration index, is the ratio of the average net thrust per cutter to the penetration per cutterhead revolution, used normally for evaluating TBM performance). Guo et al (1988) have investigated the mode of fracture in a drag bit by using strain energy release rate. It was shown that a sharp bit is able to provide much more strain energy for crack development than a blunt bit. Deliac (1988) suggested two fundamental chipping modes called Mode A and Mode B. Mode A is typical shear fracture of the rock, while Mode B is a fracture propagation mode. Mode A is predominant when the pick is wide, the rock is not very brittle, or when the depth of cut is high. Mode B is predominant when the pick is sharp and when the rock is brittle.

To measure fracture toughness Deliac (1986) used the double torsion test, short rod and three point bend test, unfortunately these three tests do not comply with the ISRM suggested methods for fracture testing.

3 PHYSICAL PROPERTIES

The bulk density and porosity of specimens were measured according to ISRM and the results for various rocks are as follows (Table 1):

Table 1. Porosity and density

Rock	Porosity %	Density (Mg/m^3)
Springwell sandstone	12	2.30
Pennant sandstone	0.14	2.66
Teesdale Whinstone	0.18	2.95
Matlock limestone	6.9	2.46
Welton chalk	1.05	2.30

4 MECHANICAL PROPERTIES

4.1 Uniaxial compressive strength and tensile strength

Most theories of rock cutting predict a correlation between rock material strength and productivity. This relationship has been used by a number of researchers to predict roadheader performance.

A suitable arrangements satisfying ISRM requirements was used for applying and measuring axial load to the specimen. The rock cores had nominal dimensions height = 150 mm and diameter = 750 mm. LVDTs (linear variable differential transformers) were used for measuring the axial and lateral deformation. Load on the specimens was applied continuously at a constant stress rate of 3 Mpa/min (according to ISRM).

Table 2. Uniaxial compressive strength and tensile strength

Rock	σ_c (MPa)	σ_t (MPa)
Springwell sandstone	43.9	3.77
Pennant sandstone	145.92	11.28
Teesdale Whinstone	145.6	20.8
Matlock limestone	122	5.23
Welton chalk	47.5	4.41

The Brazil disc test was used to measure indirectly the uniaxial tensile of rock specimen.

A cylindrical Specimen of diameter 60 mm and thickness 30 mm (ratio 2:1) was compresses diametrical between two platens. A constant loading rate was applied so that failure occurred within 20-25 s of loading.

4.2 Hardness and abrasivity

4.2.1 Shore scleroscope hardness

This method determines an index of hardness of rock minerals as an average of readings taken at random on an individual mineral grains and reflects a measure of rock mineralogy, elasticity and cementation.

To perform the test (according to ISRM) a core specimen with a flat ground surface is marked off in small squares of 10 mm sides and instrument is positioned vertically with the bottom of the barrel in firm contact with the specimen, and normal to its surface. The hammer is brought up to elevated position by squeezing a rubber bulb, and the allowed to fall to strike the test surface. The rebound height on the first bounce is observed and recorded as the hardness of the sample. Only one test at the same spot is made by spacing the indentation at least 5 mm apart. The test is repeated another 19 (R_{20}) times and the average over these 20 reading is taken as the shore scleroscope hardness.

4.2.2 CERCHAR abrasivity

This is done by a simple scratch (10 mm) on a plate of rock specimen by the CERCHAR machine. The measurement of abrasivity being in terms of the wear flat diameter in 1/10 mm units. The wear flat is measured under a microscope with a micrometer. A traveling microscope with an eyepiece micrometer is used to measure the wear flat in two orthogonal directions. They are allowed to coincide at one edge and first reading R1 recorded from the micrometer. Then the other line is moved to other edge of the wear flat and reading R2 is recorded. The difference, R2-R1 is then compared. The tests is repeated on each sample 5 times, (West, 1989).

Table 3. Hardness and abrasivity

Rock	(1)	(2)	(3)	(4)
Springwell sandstone	1.33	1.74	46.85	32
Pennant sandstone	1.58	2.82	70.3	6
Teesdale Whinstone	1.54	3.85	88	9
Matlock limestone	0.79	1.74	41.8	50
Welton chalk	0.32	1.74	38.65	49

(1) CERCHAR abrasivity (2) NCB cone indenter
(3) shore scleroscope
(4) shore plasticity (McFeat-Smith, 1975)

5 FRACTURE MECHANICS TESTS

From the methods for measuring fracture toughness, ISRM has suggested the following two :
1. three point bend test
2. short rod test

Theses two methods are based on mode I fracture mechanics and there are two levels, level I that is employing load controlled. The main difference between level II and level I is time and accuracy. The available testing facilities were not suitable for measuring fracture toughness at level II. Therefore in this investigation level I has been adopted for both types of test with a loading rate of 20.3 kN/min.

The tests were carried out on a 50 kN servo-hydraulic, closed loop testing machine. The load was read by a load ring-shaped transducer and recorded by a data logger using an LVDT. One LVDT and one clip gauge were used to measure the displacements. The LVDT was used for measurement of loading point deformation, the clip gauge was manufactured in the University of Newcastle upon Tyne the basis of the ASTM E399-78 drawings and it was calibrated to the nearest 0.0025 mm. A pair of knife edges that support the gauges'arms and serve as displacement reference points, were attached to the specimens using adhesive prior to testing. All the data, which included the load, the point displacement, and crack open displacement, were monitored using a data logger.

5.1 Three point bend test

One of the methods introduced by ISRM is the chevron notched three point bend test. Loading at three points causes the chevron-notch to start crack propagation at the crack tip and proceed to the core axis in a stable fashion until the point where the fracture toughness is evaluated. Using the geometry guidelines given in ISRM the following geometry was implemented :
D=61mm, S=200mm, a_0=9mm, a=200mm (according to ouchterlony 1988).

In this type test Springwell sandstone, Pennant sandstone, Teesdale Whinstone, Welton chalk and Matlock limestone were used characterized by the following grain size:
Springwell sandstone 0.5 mm to 0.75 mm, Pennant sandstone up to 1 mm, Teesdale Whinstone 0.2 mm to 1 mm.

Welton chalk and Matlock limestone are characterized by an amorphous mineral structure consequently grain size identification is not relevant.

The process followed requires a cylindrical specimen of diameter approximately 61mm, length 220mm notched with milling machine using a 100mm diameter diamond wheel saw cut. The notch should be placed at equal distances from the ends of the face. When the specimen is placed in the vice, first of all the distances between the saw blade and the specimen should be zeroed position, then the specimen should be cut to a depth of 0.25D, which is 15mm. After this cut, the specimen is rotated 90° with the aid of the fixture and cut again to the same depth as in the first cut.

Special alignment aids that facilitate accurate positioning of the specimen in the load fixture should be used in this method. First it is used the center the support rollers with respect to the upper (loading) roller and to give an exact support span length. Second, it is used aligning the specimen axis perpendicular to the rollers. Third the guide ensures that accurate centering of the notch between support rollers is achieved. Special fixture has been used to set up three point bend test specimens. Once the load roller and two support rollers are fixed to parallel position, this set up should be suitable for all of the tests and there is no need for any movement. another alignment aid which can be used is a metal plate which has a spirit level whose reference plane is the base of the testing machine. The direction of loading can be checked by inserting this plate into the notch. If this angle is unsatisfactory the specimen can easily be notched into the correct position. This metal plate can be used to identify the axis of span which is perpendicular to two support rollers.

5.2 Short rod test

The specimen dimensions for this method suggested by ISRM are D = 44mm, a_0 = 21mm and a = 30mm (Ingraffea, 1982)

A diamond wheel saw blade is used to cut the specimens to the required depth and to cut the required notch. The chevron notch should be straight and the holding device keeps the specimen in the position so as to ensure that chevron angle (27.3°) conforms to the given tolerances. Before making any notch the rock should be cut to a nominal length 1.45D which is approximately 44 mm. When the first notch is cut to an angle of 27.3°, because the milling machine with the saw blade is fixed, the set-up rig with the specimen is rotated to 54.6° degree to the required cutting depth. The specimen subsequently positioned in the testing machine and two aluminum plates are used to apply tensile load to the specimen. These plates are glued to the core surface and are set parallel and at equal distance from the crack position, using a space bar with a stem in the middle.

Short rod testing requires a tensile load to be applied to the specimen to be tested. A special fixture has been used to record the maximum load fixture in level I testing as described before. The clip gauge measures the relative displacement of two precisely located gauge positions spanning the notch mouth with a pair of accurately machined knife edges that support the gauge arms serve as displacement reference points. If necessary a sling is used to hold the specimen firmly in the testing rig. A stem can apply load corresponding to a displacement with a constant rate of 4.8 mm/min.

5.3 Fracture mechanics results

Results of load versus load point displacement, load versus horizontal displacement, and load versus crack open displacement from the short rod with Teesdale Whinstone are as follows : (Table 4,5)

Table 4. Short rod results

Rock	1	2	3	4	5	6
Springwell sandstone	0.86	0.65	0.86	0.65	0.65	0.73
Pennant sandstone	1.08	2.38	3.24	2.59	2.81	2.45
Teesdale Whinstone	2.12	3.46	3.46	--	--	3.12
Matlock limestone	2.59	2.16	2.16	2.38	--	2.32
Welton chalk	1.08	0.65	0.65	1.30	0.86	0.91

6 : Mean (Mpa $\sqrt{m}$)

Table 5. Chevron bend results

Rock	1	2	3	4	5	6
Springwell sandstone	0.50	0.50	0.62	--	--	0.50
Pennant sandstone	2.08	2.13	1.85	2.02	2.24	2.06
Teesdale Whinstone	2.36	2.99	2.66	--	--	2.67
Matlock limestone	1.01	0.90	0.95	--	--	0.95

9 : Mean (Mpa $\sqrt{m}$)

These are two differences between theses methods. To set up the CB method is the time consuming while short rod is very simple. Another factor is choosing the method is the number of specimens available. One specimen in the three point is equivalent to two short rod, as appeared from the ISRM's suggestion that after three point the specimen can be used for short rod testing as well. It appears therefore that use of short rod test is a better option, when limited quantity of material is available for investigations.

6 MECHANICAL CUTTING

A core grooving test developed in the University of Newcastle upon Tyne (Roxborough, 1975) is used extensively in the assessment of the machineability of rock for roadheaders and other drag tool equipped machines.

The test may be carried out on either core samples or black samples of rock. Four cuts are normally made in the rock sample at a constant depth of 5 mm with a tungsten carbide tool 12.7 mm wide, chisel-edged, with a -5 front rake and +5 back clearance angle. This tool is mounted on an instrumented shaping machine.

6.1 Data collection and analysis

The following parameters are used in assessing the performance of a cutting tool under different cutting conditions. they are as follows :

Mean cutting force (MCF)· is the average force acting on the tool in the direction of cutting. The product of the MCF and the distance cut gives the work done.

Mean normal force (MNF) is the average force acting normal to direction of cutting. This value is the thrust required to maintain the tool in place at the required depth of cut.

MPCF is the average of maximum values of the cutting force and varies greatly as chips form. This shows the lowest force that has to applied to the tool to cut the rock.

MPNF is the average of the maximum forces generated during cutting to keep the tool in the rock.

Yield, (Q) is the mass of rock cut by the pick per unit length of cut and is expressed in m3/m.

Specific energy, (SE) is defined as the work done to cut a unit volume of rock and is expressed in MJ/m3. It is obtained by dividing the mean cutting

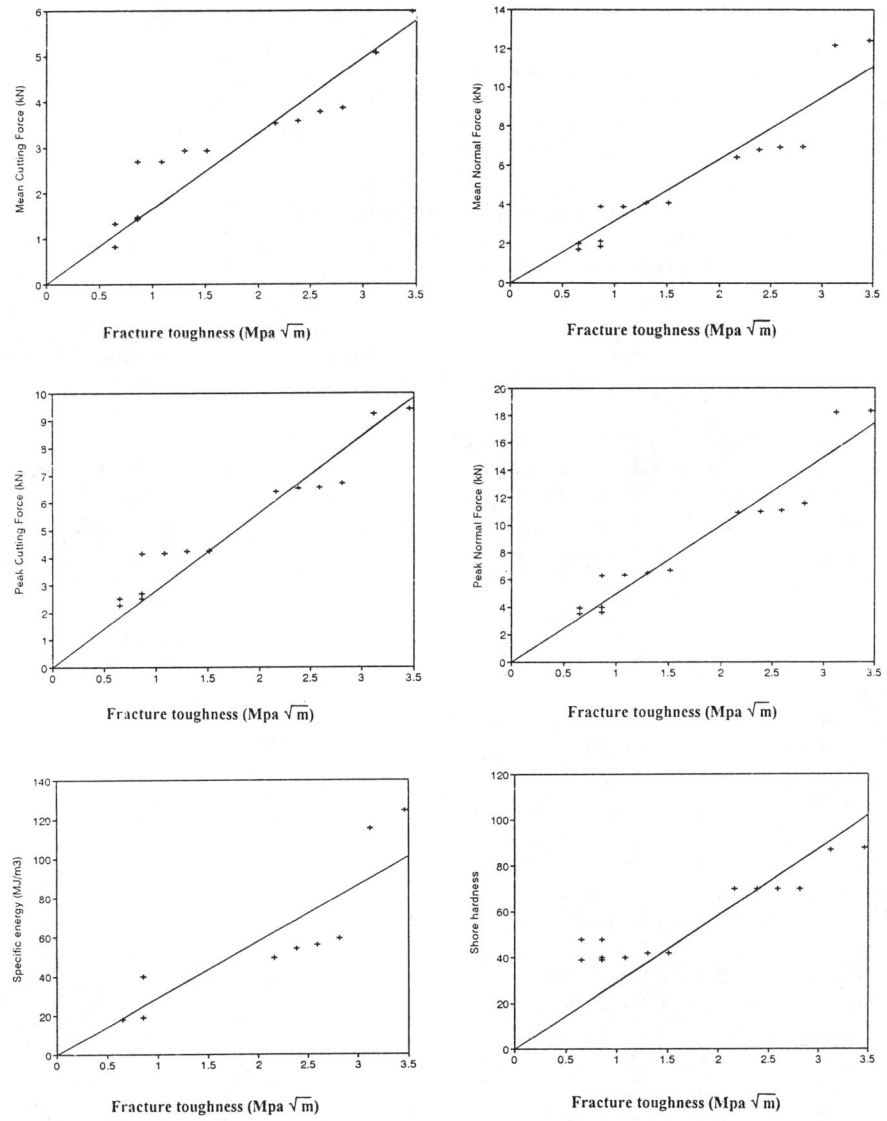

Figure 1. Cutting Parameters V Fracture toughness

force component by the yield, the latter being expressed as the volume of material cut per unit distance cut.

7 CONCLUSIONS

Rock fracture toughness is an important rock property, which indicates a measure of cuttability. The results show that the cutting forces increase as rock fracture toughness increases. Correlation between the rock fracture toughness and the cutting forces show that rock fracture toughness is a better parameter predict cutting performance than the other parameters. Rock fracture toughness is one of the best measures of rock intrinsic properties available for evaluation of cutting processes, and directly related to cutting parameters. Strength and hardness are two of the most fundamental factors which control fragmentation processes. It is reasonable to expect that the rock fracture toughness may be the best rock property which can be related to the cracking process. However additional rock property is required to assess overall fragmentation performance, i.e hardness. Various hardness tests have been proposed for the cutting process, these are: cone indenter, CERCHAR hardness, shore hardness. The other properties such as UCS, cone indenter, shore hardness are correlated to rock fracture toughness. Correlation coefficient between UCS and fracture toughness shows fracture toughness is better rock property to predict cutting parameters, Fig 1.

ACKNOWLEDGMENT

I would like to express my appreciation to Dr. E.K.S. Passaris, lecturer in geotechnical engineering department of Newcastle upon Tyne for providing the opportunity to carry out this research.

REFERENCES

ASTM E399 (1978). Standard Test Methods for Plane-Strain Fracture Toughness of Metal, ASTM Testing Standard.

Deliac, E.P. (1986) Optimization des machines d'abattage a'pics. Doctoral dissertation, Univ. of Paris VI, ed. by ENSMP/CMI, Paris/Fontainbeau, 501 p.

Deliac, E.P. (1988) Theoretical and practical investigations of improved hard rock cutting system. 29th U.S symposium, University of Minnesota. pp 553-562.

Guo, H., Standish, P., Schmidth, L.C.& Aziz, N.I. (1988) A method of mechanical efficiency analysis for rotary drag bits, second international conference on mining machinery. Brisbane, Australia, pp 322-326.

Ingraffea, A.R. and Beech, J.F. (1982) Three dimensional finite element calibration of the short rod specimen, International Journal of Fracture. Vol 18, No.3, pp 217-229.

ISRM (1988) Suggested Methods for determining the fracture toughness of rock. Int. J. Rock Mech. Min Sci & Geomech Abstr. Vol 25, No.2, pp 73-96.

ISRM (1989) On the back ground to the formulae and accuracy of rock fracture toughness measurements using ISRM standard core specimens. Int. J. Rock. Mech. Min. Sci. & Geomech Abstr. Vol 26, No.1, pp. 13-23.

McFeat-Smith, I. (1975) Correlation of rock properties and tunnelling machine performance in selected sedimentary rocks. PhD thesis, University of Newcastle upon Tyne.

NCB (1977) NCB cone indenter handbook.

Nelson, P.P., Ingrafea, A.R.& O'Rourke, T.D. (1985) TBM performance prediction using rock fracture parameters, Int. J. Rock Mech. Min. Sci. & Geomech Abstr. Vol 22, No.3, pp 189-192.

Ouchterlony, F. (1988) Fracture toughness testing of rock with core standard. The development of an ISRM standard, fracture toughens and fracture energy, pp 231-251.

Roxborough, F.F. & Phillips H.R. (1975) The mechanical properties and cutting characteristics of the Bunter sandstone, report to the Transport and Road Research laboratory, University of Newcatle upon Tyne.

Teale, R. (1965) The concept of specific energy in rock drilling, Int. J. Rock Mech. Min. Sci & Geomech. Abstr. Vol. 2, pp 57-73.

Choice of optimal operating conditions of the mining machine during tunnel drivage in geodynamically hazardous zones

A.N. Shabarov, N.V. Krotov & A.P. Zapryagaev
State Research Institute of Mining Geomechanics and Mine Surveying (VNIMI), St-Petersburg, Russia

ABSTRACT: This paper presents an information about the method of geodynamic zoning used for construction of maps of active faults and the detection of hazardous sections in rock mass along the tunnel track. An assessment of geodynamic state of rock mass within the working face section is made with the use of diagnostic systems of mechanical conditions of a tunneling machine on the registered characteristics of its members. The transition of the face section into the geodynamic hazardous state is accompanied by an abrupt change in the resonance frequencies of vibration of the tunnel-boring mechanism.

The maps on which the hazardous zones are indicated, where the vibration intensity of the machine members in its usual operating conditions differs from its normal values, will enable us to select the operation regime for tunnel drivage. An assessment of a hazard is made both by direct instrument measurements with using the geophysical equipment and also by the core separation and by the yield of boring fines out of a borehole.

The research work covers, first of all, the mechanisms and installations (units), the stoppage of which involves great expenses, or those cases when the working conditions do not allow to make direct measurements (hazardous sections).

Technological requirements to tunnel drivage dictate the specific conditions regarding the great velocity of drivage and steady trouble-free operation of driving equipment. The leading firms-producers of them driving machines equip with systems for automatic control of drivage velocity. Such systems allow to optimize the operating conditions of a machine by regulating the cutting process due to quick response to changes in rock strength within the face.

Tectonically stressed and relieved zones in rock mass are formed due to its inhomogeneous structure, the presence of various tectonic disturbances, the existence of tectonically active forces. In addition, the stress state within these zones is attributed to different strength and deformation properties of rocks. Within the tectonically stressed zones the strength and deformation properties of rocks substantially differ from those ones when rock mass is in a normal state. In the tectonically relieved zones the similar characteristics of rocks have lower values.

During tunnel drivage with tunnel boring machine the redistribution of stresses occurs in the face. Depending on rock properties, native ground pressure and the drivage velocity this redistribution may occur calmly or be accompanied by shocks and clamping of an operating member. Finally, it intensifies the vibration of machine joints, causes the wearing of equipment and reduces its service performance.

The method of geodynamic zoning enables us to perform rather reliably the prediction of native stress in rock mass along the whole track of the designed tunnel and to evaluate the technogenic effects of the tunnel boring machine.

An assessment of stress state of rocks along the tunnel track may be made by the instrumental investigations of intensity of acoustic emission in walls and face of the workings and also of geomechanical processes within the shaft zone of drilled boreholes - separation of core into disks, excess yield of boring fines.

Transitions of rock mass into ultimate stress state in face and walls of the tunnel may take place at the working sections in the direct vicinity of various tectonic disturbances. Under such conditions the stress level is evaluated by the ratio of actual vertical stresses, γH, and ultimate strength in uniaxial compression $[\sigma]$.

A high level of stress state of rocks along the tunnel track may be at great depths along the tunnel track may be at great depth and with relatively small indices of strength characteristics of rocks.

In practice, in tunnel drivage at shallow depth the problems arise, associated with rock pressure manifestations in dynamic form within the tectonically active zones.

The magnitude of γH, indicating the maximum radial stresses, $\sigma_{r.max}$, in the tunnel contour zone, may have also the critical values in tunnel drivage in the mountainous terrain.

For example, in drivage of the tunnel of the Irnagai hydroelectric station in the North Caucasus with the tunneling complex "Robbins", the values of KγH at the depth of 700-800 m exceeded the strength characteristics of rock mass at some sections of the tunnel track. In conditions such as these and yet with the presence of tectonic disturbance, the probability of rock pressure manifestations in dynamic form was not neglected. In tunnel drivage with great variations of the surface reference marks in the mountainous regions the reliable information about the tectonic disturbance of rock mass is absent.

The qualitative assessment of stress state of rocks at the tunnel sections with high stress levels - in direct vicinity of tectonic disturbances - was performed during instrumental investigations of acoustic emission in the face and walls of the working with nonoperating machine. It was stated thereby that the intensity of acoustic emission of rocks within the zone and in walls of the tunnel reduced with the decrease of the speed of machine advance.

The determination of the optimal speed of tunnel drivage in conditions like these, independent of the drivage method is the most significant factor in ensuring the safe (no-failure) operation.

Under these complicated conditions it is possible also to assess the situation in the face and walls of the tunnel in core drilling of prospect holes of 43-50 mm in dia and with the length not less than the warking's diameter.

The availability of the process of core separation in convexo- concave disks allows to assess the behaviour of rock mass and its susceptibility to rock pressure manifestations in dynamic form The methodological basis of such instrumental investigations and the substantiation of the degree of rockburst - hazard of rock mass with due account of principal factors are contained in the appropriate instructions.

These substantiations were preceded by the previously performed instrumental investigations of core disking and stability of "holes" in their boring in large-sized specimens under different types of stress state and stress levels. The investigations of quantitative indices of these processes were carried out on the specially-designed high-pressure hydraulic equipment.

This equipment provided the formation of different volumetric stress states in large-sized cylindrical specimens under various radial and axial stresses up to 200 MPa.

Figure 1 shows the sandstone specimen sawn along the axis of the bored "hole" of 22 mm in dia and in disks in a "hole" of various thickness after the tests completion. Disks were drilled in the stressed sandstone specimen, the strength of which in uniaxial compression [σ] was equal to 160 MPa. The relationship between the axial and radial stresses $\frac{\sigma_{ax}}{\sigma} = 0.4$; the determinants of the type of stress state of sandstone were constant in a hole boring. With the increase of stress level of rock, $\frac{\sigma_{ax}}{[\sigma]}$, the thickness of drilled disks, equally as the ration $\frac{1}{d}$, will decrease.

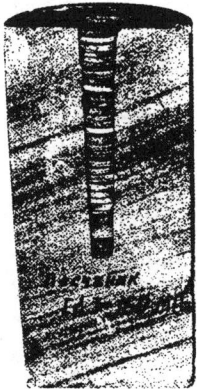 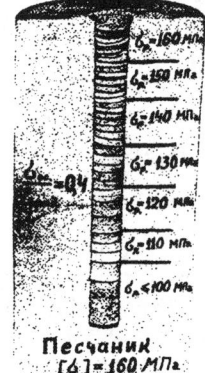

Fig.1

Continuous evaluation of geodynamic state in the tunnel face during its drivage, in our opinion, may be made on the records of vibration characteristics of the machine nodes (Krotov N.V., 1990). The tunneling machine operation within the zone of abnomal effects of properties and stress state of rocks is accompanied by a sharp change in the resonance frieguences of vibration. The prognostic map of distribution of these zones along the tunnel track allows to transfer timely to the control mode of vibration level, to change the loads on the face and to ensure safety for the interaction of a machine with stressed rocks within the near-face area.

Another important moment is the increase in reliability and conservation of a tunneling machine due to drop in the rate of accumulation of breakdowns in its nodes.

The dangerous interaction of the nature technogenic factors appears with the tunnel face entering into the zone "A" of abnormal properties of rock mass and when leaving it.

Proceeding from the formation of reference loads ahead of tunnel face, the determination of the strength and deformation abnormal zones in rock mass is made on the criteria of change in rock pressure level, σ_y, and the distance the point of maximum stresses. It follows from Fig.2 that the decrease in stress level in the epure 1 of support pressure supposes the existence ahead of face of the tectonically stressed zone and vice versa, the increase in stress level within the support pressure zone 2 evidences the oncoming meeting of the face with the tectonically relieved zone. Thereby, the transition of abnormal zone "A" in both cases induces the rise in loads within the tunnel face zone. Prediction of a type of abnomal zone (with lower or higher strength) should be performed at the distance "S" defined from the expression:

$$S = 3\beta(a+b),$$

where β is the coefficient defined on the nomogram (Fig.3); a - is the tunnel's width; b - the zone width of ultimate stress state of rock in the near-face area.

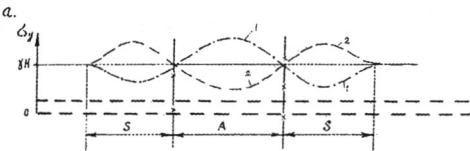

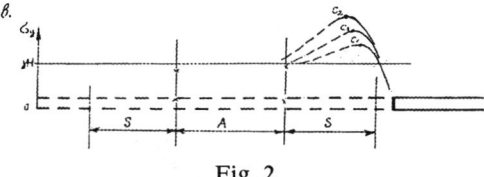

Fig. 2

In this case

$$D = \frac{\sigma}{K\gamma H}\left(\frac{a+b}{m}\right),$$

where σ is the strength of rocks; $K\gamma H$ - the stress state of intact rock mass at the depth, H, of tunnel drivage; m - is the height of a tunnel.

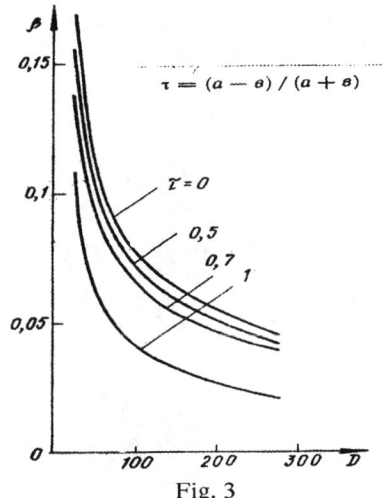

Fig. 3

An assessment of the emergence of geodynamically hazardous nature - technogenic situations in the vicinity of abnormal zones is made on the analysis of inflow of elastic energy in the near-face zone of the tunnel (3). The geodynamic hazard appears in cases when the condition is satisfied;

$$\frac{2M}{(1+V)E}\frac{X}{m} \geq 1,$$

where M is modulus of drop in the diagram of rock strength; V,E are the Poisson's ratio and average value of modulus of rock elasticity; X - the distance up to the point "C" at the maximum of reference pressure.

The predicted energy inflow ΔE per unit of tunnel length is defined from the expression

$$\Delta E = \pi \frac{(\gamma H)^2 a^2}{E}$$

Detection of abnormally deformed sections along the tunnel track and their boundaries at the stage of geological prediction allows to apply in advance the measures to ensure the geodynamic safety and to choose the optimal operating conditions of the tunneling machine. The investigation results enable to improve the safety and efficiency in service performance, to increase the service life of mining equipment and to ensure its no-failure operation.

REFERENCES

Krotov N.V. 1990. The Way of Assessment of Rockburst-Hazard of Mineral Deposit. Patent of Russia, № 1710775

Numerical analysis on the effect of pre-notches of borehole on fracture propagation

W.K.Song
Korea Institute of Geology, Mining and Materials, Taejon, Korea

ABSTRACT: This paper presents numerical analysis about the effect of two opposed pre-notches of borehole upon the fracture propagation using a FEM program. The study has been focused on the role of gas pressures in the quasi-static stage that follows the dynamic stage related to shock wave. It is assumed that a constant gas pressure of 25MPa is applied to the borehole wall and crack surfaces throughout the calculation. Stress distribution around pre-notched borehole has been analysed and compared with that for a simple hole. The most important factor between width and length of notches affecting the efficiency of fracture propagation has been determined. Two borehole model has also been analysed in order to study the effect of neighbouring boreholes on fracture process. The results of this study could be applied to blasting works not only in quarry mines but also in civil construction.

1 INTRODUCTION

When blasting a rock by explosives in quarry mines or civil workings, it is essential to yield a fracture surface in the desired direction without giving severe damage to rear part of the rock to be cut. This objective could be achieved by putting a notch on both sides of boreholes before blasting since fractures may propagate to the pre-notch direction prior to any other directions. It is known that, when a borehole containing cracks is pressurised, concentration of tangential component of stress is produced around crack tips and makes a failure of surrounding rock. Such phenomenon has been studied from laboratory tests by several researchers (Kutter, 1969, Kutter and Fairhurst, 1970 and Ito and Sassa, 1969). They showed that fracture surface is principally formed in the same direction as pre-notch, while only some cracks are generated within short distance in the other directions.

In this study, the effect of two opposed pre-notches of borehole on the fracture propagation has been analysed in numerical way using a FEM program. The analysis has been focused on gas pressures playing during the quasi-static stage that follows the dynamic stage related to shock wave, since fracture propagation is mainly controlled by gas pressures. Stress distribution around pre-notched borehole has been analysed and compared with that for a simple, not notched borehole. The study to determine the most important factor between width and length of notches affecting the efficiency of fracture propagation has also been carried out. Furthermore, analysis of a model with two boreholes has been conducted to study the effect of neighbouring borehole on fracture process.

2 FRACTURE PROCESS

The forces induced by explosion result from the detonation of a chemical charge. It is the ultra rapid chemical reaction of a thermodynamically unstable substance that produces shock waves and gases with high pressure and temperature. Generally, fracture phenomenon of a rock due to a detonation may be divided into two distinct phases. The first is dynamic phase related to shock wave. The second is related to gas pressures and called quasi-static phase since they act relatively long time inside borehole (Song, 1993).

In dynamic phase, the stresses induced by the shock wave exceed to a large extent the compressive strength of the rock so that the rock in the vicinity of the charge is intensively crushed and shattered. However, crushed zone is not enlarged so far since the induced stresses are attenuated very rapidly due to dissipation in the form of friction, heat and fracture(Fig.1.a).

Fractures created in the dynamic phase are extended by the expansion of gases emitted by

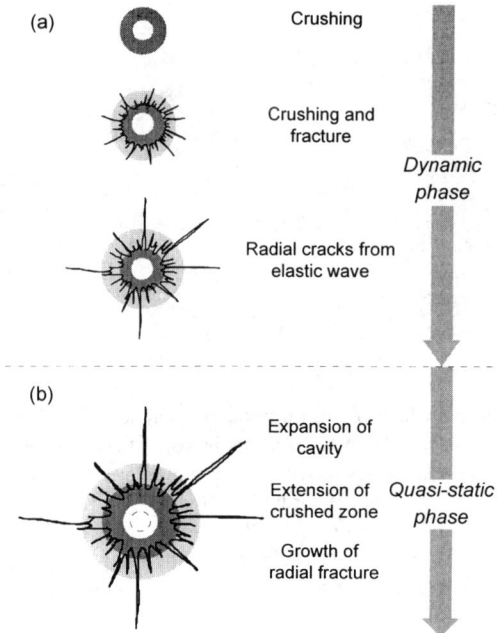

Fig.1. Two different phases of the fracture process in blasting

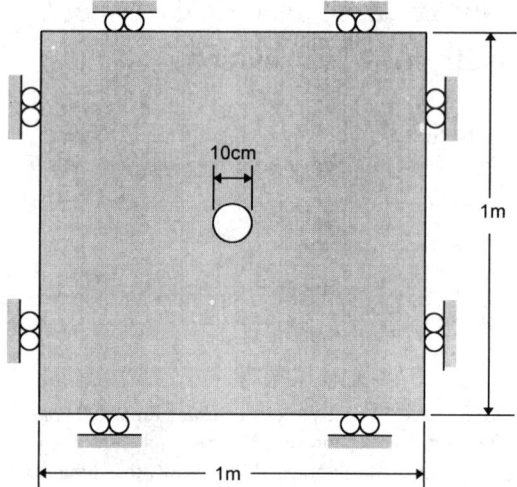

Fig.2. Geometry of the numerical model and boundary conditions

chemical reactions of explosives. As the tensile strength of rocks is considerably lower than their compressive strength, the tangential stress due to gas expansion is large enough for the readily created radial cracks to grow up(Fig.1.b).

This report concerns roles of gas pressure on fracture propagation during the quasi-static stage. In numerical modelling gas pressure applied to the inside of borehole, p_o, is assumed to be of 25MPa which corresponds to the approximate critical pressure to initiate fractures in granite and this value remains constant during calculation. This assumption of constant pressure may not be realistic because gas pressure will decrease with the expansion of cavity and, therefore, stress concentration around crack tip could be overestimated. Nevertheless, it would be acceptable to obtain qualitative information about fracture phenomenon. Such faults should be improved by assigning pressure varying with time in further studies.

3 SET-UP OF THE MODEL

Numerical model consists of a section perpendicular to a blasting hole. Typical geometry of the model is depicted in Fig. 2. Diameter of a blasting hole is of 10cm and the border lays 50cm away from the centre of hole so as not to give an influence on stress field around the borehole. The calculation is made in plane stain condition since vertical displacements to the plane must not be occurred due to overlaying materials. Displacement perpendicular to the external boundary is not permitted.

The model is supposed to be composed of elastic and isotropic materials and their physical properties are characterised by:

Young's modulus, E = 50Gpa
Poisson's ratio, $\nu = 0.25$

Fig.3 shows close-up view of typical grid around a borehole with pre-notches and cracks. Larger openings represent pre-notches and smaller ones cracks created by shock waves. It should be noted that grid density is higher around pre-notch or crack tips than other regions in order to detect rapid variation of stress concentration there.

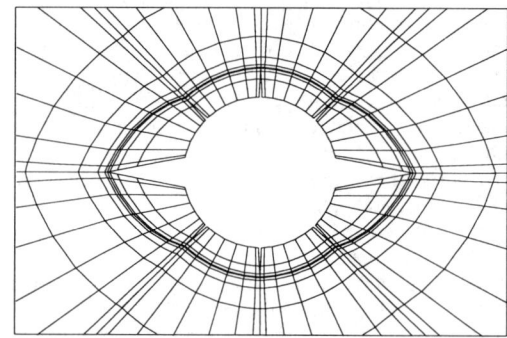

Fig.3. Typical grid around borehole with pre-notches and cracks represented by larger openings and smaller ones respectively

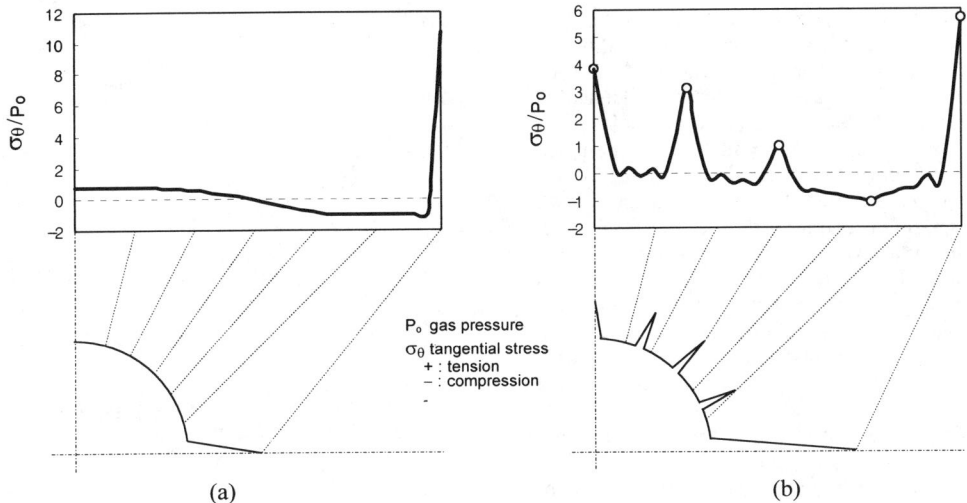

Fig.4. Tangential stress distribution on the wall surface for a borehole: (a) without cracks and (b) with several cracks

4 ONE BOREHOLE MODEL

4.1 Stress distribution around borehole

In order to analyse effects of pre-notch on fracture propagation two models have been studied: one has pre-notched borehole without any cracks and the other with several cracks. The first model represents the geometry at the moment of detonation and the last one the geometry when some cracks have been produced after the detonation. Corresponding tangential stress distribution along the borehole wall has been plotted in Fig.4.a and b, respectively.

As shown in Fig.4.a, an enormous stress is concentrated on the tip of pre-notch, while stress concentration in other zone is quite low. It can be seen that the value of tensile stress on the top of the borehole wall is much lower than that for a simple, not-notched borehole in which tensile stress is equal to the internal pressure and that even compressive stress appears in the vicinity of the pre-notch. This implies that, while pre-notch can promote the propagation of fracture in that direction, it can constrain creation of cracks in other directions.

Once some cracks have been created around borehole, stress concentration on the tip of pre-notch diminishes to a great extent but the value of tangential stress is still much higher than those on crack tips(Fig.4.b). It should be noted that the nearer a crack is to pre-notch, the lower the tangential stress concentration is, and that even an compressive stress acts on the tip of the nearest crack to the pre-notch. This implies that major fracture will advance along the line of pre-notch and minor crack may be developed in the direction perpendicular to pre-notch.

It is certain that rock materials around crack tips are affected by pressure exerted on the surface of pre-notch. This phenomenon is proven by the results of laboratory test using a acrylic plate with a pre-notched borehole in bi-direction (Ito and Sassa, 1969).

Comparison of full distribution of the tangential stress around crack tips between simple and pre-

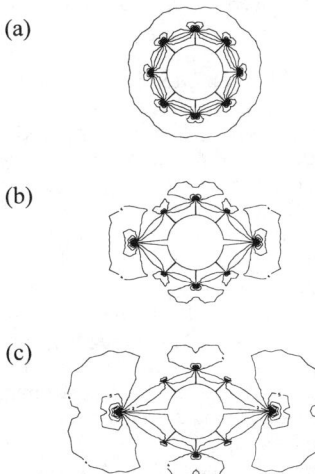

Fig.5. Tangential stress contours for a simple borehole (a) and for a pre-notched borehole according to the growth of fracture (b and c)

143

notched borehole is presented in Fig. 5 in terms of stress contours. Fig.5.a shows stress distribution for a simple borehole where star cracks have been produced, and Fig.5.b and c a variation of tangential stress contours for a pre-notched borehole when fractures grow up preferentially in pre-notch direction. When pre-notches do not exist, equal stresses are generated on the tip of fissures(Fig.5.a), whereas stress concentration on the tip of pre-notch is more important than on the tip of cracks created by detonation(Fig.5.b). The influence zone of stress concentration is also larger for pre-notched hole than for the simple hole. Particularly, stress concentration on the tip of the cracks just next to the pre-notches is largely attenuated with fracture being developed in pre-notch direction(Fig.5.c).

According to the difference of stress concentration between on the tip of pre-notches and cracks, deformation for each opening appears in different scale depending not only on the kind of openings (pre-notch or crack) but also on their location(Fig.6). The maximum deformation occurs without any doubt in pre-notch crack and the nearest crack to the pre-notch has a tendency to close.

4.2 *Effect of notch dimension*

In engineering practice, it is important to decide notch dimension to obtain the best results when blasting a rock. Two parameters are related to this subject: length and width of pre-notch. Notch length is defined by the distance between the notch mouth and its tip, and width by length of the opening at the wall side. Numerical analysis on stress concentration on pre-notch tips has been conducted varying the values of each parameter.

Variation of tangential stress on the tip of pre-notch versus its width is plotted in Fig.7 when pre-notch is of 4cm long. The curve shows a little change of tangential stress. It means that the crack propagation is not affected by the width of pre-notch. On the other hand, the length of pre-notch has a great influence on the value of tangential stress. Stress variation with respect to the length of pre-notch is given in Fig.7.b when the width of pre-notch is fixed to 1cm. The curve shows a very steep increase of tangential stress until the pre-notch length becomes equal to the radius of the borehole and then it converges to a certain value. From these results, it can be concluded that the most important factor affecting crack propagation is not the width of pre-notch but the length.

It should be noted that about 80% of the maximum stress can be obtained when the length of pre-notch is equal to the radius of the borehole. It may give a criterion to determine the dimension of pre-notch when performing blasting in situ even though the modelling has been executed under elastic condition.

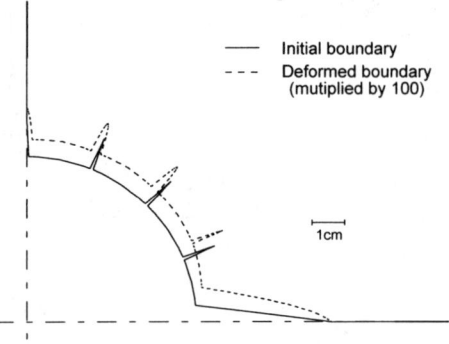

Fig.6. Differential deformation of pre-notch and cracks created by detonation

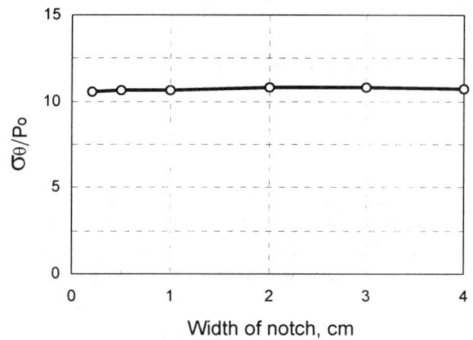

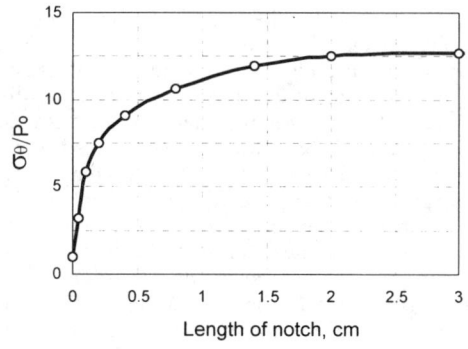

Fig.7. Variation of the tangential stress on the tip of pre-notch in function of the width (top) and the length (bottom)

5 TWO BOREHOLE MODEL

5.1 Simple boreholes

Effects of adjacent boreholes upon fracture of rock have been analysed using a model with two boreholes. Mechanical properties are the same as those used in one borehole model. A typical grid for two borehole model is presented in Fig.8.

A pressure of 25Mpa is charged simultaneously to the internal boundary of both boreholes. Stress concentration has been examined according to different distances between two boreholes. Variation of tangential stresses on the top of the borehole wall ($\sigma_{\theta,a}$) and at the nearest point to the neighbouring borehole ($\sigma_{\theta,b}$) with respect to the shortest distance between the wall surfaces is plotted in Fig.9.

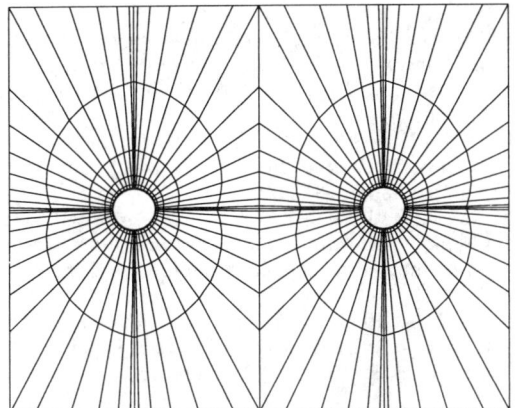

Fig.8. Typical grid for two borehole model

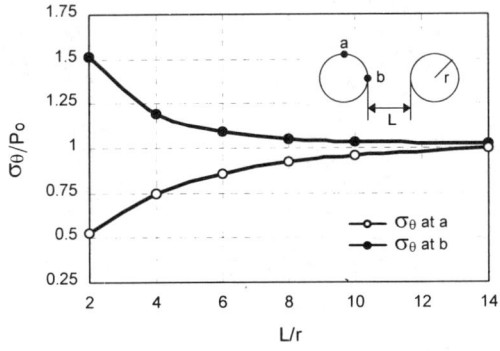

Fig.9. Variation of the tangential stress on the wall with respect to the distance between two simple boreholes

The distance between two boreholes varies from two to fourteen times of the hole radius. When the distance is two times of the radius of borehole, an increase of tangential stress equivalent to about a half of gas pressure occurs on the nearest point (point b), while a decrease as much as this increase on the top of boreholes (point a). According to the increase of the distance stress concentration on the point b is rapidly attenuated and, on the contrary, it increases on the point a. When the distance becomes more than about 10 times of the radius, the effect of adjacent borehole nearly disappears.

These results imply that growth of a fracture can be promoted in the direction of the centre line joining neighbouring boreholes due to stress concentration when a borehole is close to an another. Nevertheless, propagation of a fracture in the sense perpendicular to the centre line is constraint. These curves show clearly the influence of an adjacent borehole on the fracture propagation.

5.2 Pre-notched boreholes

In pre-notched borehole models, effects of the growth of fracture of a borehole on fracture process of the another have been examined. Two boreholes are set far away so as not to exert any influence on a neighbouring borehole when any cracks have not yet created.

Stress distributions along the line joining the centres of two boreholes with increasing the length of fracture are plotted in Fig.10 and also variation of the maximum stress concentration on the crack tip. The length of pre-notch varies from zero to six times of borehole radius. The curves have been plotted only for a half of whole line since they are symmetrical with respect to the middle point of the centre line. The increase of pre-notch length can be regarded as an evolution of a fracture due to gas pressure.

The curves show a parallel transition according to the increase of the opening length without any particular change of the pattern of stress concentration. It is common to them that they exhibit a rapid decrease within a short distance from notch tip, which implies that a fracture initiated in a borehole does not give an influence on the another one until they approach very closely each other. It should be noted that stress concentration around crack tip region decreases in a small extent when the crack length exceeds 1.5 times of hole diameter. Nevertheless, this decrease does not result from the approach of the adjacent crack but from the extension of the crack itself since the influence of opening pressure within the borehole decreases in a long distance from borehole wall.

Corresponding variation of tangential stress on the top of borehole is shown in Fig.11. Stress concentration decreases in proportion to an increase

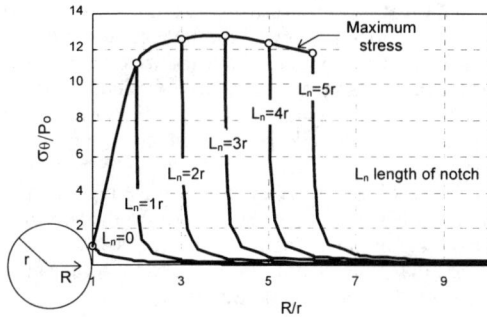
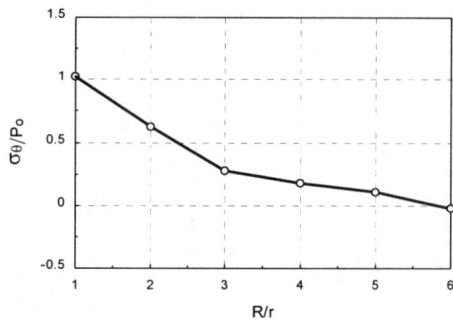

Fig.10. Variation of tangential stress along the line joining the centre of two boreholes with increasing notch length (L_n) and maximum stress concentration on the tip of notch

Fig.11. Variation of tangential stress on the top of borehole wall according to the extension of crack length (R=distance to notch tip from the centre of the borehole, r=radius of the borehole)

of crack length, even into a compression. Such decrease does not have relation with the extension of fracture generated in the adjacent borehole since the curve is the same as that for one borehole model.

CONCLUSION

Numerical analysis has been conducted to study the effect of pre-notches of borehole on fracture propagation in blasting rocks using FEM. It was supposed that the borehole is subjected to a gas pressure of 25MPa which is an approximate critical pressure to cut a granite and the model consists of elastic and isotropic materials.

The existence of pre-notches yields a great concentration in the region of notch tip and, on the contrary, a decrease of tangential stress in the other region. Even though small cracks have been produced along the wall of borehole, stress levels on their tip are still fairly lower than that on the tip of pre-notch. Much smaller stress concentration is produced on the crack near to pre-notch. This phenomenon may explain that fracture propagation is constrained more strongly in the direction near to pre-notch than in the direction away from it.

It was revealed that the parameter to determine the efficiency to fracture propagation is not the width of pre-notch but its length since stress concentration around the tip of pre-notch increases rapidly according to its length but not to its width..

In the case of simple, not notched boreholes, stress concentration is affected by the distance from adjacent boreholes, while in case of pre-notched boreholes the length of fracture itself plays more important role on fracture propagation rather than the distance between boreholes as long as concerned only with gas pressures.

REFERENCES

Ito, I. And Sassa, K., 1969, *Study on the fracture mechanism during the smooth blasting by explosive*, L'Explosif, No 3, pp 93-105 (in French).

Kutter, H.K. and Fairhurst, C., 1971, *On the fracture process in blasting*, Int. J. Rock Mech. and Min. Sci., Vol. 8, pp 181-202.

Kutter, H.K., 1970, *Stress analysis of a pressurised circular hole with radial cracks in a finite elastic plate*, Int. J. Fracture Mech., Vol. 6, No 3, pp 233-247.

Song, W.K., 1993, *Modelling of the action of gas in cutting rocks by explosives*, Master's Paper, Ecole des Mines de Paris, pp 43 (in French).

Rock fragmentation with plasma blasting method

Kyung Won Lee, Chang Ha Ryu, Joong-Ho Synn & Chulwhan Park
Korea Institute of Geology, Mining and Materials, Taejon, Korea

ABSTRACT: Rock fragmentation with plasma blasting technique has advantageous properties in contrast to the conventional blasting method in controlling of flying rocks and ground vibrations, when residents are complaining or surrounding structures stay in protection from blasting operations. The experiences show in urban construction works that the plasma blasting is the most possible method to prevent damages and minimize adverse environmental impacts. The fragmentation energy level is evaluated by numerical simulation using PFC^{2D} for various drill hole patterns and tested accordingly to get the feasibility. The energy output of plasma blasting system has been improved to a level of 1 MJ, which can break a 2-3 m^3 granite boulder or 1.5 m height bench face. Measurements are carried out to get the ground vibration level and propagation equation, so that the control of the blasting operations can be performed more precisely and safely.

1 INTRODUCTION

The conventional blasting technique, drill and blast method, has adverse environmental effects and in many cases where explosives of brisance is rather cumbersome or not possible to apply, when neighbouring structures stay in protection or residents are complaining for the possible damage and annoyance by blasting activities. To avoid such adverse effects the plasma blasting method is developed to utilize the advantageous properties. The plasma transformed from the electrolyte solution in a borehole reaches an energy of 280 kJ by electrical input of 8.5 kV and 200 A and this rapidly released energy in the borehole develops a shock wave, which in turn produces a stress field that fractures the rock without producing excessive dust and fly rocks. The range of plasma strength varies with chemical additives. The increased energy delivery covers a wide scope of applications, from breaking a boulder to bench cutting of grounds for buildings in urban area.

2 DETERMINATION OF PLASMA STRENGTH

Electrical energy stored in a capacitor bank is turned on by a high current gas piping switch; this causes the arc effect at the tip of a coaxial blasting electrode, which is tamped tightly in a blast hole drilled into the rock face. Under these conditions the electrolyte turns into a high temperature, high pressure plasma. The pressure must be high enough to fracture hard rock. Electrical energy input, an amount of the electrolyte and the oxidizer as additive surrounding the space between the electrodes and the rock determines the energy output of the system. In this case explosion energy can be interpreted analogically; it is initiated from the electrical source and chemical reaction.

From the equation of chemical reaction by aluminum powder as additive the oxygen balance reveals -0.89 gr and explosion energy is 22.3 kJ for 1 gr of aluminum. The plasma blasting system can supply 280 kJ by electrical energy for 1.5 msec as shown in Fig. 1, so the system may deliver 0.95 MJ additionally as a blasting energy when 30 gr of aluminum completed its chemical reaction. This amount becomes 0.243 of relative weight strength(RWS), which is the equivalent strength of 243 gr of ANFO, 201 gr of GD or 320 gr of Emulite. The strength of an explosive can be explained with Langefors' empirical equation by the explosive energy, e, and the gas volume, v, as follows;

$$s = \frac{5}{6}e + \frac{1}{6}v \qquad (1)$$

Table 1. Equivalent weights of explosives to the plasma strength (unit: gr)

Explosives	in RWS theoretically	in RWS by equation	in site test	final estimation
Korean				
GD HiMite 6000	185	164	199 - 217	180
GD HiMite 5000	196	172	210 - 228	200
NewMite 4000	276	236	296 - 312	280
Finex I	308	266	331 - 352	310
American				
ANFO	243	205	261 - 271	230
Gelatin Dynamite	209 - 222	205	224 - 271	210
Slurry (Ireseis)	279 - 295	250	300 - 331	280
Swedish				
Emulite (100)	312	253	330 - 340	310

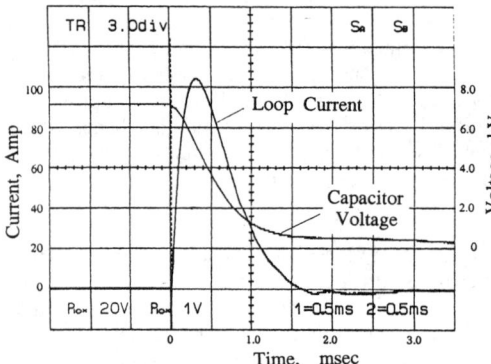

Fig. 1. The loop current and capacitor voltage history during a typical plasma discharge

Referring to this concept, the strength of the plasma system can be calculated as RWS 0.205, GD HiMite 5000 as RWS 1.19 respectively to RWS 1.0 of ANFO theoretically.

In the first developing stage of the plasma blasting technique the delivered energy did not include the chemical reaction. Tests were carried out with 1 m^3 concrete blocks by 0.28 MJ of the electrical energy only. This amount was expected to have the power as much as 66 gr of gelatine dynamite(4.27 MJ/kg). The specimen was caused three evident cracks which divided the upper half to equi-volume of 4 pieces and did not break the lower half. Simultaneously chemical explosive was applied to the same specimens which resulted in 1, 2 or 3 cracks perpendicular each other. That system was equivalent to the same power of 46~67 gr of gelatine dynamite without tamping and when sand tamping was done to the power of 25~35 gr.

Adding aluminum powder in the electrolyte, the strength of plasma was so large enough to break hard rocks by chemical reaction that the scope of the application can be extended to the blasting operations for excavation works. One test for granite, of medium hard rock was completed to compare with Emulite. It was proved that the plasma blasting system has the power of 330 gr of Emulite, which is a little more powerful than theoretically expected. The relative strength of the plasma is listed in Table 1 which can be applied in rock blasting design.

3 NUMERICAL SIMULATION OF PLASMA BENCH BLASTING

3.1 *Modelling technique*

The fragmentation phenomenon for a rock bench model by plasma blasting is simulated using PFC^{2D}(Particle Flow Code) which is based on distinct element method. The particle elements in PFC^{2D} are assumed to be rigid balls, and the behavior at all contacts is characterized using a soft-contact approach, in which finite normal and shear stiffnesses are taken to present the measurable contact stiffness. The behavior of a solid can be simulated by bonding groups of balls together at their contact points using contact bond. The existence of a contact bond precludes sliding and limits the allowable magnitudes of normal tension and shear force acting at the contact. If either the normal or shear limit is reached, then the bond breaks and the contact cannot subsequently take tension.

The model of a rock bench is shown in Fig. 2. The interval between holes and the distance from free face to the 1st row of holes are same

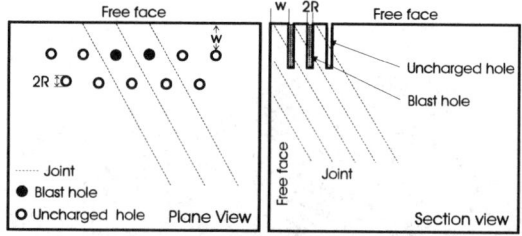

Fig. 2. The model of a rock bench blasting

as 60 cm, and the hole diameter is 10 cm. The dotted lines are pre-existing joints.

It is known that the plasma reaction induces shock wave only without gas production. So in this modelling, only the effect of shock wave is considered and the pressure by shock wave is imposed on the boundary particles of blasting hole. The history of borehole pressure with time by blasting is given as sine function of equation (2). Here, P_b is borehole pressure with time, P_o is the maximum borehole pressure, f is frequency, and t_o is the period of sine function.

$$P_b = P_o \cdot \sin(2\pi f t), \quad f = 1/t_o, \quad 0 \leq t \leq t_o/2 \quad (2)$$

Referring to the discharge history of current and voltage during plasma reaction, lasting duration of the borehole pressure by plasma reaction is set as $t_{max}(=t_o/2) = 1.5$ msec.

3.2 Fragmentation energy with blasting pattern

For the plane view model of the rock bench, input borehole pressure, P_o, and fragmentation pattern according to single hole blasting and synchronous multi-hole blasting are analyzed. The optimum fragmentation pressure, namely, the minimum level of P_o which can produce crack formation to the bench face is obtained as about 692 MPa in case of the single hole blasting. And it is obtained as 620 MPa in cases of synchronous two-hole blasting and as 604 MPa in case of synchronous four-hole blasting. It is shown from this result that the synchronous multi-hole blasting can produce similar fragmentation effect with 87~90% of the input energy compared with the single hole blasting.

The effect of delay blasting on fragmentation with varying the delay interval between 1st and 2nd row of blasting holes is simulated. Several energy terms can be calculated in PFC^{2D} analysis. Here, the ratios of the kinetic energy by particle movements and the strain energy stored at contacts to the body work by all applied forces are compared. One period of delay interval is set as 1.5 msec which is the lasting duration of the borehole pressure by plasma reaction. According to this analysis, the delay interval between 1~2 period shows good fragmentation efficiency as shown in Table 2.

This fragmentation modelling should be well related with field condition such as pre-existing joint pattern.

Table 2. Fragmentation energy ratio with delay interval

Delay interval	ratio of kinetic energy to body work	ratio of strain energy to body work
Synchronous	0.40	0.58
0.5 period	0.39	0.49
1.0 period	0.71	0.20
2.0 period	0.59	0.25
5.0 period	0.57	0.24

3.3 Effect of the pre-existing joint pattern on fragmentation

It is known that pre-existing joint patterns have effects on rock fragmentation by blasting. Three cases of joint patterns are analyzed for bench blasting simulation.

In case of the joint pattern parallel to the bench face(Fig.3), the region between blasting hole and bench face is divided into several blocks with some opening of joint to distant region. On the other hand, there is little effect of pre-existing joint on fragmentation pattern in case of pre-existing joints perpendicular to the bench face. The fragmentation pattern with inclined pre-existing joints is represented in Fig.4. In the left side of the model where shock wave propagates toward the joint plane by obtuse angle, new crack is not produced even though the energy is almost consumed in joint opening and released through joint. On the other hand, in the right side where shock wave propagates toward the joint plane by acute angle, some new cracks are produced across the joint. The section view model with a pre-existing joint set inclined inward the rock bench as shown in Fig.5 shows good fragmentation with new crack formation than the model with a pre-existing joint set inclined outward the rock bench as shown in Fig.6.

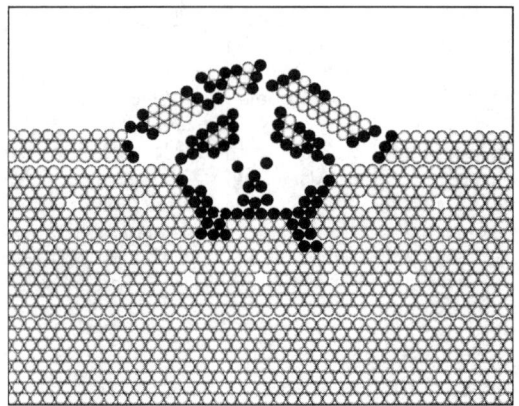

Fig. 3. Fragmentation pattern for the plane view model with a joint set parallel to bench face

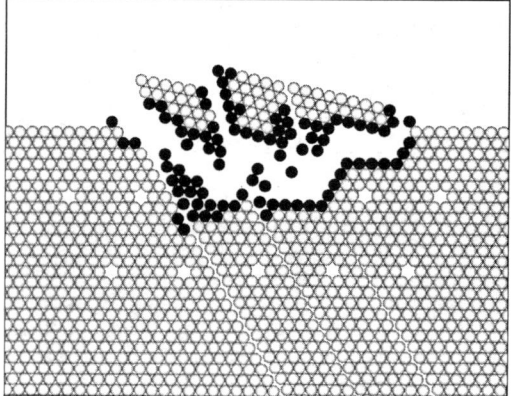

Fig. 4. Fragmentation pattern for the plane view model with a joint set inclined to bench face

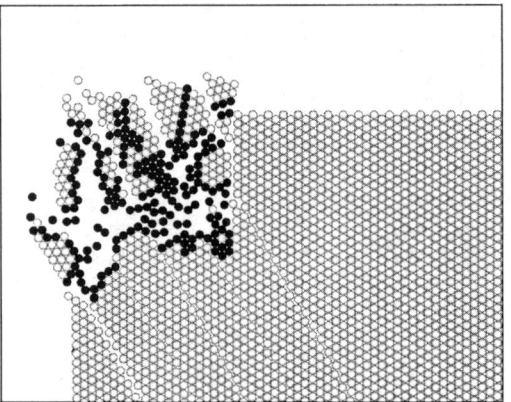

Fig. 5. Fragmentation pattern for the section view model with a joint set inclined inward the rock bench

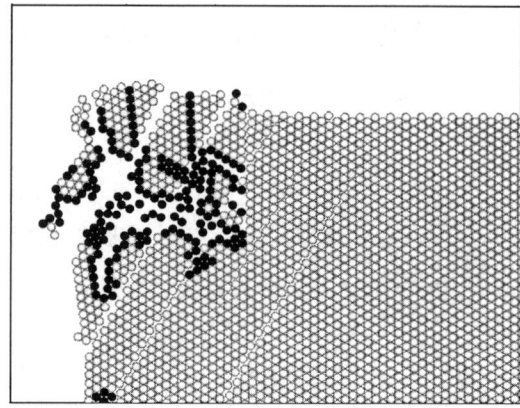

Fig. 6. Fragmentation pattern of the section view model with a joint set inclined outward the rock bench

The borehole pressure history and frequency characteristics of the shock wave by plasma reaction, which is related with damping effect, should be well evaluated as well as field rock conditions for numerical simulation of plasma blasting and its application to field.

4 CHARACTERISTICS OF GROUND VIBRATION, AIR PRESSURE AND FLY ROCK INDUCED BY PLASMA BLASTING

Blasting operations associated with the rock excavation work may have an environmental impact on nearby structures or human beings. Use of explosives is sometimes restricted in urban construction works due to the complaints for vibration and noise. The Plasma blasting system are developed as one of non-explosive demolition methods to be applicable in the residential area where the use of explosives is restricted. Because the source characteristics of plasma system is different from that of the explosive blasting, it is also expected to generate different types of ground vibration, air pressure, and fly rock. Field measurements were performed to get understandings of environmental effects of plasma blasting system.

4.1 *Measure of vibration to assess damage potential of structures*

Peak particle velocity has been suggested as the best descriptor to assess the damage potential of structures. Although peak particle velocity has been widely used, velocity itself is not sufficient to evaluate structural damage without considering tolerance of the structure. Structures respond differently to vibrations of differing frequency content. In recent years frequency content has become an increasingly important parameter in the measurement and analysis of the ground vibrations from blasting. The former U.S. Bureau of Mines and Office of Surface Mining recommended safe blasting vibration criteria for residential structures, depending on the peak particle velocity varying with respect to the frequency (Siskind et al. 1980). The criteria incorporate an important element of response spectra technique in some respects. The German vibration standard, DIN 4150, also provides similar criteria for several types of structures.

4.2 Vibration level and propagation equation

The blast-induced ground vibration decreases in amplitude with increasing distance. The ground motion can be measured as displacement, velocity or acceleration of a particle in the ground. The propagation characteristics of the vibration is influenced by rock properties, geological discontinuities and design parameters. The most general form used for the prediction of ground vibrations from explosive blasting is given by

$$PPV = K \left(\frac{D}{W^b}\right)^n \quad (3)$$

where PPV is the peak particle velocity in mm/sec, W is the charge weight per delay in kg, D is the distance from a blast source in m. The constants K, n and b are empirical and site specific. Analysis of measurement data shows that good correlation of results could be represented by square-root or cube-root scaling where the power b in D/W^b is 1/2 or 1/3, respectively (Nicholls et al. 1971; Ryu & Lee 1979). Fig.7 shows the plots of each component of measured ground vibration induced by plasma system on log-log scale. Measurements were carried out at Kimpo granite quarry. X-axis is the scaled distance which scales the distance by cube root of voltage, kV. Solid line is the regression line representing the mean trend of peak particle velocity as a function of scaled distance. The propagation equations for each component are as follows:

$$PPV(L) = 162.0 \left(\frac{D}{KV^{1/3}}\right)^{-2.130} \quad (4)$$

$$PPV(T) = 80.0 \left(\frac{D}{KV^{1/3}}\right)^{-1.887} \quad (5)$$

$$PPV(V) = 27.0 \left(\frac{D}{KV^{1/3}}\right)^{-1.437} \quad (6)$$

The correlation coefficient is 0.854 for the longitudinal component, 0.826 for transverse and 0.68 for vertical, respectively. It shows little better correlation when distance is used instead of scaled distance.

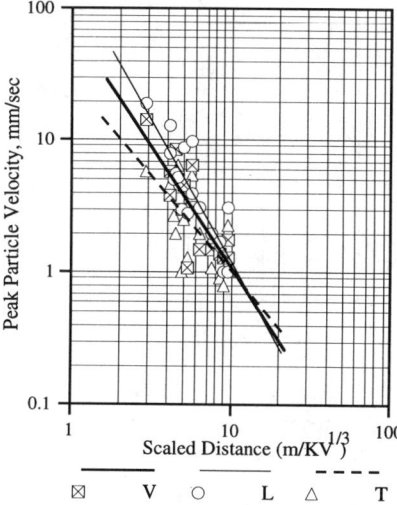

Fig. 7. Peak particle velocity vs. scaled distance

The propagation equations without scaling are as follows:

$$PPV(L) = 802.5 \, D^{-2.211} \quad (7)$$
$$PPV(T) = 338.5 \, D^{-1.968} \quad (8)$$
$$PPV(V) = 89.8 \, D^{-1.543} \quad (9)$$

The correlation coefficient is 0.865 for longitudinal component, 0.841 for transverse and 0.712 for vertical, respectively. Fig. 8 shows the results of plotting PPV with respect to the distance from source.

In explosive blasting data used for comparison, slurry explosive (KOVEX) is used in granite and distance varies between 15 to 50 m from source. The propagation equations obtained from the data are as follows:

$$PPV(L) = 189.6 \, D^{-1.152} \quad (10)$$
$$PPV(T) = 590.7 \, D^{-1.463} \quad (11)$$
$$PPV(V) = 588.2 \, D^{-1.510} \quad (12)$$

The correlation coefficient is 0.790 for

longitudinal component, 0.864 for transverse and 0.932 for vertical, respectively. The attenuation index, n, appears to be about 2.0 for blasting, and about 1.5 for explosive blasting. That is, level of ground vibration induced by plasma blasting decrease more rapidly.

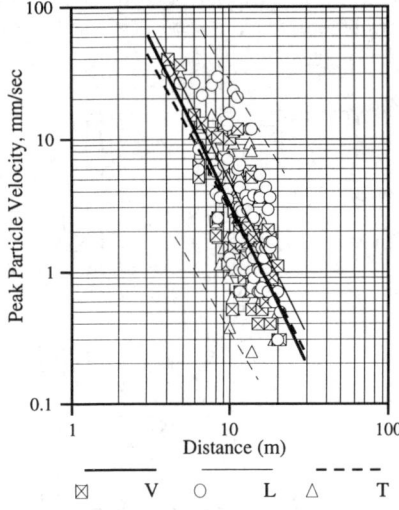

Fig. 8. Peak particle velocity vs. distance

4.3 *Frequency, noise and fly rock*

The distribution of principal frequency which is defined as that associated with peak amplitude of vibration level is ranging from 30 to 110 Hz in plasma blasting while 20-70 Hz in explosive blasting. Although field conditions are not exactly same, current measurement data show the tendency of higher frequency in plasma blasting.

Characteristics of generation of air pressure is not analysed due to lack of data. Rock breaking works, however, were carried out with satisfying the allowable limit described in the regulations. It was shown that sound level was reduced by 15 dB in A-weighting measurement by installing a fence.

The distance that fly rock is thrown by plasma blasting is very limited to surroundings of a source so as to allow nearby measurement. It is the typical difference from that by the explosive blasting. Rock fragments may be propelled by build up gas pressure in explosive blasting. Uncovered explosion sometimes yields unusual distant throw and causes hazards.

5 CONCLUSION

The plasma blasting technique opens a new field in rock breaking activities where adverse environmental impacts have to be considered. The plasma strength is improved with chemical additives that the rock fragmentation is feasible not only for boulders but also bench face. The numerical simulation shows the practicability of multi-hole and controlled blasting in jointed rocks. Analysis of ground vibration data induced by plasma blasting may be predicted using the equation similar to that used in explosively induced vibration. The peak level of ground vibration decreases rather rapidly and associates higher frequency which is more favourable considering the damage potential of structure.

REFERENCES

DIN 4150, Teil 3. 1986. Erschütterungen im Bauwesen - Einwirkungen auf bauliche Anlagen.
Nicholls, H.R., C.F. Johnson & W.I. Duvall 1971. Blasting Vibrations and Their Effects on Structures. U.S.B.M. Bulletin 656.
Ryu, C.H. & C.I. Lee 1979. A Study on the Effects of the Ground Vibration due to Blasting on the Structures. J. Korean Inst. Mineral and Mining Engineers 16: 41-50.
Siskind, D.E., M.S. Stagg, J.W. Kopp & C.H. Dowding 1980. Structure Response and Damage Produced by Ground Vibration From Surface Mine Blasting. U.S.B.M. RI 8507.
Potyondy, D.O. & P.A. Cundall 1996. Modeling of shock- and gas-driven fractures induced by a blast using bonded assemblies of spherical particles. Rock Fragmentation by Blasting: 55-62. Rotterdam: Balkema.

Blast vibration mechanism for underground blasting

T.N.Singh & C.Sawmliana
Department of Mining Engineering, Institute of Technology, Banaras Hindu University, Varanasi, India

ABSTRACT : Blasting in underground excavations generate some ill effects to safety, economy and efficiency. These ill effects are in the form of ground vibration, air over pressure, fumes, overbreaks etc. The nature of damage in underground depends upon many factors such as rock types, geological discontinuities, physico-mechanical and physico-chemical properties of rock units, explosive charge weight, blast geometry, type of explosive, initiation system etc. Though ,the ground vibration are integral part of process of blasting but it need to be predicted, monitored and controlled by the blasters. For safe and economic design it is essential to understand the mechanism of ground vibration.

An attempt is made in this paper to study the different parameters induced in blast vibration mechanism, their prediction and safe permissible limits for smooth rock-friendly excavation.

1 INTRODUCTION

Blasting is one of the major single operation which influence the production and productivity of the mine and put impact on recovery, support requirements and dilution. In addition, environmental impact is also very important and it enforced reductions in productivity or even closure to operational activity. The most important ill effects due to blasting are ground vibration, air blast, fumes, noxious gases, fly rock and over back breaks.

When an explosive energy is initiated in a blast hole, the chemical energy of an explosive is released instantaneously.. The energy released is divided into the number of processes which take place together independently and generates fractures in rockmass. The processes of crack propogations within rocks involved more than twenty variables (Da Gama, 1983). These variables are connected to different rock properties. The rock strength is the one of the most important parameter in blast mechanism but it is not clearly understood that which particular strength should be taken into account for blasting. The mechanism gets more complication due to inhomogeneity of rockmass, presence of discontinuities, joints plane, etc.

A proper understanding of different phenomena is essential in controlling the blasting parameters so as to obtain a better utilization of the available energy. In the perspective of opencast blasting little systematic research work was carried out on the effect of blasting on underground workings. Due to increasing awarness / awakening in the society, it is essential to use underground space technology to protect the surface morphology and structures.

2 GENERATION AND PROPOGATION OF GROUND VIBRATION

Detonation of explosive in a blast hole liberates the energy in a very short time and generates an enormous pressure and temperature (Gosh,1990). The rapid expanding gases impact the hole wall and produce an intense pressure pulse. These pulse which are generated due to rapid reaction rate send out stress wave radially away from the blast hole and pulverized the blast hole area. The intensity of shock wave attenuates vary rapidly as a large amount of energy is consumed in crushing and producing cracks. In elastic or semi-elastic zones located away from the source, the intensity drop significantly and thus produce no permanent deformations. The remaining energy goes directly into the surrounding rocks in the form of seismic waves. The waves propagate elastically and can be classified as body waves and surface waves. Body waves travel within the medium while surface waves are restricted to travel along free interfaces, such as the ground surface of the mine openings or wall of the excavation. Body waves comprises two components - compression or P - wave and shear or S -wave. Two type of surface waves are usually produced from normal mine blasting Rayleigh (R) waves and Love (L) waves. Other type of surface waves include coupled (C) waves, hydrodynamic (H) waves, channel waves, etc. (Persson, 1993).

3 FACTORS AFFECTING GROUND VIBRATION

The nature and intensity of blast induced vibration produced in underground excavation depend upon many factors. The most important factors may be summarized as follows,
- Rock types,
- Geological discontinuities,
- Physico-mechanical properties of rock type,
- Explosive charge weight,
- Blast geometry,
- Method of excavation and
- Explosive types.

All the above mentioned parameters are dependent upon each other and mostly interrelated. If a particular variable will be changed others also changed.

The surrounding rock type have moderate influence on ground vibration behaviour (Wiess and Linehan,1978). It is possible to mininmise ground vibration and other undesirable outputs of blasting keeping in view the characteristics like the static and dynamic strength, poissions ratio, elastic limit stress attenuation characteristics, porosity, etc. However, structural discontinuities like nature of joints, their physico-mechanical behaviour, specific gravity, hardness, lineation and foliation should be considered also (Goyal et al., 1994).

The geological discontinuity play an important role in transmission of ground vibrations. It is also dependent upon the orientation and inclination of joint sets. Gneissosity and schistosity or joint system affect the direction of vibration upto a great extent (Singh et al.,1994).

The vibration level decreases with increasing distance of structure from the source of vibration. It is due to dissipation and dispersion of shock waves at layer distance. The natural frequency of structure varies with type of construction and condition of structure, which in turn effects its response to the incoming ground - vibration.

The ground vibration when reach the surface of the earth, get reflected and refracted their by produce both vertical and horizontal motion in the rockmass. Ground vibration waves of different types will propagate at their sonic velocity. The P - wave is fastest followed by S - waves. Prediction of ground vibration is essential because large underground caverns, tunnels, drainage galleries, dam foundations and other underground headings are expected to serve several years. The main objective is to get greater pull, higher rate of advance, less explosive consumption and minimum over and under breaks and to control the vibration level also (Dupoint,1977).

4 BLAST VIBRATION PREDICTOR

A number of investigators have studied ground vibration from blasting and have developed theoretical analysis to explain empirical data. Empirical equation may be deemed on the basis of test blasts conducted at a particular site. The equation deemed for the prediction should be site specific. All the commonly used predictors are given in Table. 1.

In the above mentioned equations mostly given on the basis of surface blasting and little consideration was paid to underground excavation. The general equation proposed by Davies et al. (1964) is most widely acceptable predictor

$$V = k \cdot W^a \cdot D^b \quad \text{------------------------(1)}$$

where, k, a and b are empirical constants.

In the present paper, most general equation is used because it contains three empirical constants so it probably gives better results than other equations. Taking natural logarithms of both sides of equations we obtain,

$$\log V = \log k + a \log W + b \log D \text{-----(2)}$$

Thus, multiple regression analysis may be used with log W and log D as the independent variables and log V as the dependent variable. The dependent variable log V is denoted by Y and log W and log D by X_1 and X_2 respectively. The equation (2) becomes

$$Y = C_0 + C_1 X_1 + C_2 X_2 \text{------------------(3)}$$

Table 1. Vibration predictor commonly used by various reseachers.

S.No	Predicted Equation	Proposed with year
1	$V = k (D/Q^{1/2})^{-B}$	**USBM, 1950**
2	$V = k (D/Q^{1/3})^{-B}$	**Ambraseys Hendron, 1968**
3	$V = k (Q/D^{3/2})^{B}$	**Indian Standard, 1973**
4	$V = k W^{a} D^{b}$	**General equation, Davies 1964**
5	$V = k (D/Q^{1/3})^{-B} X e^{-pD}$	**Modified USBM, Gosh and Daeman, 1983**
6	$V = K (Q/D^{1/3})^{B} X e^{pD}$	**Modified Ambrasey and Hendron 7**
7	$V = K^{B-\beta} Q^{A} e^{-pD}$	**Modified Gosh and Daeman, 1983**

where, $C_0 = \log k$, $C_1 = a$, $C_2 = b$ are constants that must be determined from the given set of n data points. Thus C_0, $C1$ and $C2$ are to be chosen such that the sum S of the squares of the deviations of data point from the values obtained from the equation (3) is a minimum.
This implies that

$$S = \sum_{i=1}^{n} (Y_1 - C_0 - C_1 X_1 - C_2 X_2)^2 \text{ minimum} \quad (4)$$

where the subscript 'i' which varies from 1 to n, is used to denote the 'n' data points. Differentiating S with respect to the coefficients and setting the partial derivatives equal to zero yields the minimum value of S. Thus,

$$\delta S/\delta C_0 = .2 \Sigma(Y_1 - C_0 - C_1 X_{1.1} - C_2 X_{2.1}) = 0 \quad (5)$$

$$\delta S/\delta C_1 = .2\Sigma X_{1.1}(Y_1 - C_0 - C_1 X_{1.1} - C_2 X_{2.1}) = 0 \quad (6)$$

$$\delta S/\delta C_2 = .2 \Sigma X_{2.1}(Y_1 - C_0 - C_1 X_{1.1} - C_2 X_{2.1}) = 0 \quad (7)$$

where the summations are from i = 1 to i = n. The above equation yields the following system of linear equations for the unknown C_0, C_1, and C_2

$$nC_0 + C_1 \Sigma X_{1.1} + C_2 \Sigma X_{2.1} = \Sigma Y_i \quad (8)$$

$$C_0 \Sigma X_{1.1} + C_1 \Sigma (X_{1.1})^2 + C_2 \Sigma X_{1.1} X_{2.1} = \Sigma X_{1.1} Y_i \quad (9)$$

$$C_0 \Sigma X_{2.1} + C_1 \Sigma X_{1.1} X_{2.1} + C_2 \Sigma (X_{2.1})^2 = \Sigma X_{2.1} Y_i \quad (10)$$

The above system of linear equations 8 to 10 was solved by using the programme O Gaussian elimination to yield the value of C_0, C_1 and C_2. From equations (2)

$C_o = \log k$, then $k = \exp. (C_o)$, $C_1 = a$, and $C_2 = b$.

Therefore, the best fit to the given data is obtained by putting the value of k, a and b in the equation (1)

In this case, a regression plane is obtained instead of variables W and D (Table 2). The values of k, a and b were calculated for the three mines namely mine A, mine B and mine C based on the monitor data from the field.
The blast monitor data is given in Table 3

Table 2. The best fit equations for different mines as obtained as

MINE	k	a	b
Mine A	000.934	-0.023	0.558
Mine B	408.055	0.479	-1.198
Mine C	000.802	0.685	0.451

Table 3. Blast Monitoring Data

MINE	W Kg	D, m	V mm/Sec
Mine A	08.90	25	05.52
	03.83	24	04.77
	05.85	24	05.03
Mine B	39.00	60	17.00
	61.00	60	22.22
	52.68	95	11.63
Mine C	00.38	52	02.21
	00.63	50	03.24
	00.75	37	03.27

W = maximum charge / delay, D = distance from the blast site to monitoring point, V = peak particle velocity.

A comparison is made between predicted values and measured values (Table 4) and graphs were also plotted for measured and predicted values of peak particle velocity (Figs. 1, 2 & 3) in order to validate the model. The values of measured and predicted peak particle velocity shows high degree of correlation coefficients i.e. more than 98 %.

Table .4 : Comparision between predicted and measured peak particle velocity (ppv).

Mine	measured (ppv)	Predicted (ppv)	Correlation Co- efficient
Mine A	5.52	5.35	
	4.77	5.33	0.999
	5.03	5.28	
Mine B	17.00	17.22	
	22.22	21.29	0.981
	11.63	11.45	
Mine C	2.21	2.42	
	3.24	3.38	0.994
	3.27	3.35	

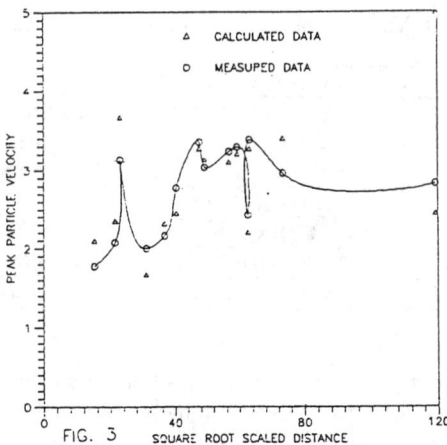

FIG. 3 SQUARE ROOT SCALED DISTANCE

Peak particcle velocity as a function
of square root scaled distance
(For Mine C)

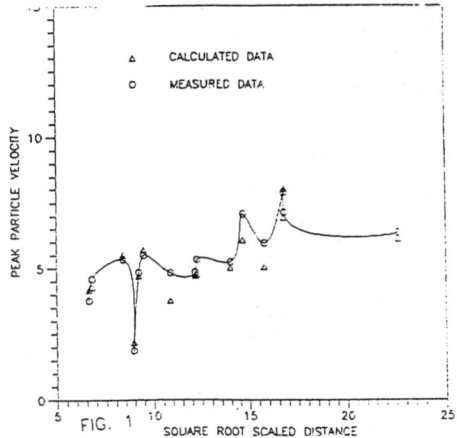

FIG. 1 SQUARE ROOT SCALED DISTANCE

Peak particcle velocity as a function
of square root scaled distance
(For Mine A)

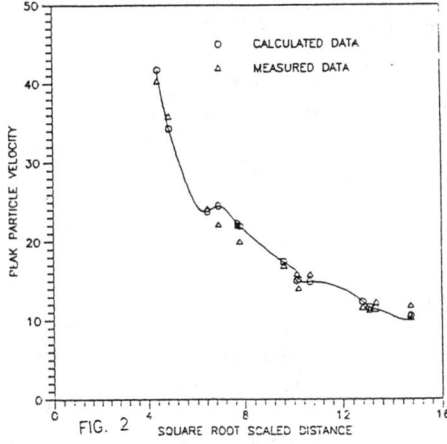

FIG. 2 SQUARE ROOT SCALED DISTANCE

Peak particcle velocity as a function
of square root scaled distance
(For Mine B)

5 CONCLUSIONS

Prediction of ground vibrations induced by blasting is becoming important. A large number of factors influence the ground vibration and it is not possible to incorporate all the factors in any of the predictor equations for calculating the safe charge. The predictor equation used for calculating the safe charge is site specific and not universal.

A study of blast induced vibrations was conducted in three different underground mines. The peak particle velocity was measured and values obtained was compared with the calculated values. The general equation was used to calculate the peak particle velocity at all the three sites because it contains three empirical constants and hence gives better result. The comparison between calculated and measured values shows a very good correlation coefficient.

ACKNOWLEDGEMENT

Authors are thankful to Mr. A. Singh, Research Scholar, BHU, Varanasi, for his help during the preparation of this paper.

REFERENCES

Ambrasey's N.N. and Hendron, J. ,1968. Dynamic Behaviour of Rockmass in Engineering Practice, K.G. Stagg and O.C. Zienkiewicz, edited, John Wiley & Sons, New York: 203-227.

Da Gama, C.D. ,1983. Use of comminution theory to predict fragmentation of jointed rockmass subjected to blasting, Int. Symp. on Rock Fragmentation by Blasting, Lulea, Sweden: 565-597.

Davis, B. , Farmer, I.W. and Attwell, P.B. , 1964. Ground vibration from shallow sub-surface blasts, The Engineers, 217:55-559.

Dupont, E.T. ,1977. Blaster Hand Book, 175th. Anniversary Edition, E.T. Dupont de Nemors, Inc. Wilmington, Delawane, 494p.

Gosh, A. , 1990. Fractals and numerical models of explosive rock fragmentation, Ph. D Thesis, The University of Arizona, U.S.A.

Gosh, A. and Daeman, J.K. , 1983. A simple new blast vibrator predictor based on wave propagation laws, Proc. 21th. US Symp. on Rock Mechanics, Texas, U.S.A.

Goyal, M. , Singh, T.N. and Singh, D.P. , 1994. Ground vibration due to blasting and its impact, ENVIRONMIN 94, IM & EJ, 289-299.

Indian Standard , 1973. Criteria for safety and design of structures subjected to underground blast, ISI Bull. No. IS-6922.

Persson , P.A., Holmberg, R. and Lee, J., 1993. Rock Blasting and Explosive Engineering, CRC Press. London, U.K.

Singh, A., Singh, T.N. and Singh, C.S., 1994. Prediction of ground vibration induced by blasting, IM & EJ, 31-34.

Singh, D.P., Goyal, M. and Singh, T.N. , 1994. Blasting technique for improving overall mining cost productivity and technology, MGMI, Calcutta, 153-166.

Singh, T.N., 1995. Prediction of ground vibration to save nearby structures due to blasting, Indian Cement Industry Deskbook, Commercial Publication, Bombay, 69-75.

Wiss, J.F. and Linehan, P.W., 1978. Control of vibration and blast noise from surface coal mining, Coal Mining I, III Reports to Mines, Bureau of Mines, OFR, 1-474.

Subsidence and ground control

Some incorrect problems of biharmonic equations arising in land subsidence mechanics

V. I. Dimova
University of Mining and Geology, Sofia, Bulgaria

ABSTRACT: The rock mass over a horizontal coal seam subjected to mining operations and lying at a depth H is considered. A problem is posed to determine the stress and strain fields in the rock mass over the seam by measured displacements of the points from the earth's surface and those of the immediate top.
The problem is reduced to a mixed problem of the biharmonic equation, which is decomposed into two separate problems: the first one is Dirichlet's problem for Laplace equation and the second one is an incorrect problem of mathematical physics for analytical continuation of a harmonic function.
The problem posed is solved for both the 3-D and 2-D cases.

1 INTRODUCTION

It is considered a lineary-elasic geomaterial layer with thickness H. The origin of the Cartesian coordinate system Oxyz is displaced over the down restricting plane (the immediate roof of the mined layer). This plane coincide with the coordinate plane Oxy. The axis z is directed conversely the gravitational force.

Our aim is to determine the displacement fields $u = u(x,y,x)$, $v = v(x,y,z)$ and $w = w(x,y,z)$ in the geomaterial band (the rock mass), by given values of these displacements over the upper (the earth surface) and down (the immediate roof) restricting planes: $u(x,y,0) = \varphi_1(x,y)$, $u(x,y,H) = \varphi_2(x,y)$, $v(x,y,0) = \psi_1(x,y)$, $v(x,y,H) = \psi_2(x,y)$, $w(x,y,0) = \chi_1(x,y)$, $w(x,y,H) = \chi_2(x,y)$. These data are obtained from the geodetic measurements.

The problem posed, has a clear geomechanical meaning and a significant practical value. Its solution will give us information for the geomechanical conditions under which we can eventually perform an underground operations (f.e. drawing a tunnel) above the mined layer.

Now we pass into the basic equations of the problem.

Under the supposition, that the mass forces are negligible (they are indirectly given in the boundary conditions), the equations of the theory of elasticity (the Lamé's equations) (Filonenko - Borodich 1959)

$$(\lambda+\mu)\frac{\partial\theta}{\partial x}+\mu\nabla^2 u+\rho X = 0$$

$$(\lambda+\mu)\frac{\partial\theta}{\partial y}+\mu\nabla^2 v+\rho Y = 0 \qquad (1)$$

$$(\lambda+\mu)\frac{\partial\theta}{\partial z}+\mu\nabla^2 w+\rho Z = 0$$

where λ, μ are Lamé's constants, ρ is density, X, Y, Z are the intensity of the mass forces, $\theta = \partial u/\partial x + \partial v/\partial y + \partial w/\partial z$, $\nabla^2(\cdot)$ is Laplace's operator, take the form (Filonenko - Borodich 1959):

$$\Delta^2 u = 0$$

$$\Delta^2 v = 0 \qquad (2)$$

$$\Delta^2 w = 0$$

where $\Delta^2(\cdot)$ is the biharmonic operator.

161

So in the considered case we are in a possession with three independent biharmonic equations, which together with the boundary conditions will allow us to solve the problem. Now we will formulate the problem precisely.

We have to underline, that we will consider one of the displacement components. Analogicaly the other two components can be also considered.

2 FORMULATION AND SOLUTION OF THE 3-D PROBLEM

In the half-space $z > 0$ we want find the solution of the biharmonic equation

$$\Delta^2 u(x,y,z) = 0 \qquad (3)$$

under the supposition that $u \to 0$ for $z \to \infty$:

$$u(x,y,0) = \varphi_1(x,y) \ ; \ u(x,y,H) = \varphi_2(x,y), \qquad (4)$$

where $\varphi_j(x,y)$ $(j=1,2)$ are given functions;

$$\varphi_j(x,y) \equiv O\left((x^2+y^2)^{-2-\varepsilon}\right), \ \varepsilon > 0$$

The problem posed is incorrect in the sense of J. Hadamard. Really, because of the representation

$$u(x,y,z) = u_1(x,y,z) + z u_2(x,y,z) \qquad (5)$$

which is valid for any biharmonic function $u(x,y,z)$, under the conditions (4), we rich to the following problem:

$$\Delta u_1(x,y,z) = 0 \ , \ z > 0 \ , \ u_1(x,y,\infty) = 0$$
$$u_1(x,y,0) = \varphi_1(x,y) \qquad (6)$$

$$\Delta u_2(x,y,z) = 0 \ , \ z > 0 \ , \ u_2(x,y,\infty) = 0$$
$$u_2(x,y,0) = \varphi_2(x,y) \qquad (7)$$

where

$$\varphi_3(x,y) = \frac{1}{H}\left[\varphi_2(x,y) - u_1(x,y,b)\right];$$

$$\varphi_j(x,y) \equiv O\left((x^2+y^2)^{-2-\varepsilon}\right), \ j=1,2,3, \ \varepsilon > 0$$

It can be immediately seen, that the problem (7) is incorrect for the harmonic continuation in the area $z > b$ in the band $0 < z < b$ (Atahodjaev 1986) (Ivanov 1965).

It can be imposed additional information about $u_2(x,y,z)$, which will ensure a conditional correctness of the problem (7), sequentially of the problem (3) - (4). In that way an approximate solution, which converges to the real solution, can be constructed.

Let first we suppose, that $\varphi_j (j=1,2)$ are given precisely. We will seek a solution of the problem (6)-(7), by applying double Fourier's transformation by x and by y. Thus

$$u_1(x,y,z) = \frac{1}{(2\pi)^2} \int\int_{-\infty}^{+\infty+\infty} \varphi_1(s,\eta) \cdot$$
$$\cdot \exp\left(-\sqrt{s^2+\eta^2}\,z + i(xs+y\eta)\right) ds d\eta \qquad (8)$$

$$u_1(x,y,z) = \frac{1}{(2\pi)^2} \int\int_{-\infty}^{+\infty+\infty} \varphi_3(s,\eta)$$
$$\cdot \exp\left(-\sqrt{s^2+\eta^2}(z-H) + i(xs+y\eta)\right) ds d\eta \qquad (9)$$

where

$$\varphi_j(s,\eta) = \int\int_{-\infty}^{+\infty+\infty} \varphi_j(x,y) \cdot$$
$$\cdot e^{-i(xs+y\eta)} dx dy, \ (j=1,3)$$

Let us underline, that for the values of z from the interval $[0,b]$, the formula (9) is insatiable, which one more time proves the incorrectness of the problem (7), sequentially the incorrectness of the problem (3) - (4).

If following V. Ivanov (Ivanov 1965), we regularize (9), we will receive

$$u_{2\alpha}(x,y,z) = \frac{1}{2\pi} \int_{-\infty}^{+\infty}\int_{-\infty}^{+\infty} R_\alpha(x-t, y-t, z) \cdot \varphi_3(t,\tau) dt d\tau \qquad (10)$$

where

$$R_\alpha(x,y,z) = \frac{1}{2\pi}\int_{-\infty}^{+\infty}\int_{-\infty}^{+\infty} \exp\left(-\alpha^2\left(s^2+\eta^2 - \sqrt{s^2+\eta^2}(z-H)\right)\right) \cdot \cos(xs+y\eta) ds d\eta$$

where α is reguralization parameter.

It can be proved (Ivanov 1965) (Atahodjaev 1986), that the function $u_{2\alpha}(x,y,z)$, determined by the equality (10) can be considered as an approximate solution of the problem (7): for any $z \in [0,H]$ is valid the equality

$$\lim_{\alpha \to 0} u_\alpha(x,y,z) = u(x,y,z)$$

monotolically in respect to x and y over $(-\infty, +\infty)$.

Let us now suppose, that instead of $\varphi_j(x,y)$ ($j=1,2$), theirs approximations $\varphi_{j\delta}(x,y)$ are given. and let these approximations are continuous, bounded and differ from $\varphi_j(x,y)$ not more, that with a given $\delta > 0$

$$|\varphi_j(x,y) - \varphi_{j\delta}(x,y)| \le \delta$$

If we now apply (10) over $\varphi_{3\delta}(x,y)$, then instead of $u_\alpha(x,y,z)$, we receive the approximation

$$u_{2\alpha\delta}(x,y,z) = \frac{1}{2\pi}\int_{-\infty}^{+\infty}\int_{-\infty}^{+\infty} R_\alpha(x-t, y-t, z) \cdot \varphi_{3\delta}(t,\tau) dt d\tau \qquad (11)$$

which can be considered as a solution to the posed problem.

On the base of the analogical reasoning we can receive the other components of the displacement vector. Then we will have the solution of the problem posed in the item 1.

3 FORMULATION AND SOLUTION OF THE 2-D PROBLEM

The reasoning made for the elastic geomaterial layer, laid out in item 1., could be easily adapted for a geomaterial elastic band. Thus the problem formulated in item 2. takes the form: to be found in the semi-plane $z > 0$ the solution of the biharmonic equation

$$\Delta^2 u(x,y) = 0 \qquad (12)$$

under the supposition that:
$u \to 0$ for $y \to \infty$; $u(x,0) = \varphi_1(x)$; $u(x,H) = \varphi_2(x)$, where $\varphi_j(x)$ ($j=1,2$) are given functions as in it

$$\varphi_j(x) \equiv O\left(|x^2|^{-2-\varepsilon}\right), \varepsilon > 0 . \qquad (13)$$

The problem (12) - (13) is incorrectly posed in the sense of J. Hadamard. Really, with the presentation

$$u(x,y) = u_1(x,y) + y u_2(x,y) \qquad (14)$$

which is valid for any biharmonic function in $y > 0$, the conditions (13) leads to the following problems:

$$\Delta u_1(x,y) = 0, \; y > 0,$$
$$u_1(x,\infty) = 0, \; u_1(x,0) = \varphi_1(x) \qquad (15)$$

$$\Delta u_2(x,y) = 0, \; y > 0,$$
$$u_2(x,\infty) = 0, \; u_2(x,H) = \varphi_3(x) \qquad (16)$$

where

$$u_{2y} = \Delta u(x,y), \; \varphi_3(x) = \varphi_2(x) - u_1(x,H)$$

$$\varphi_j(x) = O\left(|x|^{-2-\varepsilon}\right), j=1,2,3, \; \varepsilon > 0$$

It can be seen that the problem (16) is incorrectly posed problem for the harmonic continuation from $\{y > b\}$ to the stripe $0 < y < b$ (Atahodjaev 1986). Its solution is achieved by the A. Tichonov's regularization method in the variant, proposed by V. Ivanov (Tikhonov & Arsenin 1977).

Thus for approximate solution can be taken

$$u_{2\alpha}(x,y) = \frac{1}{2\pi}\int_{-\infty}^{+\infty} R_\alpha(x-t,y)\varphi_3(t)dt \qquad (17)$$

where

$$R_\alpha(x,y) = \frac{1}{2\pi}\int_{-\infty}^{+\infty} \exp(-\alpha^2\omega^2 - |\omega|(y-b))\cos\omega x\, d\omega$$

In the cases when the functions $\varphi_1(x)$ and $\varphi_3(x)$ are given approximately, we act as in the item 2.

4 CONCLUSION

The problems considered above are original as fare in the elasticity theory, as in the applied geomechanics. Theirs solutions permits by given subsidence on the earth's surface and on the immediate roof, to determine displacements field in the rock mass. This is of extreme importance for the underground human activities (drawing a tunnels, mining above mined layer). Then the solution of the posed problems give us information for the conditions under which the performances will be carried out.

REFERENCES

Atahodjaev M.A. 1986. *Incorrect problems for the biharmmonic equation*, FAN, Tashkent (in Russian).

Dimova V.I. 1990. *Some direct and inverse problems in applied geomechanics*, UM&G Press, Sofia.

Filonenko - Borodich M.M. 1959. *Theory of elasticity*, FIZMATIZ, Moscow (in Russian).

Ivanov B.K. 1965. Caushy's problem for the Laplace's equation in infinite stripe, *Differential equations*, vol. 1 (in Russian).

Tikhonov A.N. & V. Arsenin 1977. *Solutions of ill-posed problems*, V.H. Winston & Sons, Washington D.C.

The prediction of tunnelling induced settlements in weak rock

Harry Asche
Connell Wagner, Brisbane, Qld, Australia

ABSTRACT: Available methods for the prediction of the settlements above shallow tunnels in rock are reviewed. The empirical method is direct and accurate where there is a good database of similar tunnels in similar ground. Where no such database exists, the method is of dubious value. Numerical methods are promising for all cases, but are very dependant upon input parameters, which are usually only poorly known. Case histories are presented where both methods were used. In the design of the New Southern Railway Sydney, results from the back-analysed Brisbane Rail Tunnels were used to help with the selection of the support systems. Construction of the New Southern Railway in currently underway, and the predictions can now be compared with measurement.

1 SUMMARY OF METHODS USED TO PREDICT SETTLEMENT

The design of shallow tunnels in weak rock is often governed by the minimisation of tunnelling induced settlements to buildings and services above. The choice of the most appropriate support system will be that which is the least total cost combination of building and service remedial measures with excavation and support systems. To effectively assess the available options of excavation and support, accurate prediction of tunnelling induced settlement is necessary.

The methods available can be classified into two categories; empirical methods, and numerical methods.

1.1 The Empirical Method

This method originated with Peck (1969) and has reached a high degree of sophistication, particularly in the UK.

Briefly the method assumes a "loss of ground", equal to a proportion of the tunnel volume, which is called the "face loss". The face loss is predicted from considering previously measured settlements of tunnels constructed in similar ground and using similar construction methods. Associated with the volume of the face loss is the volume of the settlement trough formed on the surface above the tunnel. Normally these volumes are assumed to be equal. The method clearly originates from, and is normally used for, shield driven tunnels in soft ground, hence the nomenclature of "face loss" and the assumption of zero volumetric change - an assumption which is relevant in clays, but not necessarily in rock.

Note also, that the assumption made in most numerical models - i.e. that the ground load is released by the excavation and that the lining deflects with the ground to take up some of this load - is totally absent in the empirical method. All of this behaviour is identified as "face loss".

The settlement trough is assumed to have a Gaussian "bell curve" shape. The width of the trough is related to the depth of the tunnel and to the type of ground. Therefore, from knowledge of the tunnel diameter, the face loss and the depth of tunnel, the settlement trough can be predicted.

The weakness of the empirical method is where tunnels are to be driven in ground where a small or no database exists, or where construction methods are different from those previously employed. Without this database, the possible values of face loss, and the predicted settlement arising from these values, range over an order of magnitude.

1.2 Numerical Methods

By modelling the ground and the lining with finite element or with finite difference methods, in theory, any process of excavation and support can be modelled in any ground type, providing the actions and deflections of the tunnel support, and the settlement of the surface as well.

Practically, the numerical method is limited by three factors:

1. Lack of knowledge of "true" ground and support properties. There are many ground models available to the modeller, starting with simple elasticity and ranging up to include for plasticity or time dependent behaviour. But even for simple elasticity, the real value

of the insitu stress field or the Young's modulus is usually difficult to measure.

2. Limitations due to assumptions inherent in the method. The 2D modeller uses a number of methods to take account of face advance effects or shotcrete strength gains. But even if a three dimensional model is used, local effects such as localised yield at the base of steel ribs, or redistribution of loads due to yield and plasticity in shotcrete linings can make the model miss key features of reality.

3. Limitations due to computing power. This problem is one which is going away with time. The progress in increased computing power is steadily reducing restrictions on discretisation. However, it is still necessary to be aware of the limitations on accuracy caused by coarse meshing or by locating problem boundaries too close to the area being analysed.

Each of these factors affect the predictive abilities of the numerical method.

1.3   Practical Example

To give a simple example of the limitations of both methods applied to settlement predictions, results for some simple 2D numerical tunnel models are given in Table 1. The analysis was carried out with the program FLAC, of a 6 m diameter tunnel with 14 m of cover.

The calculated settlements in Table 1 are reported relative to a datum on the surface. This datum has been chosen to be located a distance sideways from the centreline equal to the depth from the surface to the tunnel axis. The reason for this procedure, and for not just reporting the absolute value given by the program is as follows. Any tunnel excavation produces a net upwards force, equal to the weight of the ground removed. According to the theory of elasticity, apart from the local effects of the excavation which may involve plastic or elastic deformation around the tunnel, there is a global effect caused by the upwards force, which produces an upwards deflection proportional to the logarithm of the distance to an assumed point of zero deformation. In a numerical analysis, the mesh boundaries are usually defined have zero deformation. Therefore, the global deflection due to the uplift force will be a function of the distance to the bottom boundary, and will generally be upwards. By a suitable choice of boundary distance, any arbitrary choice of upwards deflection can be produced by our model. In contrast, the deflection relative to an assumed datum point, such as the one described above, will be independent of the boundary distance once the boundary is sufficiently far away. It is important to separate global movements due to the far boundaries from local effects due to the tunnelling. Large global deflections are not observed in practice, for a variety of reasons, for example, the effective Young's modulus of a real rockmass increases, both with depth, and as strain becomes smaller. A full discussion is contained in Oteo and Sagaseta (1982). Note that the choice of a datum, relative to which the settlement is measured, exactly follows the survey procedure used to measure the actual settlements.

Table 1 Results of practical example

No	Description	Settlement
E1	Base Case: Elastic only $E=2000$ MPa, $K_o=1.0$	0.32 mm
E2	Case E1, with $E=1000$ MPa	0.64 mm
E3	Case E1, with $K_o=1.5$	-0.30 mm (heave)
E1B	Case E1, with mesh boundaries too near	0.41 mm
E1M	Case E1, with a coarse mesh	0.31 mm
P1	Case E1 with Plasticity $c=100$ kPa, $\phi=20°$	1.43 mm
P2	Case P1 with $E=1000$ MPa	2.85 mm
P3	Case P1 with $K_o=1.5$	-1.41 mm (heave)
P4	Case P1 with $c=80$ kPa	147.32 mm (collapse)
P1B	Case P1, with mesh boundaries too near	1.45 mm
P1M	Case P1, with a coarse mesh	1.12 mm

Conclusions to be drawn from the results are as follows:-

1. As is expected, the results are directly proportional to the value of Young's modulus (E), even in plasticity. (Plastic deformation is also calculated proportional to the Young's modulus.)

2. The results are very sensitive to the ratio of horizontal to vertical stress (K_o).

3. The results can be extremely sensitive to the plastic parameters. In this example, the plastic parameters were chosen so that a ring of plasticity formed around the tunnel. In this case, the effect of changing the plastic parameters slightly has the effect of allowing the plasticity to extend to the surface, and collapse ensues. This is the effect shown in P4. (Collapse also occurs when K_o is reduced to 0.5, because the reduced minor principal stresses causes more widespread plasticity than with K_o of 1.) If plastic parameters had been chosen so that the extent of plasticity was smaller, the sensitivity to these parameters would be smaller. The addition of a lining (not modelled in this simple example) also has the effect of controlling the plasticity, and reducing the sensitivity of the results.

4. The results are sensitive to using a coarse mesh, when plasticity is used, but not so much with elasticity only. The coarseness of the mesh is not so bad as to affect the elastic result. This highlights the fact that plastic analyses need finer meshes to avoid mesh dependency than elastic meshes.

5. The results are sensitive to using short boundaries with an elastic only analysis, but the local effects of plasticity are not much affected by the boundary distance.

Table 2 Results analysed for comparison with empirical method

No	Face loss volume (% face)	Trough volume (% face)	Ratio face-trough	K
E1	0.057	0.0057	10	0.44
E2	0.11	0.011	10	0.44
E3	0.071	-0.0046	-16	0.35
P1	0.13	0.023	5.7	0.43
P2	0.26	0.046	5.7	0.43
P3	0.28	-0.013	-22	0.18
P4	2.5	2.3	1.1	0.35

In Table 2 the results of Table 1 have been analysed for comparison with the assumptions of the empirical method. The actual face loss and trough volumes are calculated from the reported displacements. (Note that it is the trough volume that is usually measured and placed in the database of empirical results.) It can be seen that the assumption that these are the same value is not well reproduced by the numerical models. However, this assumption improves as the level of plasticity in the model increases.

The trough width parameter K is generally in the range reported by the empirical method (i.e. between 0.35 and 0.6).

The empirical method does not predict heave, nor does heave ever actually occur unless in the case of an EPBM overthrusting in very weak ground.

2 BRISBANE RAIL TUNNELS - RAILWAY TRAINING CENTRE

2.1 Brief Description of Project

Full descriptions of the project are contained in Chappel (1990), Asche & Baxter (1993) and Baxter (1993). The Brisbane Rail Tunnels comprise two pairs of single track tunnels, the Brunswick Street Tunnels (725 m) and the Roma Street Tunnels (300 m). These tunnels were designed in 1990, and driven in 1992-1994 by roadheader. Support comprised rockbolts, shotcrete, and where necessary, steel sets.

The cover to the tunnels ranges from 4 m to 20 m. A number of significant buildings exist along the route.

2.2 The Railway Training Centre

The first building to be encountered during construction was the Railway Training Centre. This building was a four story brick building located on Barry Parade. The tunnels passed approximately 4 m below the pad foundations. The building was located above the poorest conditions to be encountered along the Brunswick Street Tunnel Route.

2.3 Calculations Made Prior to Excavation

Settlement estimates were made using both the empirical and the numerical methods. The empirical method, adopting a face loss value of 0.5%, gave a settlement prediction of 20 mm. The face loss value was chosen as being equivalent to a well driven shield in good stiff clay.

Numerical calculations were carried out. The calculations assumed :

1. No stiffening effect due to the lining,
2. Values of ground stiffness which were reduced from those recommended by the geotechnical report,
3. In addition to item 2, a shattered zone was postulated with reduced stiffness around the excavation; due to possible blasting damage.

With these conservative assumptions, the predicted settlement was 12 mm. The support design then took place. The design was based on predicted support pressures derived from empirical analyses, as well as from simplified numerical models. However, it was the judgement of the designers that it was vital to make the tunnel support as rigid as possible to minimise settlement, despite the results of the numerical modelling described above, which showed that with no support (or just enough to make the excavation stable), the surface settlement would be small. In addition to the very stiff support in the tunnels, a raft slab was poured within the building to tie the pad foundations together.

2.4 Observations Made During Excavation

The tunnel was excavated with a roadheader. The maximum settlement measured was 13 mm. The settlement trough shape appeared to confirm that the raft slab was working as planned by tying the building together over the tunnel.

The support instrumentation showed that the steel sets behaved as expected, taking initial load quickly, while the shotcrete stresses rose slowly, as the shotcrete gained in stiffness.

2.5 Conclusions

The selection of the support was made on the judgement of experienced tunnelling engineers. Neither the empirical method or the numerical method were of assistance for designing the support. The case emphasises the role that experienced engineers must play in the design process, despite the production of sophisticated (and Quality Assured) calculations.

The main question which remains is:- what was the actual relationship between the support thickness and the settlement? Could the support could have been reduced to be just strong enough to permit safe tunnelling with the settlement increasing only marginally, say to 14 or 15 mm?

3 THE NEW SOUTHERN RAILWAY - CLEVELAND STREET

3.1 Brief Description of Project

A full description of the project (for the Hard Rock Tunnel section) is given in Asche (1996). The New Southern Railway comprises some 10 km of dual track tunnel. The Hard Rock Tunnel section was designed in 1995/1996 and construction is ongoing. The method of excavation is by roadheader.

The tunnel is shallowest at the northern end, with minimum cover of 4 m. The tunnel falls to the south, and the cover increases to 20 m. The tunnel route is beneath George Street, Redfern for the first 1500 m. The tunnel crosses beneath a number of significant services, and significant buildings exist on either side of George Street.

3.2 Cleveland Street Area

The New Southern Railway enters driven tunnel at Cleveland Street. Deeper tunnelling is prevented by the need to cross above the Eastern Suburbs Railway. This forces the 10 m wide by 7 m high New Southern Railway into shallow cover (about 4 m) in highly weathered Ashfield Shale and residual clays for approximately 200 m.

Along this section of the route, a number of services are crossed including a 750 mm water main in Cleveland Street, and the tunnel passes in front of a number of medium rise buildings in George Street.

In addition to the requirement to avoid damage to the buildings and services in the street, the Contract conditions require that settlement is limited in this area to a maximum of 40 mm.

3.3 Empirical Calculations

Information on the settlement due to tunnelling in Ashfield Shale was at that time non existent. The only significant tunnelling in Ashfield Shale had taken place during the construction of the Eastern Suburbs Railway tunnels between Central Station and Redfern, and no measurements had been made.

It was decided to adopt face loss parameters based on the measured settlements in the Brisbane Rail Tunnels at Albert Street where residual clay was encountered at shallow cover. There, 12 m long canopy tubes had been installed ahead of the face. The arch support was a shell of shotcrete. This was the method initially proposed for the Cleveland Street End of the New Southern Railway.

Back analysis of the 15 mm settlement at Albert Street gave an average face loss in residual clay of 0.25%. Where more of the face was in a more weathered material, the settlement was higher. Face loss values were assigned to various degrees of weathering. The predicted settlement above the New Southern Railway was 65 mm at Cleveland Street.

Although this was the only prediction of settlement that could be made with the available data, other information could be used in an empirical way. This was the assessment of the effect of nails drilled and grouted into the face, thus increasing the resistance of the ground to movements ahead of the face. A trial had recently taken place using face nails in London clay in a NATM tunnel in an area where settlements were critical. The results showed a reduction of settlement to about 70% of its value without the face nails.

This result allowed a simple assessment of the value of the use of face nails, which can be compared with the very difficult 3D analysis which would be needed to make a comparable assessment by the numerical method.

Therefore, the role played by the empirical method in this tunnel was:

1. to indicate that canopy tubes with a shotcrete-only shell (no sets) would not meet the settlement criteria.
2. to give a figure for comparison with the numerical models.
3. to assess the effect of face nails.

3.4 Numerical Modelling

The numerical modelling was performed with the program FLAC. In an effort to improve the results of this modelling, a totally new analysis was performed on the Brisbane Rail Tunnels, both at the Railway Training Centre and at Albert Street. The boreholes were analysed to give elastic and plastic properties using the methods of Hoek et al.(1995). The support was modelled using stiffness properties derived from the actual material properties, including allowances for creep and shrinkage in the shotcrete.

The modelling was carried out in 2D. The effect of the face was modelled by the "convergence-confinement" method. In this method, some of the excavation loads are released onto the unlined tunnel, and the remainder are released onto the lined tunnel. The excavation load split can be varied, for example if most of the excavation load is released onto the unlined tunnel, this implies an installation of the lining well back from the face.

The effect of the canopy tubes was modelled by providing springs in the ground at the canopy tube locations. These resisted movement in the unlined model. A proportion of their load was subsequently

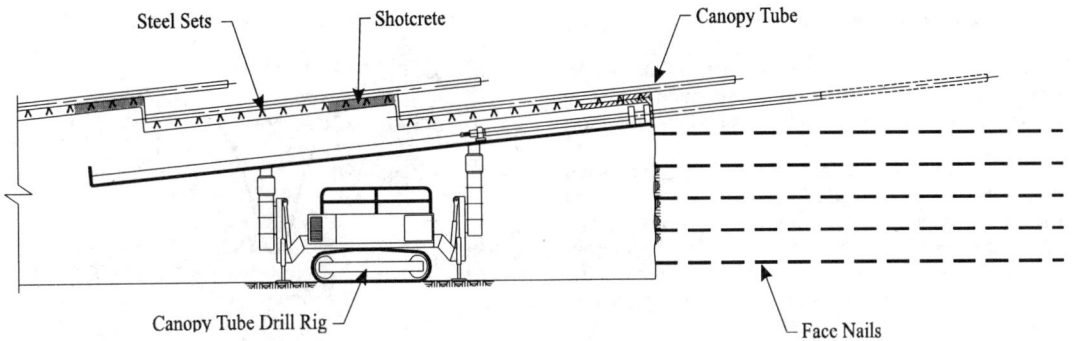

Figure 1 Installation of Canopy Tubes at Cleveland St

applied in the opposite direction (i.e. mainly downwards) for the lined case, modelling the effect of the rear ends of the canopy tubes applying load from the face to the shotcrete lined tunnel behind. This has the effect of delaying the onset of load onto the lining, thus allowing stiffening the support system, and reducing the settlement.

The predicted settlements, convergences, steel and shotcrete stresses were then compared with those actually measured in the Brisbane Rail Tunnels. In every case, the actual settlement and convergence was significantly underestimated by the numerical models while the steel and shotcrete stresses were overestimated.

Therefore, further models were run, and a crude backanalysis was carried out to get a match with the actual measured values. The result was adjusted parameters which were then applied to the modelling of the New Southern Railway. These adjusted parameters are given in Table 3.

The adjusted value for the shotcrete stiffness is of particular interest, as it implies a significant amount of creep and shrinkage of the shotcrete.

Table 3. Adjustments made to input values in settlement models

Parameter	Adjusted value
Young's modulus of ground	40% of the results derived from Hoek et al.(1995)
Mohr Coulomb plasticity parameters of ground	100% of the results derived from Hoek et al.(1995)
Excavation load split	50% on unlined tunnel, 50% on lined tunnel
Shotcrete stiffness	Young's Modulus of 400 MPa when used in second stage of excavation load split stage
Canopy tube stiffness	Equivalent to a point load at midspan of a 5 m span with fixed ends
Steel set stiffness	20% of values from section properties

Numerical calculations were now performed to evaluate different methods of excavation of the New Southern Railway at Cleveland St. The predicted settlement for support using canopy tubes was 50 mm.

Efforts were made to improve the settlement performance for the canopy tube method. The addition of steel sets was considered. This was resisted at first, because making steel sets to match the varying profile of the canopy tubes is not easy. However, by using TU

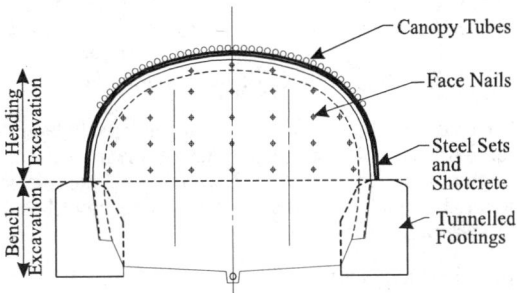

Figure 2 Designed support at Cleveland St, Cross Section

section nesting sets, providing a varying profile could be achieved, utilising a sliding joint which could be clamped and locked. The use of steel sets showed great promise in reducing settlement with the canopy tube method. Assuming a completely efficient connection between the ground and the steel sets, the numerical method predicted a settlement of 16 mm at Cleveland St.

However, backanalysis of the Brisbane Rail Tunnels at the Railway Training Centre had shown that the steel sets could be inefficient at providing the stiffening expected. In practice the sets can punch into the ground at the footings, releasing their load, and if not blocked properly, the ground can deform at the blocking points as well. Using a conservative efficiency of the connection between the sets and the ground, the predicted settlement was 30 mm at Cleveland St.

3.5 Proposed Support System

The system of support finally adopted is shown on Figures 1 & 2. Particular efforts were made to ensure the rigid connection between the ground and the steel sets.

The canopy tubes were 220 mm diameter steel tubes installed 12 m ahead of the tunnel by specialised drilling equipment designed for the purpose. The tubes were driven around the crown with a 5° lookout angle so the process can be repeated. To maintain the protection to the face, an overlap of 3 m was required, so that the cycle repeated every 9 m.

The tubes were embedded in a shotcrete arch behind. To reduce settlements in the lowest cover and weakest ground, the following additional measures were specified:
1. Steel sets in contact with the tubes to provide immediate stiff support.
2. Tunnelled concrete filled footings for the steel sets. This detail prevents the steel sets punching into the ground.
3. Preloading of the steel sets to take out any flexibility at either the footing or the blocking points.
4. Shotcrete embedment of the steel sets.
5. Face nailing.
6. Face shotcreting where necessary.

3.6 Results of Monitoring at Cleveland St

The results of the settlement monitoring are shown in comparison with the predictions of the numerical method, in Figure 3. It can be seen that the measures adopted to make the steel sets effective have been successful.

4 CONCLUSIONS

The paper has described the process of design for settlement in urban areas.

The empirical method can perform well where there is a good database of tunnelling activity. Even where there is not, the empirical method can be used as an aid to judgement, by providing comparison to other tunnelling conditions.

The numerical method is considered to have promise but requires good input data to use. The necessity for the improvement and extension of existing field data collection, particularly in relation to large scale ground deformation characteristics is highlighted.

With careful use the numerical method can now assist in support design and provide comparisons between different construction methods.

5 ACKNOWLEDGEMENTS

The author wishes to thank Queensland Railways and Mr Claudio di Berardino of Transfield Bouygues Joint Venture for permission to publish this paper.

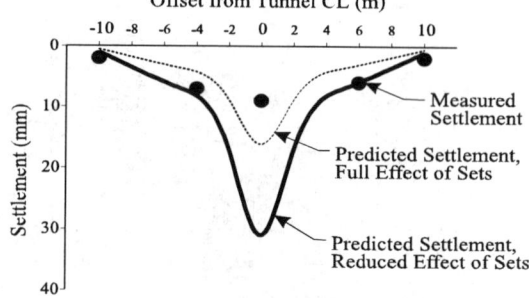

Figure 3 Predicted and Measured Settlements

REFERENCES

Asche H.R. 1996. Preliminary design of the New Southern Railway - Rock Tunnel Section. *Proceedings of the 1996 National Engineering Conference, Darwin*. The Institution of Engineers, Australia. 219-227.

Asche H.R. & Baxter D.A. 1993. Design and Ground Support Monitoring of the Brunswick Street Rail Tunnels, Brisbane. *Proceedings of the VIII Australian Tunnelling Conference, Sydney*. 233-243.

Baxter D.A. 1993. Construction of the Brunswick Street Rail Tunnels. *Proceedings of the VIII Australian Tunnelling Conference, Sydney*. 109-125.

Chappel R.A. 1990. Planning, Investigation and Preliminary Design of the Duplication of Rail Tunnels - Brisbane. *Proceedings of the VII Australian Tunnelling Conference, Sydney*. 258-263.

Hoek E., Kaiser P.K. & Bawden W.F. 1995 *Support of Underground Excavations in Hard Rock* A. A. Balkema, Rotterdam.

Oteo C.S. & Sagaseta C. 1982. Prediction of settlements due to underground openings. *Proceedings of the International Symposium on Numerical Methods in Geomechanics. Zurich*. Vol. 1. 653-659.

Peck R.B. 1969. Deep Excavations and Tunneling in Soft Ground. *Proceedings of the 7th International Conference on Soil Mechanics and Foundation Engineering, Mexico City*. Vol. 3. 225-290.

Predicting underground stability using a hangingwall stability rating

E. Villaescusa
Department of Mining Engineering and Mine Surveying, Western Australia School of Mines, Kalgoorlie, W.A., Australia (Formerly: Mount Isa Mines Limited, Kalgoorlie, W.A., Australia)

D. Tyler
Mount Isa Mines Limited, Kalgoorlie, W.A., Australia

C. Scott
Julius Kruttschnitt Mineral Research Centre, Australia

ABSTRACT: The work described here assesses the stability, performance and behaviour of unsupported openings in bench stoping. The development, application and a recent update of a locally developed bench stability method which links ground conditions and excavation geometry is presented. Experience and back analysis have been used to rate the importance of the factors controlling unsupported hangingwall behaviour.

1 INTRODUCTION

The Mount Isa Mine is one of the largest underground mines in Australia, and a major world producer of copper, silver, lead and zinc. This paper describes a recent update of a hangingwall stability method developed to link ground conditions and excavation geometry within the silver/lead/zinc operations (Villaescusa 1996).

During the 1970's and early 1980's, developments in cable support at Mount Isa in conjunction with ground behaviour monitoring, computer-based stress analysis techniques and the understanding of hangingwall behaviour allowed the introduction of large-scale open stoping in orebodies that had been previously mined using cut-and-fill methods (Bywater et al 1983; Beer et al 1983; Greenelsh 1985). During the last five years, sublevel bench stoping has been used to extract the closely spaced, narrow and steeply dipping silver/lead/zinc orebodies referred to locally as the Lead orebodies. Bench stoping has been successful largely due to advances in the understanding of unsupported hangingwall behaviour and performance, backfilling technology, ground support practices, drilling and blasting and the application of remote mucking technology (Villaescusa et al 1994).

2 BENCH STOPING

At Mount Isa, benching involves an initial mining of both a drilling and extraction drive for the entire length and width of an orebody. A bench slot is created between the two horizons at one end of the orebody by enlarging a cut-off raise (or a long hole winze) located near the footwall of the orebody. The slot is used as an expansion void into which the remainder of the bench stope is formed. In most cases, the production holes are drilled in rings parallel to the orebody dip between the two drives. Mining proceeds through the sequential firing of production rings into the advancing void (Figure 1).

Typical bench stopes are extracted to full orebody width, having an average strike length of 100m. This results in an overhanging unsupported inclined hangingwall, which ranges from 12 to 45m in length down dip. Following extraction of the orebody, the void is then filled with hydraulic sand or dryfill (aggregate) to the floor of the drill drive, which becomes the new extraction drive for the next lift.

3 ROCK MECHANICS

In order to enable the geomechanics assessment of the localized (and overall) ground behaviour resulting from current and previous mining geometries, stress analyses of mining sequences are routinely performed. These are used to avoid leaving blocks of highly stressed rock and to limit the number of openings within future pillar areas. Closely spaced orebodies are benched in a certain sequence, in order to minimize the effects that stopes might have on each other.

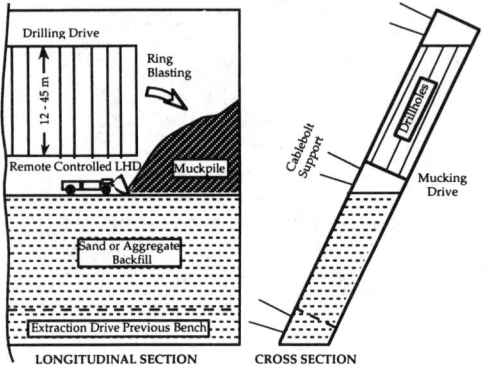

Figure 1. Schematic bench stoping geometry at Mount Isa.

Extraction sequences can generate large concentrations of stress along the strike and dip of the benching blocks. In-situ stress measurements at Mount Isa indicate that the maximum principle stress is oriented normal to bedding while the other two principal stresses have similar magnitudes and act in the plane of bedding. Numerical modelling has shown that the best mining sequence is to keep the stope backs on a line perpendicular to bedding. This minimizes the likelihood of rockfalls through slabbing and unravelling across the exposed excavation backs (Lee 1980).

4 BENCH STABILITY

The behaviour of unsupported spans is critical to performance of benching with respect to external dilution control around the ore-waste boundaries. Bench sills for longitudinal drilling and mucking operations are developed from permanent cross cuts and are located at pre-established elevations, usually defined by the bench heights. Therefore, information regarding the likely stable hangingwall dimensions is required at the early stages of planning a bench block. Some of the orebody hangingwalls are weak, and disruptive failures were experienced in some orebodies during the earlier cut and fill operations.

Initially at Mount Isa the bench heights were recommended based on the relationship between the frequency of bedding plane breaks and the performance of cut and fill operations for each of the orebodies (Baczynski 1974). However, it is clear that information on bedding plane breaks alone can not be used as a predictive or a design tool, since it does not include geometrical information regarding the typical shape and size of the bench stope openings. Furthermore, it was recognized that other factors such as the induced stresses and the blasting practices also influence the behaviour and the stability of the mine openings. A concern regarding bench stability was highlighted when no design tool was available to calculate the initial bench heights and strike lengths that could be safely exposed for the full range of orebodies at Mount Isa.

Disruptive bench hangingwall failures during the extraction of initial benches in weak rock led to the initial development of an empirical model for bench stope hangingwall stability (Villaescusa 1996). The objective was to establish a stability chart which linked the factors controlling ground behaviour and the excavation geometry for each of the orebodies. It was envisaged that such a predictive tool could be used in mine planning to optimize bench block designs. For example, greater bench heights could be planned for better quality orebodies, with significant reduction on sill development costs. Alternatively, when a particular bench height is fixed, the stability chart may be used to calculate the maximum (unsupported) stable length that can be safely exposed.

The Hangingwall Stability Rating (HSR method), assumes that the geological discontinuities, induced stresses, blast damage and the excavation geometry are the main factors controlling hangingwall stability (Villaescusa, 1996). Following the development and routine applications of the rating method, further calibration has led to an update of the factors controlling the unsupported hangingwall behaviour as follows: Geological discontinuities (70%), mining induced stresses (15%) and blast damage (15%). The maximum permissible value that HSR allows is 60. The calculated rating value can be plotted against design or open unsupported hydraulic radii producing a stability chart.

4.1 Geological discontinuities (70%)

Geological discontinuities such as bedding and jointing which are by far the most important factors controlling the dimensions of the stable openings at Mount Isa. Experience has shown that the linear frequency of bedding plane breaks is the most important parameter accounting for a total of 50% of the global ratings. If the immediate hangingwall (1-3m) is fissile, then localized or total hangingwall collapse may result. Jointing accounts for the remaining 20% that is assigned to the discontinuities. Hangingwall bulge and buckling of the unsupported mid-span where the failure surface is defined by jointing have been observed in most failures (Beer et al, 1983).

4.1.1 Bedding plane break frequency (50%)

The rock mass in the Lead Mine at Mount Isa can be generally described as highly jointed and steeply dipping bedded rock. The average dip of bedding is 65° to the west. The orebodies are tabular, stratabound and the boundaries of the hangingwalls and footwalls of all the bench stopes are essentially defined by bedding planes. Experience indicates that hangingwall behaviour and stability are mainly controlled by the frequency of bedding plane breaks within the immediate stope boundaries. Cross-cut mapping (Landmark & Villaescusa 1992) or borehole logging (Villaescusa et al 1995) data is required to define the number of bedding plane breaks per metre, within the first three meters of hangingwall, for each of the orebodies as shown below:

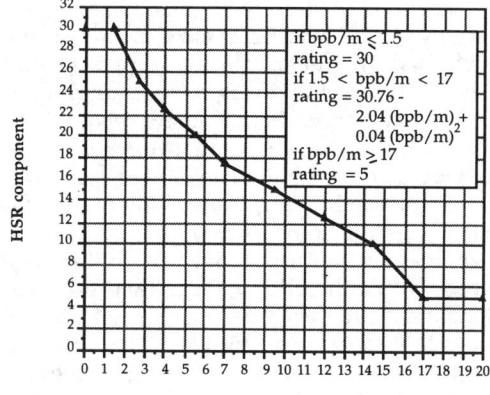

Figure 2. Rating by true linear frequency of bedding plane breaks (first 3m of hangingwall).

4.1.2 Number of joint sets and continuity (20%)

It is generally accepted that the mode of failure experienced in the hangingwalls at Mount Isa, is through mid-span buckling of individual bedding plates. These buckling failures are usually enhanced by the number and the size of the individual structures intersecting the hangingwalls. Individual structures such as faults, shears and joints normally control this behaviour. The number of joint sets can be readily defined by eye in the walls of the development drives or in the exposed bench stope hangingwall. The total number of joint sets is calculated regardless of the size of the individual joint set trace lengths.

A transecting joint set has some of its trace length actually running across the entire bench hangingwall. Information regarding joint size, as collected by the strip mapping method, can be used to predict the likely number of joint set transecting a planned bench hangingwall (Landmark & Villaescusa 1992). Alternatively, this information can easily be determined by observing the exposed hangingwall of a producing bench stope. This means that the method does not require detailed geotechnical mapping and it represents a very straight forward procedure that can be carried out routinely. The data base is updated with every visit to a producing stope. The following table has been defined:

Table 1. Rating by joint set number and continuity.

Joint set number	Joint sets transecting	HSR component
>5	>3	2
3-5	3	4
2-4	2	6
1-3	1	8
1-2	1	10
1-2	0	12

4.2 Mining induced stresses (15%)

Mining induced stresses significantly affect the stability of exposed hangingwalls and bench stope boundaries at Mount Isa. The HSR specifically refers to the predicted induced stress normal to the orebodies. Once a bench excavation is made, the hangingwall and footwall will deform with the release of stresses at the free surfaces. Experience has shown that higher pre-benching stresses will result in greater off-loading and increased hangingwall relaxation (Beer et al 1983). Furthermore, increased damage from stress re-distributions (detrimental pre-conditioning, cracking, etc.) close to an advancing bench face and brow will be experienced at higher pre-benching normal stresses.

The complex mining geometries at Mount Isa induce changes in the state of stresses both along strike and down dip of the tabular orebodies. Induced stress values can be predicted from suitably calibrated stress analysis computer programs. At Mount Isa, the computer program NFOLD is used routinely to update and model short term and long term extraction schedules. Based on the displacement discontinuity technique, the program was originally developed by Golder Associates. Since purchasing a copy from Golders, Mount Isa Mines has extensively tested and modified the program (Bywater et al 1983). In addition, pre- and post-processing programs have been written to ease data generation and output interpretation. Typically, the NFOLD program is used to model the mining geometries in 6 months, 2 and 5 year production schedules. Areas of high stress or sudden changes in stress levels can be identified. Back analysis and rock mechanics principles have shown that once the rock mass reaches stress levels of around 80 MPa, or when it experiences sudden stress changes of 10 to 15 MPa (at any absolute stress level), it will exhibit signs of failure. Alternative extraction sequences can then be devised or the rate of extraction slowed to allow for stress re-adjustment. The relevance of the induced normal stresses with respect to the bench stope hangingwall stability is summarized in Table 2. The highest rating is assigned to the interval where the induced normal stresses approach in-situ conditions for the current mining depths in the Lead Mine (approximately 1000m).

Even though the NFOLD program provides outputs other than normal stresses, such as dip and strike shear stress and convergence, dip and strike ride displacements, only the normal stress values have been used as the comparison index for bench stability. Prediction based solely on the normal stress component have been seen to be reliable with the results being well supported by observations.

4.3 Blast Damage (15%)

Excessive vibrations or gas penetration from blasting can detrimentally affect the unsupported stope walls. At the time of developing the HSR method, the majority of the blast patterns used in benching were square, with the hangingwall hole fired last on the sequence and charged with a low energy ANFO-based explosive. More recently (over the last 2 years), staggered patterns have been introduced to decrease the expected blast damage experienced by the unsupported hangingwalls. Observations of hangingwall behaviour led to the conclusion that blast damage is more significant than previously considered at the time of the initial development of the HSR method (Villaescusa, 1996).

At Mount Isa, the orebodies have been classified into three different classes. Weak (orebodies having a linear frequency of bedding plane breaks greater than 12), Medium (orebodies having a linear frequency of bedding plane breaks ranging from 4 to 12) and Strong (orebodies having a linear frequency of bedding plane breaks lower than 4).

The effects of blasting on bench hangingwall stability (in the near field) are considered to be controlled by the explosive type; the hole diameter; the ring design (hole angle); the stand-off distances from the stope boundaries as well as the attenuation properties of an orebody (and the immediate hangingwall rock). The overall underlying assumption used is that blast damage (and breakage) across the orebodies is mainly caused by vibrational energy, while the blast damage (and breakage) along the orebodies is mainly caused by the gas energy. This assumption is supported by observations of single shot breakage angles and the pervasive nature of the bedding planes. All monitoring programs to date have been directed to near field damage characterization,

Table 2. Rating for normal induced stresses.

Normal stress (MPa)	HSR component
>90	1
80-90	2
70-80	3
60-70	5
50-60	6
40-50	8
30-40	9
20-30	7
< 20	4

rather than establishing the effects of blasting in the far field.

Table 3 shows the effects of blasting. An assessment of orientation and the stand-off distance of the hangingwall holes with respect to the final bench stope boundary is required.

Table 3. Effects of blasting on hangingwall stability.

Rock type	Hangingwall hole orientation to bedding	HSR component
Weak (13/80, 13, 11 orebodies)	Across	1.5
	parallel	3
Medium (5/110, 12, 14 orebodies)	Across	4.5
	parallel	6
Strong (5/60, 5,7, 8 orebodies)	Across	7.5
	parallel	9

The analysis must also couple a consideration of explosive energy and the type of rock to be blasted. Table 3 is based on a standard 73mm diameter hangingwall hole, charged with Isanol 50, blasted last in the ring sequence and located 0.8m away from the stope boundary (Cameron & Paley 1991).

5 STABILITY OF BENCH STOPES

As stated by Laubscher (1990), an empirical classification must be based on parameter data which can be collected routinely by mine geologists or rock mechanics engineers. One of the key functions of the rock mechanics engineers at Mining Research in Mount Isa is to analyze the extraction sequences using stress analysis programs such as NFOLD. The aim is to predict the induced stress levels during the extraction of every bench in the proposed sequence. The other classification parameters of the bench stability model, such as bedding, jointing and blasting, are readily obtained during routine visits by the planning engineers, geologist or operating personnel as bench extraction progresses.

Each of the four controlling parameters is assigned a value according to its perceived importance. The overall rating for the hangingwall ground conditions is then determined by adding the individual parameter values (See Figure 3).

The relationship between the total adjusted ground condition rating and the unsupported excavation size is shown in Figure 4. In this figure the HSR is plotted against the unsupported hydraulic radii. All of the points shown in the chart were calculated for benches extracted and filled using the following methods: continuous dryfill, hydraulic fill (and pillar recovery), permanent pillars and Avoca fill pillars. The maximum permissible rating in the system is 60. To date calculated HSR values have ranged from 18 to 48 (for the Lead Mine) with the majority of benches rating between 25 and 35. Due to the geometry in the Lead Mine, the majority of these benches have been extracted with hydraulic radii ranging between 2 and 6.5. Points calculated with hydraulic radii in excess of 6.5 are for high lift benches, which are backfilled with hydraulic fill. All of the points have been calculated for steeply dipping (usually greater than 55 degrees) bedded hangingwalls.

HANGINGWALL STABILITY RATING

OREBODY: 12   DATE: 22/3/96
BENCH: 12F6   EXTRACTED FROM: MR34 TO: MR44

	DESCRIPTION	RATING
GEOLOGICAL PARAMETERS		
Bedding Plane Breaks/m: 6	Good	18.5
Number of Joint Sets: 2 - 4		
Joint Set Continuous: 2	Fair	6
INDUCED STRESS PARAMETER		
Normal Stress Levels (MPa): 70 - 80	Poor	3
BLASTING INFORMATION		
Distance of Holes to Hangingwall		
Drilling Horizon: 0.8 m		
Mucking Horizon: 0.5-0.8 m		
Hole Angle: Parallel		
Hole Diameter (mm): 73 mm		
Burden (m): 2m		
Rock Blastability Class	Medium	6
STOPE SIZE AND SHAPE INFORMATION		
Unsupported Height (m): 24 m		
Strike Length (m): 20 m		
Hangingwall Conditions: Stable		
HYDRAULIC RADIUS (m): 5.45	TOTAL	33.5

Figure 3. Hangingwall stability assessment for 12 orebody.

As described by Laubscher (1990), these type of stability charts can be described by three general zones: At Mount Isa, bench stability is described by:

1) A stable zone above the curve, where unsupported bench hangingwalls remain stable with minimum or no dilution. 2) A dilution onset zone, where the outer hangingwalls skin layers start to deteriorate and minor localized

failures of up to 1m deep are observed. These failures do not propagate along the strike length of the bench. 3) A failure zone below the curve, where substantial hangingwall failures, up to 3m deep are experienced along the entire hangingwall plane boundary.

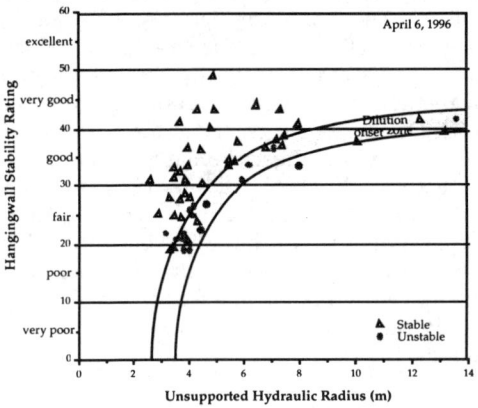

Figure 4. Bench stope stability chart.

Failures are usually arrested up-dip by the appropriate cable support provided on each of the bench sill intervals. Experience shows that permanent pillars are very effective in arresting failures along the strike of the benches.

CONCLUSIONS

An empirical method which is used to predict hangingwall stability has been developed and calibrated, (over the past four years) within the Lead Mine rockmass. The HSR method described in this paper is considered to be applicable to other stratabound steeply dipping deposits following local rating adjustments. This paper outlines recent modifications to this hangingwall stability rating method. The method is used both as a design tool, for mine planning, and as a predictive tool which is used routinely in daily operations. The rating can be routinely updated as extraction progresses on an individual bench, or block basis. Data collection and collation is relatively straight forward.

ACKNOWLEDGMENTS

The author wish to thank the management of Mount Isa Mines Limited for the permission to publish this paper.

REFERENCES

Baczynski, N.R.P. 1974 Structure and hangingwall stability in lead orebodies at the Mount Isa Mine. M.Sc. Thesis, James Cook University of North Queensland.

Beer, G., J. Meek, & R. Cowling, 1983. Prediction of the behaviour of shale hangingwalls of deep underground excavations. *Proc. Fifth ISRM Congress*, Melbourne Australia, D45-D51.

Bywater S., R. Cowling, & B. Black, 1983. Stress measurement and analysis for mine planning. *Proc. Fifth ISRM Congress*, Melbourne Australia. D29-D37.

Cameron, A. & N. Paley 1991. Assessment of blasting to reduce damage in B704 bench stope at Mount Isa Mines. *Proc. Western Australian Conference on Mining Geomechanics*. Kalgoorlie, Western Australia, 375-383.

Greenelsh, R.W. 1985. The N663 stope experiment at Mount Isa Mine. *International Journal of Mining Engineering*, 3, 183-194.

Landmark, J. & E. Villaescusa 1992. Geotechnical mapping at Mount Isa Mines. *Proc. Western Australian Conference on Mining Geomechanics*. Kalgoorlie, Western Australia, 329-333.

Laubscher, D.H. 1990. A geomechanics classification system for the rating of rock mass in mine design. *J. S. Afr. Inst. Min. Metall.*, 90 (10), 257-273.

Lee M.F. 1980. Review of MICAF rock mechanics, Mount Isa Mines Limited, *Technical Report No RES MIN 58*, (unpublished).

Villaescusa, E., L.B. Neindorf & J. Cunningham, 1994. Bench stoping of Lead/Zinc orebodies at Mount Isa Limited. *Proc. MMIJ/AusIMM Joint Symposium*, New Horizons in Resource Handling and Geo-Engineering, Yamaguchi University, Ube, Japan, 351-359.

Villaescusa, E, G. Karunatillake & T. Li 1995. An integrated approach to the extraction of the Rio Grande Silver/Lead/Zinc orebodies at Mount Isa. *Procc. 4th International Symposium on Mine Planning and Equipment Selection*, Calgary.

Villaescusa, E. 1996. Excavation design for bench stoping at Mount Isa Mine, Queensland Australia. *Trans. Instn. Min. Metall. (Sect A: Min. Industry)*, 105, A1-A10.

Reclamation and construction in abandoned mining districts in Japan: Review of practical experiences and countermeasure instruction

N. Kameda & N. Mori
Faculty of Engineering, Kyushu Kyoritu University, Japan

T. Esaki, Y. Jiang & G. Zhou
Faculty of Engineering, Kyushu University, Japan

ABSTRACT: The shortage of adequate construction sites and their difficult location has compelled many engineering structures to be planned in abandoned mining districts in Japan. This paper shows, first, the characteristic of the excavated underground environment, especially the delayed mining damages of cave-in caused by remaining cavities at shallow depth, and, next, the necessary measures against designing and prevention them during and after construction of surface structures. Finally, the risk assesment by using geographic information system for an expressway project passing through an abandoned mining district is studied as an example.

1 INTRODUCTION

In a period of one hundred years until 1960's, Japan had approximately 800 coal mines, but at present only two coal mines are in operation. In mining districts, subsidence due to mining has caused various kinds of damage to the surface affairs, and actually great deal of mining damages have grown into the severe social problem at some time in many mining districts, but the some of the mining damage has been restored at present.

Recently, the effect of old coal working on the deformation and the bearing capacity of their base grounds of newly constructed structures has become into question, because it was obliged to plan many structures in abandoned mining districts for the shortage of adequate construction sites and the difficulty of getting location in the increasing investments of social capital. That is, public buildings, dams, bridges for railroad and expressway, power plants, tunnels and storage tanks have to be built on the ground which has experienced old coal working. Under these circumstance, the ground which experienced underground working is affected considerably by the change of load of surface structure and water level due to construction activities and may suffer further subsidence, cave-in and the decrease of bearing capacity. This behavior of the ground may cause the damage to the surface structures. These phenomena also proceed in the case of earthquake.

In these case, it is necessary and important to predict the ground movement and cave-in in advance when surface structures are designed and constructed.

This paper describes the characteristics of cave-in at shallow depths, and shows some practical preventive measures against their ground movement for secure construction of the surface structures. Finally we introduce the application of geographic information system(GIS), the new technology, to evaluation of the stability of the ground and to manage the preventive measures at the construction site as a case study.

2 CHARACTERISTICS OF CAVE-IN

Fig.1 shows a probability curve of cave-in by the change of the depth of old working from statistics in the concerned authorities. The average depth of cavities which has caused cave-in is 12.5m and most of cave-in has caused at the depth of cavities within 20m. The cave-in came out not only at or just after the time of working but also even after the long elapsed period

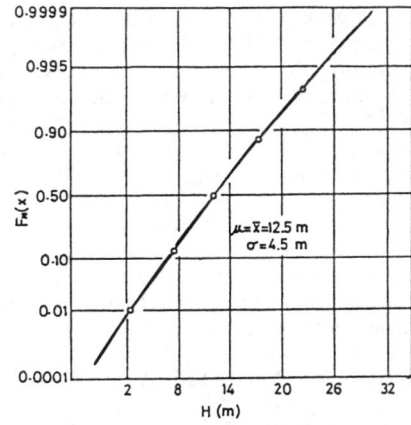

Fig. 1 A probability curve of cave-in($Fn(x)$) by the change of the depth of old working(H).

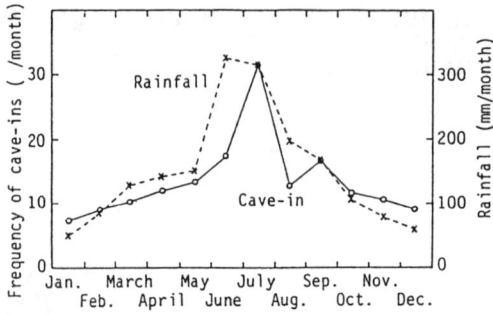

Fig.2 Average monthly frequency distributions of cave-in and rainfall in Chikuho district between 1974 and 1987

Photo.1 Cave-in due to the drainage of ground water for construction work of Sanyo Shinkansen.

from working. Cave-ins continue for a long time after working. As clear records, the lapse time such as 118 years in England are recorded. In Japan, there have been occurred many cave-ins at the unrecorded working areas which was estimated that the working had done at least one hundred ears ago. Fig.2 shows the average monthly frequency distribution of cave-ins, occurring between 1974 and 1987 in the Chikuho district of Kyushu Island together with rainfall. In the district, there existed many coal mines since about 100 years ago, and many cavities remained at shallow depths. Fifty to sixty cave-ins occurred in every year and the frequency increased especially during the rainy season. Photo. 1 shows an example of cave-in due to the drainage of ground water for construction work of Sanyo Shinkansen. In this case, remained cavities in the lignite seam were distributed about 2 m - 3 m under the ground level and cavities were filled up with ground water. When the water was drawn out from a water-filled cavity by the construction work, cave-in occurred at the field near the construction site.

An example of cave-in due to earthquake is also shown. In the Tohoku district, north-east of Japan, many lignite mines have been worked at very shallow depths by the pillar and stall mining method in Kitakami, Miyagi, Mogami, Shounai and Souma coal fields since 1890s. On June 12, 1978, an earthquake of 7.4 magnitude, named as Miyagi-ken-Oki earthquake, occurred about 70 km off the east coast of Tohoku area. Fig.3 shows the frequency of cave-ins, occurring in this district from 1974 to 1987. According to the records of authorities, about 15-20 cave-ins normally occurred every year, but in 1978, 219 cave-ins occurred. This unusual increase was attributed to the effect of earthquake. Especially at Ohhira in the Miyagi coal field, the maximum vertical acceleration was about 100 gals and 22 cave-ins occurred in farmland just after the earthquake.

3 NUMERICAL AND PHYSICAL ANALYSES

The mechanism of cave-in numerically and physically

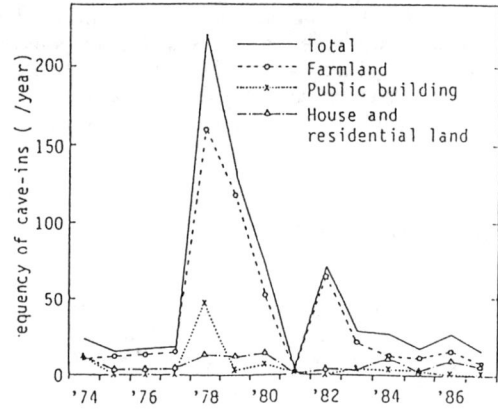

Fig.3 Frequency of cave-ins in the Tohoku district from 1974 to 1987.

have been studied. In numerical method, the incremental load procedure of the finite element method is useful for evaluating whether cave-in will occur or not. The occurrence conclusively depends on the extension of the plastic zones in the roof above the cavities to the surface. But, FEM can not be considered large deformation phenomena. Recently, the large deformation finite difference method is more effective with considering the detailed deformation and failure behavior of cave-in. Fig.4 shows an example of the developing process of failure above a rectangular cavity by using the large deformation finite difference method(Nakagawa et al. 1995).

Numerical calculation has some limits realistically to simulate a complex geotechnical conditions, such as discon-tinuities and/or stratified condition of the ground. In this case, it is convenient to study experimentally by physical model. Many kinds of physical tests have been tried, such as an ager-ager model, trapdoor model, centrifuge model and base friction model to simulate the cave-in phenomena. The base friction technique is one of the most suitable physical model test to simulate the cave-in phenomena

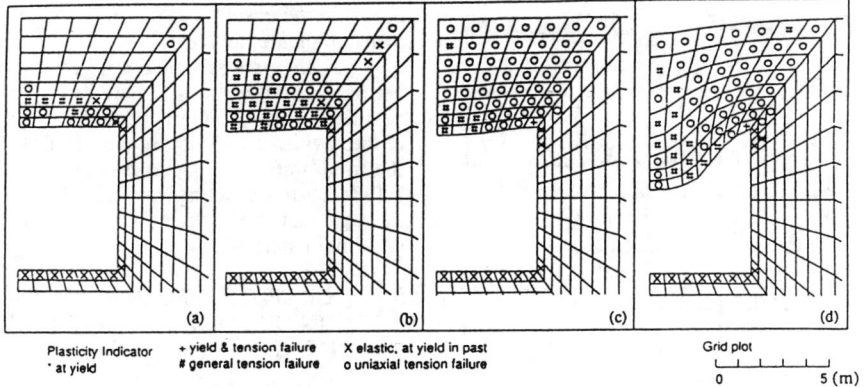

Fig.4 The developing process of failure above a rectangular cavity.

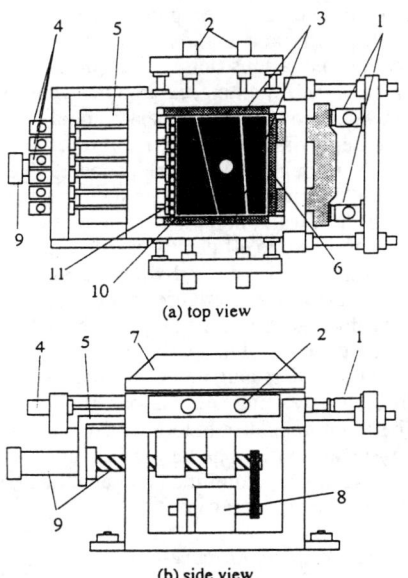

Fig.5 Schematic view of the new base friction apparatus.
1:load cells, 2:lateral loading jacks, 3:rubber bags, 4:vertical loading jacks, 5:frictional plate, 6:barrier beam, 7:cover plate, 8:geared motor, 9:screan gear system, 10:model, 11:loading plates.

which is dominated by the effect of gravity force at the near surface of the ground. The base friction apparatus supplying compressed air over the upper surface of the model can provide quantitative predictions with similitude laws, but stretching of the endless belt and indistinct boundary conditions of model have remained as unsettled problems.

We have developed a new base friction apparatus which adopts a very stiff metal plate coated with uretan rubber in place of the endless belt. The base friction shear force is provided by dragging the plate at a very slow speed across the model under test. Furthermore, this apparatus has vertical and lateral loading system

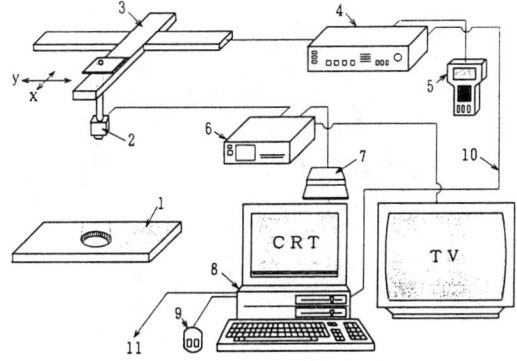

Fig.6 Newly developed picture analyzing system for the new base friction apparatus.
1:model, 2:CCD camera, 3:intelligent actuator, 4: actuator controller, 5:remote controller for actuator, 6: 2-inch video floppy disk, 7:super impose unit and frame memory, 8:personal computer, 9:mouse, 10:RS-232C cable, 11:to digital plotter and printer.

that can provide voluntary vertical and lateral forces to simulate the stress condition at depth. Fig.5 shows the schematic view of the new base friction apparatus and Fig.6 shows the attached image analyzing system. Photo. 2 shows a typical result of the experiment of cave-in using the new base friction apparatus(Nishida 1986,Kameda 1987, Esaki 1993).

4 PRACTICAL PREVENTIVE MEASURES FOR CAVE-IN

When heavy structures are built on the ground which has cavities at shallow depths, we must consider preventive measures against cave-in. The suitable preventive measures must be adopted with taking into account of the following conditions; characteristics of ground (mechanical properties, stratified condition, amount of discontinuities), distribution of cavities(depth, width, filling condition, mining style and

Photo. 2 A typical result of the experiment of cave-in using the base friction apparatus.

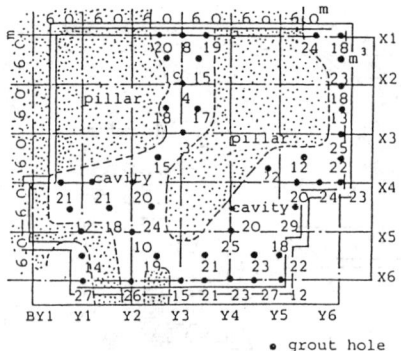

Fig.7 Distribution of cave-ins and the amount of grouting materials

records), surface affairs(amount of load, style of foundation, allowable deformation, social importance)

Our group experienced more than one hundred cases of the design and the execution of works of concrete measures, representative are the Sanyo new railways, Kyushu and Tohoku expressways, four power plants which output are more than 500 MW.

There are 4 types of preventive measures which have been selected depending on the each condition.

(1) Grouting: It will be undertaken only where the cavities are of a relatively small-scale and remain at shallow depth, that is, location and distribution of them are relatively clear, because grouting is very expensive.

(2) Excavation and filling:It is used when the cavities exist at shallower depths. This method is more reliable than grouting.

(3) Piling until below the floor of cavities:It can be used for the soft ground, but may not be used for superposing cavities which extracted multiple layers, particularly in the case of closed multi-layers. In this case, the negative friction on piles must be taken into consideration.

(4) Foundation in the form of slab: The foundation to resist cave-ins is relatively expensive but it is very useful in the case of the cavities which location and distribution are not clear.

Fig.7 shows a practical application of grouting, where a structure was planned to construct above cavities remaining at 8m depth. Although the volume of the cavities were initially estimated at about 800m^3, the total amount of grouting materials came up to 1,173m^3. After grouting, it was confirmed from check borings that the cavities were perfectly filled by grouting materials.

5 GEOGRAPHIC INFORMATION SYSTEM(GIS) TECHNIQUE FOR GEO-ENVIRONMENTAL CONTROL

5.1 GIS Solution

In the recent years, geographic information system (GIS), has attracted extensive attention, and GIS applications in various fields, like land use planning, construction facility sitting, environmental assessment, groundwater modeling, watershed modeling, facility management, disaster prevention etc., have been carried out throughout the world(Zhou 1997).

GIS is defined as " an organized collection of computer hardware, software, geographic data, and personal designed to efficiently capture, store, update, manipulate, analyze, and display all forms of geographically referenced information".

As mentioned above, the obvious characteristics of GIS is that it can solve spatial problems. It is very important to understand how GIS store spatial data.

Generally, by using GIS, various kinds of spatially distributed information are stored as different data base, then by using spatial data analysis function in GIS, different layers are combined and analyzed, and the solution of the problem can be obtained.

5.2 Risk assessment for cave-in

A study of preventive measure for cave-in with respect to a Sanyo expressway project at Onoda area in Yamaguchi-Pref. is given as a practical example. This area has been excavated by many coal mines about until 40 years ago, and many cavities remain at a shallow depths. There are two coal layers(named layer 3 and layer 5 which are valuable for extraction) under the ground. For the depth of these cavities research by GIS, we input into the GIS data base, the ground level information by the geographic feature and the distribution of coal mined area.

The basic research procedures are as follows.
(1)Terain analysis
The research area is 1 km long and 100 m wide, a rectanglar area along expressway center line. The terain countor scaled 1:1000 and benchmarks have been scanned and line coverage and point coverage have been established. Then by using TIN(triangular irregular network) function in GIS, the digital elevation

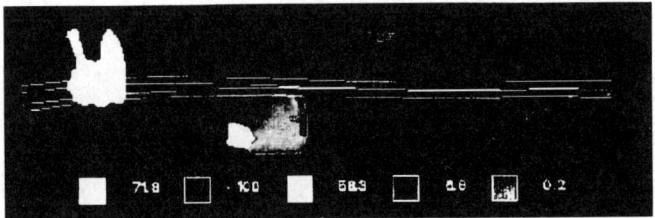

Fig.8 The distribution of cave-in probability along highway(before construction).

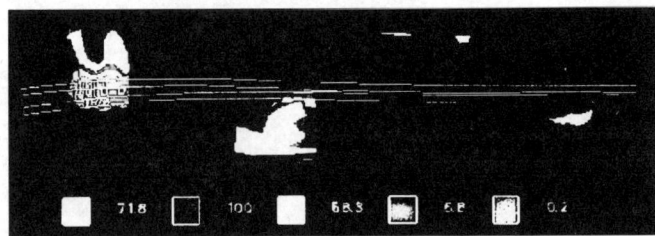

Fig.9 The distribution of cave-in probability along highway(after constructin).

model (DEM) is established and the land elevation in any point of research area can be obtained easily.

(2)The distribution of coal layers

In this area, two layers (layer 3 and 5) are existed and excavated. Firstly, the excavated coal layers are input as polygon coverage, then by using data convertion, polygon data are changed to grid data, by using grid data analysis function and terain data, the depth from surface to excavated space is calculated and stored in every cell of grid.

By above operation, the spatial distribution and depth of excavated coal layers are grasped quantatively.

(3)Cave-in risk assesment

According to the statistics data on cave-in occurrence probability and depth of shallow space, the cave-in occurrence probability in Onoda area is calculated. First, the relation of cave-in occurrence probability and depth to excavated shallow space is input to GIS, and using grid analysis function, the cave-in occurrence probability in the area is calculated and displayed, the results provide very important information to expressway construction.

Fig.8 and Fig.9 show the cave-in occurrence probability map for before and after construction of expressway against layer 3. From these figures, we will predict the variation of cave-in risk for before and after construction of expressway.

6 CONCLUSIONS

It was revealed that the mechanism and actual occurrence of cave-in phenomena and rainfall, earthquake and surcharge on the ground surface accelerate the occurrence of cave-ins. Since some preventive measures for cave-in have been already proposed, one of these technique should be adopted according to the situation of each project. Furthermore, a new technology, GIS is successfully applied in environmental impact assessment due to mining damage at a shallow depths.

REFERENCES

T.Nishida, T. Esaki and N.Kameda : A development of the base friction technique and its application to subsidence engineering, *Proc. Int. Symp. on Engineering in Complex Rock Formations*, pp.386-392, 1986.

N.Kameda, T.Esaki, T.Nishida and T.Kimura : Prediction of the behavior of roof cavities in discontinuous bedded rock using the new base friction technique, *Proc. 28th US Symp. on Rock Mechanics*, pp.789-796, 1987.

T.Esaki, A.Aikawa, Y.Jiang and Y.Mitani : Development of a new base friction technique and its application to geotechnical study, *Int. Symp. on Assessment and Prevention of Failure Phenomena in Rock Eng.*, pp.403-408,1993.

M.Nakagawa, Y.Jiang and T. Esaki : Application of large strain analysis for prediction of deformational behavior of tunnels in soft rock, *Proc. of the 26th Symp. of Rock mechanics*, pp. 515-519, 1995.

G.Zhou : A coupled groundwater flow-land subsidence simulation and quantitative evaluation of its influence to flooding by using GIS, *PhD. thesis*, Kyushu University, 1997.

Underground injection technology for protection of the Wieliczka Salt Mine in Poland

Andrzej Gonet, Jerzy Stopa, Stanislaw Stryczek & Stanislaw Rychlicki
University of Mining and Metallurgy, Krakow, Poland

ABSTRACT: Wieliczka is one of the oldest salt mines in the world and it is listed by UNESCO as an object of World Cultural Heritage. The attraction is the mine itself, offering a few kilometres long tour, 13-th century Salt Mine Museum and the Crystal Chamber with absolutely unique crystals measuring up to 40 m. The Wieliczka Mine has been producing salt continuously for over 700 years. In 1992, this exceptional salt mine was in danger of being flooded by catastrophic ground water inflow at the rate of 20 cu.m/hr. The authors of the paper were involved in the designing of the rescue operation. The method used successfully to stop the water inflow was based on the underground injection wells system. The wells number, length, locations and operating parameters were optimised using the method of mathematical modelling and computer simulation to reduce risk and cost. This paper describes some geological, technical and mathematical aspects of the rescue system designing and exploitation. The mathematical model (with time depended viscosity) and new analytical solutions describing the injected fluid movement in complicated geological conditions are presented and discussed. In the paper, both the mathematical formulae and the results of numerical computations compared with the field measurements are presented. The results of laboratory experiments to find the optimal dispensing of the injected fluid are also reported. The paper concludes with the results of field measurements showing the protection system's effectiveness. The results presented here may be of interest to the engineers and scientists working in the area of underground injection, rock mass stabilisation and ground water flow control.

1. INTRODUCTION

The Wieliczka Salt Mine is among the oldest functioning mines in the world. Exploitation of salt deposits which has been carried out for over 700 years led to the construction of about 2000 chambers of about 7.5 million m^3 total capacity and over 200 km of dog heading on 9 levels. Due to the Wieliczka Salt Mine's exceptional beauty and historical character, the Mine was added to UNESCO's World Cultural and Natural Heritage list. A saline museum can be found in one part of the mine and a sanatorium located on level V.

2. DESCRIPTION OF THE SALT DEPOSIT AND OF THE UNDERGROUND WATER THREAT

The salt deposit is Miocene background consisting of several different types of salt. Formations of green lode salt lie on the oldest salt lodes. A complex of bronze salt, which makes up a group of salt bank with packs of shales, mudstones, sandstones and anhydrites are located higher. The uppermost part of the deposit consists of marly siltstones and salt break. The salt deposit is surrounded by a silty - gypsum lag which borders on the southern side formations of Carpathian flysch of a low water saturation, and on the northern side "chodowieckie" layers made from sandstone, siltstone and mudstone. Quaternary and Miocene water-bearing layers are distinguished around the salt deposit. The latter create around the mine a huge collector of water comprising the main water threat for the existing galleries. As a result of many years and multiple types of mining activities, several times the silt - gypsum screen of the deposit was disturbed. More than once it caused the appearance of damp patches in the mine. It is estimated that starting from the XIV century as much as 5,7 mln m^3 of water, which was used for the salt exploitation of the salt deposit, have flowed into the mine. Most recently, a very serious inflow to the mine appeared on April 13, 1992. This occurred on level IV of the mine in the central part at a depth of 170 m beneath the surface level (exploitation of this gallery began around 1805 and was finished in 1912). It is estimated that the maximum inflow was about 700 m³/day at a saturation of 35-40 g NaCl/dm³. The inflow was of a pulsating character and brought with it substantial amounts of silt-sand elements, which significantly hindered pumping of water and removal of fixed parts from the end of the gallery. Given this situation in the mine, rescue measures were taken whereby the museum and sanatorium had to be closed to visitors.

3. PROTECTION EFFORTS.

Actions were taken to limit the inflow by controlling the water intake. With every decrease in the water inflow, rescue efforts proceeded in the mine. During the first phase of the work, a simple dam and two dams were built between which a silt-concrete plug was made. A pipeline to control water intake from the leakage was located inside of the plug. At the same time two rescue- exploration wells were drilled. Their purpose was to better recognize the geomechanical and hydrogeological conditions. One of them was the large-diameter well, through which water which was flowing in the direction of the gallery Mina was pumped out to the surface. Thanks to this action, a depression was made on the foreland of the mine and the inflow of water to the mine was limited. At this time at the end of the "Mina" gallery, a tri-segmental dam made of wood, silt and two cement sections was made. Unfortunately, during a test of the dam water-tightness, water leakage in the gallery in front of the dam appeared. Fissures and permeable cracks of the formation were paths for water inflow to the mine. To ensure miners' safety, a water-tight screen was built from sheet metal inside the casing. tP-1 from dam T4 to the water dam covering a length of 34 m. Next the formation was strengthened and sealed using well-injection method from dam T4 toward the water dam. Irrespective of this, two drainage wells through the water dam were drilled. The wells have retained the water on the foreland of the salt deposit and thus allowed it to be directed to the mine in a controlled manner and further used for the well-exploitation of salt. Zonal injection work (so-called "fan-shaped" injection wells) was carried out along the end of the "Mina" gallery. At present efforts to strengthen and to seal the formation are being continued from the water dam in the silt-gypsum screen of the salt deposit toward the north in the so-called "sealing dome". Various slurries were used to fill caverns located in the formation.From a technological and economic point of view, mineral hydraulic slurries possess the most advantageous parameters. Their usefulness has been confirmed in practice. Fig. 1 shows water intake rate vs. time. A mathematical model was drawn up for the slurries and injection wells to permit the logical design of technology for the strengthening and sealing of formation.

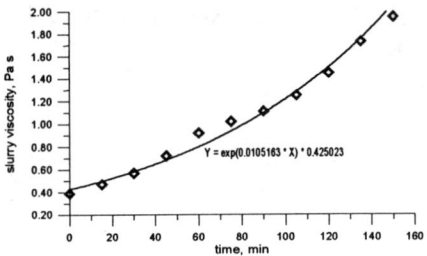

Figure2. Slurry viscosity vs. time, laboratory measurements

4. MATHEMATICAL MODEL OF SLURRY MOVEMENT DURING UNDERGROUND INJECTION.

Two phenomena may be important in the modeling of slurry movement in porous media. The first is permeability and porosity reduction by the invasion of solid particles into pores. The second is the time-dependent slurry viscosity which yields continuous increases in flow resistance and consequently stops the flow. Due to the slurry composition used at Wieliczka and specific pore structure containing fractures and fissures, the second phenomenon was more important in the presented case. Consequently, the first phenomenon was neglected in the present mathematical model. The typical dependence of slurry viscosity vs. time is shown in Figure 2.

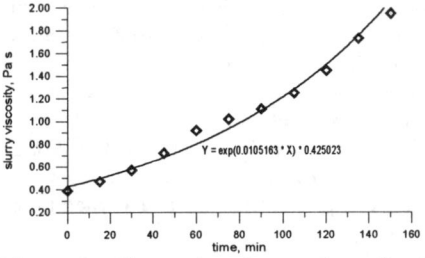

Figure. 2 - Slurry viscosity vs. time after laboratory measurements.

The empirical relationship of the slurry viscosity μ_1 vs. time t spending from the moment of slurry preparation to the time of measurement is:

$$\mu_1 = \mu_1^0 e^{at} \qquad (1)$$

Having assumed that the slurry movement through fissured porous rocks is governed by the slightly compressible fluid equation, the "abrupt interface" displacement model was adopted. The resulting equations for homogeneous infinite reservoir in radial geometry are:

$$\frac{\partial p_1}{\partial t} = \chi_1(t)\frac{1}{r}\frac{\partial}{\partial r}(r\frac{\partial p_1}{\partial r}) \text{ for } 0 < r < l(t) \quad (2)$$

$$\frac{\partial p_2}{\partial t} = \chi_2 \frac{1}{r}\frac{\partial}{\partial r}(r\frac{\partial p_2}{\partial r})$$
for $l(t) < r < \infty$ \quad (3)

where

$$\chi_1(t) = \frac{k_i}{\phi \mu_1(t) c_{f1}} = \frac{k e^{-at}}{\phi \mu_1^0 c_{f1}}, \chi_2 = \frac{k}{\phi \mu_2 c_{f2}}$$
for $l(t) < r < \infty$ \quad (4)

The $r<l(t)$ refers to the slurry injected zone, and $r>l(t)$ refers to the water saturated zone. The unknown function $l(t)$ describes the slurry/water interface position.

The boundary and initial conditions are:

$$-2\Pi h \frac{k}{\mu_1^0} \cdot \lim_{r \to 0}(r\frac{\partial p}{\partial r}) = Q \quad (5)$$

$$p_2(r,o) = p_2(\infty,t) = p_e \quad (6)$$
$$p_1(l,t) = p_2(l,t) \quad (7)$$

$$\frac{k}{\mu_1(t)}\frac{\partial p_1(l,t)}{\partial r} = \frac{k}{\mu_2}\frac{\partial p_2(l,t)}{\partial r} \quad (8)$$

$$\phi \frac{dl}{dt} = -\frac{k}{\mu_1(t)}\frac{\partial p_1(l,t)}{\partial t} \quad (9)$$

$$l(0) = 0 \quad (10)$$

Introducing the new variable:

$$\tau = \int_0^t \chi(u)_1 \, d \quad (11)$$

assuming: $\mu_1(t) >> \mu_2$
and putting $\chi_1/\chi_2 = \varepsilon \to 0$
the self-similar solution of the form:

$$p_i = p_e \cdot f_i(\sigma) \quad (12)$$

$$l(t) = \alpha \cdot f\sqrt{\tau} \quad (13)$$

where:

$$\sigma = r/\sqrt{\tau} \quad (14)$$

may be found.

Using (11) - (14), the problem is reduced to the system of ordinary differential equations which may be solved analitically.

Solving that system one may verify that α is the solution to the algebraic equation:

$$\alpha^2 e^{\alpha^2/2} = \frac{Q\mu_1^0 c_{f1}}{\Pi k h} \quad (15)$$

which may be solved numerically.

For small α:

$$\alpha \approx \sqrt{\frac{Q\mu_1^0 c_{f1}}{\Pi k h}} \quad (16)$$

From (13), the radius of the injected slurry zone is:

$$l(t) = \sqrt{\frac{Q}{\Pi h \varphi a}(1 - e^{-at})} \quad (17)$$

The maximum effective injection radius is:

$$\lim_{t \to \infty} l(t) = \sqrt{\frac{Q}{\Pi h \varphi a}} \quad (18)$$

CONCLUSION

The penetration of water to the Wieliczka Salt Mine on level IV on April 13, 1992 comprised a catastrophic threat for the mine. Well injection was used in the form of so-called water-tight ring and sealing dome allowing for the effective strengthening and sealing of the formation/orogen in the area around the leak. The correct choice of slurry for actual geological, geomechanical and hydrogeological conditions is an important factor deciding about the success of security efforts for each salt mine. Thanks to their effective implementation in the case of Wieliczka Salt Mine, a most beautiful and unique on a world scale historical monument has been preserved for future generations.

NOMENCLATURE:

- a, μ_1° — empirical constants
- c_f — fluid compressibility
- k — permeability
- p — pressure
- t — time
- r — spatial coordinate
- $l(t)$ — radius of slurry saturated zone
- μ — viscosity
- h — thickness
- Q — injection rate
- φ — porosity

SUBSCRIPTS:

- 1 — slurry (displacing fluid)
- 2 — displaced fluid
- e — well influence zone

A study on the mechanism of chimney subsidence

H.K. Moon, D.H. Huh & B.C. Kim
School of Geosystem, Environment and Civil Engineering, Hanyang University, Seoul, Korea

ABSTRACT: The ground subsidence due to underground mining, in particular, the chimney caving which forms sinkhole type subsidence at ground surface, is investigated. The methods employed in this study are the limit equilibrium and bulking factor approaches.
Analyzed and presented in this paper are the critical depth of protective coal seam, critical span of stopes, and maximum possible height of roof caving for various shapes of caved zone. The effects of bulking factor on the progression of vertical chimneying and the approximate estimate of surface subsidence are discussed.

1 INTRODUCTION

Underground excavation in mining and tunnelling induces ground movement (deformation) toward the openings. When such movement is small as in the case of elastic deformation, the rock mass surrounding the opening may reach a stable state of equilibrium. In some cases, the rock mass around openings may undergo plastic yielding and unstable (progressive) failure. Ground subsidence occurs, when such zone of progressive failure extends in the rock mass and migrates toward ground surface. In general, ground subsidence can be classified into two: One is the continuous trough-type subsidence that occurs usually in longwall mines, and the other is the discontinuous subsidence which includes the crown holes, chimney caving, sinkholes, piping, funnelling and plugging.

In Korea, the ground subsidence problem due to the increasing number of abandoned coal mines became serious. Recently, the sinkhole type subsidence occurred in many abandoned mines has raised an urgent stability question on the nearby railroads, bridges and buildings.

The procedure analyzing the trough subsidence is well investigated by the National Coal Board (1975). Comprehensive research and analysis on the ground subsidence are given by Wittaker and Reddish (1989) and Kratzsch (1983). Compared to the great research efforts concentrated on the trough subsidence, the study on the mechanism of discontinuous subsidence has not attracted much attention in the past.

This study is mainly concerned with the prediction and control of such small scale but highly hazardous subsidence. The specific goals of this study are to: (i) determine the depth of an inclined coal seam to be left to prevent plugging, (ii) analyze the effects of bulking factor on the roof caving and chimney subsidence, (iii) predict the maximum height of roof caving for different shapes of caved zone. The analyses are based on the limit equilibrium and bulking factor approaches.

2 LIMIT EQUILIBRIUM ANALYSIS

The limiting equilibrium approach is useful in estimating the caving conditions of chimney subsidence, especially the plug subsidence, in case that the dilation on the slip surface is negligibly small. The limit equilibrium analysis of a vertically sided chimney caving is described in Brady & Brown (1985). In this paper, the same method is applied to an inclined coal seam shown in Figure 1. It is assumed that the coal seam of width a, depth z and dip α is left to protect the stope underneath.

The available shear resistance mobilized by the linear Mohr-Coulomb shear strength criterion can be expressed as follows:

$$R = \int_{\Gamma_1} [c + \gamma_r z (k \sin^2\alpha + \cos^2\alpha) \tan\phi] \, dA$$

$$+ \int_{\Gamma_2} [c + \gamma_r z (k \sin^2\alpha + \cos^2\alpha) \tan\phi$$

$$+ a \gamma_c \sin\alpha \cos\alpha \tan\phi] \, dA \qquad (1)$$

where R is the shear resistance, Γ_1 and Γ_2 denote

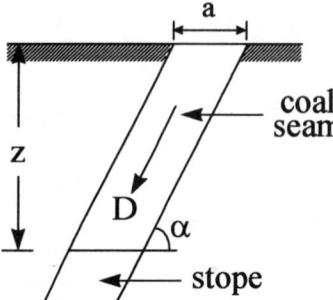

Figure 1. Schematic representation of an inclined coal seam.

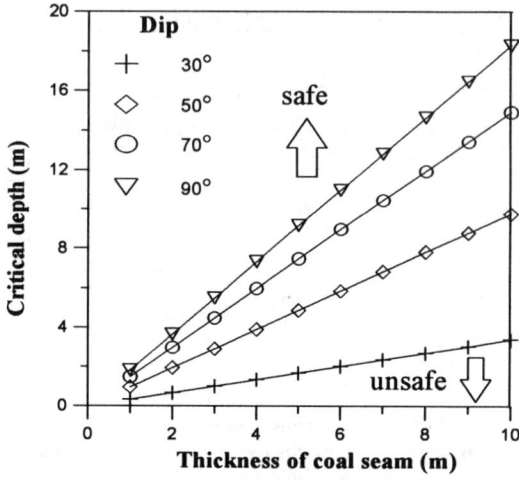

Figure 2. Variation of critical depth with coal seam thickness for various seam inclination.

the upper and lower slip surfaces, z and α are the depth and dip of the seam respectively, c and φ are the cohesion and friction angle of the slip surface respectively, k is the lateral earth pressure coefficient, i.e. the ratio of the horizontal stress to the vertical stress, γ_r and γ_c are the unit weights of the rock and coal respectively. The driving force due to the weight of the coal seam is determined by

$$D = \gamma_c\, az \sin\alpha \qquad (2)$$

where D is the driving shear force causing slip on the coal-rock interface. The limit equilibrium state of R = D results in the following expression for critical depth of protective coal seam.

$$z_{crit} = \frac{a\gamma_c \sin^2\alpha - 2c - a\gamma_c \sin\alpha \cos\alpha \tan\phi}{\gamma_r (k\sin^2\alpha + \cos^2\alpha)\tan\phi} \qquad (3)$$

Figure 2 shows the variation of critical depth with increasing seam thickness, where c = 0, φ = 20°, γ_r/γ_c = 1.5 and k = 1 are assumed. In the case of average seam thickness 4 m which is common in Korea, the relationships between critical depth, lateral earth pressure coefficient and seam inclination are given in Figure 3.

Bétournay et al. (1994) also presented the limit equilibrium equation for chimney caving due to upward progression of symmetric circular arcs. Their analysis provides the relationships between the stope span, rock mass cohesion, and factor of safety for chimneying as shown in Figure 4. The critical span at FS = 1 is approximated by

$$L_{crit} = 0.1 c_m \qquad (4)$$

where L_{crit} is critical stope span in meter, and c_m is rock mass cohesion in kPa.

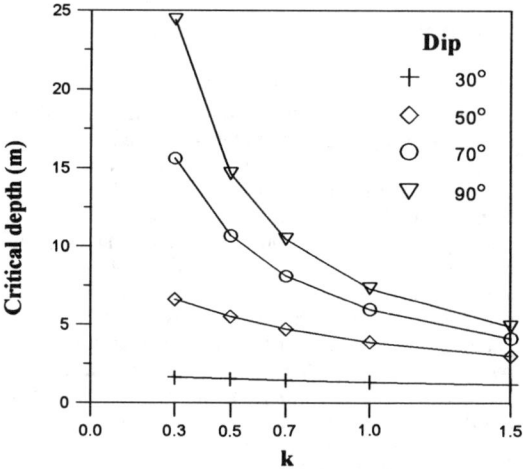

Figure 3. Relationship between critical depth and lateral earth pressure coefficient.

3 BULKING FACTOR APPROACH

It has been well recognized in subsidence mechanics that the chimney caving propagates upward until it forms a stable arch or until the disintegrated rock fills the cavity completely. The bulking associated with chimney caving consists of the volume increase due to fracturing of solid rock and void increase due to loose accumulation of fragmented rock in the mine cavity. A sample model of roof caving is shown in Figure 5.

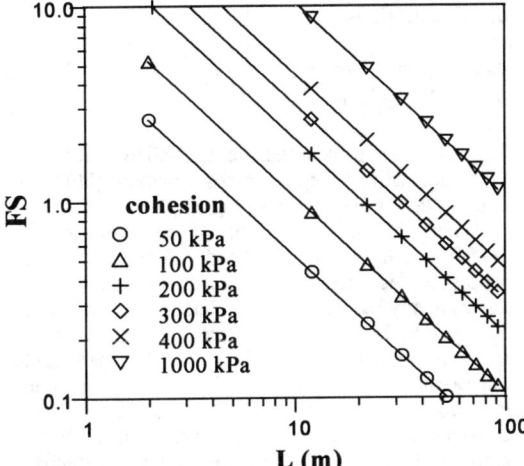

Figure 4. Relationship between stope span (L) and factor of safety (FS).

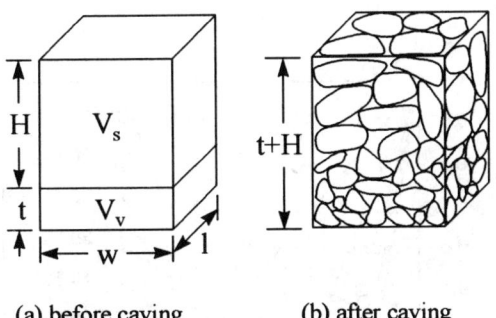

(a) before caving   (b) after caving

Figure 5. Idealized model of chimney caving.

Bulking factor is defined as follows.

$$B = \frac{V_f - V_i}{V_i} = \frac{(V_s + V_v) - V_s}{V_s} = \frac{V_v}{V_s} \quad (5)$$

where B is bulking factor, V_i is initial rock volume before caving, V_f is the final volume after caving, V_s is the volume of solid rock, V_v is void volume, i.e. the volume of mine opening before caving. In order to determine the maximum height of caving progression, one must know in advance the shape of the caved zone. Since it is difficult to predict the exact geometry of the caving, we tested in the following a variety of caving shapes for determination of the relationships between the maximum caving height (H), the height of mine opening (t), and the bulking factor (B).

H/t = 1/B ⋯ shape (1): rectangular prism; circular cylinder.

H/t = 1.5/B ⋯ shape (2): ellipsoid with circular crosssection; ellipsoid with elliptical crosssection; hemisphere.

H/t = 2/B ⋯ shape (3): wedge; paraboloid.

H/t = 3/B ⋯ shape (4): circular cone; rectangular cone; pyramid.

H/t = 1/B + r/3t ⋯ shape (5): circular cylinder with a hemisphere cap of radius r.

Figure 6 shows the variation of maximum height of caving progression with increasing height of opening for a bulking factor of 45 %. The maximum height increases as the shape of caved zone changes from a prism (or cylinder) to a cone. In view of the usual shape of stable pressure arch reported in rock mechanics, the ellipsoid, hemisphere and paraboloid grouped in shape (2) and shape (3) seem to be likely cases. The relationship between the bulking factor and the caving height normalized to opening height is shown in Figure 7. The extent of roof caving is significantly decreased by a small increase of bulking factor in its relatively low range.

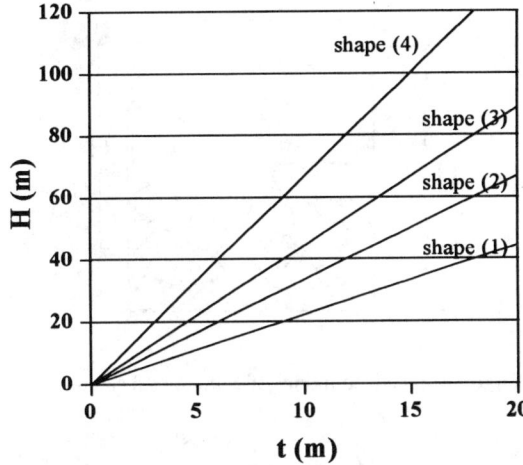

Figure 6. Caving height vs. height of the opening for a bulking factor of 45%.

Apart from the roof caving contained in the overburden by the formation of a stable pressure arch, another possibility is that the caving propagates to ground surface resulting in a chimney subsidence at the surface. The occurrence of such

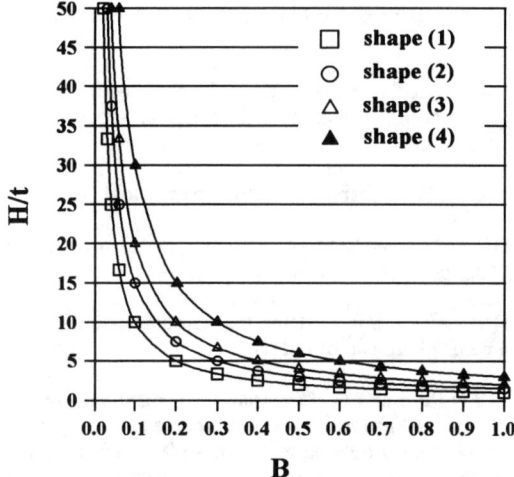

Figure 7. Variation of the maximum height of roof caving with bulking factor for four representative shapes.

subsidence can be analyzed using the bulking factor approach. Using the simple model shown in Figure 8, bulking porosity of the caved rock can be expressed as follows.

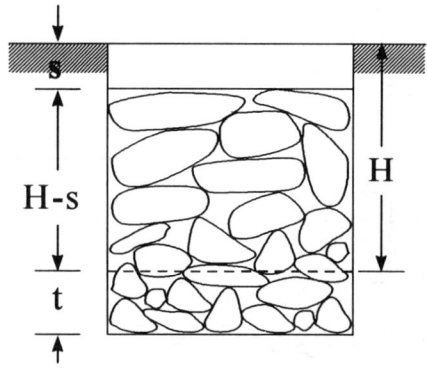

Figure 8. Idealized model of chimney subsidence.

$$n = \frac{V_v}{V_t} = \frac{A(t-s)}{A(t-s)+AH} = \frac{1-\frac{s}{t}}{1-\frac{s}{t}+\frac{H}{t}} \quad (6)$$

where n is bulking porosity, V_v is void volume, V_t is total volume of the caved zone including the opening volume, A is the area of the caved zone, t is the opening height, s is subsidence at the surface, and H is depth to the roof of the opening. The expression for the chimney subsidence normalized to the opening height is:

$$\frac{s}{t} = 1 - (\frac{n}{1-n})(\frac{H}{t}) = 1 - B(\frac{H}{t}) \quad (7)$$

The relationship between the normalized subsidence (s/t), normalized depth of the opening (H/t), and bulking factor are shown in Figure 9.

When the bulking factor of an overburden is 10 % (the lower limit), the depth to the opening should be less than 10 times the opening height to generate any subsidence at the surface. If the bulking factor is 50 %, no surface subsidence will occur as far as the opening is at a depth greater than double the opening height. But, if the opening is steeply inclined stope, such optimistic prediction is not possible. For instance, in the case of a 50 m high stope with an overburden of 50 % bulking factor, a surface subsidence as much as 5 m can occur, if the depth to the roof of the stope is 90 m. Table 1 shows the range of bulking factor for rocks and coal reported in the literature. It is difficult to determine the exact value of bulking factor for in-situ rock mass.

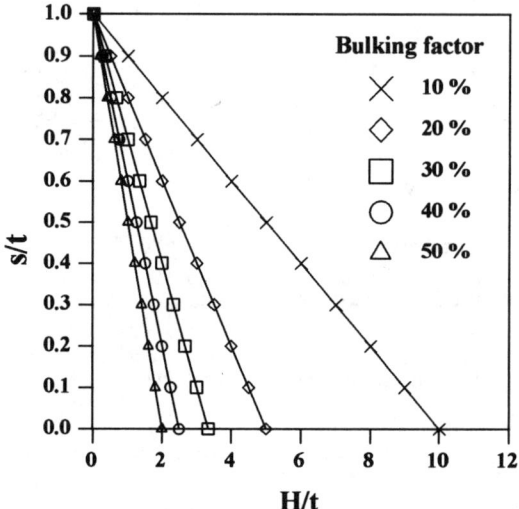

Figure 9. Relationship between surface subsidence and the depth of opening for various bulking factor.

4 SUMMARY AND CONCLUSION

Abrupt sinkhole type subsidence is frequently observed in many abandoned mine areas in Korea, where more than 300 coal mines have been shut down since 1989. The Korea Coal Industry Promotion Board has made every efforts to protect the national infrastructure (e.g. railroads and

Table 1. Bulking factor for various rocks

Rock type	Bulking factor
Laminated mudstone	10 % ~ 20 %
Massive sandstone	≥ 50 %
Weathered rock	30 %
Weak rock (shale, sandstone)	40 %
Medium strong rock (limestone, porous andesite)	60 % ~ 62 %
Strong rock (granite, andesite)	65 % ~ 70 %
Very strong rock (quartzite)	100 %
Coal	40 % ~ 50%
Shale	40 % ~ 50%
Sandy shale	50 % ~ 100 %

bridges) from ground subsidence. As part of the Board's efforts to predict and control the subsidence, the study on the mechanism of chimney caving is conducted first.

The limit equilibrium analysis and bulking factor approach are employed in this study. First, the determination of the critical (minimum) depth of inclined coal seam for preventing plugging subsidence is based on the limit equilibrium method. The principal parameters in this case include the frictional and cohesive property of coal-rock interface and the state of stresses expressed in terms of lateral earth pressure coefficient. The risk of vertical chimney caving over a stope is evaluated by calculating the factor of safety for various span. In this case, the major parameter is rock mass cohesion.

Using the bulking factor analysis, the maximum possible height of roof caving is determined for five different groups of caving shape. For this analysis, the geometry of caved zone and the size of underground openings must be known in advance. The relationships between the surface subsidence, bulking factor and depth of opening are obtained. In order to predict the possibility of sinkhole due to chimney caving, one must know the bulking factor of rock mass, the location and geometry of underground openings, and the shape of caving progression. A better understanding of subsidence mechanism is essential not only to make an efficient plan for subsidence control, but also to select reliable protective measures for subsidence prone area.

ACKNOWLEDGEMENT

Financial support for this study from the Korea Science and Engineering Foundation (KOSEF) under grant number 961-0407-022-2 is gratefully acknowledged. The authors are especially grateful for financial and technical support provided by the Korea Coal Industry Promotion Board.

REFERENCES

Bétournay, M.C., Mitri, H. S. & Hassani, F. 1994. Chimney disintegration mechanism of hard rock mines. *Proceedings 1st NARM*: 98S7-996.

Hoek, E. & Brown, E. T., 1980. *Underground excavations in rock*: 527: The Institution of Mining and Metallurgy.

Kratzsch, H. 1983. *Mining subsidence engineering*. 543: Springer-Verlag.

Karfakis, M. G., 1993. Residual subsidence over abandoned coal mines. *Comprehensive rock engineering* Vol. 5-I: 451-476.

National Coal Board, 1975. *Subsidence engineers' handbook*: 111.

Whittaker, B. N. & Reddish, D. J., 1989. *Subsidence: occurrence, prediction and control*. 528: Elsevier.

Underground constructions: Estimation of their influence on the earth surface deformations and stability of buildings in Odessa, Ukraine

K.K.Pronin, E.A.Cherkez & V.I.Shmouratko
Odessa State University, Ukraine

ABSTRACT: Under the territory of Odessa, there are two kinds of underground cavities: a labyrinth of artificial quarries, and karst cavities. The age of quarries is over 100-150 years. Therefore, recent decades have seen deterioration in their mining-technical condition. The workings stability depends on their size, depth of installation, geological structures of areas, lithological composition and physical-chemical characteristics of limestones. Data analysis on the deformations in workings and buildings on the earth surface showed that the risk of disasters especially grows when the coefficient of total amount of workings beneath the city territory is more than 0.2, while their depth is less then 15m, and their width is about 4m.

1. INTRODUCTION

Odessa is a relatively young city — 203 years of age. The construction material for the overwhelming majority of urban buildings is regularly cement the Pontian shell limestone. This limestone was extracted directly from under the city. As a result, a vast labyrinth of underground quarries was generated under Odessa. Its total extent makes about 2,500 km. Quarries are often two-level. At the time driving and research, various types of natural karst cavities were found in the Pontian limestone quarries.

The site described in the article is a site of conventional quarrying of the construction limestone. Its extraction from underground quarries started on the site in the 1840's. Quarrying was carried out most intensively in the 1870-80's. Some fragments of underground workings were attached a number of times according to different techniques. In other quarrying sites, mining stopped and restarted time after time.

As to the surface constructions, some were built in the 1870-80's and still exist; other buildings were repeatedly reconstructed; still others were pulled down and, on their locations, new buildings were constructed. Sizes of buildings, number of storeys, bases types — all these vary a lot from one location to another. The above-mentioned circumstances are the reason for complex interaction of underground workings and surface constructions. This is why the question of the influence of various types of underground cavities on the bed-rock massif stability, on deformations of buildings and constructions, has been of extreme urgency for the city.

2. GEOLOGICAL STRUCTURE

The geological structure of the site under research is typical for the whole city territory. Underground cavities (quarries and karst cavities) are located in the Pontian limestone. Therefore we shall consider the site geological section, starting from the Meotian deposits on which the Pontian limestone lies.

The Meotian deposits (N_1m) have 50-60 m thickness. They consist predominantly of dense clay of grey-green colour. They contain interlayers and lenses water-rich sand. Clay is the Pontian aquifer aquifuge.

The Pontian limestone layer (N_2pn) has a complex structure. Its properties change in the vertical direction, as well as in lateral. The Pontian limestone formation can be divided into four layers. The first layer — the bottom-most — is the flaggy limestone. Its thickness makes 0.2-1.0 m. The second layer consists of regularly cemented shell-limestone (the 'construction' limestone). Its thickness is 4.5-7.9 m. It is in this layer that quarries were built. Thus the sizes and quantities of quarries depend on the limestone quality. The third layer is formed by the heavily recrystalised shell-limestone. The fourth layer — the top-most — is mostly the flaggy-debris limestone. The general thickness of the third and fourth layers is 5.0-5.6 m. In the upper limestone formation rose marl up to 0.3 m thick can be found).

The Pontian aquifer is present in the limestone on the whole city territory. It is supplied predominantly with the upper aquifer water downflow. On the site, the second Pontian limestone layer is partially moistened. The water level has increased 1.2 m in the recent two years.

The Pontian limestone is blocked with dense red clay (N_{2-3}) with gypsum inclusion and carbonates. Clay is the aquitard for underground waters. The red clay thickness is 3-6 m.

Loesses and loess-like loams (Q_{1-3}) of various consistencies are located in the upper layers. Their thickness reaches 10-12 m. The loess rock often contains the Quaternary aquifer. It was generated at the expense of communications leakage.

Flexures and other structural heterogeneities are found in relief of the Pontian limestone bottom and roof.

The limestone is infringed by systems of fractures running in various directions. The angle of fractures' dip makes 70...90°, their main stretch — 320...360°, 20...30°, and 60...80°. The fractures are of different width (from several mm to 50 cm).

Sometimes they are filled with the limestone lumps, more often — with red clay. The fractures are especially noticeable in the limestone deposited in close vicinity from ravines or the sea coast. The distance between fractures increases towards the watershed. Statistical analysis shows that north-west directed fractures prevail.

Karst formation in the Pontian limestone occurred most intensively along the fractures directed the same way. These fractures are approximately 30-50% wider that those directed north-east. It does not depend on the degree of karst density in the limestone and sites of underground working fields.

The fractures in the limestone form fracture zones. We have carried out the analysis of distribution of the index of specific fracture density in underground workings on several sites. To determine directions and width of fractures, the mine-survey map of workings on the scale 1:500 was used.

The density of fractures was calculated using the 'sliding window' technique. The analysis of received results has allowed to establish the following facts. Within each site, fracture zone are oriented at the right angle to each other (NW-SE and NE-SW). The width of zones makes 40-60 m, and distance between them is approximately 100-120 m.

The physical-mechanical properties of the Pontian limestone depend on its jointing. The size of resistance to the limestone squeezing ranges widely. Increase in the limestone strength towards watersheds and decrease in their strength close to large ravines are the characteristic features of the limestone. On separate sites of the territory of Odessa, the limestone strength in dry condition changes from 0.5 to 2.0 MPa. In completely saturated condition the limestone strength is reduced 1.5-2 times.

Jointing and space variability of physical-mechanical properties of the limestone determine the degree of their blocking, permeability, deformability and bearing capacity of rock massifs are major factors in forming the engineering-geological conditions of the city territory.

3. QUARRIES

Theodolite survey of quarries on the site under research started in the 1940's, but it was of incidental nature; only small, unconnected fragments of mining fields were mapped. The complex mapping of the whole mining field was made from 1966 till 1973. Then, a nearly complete map of all workings was made, with the exception of heavily blocked up inaccessible sites.

Underground mapping made the researchers create a geodetic network with long-term positioning of reference points and bench-marks. The maps of workings are constructed as a right-angled coordinate system and are made as standard plates of the 1:500 scale. Based on these plates, maps in the 1:1000 scale are created on which is plotted the mining-geological and other special information. Nearly the whole site described is undermined with old underground workings.

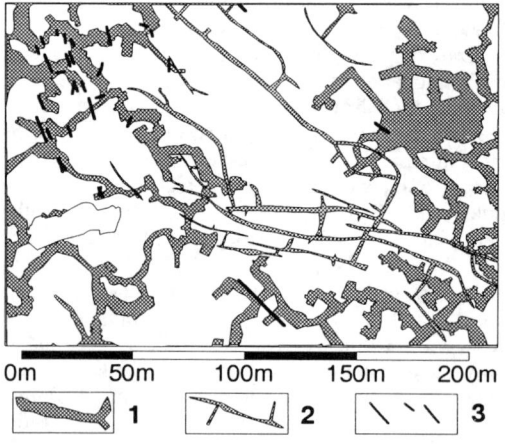

Fig.1. Fragment of the working density on the territory of Odessa.
1 — quarries, 2 — karst caves, 3 — karst fractures

A large mining field is located there, comprising numerous joined small-sized quarries and five local mines. The quarries consist of the following elements: (1) mining workings 2.5-6.0 m wide and 1.6-4.0 m high; (2) sites of joining with 1.5-2 m section, between mining workings and mines; (3) mining shafts with round/square section 4 m in diameter and 12...25 m deep from the earth surface; (4) water sumps of up to 3 m diameter; (5) inclined mining shafts and secret entrances to the quarries; (6) induced fan pit holes and wells 200-800 mm in diameter; (7) bunkers for military purposes. In some locations quarries are joined by sumps and inclined passages with building basements located in loess-like loams. In many cases quarries are uncovered by exploration and drainage wells. Total length of quarries on the site makes approximately 30 km.

4. KARST CAVITIES

Karst caves were mapped through measuring magnetic azimuths with subsequent reference of cave maps to existing fixed points of control of mine-survey mapping.

The considered site is subject to intensive karst formation. Most karst cavities are concentrated in its central and northern parts. The karst cavities are shown by: (1) pits up to 12 m deep, and up to 20 m in diameter; they are filled with red clay nowadays; (2) vertical karst channels oval in section, 5-15 cm in diameter; (3) horizontal karst channels up to 10 cm in diameter. Most widespread are karst-filled fractures, their number being over 200 on the site. The fractures are mainly vertical, rectilinear, 0.5-20 cm wide; their visible length makes 3-4 m.

Occasionally, the fractures are filled with red clay, though in most cases they are empty, covered in the bottom with gypsum crystals.

Karst caves are largest natural cavities on the site. They are all incorporated in the middle and upper parts of the second layer of the Pontian limestone. Now are revealed 26 caves, from 8 to 1292 m long. Cave channels are, basically, triangular or rhombic in section; they are less often trapezium-like, crack-like, rectangular, oval. The channels are up to 3.8 m high, up to 2.5 m wide. Main morphological elements of caves are (1) well-like or crack-like cavities in the roof, reaching 6 m of height and 1.5 m in diameter, and (2) channels up to 2 m deep and up to 60 cm in diameter.

In most cases caves are empty, but around karst funnels they are filled with red clay with the Upper Pliocene animal bones. In the bottom of empty channels, walls are covered with calcyte and gypsum crystals. Total length of caves on the site makes 3840m.

5. MINING-TECHNICAL CONDITION OF UNDERGROUND CAVITIES, AND THEIR INFLUENCE ON DEFORMATION OF CONSTRUCTIONS

The mining-technical condition of workings varies. Some of them are well-preserved. Many workings are almost completely downfallen; cavities formed above them, the heap downfall height reaches 3.5 m. All the above-mentioned relates also to other catacombs elements — mining shafts, water wells, and inclined channels. Almost all of them are completely or partially filled with earth. For the time of existence of workings, due to a variety of reasons some of them were stripped and filled.

Vertical mining shafts driven through the red clay layer serve as drains for the underground waters. Around shafts (in the discharge area) is always observed water infiltration, and through shafts, filling of quarries with diluted red clay and loams occurs. The modern concrete support is heavily leached by the underground waters; stalactytes and stalagmytes are formed on it.

The quarries have different modern appearance owing to various driving methods, distinctions in their age and size, as well as technologies of fastening. Some sites were fastened by timber, others — by stonework, still others were filled by stone or earth. Several sites have fallen down and were destroyed naturally. In some cases, the support was stripped and restored again a number of times. Since the late 1950's, grouting of workings with sandy pulp started. Some time later, working were fastened by walls made of monolithic concrete.

Estimation of influence of workings and karst density of the territory on the condition of constructions and buildings is a difficult task. It is stipulated by two groups of reasons: (1) reasons connected with general condition of constructions and buildings, and (2) reasons linked to the features of underground cavities. The most important are the following reasons:

(1) different age of constructions and buildings;
(2) different designs of buildings and their bases;
(3) different construction materials;
(4) different functional purpose of constructions and buildings (industrial and civil constructions);
(5) different conditions of operation;
(6) workings density and volume on the territory;
(7) technology and period of time since fastening of workings;
(8) space variability of physical-mechanical properties of the limestone and rock overlapping quarries;
(9) different thickness of the overlapping rock;

(10) technogenic load (the underground water filtration through hydraulic windows, and drainage of the water through wells and shafts; vibrations stipulated by industrial production, automobile and railway transport. It is obvious that the research of the dependence of deformations of constructions and buildings on each of the listed factors is extraordinarily difficult. Therefore, almost always this task has to be solved in a complex.

For example, it is established that the risk of construction failure grows considerably, with the factor of specific workings density on the territory exceeding 0.3-0.4 and their depth from the surface less than 15 m (ravine slopes). Thus the width of workings renders less influence than the change of physical-mechanical properties of the limestone. It occurs owing to the technogenic increase of the Pontian aquifer thickness and the appropriate decrease in the general strength of pillars.

The influence of karst cavities is even more complex. It is connected to the fact that the network of karst cavities (caves, fractures, etc.) was formed on the zones of tectonic dislocations. The tectonic conditionality of their space arrangement is confirmed by the regular nature of the network (with the characteristic step of 30-60 m), and by the regular stretches of the main systems (diagonal and orthogonal). The workings density of the territory, connected with karst phenomena, is 7-10 times less than the workings density due to the quarries.

At the same time, analysis of deformations of constructions and buildings on the city territory as a whole shows that the sites of strongest deformations are not chaotic in space. They are regularly grouped in linear zones oriented mainly in NW and SE directions. This allows to state the assumption that these zones — as well as the karst — can be connected with tectonic dislocations of corresponding stretches. Both factors — karst, and modern tectonic activity — are concentrated in relatively narrow zones, which are to be considered as the zone of increased construction risk.

This enables to explain, in particular, why on sites of prevalence of karst the relative increase of constructions deformation is observed.

Problems of the surface subsidence above old workings

V.N. Zemisev
State Research Institute of Mining Geomechanics and Mine Surveying (VNIMI), St-Petersburg, Russia

ABSTRACT: This paper considers the problems regarding the surface subsidence above the old workings in coal deposit mining in Russia and within the CIS, as well as the stability of pillars, the changes in hydrogeological conditions of the surface and groundwater flow. The emphasis is given to the necessity for the classification of undermined areas.

In terms of abandonment of the unprofitable mines, at present the problems still more arise with respect to further employment of the areas above the mined-out space. Major unfavorable aftereffects caused by coal extraction appear to be the deformations and downfalls of the surface occurring in time above the productive and development workings due to the movements and failure of rocks and pillars of different designation as well as the change in subsurface and ground water flow.

Upon revealing the zones of potential subsidence and deformations of the surface, change in hydrogeological conditions of subsurface water flow, the previously undermined areas may be classified according to the plausible usage and designing of the new structures.

At the deposits within Russian Federation and the CIS the earth surface subsidence occurs in mining at shallow depths both above the productive and development workings. The downfalls of the surface above the productive workings are set by several factors: minable thickness and dip angles of seams, the depth of upper boundary of productive workings and the rock properties.

In mining the flat-lying seams with dip angles of 35° the earth surface subsidence occurs, as a rule, at the mining depth, $H \leq 20 \cdot m$, where m is the working thickness of the seam.

In order to look into the formation of downfalls in the inclined and pitch strata in different fields their frequencies were analytically treated depending on the relation of the working thickness to the distance along the vertical between the upper and lower boundaries of a pillar left under sedimentaries.

Analysis of numerous statistical data on the downfall formation has shown that the dependence of downfalls number, n, on the relation of the working thickness to the vertical size of a pillar, $\dfrac{m}{h_P}$, may be expressed by the equation of regression as follows:

$$\frac{m}{h_P} = A \cdot n^2 + B \cdot n + C \qquad (1)$$

where n is a frequency of downfalls occurrence (relation of lengths of surface section disturbed with downfalls, to total length of the surface section on the strike where mining operations took place); A, B, C are the coefficients, the values of which depend on the deposit type and rock properties.

Let $n = 0$ in Eq. (1), then we'll receive the critical size of pillars, with which downfalls do not occur. Within the Kuznetsk coal basin, the vertical sizes of pillars with which downfalls do not happen, are: if $m = 2$–5 m, $h_P = 120$ m, and when $m = 10$–12 m, $h_P = 160$–200 m.

Within the Donetsk coal basin, in long-pillar mining along the strike with roof caving and with seam thickness, $m \leq 2$ m, downfalls do not happen if the pillar sizes $l_P = 60$ m.

At the Bulanash deposit downfalls do not occur if the vertical height of pillars left at upper levels (under sedimentaries) exceeds the value $30 \cdot m$. In the Cheliabinsk coal basin, with the working thickness of seams, $m = 5$–10 m, the vertical size of a pillar, with which downfalls do not occur, is: $h_P =$

120 m, and with seam thickness up to 5 m, h_p = 100 m.

The formation of downfalls is highly influenced by the mining depth of the seam under study. The critical mining depth when the downfalls may occur, is 200 m in soft rocks, 400 m in medium-hard and hard rocks.

As a rule, the boundary on the side of seam dip does not fall outside the isohypse projection drawn at the depth of 80 m. Along the strike and on the side of up-raise, assume as a downfall zone boundary a line removed from the projection of mining boundary at a space interval equal to the sediments thickness, but not less than 20 m.

Downfalls above the development workings may generate in different time intervals, both in drivage of workings and their exploitation, due to inrush of watered rocks into working and after the termination of the workings operation if they were not packed with rocks. In the latter case, rock caving and downfall outcrop may occur due to the loss of bearing capacity of the support, change in hydrogeological conditions, etc.

It was established that the surface downfalls above the development workings with no by-passing of caved rocks may happen if the distance, H_W, along the vertical from the working roof to the contact of bed rocks with sediments, satisfies the condition:

$$H_W < 15 \cdot h_w \cdot k_r \qquad (2)$$

where h_w is the height of the working in the rough; k_r is the coefficient depending on the structure and properties of rocks; with thinlayered soft rocks k_r = 1.2; with hard monolithic rocks k_r = 0.7; with presence of zones of tectonic disturbance above the working k_r = 1.4.

With the existence of inrush-hazardous water-bearing rocks in rock mass, for example, in the Podmoskovny brown coal basin, one may observe the inrushes of watered sand into mine workings under the influence of substantial head in water-bearing horizons positioned in cover rocks. Analysis of the observation results has shown that the formation of the surface downfalls under these conditions occurs with the certain (critical) volumes of watered sand rushed into the development working. These critical volumes of evacuated sand depend on the depth of the working layout.

Limiting depths, at which the downfalls of surface generate and critical volumes of the sand evacuated into the working, respectively, are presented in the Table below.

Table.

Critical volume of sand evacuated into the working, m³	Limiting depth of generation of downfalls, H_l, m
up to 50	50
250	75
1000	90

Deformations of the surface above the old workings may be caused also due failure of pillars to left in the mined-out space, the bearing capacity of which does not satisfy the requirements of long-term strength.

In particular, in mines of the Leningrad combustible shale deposit, some sections are revealed, where in time the failure of interlongwall pillars may occur, in consequence, the substantial deformations of the earth surface may happen. An evaluation of the pillar stability is made by comparison of long-term strength of coal (shale) pillars and the applied load.

One of the most important problems in assessment of the surface conditions above the old workings is the determination of potential inundation of the surface in wet liquidation of mines.

With sufficiently high level of subsurface water when the surface subsidence exceeds the distance from the surface to groundwater level with existence of the watered rock outcropping, the swamping or inundation of the surface may happen. As a rule, the surface inundation occurs with the absence of subsurface drainage when the condition is satisfied:

$$\eta_m + A > h \qquad (3)$$

where η_m is maximum surface subsidence in the section under study; A is the amplitude of seasonal fluctuations of the subsurface water level; h is the medium distance from the earth surface to the level of subsidence water.

More detailed recommendations regarding the generation of downfalls above the old workings and inundation of the surface are given in the methodological instructions and technical manuals developed in VNIMI.

REFERENCES

Safety Regulations for Protection of Structures and Natural Objects against Harmful Influence of Underground Mine Workings at Coal Deposits. 1981, Moscow, Nedra Publishers (in Russian).

Methodological Instructions for Prediction of Subsidence and Deformations of the Earth Surface and the Determination of Loads Applied on Buildings in Multiple Undermining. 1987, VNIMI, Leningrad (in Russian).

An inverse problem approach in mining subsidence

A. Constantinescu & D. Nguyen Minh
Laboratoire de Mécanique des Solides, CNRS, Ecole Polytechnique, Palaiseau, France

ABSTRACT: Subsidence due to underground mining excavation is a typical limited data problem. For a preliminary identification problem, elastic rockmass behavior is supposed, and a method based on minization error of constitutive law has been used. In this subsidence problem, joints behaviour appear to have a great importance. This inverse method allows for the moment, to take account of two limiting cases for contact between layers, corresponding to perfectly bonded interfaces and smooth interfaces.

1 INTRODUCTION

In underground mining, subsidence is much more important in intensity and in extension than in any other underground excavations (e.g. tunnels or roadways). Its prediction and control by adequate mining procedures are the task of subsidence engineering which must take into account protection of surface structures, and other important environmental issues such as surface hydrogeology.

Long term behaviour of mines is also another problem, such as the abandoned Nord Pas de Calais mining basin, a wide region of some 150×15 km, with an overburden with nearly one and half million inhabitants, which has to be refitted and managed. Subsidence may attain there in some places some metric depth, depending on the mining exploitation method. Long term evaluation should have to use various tools such as numerical models, taking account of hydromechanical aspects. The limited data problem is another aspect of subsidence evaluation.

The purpose of this paper is mainly focused on the possibility of identifying rockmass mechanical parameters from surface subsidence observation, in the framework of continuum analysis, by using an inverse method. For this purpose, a reference simplified model of a mine in the typical geological context of the Nord Pas de Calais region has been used.

The paper shares into two parts : in a first part, a mechanical analysis by Finite Elements is performed to analyse the behaviour of the model. The system appears to be strongly dependent on joint behaviour.

The second part is devoted to the identifying technique, and, as a preliminary work, within hypothesis of an elastic rockmass behavior. Identification techniques have already been proposed for the identification of parameters in geotechnics [2, 4, 8]. The authors proposed a solution by a direct minization of the distance between measurements and simulated values. In this work a new method is proposed, based on the minization of error on constitutive law, e.g. a distributed error energy. This method presents the advantage that regions containing the "bad" parameters can be localised [1, 3].

2 THE GEOLOGICAL CONTEXT AND THE MODELISATION

The plane strain model shown on figure 1. is a simplified cross-section of a mine from a site near from Valenciennes. Alternation of the lithostatic layers have been choosen according to geological information provided by the geological laboratory. The rockmass was assumed to behave as a Coulomb elastoplastic material, with "reasonable" parameters. There is no available data on joints behavior, so we shall limit analysis to two extreme interface hypotheses.

2.1 Perfectly bonded interfaces

It can be verified that the subsidence of the model remains practically elastic, since compressive plastic zones are just limited at the two front faces. The clo-

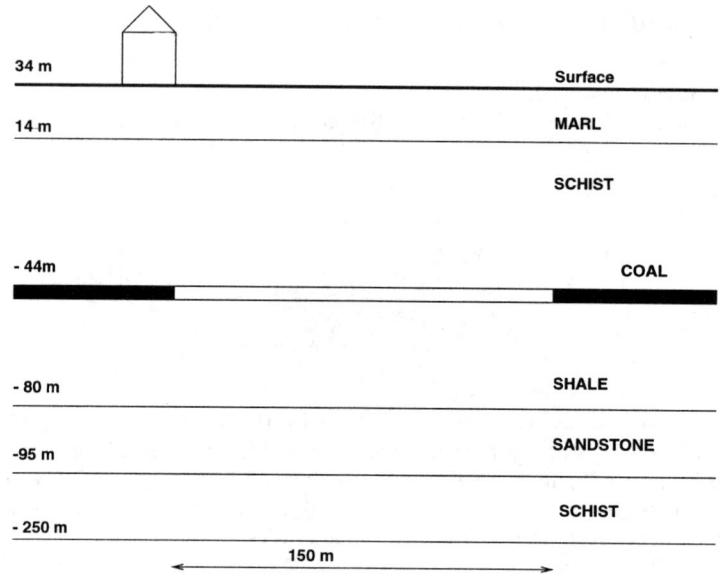

Figure 1: The reference problem of the study

sure of the mine roof is practically translated up to surface subsidence. On the roof and on the bottom, tensile stresses prevail, but remain in a limited zone. A parametric analysis, with lowering the yield compressive strenght by one half, shows no difference in subsidence (see figure 3.).

This shows, with the yield characteristics adopted for the model, that an elastic approach is sufficient when joints are supposed to be perfectly bonded.

2.2 Perfectly smooth interfaces

This analysis illustrates the great importance of joint behaviour. The surface settlement, which was initially 18 cm with fully bonded interfaces, increases to 30 cm when one joint above the opening is released into a smooth interface (schist/marl interface, figure 4.), and attains 45 cm when the coal seam/schist interface above the opening is released too (fig. 2.1). This is mainly due to bending of the overlying strata when they are liberated by joint release, inducing large development of yielding tensile zones. Finally tensile strenght appears in this problem to be a leading parameter, contrary to the compressive strength.

3 THE INDENTIFICATION OF THE ELASTIC MODULI

The identification problem which will be addressed next, will be defined in a preliminary approach, within an elastic rockmass behavior hypothesis. According to the previous study, this hypothesis is valid if joints between layers are supposed to be perfectly bonded, or eventually perfectly smooth if tensile failure is not considered. The identification problem then states as follows:

	$\rho\,[kg/m^3]$	$E\,[GPa]$	$\nu\,[adim]$	$C\,[MPa]$	$\varphi\,[deg]$
chalk	2000.	2.0	0.2	1.3	30.
coal		1.2	0.1	0.5	45.
sandstone		40.	0.15	22.	30.
tourtia		4.	0.2	0.001	28.
marl		6.	0.25	5.	3.
schist		10.	0.25	4.8	58.

Figure 2: Geotechnical characteristics of the materials

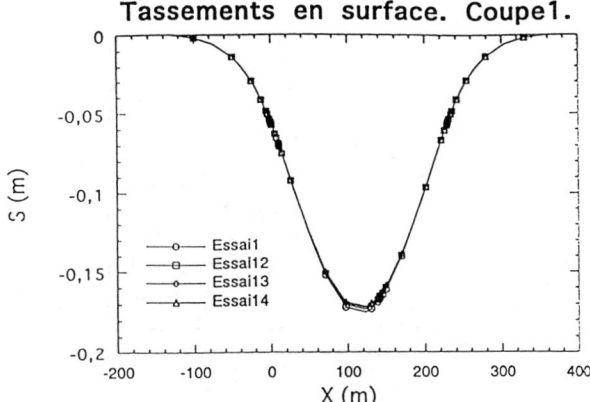

Figure 3: Vertical surface displacements from FEM comptutions with elastic and elastoplastic constitutive models

Can one find out elastic moduli of the geological layers from the knowledge of displacement measured on the surface ?

This problem is a particular case of the more general one: can a distributions of elastic moduli be found out from the knowledge of boundary displacement and force measurements ?

From a mathematical point of view, the answer is affirmative in the case of isotropic and special cases of elasticity provided one gets *all* displacement-force measured on the boundary. This has been proven by Ikehata [5] and Nakamura & Uhlmann [6]. They also showed that one can expect only a weak stability of the identified moduli to noise in the measurements.

From the practical point of view, a numerical reconstruction algorithm of a continuous distribution of elastic moduli from a *finite* number of measurements is described in [3].

However, the mathematical and the numerical approaches have been based on a large number of measurements, which are unavailable from a practical point of view in the subsidence problem. In order to balance these deficiencies some hypotheses have been stated:

1. isotropic elasticity,

2. constant elastic moduli in each geological layer,

3. identification of one layer at a time

3.1 The identification method

In the following it will be assumed that the behaviour of the body is governed by the following set of equations:

$$\epsilon = \frac{1}{2}(\nabla \mathbf{u} + \nabla^T \mathbf{u})$$
$$\sigma = \lambda \mathrm{tr}\epsilon + 2\mu\epsilon \quad (1)$$
$$\mathrm{div}\sigma = \rho g\, \mathbf{e}_z$$

and the following boundary condition on the surface:

$$\mathbf{u}|_{Surface} = \mathbf{u}^m \quad (2)$$

and blocked displacements on the boundaries in the rock mass. $\mathbf{u}, \epsilon, \sigma$ denote respectively the displacement vector, the strain and the stress tensor. λ, μ are the Lamé moduli.

The reconstruction algorithm is based on the minization of the error on the constitutive law [3]:

$$I(\epsilon, \sigma, \omega, \eta) = \int_\Omega \|\frac{1}{\eta}\mathrm{tr}\sigma - \eta\,\mathrm{tr}\sigma\|^2 dv$$
$$+ \int_\Omega \|\frac{1}{\omega}\mathrm{dev}\sigma - \omega\,\mathrm{dev}\sigma\|^2 dv$$

over kinematically admissible strain fields, statically admissible stress fields and admissible distributions of the bulk modulus η and the shear modulus ω. An *alternating direction minization method* conducts to the following algorithm:

- initialize the elastic moduli.

- with given elastic moduli do the following computations:

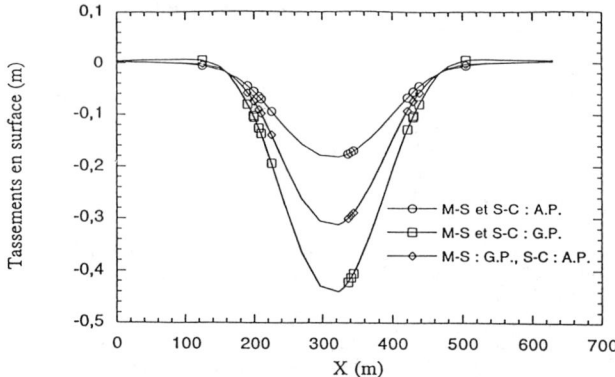

Figure 4: Vertical surface displacements from FEM comptutions for perfectly bonded (AP) and perfectly smooth (GP) interfaces between different geological layers

1. Compute the strain field ϵ from the *Dirichlet Problem*:
equations (1) and the boundary conditions

$$u|_{Surface} = u^m \quad (3)$$

2. Compute the stress field σ from the *Neumann Problem*:
equations (1) and the boundary conditions

$$\sigma n|_{Surface} = 0 \quad (4)$$

3. with ϵ and ϵ calculated before update the elastic moduli:

$$\eta^2 = \frac{(\mathrm{tr}\sigma)^2}{(\mathrm{tr}\epsilon)^2}$$
$$\omega^2 = \frac{(\mathrm{dev}\sigma)^2}{(\mathrm{dev}\epsilon)^2} \quad (5)$$

3.2 Numerical results in the subsidence problem

The numerical computations have been done on using the object oriented finite element code CASTEM 2000 on a HP 720 workstation and each iteration corresponding to two elastic computations took a couple of seconds.

The numerical computations show that the Young modulus is generally found out within 5 % of the actual value after $10 - 20$ iterations (see figure 6.). The initial values have been chosen in a range of one fifth to five times the actual Young modulus value.

Reconstruction of the Poisson's ratio gave globally poor results. When initial values where far from the actual value, the algorithm converged slowly or even diverged. This result might be explained by the prescribed rigid lateral boundary conditions. Due to these rigid borders, the lateral dilatation of the domain has been empeded, and moreover, no measurements are provided on these boundaries.

It is important to notice that the equations (5) allow for a distributed update of the elastic moduli in the whole domain. This property has not been used in this problem because of a non uniform distribution of energy over the whole domain which can lead to a bad reconstruction in regions with less energy. Using the fact that moduli are constant in every geological layer, the moduli have been computed from average values of stress and strain in zones inside the layer (see figure 5.). If the region considered in the average was sufficiently large, no difference in the results where observed. However in the case of the schist layer, results where substantially better if the average has been computed on a region including the front face. In an elastic computation one can remark that the mine front face concentrates deformation similarly to a crack tip. This energy concentration permitted a better average in the region and thus better overall elastic moduli.

4 CONCLUSION

Although a rather simplistic mining model has been used, such as tensile plastic yielding, the elastoplastic finite element analysis of the mine model has clearly shown that behaviour of joints between different layers play an important role on the response of

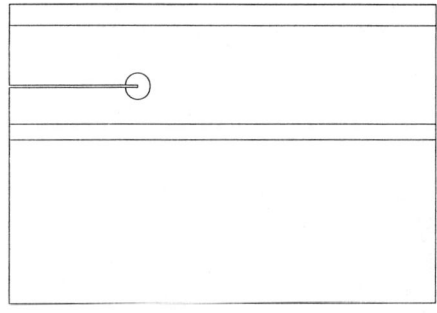

Figure 5: The contours of the different domains and layers in the inverse problem (half of the site is modelised)

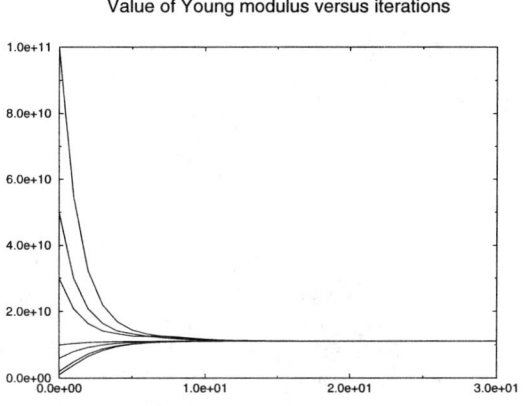

Figure 6: Converge of the young modulus for different initial values (sandstone layer)

the system. Two limiting cases have been considered for joints :

- fully bonded joints, which leads to an elastic behaviour of the system, as plastic zones are concentrated at the front face of long wall excavations.

- perfectly smooth interfaces, which leads to large development of plastic tensile zones. Tensile strength is then the leading mechanical parameter, due to bending of the seams.

In any case, the compressive strength is of secondary importance.

Assuming, for a preliminary inverse problem study, that the system behaves elastically, one has considered the problem of identification of mechanical parameters from surface displacement measurements. This method, based on the minization of the error on constitutive law, has proved to give accurate results and rapid convergence. It has been experienced on the fully bonded interface model, but it also applies to the other limiting case, for smooth joint interface. Extension of this method to non linear rockmass behaviour wil be the next step of this work.

Acknowledgement This study was conducted within the GEO's urban underground cavities workshop, including different laboratories, namely, the LG-ENSG Laboratory (Nancy), the LG-UST Laboratory (Lille), the LMS Laboratoiry (Palaiseau), the LML-EUDIL (Lille), and the Ecole Centrale (Lille).The authors would like to thank their partners

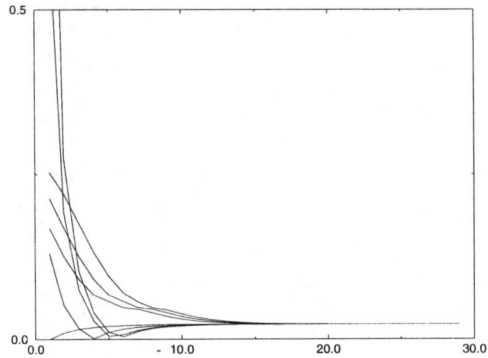

Figure 7: Signed relative error on the surface displacement during the iterations for different initial values of the young moduli

for fruitful discussion, Mr. Melliez (LG-UST) for his kind support in providing information on the studied site, and Mr.Benamar (LMS-G3S) for the elastoplastic computations of the direct problem.

References

[1] BUI H.D. - *Introduction aux problèmes inverses en mécanique des matériaux*, Editions Eyrolles, Paris, 1993 (english translation - CRC Press, Boca Raton, 1994; japanise translation - Shokabo, Tokyo, 1994)

[2] CIVIDINI A., JURINA L., GIODA G. Some Aspects of Characterisation Problems in Geomechanics *Int.J.Rock Mech. Min. Sci. & Geomechanical Abstr.*, Vol. 18, No. 7, pp 487-503, 1981

[3] CONSTANTINESCU A. - On the identification of elastic moduli from displacement-force boundary measurements *Int.J.Inv.Probl.Eng.*,vol. 1, 1995, p. 293-313

[4] GIODA G., SAKURAI S. Back Analysis Procedures for the Interpretation of Field Measurments in Geomechanics *Int.J. for Numerical and Analytical Methods in Geomechanics*, Vol. 11, pp 555-583, 1987

[5] IKEHATA M. - Inversion for the linearized problem for an inverse boundary value problem in elastic prospection *SIAM J.Appl.Math.*, Vol. 50, No. 6, dec. 1990, p. 1635-1644

[6] NAKAMURA G., UHLMANN G. - Global Uniqueness for an inverse boundary value problem arising in elasticity *Invent.Math.*, Vol. 118, 1994, p. 457-474

[7] QUINTANILHA J.E., NGUYEN MINH D. - Numerical Modelling of leached cavern fields by a mixed BEM-FEM Method *Proc. 4th Int. Conf. on Mechanical Behaviour of Salt*, Toronto (Canada), 1996

[8] SAKURAI, AKUTAGAWA S., TOKUDOME O. - Characterisation of yield function and plastic potential function by back analysis, in *Computer Methods and Advances in Geomechanics*, eds. Siriwardane & Zaman, Balkema, Rotterdam, 1994 (ISBN 90 5410 380 9)

[9] YANG G., CHUGH G.P., YU Z., SALAMON M.D.G. A Numerical Approach to Subsidence Prediction and Stress Analysis in Coal Mining Using a Laminated Model *Int.J.Rock Mech. Min. Sci. & Geomechanical Abstr.*, Vol. 30, No. 7, pp 1419-1422, 1993

Calculation of surface subsidence associated with oil and gas production from multilayered deposits with abnormal pore pressure

I.A.Garagash
United Institute of the Earth Physics, Moscow, Russia

ABSTRACT: The subsidence of surface is considered within the framework of model of bending layer, lying on the pliable bed. The coefficient of bed and the transverse load on the cover layer are expressed through the parameters of layers forming the deposit and pressure drop.

1 INTRODUCTION

It is known, that the production of gas and oil deposits results in dangerous consequences on the surface. As extraction of hydrocarbons decrease the pore pressure the equilibrium of rocks is broken and occurs subsidence of the earth surface. Vertical and horizontal tectonic loads and weight of overlapping sediments is balanced by the stresses in the rigid skeleton and fluid pressure. Decrease of pressure increases the load on the skeleton, that is the reason of deformation. It is clear, that thus many depends on those which sediments compose the reservoir. If the sediments have high rigidity (dolomites, limestones and rigid sandstone), the deformations will be insignificant (no more than 10^{-4} and the effects on the earth surface will be insignificant. If the reservoir is carried out by aleurolite or loose sand, the deformations can reach about 10^{-3} and the vertical size of the reservoir will decrease on a few decimeters (down to meter) with the appropriate consequences on a surface. The largest deformations (about 10^{-2} and more) it is necessary to expect in sediments (for example, clay sediments) with high abnormal pore pressure. Usually such sediments have lowered density, as than them porosity is higher then normal values for the given depth. Noncompaction can reach 20 % from normal values. In result the skeleton does not keep the load and fluid pressure as the minimum is equal to lithostatic pressure. It is clear, that downturn of pore pressure in this case results in fast reduction of porosity up to normal values and, as a consequence, to the large displacement on the surface.

2 BASIC RELATIONS

The development of gas and oil deposits is accompanied by decreasing of pore pressure p and increasing of effective vertical pressure in a rigid phase $\sigma_{33(eff)} = (\sigma_{33} - p)$ of layer (σ_{33} is positive at compression), that results in change of vertical deformation ε_{33}. The most simple relation, connecting these two values looks like

$$\varepsilon_{33} = -K\sigma_{33(eff)} \qquad (2.1)$$

where K is the factor of proportionality.

If layer is submitted by rigid rocks (dolomites, limestones, firm sandstones) and the pore pressure poorly differs from hydrostatic on the given depth, factor K will be correspond to elastic reaction. Starting from that points of layer at decreasing of pore pressure move only on the vertical, we shall neglect ε_{11} and ε_{22} in comparison with ε_{33} and write according to the Hookes law

$$K = \frac{1-2v^*}{2G^*(1-v^*)} \qquad (2.2)$$

where G^* and v^* are shear modulus and Poisson's ratio for layer respectively.

At drop of pressure in layer on value Δp its deformation will change on value

$$\Delta\varepsilon_{33} = -\frac{1-2v^*}{2G^*(1-v^*)}\Delta p \qquad (2.3)$$

The typical meanings of the mechanical characteristics, taken from the book, are resulted in Tab. 1.

Table 1.

Lithology	E^*, MPa	v^*
Plastic clay	266	0.44
Sandstone	11200	0.26

If the reservoir is carried out by aleorolites, clayes or loosening sandstones with abnormal high pore pressure, such sediments have increased porosity appreciably exceeding normal for the given depth. Compaction in these cases can less on 20% than normal values (Garagash et al. 1995). The normal distribution porosity with depth is usually defined by simple exponential law,

$$\phi_n = \phi_0 \exp(-cx_3) \qquad (2.4)$$

In Tab. 2 the meanings of porosity on surface ϕ_0, density ρ_0 and lithology constant c for different sediments are resulted.

Table 2

Lithology	ϕ_0	$c \times 10^{-4}, m^{-1}$	$\rho_0, gr/cm^3$
Sand	0.39	2.50	2.65
Clay	0.60	5.00	2.72
Aleurolite	0.53	3.03	2.67
Argillite	0.57	3.57	2.70
Marl	0.65	6.67	1.93
Limestone	0.70	6.33	2.71
Conglomerate	0.50	3.33	2.65
Dolomite	0.24	1.60	2.83
Anhydride	0.28	1.90	2.55

The law (4) is realized in conditions of normal compaction of sediment, when at its accumulation pressure in pores all time equal to hydrostatic $\rho_w g x_3$ on the given depth (ρ_w is density of water, g is gravitational acceleration). In this case the normal value of effective stress looks like

$$\sigma^n_{33(eff)} = (\sigma_{33} - \rho_w g x_3) \qquad (2.5)$$

where

$$\sigma_{33} = \int_0^{x_3} \rho g dz \qquad (2.6)$$

is full lithostatic pressure.

Final changes of volume of sediment, caused by reduction of the pore sizes at its subsidence, in case of normal compaction of particles result in deformation [1]

$$\varepsilon^n_{33} = \ln(1-\phi_0)/(1-\phi_n) \qquad (2.7)$$

The change of deformation of clay sediment with depth under the formula (2.7) most quickly increase on rather small depths up to 2km. Hence and greatest deviations from normal distribution, in consequence of action of abnormal pore pressure will take place on these depths. Thus the deposits on these depths are potentially dangerous from the point of view of maximum subsidence of surface.

Substituting relations (7) and (5) in (1), we shall define coefficient K as follows,

$$K = -\varepsilon^n_{33}/(\sigma_{33} - \rho_w g x_3) \qquad (2.8)$$

Thus agrees (8) and (1), if the pore pressure p differs from hydrostatic, the deformation of elementary volume ε_{33} at his movement from surface on depth will lag behind normal deformation ε^n_{33},

$$\varepsilon_{33} = \varepsilon^n_{33} \frac{\sigma_{33} - p}{\sigma_{33} - \rho_w g x_3} \qquad (2.9)$$

The state, supported by abnormal pressure p, is nonequilibrium and in case of his drop on incremental Δp porosity will decrease, that will result in deformation of layer on value

$$\Delta\varepsilon_{33} = \Delta p \frac{\ln[(1-\phi_0)/(1-\phi_n)]}{\sigma_{33} - \rho_w g x_3} \qquad (2.10)$$

We shall notice, that the greatest possible jump of deformation $\Delta\varepsilon_{33}$ corresponds to drop of pressure up to normal values $\rho_w g x_3$, i.e. at $\Delta p = p - \rho_w g x_3$. Further the drop of pressure will result only in deformation under the law (3), as porosity will reach normal values.

We compare deformations $\Delta\varepsilon_{33}$, counted up under the formulas (3) and (10), on an example of Bovanen gas deposit. Bovanen gas deposit is located in the northwest part of Yamal peninsula. The sizes of deposit 51×21 km in the plane and thickness $h = 100$ m. Average depth of deposit roof equals 550 m. The whole deposit is covered by the permafrost layer with thickness equals $200-250$ m. Deposit of gas and condensate are confined to Senoman sand collector. Productive layers of sandstone have porosity from 20% up to 34%. Is expected, that in result of development the pressure will drop with 6.9 MPa up to 2.5 MPa, i.e. drop of pressure will make $\Delta p = 4.4$ MPa.

According to the formula (3) for $\Delta p = 4.4$ MPa and sandstone from Tab. 1 we shall receive, that $\Delta \varepsilon_{33} = -3.2 \cdot 10^{-4}$. The maximum reduction of the reservoir with thickness equal 100m will make 3.2cm.

According to the formula (10) for sandstone with parameters from Tab. 2 and average density of cover layer $\rho = 2.6 \, gr/cm^3$ we shall receive, that on depth 600m $\Delta \varepsilon_{33} = -8 \cdot 10^{-3}$. The maximum reduction of the reservoir with thickness equal 100 m will make 80 cm.

In the case of clay soils the deformation will much increase. Calculations according to Tables 1 and 2 show, that reductions of the reservoir with thickness 100 m will make 51 cm and 308 cm accordingly.

Method of calculation of movements, strains and stresses in the cover layer rested on the multilayered hydrocarbon deposit as result of pressure drop $\Delta p(x_1, x_2)$. As a basis the problem about bending of layer on the soft foundation is taken.

3 DEFINITION OF THE LOADING ON THE COVER, ASSOCIATED WITH OIL AND GAS PRODUCTION

We shall consider equilibrium of the elastic layer by thickness H, rested on the soft foundation with bed coefficient k. We shall place coordinate axes $0x_1$ and $0x_2$ in the middle surface of layer, from which measure the vertical coordinate z. On Fig.1 section of the layer is shown. The layer is bent by transversal forces $Q(x_1, x_2)$. After deformation the points in the middle surface receives the deflection $w(x_1, x_2)$. The equilibrium equation of the layer have the form (Lukasiewicz 1979)

$$D\nabla^4 w + kw = Q \qquad (3.1)$$

where $D = \dfrac{Gh^3}{6(1-v)}$ is bending rigidity.

We shall examine the definition of transversal loads Q (x, x) and bed coefficient k through parameters of layers, and drop of the pore pressure.

First of all we shall consider the case of elastic reaction of reservoir, consisting from m layers. We shall designate through G_i^*, v_i^*, h_i and $\Delta\varepsilon_{33(i)}$ the modulus of shear, Poisson's ratio, thickness and increment of vertical deformation for i layer accordingly.

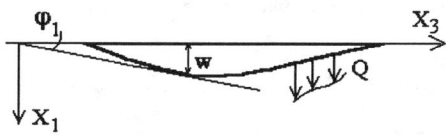

Fig.1 Section of the bending layer

It is obvious, that the deflection of cover w should be equaled to the total reduction of thickness of reservoir, i.e.

$$w = -\sum_i^m \Delta\varepsilon_{33(i)} h_i \qquad (3.2)$$

The vertical deformation of i layer $\Delta\varepsilon_{33(i)}$ agrees (2.1) and (2.2) looks like

$$\Delta\varepsilon_{33(i)} = -\Delta\sigma_{33(eff)}^{(i)} \frac{1-2v_i^*}{2G_i^*(1-v_i^*)} \qquad (3.3)$$

In the considered case

$$\Delta\sigma_{33(eff)}^{(i)} = -D\nabla^4 w + \Delta p_i \qquad (3.4)$$

where Δp_i is drop of pressure in i layer.

At the deduction of relation (4) it is necessary to take into account, that before drop of pressure the system cover-deposit is in balance. After reduction of pressure in i layer the load on its skeleton is increased on the value Δp_i except $D\nabla^4 w$, caused by bending rigidity of the cover.

Substituting (4) and (3) in (2), we shall receive

$$w = \sum_{i=1}^m [(-D\nabla^4 w + \Delta p_i) \frac{1-2v_i^*}{2G_i^*(1-v_i^*)} h_i] \qquad (3.5)$$

Comparing the equations (3.1) and (3.5), we come to the conclusion, that the coefficient of bed

$$k = 1 \Big/ \sum_{i=1}^m \frac{(1-2v_i^*)h_i}{2G_i^*(1-v_i^*)} \qquad (3.6)$$

and the transversal load

$$Q = \sum_{i=1}^{m} [\Delta p_i \frac{(1-2v_i^*)h_i}{2G^*(1-v_i^*)}] \bigg/ \sum_{i=1}^{m} \frac{(1-2v_i^*)h_i}{2G^*(1-v_i^*)} \quad (3.7)$$

Acting similarly in the case, when the deformation is caused by the reduction of abnormal pressure, according to the formula (2.9), we shall write down

$$w = \sum_{i=1}^{m} \{[(-D\nabla^4 w + \Delta p_i) \\ \times \frac{\ln[(1-\phi_{0(i)})/(1-\phi_{n(i)})]}{\sigma_{33(i)} - \rho_w g x_{3(i)}} h_i\} \quad (3.8)$$

where the greatest possible drop of pressure $\Delta p_i = p_i - \rho_w g x_{3(i)}$.

Comparing the equation (3.1) and (3.8), we come to the conclusion, that the coefficient of bed

$$k = 1 \bigg/ \sum_{i=1}^{m} \frac{\ln[(1-\phi_{0(i)})/(1-\phi_{n(i)})]}{\sigma_{33(i)} - \rho_w g x_{3(i)}} h_i \quad (3.9)$$

and transversal load

$$Q = \sum_{i=1}^{m} \frac{\Delta p_i \ln[(1-\phi_{0(i)})/(1-\phi_{n(i)})]h_i}{\sigma_{33(i)} - \rho_w g x_{3(i)}} \bigg/ \\ \sum_{i=1}^{m} \frac{\ln[(1-\phi_{0(i)})/(1-\phi_{n(i)})]h_i}{\sigma_{33(i)} - \rho_w g x_{3(i)}} \quad (3.10)$$

4 CONCLUSION

On the basis of the fundamental solution of the biharmonic equation (3.1) the program of calculation of subsidence and stresses in the cover is created. Thus in the case of elastic reaction of deposit on the drop of pressure the coefficient of bed k and transversal load Q should be calculated under the formulas (3.6) and (3.7). In the case, when porosity is supported by abnormal pore pressure, for definition of parameters k and Q it is necessary to use the formulas (3.9) and (3.10) accordingly.

The calculations are carried out on the regular grid, the step of which automatically gets out, starting from relation for the given rigidities of cover and reservoir. The program allows to choose one from two variants or case of the rigid elastic collector or collector with abnormal porosity supported by abnormal pressure. The properties of the collector can be nominated according to Tables 1 and 2. The results of calculations of deflection and stresses are kept as standard DAT-files that allows to use any graphic packages for analysis of results. Moreover the results are kept as special DEM-files, that allows to make the fast output of any parameter on the display.

REFERENCES

Garagash I.A., Volozh U.A., Lobkovsky L.I. Analysis of layer pressure distribution in Precaspian pre-salt unit during tectonic subsidence. Abstracts of 5-th Zonenshain conference on plate tectonics, Moscow, GEOMAR, 1995, 110-111.

Lukasiewicz S. Local loads in plates and shells. Leyden, 1979, 542p.

Questions and measures for safety in the design and construction of tunnels

Design of undersea and under-river tunnel linings

N. N. Fotieva & N. S. Bulychev
Tula State University, Russia

ABSTRACT: Analytical method of designing undersea and under-river circular tunnel linings upon the action of the bottom rock's own weight and water pressure on the bottom including the case of water penetration into the bottom rock is proposed in the paper presented. Examples of the design and results of the investigation of the lining stress state depending on basic influencing factors are given.

One of the basic features of undersea and under-river tunnels is the action of the water pressure uniformly distributed on the bottom surface and the possible water penetration into the bottom rock surrounding the tunnel.

With the aim of those features to be taken into account the analytical method of designing undersea and under-river tunnel linings of a circular cross-section shape upon the action of the bottom rocks own weight, water pressure on the bottom or the pressure of water penetrating through the bottom rock has been developed.

The method is based on mathematical modelling the interaction of the tunnel lining with the surrounding rock mass undergoing the pressure of water covering the bottom or with the water bearing rock mass (in the case of water penetration through the bottom rock) having an initial stress field caused by the rock own weight and by all water head counted off from the tunnel axis.

With that aim for the case when the bottom rock may be considered as non-penetrating one the analytical solution of the elasticity theory plane contact problem the design scheme of which is given in Figure 1 is being applied.

Here the S_1 ring restricted by R_0 and R_1 radii the material of which has the E_1 deformation modulus and the ν_1 Poisson ratio simulates the tunnel lining and supports the opening in a heavy linearly deformable S_0 semi-plane with the E_0, ν_0 deformation characteristics simulating the bottom rock mass loaded on the L_0 infinite straight boundary by the uniformly distributed

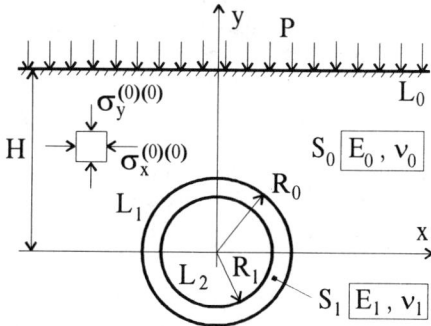

Figure 1. Design scheme.

pressure

$$P = -\gamma_w H_w \qquad (1)$$

where γ_w = water specific weight, H_w = depth of the sea or river.

The action of the rock own weight is simulated by initial gravitational stresses

$$\sigma_y^{(0)(0)} = -\gamma(H - y),$$
$$\sigma_x^{(0)(0)} = -\lambda\gamma(H - y), \qquad \tau_{x,y}^{(0)(0)} = 0 \qquad (2)$$

where γ = rock specific weight, H = depth of the tunnel under the bottom, λ = lateral pressure coefficient in an intact rock of the bottom.

Since as it is known (Muskhelishvili 1966) the action of the load uniformly distributed on the all boundary of the semi-plane results in initial stresses

$$\sigma_y^{(0)(0)} = \sigma_x^{(0)(0)} = -\gamma_w H_w, \quad \tau_{x,y}^{(0)(0)} = 0 \quad (3)$$

appearing in the S_0 medium, the problem being considered may be represented as the elasticity theory plane contact problem for a semi-plane weakened by a supported circular opening and having an initial stress state with components:

$$\sigma_y^{(0)(0)} = -\alpha^*[\gamma(H-y) + \gamma_w H_w],$$
$$\sigma_x^{(0)(0)} = -\alpha^*[\lambda\gamma(H-y) + \gamma_w H_w], \quad (4)$$
$$\tau_{x,y}^{(0)(0)} = 0$$

where α^* = correcting multiplier introduced for an approximate registration of the influence of the l distance between the lining being constructed and the tunnel face which may be determined by formula (Bulychev & Fotieva 1991):

$$\alpha^* = 0.64 e^{-1.75 l/R_0} \quad (5)$$

On representing complete stresses in the S_0 medium as sums of the initial stresses (4) and the additional stresses $\sigma_x^{(0)}, \sigma_y^{(0)}, \tau_{xy}^{(0)}$, caused by a presence of the opening (displacements are considered only as the additional ones) the boundary conditions of the problem being solved for determining the additional stresses and displacements have the form:

$$\sigma_y^{(0)} = 0, \quad \tau_{xy}^{(0)} = 0 \quad \text{on the } L_0$$
$$\sigma_r^{(1)} = \sigma_r^{(0)} + \sigma_r^{(0)(0)}, \quad \tau_{r\theta}^{(1)} = \tau_{r\theta}^{(0)} + \tau_{r\theta}^{(0)(0)},$$
$$u_x^{(1)} = u_x^{(0)}, \quad u_y^{(1)} = u_y^{(0)} \quad \text{on the } L_1 \quad (6)$$
$$\sigma_r^{(1)} = 0, \quad \tau_{r\theta}^{(1)} = 0 \quad \text{on the } L_2$$

where $\sigma_r^{(j)}, \tau_{r\theta}^{(j)}, u_x^{(j)}, u_y^{(j)}$ (j = 0, 1) are correspondingly the radial and shear additional stresses, horizontal and vertical displacements in points of the S_j (j = 0, 1) areas simulating the rock mass (j = 0) and the lining (j = 1) ; $\sigma_r^{(0)(0)}, \tau_{r\theta}^{(0)(0)}$ are initial radial and shear stresses in points of the S_0 medium being determined from the formulae (4) after passing to the polar co-ordinate system.

In case when the bottom rocks are water penetrated the same problem with the boundary conditions similar to (6) is considered at a presence of the following initial stresses in the S_0 medium:

$$\sigma_y^{(0)(0)} = -[\gamma(H-y) + \gamma_w(H_w + H - y)],$$
$$\sigma_x^{(0)(0)} = -[\lambda\gamma(H-y) + \gamma_w(H_w + H - y)] (7)$$

The last formulae may be transformed to the form:

$$\sigma_y^{(0)(0)} = -[(\gamma+\gamma_w)(H-y) + \gamma_w H_w] =$$
$$= -[\gamma'(H-y) + \gamma_w H_w], \quad (8)$$
$$\sigma_x^{(0)(0)} = -[(\lambda\gamma+\gamma_w)(H-y) + \gamma_w H_w] =$$
$$= -\{[\lambda(\gamma+\gamma_w) + (1-\lambda)\gamma_w](H-y) +$$
$$+\gamma_w H_w\} = -(\gamma+\gamma_w)[\lambda + (1-\lambda)\frac{\gamma_w}{\gamma+\gamma_w}] \times$$
$$\times (H-y) - \gamma_w H_w = -[\lambda'\gamma'(H-y) + \gamma_w H_w]$$

where

$$\gamma' = \gamma + \gamma_w, \quad \lambda' = \lambda + (1-\lambda)\frac{\gamma_w}{\gamma'} \quad (9)$$

So, in both of the cases considered we came to a single problem with boundary conditions (6) at different values of the rock specific weight, lateral pressure coefficient in an intact bottom rock and correcting multiplier α^*.

The initial stresses in S_0 area are being represented by formulae:

$$\sigma_y^{(0)(0)} = -\alpha^*[\gamma'(H-y) + \gamma_w H_w], \quad (10)$$
$$\sigma_x^{(0)(0)} = -\alpha^*[\lambda'\gamma'(H-y) + \gamma_w H_w]$$

where

$\gamma' = \gamma, \quad \lambda' = \lambda$ if water does not penetrate into the bottom rock;

$\gamma' = \gamma + \gamma_w, \quad \lambda' = \lambda + (1-\lambda)\frac{\gamma_w}{\gamma'}$ if water penetrates into the bottom rock.

The α^* correcting multiplier is being determined by formula (5) in the first case and $\alpha^* = 1$ in the second case.

The problem mentioned above has been solved with the application of the complex variable analytic functions theory by way of modification of the method proposed by I.G.Aramanovich (1955) allowing to obtain (after analytical continuation of Kolosov-Muskhelishvili complex potentials through the semi-plane boundary) the problem for circular ring in the whole plane undergoing loads represented in the form of Loran series on the external ring outline.

The computer program has been developed. As examples the results of designing underwater tunnel lining are given below. Input data are the following: $R_0 = 4.9$ m, $R_1 = 4.4$ m, $E_0/E_1 = 0.5$, $\nu_0 = \nu_1 = 0.2$, $\gamma = 25$ kN/m³, $\gamma_w = 10$ kN/m³, $H_w = 50$ m, $\lambda = 0.25$, $l = 0$ ($\alpha^* = 1$). The depth of the tunnel was assumed $H = 7.35$ m and $H = 49.0$ m. In case of water penetration into the bottom rock we have $\gamma' = 35$ kN/m³, $\lambda' = 0.464$.

The distributions of the σ_r radial contact stresses, σ_θ^{ex} and σ_θ^{in} normal tangential stresses appearing on the external and internal outlines of the lining cross-section if the bottom rock is non-penetrated for water are shown in Figure 2 (by solid lines - if $H = 7.35$ m and by dotted lines - if $H = 49$ m). The same stresses in the case of water penetration into the bottom rock are given in Figure 3.

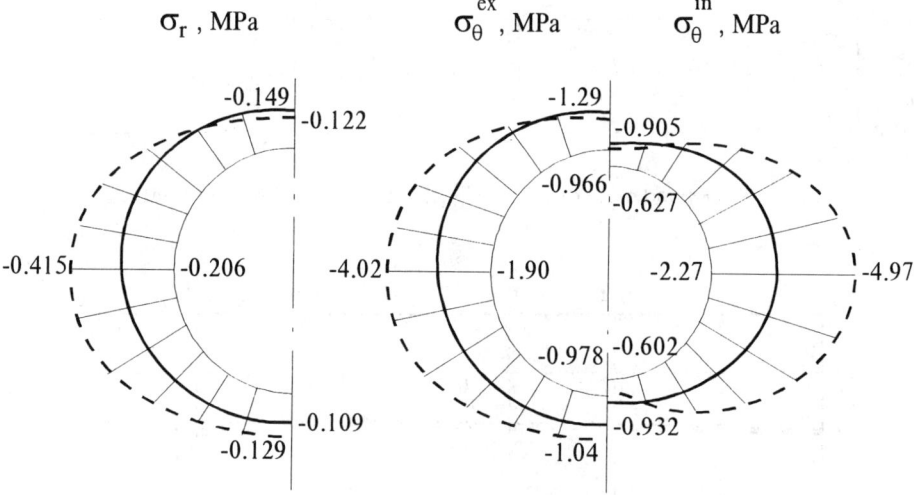

Figure 2. Distributions of stresses in the lining if water does not penetrate into the bottom rock.

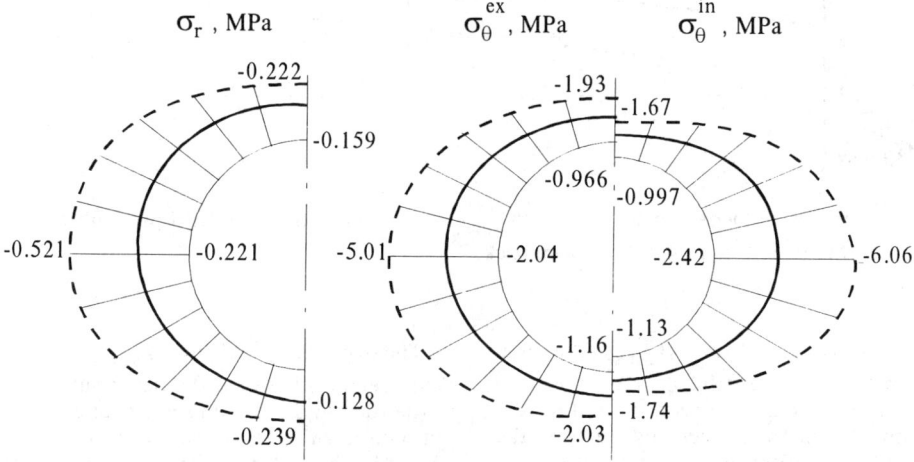

Figure 3. Distribution of stresses appearing in the lining if water penetrates into the bottom rock.

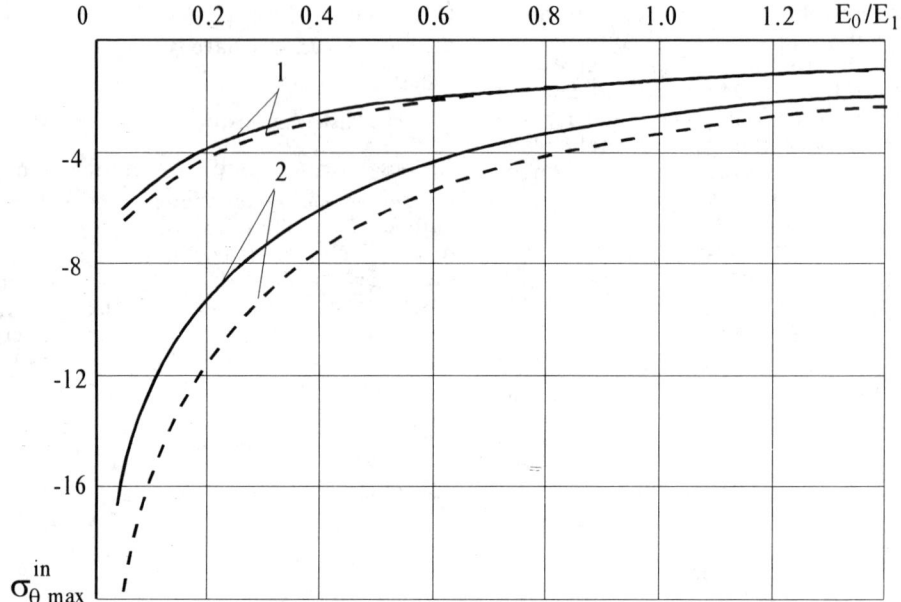

Figure 4. Dependences of the $\sigma_{\theta\,max}^{in}$ stresses on the E_0/E_1 relation.

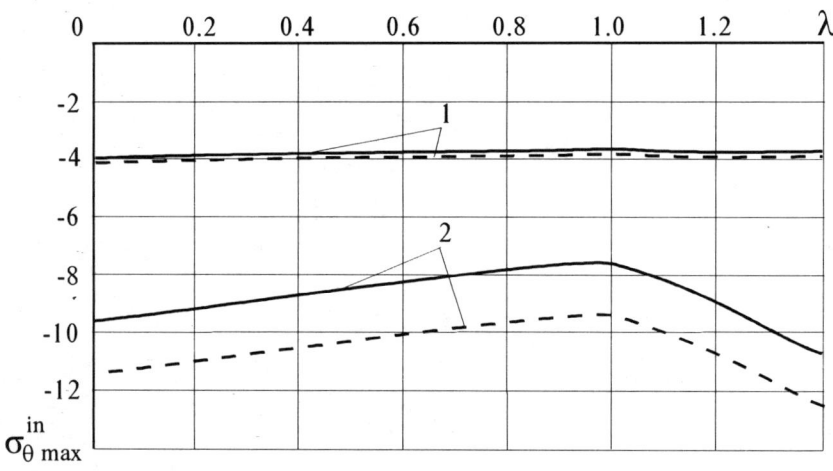

Figure 5. Dependences of the $\sigma_{\theta\,max}^{in}$ stresses on the λ lateral pressure coefficient in an intact bottom rock.

For the investigation of the $\sigma_{\theta\,max}^{in}$ maximal compressive normal tangential stresses appearing on the internal lining cross-section outline depending on the main influencing factors the multi-variant calculations have been fulfilled.

The dependences of the $\sigma_{\theta\,max}^{in}$ values on the E_0/E_1 relation between the deformation modules of the rock and the lining material at the rest input data mentioned above are shown in Figure 4 where the dependences for the tunnel located on the H = 7.35 m and H = 49 m depth are given by curves 1 and curves 2 correspondingly. The maximal stresses in the case if water does not penetrate into the bottom rock are shown by solid lines and in the case if water penetrates into the rock - by dotted lines.

The similar dependences of the $\sigma_{\theta\,max}^{in}$ maximal stresses on the λ lateral pressure coefficient in an intact rock obtained for the case when $E_0/E_1 = = 0.2$ are given in Figure 5.

Dependences of the $\sigma_{\theta\,max}^{in}$ stresses on the Δ lining thickness obtained at the $E_0/E_1 = 0.2$, $\lambda = = 0.25$ and on the H_w depth of the body of water obtained at the $E_0/E_1 = 0.2$, $\lambda = 0.25$, $\Delta = = 0.5$ m are given in Figure 6 and in Figure 7 correspondingly. Here and in Figure 5 both the curves 1, 2 and the solid and dotted lines mean the same as in Figure 4.

CONCLUSION

In conclusion we can mark that the method proposed may be generalised for designing undersea and under-river tunnel linings being constructed with the application of grouting or other technique of the bottom rock strengthening, for designing of several parallel circular tunnels and tunnel linings of a non-circular cross-section shape.

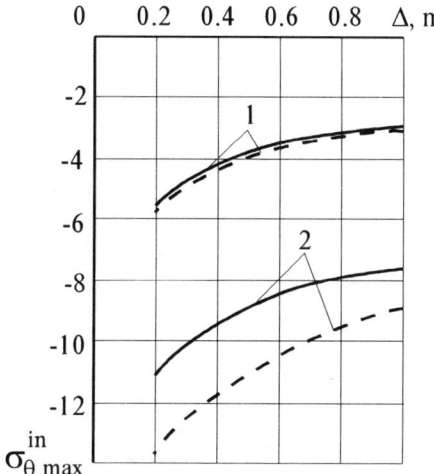

Figure 6. Dependences of the $\sigma_{\theta\,max}^{in}$ stresses on the lining thickness.

REFERENCES

Aramanovich, I.G. 1955. Distribution of stresses in the elastic semi-plane weakened by the supported circular hole. *Dokl. AN USSR*, v. 104, N 3: 372-375.

Bulychev, N.S. & N.N. Fotieva 1991. Basic problems of Underground Structures Mechanics. *Podzemnoye i shakhtnoye stroitelstvo.* 1: 8-11.

Muskhelishvili, N.I. 1966. Some basic problems of the mathematical elasticity theory. *Nauka*, Moscow.

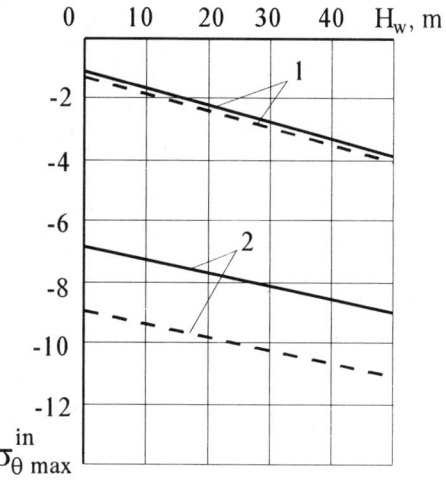

Figure 7. Dependences of the $\sigma_{\theta\,max}^{in}$ stresses on the H_w value.

Study of stability for tunnel group in alternative stratified rock

Moweng Xie & Shuqing Yan
Wuhan University of Hydraulic and Electric Engineering, People's Republic of China

Yichang Li & Jianzhong Yu
Qing Jian Hydroelectric Development Corporation, People's Republic of China

ABSTRACT : Based on the results of FEM and geomechanical model test, this paper presents the study of the stability for a group of tunnels and the possibility of tunnelling scheme. The comparison of the result of monitoring and the theoretical analysis show the stability of group of tunnels in the alternative stratified rock masses.

1 INTRODUCTION

Geheyan Hydropower station is located in the middle part of China and its installed capacity is $4 \times 40 = 120$ MW. The group is consisted of four tunnels with excavated diameter 11.3~12.5m. The thickness of rockwall between two tunnels in 12m.

From stake 0+062 to stake 0+149, the surrounding rock of tunnels is consisted of interbeded stratum of Limestone and shale, the whole strength of rock is low beeause of the fractures and the joints. This is the main point of stability of surrounding rock.

Because of the large diameter of tunnel, the thin rockwall and the complicated geological condition, in the design, it is demanded that after excavating and lining $1^\#$, $3^\#$ tunnel, then excavating $2^\#$, $4^\#$ tunnel. But this constructing procedure cann't satisfied the schedue of generate electricity. In this condition, we must study the possibility of excavating the $2^\#$, $4^\#$ tunnel before lining the $1^\#$, $3^\#$ tunnel. Based on the F.E.M. and the geomechanical model test, we have study the possibility of excavating $2^\#$, $4^\#$ tunnel before the Lining $1^\#$, $3^\#$ tunnel and the stability of surrounding rock of each possible schemes. Finally, the concrete excavating scheme is presented. In the excavating $2^\#$, $4^\#$ tunnel, first excavating a small lead hole, then the side wall, Finally full excavating, and using the New Austria Method, monitoring the rock stability.

2 BASIC INFORMATION

Selecting the 0+115 section for study, the geological section of stake 0+115 is shown in Fig.1 and the physicmechanical parameters of rock is presented in Table.1.
The program of excavating and lining as follow, is shown if Fig. 2.

(1)Full excavating $1^\#$, $3^\#$ tunnel and making bolt and shotcrete lining;

(2) Excavating the lead hole of $2^\#$, $4^\#$ tunnel and making bolt and shotcrete lining;

(3)Exacating the left side wall of $2^\#$, $4^\#$ tunnel and making bolt and shotcrete lining;

(4)Lining $1^\#$ tunnel, excavating the right side wall and make bolt and shotcrete lining;

(5)Lining $3^\#$ tunnel, excavating the left of $2^\#$ tunnel;

(6)Excavating the left of $4^\#$ tunnel.

Table 1 physical and mechanical parameters of rock

rock \ parameter	E (MPa)	u (MPa)	r (kN/m³)	R_c (MPa)	R_t (MPa)	C (MPa)	f
Limestone	14×10^3	0.25	26.5	50	1.0	1.0	1.2
shale	2×10^3	0.35	25.5	20	0.2	0.3	0.6
$401^\#$, $403^\#$, $405^\#$	200	0.4	25.0	/	0.02	0.1	0.25
F_{18}, F_{52}	500	0.4	25.5	12	0.05	0.05	0.35

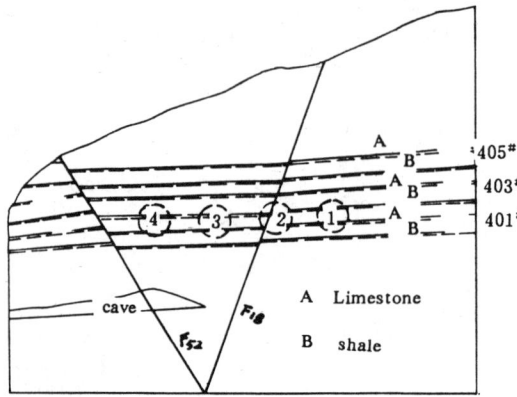

Fig. 1 0+115 section view

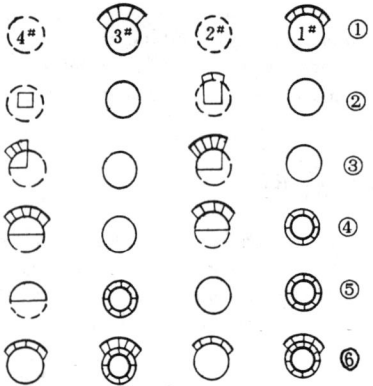

Fig. 2 construction program

We have simulated whole construction procedure under the plane strain in the F.E.M. and the geomechanical model test.

In the geomechanical model test, the simulation scale $C_L=1/160$, the dimension is length×width×thickness = $80×60×20$(cm), the stress simulation scale $C_σ=5$. The simulation material is consisted of sand、gypsum、emulsion、glycerine、citric acid and water.

3 EXCAVATING STABILITY OF TUNNEL GROUP

3.1 *Excavating* $1^#$、$3^#$ *tunnel*

Because the bad geological condition of $1^#$、$3^#$ tunnel in the alternative stratified rock, in the excavating, first excavating the upper part and making shotcrete, then making bolt, and then excavating the full tunnel. The length of bolt is 3.5～5.0(m), the bolt is friction bolt.

From the result of F.E.M., After excavated $1^#$、$3^#$ tunnel, There is convergence displacement in $1^#$ tunnel and the maximum displacement is 10mm on the top. But in $3^#$ tunnel, there are convergence displacement on top and bottom and the out—direct displacement on sides, the maximum displacement is 15mm on the top. After excavating, the damage range is limited in the loosening zone which is about 1.2m thick, and the range of stress disturbed is limited in 12m circle.

From the result of the geomechanical model test, in the top of $1^#$ tunnel, the tangential stress increased by 1.7 times of the initial stress and the radial stress loosened, and from that, we know the cave-in of top shale is unvoided. In the excavating site, After detonation 3～5 hours, we can see the partial cave-in, because of the timely bolting and shotcrete lining, the surrounding rock is become stable, Fig. 3 is the curve of displacement and time on the 0+111 section, we can see that from 4/1990 to 20/4/1991 (before excavating $2^#$ tunnel), the surrounding rock is stable.

We also know from the model test, After excavated $1^#$、$3^#$ tunnel, the limestone on side wall is become the bearing strate and the tangential stress is 1.5 times of the initial stress, and in the shale of the bottom, the radial stress loosen and appears partial tensile strain. From all above we know, making bolt and shotcrete lining is useful for protect the bearing force of surrounding rock.

In spite of the bad geological condition, because of using the bolt and shotcrete lining, the surrounding rock of $1^#$、$3^#$ tunnel is stable during excavating period.

3.2 *Excavating* $2^#$、$4^#$ *tunnel*

From the result of the F.E.M. and model test[2], if excavating fully $2^#$、$4^#$ tunnel under the condition of unlining $1^#$、$3^#$ tunnel, the surrounding rock of $1^#$、$3^#$ and $2^#$、$4^#$ tunnel may be caving in.

3.2.1 Excavating $2^#$、$4^#$ lead hole. The dimension of the lead hole is width×highty=$6×7$(m). From the result of model test, the tangitial and radial stress loosened in the top rock and appear certain damage section. In the site excavating, we also saw small irrugular caving in on top, because of the timely bolting and shotcrete lining, the top rock then became stable.

3.2.2 Excavating the left side wall of $2^#$、$4^#$ tunnel. From the result of the F.E.M. and model test, after excavated, the stress of left rock is higher than that of right rock, the tangitial and radial stress of the left top of F_{18} is become more loose, the stress concentration of the top is shifted to right. At the same time, the excavating do little influence on the stress of $1^#$、$3^#$ tunnel rock.

3.2.3 Excavating the right side wall of $2^#$、$4^#$ tunnel. Fig. 4 shows the displacement of the surrounding rock, the sink displacement of the top of $2^#$ tunnel is 18mm,

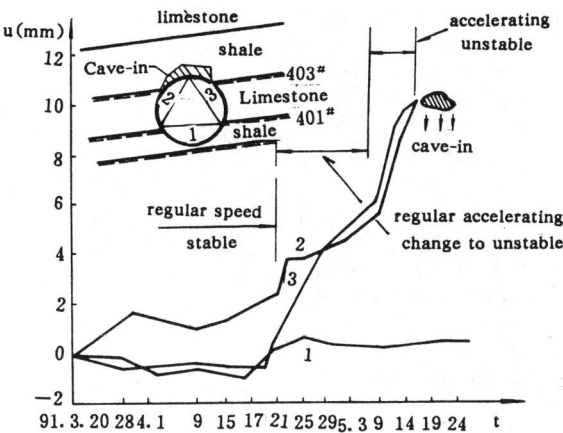

Fig. 3 Curve of displacement and time on the 0+111 section

Table. 2 Concentrated coeffcients of strains at rockwall (k)

order of excavation \ rockwall \ rock \ K	$1^\#\sim2^\#$		$2^\#\sim3^\#$		$3^\#\sim4^\#$		Notes
	ε_2 K	ε_1 K	ε_2 K	ε_1 K	ε_2 K	ε_1 K	
$1^\#$、$3^\#$ tunnels	1.58	1.19	0.87	1.16	1.00	1.28	ε_1—shale
$2^\#$、$4^\#$ leading opening	1.72	1.20	0.92	1.19	1.14	1.30	ε_2—Limestone
Leftwall of $2^\#$、$4^\#$ tunnels	1.83	1.24	1.25	1.22	1.43	1.36	$K=\dfrac{\mu\varepsilon}{\mu\varepsilon_0}$
rigntwall of $2^\#$、$4^\#$ tunnels	1.92	1.19	1.43	1.19	1.58	1.35	$\mu\varepsilon$—The second strain
below half of $2^\#$ tunnel	2.08	1.07	1.72	1.1	1.64	1.26	$\mu\varepsilon_0$—initiol strain
below half of $4^\#$ tunnel	2.14	0.97	2.02	1.1	1.76	1.18	

and the displacement of the side wall is directed out. From the result of model test, the tangitial and radial stress of the top rock is obviously become loosen.

3.2.4 Full excavating $2^\#$ tunnel. From Fig. 4 we know that the top rock upheaves $2\sim 3$mm; From the model test, we also get that the tensile stress of the top rock decreases. Because of the circle shape, the stress condition of the top rock improved.

From the result of model test, an certain radial tensile stress section appears in the bottom rock of $2^\#$ tunnel, the increasements of tangitial strain in the top and bottom shale rock and in the shale of rock wall are both tensile strain, so the whole section bolting and shotcrete lining is necessary. At the same time, certain damage appear in the top rock of $3^\#$ tunnel near the F_{18}.

3.2.5 Full excavating $4^\#$ tunnel. Because the excavating $4^\#$ tunnel is fater the $3^\#$ lining, the $3^\#$ tunnel get a little influence. During the excavating of $4^\#$ tunnel, the smooth blasting is bestly used, the relaxed rock zone is only 0.5m.

3.2.6 Stress of rockwall. Limited by the topography, the inter space of adjant tuunel is very small, and plus the complicated geological condition, the stability of rockwall is very important. Table 2 is the strain concentration factor of rock pillar, the change tendency of tangitial stress of rock pillar as follow:

(1) With the excavating, the stress of Limestone of rockwall gets increased, and the change of $2^\#\sim3^\#$ rock

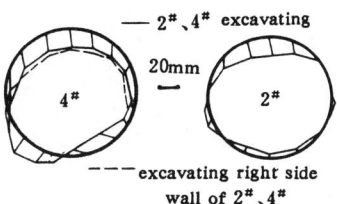

Fig. 4 Displacement curve of $2^\#$、$4^\#$ tunnel

pillar is serious which is obviously influenced by F_{18} fault. The vertical stress concentration factor is about 2.0 in the limestone, the stress is about -4.0 MPa. The maximum vertical stress is in the border of tunnel but not the midest of the rock pillar, from the F.E.M., $1^{\#}$ tunnel is -7.0 MPa, the right of $2^{\#}$ tunnel is -10.0 MPa, the right of $3^{\#}$ tunnel is -4.3 MPa and the left is -12.8 MPa. From this, the reinforced surrourding rock is necessary for protect the damage of rock pillar.

(2) The change of stress of rock pillar in shale is, with the excavating, the compress stress is increased at first, but decreased during the excavating of $2^{\#}$、$4^{\#}$ bottom semicircle, this means the shale is now in the plastic condition, the bearing capacity has decreased and it is necessary for reinforced.

4 CONCLUSIONS

1. During the excavating procedure the tangential stress in limestone layer is gradually increased, the maximum value of stress is about 2.5 times of the initial stress. But in shale layer appeare the contrariant tangential stress

2. The rock wall bettween tunnels becomes very important part for stability of surrounding rock during excavating.

3. The failure of rook mainly occured on the top of tunnel of shale layer, its failure mechanisum is shear and tensile failure along the stratification and joint.

REFERENCES

Zhao Daisheng, Yan Shuqing, Stability of Surrounding Rock of Underground house of an certain Hydropower station, Conservancy Journal, 1988; 6

Xie Moweny, Excavating Stability of tunnels of Geheyan Hydropower station, Journal of WUHEE, 1994 (8):35~39

Effect of mud slurry of surrounding soil in using pipe jacking

H. Shimada & K. Matsui
Department of Mining Engineering, Kyushu University, Fukuoka, Japan

ABSTRACT : Recently, small-diameter shallow tunnels have been often constructed by using pipe jacking. This is a sewage tunnel drivage method. This system involves the pushing or thrusting of a drivage machine through concrete pipes ahead of jacks. The method utilizes mud slurry which is formed around the pipes in order to stabilize the surrounding soil. However, the behavior of the soil and the mud slurry around the pipes is not clearly understood.

From this perspective, this paper discusses the performance of mud slurry around drivage pipes by means of laboratory testing and two-dimensional Eulerian Lagragian seepage analysis.

1 INTRODUCTION

In order to protect workers' safety during construction, as well as for environmental and cost reasons, efficient small-diameter shallow tunneling methods have become increasingly important in regards to outside plant engineering such as for water supplies, electricity, telecommunications and gas recently. The effects of the above projects, in overcrowded urban areas is significant and often results in substantial impact and traffic delays with associated loss of travel time. Clearly the solution to these utility placement problems, if the full impact of trench excavation is to be avoided, is trenchless technology (McFeat-Smith and Herath, 1994). In particular, for construction work near existing facilities, an underground tunnel that is excavated by pipe jacking is being increasingly employed in order to avoid problems.

Pipe jacking is firmly established as a special method for the non-disruptive construction of the underground pipelines of sewage systems. Pipe jacking, in its traditional form, has occasionally been used for short railways, roads, rivers, and other projects (Hunt, 1978). Basically the system involves the pushing or thrusting of a drivage machine through concrete pipes ahead of jacks. This method utilizes mud slurry which is formed around the pipes in order to stabilize the surrounding soil (Shimada and Matsui, 1995, 1996). However, the behavior of the soil and mud slurry around the pipes is not clearly understood.

From these perspectives, this paper discusses the performance of the mud slurry around the drivage pipes by means of laboratory testing and two-dimensional Eulerian Lagragian seepage analysis.

2 CONSTRUCTION OUTLINE

Pipe jacking, like many other methods of below ground construction, is best employed in stable, water free soil conditions. Unfortunately, with the demands on available space and the need to provide more services, it is not always possible to select stable strata, which means that contractors have to contend with difficult ground below the water table (Cole, 1977, Hough, 1978). The pipe jacking system is applicable to the above situations.

Fig.1 shows the pipe jacking system scheme. This system is particularly suited to both cohesive and sandy soil, and can be used to construct pipe tunnels up to 2,000mm in diameter. The features of this system are as follows:

1) It can be applied to sand with high-pressure or gravel-rich soil by using an excavating method.

2) It can construct both long (maximum length: 600m) and curved tunnels (minimum curvature radius: 50m).

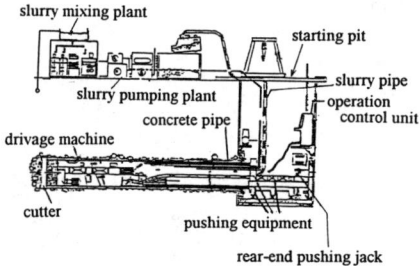

Fig.1　Pipe jacking system scheme.

Recent technological developments have led to successful methods for stabilizing unstable strata by excluding water from excavations by means of the mud slurry around the pipes. This system meets modern environmental standards. This type of mud slurry is formed by water, decomposed granite soil, carboxyl methyl cellulose, lost circulation control material, bentonite, fine aggregate, and other materials. Fig.2 shows an illustration of a concrete pipe and the soil in this system. The proportional mixes vary the viscosity and the seepage capacity of the mud slurry and will be determined by whether mud slurry is required to strengthen the soil.
This system is used as follows:
  1) Install the pushing equipment in a starting pit, then set the drivage machine.
  2) Extend the rear-end pushing jack and push the drivage machine.
  3) By repetition of the pushing process, if the rear-end driving after one concrete pipe length (2.5m) is driving in, connect another pipe.
  4) After driving is completed, recover the drivage machine from the arrival pit.

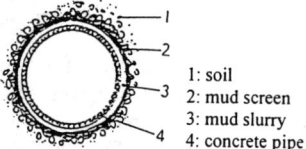

1: soil
2: mud screen
3: mud slurry
4: concrete pipe

Fig.2　Illustration of a concrete pipe and soil.

More widely used techniques that are now being used elsewhere for micro-tunneling, use remote control drivage methods in conjunction with pipe jacking. During the pushing processes, the mud slurry is injected into the face and the over-cutting area which is between the concrete pipe and the soil.

After the soil voids are filled by the slurry, soil stabilizes due to slurry pressure. To minimize ground deformation in the pushing process, it was necessary to maintain the slurry pressure that was kept on underground water pressure at plus $0.2 kgf/cm^2$ (Katano and Ogawa, 1994). The mud slurry plays an important role in reducing the friction resistance in pushing the pipes. However, the behavior of the soil and the mud slurry around the concrete pipes is not clearly understood.

3　LABORATORY TESTING

3.1　Outline of method and apparatus

Mud slurry plays a significant role in the pushing process. In order to discuss the performance of the mud slurry around the pipes, a laboratory test was performed.
Fig.3 shows the laboratory testing apparatus. Our testing with dimensions of 600mm in width, 200mm in length, and 1,000mm in height was employed in a model of the pipe jacking. Graded sand which had been hand-shaken on a 2.0mm sieve was chosen for the ingredients of the model material. The density of the ingredients was determined to be $1.47 g/cm^3$ by the dropping the graded sand 120mm through the mold.
Slurry pressure was simulated as the water level difference between the container which was connected with the inflow pipe and a model of the concrete pipe. The container which was made of plastic was 300mm in width, 400mm in length, and 250mm in height. An acrylic pipe 100mm in diameter which was connected with the inflow pipe included 165 holes which were each 4mm in diameter. The holes were utilized in order to supply the slurry pressure to the soil ingredients. A piston was inserted into the end of the acrylic pipe in order to ensure gentle slurry pressure. After this piston was removed from the end of the acrylic pipe, the slurry pressure

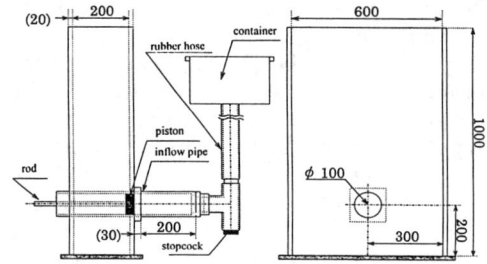

Fig.3　Laboratory testing apparatus.

222

was loaded into the soil.

The mud slurry was composed of water, clay, lost circulation control material, and carboxyl methyl cellulose, as listed in Table 1. The geometric scale factor was selected to be 1/10.

Three different parameters, such as the overburden depth, the slurry pressure, and the density of the mud slurry were tested for our study on small-diameter tunnel behavior. Table 2 lists the six models of different parameters which were denoted as ① to ⑥.

3.2 Laboratory testing results

Fig.4(a) shows the relationship between the elapsed time after the inflow of the mud slurry to the acrylic pipe and the vertical length of the slurry permeation from the top of the crown. This graph indicates when the vertical length of the permeation with high slurry pressure increased more significantly than with low slurry pressure.

Fig.4(b) shows the relationship between the elapsed time after the inflow of the mud slurry into the acrylic pipe and the surface settlement along the vertical symmetric axis of the pipe. Surface settlement decreased with an increase in slurry pressure. Comparing the results of (a) and (b), one can see that the larger slurry permeation zone tends to develop less surface settlement due to the grouting effects of the mud slurry ingredients on the soil voids.

Figs.5(a) and (b) give the above comparison in regards to two types of mud slurry density. This shows that the slurry permeation zone decreases with an increase in the density of the mud slurry. Here, more solid bodies were included in the high density of the mud slurry than in the low density. The slurry screen between the soil and the acrylic pipe was easily formed by high density mud slurry under pressure. Consequently, the slurry screen prevented the mud slurry ingredients from permeating into the soil.

The effects of the overburden on the slurry permeation zone and the surface settlement are illustrated, as shown in Figs.6 (a) and (b). When the ratio of pipe diameter D to overburden depth H was 1/2, the results of the surface settlement could not be

Table 1 Mixture ratio of the mud slurry.

clay	lost circulation control material	carboxyl methyl cellulose	water	slurry density
224.5g	10.00g	1.80g	904.50g	1.15g/cm³

Table 2 Laboratory test parameters.

number	overburden D/H	slurry pressure (kgf/cm²)	slurry density (g/cm³)
①	1/4	0.100	1.15
②	1/3	0.100	1.15
③	1/2	0.100	1.15
④	1/4	0.075	1.15
⑤	1/4	0.150	1.15
⑥	1/4	0.100	1.28

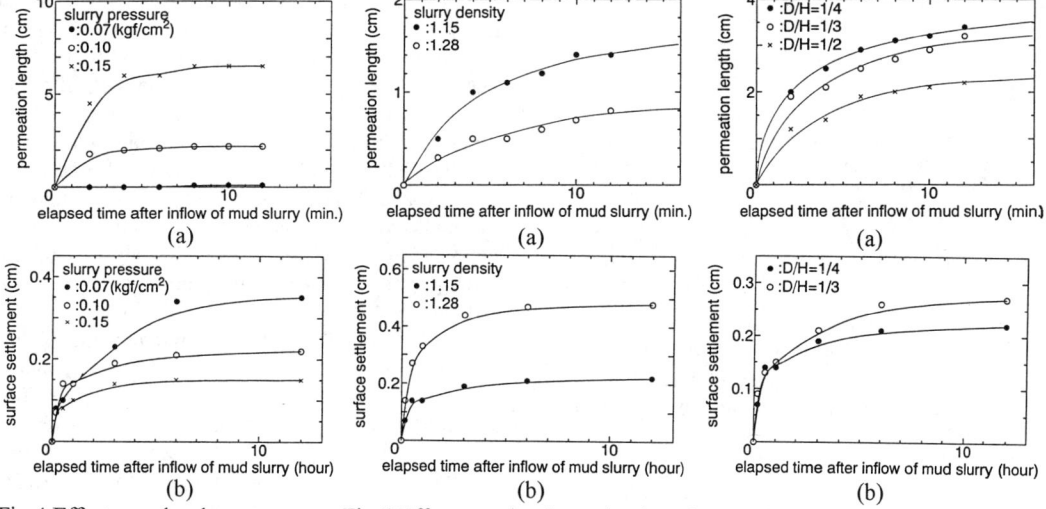

Fig.4 Effects on the slurry pressure. Fig.5 Effects on the slurry density. Fig.6 Effects on the overburden depth.

obtained due to the collapse of the soil (see Fig.6 (b)). Enhancing surface settlement leads to the collapse of the soil. Therefore, this indicates that surface settlement is caused by the development of excessive deformation around the pipes.

Fig.7 shows the relationship between the elapsed time after the inflow of the mud slurry to the acrylic pipe and the vertical lengths of the slurry permeation from the top of the crown and the bottom, respectively. When the ratio of pipe diameter D to overburden depth H was 1/4, the slurry pressure was 0.075kgf/cm^2 and the density of the mud slurry was 1.15g/cm^3, as denoted in ④ in Table 2. The vertical length of the slurry permeation is defined as the distance from the top or the bottom of the pipe to the point where the mud slurry permeated into the soil. This reveals that the permeation length of the upper side was larger than the lower side. The thickness of the mud screen was measured by using an acrylic pipe 8mm in diameter in all directions after the acrylic pipe was withdrawn from the apparatus at the end of each testing, as listed in Table 3. The results in this table contained the same tendency. This phenomenon is because of the differences in the density of each of the ingredients in the mud slurry. In other words, the mud slurry separated the high density from the low density ingredients. As a result, the permeation length of the upper side was larger than the lower side. In addition, once the mud screen was formed, the thickness of the mud screen increased gradually because it behaves like a filter paper under pressure. Thick mud screen prevents the mud slurry from permeating into the soil.

In light of the above results, it is clear that the performance of mud slurry plays an important role in the pushing process.

Table 3 Thickness of the mud screen.

number	measurement direction(mm)			
	upper	lower	left	right
①	2.0	3.0	1.0	2.0
②	1.0	3.0	2.0	2.0
③	(-)	(-)	(-)	(-)
④	1.0	1.0	1.0	0.0
⑤	2.0	5.0	3.0	2.0
⑥	2.0	3.0	1.0	1.0

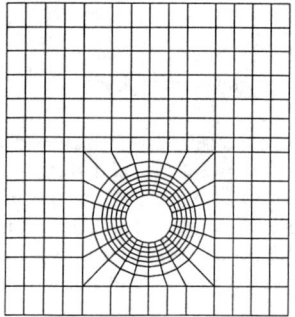

Fig.8 Eulerian Lagragian model.

4 EULERIAN LAGRAGIAN SEEPAGE ANALYSIS

4.1 Simulation outline

In an effort to identify the effects of the mud slurry on the surrounding soil, two-dimensional Eulerian Lagragian seepage analysis was performed.

In order to prevent the collapse of the soil around the pipe, it was necessary to maintain the slurry pressure that was constantly kept on the underground water pressure at plus 0.2kgf/cm^2. Therefore, seepage flow entered the soil, after the mud slurry flowed to the area which had been driven by the acrylic pipe. Once the soil voids were filled by the slurry, the soil was stabilized under the pressure and a mud screen was formed around the soil immediately. Stable slurry pressure allowed the surrounding soil to seal off the ingress of the underground water and to prevent the collapse of the surface.

Fig.8 shows the Eulerian Lagragian model that was used in this analysis. In this model, previous laboratory testing was considered. A pipe 100mm in diameter was driven into the soil with a width of 600mm and a height of 1,000mm. The soil was assumed to dry sand which has the same properties as the soil in the laboratory test. In seepage analysis,

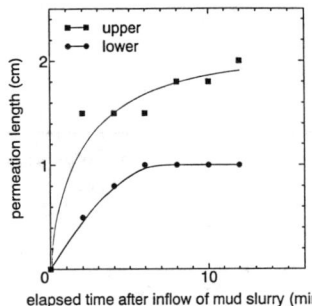

Fig.7 Relationship between elapsed time after the inflow of the mud slurry and the vertical length of the slurry permeation.

the model of the bottom and the side boundaries were not permeable. The pressure of the outer periphery of the area which was driven by the pipe was constant.

In dispersion analysis, the concentration of the mud slurry was constant. The particle elements which were necessary to calculate the seepage concentration of the mud slurry in each section of the model were located as shown in Fig.9. The black circles were particle elements which were in the mud slurry concentration and the concentration of the white circles were zero. In this seepage and dispersion analysis, the following assumptions were used.

1) Black circles were arranged for the known slurry concentration in t=0.
2) Other particles in the slurry concentration were not provided from the periphery of the drivage area because of the formation of the mud screen.
3) The dispersion of the mud slurry was occurred due to the black circles in t>0.
4) Slurry concentration was defined as the percentage of the weight of the ingredients to the weight of the solution.

The ratio of the diameter of the pipes to the overburden depth, the slurry pressure, and the soil characteristics were simulated in the model, as listed in Table 4.

4.2 Simulation results

Figs.10(a), (b) show one of the results obtained by

Table 4 Eulerian Lagragian analysis parameters.

overburden (D/H)	1/4
slurry pressure(kgf/cm²)	0.075
slurry density(g/cm³)	1.15
soil density(g/cm³)	1.50
porosity(%)	30
permeability(cm/s)	3×10^{-3}

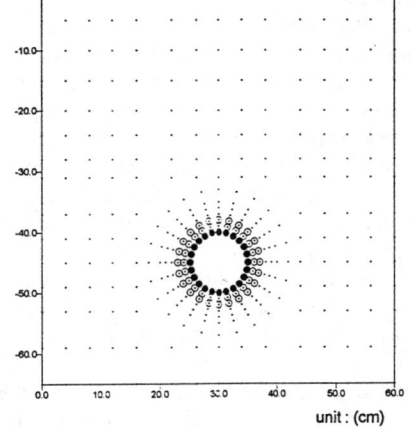

Fig.9 Particle elements which are needed in order to calculate dispersion analysis.

using Eulerian Lagragian analysis. The contours of the mud slurry concentrations from 10% to 50% are given in Fig.10(a), (b). The permeation zones were formed around the pipes immediately (as shown Fig.10(a)). This phenomenon indicates that the ingredients of the mud slurry were permeated and dispersed to the soil after the inflow of the mud

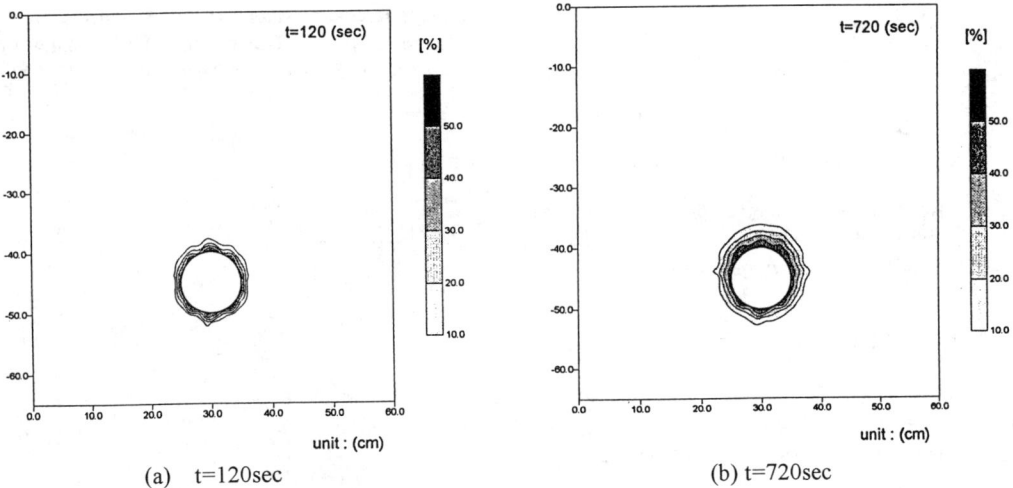

(a) t=120sec
(b) t=720sec

Fig.10 Mud slurry concentration.

slurry around the pipes. High concentrations of mud slurry were distributed around the pipes; however the permeation zone surrounded the pipes widely. Consequently, a mud screen was formed due to the above. Moreover, a thick mud screen and a permeation zone prevented mud slurry from permeating the soil and kept the soil stable.

Fig.11 shows the relationship between the elapsed time after the inflow of the mud slurry to the pipe and the vertical lengths of the slurry permeation from the top of the crown and the bottom, respectively. Where the vertical length of the slurry permeation was defined as the distance from the top or the bottom of the pipe to the point where a mud slurry of 10% concentration permeated into the soil. This shows that the permeation length of the upper side is larger than the lower side. The results in this figure had the same tendency as did the laboratory test. This is because of the differences in the density of each of the ingredients of the mud slurry, as mentioned above. The mud slurry separated the high density from the low density ingredients. Accordingly, the permeation length of the upper side was larger than the lower side. Moreover, once the mud screen was formed, the thickness of the mud screen gradually increased because it behaves like a filter paper under pressure. A thick mud screen prevents mud slurry from permeating into the soil.

These results show that the performance of mud slurry plays an important role in the pushing process.

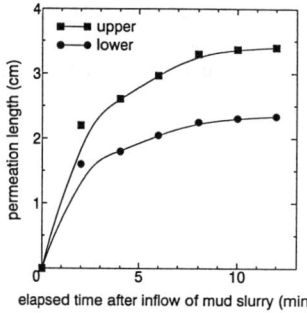

Fig.11 Relationship between elapsed time after the inflow of the mud slurry and the vertical length of the slurry permeation.

5 CONCLUSIONS

This paper discussed the performance of mud slurry around drivage pipes by means of laboratory testing and two-dimensional Eulerian Lagragian seepage analysis.

From our research, it is clear that the performance of mud slurry plays an important role in the pushing process. Moreover, it is necessary to stabilize the soil by applying suitable slurry pressure. However, in order to grasp the more detailed behavior of pipe jacking, more research and field data are needed.

ACKNOWLEDGMENTS

The authors would like to thank the many civil engineers and employees of Showa-Doboku Co. Ltd. for their assistance in this research project.

REFERENCES

Cole, J.M. 1977. Pipe Jacking Case Histories. Tunnels and Tunneling, July : 91-93.

Hough, C.M. 1978. Pipe Jacking Case Histories. Tunnels and Tunneling, April : 51-52.

Hunt, M. 1978. Pipe Jacking the Harburg Sewer. Tunnels and Tunneling, July : 19-21.

Katano, S. and T.Ogawa. 1994. Effect of Slurry Shield Tunneling in Soft Alluvial Clay on an Adjacent Underground Subway Structure. Proc. of Int. Congress on Tunneling and Ground Conditions : 151-156.

McFeat-Smith,I. and P.S.Herath. 1994. Construction of Large Sized Sewers by Slurry Machines in Hong Kong's Bouldery Soils. Proc. of Tunneling and Ground Conditions : 41-47

Shimada, H. and K.Matsui. 1995. Shallow Tunnel Drivage by Using Long Distance Curve Method. Proc. of the 26th Sympo. on Rock Mechanics : 216-220 (in Japanese).

Shimada,H. and K.Matsui. 1996. Stability of Soil around Shallow Tunnel Using Pipe Jacking. Proc. of Mining Science and Technology : 287-292.

Technical analysis on the cost-saving in Norwegian rock excavation

Min-Kyu Kim
Korea Institute of Geology, Mining and Materials, Taejon, Korea

Einar Broch & Bjørn Nilsen
Norwegian University of Science and Technology, Trondheim, Norway

ABSTRACT: Norway is characterized by hard bedrocks, high mountains, deep valleys and fjords. In this setting a wide variety of tunnels and rock caverns have been constructed. In spite of high salaries to the tunnel workers, Norwegian contractors are probably producing the cheapest tunnels and rock caverns in the world. Besides benefit of hard-rock geology, Norwegian cost-saving is explained by the excavation technique, reasonable contract system and organization of workers developed from the accumulated experience. Brief analytical descriptions of these items are given in this paper in order to stimulate to the utilization of underground space.

1. INTRODUCTION

In Norway underground space has been taken into use for a wide variety of purposes such as oil, gas, frozen food storage, and sewage treatment plants, etc. Every year, about 100 km of hydropower tunnels and 40 km of road tunnels are added to 4,000 km of hydropower tunnels of 200 underground powerhouses, over 800 road tunnels, and more than 750 railway tunnels. This competitive market has led to very cost-saving construction methods. In spite of high salaries to the tunnel workers, Norwegian contractors are probably producing the cheapest tunnels and rock caverns in the world. The cost-saving in Norwegian rock excavation owes much to the hard bedrock, but has mainly also been achieved through technical innovations such as unlined pressure tunnels, unlined air cushion chambers, underwater piercing, reasonable contract system, effective prognosis medel, and optimized organization of the works. Brief presentations of those will be given in the following chapters.

2. THE CHEAPEST TUNNELLING COSTS IN THE WORLD

The direct comparison of tunnelling costs between different countries is not simple because of the diversity of many parameters that affect construction costs. The main parameters are geological conditions, construction procedures,

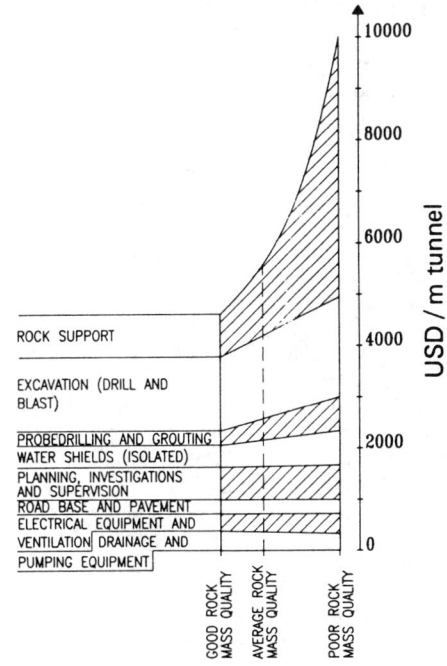

Fig. 1. Appoximate distrubution of cost for two lane 50 m2 subsea road tunnel, from Palmstrøm (1992).

construction materials, and geometrical variables of tunnel. Labour conditions and tax system are different from country to country and also have to be considered in comparing construction

Table 1. Construction costs of road tunnels (in 1996 prices)

Country	Tunnel, Type	Cost in 1996 US$/m	Cost in 1996 US$/m³	Cost in const. year() local currency /m	Cross-Section (m²)	Ground Condition	Source
Norway	subsea tn.	7,020	140 average	5,600 US$, (1990)	50	Norwegian Bedrocks	Palmstøm A, 1991
	road tn.	4,160-16,640	92-370	15,000-60,000 NOK, (1984)	40-50		Øvstedal E, 1986
Austria	ARLBERG ht	47,700	463	277,610 SCH, (1978)	103	gn to ms with shear zones	Steiner W, 1979
	GLEINALM ht	15,500	184	90,000 SCH, (1978)	84	gr - gn	
	EISEN-HOWER ht	87,500	650	37,570 US$, (1978)	135	gr - gn to fault zone	
U.S.A	TBM tn.	-	73-522	50-350 US$, (1990)	-	-	Done Rose, 1991
	B&D tn.	-	160-657	110-450 US$, (1990)	-		
Vietnam	road tn.	5,000-13,500	172-255	5,000-13,500 US$, (1996)	20-80	Vietnamese hard-rocks	Interv. with a Engr. of PIDC, 1997
Korea	road tn.	12,000-20,000	150-230	120-180thous. WON, (1996)	80-85	Korean hard-rocks	Interv. with managers of three Korean Const. Co., 1977

tn:tunnels SCH:Schillings, NOK:Norwegian Krone, WON:Korean Won, ht:highway tunnel, gr:granite, gn:gneiss, ms:mica schist

costs. Among these parameters it is the geological conditions given from nature that mostly make it difficult to compare the construction costs. Here, some information about published tunnelling costs are presented together with some costs collected from interviews. Table 1 gives a rough impression of the cost differences. The costs in the table are given in U. S. dollars (1996 price level) based on the exchange rate and escalation factors shown in Table 2. Average escalation factors are given from the consumer price index numbers during the period 1990-1993. The incurred years of construction costs are assumed as one year before publishing the data when not clearly represented. The total costs of Norwegian road tunnels vary from 15,000 to 60,000 NOK per route meter (Øvstedal, 1986), and subsea tunnels from about 4,600 to 10,000 US$ per route meter, where the lower excavation costs were a result of rapid tunnelling (Palmstrøm, 1992). Fig. 1 shows the varying total costs of subsea tunnels in Norway according to a survey of the various items of expenditure.

Steiner (1979) compared the construction costs for subways in Germany and U. S. and gave a conclusion that costs per cubic meter tent to be 50 to 100% higher in U. S. than in Germany. It was also presented that the cost of the Eisenhower Tunnel was about 2.5 times that of the Arlberg Tunnel in Austria and the conditions were in many aspects more favorable in the Eisenhower Tunnel. This comparison is also listed in Table 1. NATM (New Austrian Tunnelling Method) is based on the concept of flexible and self-supporting rock. But geological conditions in Austria are generally worse than in Norway. The cost per cubic meter of tunnel is highest in the Arlberg tunnel and lowest in the Gleinalm tunnel among the 6 highway tunnels constructed in 1970-1981.

The supporting in U.S. tunnels is acknowledged as very conservative. Hydropower tunnels in U.S. are always supported with continuous concrete linings. In these cases the costs are thought to be almost three times higher than those of unlined pressure tunnels in Norway. Korea has many similarities in bedrock conditions with Norway, but tunnelling costs of Korea seems to be more than 50% higher than those of Norway. The bedrocks in many parts of Vietnam is said to be Precambrian hard rocks. But the supporting in Vietnam are also very conservative with continuous concrete linings. The construction cost of tunnels in Vietnam seems to be almost the same as in Korea. Vietnames conservativeness in supporting may offset their merits of low wages and low prices of goods. In conclusion we are able to say that Norwegian tunnelling costs is one of the cheapest in the world in spite of its highest salary level as listed in Table 2.

Table 2. Consumer price index and wages.

Year[1]				A.E.[3] '90-'93	E.R.[4] 1996	M.E.[5] US$ 1996[2]
78	84	90	96[2]			
U.S.A. 79	126	159	184	4.0	1 US$	3,250
Austria 91	123	141	163	4.0	10.42 SCH	2,637
Norway 86	146	209	262	9.0	7.06 NOK	4,487
Korea -	138	184	282	9.8	800 WON	2,203

1) from IMF, 2) calculated value, 3) A.E.:Average Escalation, 4) E.R.:Exchange Ratio per US$, 5) M.E.:Monthly Earnings

3. FAVORABLE GEOLOGICAL CONDITIONS

The geological and topographical conditions in Norway explains to some extent why the comprehensive use of the subsurface has taken place in this country. Norway forms part of the Fenno-Scandian Precambrian shield. Approximately two thirds of the country is Precambrian rocks, with different types of gneiss dominating. Other major types of rocks from this era are granites, gabbros and quartzites. The remaining patrs are mostly rocks of Cambrian, Ordovician and Silurian age. As a result of the Caledonian movements the greater part of these rocks are metamorphic, but to very varying degrees. Through the later Quaternary glaciations, Norway today may be topographically characterized as a mountainous country, with young soils (less than 12,000 years) on the top of almost unweathered rocks. From the engineering geology point of view, one may generally describe Norway as a typical hard rock province but the faulting may have a great influence on the stability in underground openings. The Norwegian bedrock, often exposed in outcrops, makes mapping and sampling fairly easy. Due to glacial erosion, weakness zones, faults and gouges are in general well exposed.

4. FLEXIBLE GEOLOGICAL INVESTIGATION

The main goals of Norwegian geological investigation are to obtain: (a) The necessary input for the evaluation and alignment alternatives and for the overall planning of the scheme. (b) A basis for evaluation of potential stability problems and the necessary input parameters for stability analyses and planning of rock support. (c) A basis for cost evaluation and for preparation of tender documents.

Norwegian geological investigations seem to thoroughly pursue these aims but show flexibility according to varying conditons. Norwegian geological investigations can divided into two main stages with two substages each as shown in Table 3. The characteristic investigations for each of the four substages are briefly listed in the table. Not all kind of investigations will be carried out for all tunnels. A short road tunnel through rocks which can easily be mapped on the surface does not necessarily need a two-stage pre-investigation phases. On the other hand, for a hydropower scheme with several alternative tunnel alignments, or for a complicated subsea tunnel, a subdivision of the pre-investigations into more than two stages may be considered. Reports of the various investigation stages are required and represent as very important documents for the planning and operation of underground facilities.

Table 3. Geological site investigation stages, from Broch (1988).

PRE-INVESTIGATIONS (PRECONSTRUCTION PHASE)	
Feasibility study exploration	*Definite plan study investigation*
· Desk study · Walk-over survey · Geophysical investigation	· Eng. geol. mapping · Special investigation. · Sampling and laboratory testing

POST-INVESTIGATIONS (CONSTRUCTION PHASE)	
Detailed subsurface investigation	*Tunnel mapping*
· Sampling and testing · Supplementary investigation. · Control and revision of pre-investigation reports · Recommendation. of rock support.	· Mapping in tunnel · Registration of all rock support · Evaluation of excavation performance

5. EFFECTIVE PROGNOSIS MODEL

The capacity of tunnelling may vary widely depending on engineering geological parameters like rock mass quality, rock strength, degree of jointing, etc., and also the type of equipment being used. In favourable rock conditions advance rates per hour of more than 10 m have been achieved, but poor conditions the advance rate may drop to few tens of centimeters per hour. This, together with several non-geological factors like site location, labour costs, costs of electricity, etc. cause great cost variations.

Prognosis models have been developed for drill and blast- as well as TBM tunnelling, and are based primarily on correlations which have been established between tunnelling results and geological parameters. By systemizing and analyzing a large amount of tunnelling data, time- and cost- curves for the various operation have been established. When the conditions of a planned tunnel are known, the prognosis models make it possible to evaluate the cost and time schedule for the project. Since the introduction of the Norwegian tunnelling prognosis models in the mid 1970s, continuous revision and refinement have taken place. The models are widely used today as valuable tools at the planning stage for evaluating the cost and capacity of hard rock tunnelling.

For a TBM, the total cost of fullface tunnel is a result of a number of factors. In the prognosis model several empirical diagrams are used for defining the so-called "normalized cost". This parameter represents a summary of the various cost elements such as TBM-cost, cost of assembling/ dissembling, cost of back-up operation, cutter costs, hauling costs, labour costs, etc. The main purpose of calculating normalized costs is in most cases to evaluate alternatives cross sections and tunnelling route (Nilsen and Thidemann, 1993).

6. REASONABLE CONTRACTUAL ASPECTS

No matter how detailed the engineering investigation are carried out, there will always remain some degree of uncertainty relating of the rock conditions and to what extent rock support will be necessary. Therefore contractual aspects represent important factors in rock engineering.

The basic principle of the Norwegian contract system is risk-sharing in order to achieve a safe and cost-effective project completion. This concept was first introduced about 25 to 30 years ago. Fig. 2 shows risk sharing according to type of contract and its influence on the project cost. Norwegian contracts are, as a rule, unit price contracts with lump sums for initial costs (transport, outfittings etc) and for running costs that are largely independent of the volumes produced. Water infiltration problems and the cost of pumping, for example, are covered by lump sums which vary stepwise with infiltration rate measured at the end of each tunnel branch. Depending on local conditions, the combined amount for these lump sums may be from 10 to 20% of the total contact figure. Thus unit prices are reasonably suitable for dealing with variable quantities. The lump sums for running cost will be kept constant in most cases if the total unit price does not deviate more than 15-20% from the corresponding contract figure. If it exceeds that limit, it will be adjusted proportionally to the additional bill. The lump sums may also be subject to claims, but only if major change beyond the contractor's control are introduced.

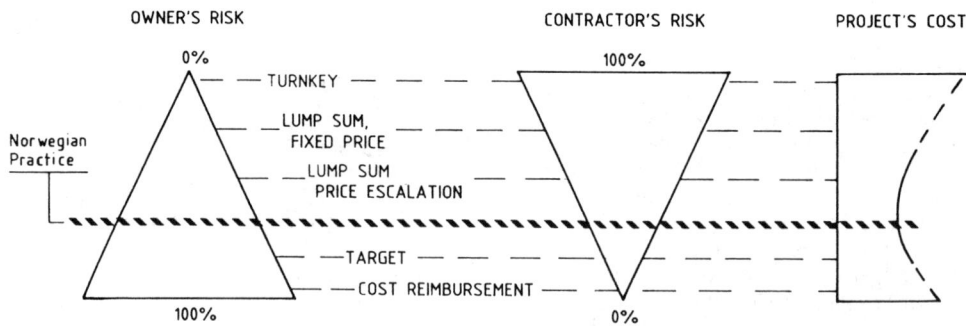

Fig. 2. Risk sharing according to type of contract, and assumed influence on project costs, from Kleivan (1988).

To compensate for deviations from the predicted quantities due to unexpected rock conditions, unit price time equivalents are defined for all the types of rock support. The unit price time equivalent system makes it easy for both the client and the contractor to recalculate the final cost according to the actual rock conditions and to reconsider the time schedule. Some examples of time equivalents for a recent Norwegian hydropower plant are shown in Table 4.

Table 4. Typical time equivalents for medium size hydropower tunnels, from Nilsen and Thidemann (1993).

ACTIVITY	TIME EQUIVALENT
Spot bolting	0.20 shift per 10 bolts
Systematic bolting	0.15 shift per 10 bolts
Shotcreting (reinforced)	0.30 shift per 10m^2
Rigging for shotcreting a new location	0.15 shift per time
Concrete lining (arch)	6 shift per 10m section
Rigging for concrete lining at new location	0.25 shift per location

Since the risk-sharing provisions became common in Norwegian tunnelling contracts, a total tunnel length of 2000 km has have been implemented with this provision of time equivalent risk-sharing in one form or another built into the contract. But very few suits or arbitrations are reported due to deviations from the assumed rock conditions from this period (Kleivan, 1988 and Dahl, 1986).

7. OPTIMIZED ORGANIZATION OF WORK

Norwegian tunnelling for traffic purposes has so far been based almost entirely upon drilling and blasting. The normal routine is as follows: (a) At the heading, 2-3 men on each shift do the drilling, charging, detonating, scaling, mucking out, placing of bolts, nets and ties, and shotcreting and pregrouting. If there are two headings within about 4 km of each other, which will normally be the case when starting out from a new adit, one 3-men crew per shift will do the same work on both faces, except scaling and mucking out, which will be dealt with by different crew. When distance between headings is more than about 1.5 km, there will be a drilling rig at each heading. (b) The day-shift behind the heading will normally consist of one mechanic, one electrician, one operator to install ventilation, piping etc., in addition to one foreman, one engineer/surveyor and if necessary, one operator on the concrete and crushing plant.

Working hours in underground works in Norway have gradually been reduced to 36 per week, and the ordinary night shift was abolished in 1977. This leaves 10 ordinary shifts of 72 working hours per week. Advance rates vary from 1.5 rounds per shift, or about 60 meters per week in stable rock, down to 20 meters per week where a full concrete lining has to be applied after blasting. When the time factor is critical, working hours can be extended up to about 108 per week by bringing in a third crew, which generally means more expensive tunnelling. Lodging expenses increase by about 50%, and the longer shifts tend to be less productive than the traditional 7.5 hours. Professional tunnellers are thus specialized, but also very versatile people, invariably paid in proportion to production. This leads to high productivity, but also to earnings up to twice the normal wages for outdoor work (Dahl, 1986).

8. UNDERGROUND POWERHOUSES WITH UNLINED PRESSURE TUNNELS AND AIR CUSHION SURGE CHAMBERS

In Norway more than 99% of the annual electricity production of approx. 100 TWh comes from hydropower. Putting the powerhouse underground is regarded as conventional solution. The reason for putting the hydropower stations underground is that it gives the most economic solution (Broch, 1989).

Penstocks were first substituted by steetlined pressure shafts and then later by unlined pressure shafts. Today, more than 80 unlined shafts and tunnels with water heads higher than 150m are in operation. As many as 15 have static head exceeding 500m. Recently, an unlined high-pressure tunnel with a static head of 1000 m was put into operation. Since no continuous lining of steel or concrete is used in the unlined concept, the rock mass itself must be able to resist very high water pressure.

The unlined air cushion surge chambers represent the most recent development in the layout in Norwegian underground hydropower plant. Since the first such chamber was successfully put into operation at Driva Hydropower Plant in 1973, ten unlined air cushion surge chambers have been completed. With an air cushion surge chamber, the usual surge chamber vented to the atmosphere at the top of a pressure shaft or a surge tower, can be omitted. The compressed air occupies from 40% to 80% of the chamber volume, and acts like a "cushion" against water hammer effects due to stops and starts or rapid changes in power production. The air cushion surge chamber concept has been used in Norway only. This is probably largely due to the Norwegian rough topography, for which the alternative with air cushion surge chamber in the unlined headrace tunnel often offers considerable cost savings. To illustrate the order of magnitude of potential cost savings, it can be mentioned that for the Kvilldal Power Plant, the air cushion concept represented a cost reduction of approximately NOK 30 millions (1987) compared with a conventional design of surge chamber (Dahlø, 1988).

In the design of unlined air cushion chambers, one main criterion is that the minor principal stress has to be higher than the air/water pressure to avoid hydraulic jacking, and the water pressure in the surrounding rock mass higher than the cushion pressure. In some cases, however, the hydrogeological condition of the area do not make this possible, and the necessary water pressure has to be created artificially with water curtain.

9. UNDERWATER PIERCING

Underwater piercing is the process of piercing the bottom of a lake with a tunnel. The main use of underwater piercing are lake taps, but the principle has found application for some tailrace tunnel and for sewer outfalls. Underwater piercing has been a regular practice in Norway for more than 75 years. During this period it have been completed more than 500, and about 70 since 1980. Piercing of a lake bottom at a depth of 105m with a tunnel of a cross-section 95 m^2 has taken place. The lake taps in Norway relate to the combination of topography and economics. The economic part is the possibility of creating reservoir by drawdown of a natural lake or by combining a low dam and a drawdown. The topographic element is that the many small lakes in high valleys and in the mountains lend themselves to such a solution. The process of tapping a lake can be divided into the design of the tunnel system, the excavation of the tunnel, and the design and execution of the final blast (Berdal et. al., 1985). Lake taps may be descrived as open or closed system according to the allowance of blasting air pressure through the open shaft or not.

10. CONCLUSION

Comparison of construction costs of tunnels and rock caverns among countries is difficult because of the diversity of parameters to be considered. However, despite of the high salary of Norwegian workers this paper shows that the cheapest constructions have been implemented in Norway as result of technologies being applied there. In countries with different geological conditions and construction systems, the Norwegian technology may have its limitaions. The authors believe, however, that many aspects of the Norwegian examples presented here could be applied in other countries directly or with some modification to increase the economical utilization of the subsurface.

REFERENCE

Berdal B. (1985): "Lake tap-Norwegian method". In: "Norwegian Hydropower Tunneling". Publ. No.5, NSREA(Norwegian Soil and Rock Engineering Association), Tapir, pp.115-119.

Broch E. (1988): "Site Investigations". In: "Norwegian Tunnelling Today". Publ. No.5, NSREA, Tapir, p.51.

Broch E. (1989):"Use of the Underground in Norway". In: Proc. of Int. Conf. "Storage of Gases in Rock Caverns" Trondheim, 1989. pp.3-5.

Dahl K. O., (1986): "Norwegian Tunnelling Experience as a Background to the Low-Cost Concept of Road." In: "Norwegian Road Tunneling". Publ. No.4, NSREA Tapir, pp.33-43.

Dahlø, T. (1988): "Experience from air cushion surge chamber". In: "Norwegian Tunnelling Today". Publ. No.5, NSREA, Tapir, pp.77-78.

Don Rose (1991): "Cost Estimating for Underground Structures". In: "Underground Structures Design and Construction". Elsevier, pp.480-515.

Kleivan, E. (1988): "NoTCoS - Norwegian Tunnelling Contract System". In: "Norwegian Tunnelling Today". Publ. No.5, NSREA, Tapir, pp.67-72.

Nilsen B. and Thidemann A. (1993): "Rock Engineering" Norwegian Institute of Technology Division of Hydraulic Engineering. pp. 115-118, 135.

Palmstrøm A. (1992): "Introduction". In: "Norwegian Subsea Tunnelling". Publ. No.8, NSREA, Tapir, pp. 9-10.

Steiner W., et. al. (1980): "Improved Design of Tunnel Support". In: "Tunneling Practices in Austria and Germany". Vol.4, pp.17-36

Øvstedal E. (1986): "Standard and Costs of Norwegian Road and Tunnels". In: "Norwegian Road Tunnelling". Publ. No.4, NSREA, Tapir, pp.19-26.

Spatial discreteness of geological environment and of underground drainage constructions in Odessa, Ukraine

E.A.Cherkez, T.V.Kozlova & V.I.Shmouratko
Odessa State University, Ukraine

ABSTRACT: The underground drainage constructions of the antilandslide complex in Odessa are used as a tool for revealing and study of high-frequency tectonic breaking. It is found out that the upper earth crust on the territory of Odessa is tectonically discontinuous. Five hierarchical levels of tectonic breaking of geological space are distinguished: 30-60m, 100-200m, 400-600m, 800-1,200m, and approx. 2,400m. Vertical and horizontal relative movements of microblocks make few mm per year. Nevertheless, the constructions preserve high engineering-geological and technical efficiency.

1. INTRODUCTION

Modern exogenous processes (subsidense, landslide, karst phenomena, etc.) are traditionally recognised as the main reasons for failures and deformations of engineering constructions. Researchers often oversee the factor of tectonic high-frequency step-type behaviour of geological space. However, recent insights in the problem have shown that in Odessa region (the North-Western coast of the Black sea), tectonic breaking of rock massifs is quite clearly seen (Voskoboynikov & Kozlova 1992; Shmouratko 1993; Zelinsky & Kuzmenko 1994; Cherkez 1996; Kozlova 1996).

Characteristic block sizes form the line spectrum. It is significant that sizes of smallest blocks are comparable with sizes of buildings and constructions. This fact formulates the necessity of making considerable corrective amendments in engineering-geological research techniques. It is obvious that the high-frequency tectonic step-type behaviour cannot but affect the features of rock massifs stressed condition, the features and nature of space distribution and dynamics of engineering-geological processes, formation of underground water flow structure and regime, and many other parameters of nature-technogenic systems.

As a tool for revealing and study of high-frequency tectonic breaking, it is appropriate to use relatively long linear constructions for lithological monitoring. On the territory of Odessa, the underground drainage constructions of the antilandslide complex have been working since 1964 as a tool of this kind.

The Odessa coast antilandslide measures called for elimination of negative influence of basic landslide-forming factors and included: (1) liquidation of coastal cliff wave wash-out, and artificial beaches creation; (2) slope cutting and lay-out; (3) interception and removal of underground waters from the Quaternary and The Pontian aquifers to the sea; (4) regulation of surface runoff.

The drainage constructions consist of three main elements:

(a) a straight row (approx. 3,300m long) of 143 drainage wells located at a distance of 15-25m from one another. They are meant for the reduction of the Quaternary aquifer level and for water dumping into the drainage gallery;

(b) the accomplished drainage gallery 5,200m long (the 1st section), and 5,700m long (the 2nd section), which is driven in bed-rock at the borderline of the Meotian clay and the Pontian limestone and is oriented parallelly the coast-line. The gallery is located at the distance of 100-180m from landslide slope (fig.1). It is intended for the Pontian aquifer draining;

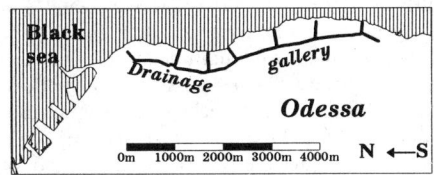

Fig.1. Plan of the drainage gallery location along the Odessa sea coastline.

(c) derivation adits oriented normally to the coast-line and connected with the gallery in the bed-rock massif. The distance between the adits makes approximately 1 km. They are intended for dumping drainage waters from the two aquifers into the sea. The adits are equipped with the geodetic benchmark network. The drainage gallery and derivation adits have the round section 220cm in diameter. They were attached in sections made of four re-inforced concrete tubings.

The geological section of the Odessa coast deep into the zone of probable influence of large landslides (60-70m from the surface plateau whose absolute marks make 40-50m) is displayed by stratigraphic-genetic complexes of the Meotian, Pontian, Middle and Upper Pliocene and Pleistocene ages. General bedding of rocks is seen as a monoclinal fall slightly inclined with azimuths to the South. In the region of the Odessa Bay above the coastal cliff are exposed clays of the Meotian stage, limestones of the Pontian stage, sand-clay deposits of the Upper Pliocene and loess-like rock of the Pleistocene (fig.2).

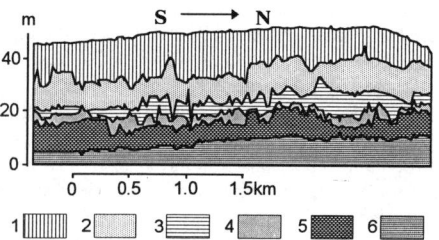

Fig.2. Fragment of a geological section along the drainage gallery.
1 - Pleistocene loess; 2 - Pleistocene loess-like loam; 3 - Upper Pliocene red clay; 4 - Alluvial sediments on Pontian limestone; 5 - Pontian limestone; 6 - Meotian clay.

The lithological monitoring for evaluation of the engineering-geological and technical efficiency of the complex and its separate elements is performed by the Odessa Antilandslide Department. It includes the following kinds of monitoring: 1) quarterly measurement of drainage well yields; 2) annual measurement of specific water inflows from the Pontian aquifer into the drainage gallery, 3) monthly measurement of adit yields; 4) geodetic monitoring of the surface benchmark network on plateaux and slopes; 5) geodetic monitoring in adits: benchmark levelling, measurement of longitudinal deformations of adits on sites between benchmarks; 6) geodetic monitoring for hydrotechnic engineering constructions (wavebreakers, traverses); 7) measurement of artificial beaches dynamics and of the Bay depths soundings.

The significant part of data received as a result of the above-mentioned monitoring can be used effectively enough for solving applied as well as research tasks. In particular, due to the fact that a distance between drainage wells is short (15-25m on the average), one can solve problems related to the study of high-frequency space variability of hydrogeological, geodeformational, geophysical and other parameters of the geological environment. In addition we performed the following kinds of measurement: 1) one-time measurement of general salt percentage in the underground waters run off through drainage wells into the gallery (data for 1996); 2) measurement of diameters of the gallery and adits in vertical and horizontal sections in 1990 (the distance between measurement points made 10 m); 3) one-time measurement of the natural impulse electromagnetic field of the Earth inside the gallery and on the day time surface along the gallery (the distance between monitoring points made 10m) (data for 1990).

2. YIELD OF THE DRAINAGE GALLERY WELLS

Drainage wells are absorbing wells that dump underground waters into the drainage gallery. The average distance between wells is equal to 23.6m. In the given work, we analyse the temporal series of average annual yields of drainage wells for the period from 1966 to 1994. Well yields were measured in the gallery according to the volume method.

The analysis of data received has shown the annual average yields change over a wide range — from 0.1 to 22.3 m^3/day. The sharp change of yields in adjacent wells located in 15-30m from one another is often observed. Approximately 50% of wells have the yield of less 2 m^3/day, about 25-27% — 2-4 m^3/day, about 10% — 4-6 m^3/day, approximately 15% are characterized by the yield of over 6 m^3/day.

Fig. 3 shows curves of the annual space changes of well yield for 28 years. As illustrated, the space arrangement of zones concerning to high and low figures of well yields are preserved through all 30 years of monitoring. It is important to emphasize that sites of high and low water abundance in wells are comparable in size and are alternated in space. The sites of space with relatively high yield are designated in fig. 3 with lowercase letters of the latin alphabet. Characteristic distances between centres of increased yield zones vary from 400 to 600 metres.

Minimum (D_{min}) and maximum (D_{max}) long-term yield amounts, as well as range of yield change for the whole period of monitoring (D_{range}) within the limits of sites with increased water abundance of wells always exceed corresponding figures on sites with lower yield amounts. Each of these parameters can serve as a criterion for geological space typification. The form of the bending around maximum long-term yield amounts (D_{max}) allows to assume that the zones a-e conform to one hierarchical level, a higher one. As will be shown below, the characteristic width of zones of this hierarchical level is approx. 2400m.

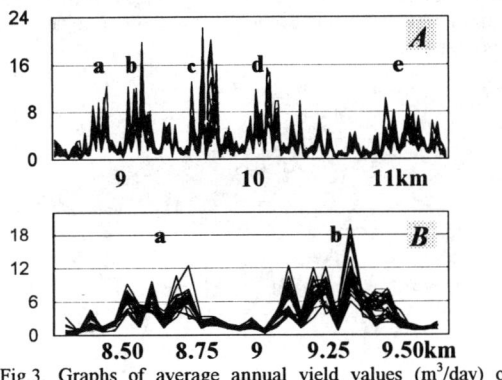

Fig.3. Graphs of average annual yield values (m³/day) of drainage wells, from 1966 to 1994:
(A) - along the whole drainage gallery;
(B) - on sites of zones (a), (b) with increased yield values.

Fig.3 designates zones of this level with letters D and E. Besides these two levels, data on drainage well yields allow to allocate the third hierarchical level of the geological space structurisation. In fig.3b the drainage well fragment is shown on the site of placement of zones a and b. In each of them, stand out 3 to 4 increased yield zones, characteristic distance between them ranging from 30 to 60m.

As long as the size of the latter discontinuity is comparable to inter-well distances, it is necessary to consider the technical aspects of well performance. As for their design, all wells are of the same type. It allows to argue that the change of yield amount between adjacent wells is not caused by distinctions in their design. Besides, the experience of regime monitoring for underground water levels near to the drainage gallery shows that the radius of sharp flow deformation zone is under 10m.

It makes less than half of size of an average distance between drainage wells. This permits to argue that hydro-geologically revealed nonuniformities (30-60m) are really existing nonuniformities of geological space. Thus, the analysis of time series of annual average well yield in the drainage gallery reveeealed three hierarchical levels of the step-type behaviour of geological space, characteristic sizes of which are 2400m, 400-600m and 30-60m.

3. GENERAL SALT PERCENTAGE OF WATER IN THE QUATERNARY AQUIFER

It is known that the intensity of water exchange and time of interaction of water with host rock largely define the structure component and general salt percentage of underground waters. In zones of increased permeability, the period of interaction of the technogenic underground water) aquifer with host rock is small. Therefore it is reasonable to expect that the water from high-yield drainage wells will have low salt percentage.

In 1996 the selection of water tests was made for determination of general salt percentage of underground waters drainaged by the gallery wells. The analysis results are shown in fig. 4. The salt percentage varies from 0.1 to 9.3 g/dm³. It is important to draw attention to the fact that the variability of salt percentage in space also forms at least three hierarchical levels which fully coincide with similar levels of yield variability. On all levels of hierarchy the dependence between general salt percentage and yield is nearly always an inverse proportion. Besides yields of wells draining the Quaternary aquifer, were analyzed the data on the Pontian aquifer water inflows into the drainage gallery.

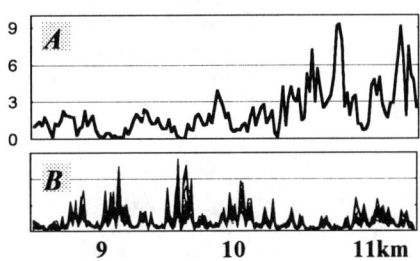

Fig.4. Comparison of general salt percentage (g/dm³) of drainage water (A) and well yields (B).

4. WATER INFLOWS FROM THE PONTIAN AQUIFER INTO THE DRAINAGE GALLERY

On sites of the first and second sections of the antilandslide constructions in Odessa, the total length of the drainage gallery makes 11,600m. Measurements of specific water inflows were made with the interval of 50m. Fig. 5 shows the curve of

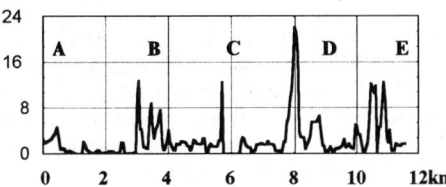

Fig.5. Specific inflow of the Pontian aquifer (m³/day*1 linear metre) on sites of the first and second section of the antilandslide complex of the Odessa sea coast.

space change of specific water inflows into the gallery. According to this parameter, two hierarchical levels of the step-type behaviour of geological space emerge. The characteristic step of the first one ranges from 2200 to 2600m, and the second one — from 800 to 1200m.

5. RADIAL DEFORMATIONS OF THE DRAINAGE GALLERY

The data about radial deformations of the drainage gallery are received as a result of measurement of its vertical and horizontal diameter using a telescopic rack. The measurements were made every 10m along the gallery site 3.9km long. While in construction (1965), the gallery diameter on the internal section made 220cm. We accept this size as the reference value. According to technological conditions of the work performance, the deviation from the given diameter should not exceed ±1cm.

We measured the internal diameter of the gallery in 1990. The data received characterise the accumulated deformations of the gallery for the recent 25 years. The sizes of deformations range from +14 to -17cm. The functional correlation between sizes of horizontal and vertical deformations is an inverse proportion. Fig. 6 shows the size of correlation of the drainage gallery horizontal deformations to vertical deformations.

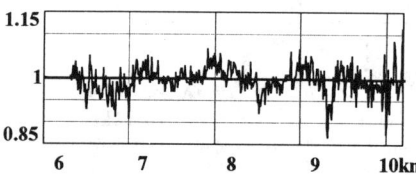

Fig.6. Factor of radial deformation of the drainage gallery (ratio of horizontal to vertical gallery diameter).

Spectral analysis of sizes of the gallery deformations revealed three characteristic space periods of their dynamics. The periods can be interpreted as three hierarchical levels of geological space discontinuities. The characteristic sizes of levels determined by this method are 800-1200m, 200-300m, and 30-60m.

6. GEODETIC MONITORING IN ADITS

Adits are subhorizontal workings 350-400m long driven in landslide deposits and bed-rock at a distance of 150-200m from slope edge (see fig.1). Adit deformations were defined via measurement of vertical and horizontal bench-mark displacement. Generally, for each adit in the monitoring period, 18-20 measurement cycles were made.

Judging from geodetic monitoring experience, all bench-marks (even the remotest from the landslide slope) are displaced. Taking this into account, each measurement cycle while processing levelling data included estimation of the relative changes of absolute marks for each bench-mark with regard to the nearest one from the bed-rock side. This allowed to reveal the adit 'body' vertical deformations accumulated for the whole period of the constructions operation. The accumulated deformations are shown as wavy bends with wave-length ranging from 60 to 120m from one adit to another (fig.7).

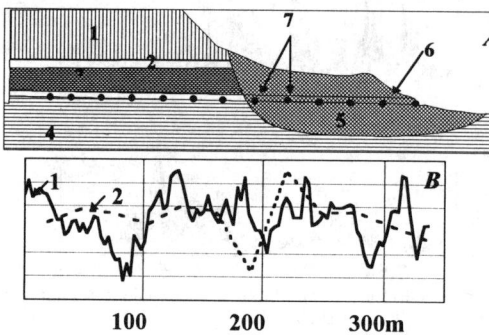

Fig.7. A - Geological cross-section along adit #3:
1 - Pleistocene loess; 2 - Upper Pliocene red clay; 3 - Pontian limestone, 4 - Meotian clay, 5- landslide deposits, 6 - adit #3, 7 - bench-marks.
B - Graphs of normalized values:
1 - factor of radial deformation in adit #3; 2 - accumulated deformation of the 'body' of adit #3 in the vertical plane.

It should be emphasized that the geological composition of a landslide slope on different sites at the Odessa coast is characterized by 'hills' and 'subsidences' in the Meotian clay. Clay relief deformations are distinguished by a relatively regular step (50-100m) in alternations of rising and sinking sites, difference of marks between them being 10-20m. The wavy character of the clay surface relief is stipulated by the bench-mark plastic deformations, discovered both in landslide preparation and on the stage of main displacement.

The wavy bends of the Meotian clay surface and adit bodies are comparable in wave-length. Like in the gallery, measurements were made in adits of their horizontal and vertical diameters. It has been found out that adit deformations are nearly analogous to the gallery deformations. In particular, spectral analysis reveals two hierarchical levels of step-type behaviour — 20-30m, and approx. 100m.

7. DISCUSSION

As far as the above-mentioned data characterise basically properties of two section elements — the loess formation and the Pontian limestone layer, — we should go into detail in their description. The Quaternary loess formation is represented by alternation of loesses and loess-like loams (the 'buried' soils), whose total thickness in Odessa region equals 15-20m on the average.

Formation and distribution of the Quaternary deposits underground waters depend on natural and technogenic factors. Owing to leakages from the utilities, by the 1970's the Quaternary deposits underground waters had appeared on the whole territory of Odessa. Now, thickness of this aquifer makes 15-18m. Its aquifuge is the red clay. Where they are absent (hydraulic 'windows'), underground waters flow in the underlying Pontian limestone layer. Loess rock has low filtration properties and weak water output.

The loess formation permeability decreases in the downward direction; the variability of filtration coefficients of the adjacent layers reaches 10-20 times, and within a separate layer in vertical and horizontal directions — 2-10 times.

The Pontian deposits on the Odessa territory are predominantly represented by two lithologic limestone types: the oolite and the shelly limestone overlapping it. The limestone thickness ranges from 2 to 15m. Zones of increased jointing in limestone form the system of the north-west and north-east strikes. There are more north-west-directed zones on which more intensive karst formation took place.

Most of the Pontian aquifer water is formed locally, on the city territory, owing mainly to water downflow from the top aquifer. The filtration properties of limestone vary a lot and on the average exceed 100-fold the loess rock filtration coefficients. It is traditionally considered that for loess-like rock in comparison with other types of sediment rock of continental genesis, rather weak lateral variability of properties is characteristic. This is connected their genesis peculiarity. At the same time, the analysis of space variability of hydrogeodynamic parameters shows high-frequency and multilevel lateral discontinuity of the Upper Pontian deposits. The generally assumed lithologic uniformity of loess properties seems to contradict these facts.

This contradiction, however, is easily enough eliminated, if we accept a new model of hydrogeo-deformation multilevel step-type behaviour of geological space, which is being developed by the authors.

Pursuant to this model, and taking into account that the space variability of well yields does not depend on the aquifer thickness, zones of increased yield amounts, — which assume stable position in space for at least 28 years — should probably be considered as zones of increased jointing in loess rocks. Such interpretation is confirmed, in our opinion, by the fact that the range of interannual yield changes takes maximum values.

We connect the interannual variability of yield with change the jointing space size. The dynamics is caused by vertical and horizontal tectonic movements. We are speaking about tectonic movements of block in the upper earth crust.

Characteristic sizes of the blocks form the hierarchical series consisting of five levels — 30...60m, 100...200m, 400...600m, 800...1200m, 2200...2600m.

Only the hierarchical system of zones of increased jointing in the loess formation satisfactorily explains the vividly expressed back proportion between the drainage well yield amounts, and salt percentage of the drainage waters. The dramatically distinct filtration properties of rock in zones of increased and decreased jointing also cause different speed of carbonate minerals dissolution. In zones of increased jointing, the duration of water exchange is insignificant (the 'flow' regime). Due to that, underground water saturation by soluble loess components does not occur.

Therefore, the salt percentage of drainage waters in these cases is close in amount to salt percentage of waters from municipal utilities.

With all this in mind, a natural explanation is given also to the fact of radial deformations of underground drainage constructions. The relative movement of tectonic blocks of the considered hierarchical levels lead to the underground constructions deformations on the borderline between blocks. This is why the deformations are of vividly expressed discrete nature and, therefore, for radial deformations of the gallery and adits (figs. 6 and 7) are characteristic such hierarchical levels of step-type behaviour, as well as for the drainage well yields (fig.3).

Geodetic monitoring in adits testifies that the relative movements of blocks of 60 m size eventually lead to formation of wave profile of the Meotian clay roof not only in landslide slope (Zelinsky *et al.* 1993), but also in rock massif. It is possible to assume that due to tectonic movements of small-sized blocks, formation of zones of local plastic deformations occurs, i. e. tectonic processes create and constantly keep in action a regular multilevel lineament network which serves as the structural-geological base for most exogenous processes (landslides, karst, abrasion, etc.).

8. RESULTS

The results obtained permit to make the following conclusions.

1). The upper earth crust on the territory of Odessa is tectonically discontinuous. Five hierarchical levels of tectonic breaking of geological space are distinguished: 30-60m, 100-200m, 400-600m, 800-1,200m, and approx. 2,400m.

2). Vertical and horizontal relative movements of microblocks make few mm per year. They deform noticeably engineering constructions.

3). It is effective to use linear constructions as a tool for revealing high-frequency space-time variability of parameters of the geological environment.

4) The antilandslide complex constructions on the Odessa coast for the 30-years operation period have been subject to deformations stipulated by microblock movements. Nevertheless, the constructions preserve high engineering-geological and technical efficiency.

REFERENCES

Cherkez, E.A. 1996. Geological and structural-tectonic factors of landslides formation and development of the North-Western Black Sea coast. In K.Senneset (ed.), *Landslides — Proc. 7th Int. Symp. on landslides, Trondheim, 17-21 June 1996*: 509-513.Rotterdam: Balkema.

Kozlova, T.V 1996. Structural-tectonic and lithogenetic features of a rock massif as factors of landslide processes In K.Senneset (ed.), *Landslides — Proc. 7th Int. Symp.on landslides, Trondheim, 17-21 June 1996*: 245-249.Rotterdam: Balkema.

Shmouratko, V.I. 1993. Role of the multi-storey tectonics at the engineering geological estimate of an area. *Geoecology*, 2: 79-93 (in Russian).

Voskoboynikov, V.M. & Kozlova, T.V. 1992. Use of the geodynamic analysis and method of the generalized variables for estimating and predicting the stability of landslide slopes (by the example of the Northern Black Sea region). *Engineering geology*, 6: 34-49 (in Russian).

Zelinsky, I.P. & Kuzmenko, G. I. 1994. On fractures and block structure of the earth crust. *Reports for the Academy of Sciences of Ukraine*, 12: 93-96 (in Russian).

Zelinsky, I.P., Moiseev, L.M., Khanonkin, A.A 1993. On the nature of plastic clay soil deformation at the landslide slopes of Odessa coast. *Geoecology*, 2: 55-66 (in Russian).

Deformations of the shallow tunnels in flysch rock mass

K.Thiel
Polish Academy of Sciences, Warsaw, Poland

L.Zabuski
Institute of Hydroengineering of the Polish Academy of Sciences, Gdansk, Poland

ABSTRACT: The paper presents selected results of analysis of deformation processes, observed during excavation of hydrotechnical tunnels, each of 9m diameter. Tunnels are located at the depth of about 55m below the terrain, in the flysch rock mass. Systematic measurements of the roof settlements, convergence as well as settlements of the terrain surface above the tunnels allowed for the collection of comprehensive results. This was the basis for elaboration some general rules for the dependences between above mentioned deformations. The paper presents elaborated models for approximation and prognosis of the deformation processes as well as the relationships between tunnel roof and terrain surface settlements.

1 INTRODUCTION

There is little number of geotechnical problems, in which deformation measuremets play such an important role as in case of underground excavations. Deformations are visible result of many processes - such as changes of natural state of stress, stress redistribution - developing in the rock mass, surrounding the excavation. The knowledge about deformation processes makes possible the recognition of these phenomena. Thanks to this, stability and security state could be relatively easy evaluated. It is the main reason, why the distinct development of deformations monitoring in the last years is noted. In many cases the results of measurements decide about the method as well as about possible corrections during excavation. Independently on the above described, purely practical scope of the measurements, their results allow for investigation of the processes around the opening and for evaluation of the influence of rock mass characteristics on the construction behaviour, etc. Generalization of the results could be important contribution in the future stability analyses. The example of the analysis of deformations measured during the hydrotechnical tunnel excavation is presented in the paper. The mathematic models are described and some examples are shown, which could be helpful in the approximation of the deformation curves as well as in the prediction of the futural deformations in the tunnels on the base of deformations actual. Additionally, dependences between terrain surface and roof settlements are analysed.

2 DESCRIPTION OF THE ROCK MASS AND THE TUNNELS

Water reservoir is constructed in the South of Poland, in Carpathy Mountains. The reservoir is created by earth dam construction. Two hydrotechnical tunnels are excavated inside the slope of the river valley, each of internal diameter of 6.5m. The tunnels allow for the outflow of water in the construction and exploitation stage of the dam and reservoir. Rock mass is built of sedimentary flysch formation. It is composed of strong and hard sandstone beds, interbedded with weak and soft clay shales. The average content of the sandstone is equal to about 70%. The rock mass is intersected with two systems of joints, approximately perpendicular to the bedding planes. Geological picture is generally very differentiated. Rock mass is heterogeneous, anisotropic, discontinuous and tectonically disturbed.

Tunnels are located at the depth up to 55m, they could be therefore qualified as "shallow excavations". In such a case, geological features of the rock mass play an important role during the excavation works. The tunnels were performed using the procedures based on the principles of New Austrian Tunneling Method. Construction process was divided into two phases. Preliminary supported tunnels were constructed in the first phase and permanent reinforced concrete lining was installed in the second. The support was composed of rockbolts, shotcrete and steel ribs. The first phase was divided into two stages. Upper half of the tunnel (kalota) was first performed on its whole length and lower part was next excavated. The tunneling procedure was accompanied by rock mass classifying. Two original classifications, KF and KFG were elaborated (Bestyński et al.,1990). Both of them are correlated with RMR classification (Bienawski,1984). Support characteristics were generally designed, basing on the rock mass class.

Careful monitoring was carried out. Deformations were mainly measured and analysed simultaneously with the excavation progress. The results allowed for the direct estimation of excavation and support appropiateness. The deformation monitoring programme comprised measurements of the tunnel shape changes (convergence, vertical displacements of roof and side-walls), extensometric measurements and mesurements of terrain surface settlements.

3 APPROXIMATION AND PREDICTION MODELS FOR DEFORMATION PROCESSES

The aim of the analysis was to determine the factors influencing deformation processes and to find a mathematic models, which approximate these phenomena and allow to predict future behaviour.

3.1 Displacements in the tunnel

Vertical roof displacements as well as convergence of the side-walls were considered. The equation was defined, which appropriately approximate these deformations, having a form:

$$U = A[1 - \exp(-B \cdot L)] \cdot t^C \qquad (1)$$

where:
t - time in days from the "zero" measurement,
L - distance between tunnel face and measuring benchmark,
A, B, C - empirical coefficients.

The coefficient A depends on the final displacements, measured after very long time -

practically after 1÷2 years. Irregularities in the tunnel advance have an influence on the value of B coefficient. Coefficient C first of all depends on the velocity of the displacements stabilization.

Figure 1 presents the example of above model application for the approximation of the roof vertical displacements. The tunnel advance was stopped for about 60 days, and the inclination of the approximation curve was significantly smaller during this period.

The model described allows also for the prediction of the roof displacements. The more regular is tunnel advance, the better prognosis. Greater number of the results gives more accurate prediction. Example is shown in figure 2. Values of the coefficients in equation (2) were calculated on the base 593 days and 1128 days measurements. One can see, that these values as well as the approximating curves are very similar. It proves, that the prognosis for a long time ahead is possible and accurate under the assumption, that the external conditions are not changed.

It should be mentioned, that the same equation was used for the prediction of the side-walls convergence. It could be thus said, that it has an "universal" sense and that the described deformation processes have the same character.

3.2 Vertical displacement on the terrain surface

The benchmarks are located on the terrain surface above the axes of the tunnels. Vertical displacements of some selected benchmark were measured, before the tunnel face was under it. Displacements were most intensive in the moment, in which the tunnel face and benchmark were on the same vertical plane. In the next stage their velocity slowly decreased. The greater distance between the face and benchmark, the smaller settlement velocity.

Equation describing this process was found, having a following form:

$$U_p = D \cdot \{1 + E \cdot \log(L^2 + 1)^{0.5}\} \cdot t^F \qquad (2)$$

where:
t, L - see eq.(1),
D, E, F - empirical coefficients.

Example of settlement approximation, by application of above equation is shown in figure 3. It could be seen, that the benchmark begins to displace for a long time before the tunnel face is

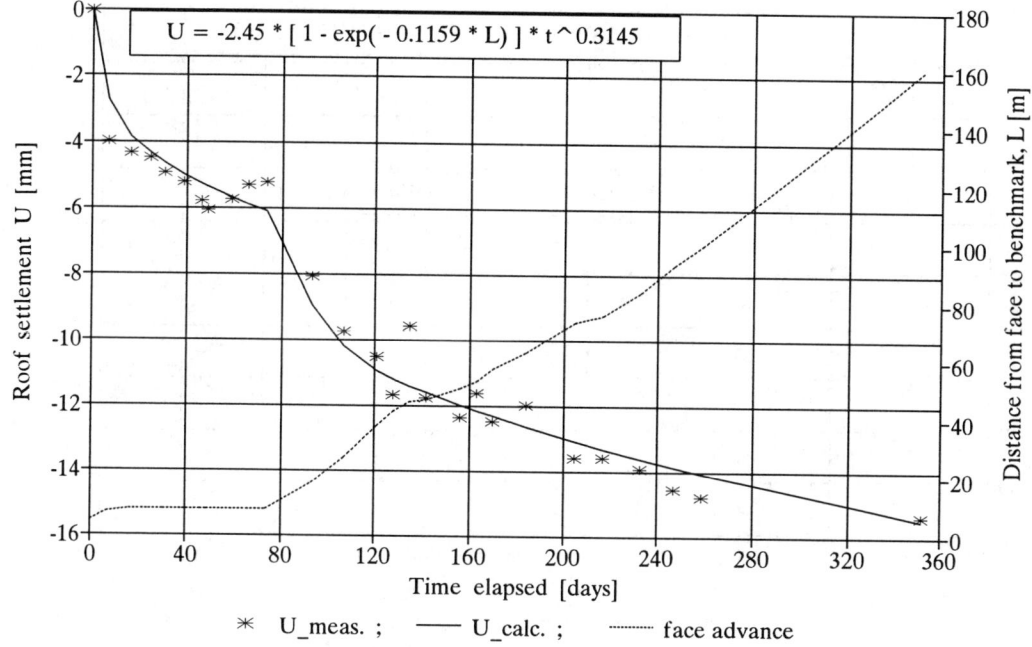

Fig.1. Approximation of the roof benchmark settlements

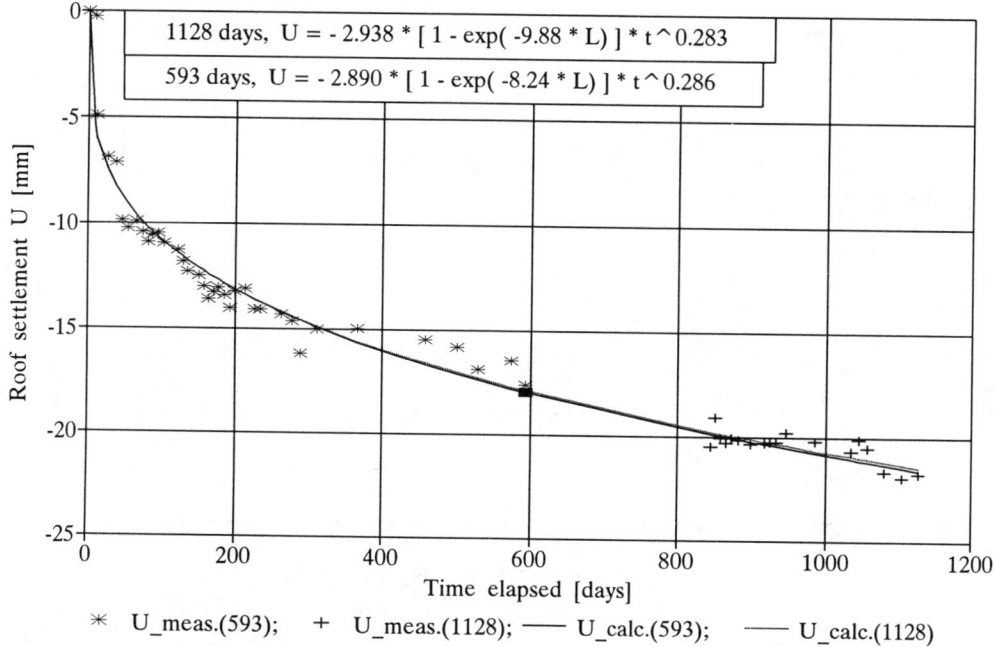

Fig.2. Prognosis of the roof benchmark settlements

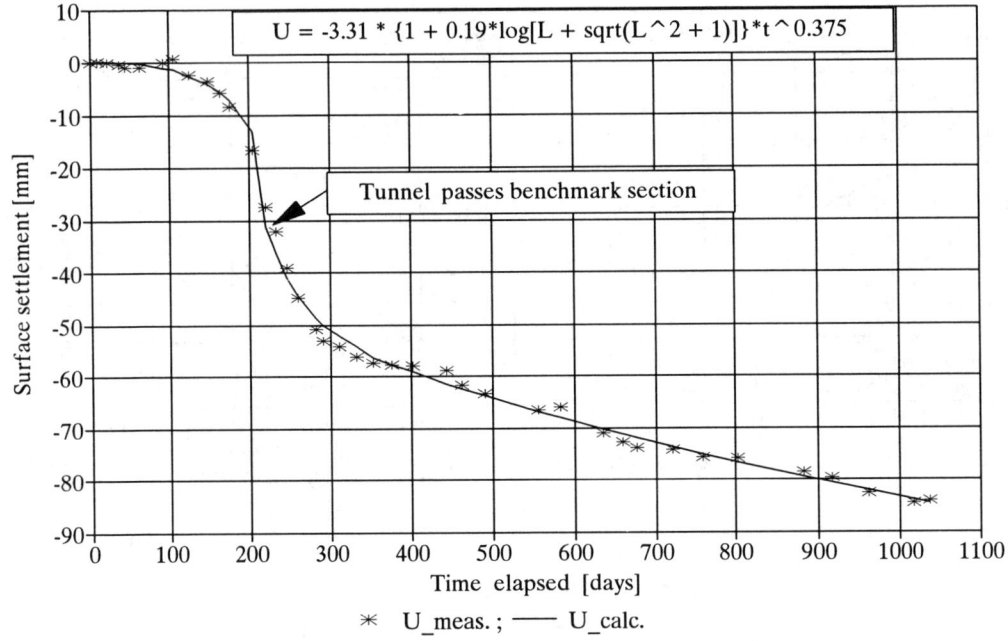

Fig.3. Approximation of the superficial benchmark settlements

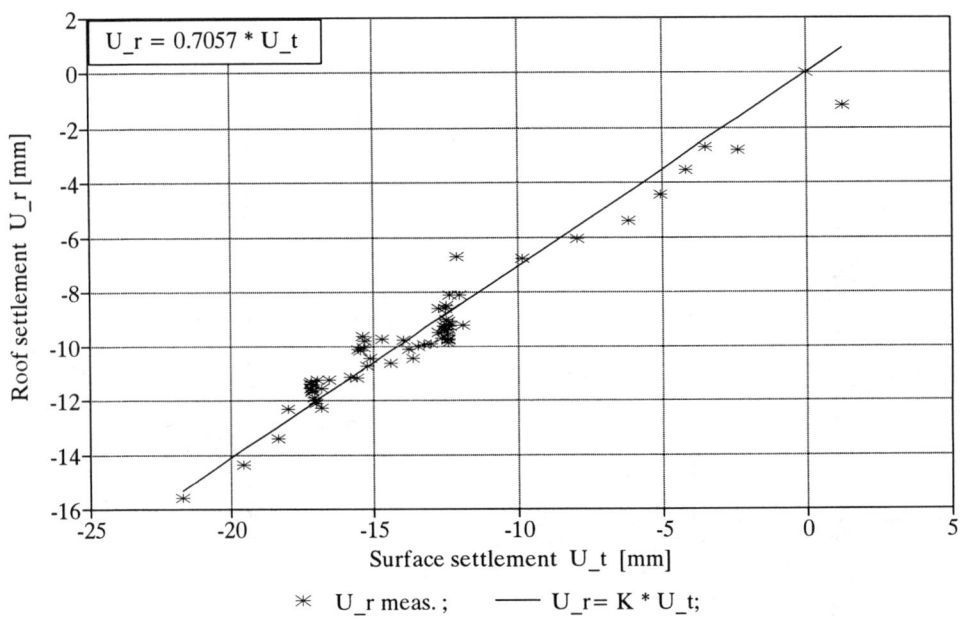

Fig.4. Correlation of the tunnel roof and superficial benchmarks settlements

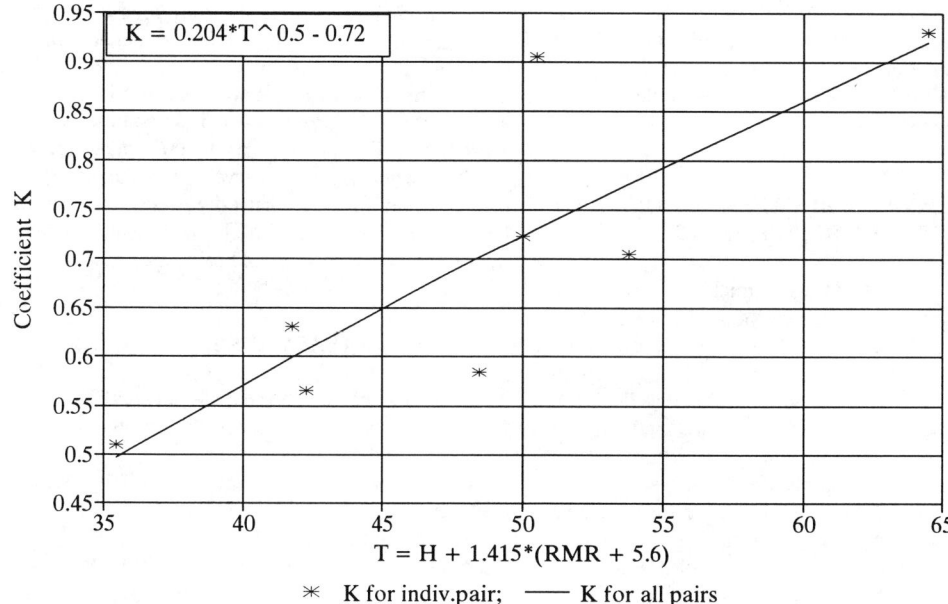

Fig.5. Coefficient K versus T

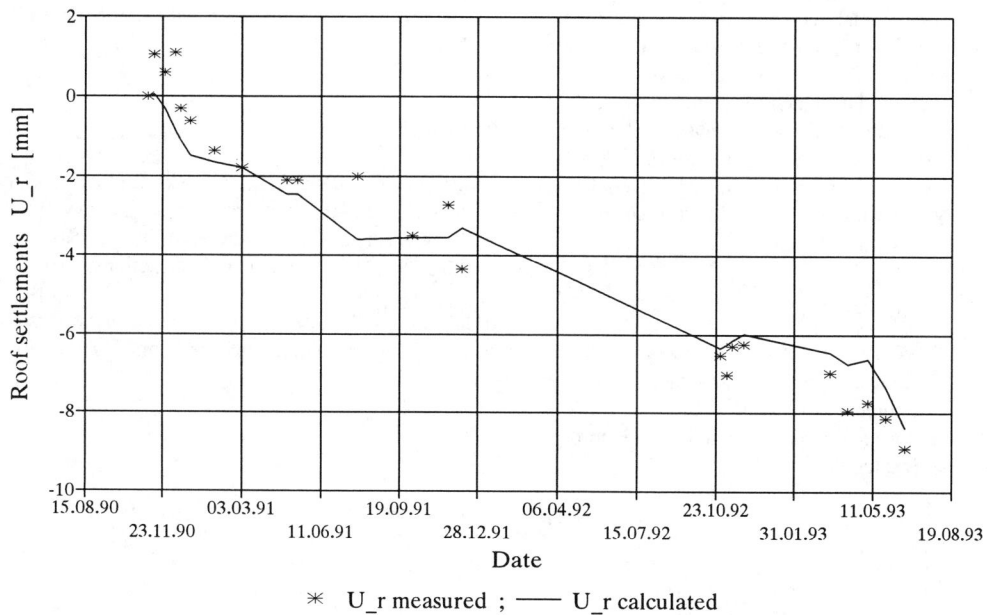

Fig.6. Measured and calculated settlements of the roof benchmark K10

under it. Stabilization of the settlements is very long and takes about three years. The meaning of the equation coefficients is similar to this for A, B, C coefficiens in eq.(1). Prediction of the settlements is also possible in this case.

4 CORRELATION BETWEEN ROOF AND TERRAIN SURFACE SETTLEMENTS

Relatively small oberburden thickness over the tunnels suggests, that the relations between roof and superficial settlements exist. Quantitative description could be important, both for practice and for theory. If such relations would be at disposal, the surface settlements could be calculated on the base of these, measured in the roof. Inverse procedure, i.e. determination of the roof settlements U_r basing on the settlements of the terrain U_t seems to be more important. Monitoring in the tunnels is difficult in some cases. Moreover, measurements in some point inside the tunnel start after excavation reaches this point and the benchmark is installed. On the contrary, the measurements on the surface can begin at any time before the tunnel face. Thanks to this possibility, superficial settlements are recorded, resulting from the excavation works in the surrounding. Basing on the possible correlation between deformational behaviour of the surface and the tunnel roof, one could approximately estimate total settlements of the roof. Such a case is considered below. The measurements results for eight pairs of benchmarks are analysed. The "pair" means the roof benchmark and the benchmark, installed above it, on the terrain surface. The linear relation between settlements was determined in form (see fig.4):

$$U_r = K \cdot U_t \qquad (3)$$

Empirical coefficient K depends on tunnel depth and on rock mass quality (expressed as KF classification number). Taking into account correlation between KF and RMR one can formulate the following equation:

$$T = H + 1.415 \cdot (RMR + 5.6) \qquad (4)$$

Function $K = f(T)$ has a form:

$$K = X1 \cdot T^{0.5} + X2 \qquad (5)$$

X1 and X2 for the tunnels described in the paper have a values: $X1 = 0.204$; $X2 = -0.72$

Figure 5 shows the curve determined on the base of equation (5); the K values calculated for the individual pairs of benchmarks are also shown. As it is seen, the dependance between U_r and U_t is almost linear. Figure 6 presents two examples of roof settlements. The points mean the measurements results, whereas the curves are calculated by using equation (5), on the base of surface benchmarks settlements. The accordance between the settlements measured and calculated is seen.

5 CONCLUDING REMARKS

The validity of the presented results is limited to the geological and technical conditions similar to these, described in the paper, i.e.:
- rock mass quality poor or medium (III÷V class in RMR),
- tunnel depth up to 50÷60m,
- support composed of rockbolts, shotcrete, steel arches or ribs,
- tunnel diameter between 6 and 10m.

It should be mentioned, that the best method of monitoring are direct measurements. However, there are sometimes difficulties in realization of the neccessary monitoring programme. In such a cases proposed methods help to obtain the informations about construction behaviour with some degree of probability.

REFERENCES

Bestyński Z., Thiel K., Zabuski L.(1990), Geotechniczne klasyfikacje masywów fliszowych (Geotechnical Classifications of Flysch Rock Masses), Hydrotechnical Transactions, Vol.52, pp.143-164

Bieniawski Z.T. (1984), Rock Mechanics Design in Mining and Tunneling, Balkema, Rotterdam-Boston

… *Environmental and Safety Concerns in Underground Construction, Lee, Yang & Chung (eds)*
© 1997 Balkema, Rotterdam, ISBN 90 5410 910 6

Effective face distance in an Indian tunnel through jointed and weak rock masses

R.K.Goel, A.Swarup & A.K.Dube
CMRI Regional Centre, CBRI Campus, Roorkee, India

ABSTRACT: The distance of tunnel face that affects the stress re-distribution is useful for the planners and the designers for designing the tunnel supports and to know the time for permanent supports.
The authors have attempted to study this aspect in an Indian tunnel through jointed and weak rock masses where they have actually measured the support pressures and closures. In the paper a correlation has been proposed to estimate the effective face distance.

1 INTRODUCTION

The insitu stress gets disturbed as soon as an underground opening is excavated. This results into the re-distribution of stresses to attain a new equilibrium state. The equilibrium state of stress at a given location, generally, is achieved only when the working tunnel face is sufficiently away from this location. The distance of tunnel face that affects the stress re-distribution may be termed as 'Effective Face Distance'. How much would be the effective face distance in a given geomining condition? This is still debatable. Some researchers are of the opinion that the effective face distance is almost three times the diameter or span of the tunnel, whereas others argue that this concept may be conservative specially in good rock masses.

The knowledge of effective face distance is useful for the planners and the designers for designing the tunnel supports and to know the time for permanent supports.

The authors have attempted to study this aspect in an Indian tunnel where they have actually measured the support pressures and closures.

2 THE TUNNEL

12 m excavated diameter horse-shoe shaped tunnel is in the final stage of construction to discharge the tail water of Salal Hydroelectric Project Stage - II. The project site is about 120 km North of Jammu city of Jammu & Kashmir state of India. The 2.6 km long tail race tunnel (TRT) is almost parallel to the TRT of Stage - I project and is 110 m apart Centre to Centre. The construction of the tunnel (TRT-II) was started in May 1990 through the jointed and sheared dolomites of Lower Himalaya. The excavation was started from three working faces with heading and benching method of excavation. One face was obtained from outlet side of the tunnel, whereas two faces have been developed by an intermediate approach adit intersecting the main tunnel at chainage (ch.) 650m from inlet side. These two faces have been named as inlet upstream and inlet downstream. Excavation and primary supporting work are almost completed, whereas the lining is going to be completed shortly.

3 GEOLOGY

The tunnel is aligned through single litho-unit of dolomitic rocks (Table 1). Since the site is located in close proximity of the 'Main Boundary Fault (MBF)', the dolomites are highly jointed. The MBF separates the younger tertiaries from the older rocks. The geological X-section is presented in Fig. 1 which shows the presence of a anticlinal fold with its axis trending NNW-SSE. The anticlinal fold is probably responsible for the change in the strike direction of the bedding planes from inlet side of the tunnel to outlet side. At inlet side, the dolomite generally strike N80°E - S80°W to E-W with dip 50°-60°

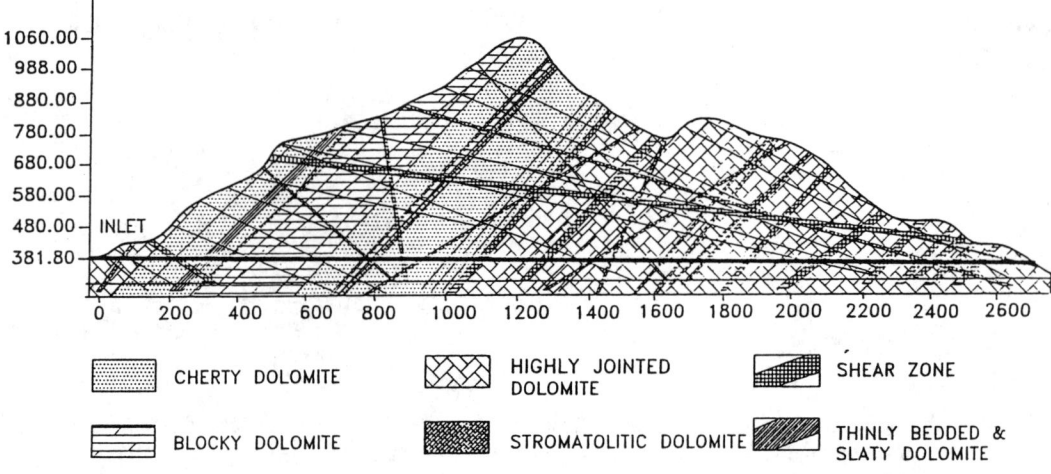

Fig. 1. Longitudinal Cross-section along TRT - II, Salal Hydel Project (Supplied by Project Authorities)

Table I - Classification of dolomites (from project reports)

S. No.	Dolomite Class	Estimated Extent in Tunnel (%)	Uniaxial Compressive Strength (MPa)
1.	**Cherty dolomites (Grade I)** - Greyish white in colour & quartz/chert bands occurred along the bedding joints spaced at 0.3 to 1.5m. Occasional occurrence of cross & transverse joints (Q = 1-2.3, N = 4.9 - 5.8 & RMR = 47)	20-25	80-120
2.	**Blocky dolomites (Grade I)** - Greyish in colour and massive in appearance with widely spaced bedding joints. Occurrences of cross & transverse joints are occasional (Q = 1-2.3, N = 4.9-5.8 & RMR = 47)		80-120
3.	**Highly jointed dolomites (Grade II)** - Dark to greyish in colour three prominent joint sets. Bedding joint spaced at 0.1 to 0.3m, cross & transverse joints spaced at 1 to 5cm (Q= 0.17-0.22, N = 1.1 & RMR = 32)	65	40-80
4.	**Crumbly & sheared dolomites (Grade III)** Closely spaced joints. Shear zones are completely crushed with gouge material (Q = 0.02, N = 0.2 & RMR = 15)	10-15	20-40

towards NNW-North and at outlet strike NE-SW with dip of 45- 60°. The orientation of tunnel axis is N20°. There are three prominent joint sets in the dolomites with the following orientation (after project reports):

Bedding joints N 80-90°, 45-60°, N 320-0°
Cross joints N 80-90°, 20-30°, N 90°
Transverse joints N 180°, 70-80°, N 90°

The dolomites exposed in the area have been divided in various categories based on their physical behaviour, extent of crushing and shearing and number of joints and their spacing as given in Table - 1 and Fig. 1. It may be noted that highly jointed dolomites are also affected by the presence of shear zones. In Table 1, Q is Barton's rock mass quality, N is rock mass number of Goel et al., (1995) and RMR is Bieniawski's rock mass rating.

As per geological details, highly jointed dolomites and the crumbly and sheared dolomites have created some type of tunnelling problems during the tunnel excavation. The tunneling problem encountered during the excavation is discussed by Goel et al., (1996).

4 TUNNEL MONITORING

Rock mass behaviour around the tunnel was monitored for evaluating and designing the supports. Load cells and closure studs have been installed in various grades of rocks for monitoring the support pressures and tunnel deformations respectively. Load cells have been installed in three consecutive ribs, whereas closure studs are installed on rib and in rock to monitor the rib and the rock closures at seven test sections. These instruments were installed close to the working face and the monitoring remainscontinued with the advancement of the tunnel face. In addition to these test sections, closure studs have also been installed at a number of locations since it does not add much cost to instrumentation and at the same time it provides a useful information on the behaviour of the rock mass. A typical load-time plot with the face advance is shown in Fig. 2 for pictorial presentation of face advance effect on support pressure. Similar effect of the face advance has been observed in case of tunnel closures.

It may be noted that no load cell has been installed in Grade I dolomites because no steel rib support has been erected in this grade of rock mass. Measured values of support pressures and tunnel closures are given in Table 2.

5 EFFECTIVE FACE DISTANCE

For developing the correlation for predicting the effective face distance, first task was to select the influencing parameters. Emphasis has been given to those parameters which were found effective theoretically and are easy to obtain in the field in order to make the correlation users' friendly and reliable. Main influencing parameters could be the insitu stress, the rock mass strength and the tunnel span or diameter. The insitu stress has been taken into account in terms of the tunnel depth, the rock mass strength in terms of the rock mass number N (defined as Barton's Q with SRF =1) and the tunnel span has been considered in D while calculating the effective face distance in terms of tunnel span.

The values for tunnel depth H in metres, rock

Table 2. Measured support pressure and tunnel closure in TRT - II, Salal Hydroelectric Project (after Goel et al., 1996)

S. No.	Location Ch. (m)	Rib spacing (m)	Type of Rock Mass	Tunnel Depth (m)	Measured Values of	
					Support Pressure (MPa)	Rock Closure (mm)
A. OUTLET END						
1.	303.6 - 305.6	1.0	Grade II	110	0.05	10.35
2.	381.4 - 382.4	0.5	Grade III	170	0.18	36.8
3.	718.05 - 720.05	1.0	Grade II	350	0.12	13.00
4.	939.8 - 9418	1.0	Grade II	370	0.124	13.15
5.	1173.0 - 1174.0	1.0	Grade II	480	0.078	11.10
B. INLET UPSTREAM						
6.	225.5 - 226.5	0.5	Grade II	70	0.058	10.55
C. INLET DOWNSTREAM						
7.	Around ch.908m	0.5	Grade III	380	> 0.47*	tunnel collapsed
8.	1252.22- 1253.52	1.3	Grade I & II	400	0.028	not recorded

pressure value inferred from the capacity of steel rib supports which has buckled

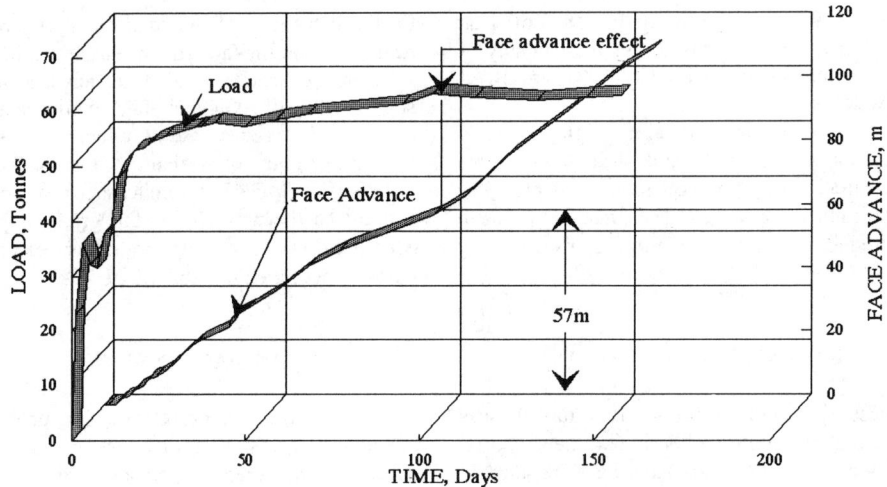

Fig.2. Load-time and face advance plots at a tunnel section

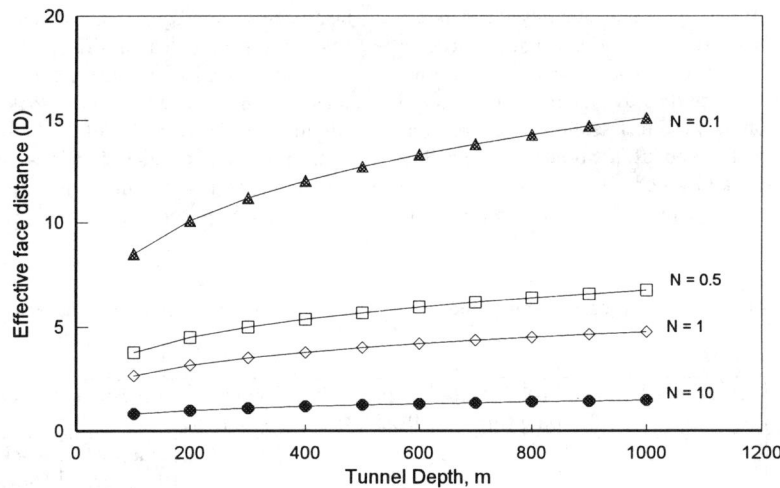

Fig.3. Plot between tunnel depth and effective face distance for various values of N

mass number N and the effective face distance D has been collected from eight tunnel sections and the following equation of the best fit line using the computer program of regression analysis has been obtained.

$$D = (0.85 \cdot H^{0.25})/N^{0.5} \quad (1)$$
$\geqslant 2.5$ times the diameter of the tunnel opening

where D = effective face distance in terms of tunnel diameter; H = tunnel depth in metres; and N = rock mass number, i.e., Barton's Q with SRF=1.

Equation 1 is tested in three tunnel sections of an Indian tunnel and it show good results, which also has generated confidence in the approach under the conditions mentioned above. The equation will further be tested as more and more data of face advance would be available.

5.1 *Discussions*

It can be seen from Eq. 1 that the effective face distance is inversely proportional to on the rock quality represented by rock mass number N, i.e., more is the N, less will be the effective face distance D, whereas it is directly proportional to the tunnel depth.

Figure 3 shows plot between tunnel depth and effective face distance for various values of N. This figure is prepared using Eq.1. It is clear from the figure that the effect of tunnel depth on effective face distance (D) is more pronounced in case of poor rock (N = 0.1) where D increases sharply up to tunnel depth around 400m and beyond this depth the rate of increase of D decreases. This observation indicates that while planning a tunnel through poor rock mass even minor difference in tunnel depth, up to 400m, has significant influence on effective face distance. Therefore, in order to reduce the effective face distance, the planners can select a different tunnel alignment having comparatively less overburden or tunnel depth.

5.2 Applicability

(i) The above correlation is applicable in both non-squeezing and squeezing ground conditions. The creep effect, generally associated with the squeezing conditions, is not considered in the correlation because of lack of data.

(ii) It is understood that, more the rate of face advance, more will be the effective face distance, though marginally. Equation 1 is developed for the tunnels where the rate of face advance is less than or equal to 2m per day. When the rate of face advance is more than 2m per day, as in the case of excavation by tunnel boring machine (TBM), the effective face distance may be more than that obtained from the above equation.

6 CONCLUSIONS

Using Eq. 1 one can estimate the effective face distance in terms of tunnel diameter for tunnels through jointed and fractured rock masses. Equation 1 is yet to be tested for squeezing conditions which have creep effect also.

Effective face distance D is inversely proportional to rock mass number N and directly proportional to the tunnel depth H (Eq. 1).

The tunnel depth up to 400m has significant influence on the effective face distance in poor rock masses (Fig. 3).

REFERENCES

Barton, N., Lien, R. and J. Lunde. 1974, Classification of rock masses for the design of tunnel support, Rock Mechanics, Springer-Verlag, Vol. 6, pp. 189 - 236.

Bieniawski, Z. T. (1973). Engineering classification of jointed rock masses. Transactions, South African Institution of Civil Engineers, 15: 335-344.

Dube, A. K., Saini, G.S., Goel, R.K., Swarup, A. and Kumar, P. (1995). Geotechnical instrumentation for monitoring the rock mass behaviour in tail race tunnel-II, Salal hydroelectric project, J&K, September.

Goel, R. K., Swarup, Anil, Dube, A.K. and Kumar, Pramod. (1996). Tunneling through the blocky rock masses of Lower Himalayas - A Case history, Proc. Conf. Recent Advances in Tunneling Technology, New Delhi, India, pp. 241-249.

Goel, R.K., Jethwa, J.L. and Paithankar, A.G. (1995). Indian experiences with Q and RMR systems, Tunneling and Underground Space Technology, Vol. 10, No. 1, pp. 97-109.

Unal, E. (1983). Design guidelines and roof control standards for coal mine roofs. Ph. D. Thesis, Pennsylvania State University (Reference Bieniawski, Z. T., 1984, Rock Mechanics in Mining and Tunneling, p.113, Rotterdam: A.A. Balkema).

The numerical simulation to the construction with double shields driving in the opposite directions

Xiao-qing Zeng & Qing-he Zhang
Department of Geotechnical Engineering of Tongji University, Shanghai, People's Republic of China

ABSTRACT: To build double tube-tunnels with both shields driving in the opposite directions is one of the construction method taking the shortest engineering time. In order to research the mutual effects produced by the two shields driving in the opposite directions on the soft soil and their effects between the shields and the earth body, the author applies the three-dimension time and space dynamic simulation to the construction process with a computer. The construction mechanic analytic method put forward in this article embodies properly the effects of time and space in the construction process of the double-tube tunnel.

1. INTRODUCTION

In China, in urban subways a horizontal double-tube parallel tunnels is often to be adopted, and in Shanghai, as to its soft soil ground, shields are usually used in construction. Their construction method may be summarized into three types: (1) single shield driving to and fro, (2) double shields driving in the same direction (3) double shields driving in the opposite directions. Since constructing with method(1) produces the minimum mutual effects between the double tubes in tunnel, this method is mostly adopted in the Shanghai Subway No.1. However, sometimes speeding up the schedule is necessary, and method(2) or (3) is to be adopted when constructing.

Regarding method(2) the effects between the two tunnels are difficult to be controlled if the two shields drive simultaneously in the same direction. So it is usually adopted in the way of asynchronously driving instead of simultaneously to speed somewhat up the engineering schedule. As to method(3), being simulated in this article , it has the shortest engineering schedule. However if the degree of their mutual effects is estimated incorrectly previously and the measures are taken unreasonably, serious damages will be produced at the place where one shield drive close to the other or both shields get together, and serious deformation of the tunnel will happen. For this reason, seeking the law of interactions produced by the double shields while driving in the opposite directions obviously has substantial and practical meaning for improving the construction, speeding up the engineering schedule and guarantee the quality and safety in construction. Nevertheless, to handle the mutual effects between tunnels, and control the earth surface stratum deformation as well as the surface subsidence properly will bring the subway construction even greater economic benefit.

The process of tunneling is a problem of three-dimensions which varies constantly with time and space. As the excavation goes on, the original balance of physics and mechanics in the underground soft soil in the urban district will vary. Hence, it is certain that changes and variations in mechanic state and deformations of the surrounding soil medium. It will take place as well between the soil medium and the tunnel, including other building. In order to research the mutual effects being produced in the process of two shields driving in the opposite directions, an analytic method of construction mechanics is introduced into this paper. In the adopted numerical simulation of construction mechanics, the loading system and the geometry-physical parameters are quite related with time. Taking the object in this research, the author uses the theoretical research method and tactics that conform with the construction reality and in this way to analyze the mutual effects in the process of the shield tunnel construction.

2 THEORETICAL MODEL

This author researches the mutual effects between the two shields both driving in the opposite directions in the horizontal double-tube tunnels, taking the shield construction at Shanghai Subway as the practical background.

The load choice : the machine load and the excavation load. The former includes the shield thrust and the friction force of the shield produced around the surrounding soil when driving forward. The latter is the excavating release force.

In order to reflect the problem of the three-dimension time & space in tunnel construction, a semi-analytic numerical method is adopted in this paper. That is to reduce the problem of the mutual actions between the three dimension medium and the structural system into the one-dimension finite numerical problem.

Regarding to the geometry model of calculation elements, the semi-analytical infinite ring element is adopted. A plane which is perpendicular to the main axis of the tunnel separates the structure and the finite media of soil into a certain amount of infinite layer element. There are four types of elements: single-tube element(I_A), single-tube element(I_B), double-tube element(II), and non-tube element(III), as figure 1 shows.

As the tunnel construction goes on, single-tube element(I_A) and single-tube element(I_B) are increasing and non-tube elements decreasing. When and after the two shields arrive to each other, double-tube elements increase gradually until the construction to be completed and to be entire double-tube elements. As the geometrical shape changes with time goes by, the general rigidity matrix also changes in various stages of time.

When composing the semi-analytical displacement function, a column coordinate system(as shown in Figure 1) is adopted, analyzing along the circular θ and in the radial r direction while discreting along the axial X. The formula of semi-analytical function is as follows:

$$u(x,r,\theta,t)=\sum_{m=1}^{P}\sum_{n=1}^{Q} H_{mn}(r,\theta) \sum_{k=1}^{S} N_k(x) u_{mn}^{(k)}(t)$$

Where $N_k(x)$ stands for polynomial interpolation function, $u_{mn}^{(k)}(t)$ stands for generalized valuable, $H_{mn}(r,\theta)$ stands for analytical function in the direction r and direction θ . The region of one of the analytical directions is infinite. Further more, the integration region of variation functional infinite. It makes up the semi-analytical infinite element. In this way, the limitation of the finite element method in solving infinite problems has been somewhat gotten rid of. Though the provided analytical function would not be able to reflect the exact solution in the direction of the problem perfectly. The approaching to the exact solution of the analytical function family could be achieved properly owing to combining with the generalized variable of the discrete directions and giving satisfaction to the boundary condition as well as to the variation equation.

This paper takes the earth surrounding as the medium of the same quality in different directions with linear and viscoelastical properties. The constitutive mode is considered as a Kelvin-Voigt mode. The lining of tunnel is taken as a linear elastical body.

3. THE NUMERICAL SIMULATION OF CONSTRUCTION MECHANICS

3.1 A brief on calculation method of construction mechanics

A construction of building double tube parallel tunnels with two shields driving in the opposite directions is taken as a sample for studying the mutual effects of the two tunnels produced from the shield construction. In order to explain the analytical method of construction mechanics in a concise way, we take a concrete and simplified case here. Provided the excavated region to be 200 meters, each time-pace takes 1 day, each day one shield drives for 10 meters. The region is divided into 20 elements with

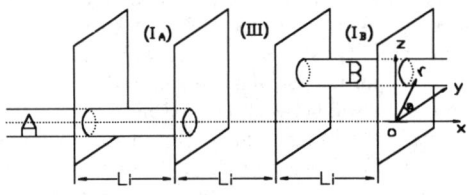

(a) Before two shields getting together

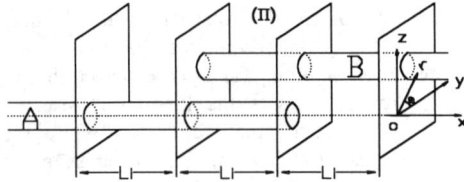

(b) After two shields getting together

Fig.1 three kinds of calculation element

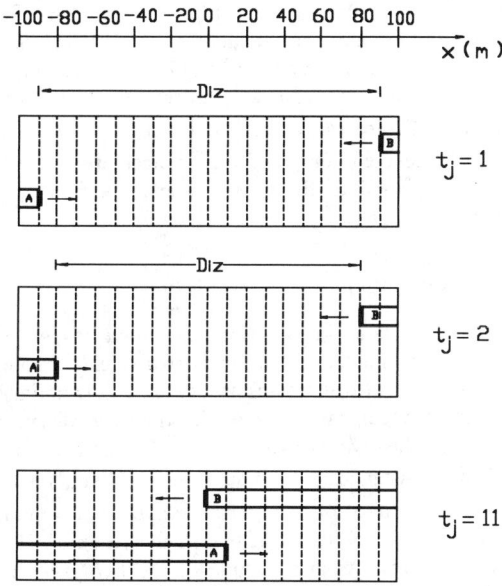

Fig.2 Construction progress of simulation calculation

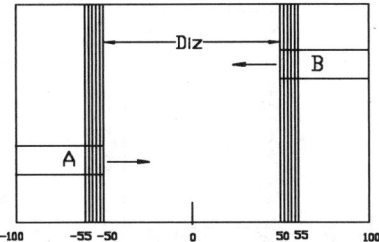

Fig.3 Schematic Drawing of Calculation Elements

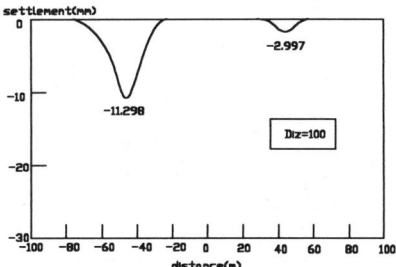

Fig.4 Longitudinal settlement when the Distance of Two Shields to be 100m

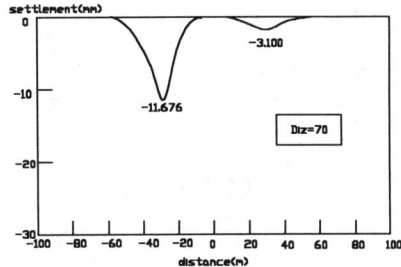

Fig.5 Longitudinal settlement when the Distance of Two Shields to be 70m

Table 1

Time (t_j) (day)	Variations of elements	Mechanics state		
		Moment	Time effect	Mechanical state at that time
$t_0=0$	20 elements of non-tube			
$t_1=1$	18 elements of non-tube 1 element of single-tube(I_A) 1 element of single-tube(I_B)	$P_1(t_1)$		
$t_2=2$	16 elements of non-tube 2 element of single-tube(I_A) 2 element of single-tube(I_B)	$P_2(t_2)$	$P_1(t_2)$	$P_2(t_2)+P_1(t_2)$
$1 \leq t_j \leq 10$	20-2j elements of non-tube j element of single-tube(I_A) j element of single-tube(I_B)	$P_j(t_j)$	$\sum_{i=1}^{j-1} P_{i(t_j)}$	$\sum_{i=1}^{j} P_{i(t_j)}$
$t_{11}=11$	9 element of single-tube(I_A) 9 element of single-tube(I_B) 2 elements of double-tube	$P_{11}(t_{11})$	$\sum_{i=1}^{10} P_{i(t_{11})}$	$\sum_{i=1}^{11} P_{i(t_{11})}$
$11 \leq t_j \leq 20$	2j-20 elements of double-tube 20-j element of single-tube(I_A) 20-j element of single-tube(I_B)	$P_j(t_j)$	$\sum_{i=1}^{j-1} P_{i(t_j)}$	$\sum_{i=1}^{j} P_{i(t_j)}$

the plane which is perpendicular to the axis. Thus, the whole time to build two tunnels in this entire region with double shields driving in the opposite directions is 20 days. Then, the calculating progress of the construction process within the region can be obtained according to Fig.2 and Table 1. In Table 1, Pi(t) stands for the mechanic state of time effect caused by excavation at the day i.

4. CALCULATION PARAMETERS

Considering the limitations of the capacity of computer and the time of calculation etc., the following hypotheses are adopted in this paper.

(1) Calculation region x ∈ [-100 , 100], displacement = 0 on the boundary line x=-100m and x= 100m.

(2) Take outside diameter of tunnel as 6.2m, inside diameter of tunnel as 5.5m, length of element as 1m, and center to center distance between two tunnels as 12 m; distance between tube center and ground surface H_0 = 10 m.

(3) Suppose adhesive of rock and soil media C=13.6 kPa, inner friction angle φ =10°, unit weight γ = 17.6 k N/ m³, elastic modules E_R= 4.2 × 10³ K p a, Poisson ratio μ_r=0.25, lateral pressure

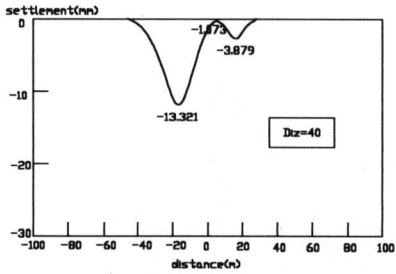

Fig.6 Longitudinal settlement when the Distance of Two Shields to be 40m

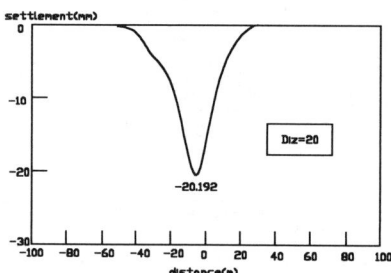

Fig.7 Longitudinal settlement when the Distance of Two Shields to be 20m

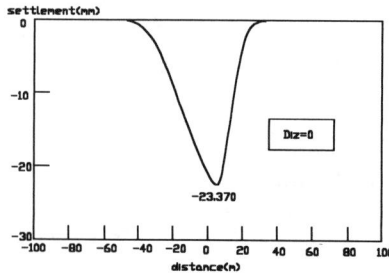

Fig.8 Settlement above Tube A when the Two Shields Reach to each other

coefficient $\lambda = 0.75$, Kelvin-Voigt model parameter $E_H = 69 \times 10^3$ kPa, $E_k = 6.0 \times 10^3$ kPa, $\eta_k = 1.2$, elastic modules of lining structure $Es = 3.5 \times 10^7$ kPa, Poisson ratio $\mu_s = 0.17$.

5. CALCULATION RESULTS

Provided Diz stands for the distance between the faces of shield A and shield B. This paper calculate the results of the surface subsidence when each shield drives forward for 5m at separately time while Diz equals 100m, 70m, 40m, 20m, and 0m. As for making the numeric simulation, the calculation elements are adopted as shown in Figure 3 when Diz equals 100m. the rest four operating modes are similar. Figure 4 ~ 8 show the longitudinal subsidence distribution of the vertical upper earth surface ground of the tunnel A and the front soil body.

6. CONCLUSION

It has been known from the above results of calculation simulation that when two shields driving in the opposite directions in soft soil, the mutual effects are comparatively weak while the distance between the two shields is over 100 metres. And when the distance between them reaches 70 metres ~ 40 metres, there effects in certain degree appear, until to the distance between the two shields reaches as far as 40 metres. As approaching of the two faces of shields, the surface settlement starts increasing, and the soil underground deformation becomes violent.

Hence we suggest that when the distance between the two shields is under 70 metres, the driving of the shields should be slow down. When it reaches as far as 40 metres, one of the shields should be stopped driving for keeping safety in construction, thus to decrease the effects against the surrounding buildings, underground pipelines, roads and other facilities.

REFERENCE

1, Cao zhiyuan, Zhang youqi, Semi-analytical Numerical Method, National Defense Industry Press, 1992,8.

2, E. soliman, H. Duddeck and H. Ahrens, Two-and Three-dimensional Analyses of Closely Spaced Double-tube Tunnels, Tunneling and Underground space Technology, 1993, Vol. 8, No. 1, 13-18

3D FEA on effects of shield tunnelling on the adjacent deep piles

Linwang Ruan & Yongsheng Li
Department of Geotechnical Engineering, Tongji University, People's Republic of China

ABSTRACT: This paper describes the effects on the adjacent piles due to shield tunnelling by using 3D FEA. The quantitative simulation on the axial deflection and the additional moment of the deep piles caused by the forward thrust of shield and the ground loss is presented. The rules of deformation and loading of piles, as well as the relation with ground movement are predicted along with the shield driving. Finally, certain factors, such as the strata properties, the upper and of pile, are studied to analyse their effects on the computational results.

1 INTRODUCTION

In urban areas, the shield tunnelling in soft and saturated clay will no doubt cause ground movement, and consequently the potential damage to the adjacent substructures. Although empirical formulas of calculating ground movements based on field measurements are available (Peck, 1969; Attewell et al, 1986), very limit theoretical research has been conducted, especially about into the interaction between substructures and the soil during the shield driving.

In Shanghai No.1 Metro Line, the adjacent bored piles as the foundation of the intercrossing bridge were affected in some degree by shield tunnelling. The detail field monitoring data was reported by Li et al (1997). In this paper, the shield tunnelling effects are divided into two parts, the axial thrust forces and the ground loss, which are modeled by a 3D elastic FEM, in which: 1) the multi-layer ground covering a large range in their mechanical properties; 2) the shield and the adjacent piles; 3) the axial thrust forces and the ground loss are taken into account. The effects of shield tunnelling in Shanghai soft clay on deep piles are analysed, and verified by the field measurements.

2 CASE HISTORY

The Shanghai No.1 Metro Line was excavated by the Earth Pressure Balance (EPB) shield with 6.34m in its outer diameter, 6.54m long and 10m deep, equipped with a rotary cutter, the outer diameter of which is 6.38m. The earth pressure in the cabin was 0.15-0.17MPa. The tunnel is assembled with the precast concrete segments. The outer diameter of the completed lining is 6.2m. During the driving, a clay mixture grout is grouted synchronously into the void

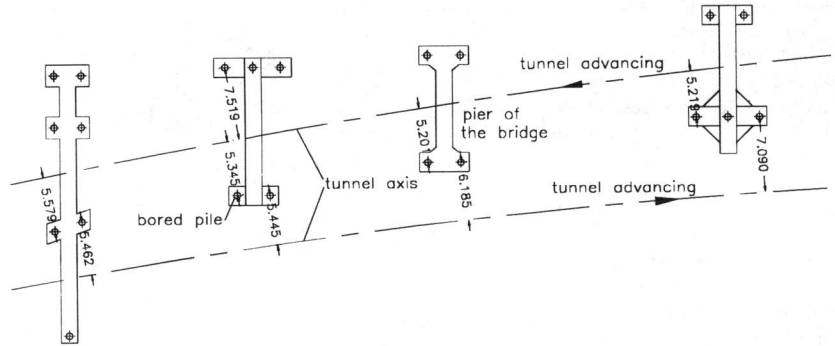

Figure 1 Distribution of the tunnel and the bored piles

Table 1 Properties of the soils

	Thickness (m)	Compressive modulus $E_{s,1-2}$(MPa)	Poisson's ratio	Cohesion (kPa)	Internal degree of friction (°)	Gravity (kN/m³)
Brown silty clay	2	5.85	0.27	20	15.2	19.4
Very soft silty clay	4	2.59	0.35	7	14.8	17.7
Very soft clay	8	2.46	0.35	9	8.5	17.1
Clay	7	5.98	0.25	3.6	24.4	18.3
Sandy silt	4	7.2	0.25	7	19.3	18.1
Black silt	5	7.8	0.22	24	11	
Sandy silt	15	8.4	0.21	7	17	
Silty sand / Silty clay	20	10	0.2	18	22	

between shield tail and segment lining.

The adjacent intercrossing bridge is supported by the bored piles which is 1m in diameter, 45m in length, and E_c=2.55 × 10⁴MPa in elastic modulus. The horizontal distance between the piles and the tunnel axis centerline of the tunnel was about 20m. Each pile is fixed with a reinforced concrete rigid beam at its upper end. The distribution of the tunnel and the piles is shown in Figure 1.

Based on the geologic investigation in situ, the properties of the soil are shown in Table 1. The shield is located in very soft silty clay and very soft clay.

3 THE AXIAL THRUST FORCES AND THE GROUND LOSS

The axial thrust forces, which squeeze the soil of the opening, are often ignored in the former research. The ground ahead the shield, however, will move upward and forward under the action of the forces. In EPB shield tunnelling, the thrust forces can be defined as:

$$P = P_i - P_z - P_w \tag{1}$$

where P_i is the soil pressure inside the cabin; P_z and P_w are the earth pressure at rest and the hydrostatic pressure on the opening face. Theoretically, P can be assumed as zero. As a matter of fact it is impossible. In this paper P is assumed as 0.14MPa.

The ground loss GAP is a measure of the volume of materiel that has been excavated in excess of the theoretical volume within the outer diameter of the tunnel lining (Lee and Rowe, 1990). It can be expressed as:

$$GAP = G_P + U_{3D}^* + \omega \tag{2}$$

where G_P is the net geometry void between the tail of the shield and the lining; U_{3D}^* represents the amount of over excavation owing to the 3D movements of the soil ahead of the tunnel face; ω is the ground loss due to workmanship reason. In EPB shield tunnelling,

the earth and water pressure on the opening face are balanced, so U_{3D}^* can be zero. Because of the development the synchronous grouting technique the ground loss is reduced significantly. For this reason, the grouting ratio must be taken into account. That means the ratio between the actual void filled with grout and the total ground loss has to be considered. In this paper, the grouting ratio is 80% and the ground loss GAP is 36mm.

4 COMPUTATIONAL ASSUMPTIONS AND ANALYSIS METHOD

The soil, the shield, the lining and the piles are all assumed to be elastic materials. The layer of soil is considered as homogeneous and isotropic. The structures and the around soil are assumed as uniform

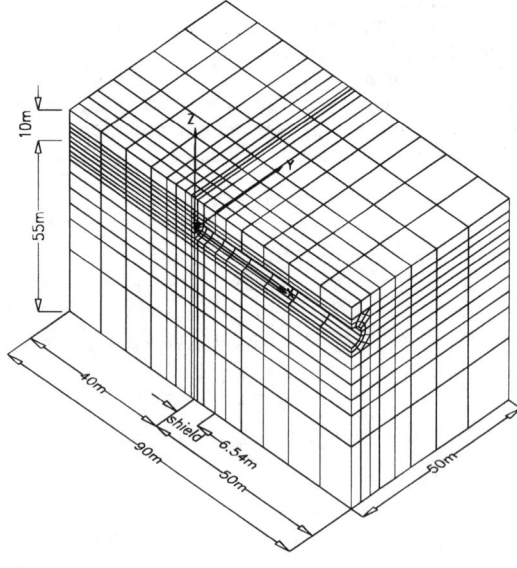

Figure 2 Finite-element mesh used in the analysis

in their deformations, and there is no relative displacement happened between them. The initial subsidence and stress due to gravity of the ground are not taken into account as both are completed before the excavation. The centerline is assumed to be a straight line. As the shield drives in step-by-step procedure, it is difficult to simulate the excavation exactly, so simplified single-step excavation theory proposed by Rowe and Lee (1991) has been employed. The tunnel is assumed to be excavated instaneously from a distance far behind the tunnel face to the designed position.

In practice, the spacing between the piles is long enough to assume there is no other structures but a single pile near the tunnel in the analysis. The distance between the piles and the spring line is 6m in the project analysed. Three situations have been analysed according to the location of the pile away from the tunnel face: -9m, 0m and 22m.

Figure 2 shows the finite element mesh used in the three-dimensional analysis. Taken the advantage of the symmetry, only half of the tunnel is considered. A total of 2076 elements and 2436 nodes are employed to represent the structures and the soil. The strata are simulated by 1828 eight-node isoparametric brick elements including 8 layers, each with different physical properties (see Table 1). The structural behaviour of the tunnel shield and the lining are simulated by 3D thin plane elements. Considering the restraint at the upper end of the pile, the 10 3D beam elements are used to simulate the pile with elastic support in horizontal direction at its top. Additionally, in this study, 88 boundary elements are used to demonstrate both the boundary conditions and ground loss.

The boundary conditions are assumed as that the surface is free; plane y=0 is not allowed to deform in y direction but is free in the other two directions; plane x=50m (the initial location of the tunnel) is fixed in x direction but is free to move in the other two directions; others are fixed.

5 ANALYSIS RESULTS

5.1 *Deflection of the pile*

The observed and calculated forward deflection and lateral deflection of the pile owing to the axial thrust forces and the ground loss are shown in Figures 3-4. For comparison, Figures 3 and 4 also give curves for the horizontal movement of the ground when there are no piles in the soil in the same place where the pile locates. Compared with the displacement of the shallow ground, the deflection of the pile decreases rapidly. As the depth increases, the deflection of the pile is gradually close to, or even larger than the displacement of the ground.

The considerable large movement of the shallow ground is due to the effects of the shield driving and the unsatisfied properties of the soil. The maximum ground displacement occurs just at the depth as same as the centerline of the shield. Below the excavation

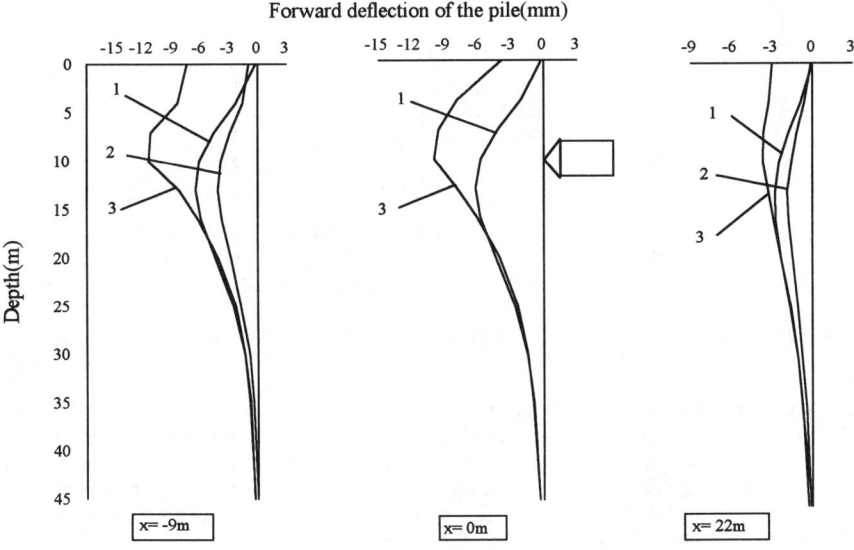

1 — FE calculated deflection of the pile; 2 — Field measurement; 3 — Horizontal movement of the ground
Figure 3. Observed and calculated forward deflection of the pile

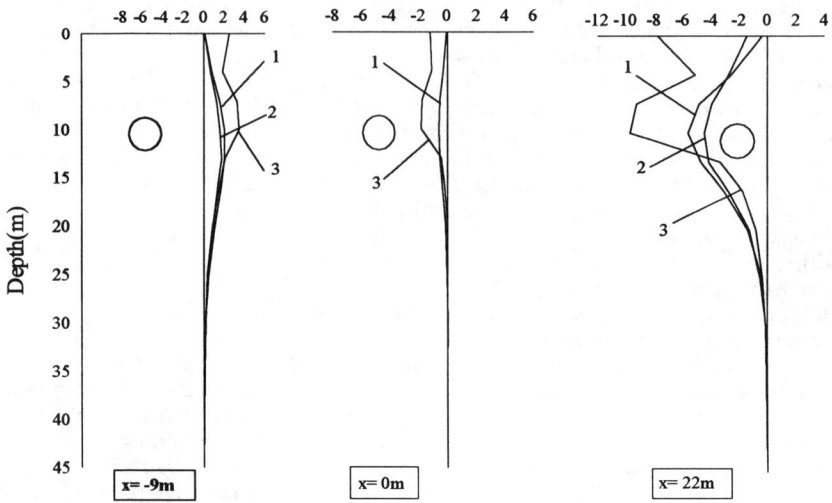

1 — FE calculated deflection of the pile; 2 — Field measurement; 3 — Horizontal movement of the ground

Figure 4. Observed and calculated lateral deflection of the pile

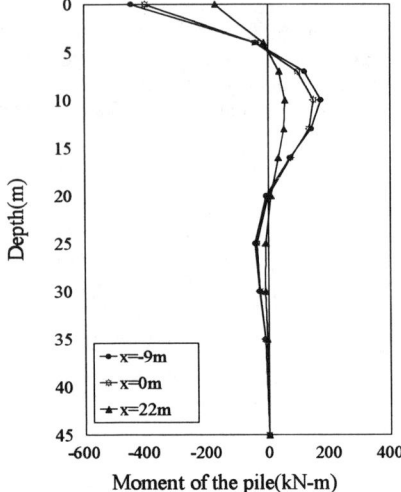

Figure 5. Forward moment of the pile

Figure 6. Lateral moment of the pile

level, the ground displacement decreases rapidly along with the depth. Different from the movement of the ground, the deformation of the pile appears uniform and is small at the upper end. This reflects the stiffness of pile and the restraint at the top of the pile. The maximum deflection occurs at 5m below the centerline.

The forward deflection of the pile, which is mainly caused by the axial thrust forces, reaches to maximum at a distance of 9m ahead of the tunnel face and gradually decreases as the shield further driving. At 22m behind the tunnel face, the forward pile deflection is about 60% of the maximum value. The affected length of pile ranges from 0 to 30m below the ground surface.

The lateral deflection of the pile is caused by the axial thrust forces and the ground loss. When ahead of the tunnel face, the pile is deflected mainly by the

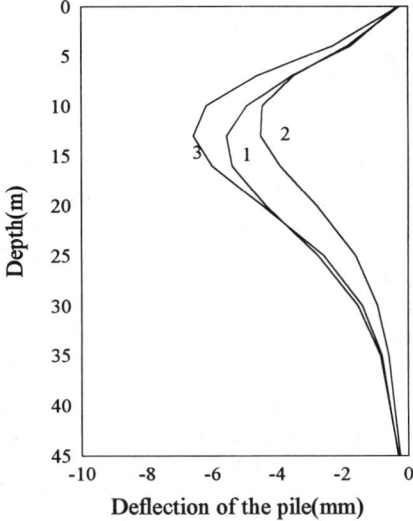 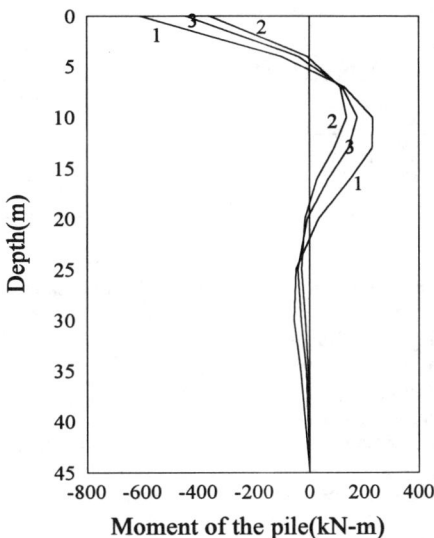

(a) (b)

Figures 7a-7b. Deflection and moment with variation of the pile elastic modulus E_p and the soil elastic modulus E_g

axial thrust forces. Once the shield passes, however, the pile is deflected toward the opposite direction owing to the ground loss. The lateral deflection reaches maximum at a distance of 22m behind the tunnel face.

The variation tendency of the calculated deflection of the pile is similar to that of the observed results, but latter is consistently less than the former by 20~30%. This could be explained as an overestimation of the axial thrust forces and underestimation of the grouting ratio.

5.2 Additional moment of the pile

Different from the deflection, the additional moment of the pile reaches to the maximum at the top and varies rapidly with the depth. Here, the restraint at the upper end of the pile processes a dominant rule to the computation results.

Figures 5 and 6 show the additional moment of the pile, in which the variation rule of the moment of the pile has a rather good agreement with that of the deflection with relation to the shield advancing. The points of contrabending locate at 5m and 20m below surface. The moment is about 5-10% of the maximum moment at 30m below the ground surface. As a conclusion, the affected part of the pile by shield driving is 30m below the ground surface in this case.

5.3 Bearing capacity of the pile

In shield tunnelling, the maximum deflection and the maximum stress of the reinforcement must be considered when discuss the bearing capacity of the pile. From figures 3 and 4, the maximum deflection of the pile is about 7mm and 0.015% of the total length, which is much less than the limiting values of l/200 (l is the length of the pile).

By additional moment, the normal stress on the cross-section of the reinforcement can be calculated. In this case the maximum normal stress 7Mpa occurs at the upper end of the pile, which is considerable to the bored pile. It is shown that the most dangerous position in this case is the upper end where the reinforcement has to be considered especially.

5.4 Factors affecting the pile deformation

There are many factors affecting the pile deformation such as the stiffness of the pile, the soil properties and the restraint at the top of the pile etc. Figures 7a and 7b show the variation of the deflection and the moment of the pile when doubling the pile elastic modulus E_p and the soil elastic modulus E_g respectively. Curves 1, 2 and 3 refer to the result when doubling the E_p, doubling the E_g and no variation of them respectively.

When doubling the value E_p, the deflection of the pile decreases while the moment increases, the maximum deflection and moment are 84.3% and 137% of the original data. The deflection reduction leads to the increasing moment, consequently the stress increment of the reinforcing steel of the pile, reduction the bearing capacity of the pile as well. So

Table 2. Deflection and moment with variation of bearing stiffness

Stiffness of the bearing (kN/m)	Maximum deflection of the pile (mm)	Displacement on the top (mm)	Maximum additional moment (kN-m)
1E5	2.916	1.937	-267.48
3E5	2.412	0.89	-389.51
5E5	2.263	0.57	-425.65
1E6	2.314	0.3	-456.8
1E11	1.988	0	-492.2

there is no significant improve to the bearing capacity of the bored pile as mentioned in this paper by increasing the stiffness of the pile.

When doubling the value E_g, both the deflection and the additional moment of the pile decrease considerably, about 68.4% and 81% of the original data. It is evident that the improvement in the soil properties affects greatly on the deformation and additional stress of the pile, the stress distribution of shallow soil stratum as well. In shield tunnelling, the upper portion of the pile is the key part, so the properties of the shallow soil play an importance role to avoid the pile damage. As an effective way to protect the underground structures, it is widely used in Shanghai to reinforce the soil around the structures.

The restraint at the top of the pile affects the pile seriously as shown in Table 2. Increasing the bearing stiffness can cause the reduce of the deflection and the increasing of the moment.

6 CONCLUSIONS

The main conclusions can be summarized as follows:

1. In shield tunnelling, the axial thrust force is a significant factor to affect the movement adjacent ground and the deformation of structures.

2. The accuracy of the computing result depends on the input parameters. When the deformations of the ground or the piles are available the axial thrust forces and the ground loss can be get through the 3D FEM proposed in this paper.

3. In the case of shallow tunnel, the top of the pile is the most dangerous section by shield tunnelling, which has to be focused on.

4. One effective way to protect the underground structures is to reinforce the soil around the structures.

5. The comparison between the field monitor data and the computation shows that with rational input parameters and the 3D FEM boundary conditions, in analyzing the tunnelling in soft soil and estimating the ground movement and effects on adjacent structures by shield tunnelling, the 3D FEM is a rather effective and practical method.

REFERENCES

Attewell, P.B. et al 1986. Soil movements induced by tunnelling and their effects on pipe lines and structures. Blackie and Sons Ltd, London.

Lee, K.M. and Rowe, R.K. 1990. An analysis of three-dimensional ground movements: the Thunder Bay tunnel. *Canadian Geotechnical Journal*. 28:25-41.

Li, Yongsheng and Huang, Haiying. Computation of mechanical effects on adjacent pile by shield excavation. *Journal of Tongji University*. 1997, (3).

Peck, R.B. 1969. Deep excavation and tunnelling in soft ground. *state of-the-art-report*. 7th ICSMFE, Mexico:225-281.

Rowe, R.K. and Lee, K.M. An evaluation of simplified techniques for estimating three-dimensional undrained ground movements due to tunnelling in soft soils. *Canadian Geotechnical Journal*, 29:39-52.

Measurement control method and expert system for tunneling by fuzzy set theory

H. Chikahisa, K. Matsumoto, H. Nakahara & M. Tsutsui
Technological Research Institute, Tobishima Corporation, Chiba, Japan

S. Sakurai
Department of Civil Engineering, Kobe University, Japan

ABSTRACT: In tunneling, it is often faced that measures are obliged to be taken without confirmation for such abnormality as unconverged movement of surrounding rock mass, growing crack of shotcrete and yield and break-off of rockbolts. In this case, it is usually said that experienced engineer's judgments for the selection of measures are importance and allowed us to get over the situations in many construction sites. But decrease of such experienced engineers need us to develop the new system to assist the selection of measures for the abnormality without any experiences of similar construction sites.

To solve this problem, the authors developed a new measurement control method and expert system by fussy set theory in tunnel construction sites. In this paper, the practical use of the method and the system are demonstrated based on the results of application to tunnel construction sites.

1 INTRODUCTION

In case of an encounter with collapse or harmful deformation of surrounding ground and support members during tunneling, it is important to define the cause of the phenomenon and then to carry out counter measures against the major cause. On actual sites, however, the engineers often have to begin to cope with the problem while the cause is still unknown. Also, when selecting the counter measures, the selection tends to be made according to personal experience and intuition of the engineers on site, but the judgment criteria depending on experience and intuition tend to be inaccurate and unclear for the unexperienced phenomenon.

The authors have been conducting research and development of a tunnel measurement control system (Yamagata et al. 1989) and an expert system whereby engineers with limited experience can utilize the expertise of experienced engineers (Chikahisa et al. 1991). This paper discusses the validity of applying the fuzzy quantification theory type II (Watada et al. 1982) to the judgment criteria for selecting the measures to increase the self-supportability of the tunnel face among a series of studies concerning tunnel measurement control.

2 MEASUREMENT CONTROL OF TUNNELS

The support pattern for tunnels in Japan is usually selected according to the elastic wave velocity and geological distribution in consideration of construction experience of similar tunnels. For this reason, the support pattern specified in the design is naturally a rough estimation, and has to be changed to adapt to the situation of the site by investigating the ground conditions and abnormal phenomena encountered during tunneling.

Here we explain the general flow of measurement control of tunnels carried out with this as a background (Figure 1).

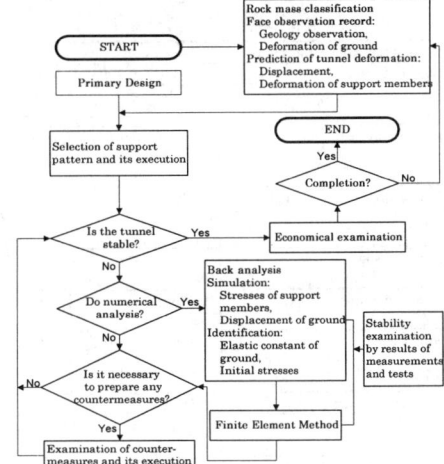

Figure 1. Flow chart of measurement control of tunnels.

Immediately after excavating, observe and record the ground of which emerges at the crown and face to grasp the proprieties. Select the support pattern and auxiliary methods and apply them based on the design, existing observation records and tunnel behavior data such as measurement results, as well as construction examples of similar types. Along with the progress of tunneling, make various measurements and observe abnormalities frequently, to investigate accordingly the current and future stability of the tunnel. In the case where these investigations suggest any problem in regard to the stability of the tunnel, increase the frequency of measurements and abnormality observations, investigate counter measures to be taken, and prepare the required equipment for the measures. When the caution level of surrounding ground or support members increases or is supposed to increase significantly, counter measures should be taken promptly.

When abnormality or harmful deformation of surrounding ground or support members is encountered, it is generally necessary to find the cause of the phenomenon. Such phenomenon is in most cases caused by multiple factors entangled together, rather than one outstanding factor. Also, abnormality is often noticed after substantial construction has been completed, by which time the state of the ground and the construction conditions on the way tend to become indefinite. For this reason, the data obtained during construction should be retained as indices within minimum required items that can easily be filled in. In this sense, the face observation record sheets (JSCE 1996) are arranged as given in Table 1 so that they can be used easily on site, and are actually used on a number of construction sites. It is also important to retain the photographs of the face immediately after excavation to supplement these data.

If it is judged that that the face is not stable or will be collapsed during tunneling, the main cause of the abnormality and effective counter measures are required to be investigated and carried out in consideration of the data collected on site. However, on tunneling sites where a number of related engineers manage to carry out various types of work, identification of the cause of abnormalities and harmful deformation of surrounding ground and the support members tends to be delayed. For this reason, abnormalities and harmful deformation during routine inspection and measurement on site try to be found out early using a measurement control table as shown in Table 2. The data obtained in this way are arranged in lists as given in Table 3 so as to be easily utilized later.

Table 1. Face observation record.

Tunnel name			Station No.		Distance from the starting point		
					Distance from the entrance		
Overburden			Rock mass classification				
A kind of rock			Rock name				
			Geological period				
Special conditions	swelling ground, leaned load, liquefaction, shallow overburden ( )m, close to important structures, under valley, others ( )						
Special geology	1.alternation of strata	2.unconformity	3.dike		4.folds	5.fault	6.others
Ground condition after excavation							
A	State of the face		1. Stable	2. Partially fall down	3.Squeezing, swelling	4. Collapse	5. Others
B	State of the crown and side wall		1. Stable	2. Partially fall down (normal supporting)	3. Unstable (immediate supporting)	4. Forepoling is necessary	5. Others
C	Compressive strength		1. σ_c>100Mpa Hammering with metallic sound	2. 100>σ_c>20MPa Broken by hammering	3. 20>σ_c>5Mpa Broken by slightly hammering	4. 5Mpa>σ_c Sunk by hammering	5. Others
D	Weathering		1. None	2. Discolored along joints, a little decreasing of strength	3. Discolored in whole, considerably decreasing of strength	4. not solidified	5. Others
E	Joint spacing		1. d>1m	2. 1m>d>20cm	3. 20cm>d>5cm	4. 5cm>d, crushed not solidified	5. Others
F	Joint conditions		1. Close	2. Partially open	3. Open	4. Filled with clay not solidified	5. Others
G	Joint shape		1. Random, square	2. Column	3. Schistose	4. Soft soil not solidified	5. Others
H	Inflow water		1. None	2. Blob	3. Partially	4. In whole	5. Others
I	Weaken by water		1. None	2. loosened	3. Weaken	4. Collapse, outflow	5. Others
Joint direction	Longitudinal direction to the face		1. Horizontal(10˚>θ>0˚), 2. Stable dip(30˚>θ>10˚, 80˚>θ>60˚), 3. Stable dip(60˚>θ>30˚) 4. Sliding dip(60˚>θ>30˚), 5. Sliding dip(30˚>θ>10˚, 80˚>θ>60˚), 6. Vertical(θ>80˚)				
	Cross direction to the face		1. Horizontal(10˚>θ>0˚), 2. From right to left(30˚>θ>10˚, 80˚>θ>60˚), 3. From right to left(60˚>θ>30˚), 4. From left to right(60˚>θ>30˚), 5. From left to right(30˚>θ>10˚, 80˚>θ>60˚), 6. Vertical(θ>80˚)				

Table 2. Measurement control table.

Control item \ Caution level	Level 1 (Stable)	Level 2 (Cautionary)	Level 3 (Abnormal)	Level 4 (Limit)
Ground				
A. State of face	Stable	Partially fall down	Partially Collapse	Collapse spontaneously wholly
B. State of crown and side wall	Stable	Partially fall down	Fall down ($<9m^3$)	Fall down ($>9m^3$)
C. Inflow water	None	Blob	Soak through	Outflow, Gushing
C'. Weaken by water	None	Partially	Wholly	Collapse, Outflow
Measurement				
D. Convergence(mm)	D<19.6	19.6<D<57.8	57.8<D<170.0	170.0<D
D'. Convergence speed(mm/day)	D'<2.4	2.4<D'<7.0	7.0<D'<20.7	20.7<D'
E. Crown settlement(mm)	E<19.6	19.6<E<57.8	57.8<E<170.0	170.0<E
F. Strain of surrounding ground(%)	F<0.179	0.179<F<0.525	0.525<F<1.546	1.546<F
G. Normal force of steel ribs(kN)	G<510	510<G<911	911<G<1264	1264<G
G'. Normal strain of steel ribs(%)	G'<0.08	0.08<G'<0.15	0.15<G'<22.5	22.5<G
H. Normal force of rock bolts(kN)	H<98	98<H<176	176<H<245	245<G
H'. Normal strain of rock bolts(%)	H'<0.10	0.10<H'<0.17	0.17<H'<20.0	20.0<G'
I. Stress of shotcrete(MPa)	I<5.9	5.9<I<11.7	11.7<I<17.6	17.6<I
Support members				
J. Mired	None	A little	Exist	Wholly
K. Steel ribs	None	Slightly separated from shotcrete	Deformed, Separated from shotcrete	Buckling
L. Rock bolts	None	Slight deformation of plates	Large deformation of plates	Punching share of plates, Broken-out of threaded portion
M. Shotcrete	None	Slight crack	Large crack, Punching share by rock bolts, Separation	Fall down

Table 3. Construction and measurement results table.

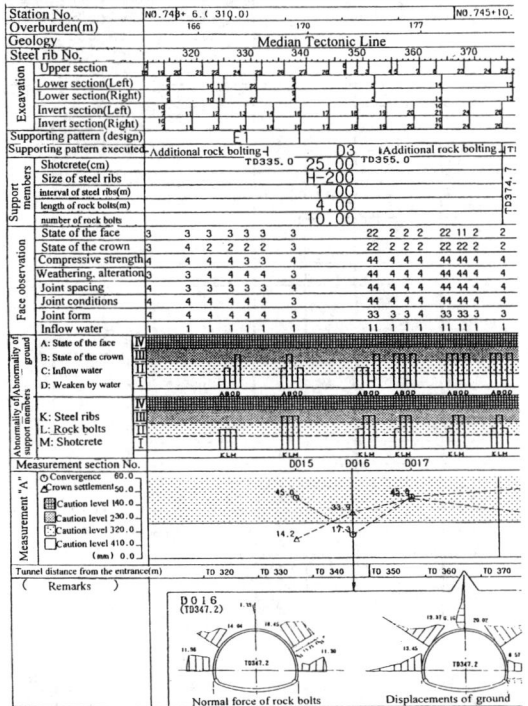

the required support patterns and auxiliary methods can be designed using numerical analysis techniques such as sequential analysis and back analysis based on the finite element method. However, on tunneling sites, a prompt decision is required in most cases without taking sufficient time for such investigation and design. In addition, as the causes of the abnormality and harmful deformation tend to be complicated and indefinite, counter measures are often obliged to be implemented before identifying them. With this as a background, we listed counter measures expected in advance based on past results as shown in Table 4. This table rearranges the counter measures taken or planned against abnormal phenomena encountered in past tunneling.

Regardless of whether or not the measures were taken, measurements should be continued to be carried out until the measurement results and caution level of the abnormality are settled. The data are later rearranged for review from the general standpoint to assess if the adopted support patterns and construction methods were economical. These data including the assessment data are utilized for subsequent ground judgment and investigation of support patterns and construction methods.

When abnormality and harmful deformation of the ground or support members are found out or supposed, the causes of the encountered abnormality should be defined and counter measures should be investigated and carried out against the causes. In such a case, geological surveys and tests can be conducted to find the causes of the abnormality and

We have studied how the above-explained measurement control technique is applied to various phenomena encountered on tunneling sites, and have developed analysis programs and test equipment as the needs arose. Therefore, the method of application varies depending on the type of the abnormal phenomenon. Among these methods of application, the following sections discuss the application in regard to the selection of auxiliary methods of supporting the tunnel face.

Table 4. Explanatory table of counter measures.

Mark	Counter measures
EX 1	Ring cutting.
EX 2	Repeat of partial excavation with temporary supports.
EX 3	Shorten bench length.
EX 4	Shorten excavating interval.
EX 5	Change excavation method (Ex. Side pilot excavation method.)
EX 6	Decrease explosives, mechanically excavating mainly
SR 1	Steel ribs (of higher grade) use.
SR 2	Over-setting upper and outer.
SR 3	Sheet piles under bearing plates.
SR 4	Foot reinforcement by Rock bolting.
SR 5	Holed web for forepoling.
SC 1	Immediately shotcreting.
SC 2	Increase of accelerator.
SC 3	Structural continuity by shotcreting.
SC 4	Face shotcreting.
SC 5	Increase of thickness.
SC 6	Additional shotcreting.
SC 7	Foot shotcreting for structural continuity.
SC 8	Foot shotcreting for increase of bearing capacity.
SC 9	Temporary strut by shotcreting.
SC 10	Reinforcement by wire mesh.
SC 11	Small wire mesh for well bond.
SC 12	Immediate ring closure by shotcreting.
RB 1	Immediately rock bolting.
RB 2	Increase of rock bolts.
RB 3	Longer rock bolts.
RB 4	Additional rock bolting.
RB 5	Face rock bolting.
RB 6	Inclined rock bolting.
RB 7	Forepoling.
RB 8	Larger bearing plates.
RB 9	Bearing plates on steel ribs.
RB 10	Increase of thickness of bearing plates.
RB 11	Accelerator in mortar for anchorage.
RB 12	Resin for anchorage.
RB 13	Rock bolts of injection type.
RB 14	Rock bolts of self-drilling type.
AM 1	Forepoling.
AM 2	Catchment work (Ex. catchment boring, well point.)
AM 3	Drainage (Ex. drip hole, drain hole.)
AM 4	Seals, watertight wall method.
AM 5	Ground-improvement method.
AM 6	Invert concrete for immediate ring closure.
AM 7	Prevent the mired.
AM 8	Enough space for deformation.
AM 9	Rank up of support pattern.
AM 10	Rebuilding.
AM 11	Seal and harden of the face.
AM 12	Vertical spiling from ground surface.

EX: For excavation, SR: For steel ribs, SC: For Shotcrete, RB: For rock bolts, AM: The other auxiliary method

3 FACE OBSERVATION RECORD

The tunnel used for the present data analysis is planned for the 2-lane highway standard with an excavation section of 90 m^2. It was constructed through tertiary sedimentary rocks by the shotcrete/rock bolt method. The tunnel encountered a fault fracture zone where the face had poor self-supportability, and had to undergo various counter measures to be supported.

Observation records as shown in Table 1 were written down on this site to be utilized for various measurement controls. The process of selecting auxiliary methods to support the face based on these observation record tables is analyzed.

4 FUZZY QUANTIFICATION TYPE II

4.1 *Membership function for pre-condition*

To establish the membership function for the pre-condition, the face observation record for 38 sections in the tunnel are selected arbitrarily from the data of 131 sections recorded for a zone of 500m (TD 500m) from the entrance of the tunnel 1310m in length. The record is arranged as a response table as shown in Table 5. This table is then analyzed with the quantification theory type II by assuming the "state of the face" to be the external criterion and other items to be the predictor variables. As a result, the following 4 items were found significant:
- External criterion: "state of the face"
- Predictor variables: "compressive strength," "joint spacing," "joint shape," and "inflow water"

It was also found that regression coefficients and correlation coefficients are as given in Table 6.

Table 5. Response table of face observation records.

No.	TD(m)	C	D	E	F	G	H	A
1	35	2	2	3	3	3	1	1
2	45	3	3	2	3	3	1	1
3	55	3	3	3	3	3	1	1
4	68	3	3	3	2	2	1	1
5	80	3	3	3	2	2	1	1
6	97	3	2	2	2	3	1	1
7	110	3	3	2	2	3	1	1
8	127	3	3	3	3	3	2	2
9	144	3	3	3	3	3	1	1
10	161	3	3	3	2	3	1	1
11	188	3	3	3	2	3	1	1
12	201	3	3	3	3	3	1	2
13	218	3	3	3	3	3	1	2
14	238	3	3	3	3	3	1	2
15	256	3	3	2	3	3	2	1
16	276	3	4	4	4	2	1	1
17	286	3	3	3	4	4	1	1
18	295	4	4	2	4	4	1	1
19	307	4	3	3	3	3	1	2
20	320	4	4	3	4	3	1	2
21	337	4	4	3	4	3	1	3
22	346	4	4	3	4	3	1	3
23	354	4	4	3	4	3	1	3
24	360	4	4	3	4	3	1	3
25	376	4	4	3	4	3	2	3
26	389	4	4	3	4	4	1	2
27	398	4	4	3	4	4	1	2
28	409	3	4	3	4	3	1	3
29	416	4	4	4	4	3	2	3
30	428	4	4	4	4	4	2	4
31	438	3	3	3	3	1	1	3
32	446	4	3	3	4	4	1	2
33	459	4	4	3	4	4	2	4
34	463	4	4	4	4	4	3	3
35	473	1	2	2	2	1	2	1
36	488	3	3	2	2	3	3	2
37	492	3	3	3	2	3	2	3
38	500	2	2	2	2	1	1	2

(C: Compressive strength, D: Weathering, alteration, E: Joint spacing, F: Joint condition, G: Joint shape, H: Inflow water, A: State of the face)

The distributions of frequencies for the sample values are as shown in Figure 2. Figure 3 shows the occurrence ratio of each level of "state of the face" for each sample value defined as the membership function for the pre-condition.

Table 6. Regression coefficients and correlation coefficients.

Category \ Item	1	2	3	4	Correlation coefficients	Partial correlation coefficients
C. Compressive strength	-3.127	-0.830	-0.101	0.453	0.430	0.730*
D. Weathering	-	0.084	-0.253	0.298	0.539	0.350
E. Joint spacing	-	-0.744	0.204	0.160	0.381	0.628*
F. Joint condition	-	-0.033	-0.249	0.180	0.462	0.238
G. Joint form	0.763	-0.789	0.062	-0.552	0.026	0.723*
H. Inflow water	-0.335	0.991	0.733	-	0.476	0.788*

*'marks indicate items which are found significant by 5% level.

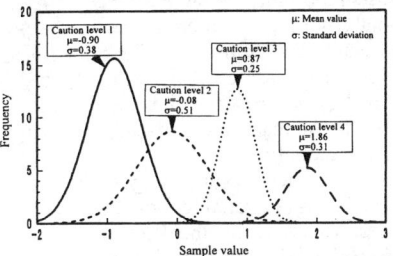

Figure 2. Frequency distribution corresponding to the sample value.

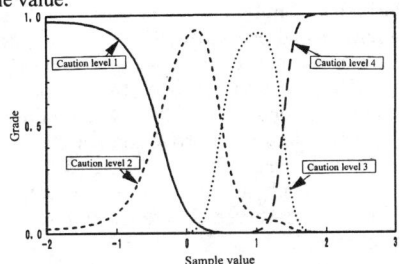

Figure 3. Membership function for pre-condition.

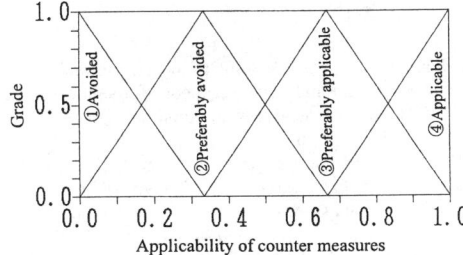

Figure 4. Membership function for post condition.

4.2 Membership function for post-condition

The applicability and grade of counter measures are established as shown in Figure 4 and defined as the membership function for the post condition. Table 7 shows the evaluation in the four levels according to this definition of the applicability of four counter measures adopted or planned to improve the self-supportability of the face on the other sites encountered the similar abnormality. The counter measures are "face shotcreting," "face bolting," "ring cutting," and "mechanically excavating,".

Table 7. Fuzzy inference rule.

Caution level	1	2	3	4	Legend
Face shotcreting	①	②	④	④	①Avoided
Face bolting	①	②	③	④	②Preferably avoided
Ring cutting	①	①	②	③	③Preferably applicable
Mechanically excavating	①	①	③	③	④Applicable

4.3 Fuzzy inference

The self-supportability of the face of this tunnel was rich after TD 580m, as the surrounding ground changed to hard granite, and no counter measures were taken. Accordingly, the data of 6 sections, which are not used in the above analysis, are arbitrarily selected from the face observation record of the zone from TD 500m to TD 600m, whose ground properties are similar to the zone used in the above analysis. These data observed are written in the upper part of Table 8. The following is the inference process of selecting the counter measures by the above-mentioned membership functions using Section 2 of this tunnel:

1. Determine the sample value S for the section from the face observation record on the basis of the regression coefficient in Table 6 as follows:

$$S = \Sigma f(4,4,3,4,3,1) = 0.862 \qquad (1)$$

2. The fitting grade of the sample value S (0.862) to each level of the "state of the face" determined from the membership functions for the pre-condition shown in Figure 3 is:

$$\omega_i = (0.0, 0.11, 0.89, 0.0) \qquad (2)$$

3. From the membership function for the post-condition shown in Figure 3 and the fuzzy inference rule given in Table 7, the inference result with the "state of the face" at Section 2 is as shown in Figure 5.

4. Assuming that the centroid of the resulting function is the result of inference, the grade of applicability of "face shotcreting," "face bolting," and "mechanically excavating" at Section 2 is 0.72, 0.62, and 0.62, respectively. Accordingly, these measures are concluded as being "③ preferably applicable." Meanwhile, the grade of applicability of "ring cutting" is 0.33, which is concluded as being "② preferably avoided."

By the same procedure, the inference results at

each section are summarized as given in the lower part of Table 8. Figure 6 shows the inference results for the sections under analysis expressed as a bar graph, in which ○ marks indicate the sections to which measures were actually applied during tunneling in this site. Consequently, the selection of counter measures for the zone of TD 500m to TD 600m can be relatively accurately predicted by using the data analysis results for the zone from the entrance to TD 500m, whose geological properties are similar.

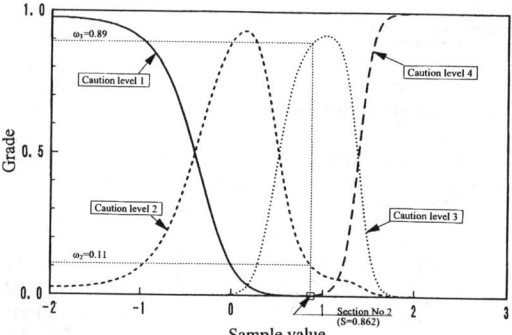

Figure 5. The Fitting grade from the membership function for pre-condition.

Table 8. Fuzzy inference results of each section.

Section No.	1	2	3	4	5	6
Face observation records						
C Compressive strength	4	4	3	3	3	3
D Weathering	4	4	3	4	3	3
E Joint spacing	3	3	3	3	2	3
F Joint condition	4	4	3	4	3	3
G. Joint form	3	3	3	3	3	3
H Inflow water	2	1	2	1	2	1
The fitting degree from membership function for pre-condition						
Caution level 1	0	0	0	0.01	0.36	0.75
Caution level 2	0	0.11	0.25	0.85	0.64	0.25
Caution level 3	0	0.89	0.75	0.14	0	0
Caution level 4	1.00	0	0	0	0	0
Fuzzy influence results						
Face shotcreting	0.89	0.72	0.62	0.54	0.31	0.24
Face bolting	0.89	0.62	0.57	0.39	0.31	0.24
Ring cutting	0.67	0.33	0.32	0.30	0.12	0.11
Mechanically excavating	0.67	0.62	0.56	0.31	0.12	0.11

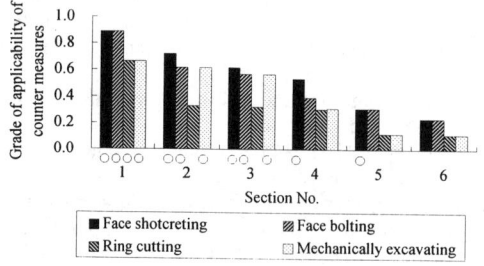

Figure 6. Fuzzy inference results of each section.

5 CONCLUSIONS

In this paper, the authors have introduced a support system for selecting counter measures to sustain the tunnel face at the actual construction site by the use of the tunnel measurement control systems that have been under research and development.

In the tunnel used for the present analysis, the "self-supportability of the face" can be judged by 4 items, i.e., the "inflow water," "compressive strength," "joint shapes," and "joint spacing" with the multiple correlation coefficient being 0.919. In regard to this tunnel, it is also found that these face observation items are also sufficient as the judgment criteria for selecting counter measures to increase the self-supportability of the face. Though the judgment criteria of field engineers in charge of on-site measurement control contains fuzziness, it is understood that these criteria eventually use the four items as key indices when selecting the counter measures to increase the self-supportability of the face during tunneling.

Also, the present results suggest that, if data for up to TD 500m is available, any engineer can make a similar selection of the counter measures for increasing the self-supportability of the face at TD 500m to TD 600m having similar geological features. Accordingly, it is considered that the unexperienced engineers can utilize the experience of the other tunnel engineers by collecting and analyzing data, such as the face observation record. The authors intend to further accumulate and analyze similar data to improve the accuracy of the inference process, as well as to improve the system so as to cope with a wide variety of deleterious ground behavior observed on site and abnormality of support members.

REFERENCES

Chikahisa, H., Y.Arai, M.Tsutsui & N. Shimizu 1991. Database system and fuzzy set theory for support system to select counter measures in tunneling. *Proc. Tunnel Eng., JSCE*, 1: 71-76.

JSCE 1996. *Tunnel standard specifications of NATM edition and its explanatory notes*. Tokyo: JSCE.

Watada, J., H. Tanaka & K.Asai 1982. Fuzzy quantification theory type II. *J. Measurement of Actions*, 9(2): 24-32.

Yamagata, S., H. Chikahisa, S. Kurosaka, Y. Arai & H. Nakahara 1989. Development of measurement control system for tunneling. *Tobishima Eng. Report*, 40: 103-117.

Reinforcement of railway tunnel in fault zone

Yung-Dae Kwon
Korean National Railroad, Civil Engineering Bureau, Seoul, Korea

Nam-Seo Park, Eun-Ryong Ha & Dae-Youl Oh
Daeduk Consulting and Construction Co., Seoul, Korea

ABSTRACT: A inner section of tunnel on Young-Dong line has been contracted due to an imbricated thrust zone which run across the tunnel. A series of reinforcement measures has been performed to decrease the inner section contraction. The applied measures were invert bridgings, poly-urethane groutings and lattice girder installations. Measurements of ground behavior during the reinforcement works show that the inner section contraction has been converged by the effect of reinforcement.

1. INTRODUCTION

A series of reinforcement measures in the railway tunnel of Youngdong line located at Taeback, Kangwon Province in Korea was performed to increase the safety of train operations.

The tunnel reinforcement measures were recommended after the safety diagnosis of tunnel by KISTEC(Korea Infrastructure Safety & Technology Corp.).

1.1 History of the tunnel

The tunnel was constructed in May, 1963 with the construction periods of only 10 months. It has a horseshoe-shape with the smallest curvature radius(R) of 250 m, 1,080 m in length and 30/1000 of longitudinal gradient. The axis of tunnel starts with the orientation of NW-SE and ends with the same direction, indicating 360° rotation of orientation within the tunnel, in order to compensate the difference of altitudes between the Tongri station and the Shimpori station(Fig. 1). The geological conditions around the tunnel are not known. However, they are determined to be very poor on the basis that the thickness of lining ranges from 30 to 60 cm depending on the geological condition and that the invert concretes were installed at 5 spots. Since tunnelling equipments and material at the time of construction were not well developed, a construction period of 10 months are considered to be too short for good construction of tunnel. After the opening of the railway tunnel, a lot of reinforcement measures and repair works, such as eleven times of waterproofing works, three times of concrete lining reinforcement and two times of drainage hole installation, have been carried out.

1.2 Deformation of the tunnel

During the installation of overhead electricity line for the train powered by electricity in 1975, the convergence measurements for inner diameter of the tunnel were performed. It was found that the spring line of inner section decreased about 300 mm compared with the as-built drawing at Sta. 98k100-110. The railway maintenance office considered the decrease of spring line as a serious problem. Therefore, the inspection for the safety of tunnel was performed by KISTEC. During the safety inspection, the changes of spring line have been measured for 4 months by the convergence measurements. An additional de crease of 3.8mm in the inner diameter were measured during that period, which indicates that an immediate reinforcement is necessary. The shape of inner section was measured precisely by the Cavity Monitoring System which uses a laser and was overlapped over that of as-built drawing(Fig. 2). The present shape of tunnel inner section is so irregular and decreased so seriously that the reinforcement measures and widening of inner section is essential for the safety of railway operation.

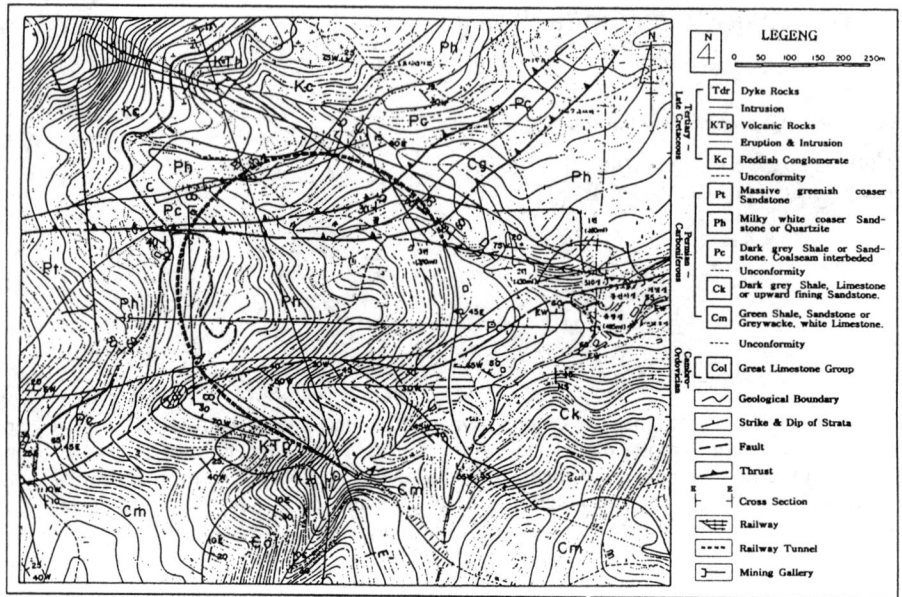

Fig. 1. Geological map

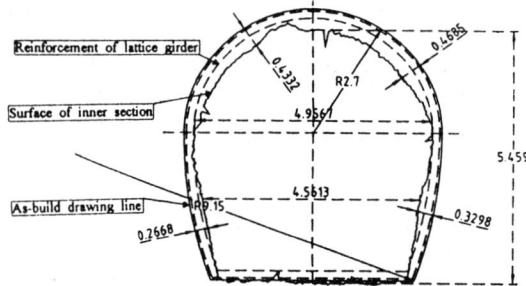

Fig. 2. Result of scanning in tunnel by C.M.S

2. TOPOGRAPHY AND GEOLOGY

2.1 Topography

A large valley oriented to NW-SE, named the Juchu valley, passes above the central part of the tunnel and one valley at the NE side and three valleys at the SW side of tunnel are distributed. Several ridgelines were made by small scale valleys which came from the larger ones. The thickness of overburden ranges from 30 to 160 m and is the maximum at the reinforced region of tunnel (Fig. 3). The southeastern and southwestern slopes of mountain above tunnel are steep. Several pits for coal mining (Youchang pit, Dongshin inclined-pit, 510ml pit) which stopped operations recently are distributed at the northwestern part from the tunnel exit. The distances from these pits to the tunnel are 57-111m and 140-370m in vertical and horizontal directions, respectively. It is considered that they do not affect the settlement within tunnel.

2.2 Geological Setting

The Permian meta-sedimentary rocks overlies the basement composed of the Cambro-Ordovician Great Limestone Series disconformably and is overlain by Cretaceous, Paleogene sedimentary rocks unconformably. Igneous rocks intruded the Permian meta-sedimentary rocks several times. These formations were folded by several tectonic events and show diverse strikes and dips.

However, strike of E-W~N70° W and dip of 30~70° NE are the most dominant. The tunnel was excavated through the upper Permian sandstone and coal-bearing shale which are relatively unstable layers. The geological setting within the tunnel is as follow.

A. the entrance~230m (Sta.98k490~98k720)
Shale, sandstone and limestone are distributed and conglomerate is embedded partly. The geological condition in this section is very poor due to many faults and joints. Although Youchang coal mine located nearby the tunnel, the distance is over 280 m from the base of this section.

B. 230m~530m (Sta.98k720~99k020)
Dark-grey shale, sandstone, limestone and coal seams are distributed. Grain sizes of sandstone vary coarse to fine upwardly. Well rounded

leucocratic, coarse sandstone is distributed after 400m. Poor geological condition around 400 m is expected because of thrust.

C. 530~830m (Sta.99k020~99k320 including the reinforced region)
Dark-grey or black sandstone and shale are imbricated because of thrust and coal seams are embedded within the shale layer. The poor geological condition is expected at the starting point of the reinforced region.

D. 830~1080m (Sta.99k320~99k570)
Dark-grey shale, limestone and massive leucocratic sandstone with embedded coal seam are distributed.

2.3 Geological structure

Two large thrust faults with strike oriented to E-W~N70°E and steep dip to N~NE are observed within tunnel. Fault breccia, fault gouge which was probably produced by thrusts fault and several coal-bearing shale layers are identified at sidewall of the tunnel by horizontal drillings. The detachment of thrust fault at several hundred meters below the tunnel and thrust ramp around the reinforced region were also identified. The tip line of thrust fault which across the tunnel appears at a surface as shown in fig. 3. Imbricate fan structures made by several faults below surface are located in the reinforcement section and disturbed the ground. It is well known that thrust is formed by high lateral pressure. Thus a high coefficient of lateral earth pressure(k_0) is expected in this region and is 2.0~2.4 by pressuremeter test.

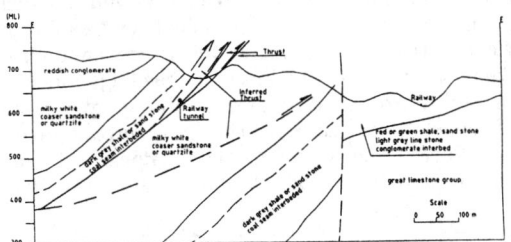

Fig. 3. Geologic cross-section

3. IN-SITU TEST AND INVESTIGATION

Three boreholes were drilled to investigate the geological profile, the geological structure, the engineering properties of the ground, the state of looseness in the ground in depth and to obtain samples for rock property test. Drill holes were located at 600 m in altitude and 150 m eastward from the reinforced section. Depths were 40-45m to penetrate the bottom of tunnel. Groundwater level observation, standard penetration test and rock test in the laboratory were also performed. Chemical analysis and X.R.D test were made for some clay samples which are intercalated in rock strata to examine the swelling characteristics of clay. Gypsiferrous minerals or smectite group of clay mineral were not detected.

3.1 Test boring

Land fill composed of coarse sand and rock fragments are distributed up to 6.5-10.1m below the surface. Shale is highly fractured and moderately weathered with low R.Q.D. value. Core recoveries vary with the depth. Sandstone is moderately weathered to fresh with well developed foliation and joints. R.Q.D values are higher than those of shale, but core recoveries vary with depth. Groundwater level is around 13.1-25.0 m below the surface. However, Since chemicals and clay were used to protect the boreholes from highly fractured zone due to fault and no flushing processes were made after drilling, the groundwater level measured at that time should be higher than actual level.

Table. 1. The results of pressuremeter test

No	Depth(m)	Coeff. of Subgr.Reaction	Elastic Modulus	Rocks	Remark
BH-1	11.0	128(kg/cm^3)	707(kg/cm^2)	Shale	sheared
	17.0	7,291	35,152	S.S	sl. weath.
BH-2	8.5	8,928	43,034	S.S	sl. weath.
	11.5	2,295	11,273	S.S	mod. weath.
BH-3	15.0	1,705	8,189	Shale	mod. weath.
	23.0	5,005	24,078	S.S	mod. weath.

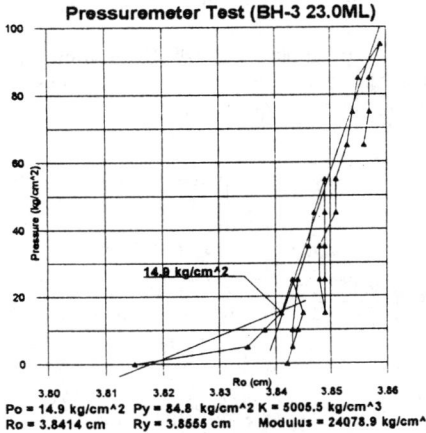

Fig. 4. The result of P.M.T

3.2 Lugeon test

Twelve lugeon tests were performed. Shale is more permeable than sandstone. The permeabilities of each rock type range 5.02×10^{-5}-4.61×10^{-4} cm/sec in shale and 1.64×10^{-5}-2.62×10^{-4} cm/sec in sandstone, indicating that shale is more fractured than sandstone.

3.3 Pressuremeter test

The moduli of elasticity calculated from pressuremeter test are shown in table 1.
The elastic moduli range 35,152-43,034 kg/cm² for slightly weathered sandstone, 11,273-24,078 kg/cm² for moderately weathered sandstone and 707-8,1889 kg/cm² for shale. The elastic moduli measured at cataclastic zone show a very low value.

3.4 Mechanical Properties of Rock

The mechanical properties were measured for only sandstone because shale is so fractured that test specimen could not be obtained. The results are shown in table 2.

Table 2. The results of the mechanical properties test

No	Spec Gra.	Vel. of Seismic Wave (cm/sec)		Uni Ax. Comp Str. kg/cm²	Ten. Str. (kg/cm²)	Er (kg/cm²)	Pois Rat. (ν)	ϕ (°)	Coh (kg/cm²)	Dep. (m)
		P	S							
BH 1-1	2.68	4800	3070	1070	80	6.78	0.26	51	180	11.7~12.1
BH 1-2	2.77	5440	3280	1130	110	8.94	0.25	49	200	15.3~18.4
BH 2-1	2.73	4750	3190	1220	130	7.34	0.25	50	210	7.7~8.4
BH 2-2	2.68	4830	2950	940	110	7.06	0.22	51	190	27.5~33.5

3.5 Chemical & X.R.D analysis of clay minerals

A horizontal borehole was drilled within the tunnel and the chemical analysis was done for fault clay obtained by drilling (table 3).

Table 3. The results of chemical analysis of clay minerals

Element (%) Sta. No.	SiO₂	Al₂O₃	Fe₂O₃	CaO	MgO	K₂O
99k 078.5	46.2	27.3	14.0	0.54	0.45	2.66
99k 083.5	50.8	25.6	8.04	0.30	0.60	4.52
Element (%) Sta. No.	Na₂O	TiO₂	H₂O	Cl (ppm)	S	Sum.
99k 078.5	0.36	1.01	1.60	57	0.03	94.15
99k 083.5	0.42	0.65	1.90	31	0.03	92.86

The contents of silica and alumina are high, which is a common characteristic of clay minerals. The total amount is less than 100% because the content of interlayer water(H_2O^+) is not included. Table 3 shows 300 ppm of sulfer, indicating existence of sulfides underground. Sulfer can corrode the concrete lining when reacted with groundwater. The results of X.R.D analysis for clay minerals are shown in table 4. The clay minerals belong to kaolinite group. This means that the possibility of deformation due to swelling characteristics of the fault clay is negligible.

Table 4. The results of X.R.D analysis

Samp. No.	Mineral
99k078.5	Pyrophyllite, Kaolinite, Mica, Chlorite
99k083.5	Quartz, Mica, Pyrophyllite, Kaolinite,

4. REINFORCEMENT METHOD

The deformation at the section of reinforcement took a place because of a high pressure in cataclastic zone caused by several faults and by yield of ground due to washing-out of fault gouge and clay mineral in the cataclastic zone by the groundwater outflow. The inner section of the tunnel has already contracted so much that the reinforcement using steel-rib or shotcrete is impossible. Therefore, the stabilization method using grouting with pressurized bolt is selected to increase the strength of ground.

4.1 Installation of invert concrete strut

Thirty one in-situ invert concretes with interval of C.T.C. 2.5 m were installed at reinforced section (L=72 m) where invert concrete had been skipped, to diminish a bending moment and to retard the distortion of lining by closure of tunnel's cross section. The procedure of invert installment consisted of 1) removal of the gravels at the railroad 2) making a groove(300×

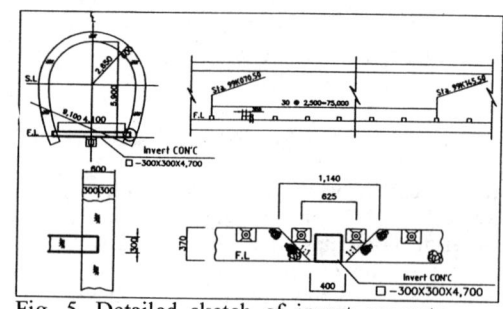

Fig. 5. Detailed sketch of invert concrete

300×300 mm) at the existing lining and 3) installation of the concrete invert with dimension of 300×300×4700 mm (Fig. 5).

4.2 Poly-urethane grouting with pressurized bolts

Two types of poly-urethane liquid with pressurized bolts were injected 1) to increase the ground strength beyond lining concrete (1.5 shot method), 2) to make earth pressure uniform 3) to fill cavities in ground beyond the lining concrete. Two types of injection pattern was used depending on the soundness of lining concrete and the degree of inner section contraction. A-type grouting was adopted at Sta. 99k082-99k111 section where groundwater flow is severe and contraction is large. B-type grouting was used at other sections where relatively minor deformation of concrete lining took a place. The locations of injection holes are shown in Fig. 6, and Fig. 7. Numerical analysis was performed to determine the injection length. The numerical analysis showed that the stress on lining is within a allowable limit and crown settlement converged into the allowable range when the injection length was 2.5 m, even though inverts were not installed. However, for A-type grouting region, the injection length of 4.5 m was used at the lowest hole of both side to increase the confining effect. Total number of hole for grouting injection are 570 and total amount of grouting liquid was 85,000 kg. The amount of injected poly-urethane per each hole was about 150 kg and applied injection pressure was 5-25 kg/cm^2 per each hole. The results of injection indicate that lots of cavities existed between the concrete lining and the ground and that the cavities weakened the strength of the tunnel ground.

4.3 Installation of lattice girder

After getting rid of the added reinforced lining (t=150 mm), Pantex lattice girder was installed to strengthen the original concrete lining (t=600 mm). Installed members are 70×20×30 mm in size. The procedure of installation consists of 1) making grooves at the existing lining(C.T.C. 1 m intervals), 2) filling up the inside of girder with mortar(1:1). The high-early-strength portland cement was used to shorten the hardening time in order to avoid a damage cause by the vibration of train operation. It's difficult to set a standard section of the reinforced region because of the deformation. Thus, six typical sections were established using C.M.S to survey the real shape of sections and dimension which need for assembly of lattice girder(Fig. 8).

5. MEASUREMENT INVESTIGATION

Eight sets of reference anchors at side-wall and crown of the reinforced region and a ground extensometer at both side-walls(Sta.99k094) where a lot of groundwater flow out were installed to analyse of the behavior of the ground and lining, which will be used in the design and execution of reinforcement. Inner section contraction was measured with a steel tape extensometer by an accuracy of 1/100 mm. Crown settlement was measured with a level by an accuracy of 1 mm. Two bench marks were installed outside and inside of the tunnel for measurement of crown settlement. The ground extensometer was multi-point type and were installed at 6, 4.8, 3.6, 2.4 and 1.2 m position. Deformation was measured using dial gauges by 1/100 mm unit.

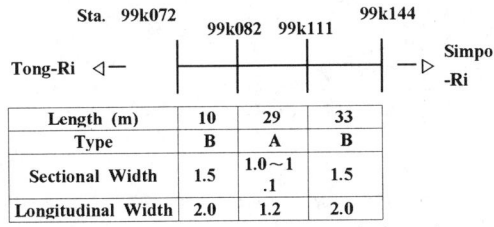

Fig. 6. Sketch of poly-urethane reinforced type

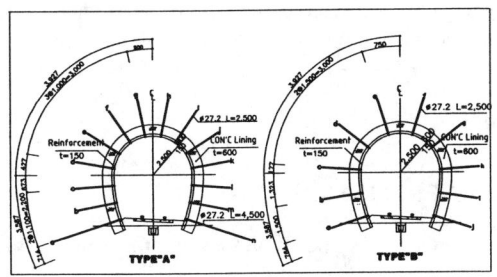

Fig. 7. Detail sketch of poly-urethane grouting

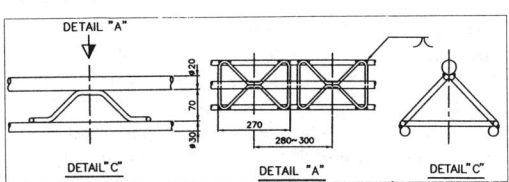

Fig. 8. Detail sketch of lattice girder installation

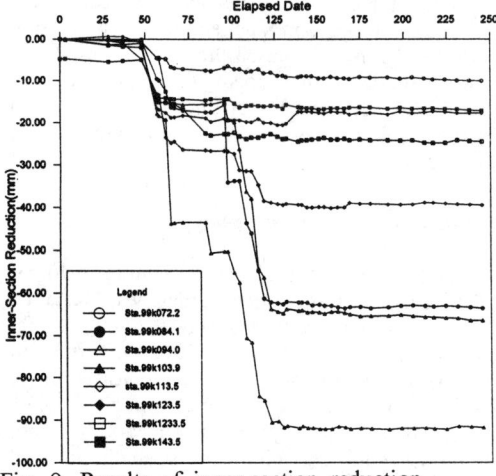

Fig. 9. Results of inner-section reduction

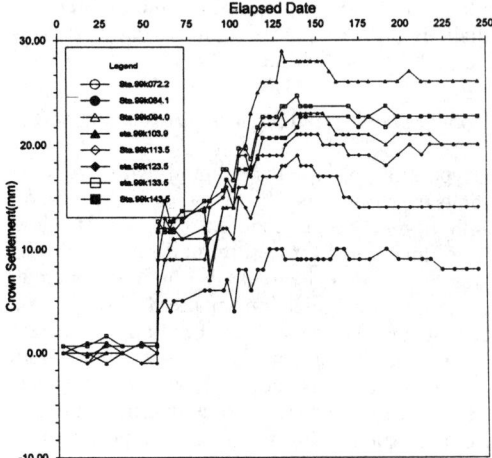

Fig. 10. Results of crown settlement

5.1 Measurement of inner section contraction

The inner section contraction ranges 90~10 mm. An abrupt contraction was measured at Sta. 99k 103.9 on August 24, September 7 and September 17~October 24, 1996. Coal-bearing shale, fault gouge, fault breccia and other weak structures were identified at this location during test boring. Inner section contraction at sections of Sta. 99k084.1, 99k094, 99k103.9 and 99k113.5 are remarkable, where A-type injections were carried out because of a weak ground and high ground water flowing. These deformations may took a place because of expansion of urethane as well as contraction of reinforced lining caused by expansion of the urethane which was injected at constructional discontinuity between existing lining and reinforced overlying lining. However, inner section contraction was converged when urethane grouting was finished. The results shows that the ground was stabilized by the grouting with pressurized bolts.

5.2 Measurement of Crown Settlement

The crown uprising of 9~26 mm was measured during urethane injection which had been done from September 3 to early November in 1996(Fig .10). After urethane injection was completed, the degree of crown uprising has converged. The result indicates that many cavities existed between concrete lining and ground at the crown of tunnel. The urethane was injected into side-wall first and into crown last. The sequence might cause the inner section contraction and the crown uprising.

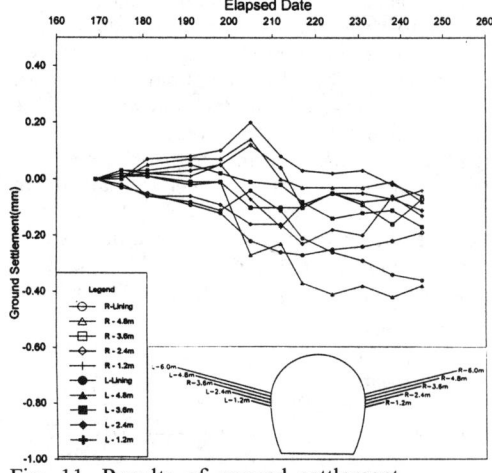

Fig. 11. Results of ground settlement

5.3 Measurement of Ground Settlement

Contraction of the right side (convex part at tunnel direction) and expansion of the left side (concave part at tunnel direction) were measured after urethane injection was finished (Fig. 10). However, the absolute deformation is 0.27~ 0.03mm which is negligible value, indicating that deformation of the ground has converged.

6. CONCLUSIONS

The tunnel has been deformed for about 30

years after the completion because of two large scale thrusts with fault breccia, fault gouge and several coal-bearing shale in cataclastic zone which run across the tunnel. The imbricate fan thrusts might also disturb the tunnel ground. The high lateral earth pressure ($k_0 \fallingdotseq 2.4$) has been measured by P.M.T. in the tunnel area. Many cavities between concrete lining and ground at the tunnel crown were identified during the reinforcement works. Therefore, refilling of cavities and restraining of cataclasite zones by urethane grouting with pressurized bolts are performed. Invert concretes and lattice girders were also installed to diminish the bending moment and to strengthen the weak lining. Measurements after reinforcement work indicate that the applied reinforcement were effective in stabilization of this tunnel.

REFERENCES

Boyer, S.E., & Elliot, D., 1982, Thrust system. Bull. Am. Assoc. Petrol. Geol., 66, 1196-1230.

Briaud, J.L., 1992, The Pressuremeter, Rotterdam : Balkema, 30-61.

KISTEC, 1996, Report of Safety diagnosis in San-Kol tunnel. Korea.

Korea Research Institute of Geoscience and Mineral Resources, 1979, The report of geological survey in Sam-Cheog coalfield, Korea.

Korean National Railroad & Daeduk Co., 1996. Report of detail design for urgent reinforcement construction to San-Kol tunnel. Korea.

財團法人 鐵道總合技術研究所, 1990, トンネル補強 補修マニュアル, Japan.

River diversion tunnelling works of Bakun Hydroelectric Project in Sarawak, Malaysia

Wang Ruel Jee
Bakun Tunnel Design Project, Dong-Ah Const., Seoul, Korea

Jeong Hoi Koo
Dong-Ah Constr., Seoul, Korea

ABSTRACT: The Bakun Hydroelectric Project is involving the construction of a hydroelectric power plant with an installed capacity of 2520MW and a power transmission system connecting to the existing transmission networks in Sarawak and Peninsular Malaysia. Construction of the river diversion tunnels and related site installations are constituted the preliminary phase of the works.

In order to allow the dam and powerhouse of the Bakun Hydroelectric Project to be constructed in dry surface conditions during the dry season, the Balui River will have to be diverted through tunnels throughout the construction period. Among the Project, three concrete lined diversion tunnels each with an internal finished diameter of 12m and a length of approximately 1,400m, including inlet and outlet structures are contracted by Dong-Ah Construction Industrial Co. Ltd. in 1995.

The geomechanical condition in the Project area have been classified into four rock mass types. As the assignment of the different rock types to rock mass types depend mainly on the degree of weathering and the occurrence of jointing, Greywacke(sandstone) as well as Shale can occur together in the same rock mass type if their rock mass properties are similar. This description deals with the construction design of underground structures, among others underground structures in view of rock mechanics.

For structures stability calculations have been performed and required rock support has been evaluated to provide sufficient stability of the construction during river diversion. It was necessary to carry out further stability calculations during the excavation progress to adjust the rock support according to the actual geological conditions.

1. INTRODUCTION

Dong Ah Construction Industrial Co., Ltd., Seoul has got the contract for the river diversion works of the Bakun Hydroelectric Project, located in Sarawak, Malaysia (see Fig.1). The power station will consist of a 205m height concrete face rockfill dam(CFRD). The installed capacity of the power scheme will be 2520MW. During construction of the dam and the power facilities, the Balui river has to be diverted by three diversion tunnels with a length of some 1400m each. The excavation diameter of the tunnels are designed between 13m and 13.4m, the inner diameter is 12m. In the portal area the tunnel width is 16m.

This report deals with the construction design of all excavation works, underground structures. In view of rock mechanics, excavation works and rock support, the river diversion is composed of the following sub-structures : diversion intake structures, vertical sidewalls at the inlet structure, diversion outlet slopes, vertical sidewalls at the outlet structure, portal area of a construction adit which crosses the diversion tunnels, construction adit, diversion tunnels and intersection between diversion tunnels and construction adit.

For all sub-structures stability calculations have been performed and the required rock support has been evaluated to provide sufficient stability of the construction during river diversion. It might be necessary to carry out further stability calculations during the excavation progress to adjust the rock support according to the actual geological conditions.

The calculations and the rock support base on the data given in the Principal Design Criteria and the Technical Specifications of the Bakun Tender Documents as well as internationally recognised standards mentioned herein such as DIN-ISO or ASTM.

Generally different rock mass types as defined in Principal Design Criteria have been considered for all calculations. Following the evaluated rock support is expressed in dependence of these different rock mass types. During excavation geotechnical engineers or geologists on site who are in charge of the different sub-structures have to decide and assign the present rock mass to the equivalent rock mass type with its rock support.

2. GEOTECHNICAL PARAMETERS OF THE STRUCTURAL ANALYSIS

2.1 Rock types

Following groups of predominant rock types corresponding to their material properties are

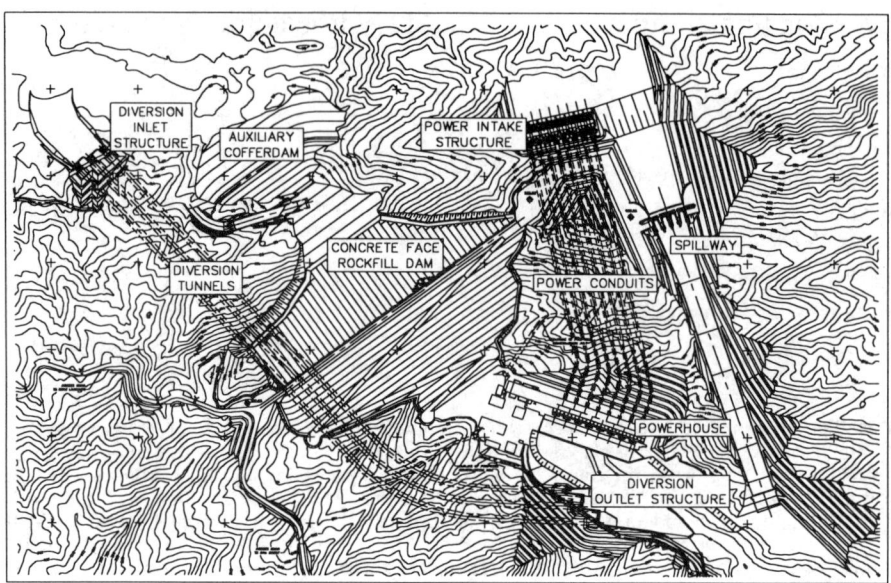

Fig. 1 Layout of the Bakun River Diversion Tunnelling Works

written. They can be summarised in below :
- Predominant greywacke. The greywacke is mostly sandstone consisting of angular and poorly sorted quartz and feldspar grains. Conglomerates occur only locally and consist of subrounded quartz grains and well rounded and flat shale fragments.
- Predominant shale mostly with a high percentage of silt. The siltstone as well as the fine grained sandstones consist mainly of quartz and have a silica matrix.
- Thinly interceded shale, siltstone and greywacke alternations. The shales are mostly silty. They consist of different clay minerals, quartz, calcite and small amounts of mica. The cement is mostly calcitic.

2.2 Rock mass types

The rock mass in the Bakun project area have been divided into four rock mass types(RMT). As the assignment of the different rock types to rock mass types depend mainly on the degree of weathering and the occurrence of jointing, greywacke as well as shale can occur together in the same rock mass type if their rock mass properties are similar.

Most parts of the tunnels will belong to RMT I to III while RMT IV is less present. For the massive greywacke banks along the diversion tunnels the assignment to RMT I instead of RMT II is most probable. The final classification of the rock into RMT I and RMT II and the assigned rock support has to be done on site after blasting. For the stability calculations it has been overtaken from the design criteria that 1% of the total tunnel length belong to RMT IV. RMT IV might also be given in the tunnel portals.

In Table 1 the geotechnical parameters depending on the rock mass types are listed which are used in stability calculations.

After shearing, the rock mass is capable to bear shear forces up to their residual values as given in Table 1. It was estimated that except for RMT I, the residual friction angles shall be

Table 1. Parameters and Safety Factors for the River Diversion Works

Parameter	Unit	Rock Mass Type				Safety Factor
		I	II	III	IV	
Rock:						
Density	ton/m^3	2.60	2.60	2.60	2.60	1.0
Uniaxial Compressive Strength (UCS$_{Rock}$)	MPa	200	100	60	35	1.0
Tensile Strength (TC$_{Rock}$)	MPa	10.0	5.0	3.0	1.8	1.0
Poisson's Ratio (ν_{Rock})		0.20	0.2	0.25	0.30	1.0
Rock Mass:						
Rating RMR	-	80	65	50	35	-
Deformation Mod(E$_D$)	MPa	1200	8000	6000	2500	1.0
In-Situ Stresses(k$_0$)	-	0	0.3	0.3	0.5	-
Peak Cohesion	MPa	0.3	3.5	2.0	0.7	1.5,2.0
Residual Cohesion	MPa	8.0	1.2	0.35	0.1	1.5,2.0
Peak Friction Angle	°	2.67	40	35	25	1.3,1.1
Res Friction Angle	°	45 40	40	35	25	1.3,1.1

the same as the peak friction angles
whereas the residual cohesion shall be reduced as follows:
- RMT I & II : residual cohesion is 1/3 of peak cohesion
- RMT III : residual cohesion is 1/6 of peak cohesion
- RMT IV : residual cohesion is zero.

The assumption to set the residual cohesion of RMT IV to zero is conservative as usually in the rock mass (not on discontinuities) a cohesion of

100kPa is always given. For numerical calculations two sets of partial safety factors are given Table 1.

2.3 Discontinuities

Three sets of discontinuities characterise the geotechnical conditions of the project area:
- J1 bedding[Dip Direction/Dip=150/55] as the dominant and most important plane of separation. In the massive greywacke it has a low degree of separation. Here, the bedding planes are mostly closed and planar.
- J2(ⅰ)[265/60] and J2(ⅱ)[065/55] as the main set of transverse joints. J2(ⅱ) is the less developed complementary set of transverse joints J2(ⅰ).
- J3[040/30] a further joint set which is directed more or less perpendicular to bedding. J3 of greywacke is predominantly rough and planar.

The peak and residual shear parameters ϕ and c applied in underground wedge stability analyses are shown in Table 2.

Table 2. Peak and Residual Shear Parameters on Discontinuities.

	Rock Mass Type	Safety Factors $\eta_{\tan\varphi}$	Safety Factors η_c	J1 (Bedding) φ [°]	J1 (Bedding) Cohes. [MPa]	J2 φ [°]	J2 Cohes. [MPa]	J3 φ [°]	J3 Cohes. [MPa]
Peak	I	1.0	1.0	40	0.60	43	0.55	40	0.60
Peak	II	1.0	1.0	35	0.40	38	0.36	35	0.40
Peak	III	1.0	1.0	30	0.25	32	0.22	30	0.25
Peak	IV	1.0	1.0	20	0.05	20	0.05	20	0.05
Resi-dual	I	1.0	1.0	35	0.20	38	0.18	35	0.20
Resi-dual	II	1.0	1.0	35	0.13	38	0.11	35	0.13
Resi-dual	III	1.0	1.0	30	0.05	32	0.04	30	0.05
Resi-dual	IV	1.0	1.0	20	0.00	20	0.00	20	0.00

3. A ROCK SUPPORTING MATERIALS

3.1 Rock bolts

From the Tender Design fully grouted SN-rock bolts are proposed to be used for the excavation works. Then a rock bolt capacity of F_{RB}=150kN =15tons is used in stability calculations. In order to save construction period it is foreseen to replace the fully grouted rock bolts with swellex rock bolts. The technical properties of the swellex system and fully grouted rock bolts are similar and in addition to that the swellex bolts provide some more advantages over fully grouted rock bolts.

The swellex rock bolts belong to the type of friction anchored rock bolts and have been developed by Atlas Copco between 1977 and 1980. This bolting system consist of a steel tubular bolt and a high-pressure water pump with installation equipment for the bolt setting.

The swellex bolt is made from a circular steel tube, which has been folded to reduce its diameter. Bushings are pressed on to both ends of the bolt which are then sealed by welding. Water is injected into the bolt at high pressure so that the bolt expand. During this expansion process the swellex bolt compacts the material surrounding the borehole and adapts its shape to fit the irregularities of the borehole. During the expansion process the bolt shortens in its longitudinal direction, pulls the faceplate firmly against the rock and causes a pre-tension of about 2 tones. After installation a powerful mechanical interlock between the bolt and the irregularities in the borehole is given and a combination of friction and mechanical interlock is generated throughout the entire bolt length.

The shear resistance of swellex rock bolts is equal or greater than that of fully grouted rock bolts due to the hollow cross section of the bolt which provides a combination of shear and tensile stresses even if the bolt is loaded with pure shear forces. In average the failure shear load is about 93% of the tensile failure load which is more compared to fully grouted rock bolts. The ductility of the swellex bolts which is mainly controlled by steel type is equal to fully grouted rock bolts and provide an elongation of 20% or 30%. The pull-out resistance can be estimated for the given greywacke to 50 to 80kN/m bolt length.

The required bond length of swellex bolts depend on the rock mass quality and has to be provided with pull-out tests. In most cases and especially with the given rock conditions a bond length of approximately 0.50m should be sufficient to transmit the full load capacity by friction into the rock mass.

The required borehole diameter should be between 32 and 39mm. It is planned to use this type instead of fully grouted rock bolts since no technical disadvantages exist and the installation time is much quicker.

The low bolt capacity of 100kN instead of 150kN, which has been taken into consideration when designing the rock support, will be regarded by reducing the rock bolt pattern. The final pattern will be determined by improving the bolt distance in longitudinal direction while the distance in radial direction is kept constant.

The main advantage of swellex rock bolts is the rapid and simple installation which is most important in weak rock. Also the pre-tensioning as a result of the contraction during installation is an advantage over fully grouted rock bolts and, in addition the full load capacity is available immediately after installation.

This swellex bolt was sellected as a rock support of the diversion tunnels, significant time and cost.

3.2 Shotcrete

In Principal Design Criteria and Technical Specification ordinary shotcrete(SC) and reinforced shotcrete(RS) is proposed as surface protection and for stability measures for the underground excavation works.

The installation of RS is usually very time consuming. To avoid delays in the construction time of the underground works Steel Fibre Reinforced Shotcrete(SFRS) will be used to accelerate the construction progress. SFRS is defined as a concrete containing steel fibres which are sprayed with a high velocity on to the rock surface. Steel fibres are incorporated in the shotcrete to improve its crack resistance, ductility, energy absorption and impact resistance characteristics. The material properties of SFRS depend mainly on the geometrical shape of the fibres, the size of the fibres, the steel quality, the weight of fibres per cubic meter shotcrete and the grain size distribution, water and cement content as well as additives.

In addition SFRS provides some technical advantages over RS. The main point is that SFRS follows exactly the contours of the rock. Wire mesh pinned on a rough rock surface can cause voids behind the mesh. Then the mesh is not embedded in the shotcrete and cannot impart any ductility in the material. Also no corrosion protection is given. On the other hand the toughness and flexural strength of SFRS is even also at rough surfaces.

The main reason for incorporating reinforcement or steel fibres in shotcrete is to impart ductility to a brittle material. Wire mesh and steel fibres in shotcrete improve the crack resistance, impact resistance and energy absorption. In addition RS and SFRS enables the post crack behaviour of the shotcrete that means a continued load carrying after cracking.

Two test methods are internationally established and will be used for the diversion works to investigate the ductility and the toughness SFRS and RS
- the standard test method for flexural toughness and first crack strength of fibre reinforced concrete according ASTM C 1018-89 or the recommendations for design and construction of steel fibre reinforced concrete of the Japan Society of Civil Engineers (JSCE) and
- the slab test method of the French Railway Company (SNCF). This method is very suitable to simulate the behaviour of the combination of rock bolts and shotcrete.

For the general approval of SFRS slab tests and three-point loading tests with test beams are performed to achieve a relationship between both tests. The evaluation of the tests will offer load-deflection and energy-deflection curves for RS and SFRS which can be compared. For the three point loading tests it is more favourable to use the Japanese evaluation methods instead of the method given in ASTM C 1018 especially for the quality control. We use the Japanese evaluation methods.

4. METHODS OF CALCULATION AND SUPPORT PHILOSOPHY

The rock mass can be treated as homogeneous and the overall stability of the diversion tunnels is controlled by the geometry and the stress state in depth. Wedge failure might only be spotwise possible if all three discontinuities are present at the same place.

The parameters for rock mass have been used in the finite element analyses for the tunnels while for local wedge stability analysis of underground wedges the parameters for discontinuities have been applied. The rock mass parameters for the different rock mass types RMT have been used in the calculations as listed in Table 1. The influence of the changing tunnel geometry on to the stress state during the excavation process has been considered with two different geometries, roof excavation and full excavation. The decisive rock mass parameters for numerical analysis and the shear parameters of discontinuities for wedge stability analysis are listed in Table 2.

4.1 Principal Method of Rock Support Evaluation

In deep underground excavations such as the Bakun diversion tunnels it is impossible to design the rock support for the given overburden. Instead the rock support is based on the so called NATM - New Austrian Tunnelling Method.

If a tunnel is under construction, the stress-strain state will be concentrated around the excavation and areas can exist where the shear strength of the rock mass or joints is reached. The depth of these overstressed areas depend on some major parameters such as the in-situ stress state with its parameter, the geometry of the excavation that means the ratio width/height, the overburden above the excavation roof, the deformation behaviour of the rock mass, the unconfined compression strength of the rock mass and the shear strength of rock mass and discontinuities.

Also wedge failure is possible if a marked discontinuity system is present which enlarges the tunnel diameter and change the stress field. Furthermore the loosened area around the tunnel after blasting influences the surrounding stress state. As mentioned before the task of the rock support is to stabilize these loosened areas in a way that the self bearing capacity of the excavated tunnel can be assured.

As a result of numerical calculations critical areas around the tunnels, depending on the rock mass type can be defined and the required rock support consisting of rock bolts, wire mesh, shotcrete or steel fibre reinforced shotcrete can be evaluated.

At the walls of tunnels or caverns wedges can occur as a result of overstressed rock mass in combination with present discontinuities. Then plane wedge failure or sliding on two planes can occur. In the roof area of such excavations two principal failure modes are possible : wedge failure as a result of a marked discontinuity system, and roof failure of overstressed rock mass as a result of the stress concentration after excavation.

Especially in a more or less homogeneous rock mass where the stability of the excavation is mainly stress controlled, the rock bolts and shotcrete in the roof area shall provide an artificial arch in the loosened zone similar to a natural arch to stabilize the excavated structure.

The task of numerical stability calculations is to determine areas in underground excavations where wedges can fail. In opposite to traffic tunnels which are located close to the surface it is in deep underground excavations impossible to show explicit with a simulated rock support in a numerical calculation that the support is sufficient to provide the stability of an underground excavation. Instead the decisive depths of the rock mass in the roof area and the walls of the diversion tunnels which have to be supported is determined 'by hand' and base on the results of the finite or boundary element calculations, that means mainly calculated overstressed areas, and on the experience of similar projects in combination with the constructed support when using that design method and on the results of wedge stability calculations in the roof and at the walls.

The amount and geometry of rock mass to be supported is then chosen individually for each underground structure and has to be adjusted during tunnel excavation.

4.2 Rock Support of Tunnel Walls

For the Bakun Hydroelectric Project at the tunnel walls plane wedge failure has been assumed as a result of the often missing discontinuities and their location related to the tunnel centerline.

In those cases where the sliding surface is more or less equal to dip and dip direction of one of the discontinuities J1, J2(i,ii), J3, the residual friction angle ϕ and the residual cohesion c has been used according to the values for discontinuities given in Table 2. In cases where no discontinuity is present the residual values for the rock mass friction angle and rock mass cohesion as listed in Table 1 has been used to evaluate the support.

The number of required rock bolts in dependence of a given safety factor can be calculated by

$$n \geq \frac{W \cdot (SF \cdot \sin\beta - \cos\beta \cdot \tan\phi) - c \cdot A}{RB \cdot (\cos\alpha \cdot \tan\phi + SF \cdot \sin\alpha)}$$

where
- n   number of rock bolts over the total wedge height and a stripe of 1m [-],
- SF   safety factor [-],
- RB   bearing capacity of rock bolts [kN/m^2],
- W   weight of the wedge [kN],
- β dip of the sliding surface [°],
- α angle between the plunge of the bolt and the normal to the sliding surface [°],
- ϕ friction angle of the sliding surface [°],
- c   cohesion of the sliding surface [kN/m^2],
- A   base area of the sliding surface [m^2].

The value of n describes the number of rock bolts which are necessary to stabilize the wedge over its total height with a unit width of one meter and the chosen safety factor.

4.3 Rock Support of the Tunnel Roofs

All types of possible failure modes have to be investigated and the worst is decisive for the required rock support and has to be designed against failure. For support considerations the behaviour of all types is assumed to be equal. Since the failure mode for those wedges is punching through the shotcrete an failing from the roof no shear strength must be considered as a restraining force.

The principle that the complete weight of such a wedge has to be supported separately by the rock bolts and by the shotcrete with a defined safety factor is used to evaluate the rock support in the roof area. When using results of finite element or boundary element calculations the first step is to define the depth of the plastified area to be supported. The next step is to determine the required shotcrete thickness. The evaluation of the rock support if a geometrical roof wedge caused by discontinuities is decisive for the stability is very similar to the above shown method.

The number of required rock bolts and the shotcrete thickness can be determined to

$$n_w \geq \frac{\eta_{RB} \cdot w \cdot V}{RB}, \quad d_{SC} \geq \frac{\eta_{SC} \cdot w \cdot V}{\tau_{per} \cdot (J_1 + J_2 + J_3)}$$

where
- n_w number of rock bolts required for a given wedge [-],
- d_{sc} shotcrete thickness [m],
- η_{RB} safety factor applied on rock bolts [-],
- η_{SC} safety factor applied on shotcrete [-],
- w   density of rock mass [MN/m^3],
- V   volume of the wedge [m^3],
- τ_{per} shear strength of the shotcrete [MN/m^2]
- J_i trace length of the 3 discontinuities [m].

At the walls the required shotcrete thickness is difficult to estimate as with the given shotcrete strength τ_{per} and realistic assumptions of rock pressures on to the shotcrete only few centimeters of shotcrete thickness will be calculated to be necessary. But this is contrary to many constructed tunnels where more shotcrete was necessary to stabilize the excavation. Therefore the shotcrete thickness for the walls is mainly based on the experience of former and similar projects and has to be adjusted during excavation considering the results of convergence or extensometer measurements.

4.4 Safety Factors

When designing a rock support for underground structures, safety factors are required to cover uncertainties lying on the shotcrete as a result of its construction, on the rock bolting since the force transmission rock bolt into mortar and mortar into rock mass can not be guaranteed to be constant over the whole borehole, on the numerical analysis when determine the overstressed areas and possible wedges in the underground excavations and on the validity of the rock support model with respect to the conditions in reality.

For the Bakun diversion tunnels the overstressed areas as a result of BEM or FEM calculations as well as the wedge heights have been determined under consideration of reduced shear parameters as given in Table 1 such as $\eta_{\tan\varphi} = 1.3$ and $\eta_c = 1.5$ for BEM or FEM calculations and $\eta_{\tan\varphi} = 1.1$ and $\eta_c = 2.0$ for the wedge stability calculation.

An additional global safety factor η_{RS} in the

area 0 to ±90° to the tunnel centerline is given which shall be valid for the construction itself. According to Principal Design Criteria η_{RS} shall determined to

$$\eta_{RS} = \eta_{RB} + \eta_{SC} \geq \frac{S_B}{W_{RM}}$$

where
- η_{RS} global safety factor of the rock support for the construction,
- η_{RB} safety factor of the rock bolt pattern of the construction,
- η_{SC} safety factor of the shotcrete of the construction,
- S_B load capacity of the chosen rock support per square meter and
- W_{RM} weight of rock mass to be supported per square meter.

It is given in Principal Design Criteria that in the areas with a denser grid of rock bolts the bolts shall be installed in a staggered arrangement to guarantee a uniform stress distribution over the rock mass. Also an economic stabilization of overstressed areas cannot be given only by rock bolting as the required grid would be too dense. Therefore, a combination of rock bolts, shotcrete and wire mesh shall stabilize the rock mass so that its self support is possible. In the roof area the safety factors shall be $\eta_{RB} \geq 0.5$ for the rock bolts and $\eta_{SC} \geq 1.5$ for the shotcrete so that the global safety factor of the rock support for the construction is given below.

$$\eta_{RS} = \eta_{RB} + \eta_{SC} \geq 2.0$$

5. STABILITY ANALYSIS OF DIVERSION TUNNELS

The stability analysis and the evaluated rock supports for the diversion tunnel considering all four rock mass types and the different overburden are presented. The determination of the rock support for the tunnels base on finite element analysis as well as on wedge stability analysis. The worst result of both methods is decisive for the rock support but after performing of a parameter variation it has been turned out that for Bakun Hydroelectric Project the stability of the diversion tunnels is mainly controlled by the stress-strain behaviour as a result of the excavation process. Due to the given discontinuity system in relation to the tunnel azimuth the calculated possible wedges are not decisive for the design of the rock support.

5.1 Finite Element Analysis

In the finite element calculations the following parameter variations have been considered for the evaluation of the rock support of the diversion tunnels :
- two geometries, roof excavation and full excavation,
- different in-situ stress states, $k_{o,x}/k_{o,y} = 0.3/0.3$ and 0.5/0.5,
- four rock mass types, RMT I to IV,
- two heights of overburden 100m and 200m
- two different sets of safety factors $\eta_{\tan\phi}/\eta_c = 1.3/1.5$ and 1.1/2.0.

Six rock support types have been defined RST A to F depending on the rock mass types RMT and the overburden. For rock mass types I and II an overburden of 200m has been considered in the finite element calculations due to the given topography in the project area and according to [1] while for the rock mass types III and IV to different overburden 100 and 200m have been applied in the finite element calculations. Following the results are explained in more detail.

Decisive for the evaluation of the rock support is following data:
- the depth of the plastified area as this is the main weight of rock mass to be supported,
- tension stresses in the roof, especially after the first excavation step(roof excavation) since cracks can activate roof wedges,
- the calculated deformations, as too large deformations can lead to cracks in the rock mass and the shotcrete which could cause serious stability problems if the increase of deformation does not stop after some time and
- the main principal stress distribution $\sigma_{1,max}$ as rock bursts could happen if the stress concentration as a result of the tunnel geometry lead to an overstressing of the rock.

Therefore the above explained results of the finite element calculations are summarised in Table 3. The comparison of the rock strength UCS_{ROCK} with the calculated principle stresses a, gives a hint of the bearing capacity of the rock. For RMT I and II generally no stability problems should arise, while for RMT IV the relation $UCS_{ROCK}/\sigma_{1,max}$ is small. Considering an usually lower rock mass strength UCS_{ROCK} than UCS_{ROCK} could arise stability problems for RMT IV.

5.2 Wedge Analysis

In addition to the finite element calculations wedge analyses have been performed to investigate whether the required rock support is stress controlled as a result of the excavation or controlled by the discontinuity system and geometrical wedges. Here it has to be mentioned again that the wedge analyses performed with the program UNWEDGE is based on the assumption that all three required discontinuities to form a wedge are present. For the Bakun diversion tunnels this a very conservative assumption as the joint sets J2(i,ii) and J3 are not very distinctive. But nevertheless a spotwise presence of all discontinuities at the same place is possible. Then the risk of wedge failure is given.

The following variations have been performed for the wedge stability analysis :
- Two azimuths 317° and 272° as the diversion tunnels change their direction after a length of about 2/3 of the tunnels.
- Three different wedge sizes have been investigated. First the biggest possible wedge due to the dip and dip direction of the given discontinuity system. Then a variation of the

Table 3. Result of Finite Element Calculations for the Diversion Tunnels

FE-Calculation	Excavation	Plastic Area d [m]	Safety Factor $\eta_{\tan\phi}/\eta_c$	In-situ stresses $k_{o,x}/k_{o,y}$	Calc. Deflections			Maximum stresses		
					Roof [mm]	Wall [mm]	Floor [mm]	max σ_1 [MPa]	UCS$_{ROCK}$ [MPa]	UCS/σ_1 [MPa]
RMT I Overburden 200m	full excavation	–	1.1/2.0	0.5/0.5	5	0	5	10-12 at wall	200	16.7
RMT II Overburden 200m	roof excavation	1.5	1.1/2.0	0.3/0.3	8	2	10	15-18 edge roof	100 to floor	5.6 1.step
RMT III Overburden 100m	roof excavation	1.5	1.3/1.5	0.3/0.3	6	2	6	8 edge roof	60 to floor	7.5 1.step
RMT III Overburden 200m	roof & full excavation	2.5	1.3/1.5	0.3/0.3 0.5/0.5	12 12	8 8	12 12	12 edge roof	60 to floor	5.0 1.step
RMT IV Overburden 100m	roof & full excavation	3.0	1.3/1.5	0.3/0.3 0.5/0.5	20 20	20 20	20 20	9 edge roof	35 to floor	3.9 1.step
RMT IV Overburden 200m	roof & full excavation	4.0	1.3/1.5	0.3/0.3 0.5/0.5	25 60	5 100	25 25	18 edge roof	35 to floor	1.9 1.step

trace length of J2(i,ii) to 5.0 and 2.5m as this is the joint set where most information on trace length and spacing is given in Principal Design Criteria.

- A variation of the different rock mass types RMT I to IV with different shear parameters as given in Table 2.

The results can be summarized as follows:

- The azimuth of the diversion tunnels of 317° lead to bigger possible wedges in the roof compared to the tunnel azimuth 272°.
- The relation wedge weight to face area is the less the smaller the wedge is. Most unfavourable is the biggest possible wedge.
- The wedge weights in the roof area are independent from the rock mass type RMT, as the failure mode of the roof wedges is not sliding but falling.
- The possibility of wall wedges is small. Most unfavourable is the combination with azimuth 317°, the biggest possible wedge and rock mass type IV. Here a safety factor of 1.47 without any support has been calculated.

In Table 4 the given safety factors for a chosen rock bolt grid and shotcrete thickness are listed. The given safety factors show that possible wedges are not decisive for the stability of the diversion tunnels but that the stability of the tunnels is stress controlled as a result of the excavation process.

5.3 Summary of the Rock Support

In the Table 5 the complete rock support for the diversion tunnels is listed. For the worse rock mass types a certain proportion of 6m bolts will be installed to improve the stress distribution inside the overstressed rock mass arch. This will also provide a better constraint in the rock mass when moving inwards.

Wire mesh or SFRS is necessary where the deformation of the tunnels increase as the steel provide a higher toughness and flexural strength. For the good rock mass type I and II no reinforcement is required as the deformation will be small, see Table 3. Here shotcrete acts mainly as a sealing or as an overhead protection. One mat of wire mesh in the middle of 10 or 15cm shotcrete will be installed to increase the toughness.

6. DESIGN OF CONSTRUCTION ADIT

The foreseen construction adit to provide access to the diversion tunnels at up to 6 faces is located in figure 1. The adit is located in shale/mudstone until station 75m following by greywacke until the end.

The rock mass can be assigned to the types RMT III to IV for the shale/mudstone and RMT I to II for the greywacke. The overburden varies from 10 to 45m in shale/mudstone and up to

Table 4. Required Shotcrete and Rock Bolts as a result of wedge stability analysis

No. [-]	Azimuth [°]	ΣJi [m]	weight [ton]	Min Required		Chosen		Safety Factor	
				d_{SC} [m]	n_w [-]	d_{SC} [m]	n_w [-]	$\eta_{SC, given}$ [-]	$\eta_{RB, given}$ [-]
1	317	8.73	0.81	0.003	0.05	>0.05	1	>1.5	>1.5
2	317	17.46	6.42	0.010	0.43	>0.05	1	>1.5	2.3
3	317	23.78	16.20	0.020	1.08	>0.05	2	>1.5	1.9
1	272	5.90	0.21	0.001	0.01	>0.05	1	>1.5	>1.5
2	272	11.80	1.72	0.005	0.11	>0.05	1	>1.5	>1.5
3	272	17.14	5.25	0.009	0.35	>0.05	1	>1.5	>1.5

Table 5. Rock Support for the Bakun Diversion Tunnels

RST [-]	RMT [-]	Overburden [m]	in the roof Area ±60° to the Centerline				in the wall Area ±60° to ±135°			
			Shotcrete [m]	No. of Wire Mesh [-]	Rock Bolt Grid [m²]	Length /Percent [m/%]	Shotcrete [m]	No. of Wire Mesh [-]	Rock Bolt Grid [m²]	Length /Percent [m/%]
A	I	all	0.05	–	SB	4/100	–	–	SB	4/100
B	II	all	0.10	1	6.25	4/100	0.05	–	16.00	4/100
C	III	≤100	0.10	1	6.25	4/100	0.05	–	9.00	4/100
D	III	>100	0.15	1	4.00	6/20 4/80	0.10	1	6.25 4.00	4/100 6/10 4/90
E	IV	≤100	0.15	1	3.06	6/30 4/70	0.10	1	2.56	6/20 4/80
F	IV	>100	0.20	2	2.25	6/40 4/60	0.15	1		

100m in greywacke in the intersection area with the diversion tunnels. For the stability investigation following characteristic parameter sets can be summarized:
- The portal area from station 12 to 28m is located in weathered shale 1 mudstone. The rock mass can be assigned to RMT IV. The overburden is small in the range of not more than 10m. Here the excavation progress might be difficult due to the fact that no rock mass arch will develop after excavation and the weight must be carried from the rock support and steel arches.
- The rock mass from station 28 to 75m can be assigned to RMT III or IV and the overburden varies between 10 and 50m. The stress-strain behaviour of the rock mass in that depth can be assigned to the class deep located excavations. Here usually an overburden of 1-2 times the tunnel radius is required to activate a rock mass arch and a self bearing capacity of the rock mass.
- The remaining part of the adit from station 75 to 287m. Here the adit is located in greywacke of RMT I or II. The overburden varies between 45m and 100m in the intersection area with the diversion tunnels.

For the evaluation of the rock support two different analyses have been performed. Behind station 28m the adit was treated as a deep located. Since the given rock mass both shale/mudstone and greywacke do not belong to the class of hard rock, finite element analyses have been performed to investigate the stress state around the adit.

For the first part of the adit where no self bearing capacity of the rock mass can be expected, the full overburden has to be supported by reinforced shotcrete and steel arches.

6.1 Finite Element Analysis and Rock Support

Following parameters have been applied in the finite element analyses as the worst case for the greywacke :
- one geometry for full adit excavation,
- two different in-situ stress states, $k_{o,x}/k_{o,y}=0.3/0.3$ and $0.5/0.5$,
- rock mass type, RMT II,
- overburden 100 m and
- safety factors $\eta_{\tan \varphi}/\eta_c = 1.3/1.5$.

The calculated maximum principle stresses in the range of $\sigma_1=7.5$MPa are far below the unconfined compressive strength of the rock mass in the range of $UCS_{RM} \approx 60$MPa.

No plastification in the rock mass close to the tunnel have been calculated. Even in the edges wall to bottom here usually the stress concentration is at the most no overstressing of the material has been calculated. The deformations into the adit for both investigated in-situ stress states have been calculated to approximately 2.5mm which is very little.

For the worst case, that means considering an overburden of 100m and the rock mass parameters for RMT II, the required rock support is very small. Spot bolting with bolt lengths of 4m are sufficient for the stability. Shotcrete with a thickness of 50mm±60° from the centerline is adequate and acts mainly as overhead protection. Depending on possible wedges only wire mesh as overhead protection might be sufficient which has to be decided on site during excavation.

In the intersection areas a systematic rock support of both roof and walls shall be installed due to the large excavation diameters of the diversion tunnels. In the case that the conditions are much better and the rock mass can be assigned to RMT I a reduction of the rock support is possible.

In shale/mudstone the worst case is same condition in greywacke bur rock mass type RMT IV, and overburden 50m is different.

From the result of analysis the maximum principle stresses have been calculated to approximately 2.0MPa along the excavation surface. Assuming a rock mass strength UCS_{RM} in the range of 5-10MPa for RMT IV, the unconfined compression strength is sufficient to withstand the stress concentration.

The shear strength is reached mainly in the walls for both simulated in-situ stress variations. As the overstressed zone is the main criteria for the determination of the rock support, a stripe of 1.25m thickness has been chosen be decisive for the rock support. In the roof ±60° to the centerline no overstressed zone has been calculated.

The calculated deformations are in a range of 50mm which is compared to the adit width of 10m not critical. According to the calculated plastic depth d=1.25m, the rock bolt length is chosen to 4m with a bolt capacity of 15ton. For shotcrete thickness of 10cm and more at least 1 layer of wire mesh shall be installed.

6.2 Wedge Stability Analysis

In addition to the finite element analyses wedge stability analyses have been performed. The construction adit has two main directions 317°(more or less parallel to the D/S part of the diversion tunnels) and 060° in the intersection area with the diversion tunnels. As shown in the wedge stability analysis for the diversion tunnels, the direction 317° is not critical for potential wedges.

The wedge stability analysis showed that sliding on J1 might be possible although the calculated safety factor with reduced shear parameters has been calculated to SF=2.72.

7. CONCLUSION

The tunnelling method of Bakun River Diversion Project were designed by NATM-Drill and Blasting methods. This tunnelling method has been defined as a concept that beneficially alters the stress pattern an underground openings.

The structural calculation and the rock support measures base on the international standards such as DIN-ISO and ASTM. Through the excavation we found that most part of tunnels are belongs to RMT II and three sets of discontinuities developed in the Bakun site. We applied to design the swellex rock bolt and steel fibre shotcrete in order to save the construction times.

It was proved that most of the rock conditions at the tunnel site are better than expected at the design stage through the geological mapping and monitoring instrumentation. We could reduce the reinforcement of concrete lining through the back analysis for the optimal design to save the construction time and cost. However, the loosening rock masses after blasting, we conducted controlled blasting with special care to increase the safety.

REFERENCE

1. Bakun Management SDN. BHD.,
   Privatization of the Bakun Hydroelectric Project. Principle Design Criteria & Technical specifications, KUCHING. 1995.
2. E. Hoek, and J. W. Bray,
   Rock Slope Engineering, Institute of Mining and Metallurgy, London, 1981.
3. E. Hoek, and E. T. Brown,
   Underground Excavation in Rock, Institution of Mining and Metallurgy, London, 1980.
4. J. L. Carvalho, E. Hoek, and B. Li,
   UNWEDGE2.2 User's Manual, Rock Engineering Group, University of Toronto, 1991.
5. W. Wittke,
   Felsmechanik. Springer Verlag, 1984.
6. B. Maidl,
   Handbuch des Tunnel-und Stollenbaus Band ll, Verlag Glückauf GmbH, Essen, 1988.

Questions and measures for safety in the design and construction of caverns and openings

Cusp catastrophe of the rockbursts induced by ore pillars and its forewarning

Hong Li, Zenghe Xu, Xiaohe Xu & Yongjia Wang
Center for Rockbursts and Induced Seismicity Research, Northeastern University, Shenyang, People's Republic of China

ABSTRACT: The rockbursts of ore pillar under a hard rock stratum are regarded as the instability failure problem of a ore pillar under a stratum beam subject to elastic supports. The cusp catastrophic theory is applied to discuss the instability mechanism of the rockbursts. It is explicate that rockbursts only occur on condition that the ore pillar presents strain-weakening property, the stiffness ratio of the system $K/\lambda_1 < 1$ and loads imposed on the roof stratum is large enough to make the deformation of ore pillar reaches the post-peak strength phase. The factors affecting rockbursts are also discussed. Based on them, the evolution process, forewarning regular pattern and forewarning signs of rockbursts are studied. It is indicated that the subsidence velocity of roof stratum, which increases dramatically and tends to infinity, is the measurable forewarning signs of the rockbursts of ore pillar.

1 INTRODUCTION

Rockbursts are the phenomena which rockmass failures suddenly and releases lots of energy instantaneously. Because they happen much quickly and there is no apparent forewarning signs, they are one of the hazards which result in the most serious damage during the mining course. Along with the mines go deep continuously, rockbursts will take place more frequently and more violently. So, it is important to study the criterion, evolution process and forewarning pattern and signs of rockbursts. Since rockbursts occur only when some portion of media in a mechanical system consisting of ore body and surrounding rock presents strain-weakening due to its rupture, meanwhile ,accompanying with that, its surrounding rockmasses restore elastically and release their accumulated energy to the rupture body rapidly, rockbursts are related to the stability of the equilibrium of deformation mechanical system (Yin Youquan et al. 1984). When the system is in unstable equilibrium state and triggered by a minute perturbation, rockbursts will occur. For the majority of mine rockbursts are induced by ore pillar failure, in this paper, rockbursts under hard roof are regarded as a instablility problem of the mechanical system consisting of hard rock roof, ore pillar and unexcavated ore strata when it perturbed by external world; the unstable mechanism of the rockbursts in this case is studied through applying the cusp catastrophe theory; the factors affecting rockbursts are discussed. In theory, it can be thought that the reasonable design of mining layout to avoid the occurrence of rockbursts will be given once the criterion for rockbursts occurrence is made clear, in practice, nevertheless, the full ascertainment of the geological environments of a mine or a stope and of the mechanical properties of rock mass is actually impossible. So it is more significant in practice to study precursor regular pattern and signs in order to predict rockbursts. In this paper, the evolution process and the precursor regular pattern of rockbursts are studied based on the discussions on the instable mechanism of rockbursts; a forewarning signs which can be monitored are proposed.

2 MECHANICAL MODEL FOR ROCKBURST OF ORE PILLAR

Pillar mining methods which the roof is supported by leaving ore pillars shown as Fig.1(a) are often employed in underground mining. If the width of coal pillar is far narrow compared with the span of stopes

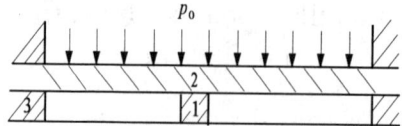

1--coal pillar 2--roof stratum
3--unmined coal stratum

(a)

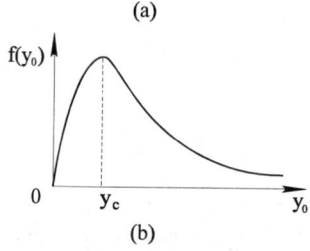

(b)

Fig. 1 The pillar mining method and the constitutive relation of ore pillar

advancing, the resistance of coal pillar to roof stratum can be approximately regarded as a concentration force R. The roof stratum can be regarded as a beam. The weight of roof stratum itself and the loads applied on strata can be regarded as the distribution of forces with the density of q. Because of its less deformation, unmined coal stratum can be regarded as elastic foundation. The ore pillar is the medium possessing nonlinear weakening behavior (Tang Chunan 1993), its constitutive curve is shown in Fig.1(b), and its constitutive equation is as follows

$$f(y_0) = \lambda y_0 e^{-y_0/y_c} \quad (1)$$

where λ -- the initial stiffness of ore pillar (i.e. the slope of the constitutive curve at $y_0 = 0$), y_0 -- the subsidence of the top end of ore pillar, y_c --the subsidence of or pillar corresponding to the peak-value strength. Based on the above simplification, a mechanical model shown as Fig. 3 is obtained. Because of its

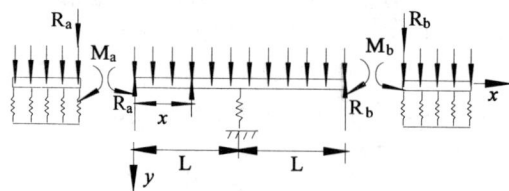

Fig.2 The mechanical model

symmetry, only the half of the model needs studying. If the non-dimensional variables $\bar{y} = y/L, \bar{x} = x/L$ are introduce, then according to the material mechanics, the equations for the beam deflection are

$$\begin{cases} \bar{y}_1^{(4)} + 4\beta^4 \bar{y}_1 = \dfrac{qL^3}{EI} &, \bar{x} \leq 0 \\ \bar{y}_2'' = \dfrac{qL^3}{2EI}\bar{x}^2 + \dfrac{M_a L}{EI} - \dfrac{R_a L^2}{EI}\bar{x}, 0 \leq \bar{x} \leq 1 \end{cases} \quad (2.1\text{-}2)$$

where y_1 and y_2 are the beam deflections in the section $x \leq 0$ and $0 \leq x \leq L$ respectively; the resistance force and the bending moment of the beam at $x = 0$; $\beta = (kL^4/4EI)^{1/4}$; EI is the curved stiffness; k -- the elastic coefficient of unexcavated ore stratum. The boundary and continuity conditions of beam are

$$\begin{cases} \bar{y}_1(-\infty) = q/kL, \; \bar{y}_1'(-\infty) = 0, \; \bar{y}_2'(1) = 0 \\ \bar{y}_1(0) = \bar{y}_2(0), \; \bar{y}_1'(0) = \bar{y}_2'(0), \\ \bar{y}_1''(0) = \bar{y}_2''(0), \; \bar{y}_1'''(0) = \bar{y}_2'''(0) \end{cases} \quad (3.1\text{-}7)$$

Having solved out (2.2) under the solution-determining conditions (3.1-7), the subsidence of the middle point of beam, i.e., the compression of ore pillar can be obtained as (4) by making $\bar{x} = 1$

$$y_0 = \dfrac{qL^4}{EI}\left(\dfrac{1}{24} - \beta_a\right) + \dfrac{R_a L^3}{EI}\left(\beta_b - \dfrac{1}{6}\right) + \dfrac{q}{k} \quad (4)$$

where $\beta_a = (1+\beta)/12\beta$, $\beta_b = (1+\beta)(1+\beta^2)/4\beta^3$. From Fig.2 we know $R_a = qL - f(y_0)/2$. Substituting the expression of R_a into (4), rearranging it produces

$$R = f(y_0) = qL[\dfrac{24(\beta_b - \beta_a) - 3}{12\beta_b - 2} \\ + \dfrac{12EI}{kL^4(6\beta_b - 1)}] - \dfrac{12EI}{L^3(6\beta_b - 1)}y_0 \quad (5)$$

Obviously, Eq. (5) is also the equilibrium equation of ore pillar.

3 THEORETICAL ANALYSIS OF THE CUSP CATASTROPHE ON THE ROCKBURST OF ORE PILLAR

If $U(y_0)$ is potential function, then Eq. (11) is equivalent to

$$U'(y_0) = f(y_0) + \dfrac{12EI}{L^3(6\beta_b - 1)}y_0 - \\ -qL\left\{\dfrac{24(\beta_b - \beta_a) - 3}{12\beta_b - 2} + \dfrac{12EI}{kL^4(6\beta_b - 1)}\right\} = 0 \quad (6)$$

According to cusp catastrophic theory (Xu Zenghe et. al. 1995), the cuspidal point of the system can be derived from formula $U'''(y_0) = 0$, so the cuspidal point $y_{01} = 2y_c$ can be found by differentiating Eq. (12) twice after substituting (1) into (6). Expanding (6) as Taylor's series at cuspidal point y_{01}. Cusp catastrophe theory assures that the qualitative properties of equilibrium equation (6) in the vicinity of cuspidal point y_{01} aren't affected after the items higher than three orders in the series are left out, and the quantitative analysis is also accurate enough. Thus, expanding (6), substituting (1) and the value of cuspidal point y_{01} into it produces

$$U'(y_0) = \{\lambda y_{01} e^{-2} + \frac{12EL}{L^3(6\beta_b - 1)} y_{01} -$$
$$- \frac{24(\beta_b - \beta_a) - 3}{(12\beta_b - 2)} qL -$$
$$- \frac{12EI}{kL^4(6\beta_b - 1)} qL\} + \{\frac{12EL}{L^3(6\beta_b - 1)} - \lambda e^{-2}\} \times \quad (7)$$
$$\times (y_0 - y_{01}) + \frac{1}{3!}\lambda e^{-2} y_c^{-2} (y_0 - y_{01})^3 = 0$$

Introducing non-dimensional variable $\tilde{x} = (y_0 - y_{01})/y_{01}$, then rearranging (7) yields the state equation and the bifurcation set

$$\begin{cases} \tilde{x}^3 + u\tilde{x} + v = 0 \\ u = \frac{3}{2}[\frac{12EI}{L^3(6\beta_b - 1)\lambda_1} - 1] \\ v = \frac{3}{2}[1 + \frac{12EI}{L^3(6\beta_b - 1)\lambda_1} - \\ \quad - \frac{24(\beta_b - \beta_a) - 3}{(24\beta_b - 4)\lambda_1 y_c} qL - \\ \quad - \frac{12EI}{kL^4(12\beta_b - 2)\lambda_1 y_c} qL] \\ \lambda_1 = |f'(y_{01})| = \lambda e^{-2} \end{cases} \quad (8.1\text{-}4)$$

$$27v^2 + 4u^3 = 0 \quad (9)$$

According to (8.1) and (9), the equilibrium surface and the bifurcation set is shown as Fig. 3(a). Cusp catastrophic theory points out that the jump of the system as a results of crossing the bifurcation set due to any mini-perturbation happens only when the system is situated in the equilibrium state on the bifurcation set. Obviously, Eq. (9) is satisfied only when $u < 0$, according to formula (8.2), it is equivalent to

$$K/\lambda_1 < 1 \quad (10)$$

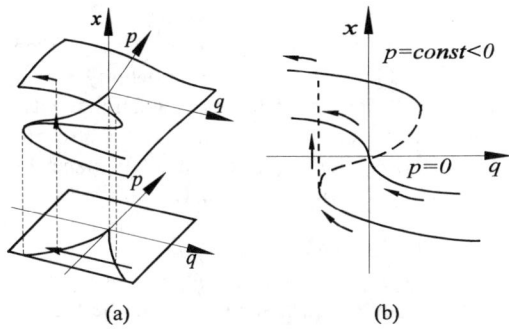

Fig.3 Cusp catastrophe and jump of state variable of rockbursts

Where $K = 12EI/L^3(6\beta_b - 1) > 0$. Formula (10) is the necessary condition of which the system loses its equilibrium stability. It can be observed from (10) and (8.4) that K/λ_1 only reflects the mechanical behaviors of the media constituting system and the geometric scale of the system. So the necessary condition of the system instability only depends on its inherent characters. Because $\tilde{x} < 0$ when $y < y_{01}$, and y_0 increases as the increase of load qL, the system is situated in the equilibrium state on the low leaf of equilibrium curved surface initially. If the curved surface is sectioned by the plane of which u is negative constant, the curve shown as Fig. 3 (b) will be obtained. It can be seen from formula (8.3) and Fig. 4 that, as the increase of load qL, the equilibrium state of the system evolves towards the direction of v descending. So it can be directly perceived through the sense from Fig. 4 that the jump of the system owing to any mini-perturbation happens only when the equilibrium state of the system crosses the left branch of bifurcation set. According to catastrophic theory, on the bifurcation set, the one order differential of equilibrium equation is always zero. So differentiating equation (7) leads to

$$\frac{dU''}{dy_0} = f'(y_0) + K = 0 \quad (11)$$

Because K is a constant always greater than zero, apparently formula (11) is satisfied only when $f'(y_0) < 0$. According to equation (1) and Fig. 1(b), $f'(y_0) < 0$ means coal pillar is situated in the deformation phase of post peak-value strength.. In accordance with the above-mentioned discussions, the conditions of the system instability are: (1) the necessary conditions of the system instability are that the coal pillar possesses the strain weakening behavior, meanwhile the ratio $K/\lambda_1 < 1$ is satisfied; (2) the sufficient conditions of the system instability is that load qL is large enough to make coal pillar enter the deformation phase of post peak-value strength and reach the equilibrium state on the left branch of bifurcation set [i.e. ,formula (9) and (11) are satisfied].

Obviously, the conditions (1) are the inherent conditions of the system, and the conditions (2) are the external conditions of the system. In the case of which $u < 0$, substituting Eq. (9) into (8.1) and noticing formula (8.2), the jump $\Delta \tilde{x}$ and the sudden compression of coal pillar Δy_0 when the system crosses the bifurcation set are

$$\begin{cases} \tilde{x}_1 = \tilde{x}_2 = -(-\frac{u}{3})^{1/2} = -\frac{\sqrt{2}}{2}(1-\frac{K}{\lambda_1})^{1/2}, \\ \tilde{x}_3 = 2(-\frac{u}{3})^{1/2} = \sqrt{2}(1-\frac{K}{\lambda_1}) \end{cases} \quad (12.1\text{-}2)$$

$$\begin{cases} \Delta \tilde{x} = \tilde{x}_3 - \tilde{x}_1 = \frac{3\sqrt{2}}{2}(1-\frac{K}{\lambda_1})^{1/2}, \\ \Delta y_0 = \frac{3}{\sqrt{2}}(1-\frac{K}{\lambda_1})^{1/2} y_{01} \end{cases} \quad (13.1\text{-}2)$$

Now we discuss the factors affecting the instability of the system. Noticing the expressions of β_b and beam's inertial moment I, Eq. (10) can be rearranged as follows

$$\begin{cases} 24EI / [L^3(1+\frac{3}{\beta}+\frac{3}{\beta^2}+\frac{3}{\beta^3})\lambda_1] < 1 \\ \beta = (kL^4/4EI)^{1/4} \\ I = \frac{bh^3}{12} = \frac{h^3}{12} \end{cases} \quad (14.1\text{-}3)$$

where h is the thickness of roof stratum, b is its width. Because of the assumption of the plane-strain state, $b = 1$. It is observed easily from (14.1-3) that, (1) the larger the curved stiffness of beam EI is, the less the possibility which the system loses its equilibrium stability (i.e. coal pillar rockbursts) is; (2) the larger the absolute value of the slope of constitutive curve of coal pillar at cuspidal point y_{01} i.e. λ_1 is, the larger the possibility which rockbursts of system happens is; (3) because the possibility of the rockbursts is associated with the third power of h and L, the possibility of rockbursts decreases extremely quickly as the raise of beam's thickness h and increases awfully quickly as the raise of the half of beam's span L; (4) the possibility of rockbursts decreases as the raise of the elastic constant k of unmined ore stratum.

4 PRECURSOR AND PROCESS OF THE ROCKBURSTS OF ORE PILLAR

In practical mining, the loads applied to roof stratum usually increase gradually, so does the deflection of roof stratum beam at mid-point $\bar{x} = 1$ (i.e. the compression of ore pillar y_0). At the place of the roof in front of the working face where the bending moment is maximum, the roof beam is the most easy to fail. To find the precursors the rockbursts of the system, now we study the evolution trends of the deformation velocity $\dot{y}_1$ in that place. Differentiating Eq. (5) with respect to time t and rearranging the derived formula, the deformation velocity $\dot{y}_0$ of ore pillar can be found as follows

$$\dot{y}_0 = \dot{q}L \frac{[\frac{24(\beta_b - \beta_a) - 3}{12\beta_b - 2} + \frac{12EI}{kL^4(6\beta_b - 1)}]}{[f'(y_0) + K]} \quad (15)$$

Solving (2.1) under the solution-determining conditions (3.1-7) and differentiating the solution with respect to time t, noticing (15) and $R_a = qL - f(y_0)/2$, furthermore, making $x = x_1$ (x_1 is the place of the roof in front of the working face where the bending moment is maximum) yields the deformation velocity $\dot{y}_1(x_1)$ in that place as following

$$\dot{y}_1 = \frac{\dot{q}}{k} + \frac{\dot{q}L^4}{EI} \exp(\beta \bar{x}) \{[\frac{(\beta+1)\cos\beta\bar{x}_1}{4\beta^3} + \frac{(\beta-1)\sin\beta\bar{x}_1}{4\beta^3}][1 - \frac{1}{\beta_b - 1/6} \times \\ + (\beta_b - \beta_a + \frac{EI}{kL^4})\frac{f'(y_0)}{f'(y_0) + K}] - \\ - \frac{\cos\beta\bar{x}_1 + \sin\beta\bar{x}_1}{12\beta(1+\beta)}\} \quad (16)$$

Making figure according to Formula (16) yields Fig. 4. In Fig.4, curve 1 indicates that when $K < \lambda_1$, rockbursts occur in the system, while 2, 3, 4, 5 indicate the states that rockbursts don't occur. It can be seen from the figure that along with the increase of y_0, $\dot{y}_1$ increases slowly in the beginning. When the deformation of ore pillar reaches the phase of post peak-value strength, $f'(y_0) < 0$ and $\dot{y}_1$ increase quickly. When the system reaches the instable equilibrium state on the left branch of bifurcation set, $f'(y_0^*) + K = 0$, where y_0^* is the compression of ore

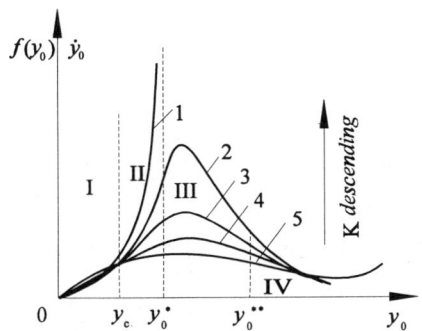

Fig.4 Deformation velocity of roof and complete process of rockbursts

pillar when the system reaches instable state, $\dot{y}_1$ trends towards infiniteness. When $K > \lambda_1$, the system can't lose its stability. This is indicated by curve 2, 3, 4, 5 when $\dot{y}_1$ can't trend towards infiniteness. According to Fig.5, the complete process that the system evolves from stability to instability can be divided as four phase by the deformation of ore pillar of peak-value strength y_c, the unstable equilibrium point y_0^* and the stable equilibrium point y_0^{**}. These four phase can be called as the stable phase I, the precursor phase II which is closed immediately to instable equilibrium state, the instable phase III in which the system loses its stability, and the post rockbursts phase IV in which the system reaches a new stable equilibrium state, respectively. Because in the precursor phase $\dot{y}_1$ becomes greater and greater, that $\dot{y}_1$ increases quickly and tends to infiniteness can be used as the forewarning sign of system rockbursts. To monitor $\dot{y}_1$ can predict the rockbursts.

5 SUMMARY

Through the discussion made in this paper by applying cusp catastrophic theory on the instable mechanism of the ore pillar rockbursts under a hard roof stratum subject to elastic supports, not only the criterions of system instability, i.e. rockbursts, are proposed, but also, based on above discussions, the relation between deformation velocity of roof $\dot{y}_1$ and the evolution process and the phases of rockbursts. The study indicates that the quick increasing of $\dot{y}_1$ and the evolution tendency of its going towards infinity are indeed the forewarning signs of ore to show quantitatively how close is the occurrence of rockbursts.

REFERENCE

Yin Youquan & Zhang Hong 1984. The Softening Behavior of Fault Zone Medium and An Instability Model of Earthquake. *Acta Seismologia Sinica.* 6(2): 1-17.

Tang Chunan 1993. *Catastrophe in Rock Unstable Failure.* Beijing, Press of Coal Engineering.1993, 10-30.

Xu Zenghe & Xu Xiaohe & Tang Chunan 1995. Theoretical Analysis of A Cusp Catastrophe of Coal Pillar Rockbursts Under Hard Rocks. *Journal of China Coal Society.* 20(5): 485-491.

Numerical stability analysis of a large cavern in weak rock

R.J. Fowell
Mining and Mineral Engineering Department, University of Leeds, UK

S.J. Ma
Geotechnical Engineering Division, Korea Institute of Construction Technology, Seoul, Korea
(Formerly: Mining and Mineral Engineering Department, University of Leeds, UK)

ABSTRACT: The extensometer results from a large hydro-pumped storage cavern were used to compare two commonly used types of numerical analysis programs, namely FLAC and UDEC for stability analysis. The geological structure was complex with over ten faults within 35° dipping bedded sandstones and siltstones along with clay seams. The modified Hoek-Brown failure criteria was applied for the prediction of rock mass properties for the numerical modelling. Cablebolts and rock bolts were installed throughout the cavern. Pre-reinforcement, pre-treatment, 'dental treatment' and steel reinforced shotcrete were used as support systems. The fault planes were converted to a Joint Command in the UDEC model. For the determination of the shear and normal stiffness of the faults, computer modelling of a shearbox test for filled joints using FLAC was used. The paper will describe the problems encountered and the solutions adopted.

1 INTRODUCTION

The Mingtan hydro-pumped power storage cavern, Taiwan is located in weak rock, containing faults and clay seams. This cavern was selected due to the amount of detailed published data available and the challenging geological setting. During the construction extensometers were installed to record ground deformations and hence provide comparisons for the analysis performed by the authors, using both UDEC and FLAC as an initial exercise in their use.

The Mingtan cavern is 22.4m-wide x 46.5m-high x 158.4m-long, lies among more than ten faults within bedding planes characterised by folds dipping at 35° and sheared zones. The underground stability at the site was determined by the shear strength of these bedding planes and faulted strata, which commonly gives rise to stability problems. The effect on stability of these features is important for the design of an underground facility and thus a suitable analytical method based on the correct geological and geomechanical information, should be chosen. Brady (1986) proposed several conceptual models for a rock mass which is a different approach to rock mechanics modelling. The commonly used numerical methods can be divided into two categories according to the discretization method, the continuum and the discontinuum method. FLAC using the continuum method and UDEC, the discontinuum method, have been used for the stability analysis of the Mingtan cavern which is characterised by weak zones.

2 GEOLOGICAL STRUCTURE AND ROCK PROPERTIES AT THE MINGTAN SITE

There are faults within the 35° dipping bedded sandstones and siltstones of the Waichecheng Series, an alternating sequence of sandstones and siltstones including joint sets which could have caused problems with cavern stability.

The Waichecheng series is composed mainly of fine-grained to conglomeratic and silty to quarzitic sandstone with interbeds of light grey sandstones and dark grey siltstones, including fault zones. The Waichecheng series form a syncline with many small folds and are further cut by several local minor faults. The fault zones are composed of multiple clay seams and shattered, softened or decomposed rock which caused a concentration of shear movements during tectonic activity, so that faulted strata has developed parallel to the bedding planes. The clay seams vary from a few millimetres to 20cm

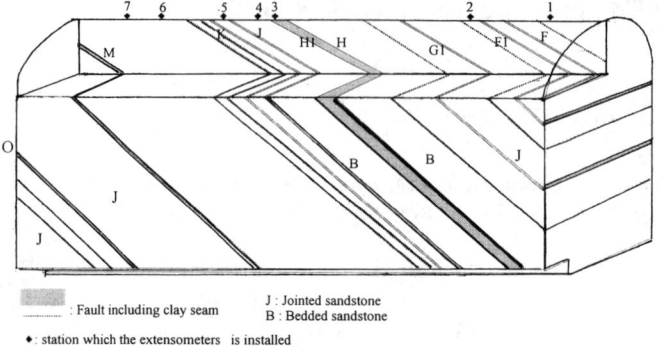

Figure-1. Fault profile along the cavern axis (Kai and Lee 1990).

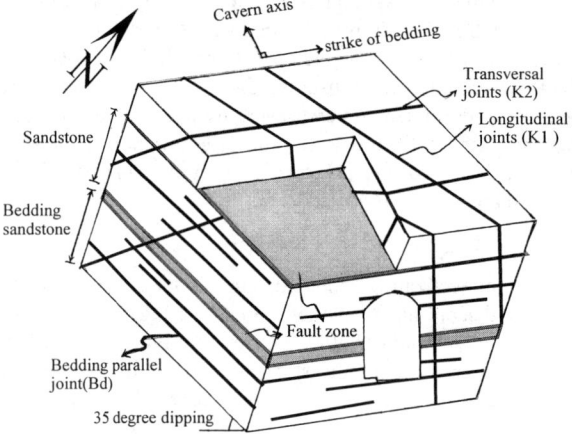

Figure-2. Geological structure at the Mingtan cavern site.

Table-1. Prediction of rock properties using the modified Hoek-Brown failure criterion at the Mingtan site.

Rock type	Jointed Sandstone		Bedded Sandstone		Faults
	Intact rock	Jointed rock	Intact rock	Jointed rock	Intact rock
Elasticity modulus (GPa)	21		14		2
Deformation modulus (GPa)		5		3	
Poisson's ratio	0.25	0.25	0.25	0.25	0.3
Friction angle (degree)	47	44.3	43	33	20
Cohesion (MPa)	32.6	5.8	14.3	0.96	0.44
UCS (MPa)	116	27.6	66	3.5	2.1
Tensile strength (MPa)	11	5.3~4.9	6	1.47~1.04	0
Q value		1.7-20		1.1-18.8	0.02-0.4

Table-2. Joint properties for UDEC-BB model.

Joint type	Jkn (GPa/m)	Jks (GPa/m)	JCS (MPa)	JRC	Ln (m)	ϕ_r (degree)	σ_c (MPa)
Jointed sandstone	12	5	35	6	0.6	30	166
Bedded sandstone	10.7	4.3	25	4.5	0.6	27.5	66

Jkn normal joint stiffness, *ks* shear joint stiffness, L_n field scale joints length (metres)

in the fault zone whilst the range of thickness of the fault zone range from a few centimetres to more than a metre, varying from place to place and undulating greatly around the powerhouse site. These fault zones gave rise to stability problems in the excavation procedure (Figure-1).

Three sets of discontinuities are identified in the cavern area, namely, the bedding planes(Bd), longitudinal joints (K1) and transversal joints (K2) as illustrated in Figure-2. The bedding planes have a major influence on the stability of the power cavern. The condition of the rock mass in the power cavern varied from good to very poor.

3 INPUT DATA FOR NUMERICAL MODELLING

1. Rock property parameters

There are three rock types to be identified, jointed sandstone, bedded sandstone with siltstone and fault zones. Barton (1994) applied his *Q*-system for rock mass classification to the Mingtan site. *Q*-system values were used for prediction of rock properties including strength using the Hoek-Brown Failure Criterion (Hoek et al. 1994) as shown Table-1.

2. Shear strength parameters for the joint

The joint properties were provided by Dr. N.Barton (1995) who took part in the Mingtan project for the adjustment of S(fr)+B design using the *Q*-system. These properties (*JRC, JCS* and L_n) are estimated from Barton's experience. The other properties, shear stiffness, normal stiffness, cohesion and friction angle, are obtained from the UDEC-BB model based on Equation (1). These properties are shown in Table-2.

$$\tau = \sigma_n \tan\left[JRC \log_{10}\left(\frac{JCS}{\sigma_n}\right) + \phi_r \right] \quad (1)$$

ϕ_r residual friction angle
JRC joint roughness coefficient
JCS joint compressive strength

3. The prediction of fault properties

In the UDEC model, faults which range in thickness from several millimetres to four metres, cannot be modelled as blocks because these fault blocks have extremely high aspect ratios and long thin zones will not bend correctly during UDEC modelling. If the bending movement of these thin zones is to be modelled correctly, they need at least 5 zones across the thickness of the fault. It is, however, impossible to subdivide thin fault zones into small zones in the Mingtan model due to limitations in the computer's memory and calculation time required plus the difference of the zone size ratio between the fault zone and the surrounding rock mass.

These faults were converted into Joint Commands in the UDEC model. For this, the shear and normal stiffness of faults, including the material and the interface deformation mechanism, should be predicted correctly. Simulated filling shear box tests (Ma 1996) were used by FLAC to predict the stiffness properties of the faults. Goodman (1970) proposed that, if the ratio of T/A is over 1, the filled joints are governed by the properties of the fill. The filling FLAC model therefore only presents the filling material's own deformation. The joint stiffness results for the filling shear box test with the FLAC analysis are shown in Table-3.

Table-3. Normal and shear stiffnesses predicted from the filling shear test by FLAC for various thicknesses.

Filling thickness (cm)	Normal joint stiffness (GPa/m)	Shear joint stiffness (GPa/m)
5	13	2.17
20	4.15	0.66
50	1.3	0.21

Normal load is 7MPa.

4 SUPPORT AND MONITORING SYSTEMS

The Mingtan cavern used special supporting systems, a pre-reinforcement supporting system in the cavern roof and fault seam treatment and steel fibre-reinforced shotcrete for poorer geological conditions.

For the pre-reinforcement supporting system in the cavern roof, two longitudinal working galleries, each with a cross section of 2m x 2m, were excavated along the entire length of the cavern at the midheight of each haunch. Using a cross-cut as access, clay seams and very weak fault zone material were washed out to a depth of approximately 4m with high or low pressure water jets or physical mining methods at faults F1, G, H, H1, J, K, M and O (Figure-1) and then was backfilled with non-shrink cement mortar as the pre-treatment support (Kao and Lee 1990).

The pre-reinforcement cables, 23T or 46T (15.2mmϕ 7 by 1 or 2 strands) according to the geological structure, were installed from the drainage and the working galleries in the entire roof area of the power cavern and transformer hall before the excavation of the main chamber. Other supporting systems were applied 50T, 60T cables (12.7mmϕ 7 by 6 or 7 strands) and 25T rock bolts geological structure. For the sidewall supporting systems 30T, 50T and 60T cables and 25T rock bolts were applied immediately after successive excavation steps (Moy and Hoek 1989, Cheng and Liu 1994). The Steel Fibre-Reinforced Shotcrete which has a higher tensile strength than normal concrete, was also applied 5 to 20cm thick depending on the geological structure around the cavern (Vandevalle 1994).

In order to monitor the behaviour of the cavern in response to excavation, borehole extensometers were provided at seven cross-section locations in the power cavern through various geological conditions (Figure-1). At each cross-section, extensometers were pre-installed for monitoring the behaviour of the arch area from the drainage gallery and, in some cases, the two working galleries for seam pre-treatment before the commencement of the cavern excavation. These extensometers installed in the arch, haunch and sidewall could measure the actual movement of the rock mass surrounding the cavern due to excavation. The real deformation obtained from the extensometers was compared with the deformation predicted by FLAC and UDEC.

extensometer and the predicted deformations of FLAC and UDEC are presented with the excavation steps at the arch and haunch in the main cavern area in Figure-3 and 4. In the two dimensional model, faults, dipping at 35° against the cavern cross section can only be shown as horizontal strata. In Figure-3, the FLAC displacement, therefore, shows only 50% of the total deformation in the Mingtan cavern because FLAC can analyse only the displacement of the rock mass. However UDEC's results show close agreement with the monitored displacement because UDEC can effectively analyse not only the displacement of rock mass but also the behaviour of the discontinuities causing the large displacement. However the FLAC and UDEC deformations in the haunch and sidewall match with the monitored deformation because major deformation occurred in the direction perpendicular to the cavern sidewall.

Cheng and Liu(1993) defined the blast damaged zone in their Mingtan FLAC model. The properties of the blast damaged zone involved a considerably weaker rock with properties similar to those of the fault material for anisotropic deformation at the arch area. The displacement then increased constantly around the powerhouse according to the successive excavation steps and the error was 50~100%. Barton (1994) suggested that different sets of data would be needed in the arches and walls for non-isotropic behaviour patterns in the arch and walls to match with the prediction of the numerical methods. However, for anisotropic behaviour at the arch area, several extra joints are placed in the arch area in this UDEC modelling. The reasons why UDEC has a high ability in these geological circumstances rather than FLAC, involves the distinctive features of the discontinuum method as follows:

• It can model slip and separation at joints and rigid-body translation and rotation in a jointed medium, including complete detachment or plastic flow in a continuous medium.

• It can allow the small displacement of discrete blocks such as in the continuum method.

• It recognises new contacts automatically as the calculation progresses.

5 RESULTS OF NUMERICAL ANALYSIS

Major deformation occurs in the arch area and fault zone in the haunch and wall area even if the comprehensive supporting systems such as pre-reinforcement and pre-treatment were applied. The variations between the deformation monitored by the

6 FAILURE MECHANISMS IN THE CAVERN

The physical fault behaviour, as mentioned above, may give rise to problems in the stability of the underground structure. Therefore the fault zone dipping at 35° throughout the whole of the Mingtan site and causing a major deformation in the roof

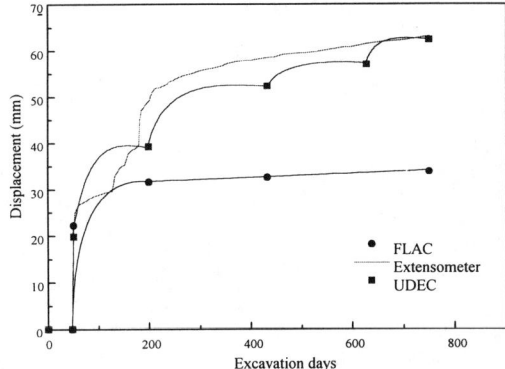

Figure-3. Comparison of displacement monitored by the extensometer and predicted displacement by UDEC & FLAC at roof area of station-2 (42m).

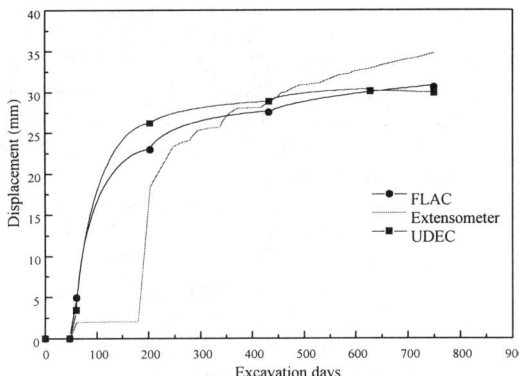

Figure-4. Comparison of displacement monitored by extensometer and predicted displacement by UDEC & FLAC at up-stream haunch area of station-2-2 (42m).

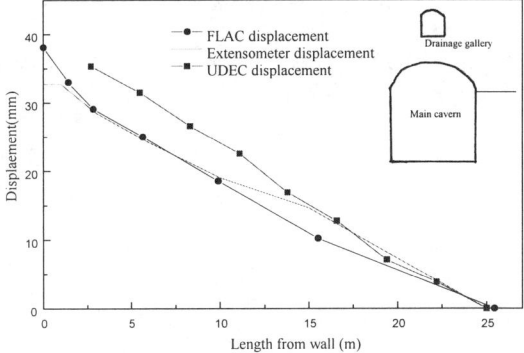

Figure-5. Actual measured displacement in the power cavern sidewall area at St 2-6 EL 320 with predictions from FLAC and UDEC models.

area, and bedded sandstone, parallel to the fault strata, must be analysed correctly when computer modelling.

Figure 6 shows a deformation vector diagram analysing the deformation in the Mingtan roof area. There are two displacement vectors occurring, firstly the displacement of the rock mass alone, which can be effectively analysed by two-dimensional modelling, secondly the displacement of the 35° dipping fault zone or bedding planes when the normal load exceeds the shear strength of the fault zone or bedding planes alone or bond strength between the jointed sandstone and the bedded sandstone or faults. The extensometers installed in seven places throughout the Mingtan site may involve both deformation vectors. In the roof area of the main cavern, there is another potential failure mechanism responsible for a large deformation besides fault slip. This shows that a 50° dipping transversal joint set (K1) passes through the fault strata almost perpendicularly. It may cause a potential failure in the rock block. Displacement analysis in the cavern wall area also shows that 35° dipping fault strata may be squeezed out almost perpendicular to the cavern wall due to the horizontal load, whereas the deformation in the direction of the 35° dip is small in the wall area due to the excavation procedures. Thus the FLAC and UDEC models can effectively analyse the cavern sidewall area as opposed to the computer modelling of the roof area.

7 CONCLUSIONS

The numerical modelling of the Mingtan site which includes high density joint sets and over ten faults must consider the behaviour of the 35° dipping faulted strata and bedded sandstone. Using two-dimensional computer models, such as FLAC and UDEC, the 35° dipping faulted strata and bedding planes could not be modelled effectively. The FLAC model can only analyse the fault material deformation due to convergence. The amount of this deformation could be less than half of the total deformation measured by the extensometers. However, the UDEC model can analyse the resultant deformation of the faulted strata toward a 35° dip because the UDEC model can model the joint and fault shear properties and can accommodate large deformations on the joint and fault surfaces.

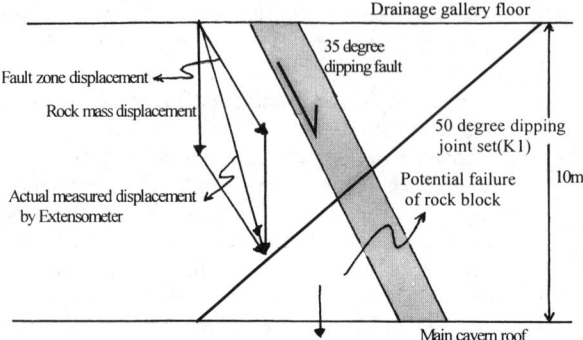
Figure-6. Displacement analysis and failure mechanics in roof area.

From the Mingtan computer modelling, the most important factor in the determination by numerical methods, is the geological conditions at the expected site. It is suggested that the Mingtan cavern could be analysed by three-dimensional software for analysing the behaviour of a 35° dipping fault and bedding planes effectively.

ACKNOWLEDGEMENTS

The authors of this paper should like to thank Dr. D.Moy, Golder Associates (UK) for assistance with this project. Dr. N.Barton, NGI, who provided the joint properties for this site and Dr. S.R. Hencher, formerly of the Department of Earth Sciences, University of Leeds, and presently at Bechtel, Seoul office, for suggesting this research project. The results of the analysis and the opinions expressed are those of the authors and not those of the organisations and individuals acknowledged above.

REFERENCES

Barton.N, 1994, A *Q*-system Case Record of Cavern Design in Faulted Rock, Torino. Italy, pp 16-1~14. Barton.N, 1995, Private Communication.

Brady.B.H.G, 1986, Boundary Element and Linked Methods for Underground Excavation Design, in *Analytical and Computational Methods in Engineering Rock Mechanics*, Editor. Brown, pp 164~204, Allen & Unwin.

Goodman.R.E, 1970, The Deformability of Joints, Special Technical Publication No. 477, Philadelphia; American Society for Testing and Materials, pp 174~196.

Ho.S.C, 1988, An Introduction to the Geology of Taiwan Explanatory Test of the Geologic Map of Taiwan, Central Geological Survey. The Ministry of Economic Affairs. Taiwan. 2nd Edition.

Hoek.E, Grabinsky.M.W, Diederichs.M.S, 1991, Numerical modelling for Underground Excavation Design, *Transaction of the Institution of Mining and Metallurgy*- Section A Mining Industry, Vol. 100, pp A22~30.

Hoek.E, Kaiser.P.K, Bawden.W.F, 1995, *Support of Underground Excavations in Hard Rock*, A.A.Balkema.

Kao.C.Y, Lee.H.J, 1991, The Influence of Geology on the Design of the Mingtan Power Cavern, *Neotetonics and Resources*. Belhaven Press. Editor Cosgrove and Mervyn, pp 353-369.

Ma.S.J, 1996, Design of Underground Storage Caverns in Weak Rock, Unpublished PhD Thesis, University of Leeds.

Moy.D, Hoek.E, 1989, Progress with the Excavation and Support of the Mingtan Power Cavern Roof, *Rock Cavern-Hong Kong*. Editor. Malone & Whiteside, pp 235~245

Vandewalle.M, 1994, *Tunnelling the World*, N.V.Bekaert S.A. 3rd Edition.

Numerical analysis of gas-liquid 2-phase fluid behavior around underground rock caverns storing pressurized gas

Hiroshi Suenaga
Geology Department, Abiko Laboratory, Central Research Institute of Electric Power Industry (CRIEPI), Chiba, Japan

Hiroyuki Tosaka & Keiji Kojima
Department of Geosystem Engineering, University of Tokyo, Japan

ABSTRACT: A numerical simulator for treating gas-liquid two phase fluid flow with solution and evaporation is developed. Using it, the case studies about underground rock caverns storing pressurized gas are done. From the results, important factors for evaluating the effectiveness and safety of natural or artificial water sealing around cavern are suggested.

1 INTRODUCTION

Construction of large underground caverns in deep hard rock site for storing pressurized gaseous fluids, such as air or light hydrocarbons, is under survey in Japan. Considering leakage of fluid from rock cavern, key issues are how to make the rock cavern wall gastight economically, how to design and operate the artificial sealing system by water injection through tunnels or boring placed above caverns, and how to evaluate reasonably the behavior of gaseous fluids in rock for the possible accidental leakage.

In case of leakage, the stored fluids will intrude into the rock pores or fractures and form two phase fluid system with existing groundwater. For numerical analysis of this type of fluid flow, the phase equilibrium, which is accomplished by gas solution to liquid phase or gas evaporation from liquid phase, has to be considered. According to Henry's law, the higher rises gas phase pressure, the larger amount of gas be dissolved in liquid. Taking into account such phase equilibrium, the free gas might move upward by the buoyancy force if gas saturation becomes higher than the movable value, while the dissolved gas might be transported with the groundwater flow.

For the purpose of evaluating the above phenomena, the authors developed a numerical simulator that treats gas-liquid two phase fluid movement with solution/evaporation, making use of the techniques historically developed in petroleum reservoir engineering.

2 NUMERICAL METHOD

2.1 GAS-LIQUID 2-PHASE FLOW

Fluid flow around rock cavern storing pressurized gas can be described as two composition (gas + groundwater) and two phase (gas + liquid). The gas composition flows as a gas in the gas phase or as a solution in the liquid phase.

The interfacial effect between two phases and the phase equilibrium have an influence on the flow as mentioned above. The former means capillary pressure, threshold pressure, and relative permeability, the latter gas dissolution to liquid phase and vaporization from it. Threshold pressure means capillary pressure on water saturated condition. Considering the leakage near the rock cavern wall, both effects have important roles. Furthermore, thinking about Henry's law for pressurized gas, the phase equilibrium is expected to be effective.

A state of gas solution/vaporization equilibrium in gas-liquid 2-phase system classifies two conditions. One is called the 'saturated' condition that is composed of gas and liquid phase when gas is fully dissolved in the liquid phase and superfluous gas exists. Another is the 'undersaturated' condition of only liquid phase including solution gas or not. The 'saturated' condition which is used here is the contrary meaning to the term used in the common groundwater analyses. To describe this phenomenon in the numerical analysis, two assumptions are necessary.

1. Gas solution into liquid and phase equilibrium is achieved instantaneously.

2. Existence or nonexistence of gas phase depends on three parameters, i.e. solubility, saturation pressure, and liquid phase pressure. Saturation pressure can be defined as liquid phase pressure on saturated condition.

2.2 FLOW EQUATIONS

First of all, we derive flow equations. Fluid flow in porus medium is expressed by Darcy's law

$$\vec{u} = -\frac{K}{\mu}\nabla\Phi \quad (1)$$

where $\vec{u}$ is a vector of flow velocity, K is absolute permeability of rock, and μ is viscosity. Φ is the fluid's potential function described as

$$\Phi = P - \rho g z \quad (2)$$

where P is pressure, ρ is density, g is gravitational acceleration, and z is depth. The conservation law of mass can be described as

$$-\nabla \cdot (\rho\vec{u}) - \rho q_s = \frac{\partial(\rho\phi S)}{\partial t} \quad (3)$$

where q_s is production rate of fluid in standard condition, ϕ is porosity of the medium, and S is saturation. In the eq. (3), the first term of the left hand side is the mass flowing into a minute region, the second is extracted mass, and the right hand side term is the rate of mass accumulation. Density of the fluid can be expressed by using formation volume factor B as

$$\rho = \frac{\rho_s}{B} \quad (4)$$

where ρ_s is density in standard condition. We finally obtain the flow equation

$$\nabla \cdot \frac{K}{\mu B}\nabla\Phi - q_s = \frac{\partial}{\partial t}\left(\frac{\phi S}{B}\right) \quad (5)$$

Next, we apply eq. (5) to two phases and expand the equation with phase equilibrium. Formulation of two phase fluid flow including solution/vaporization has been studied in petroleum engineering field concerning water-oil-natural gas system. Taking assumptions as mentioned above into account, the formulation is done.

Flow equation for water phase can be expressed as

$$\nabla \cdot \frac{Kk_{rw}}{\mu_w B_w}\nabla\Phi_w - q_{ws} = \frac{\partial}{\partial t}\left(\frac{\phi S_w}{B_w}\right) \quad (6)$$

where suffix w means water composition, and k_r is relative permeability. On the other hand, flow equation of gas composition can be described as

$$\nabla \cdot \frac{Kk_{rw}R_s}{\mu_w B_w}\nabla\Phi_w + \nabla \cdot \frac{Kk_{rg}}{\mu_g B_g}\nabla\Phi_g$$
$$-q_{ws}R_s - q_{gs} = \frac{\partial}{\partial t}\left(\frac{\phi S_w R_s}{B_w} + \frac{\phi S_g}{B_g}\right) \quad (7)$$

where gas saturation $S_g = 1.0 - S_w$, suffix g means gas composition, and R_s is solubility which is the value of solution gas volume per unit volume of water.

2.3 SOLUTION PROCEDURE

In this study, eq. (6) and (7) are solved numerically by FDM (Finite Difference Method) and Newton-Raphson's iterative calculation.

To get the solution for the equations, there are three unknown variables, i.e. water pressure P_w, saturation pressure P_s, and water saturation S_w. Stright et al. (1977) and Forsyth et al. (1984) suggested how to solve the equations.

The former method is used in this study, in which the phase condition is checked using the solution from the first Newton iteration according to the following rules ;

- If water phase pressure is greater than saturation pressure, only liquid phase exists ($S_g = 0$) and P_w and P_s are treated as unknowns.

- If there is two-phase-condition ($S_g > 0$), water phase pressure is equal to saturation pressure ($P_w = P_s$), so that P_s and S_w become independent unknowns.

Finally, by fixing two unknown variables in the second and the later iteration, we can get the convergent solution with Newton-Raphson method.

3 CASE STUDY

By the method above, several case studies were done. The objective region is assumed as the cavern storing highly pressurized gas in the rock mass which is homogeneous and isotropic porous medium and is saturated with groundwater. Fig. 1 shows the grid system for case studies. It is situated from 500m to 910m under water table vertically and is 200m long horizontally. In the 6 × 17 gridblocks, one with hatching is for the tank storing gas at 80atm (8.1×10^6 Pa). The condition on the upper side of boundary is fixed pressure and others are no flow boundaries. Properties are given in Figs. 2 through 6. Solubility, water formation volume factor and viscosity table are derived from Amix et al. (1960). Capillary pressure table is made by the empirical equation proposed by van Genuchten (1980) which describes

$$\mid \alpha P_c \mid^n + 1 = \left(\frac{S_w - S_{wirr}}{S_{wres} - S_{wirr}} \right)^{1+\frac{1}{n}} \quad (8)$$

where α and n are the parameters given by Table 1, and suffix irr and res mean irreducible and residual water saturation respectively. Those parameters are measured for granite. Relative permeability tables for gas and water are given as general curves used in petroleum engineering.

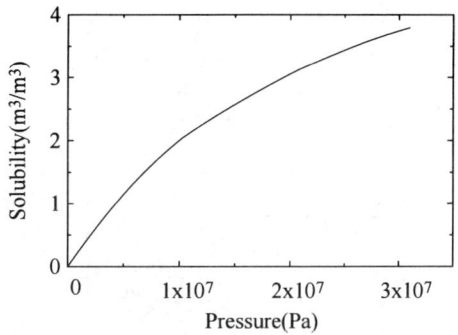

Figure 2: Solubility Curve

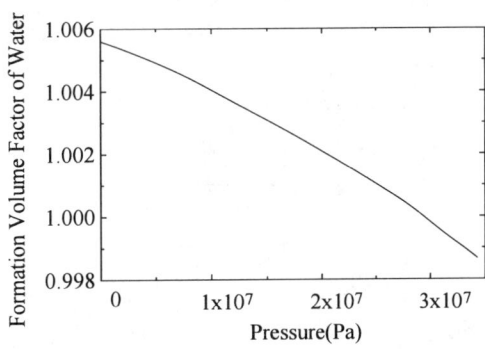

Figure 3: Formation Volume Factor Curve

Table 1: Parameters for Eq.(8)

Curve Type(in Fig.5)	α	n
Curve A	1.4×10^{-4}	2.09
Curve B	8.0×10^{-5}	2.44

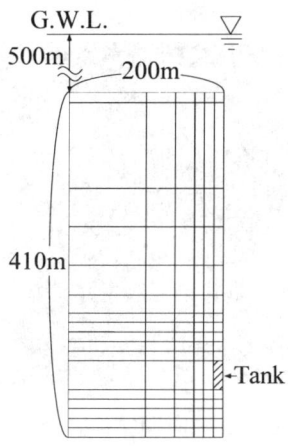

Figure 1: Grid System

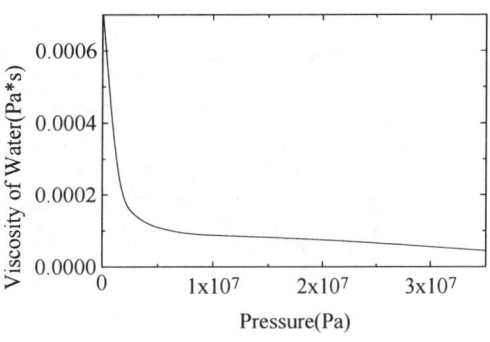

Figure 4: Viscosity Curve

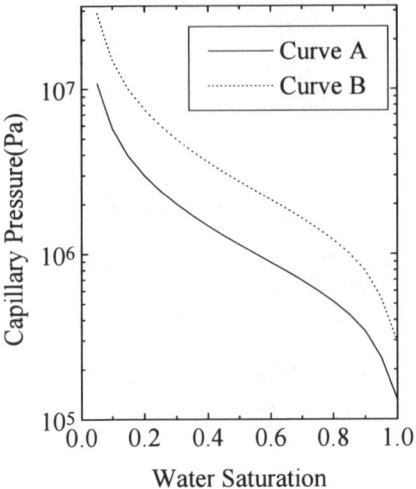

Figure 5: Capillary Pressure Curve

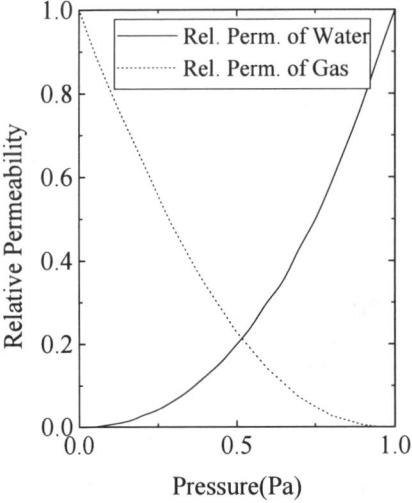

Figure 6: Relative Permeability Curve

We investigated whether gas can leak or not by varying threshold pressure (P_{th}) and the depth of the tank (D). The results are presented in Table 2. It shows the followings.

Table 2: Results of Parameter Study

P_{th} (Pa)	D (m)	Leakage or Gastight
1.3×10^5	900	Gastight
1.3×10^5	890	Leakage
1.3×10^5	830	Leakage
2.9×10^5	830	Gastight
2.9×10^5	800	Gastight
2.9×10^5	790	Leakage

When threshold pressure of rock is estimated at 1.3×10^5 Pa, pressurized gas cannot suppress by groundwater head at 830m and tank should be placed deeper than 900m. On the other hand, when evaluated 2.9×10^5 Pa, tank might be set upper to 800m.

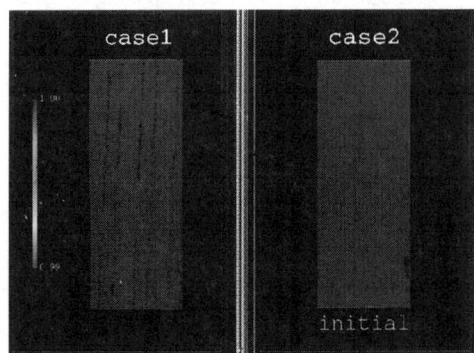

Figure 7: Gas Spread(Initial)

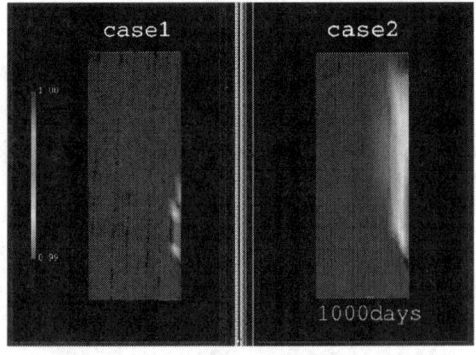

Figure 8: Gas Spread(After 1000days)

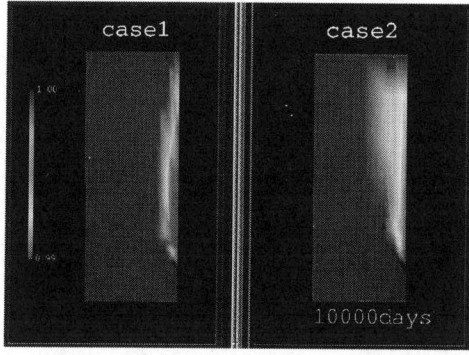

Figure 9: Gas Spread(After 10000days)

Next, we studied to examine the differences of leaked gas behavior with or without considering solution phenomenon. The results depicted in Figs. 7 through 9 show us that gas might spread wider in the case with solution phenomenon(case2) than that case without(case1).

4 CONCLUSIONS

Followings are the conclusions of this study.

1. For the purpose of modeling fluid flow around rock caverns storing pressurized gas, numerical technique to treat 2-phase, 2-composition flow with gas solution/vaporization has been studied.

2. From the results of case studies, it is suggested that for evaluating the effectiveness and safety of natural/artificial water sealing, the threshold pressure of gas phase at the boundary of rock cavern wall and the phenomenon of the gas solution to the liquid phase at the gas-liquid interface will play important role.

REFERENCES

[1] Amix, J.W., Bass, D.M., and Whiting, R.L. (1960): Petroleum Reservoir Engineering - Physical Properties, McGraw-Hill.

[2] Aziz, K. and Settari, A. (1979): Petroleum Reservoir Simulation, Elservier Applied Science Publishers.

[3] Forsyth, P., Jr. and Sammon, P.H. (1984): Gas Phase Apperance and Disapperance in Fully Implicit Black Oil Simulation, SPEJ, pp.505-507.

[4] Stright, D.H., Jr., Aziz, K., Settari, A., and Starratt, F. (1977): Carbon Deoxide Injection into Bottom-Water, Undersaturated Viscous Oil Reservoirs, J. Petrol. Technol., 29, pp.1248-1258.

[5] van Genuchten, M.Th. (1980): A Closed-Form Equation for Predicting the Hydraulic Conductivity of Undersaturated Soils, Sci. Soc. Am. J. 44, pp.892-898.

Utility of fly ash as mine fill – A geotechnical study

J.M. Kate
Civil Engineering Department, Indian Institute of Technology, New Delhi, India

ABSTRACT : The present study is conducted to investigate the most appropriate way of utilizing fly ash as mine fill taking into account certain important and inter-related aspects of strength & deformation, practical problems of stowing, environmental conditions etc. Trial mixes with different proportion of fly ash, sand and binder (cement or lime) have been prepared and tested for their flowability. The uniaxial compressive strengths and stress - strain relationship of the specimens casted out of such mixes using flowability water contents are determined. In the light of relevant characteristics of mine and mine fill material, a proposal of suitable design mix has been discussed.

1 INTRODUCTION

The Thermal power plants in India are facing an enormous fly ash disposal problem. The present production of fly ash at 72 thermal power stations in the country is estimated to be around 50 million tonnes per annum. Presently, it is being dumped on open grounds and/or stored in ponds which is leading to environmental problems e.g. pollutions of air, groundwater and soil. On the other hand, the underground and open cast coal mines are facing an acute shortage of stowing material required for filling up the worked coal seams to keep the roof of cavity intact and prevent subsidence. In India, sand has been the traditional stowing material for mines. However, due to over - exploitation of sand from nearby river bed, the river sand has become scarce and the high transport cost rules out the possibility of large quantities of sand being hauled over long distances.

Interestingly, on economic considerations and rapidity of coal transport, most of the Thermal power stations in the country are located in the vicinity of coal mines. In the above context, use of fly ash as mine fill appears to be natural choice. Thus, it becomes imperative to understand relevant characteristics of fly ash alone or mix with other materials in relation to geotechnical engineering and environmental aspects of mine fill.

The present study has been planned to investigate the most appropriate way of using fly ash as a mine fill by considering various relevant aspects. In view of this, herein, experiments have been conducted on pulverized fly ash (PFA) to assess its suitability as a mine fill material alone or in combination with other materials such as river sand and binder like cement or lime.

2 OVERVIEW

2.1 Fly ash as Pozzolana

Most of the Thermal power plants in India use crude coal and middlings, after burning which produces large percentage of fly ash. The pozzolanic property possessed by fly ash makes it very useful material in civil engineering industry. Realising the need to find a suitable substitute for sand as a stowing material in mines and at the same time availability of fly ash in large quantities around coal fields, there is a growing awareness amongst mining industries about the utility of fly ash as prospective mine fill material.

2.2 Fly ash as mine fill - some case studies

The use of fly ash alone as a mine fill material suffers from number of drawbacks predominant amongst these are inadequate strength to prevent mine subsidence or caving and ground water pollution due to leaching. However, fly ash when mixed with materials like sand, rock powder and cement or lime offers considerable improvement in the strength of the mix and resistance to aggressive waters.

An investigation carried out on stabilization of mine waste at Wardley Colliery mine (U.K.) indicated that on addition of 10% of PFA alongwith 8% OPC, the strength of the mix was increased by about 70% compared to normal cement stabilized mine stone. The galleries in an old underground Castle field limestone mine in Dudley (U.K.) were stabilized using rock paste made out of unburnt colliery spoil, 4% PFA, 0.7 to 1% hydrated lime water and small percentage of quaternary ammonium salt.

The largest use of pulverized fly ash for consolidated backfilling has been at kidd creek mine, Ontario, Canada in 1984 wherein, the blast furnace slag was completely replaced by fly ash to achieve savings in cost. Fly ash has also been used as an extender for cement for cemented fillings in mine stopes in one of the mines in Australia.

2.3 Pumpability and flowability of slurries

Hydraulic transport of fly ash slurries at weight concentration of 25 to 40% poses no problem and for such slurries use of centrifugal pump is adequate. Fly ash slurries at higher concentrations from 65 to 75% demands use of reciprocal pump to increase the overall efficiency and economy of pumping process. The frictional head loss in transporting fly ash slurries in pipe lines under gravity is much less compared to sand entailing higher concentration of fly ash than sand for the same hydraulic conditions. In India, at one of the coal mines of BCCL company, a mixture of 40% fly ash and 60% sand by volume was hydraulically stowed quite successfully without any pipe blockage or serious drainage problems. The efficient pumping system required for hydraulically transporting the non - Newtonian viscous slurries such as flyash, sand and binder mixture needs to be specially designed.

3 EXPERIMENTAL PROGRAMME

3.1 Materials

The fly ash used in this study has been procured from the electrostatic precipitators of Badarpur Thermal power plant at Delhi. The magnitudes of various properties of this pulverized fly ash were determined following the Indian Standard Codes of practice and are presented in Table 1.

Table 1. Properties of Badarpur fly ash

Property	Value
In - situ unit weight(kN/m^3)	11.2
Standard Proctor Maximum dry unit weight (kN/m^3)	12.8
Optimum moisture content(%)	32.0
Specific gravity	2.14
Specific surface (cm^2/g)	2455
Lime reactivity strength (MPa)	3.1
Setting time (minutes)	
Initial	12 to 18
Final	15 to 28
Chemical Analysis (percentages)	
Loss on ignition	5.1
SiO_2	59.3
Fe_2O_3	7.2
Al_2O_3	17.7
CaO	6.8
MgO	3.2
Other	0.7
pH of leached water	8.5 to 10

In general, this PFA satisfies specifications for fly ash to be used as Pozzolana and Admixture (IS - 3812, 1966). The particle size distribution curve shown in Fig. 1 indicates that the fly ash falls mostly within the silt size range with coefficient of uniformity of 2.8.

The sand was collected from the river Yamuna which flows down in the close vicinity of Badarpur Thermal Power Station. The curve for Yamuna sand in Fig. 1 classifies it as uniformly graded fine sand with coefficient of uniformity of 1.7. The cement used is Portland cement whereas the lime is powdered good quality hydraulic lime with CaO of around 70 percent.

3.2 Mixes and specimen preparation

The fly ash was mixed with sand in different proportions and small quantity of binder (cement, C or lime, L) was added to it. Various mixes thus prepared were fly ash to sand (F:S) mix in proportion by weight ranging from 1:0 (i.e. only fly ash) to 4:1 alongwith a binder as a percentage of this mix ranging from 1 to 5%. The cylindrical

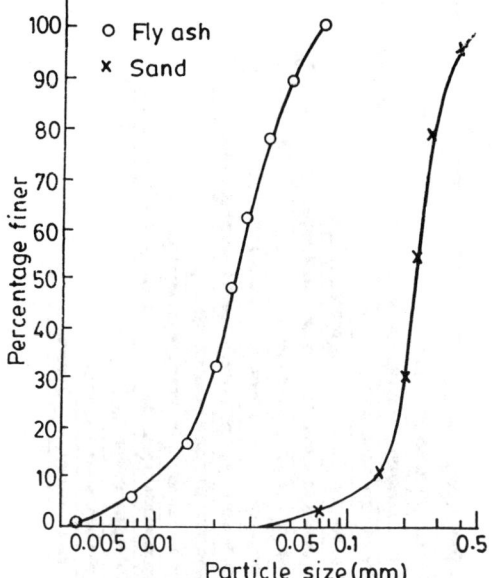

Fig. 1. Particle size distribution curves for Badarpur fly ash and Yamuna sand

specimens of 38 mm dia. and 76 mm height were casted from the slurry prepared by mixing predetermined quantity (based on flowability test results) of water with the chosen proportion of fly ash, sand and binder. Such specimens were then subjected to curing in humidity chamber for 7 days and 28 days.

3.3 *Parameters studied*

Various proportions of fly ash and sand by weight studied are 1:0, 1:1, 2:1, 3:1 and 4:1 and the percentages of binder (either cement or lime) in each of the above mix used are 1%, 3% and 5% by weight. The total number of trial mixes thus studied becomes 30. To simulate the field conditions of hydraulic stowing, flowability tests were conducted to determine the water contents required for preparing the flowable slurries of these mixes. The cylindrical specimens prepared out of these mixes and cured for 7 & 28 days have been tested for their uniaxial compressive strengths (IS : 9143 - 1979) and stress strain behaviour.

4 RESULTS AND DISCUSSION

4.1 *Flowability water contents and unit weights*

The optimum water content required to make 100% fly ash as flowable is 72%. Addition of 1:1 sand reduces flowability water content to 41%. The results of flowability tests for different mixes with lime or cement addition are illustrated in Fig. 2. It is clear that with increase in the proportion of both the fly ash as well as binder in the mix, the flowability water content also increases. For the similar mix lime requires more quantity of water compared with cement to make the slurry flowable.

The bulk unit weights of the specimens prepared out of these mixes by addition of flowability water contents decrease both with increase in fly ash as well as binder. The bulk unit weights of these mixes ranges from 13.2 to 14.8 kN/m^3 for lime and 13.6 to 15.1 kN/m^3 for cement.

4.2 *Uniaxial compressive strengths*

The specimens prepared out of mixes without any binder exhibit very low strength. However, the addition of even a small quantity of binder increases the strengths significantly. The bar diagrams of specimen strengths are illustrated in Figures 3 and 4

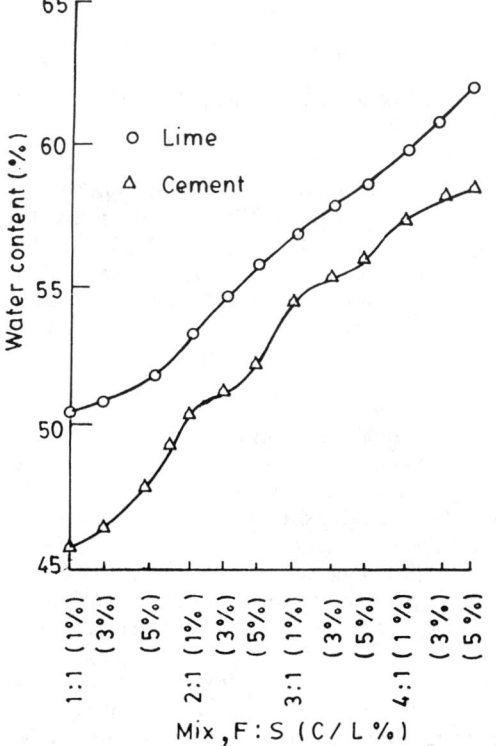

Fig. 2. Flowability water content for different mixes

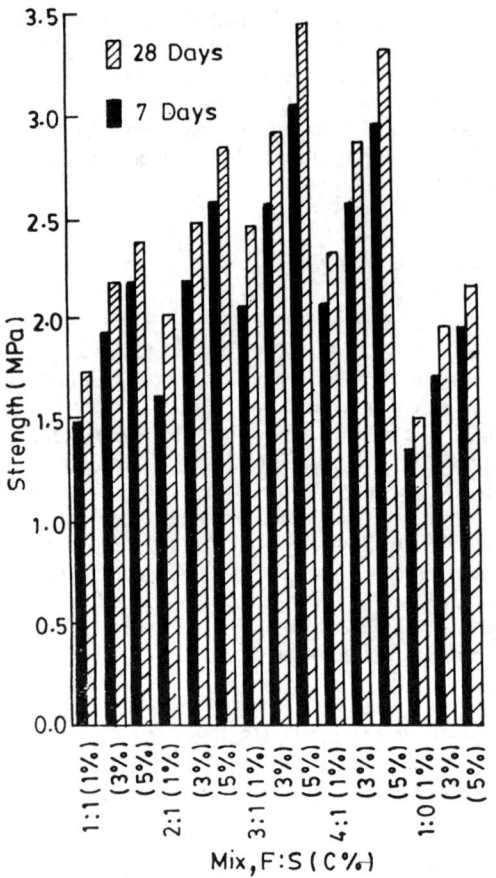

Fig. 3. Uniaxial compressive strengths for various mixes with cement

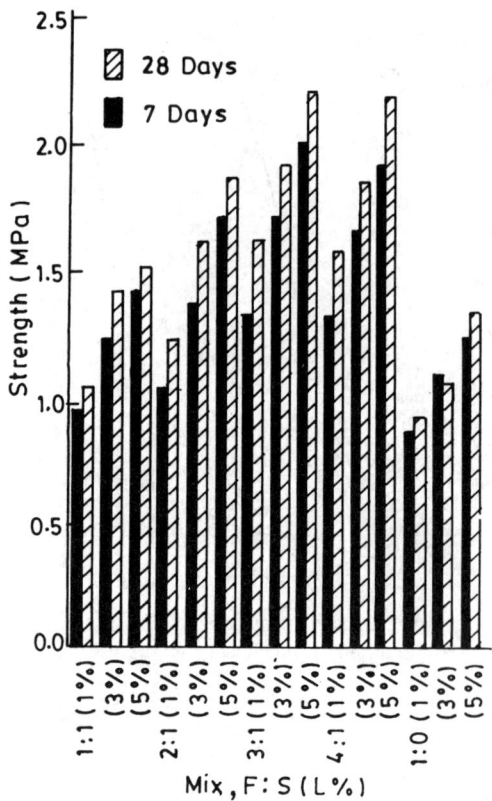

Fig. 4. Uniaxial compressive strengths for different mixes with lime

for cement and lime as binders respectively. In general, it is observed that with increasing percentage of binder, both the 7 days as well as 28 days strength increases. It is interesting to note that, with increase in proportion of fly ash in the mix, the strength initially increased and beyond certain proportion further addition of fly ash starts decreasing. This trend has been noticed for both the binders and curing periods.

On the strength considertions, the optimum mix F:S (binder %) appers to be 3:1 (5%) which gives maximum strength of the order of 3.45 MPa with cement and 2.25 MPa with lime as binder. The second highest strengths are exhibited by the mix 4:1 (5%) and the values are 3.3 MPa and 2.2 MPa respectively for cement and lime. It may be mentioned here that, the same mix when compacted at its maximum dry density and optimum water content is certain to exhibit considerably higher strength. However, flowability being one of the prime condition of mine fill material, herein the strengths have been studied only of the flowable mixes. Such a mix with flowability water content when pumped inside the underground mine cavity may find the surplus water for satisfying its curing requirements to build up the strength.

4.3 *Stress - strain relationships*

Typical stress-strain variations observed during uniaxial compressive strength tests are shown in Fig. 5 for specimens with lime and in Fig. 6 for specimens with lime and also with cement for 7 days and 28 days curing periods. In general, the curves show that, the specimens cured for 28 days are more brittle than those with 7 days curing period. The failure of specimens shifts from brittle to ductile region with decreasing binder percentage. The specimens with cement binder exhibit more brittleness compared with lime for similar mix. It

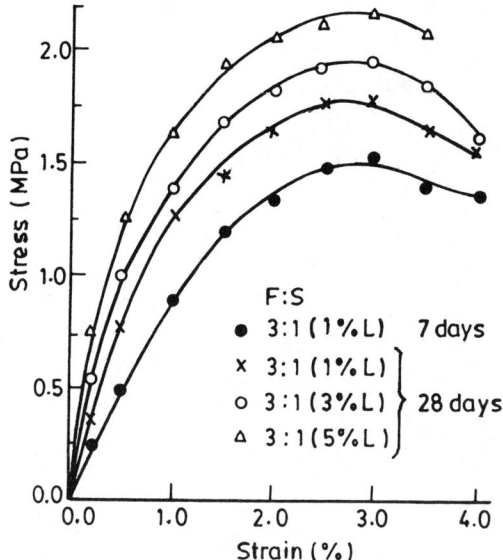

Fig. 5. Typical stress - strain variation for some of the mixes with lime

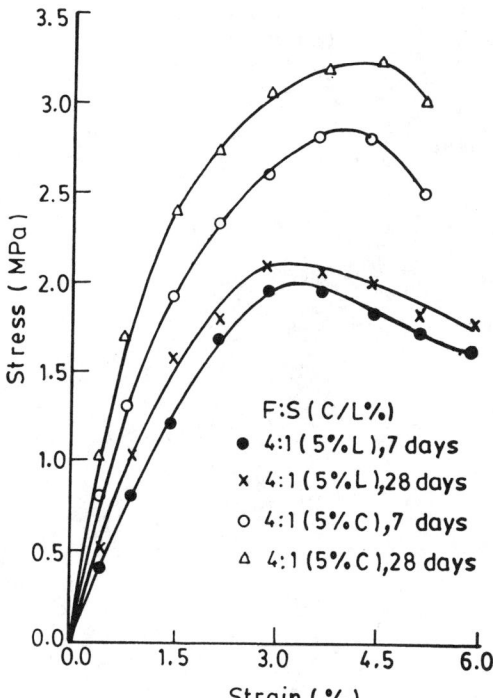

Fig. 6. Stress-strain relationship for selected mixes with cement and lime

may be inferred from these stress - strain curves that, the specimens with lime might not have gained its full strength even after 28 days of curing.

4.4 Design mix proposal

The strength and deformability behaviour of the mine fill materials to withstand the overburden stress are of prime importance while making a choice. The choice would further depend upon number of factors, some of these are geometry of mine cavity, process of stowing, environmental aspects etc. Each mine cavity or abandoned mine or open cast mine always possesses unique features. Thus the fill material/design mix to be chosen to satisfy the requirements only for that particular mine rather than generalizing its suitability.

The scope of the present investigation was limited to understand possibility of using fly ash as a major constituent of the design mix for filling up underground or open cast mines taking into account various parctical aspects. In the present study, the mixes with cement provide higher strengths compared to those with lime for similar proportions of fly ash and sand. Considering the cost aspect, cement as a binder would suit for a mine cavity of small volume but large overburden cover to support. On the other hand, a mix with lime as a binder would be ideal for large volume cavity with small overburden cover (shallow depth).

The perusal of various aspects in the present study leads to identify a mix of fly ash to sand in proportion of 4:1 by weight with 5% lime as most ideal in terms of utilizing more quantity of fly ash for nearly the same strength as given by 3:1 (5% L) mix. Addition of further proportion of fly ash (beyond 4) would reduce the strength of mix considerably as it appears from the trends in Fig. 4. The proposed design mix of 4:1 (5% L) possesses a strength to support around 125 m cover of rock overburden.

5 CONCLUSIONS

The following conclusions have been arrived at from the present experimental study.

1. With increase in proportion of fly ash in the fly ash : sand : binder mix, the uniaxial compresssive strength initially increases and beyond certain proportion of fly ash it then starts decreasing.

2. Although specifications of suitable mix depend upon several factors namely, overburden

cover, geometry of mine cavity, environmental conditions etc., the design mix of fly ash to sand of 4:1 with 5% lime is found to be satisfactory from strength, flowability and economic considerations for filling up underground mine cavity of relatively large size with shallow overburden cover to support.

3. The study substantiates the utility of fly ash as a major constituent of design mix for filling up underground/open cast mines and mine cavities.

REFERENCES

Das, D. 1994. A feasibility study of utilizing fly ash as mine fill, *UG project report*, CE Deptt., I.I.T. Delhi, New Delhi

IS : 1727 - 1967. Methods of Test for Pozzolanic Materials, *Bureau of Indian Standards*, Manak Bhavan, New Delhi.

IS : 3812 - 1966. Specification for fly ash for use as Pozzolana (Part I), *Bureau of Indian Standards*, Manak Bhavan, New Delhi.

IS : 9143 - 1979. Method for the determination of unconfined compressive strength of rock materials, *Bureau of Indian Standards*, Manak Bhavan, New Delhi.

Meena, B.S. 1996. Studies on the strength behaviour of lime treated fly ash - sand mix, *UG project report*, CE Deptt., IIT Delhi, New Delhi.

Parida, A. et al. 1996. Hydraulic transportation of fly ash at higer concentration, *Ash Ponds and Ash disposal Systems* (Ed. Raju, V.S. et al), Narosa Publishing House, PP. 17 - 28.

Excavation analysis of a large-scale power station cavern by micromechanics-based continuum model for jointed rock mass

H. Yoshida
Department of Civil Engineering, Chiba Institute of Technology, Japan

H. Horii
Department of Civil Engineering, The University of Tokyo, Japan

K. Kudo
Tokyo Electric Power Co., Ltd, Japan

ABSTRACT: The mechanical behaviors of jointed rock mass are strongly affected by the property and geometry of joints. The authors proposed a numerical method based on micromechanics assuming that the sliding and opening of joints are the governing mechanism of the jointed rock mass. It reflects the effect of orientation and density of joints in a direct manner. In this study, a problem of the excavation of a power station cavern is analyzed to examine its performance. Analytical results, such as the displacement of rock mass, are compared with measurement data showing good agreements.

1 INTRODUCTION

It is well recognized that the mechanical behaviors of jointed rock mass are strongly affected by the property and geometry of its joints. The complicated behaviors of the rock mass during a cavern excavation are considered to be caused by the sliding and opening of the joints. Thus, it is needed to pay attention to not only the imposed loading condition and the material properties of intact rock mass but also the property and geometry of the joints. When the construction of a large-scale cavern, such as a underground power station, is carried out, a numerical model reflecting the behaviors of the joints is needed to predict the behavior of jointed rock mass with reasonable accuracy.

The authors proposed a micromechanics-based continuum model (MBC model) for the behavior of the jointed rock mass. In the proposed model, the joint is assumed to have a macroscopic undulation and sliding along one part of joint surface leads to opening along the other part, which is considered to be the governing mechanism of the mechanical behavior of the jointed rock mass.

To examine the performance of the proposed method, the excavation analyses of Shiobara power station cavern constructed by the Tokyo Electric Power Co., Ltd. (Yoshida & Horii 1996), Okawachi power station cavern constructed by the Kansai Electric Power Co., Ltd. (Yoshida et al. 1996) and the trial reduced-scale cavern preceding the construction of Kazunogawa power station cavern constructed by the Tokyo Electric Power Co., Ltd. (Yoshida et al. 1995) are carried out and their results are compared with measurement data showing good agreement.

As for the cavern shape of underground power station, there are mainly three types; mushroom-type, bullet-type, and egg-type. The cavern shapes of Shiobara power station and Okawachi power station are the mushroom-type and the bullet-type, respectively. In this paper, the excavation problem of Kazunogawa power station cavern constructed by the Tokyo Electric Power Co., Ltd. having an egg-type shape is analyzed and analytical results, such as the displacement of rock mass, are compared with measurement data.

2 MICROMECHANICS-BASED CONTINUUM MODEL

The Micromechanics-based continuum theory is for the mechanical behavior of the material which is governed by the existence of initial defects and microstructures or their growth and propagation. The behaviors of microstructures are modeled and the constitutive equation of an equivalent continuum is derived from a relationship between average stress and average strain over a representative volume element (R.V.E) which contains a lot of microstructures. This theory is suitable for the analysis of rock mass containing a number of joints.

2.1 Average strain - average stress relation

Mechanical behaviors of jointed rock mass are considered on a representative volume element (RVE). Although the size of the RVE is much smaller than that of the analytical are, it is much larger than that of all characteristics of the jointed rock mass. In the RVE, the average stress and average strain are defined as

$$\overline{\sigma}_{ij} = \frac{1}{V} \int_V \sigma_{ij} dV, \quad (1)$$

$$\overline{\varepsilon}_{ij} = \frac{1}{V} \int_V \varepsilon_{ij} dV = \frac{1}{V} \int_V \left\{ \frac{1}{2}(u_{i,j} + u_{j,i}) \right\} dV, \quad (2)$$

where V is the volume of the RVE and u_i is displacement vector. Horii and Nemat-Nesser derived the average stress - average strain relationship over a body containing discontinuities in the following form (Horii & Nemat-Nesser, 1983). In the derivation, the divergence theorem is employed and no approximations are introduced. The intact rock surrounding joints is assumed to be elastic.

$$\overline{\varepsilon}_{ij} = D^R_{ijkl} \overline{\sigma}_{kl} + \frac{1}{2V} \int_\Omega ([u_i]n_j + [u_j]n_i) dS, \quad (3)$$

where D^R_{ijkl} is the elastic compliance tensor of intact rock, Ω is the exterior surface of the RVE, $[u_i] = u_i^+ - u_i^-$ is the displacement jump across the joint, and n_i is the unit vector normal to Ω. The integration is taken over the upper surface of all joints inside of RVE. Since the studied problem is non-liner, Equation (3) is rewritten as

$$\begin{aligned}\Delta \overline{\varepsilon}_{ij} &= D^R_{ijkl} \Delta \overline{\sigma}_{kl} \\&+ \frac{1}{2V} \sum_\alpha \int_{\Omega^\alpha} (\Delta[u_i]n_j + \Delta[u_j]n_i)_\alpha dS,\end{aligned} \quad (4)$$

where Ω^α is the exterior surface of joint α. If the incremental displacement jump across a joint, $\Delta[u_i]$, is expressed as a function of the incremental average stress $\Delta \overline{\sigma}_{ij}$, the overall constitutive equation of the jointed rock mass is obtained from Equation (4) as

$$\Delta \overline{\varepsilon}_{ij} = \overline{D}_{ijkl} \Delta \overline{\sigma}_{kl}, \quad (5)$$

where $\overline{D}_{ijkl}$ is the tangential compliance of the jointed rock mass.

2.2 Behavior of joints in rock mass

To express the incremental displacement jump across a particular joint as a function of the incremental average stress, the isolation of the joint by cutting it out is assumed and (a) the original problem of rock mass with embedded joints is decomposed into (b) homogeneous intact rock, (c) rock with traction on acting an open slit, and (d) a cut-out joint (Figure 1).

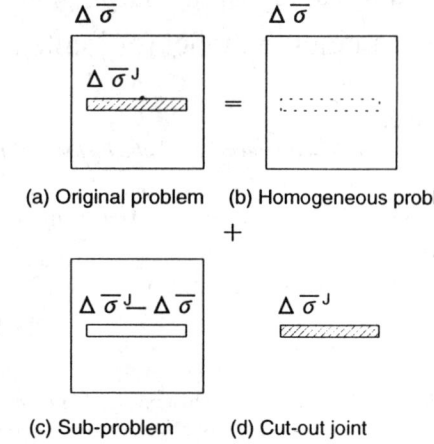

Figure 1. Decomposition of the problem of jointed rock mass

The compatibility condition of displacement jumps on the shared boundary between the open slit and the cut-out joint leads to;

$$\Delta[\overline{u}_n^S] = \Delta[\overline{u}_n^J], \quad \Delta[\overline{u}_s^S] = \Delta[\overline{u}_s^J], \quad (6)$$

where $\Delta[\overline{u}_n^S]$ and $\Delta[\overline{u}_s^S]$ are normal and shear components of average incremental displacement jump across the slit, and $\Delta[\overline{u}_n^J]$ and $\Delta[\overline{u}_s^J]$ are normal and shear components of average incremental displacement jump across the joint, respectively.

The restriction of the joint deformation by the surrounding rock mass is expressed in terms of the system stiffness introduced by Cai and Horii (1992). It regards the reduction of the overall stiffness due to the existence of embedded joints. Following relationship between the average incremental displacement jump across the slit and the average incremental stress acting on the slit;

$$\Delta \overline{\sigma}_n - \Delta \overline{\sigma}_n^J = \overline{K}_n \Delta[\overline{u}_n^S], \quad \Delta \overline{\sigma}_s - \Delta \overline{\sigma}_s^J = \overline{K}_s \Delta[\overline{u}_s^S], \quad (7)$$

where $\Delta[\overline{u}_n^S]$ and $\Delta[\overline{u}_s^S]$ are normal and shear components of the average incremental displacement jump across the slit on which the incremental normal stress $\Delta \overline{\sigma}_n - \Delta \overline{\sigma}_n^J$ and shear stress $\Delta \overline{\sigma}_s - \Delta \overline{\sigma}_s^J$ are acting, respectively. $\overline{K}_n$ and $\overline{K}_s$ are normal and shear system stiffness. Expressions of $\overline{K}_n$ and $\overline{K}_s$ are derived from the solution of a penny shape crack in an finite elastic material as;

$$\overline{K}_n = \frac{\overline{E}}{\lambda_n^o \overline{L}^J}, \quad \overline{K}_s = \frac{\overline{G}}{\lambda_s^o \overline{L}^J}. \quad (8)$$

The effective Young's modulus for normal and shear, $\overline{E}$ and $\overline{G}$, reflect reduction of the overall stiffness due to to presence of embedded discontinuities. The values of $\overline{E}$ and $\overline{G}$ are smaller than the effective Young's modulus of an intact rock and evaluated by determining the overall constitutive relationship of the jointed rock mass. The normal and shear coefficients of joint shape λ_n^o and λ_s^o are adopted as $\lambda_n^o = \frac{16(1-\nu^2)}{3\pi}$ and $\lambda_s^o = \frac{16(1-\nu)}{3\pi(2-\nu)}$ for three dimensional penny-shape joint. The effective length $\overline{L}^J$ regards the increase of joints and it can be expressed as the product of joint representative length and joint connectivity.

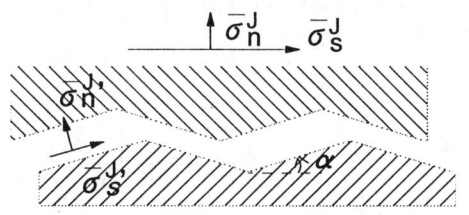

Figure 2. Joint with undulation

In this study, the saw-teeth shape shown in Figure 2 is considered to model the large-scale undulations of the joint. It is assumed that deformation of the joint is caused by sliding along one part of its surfaces and the associated opening of another part. At this moment, the opening part of joint surface is assumed to be traction free surface. The relationship between the nominal joint stresses on global joint plane and those on the sliding joint surface (see, Figure 2) is given by

$$\Delta \overline{\sigma}_n^J = \frac{1}{2}(\Delta \overline{\sigma}_s^{J'} \tan \alpha + \Delta \overline{\sigma}_n^{J'}),$$
$$\Delta \overline{\sigma}_s^J = \frac{1}{2}(\Delta \overline{\sigma}_s^{J'} - \Delta \overline{\sigma}_n^{J'} \tan \alpha), \quad (9)$$

where α is the undulation angle of the joint.

The Coulumb fracture criterion is employed, i.e. The deformation of a joint is initiated when the following condition is satisfied;

$$|\overline{\sigma}_s^{J'}| = -\overline{\sigma}_n^{J'} \tan \phi, \quad \overline{\sigma}_n^{J'} < 0, \quad (10)$$

where ϕ is the frictional angle on joint surface, $\overline{\sigma}_n^{J'}$ and $\overline{\sigma}_s^{J'}$ are the average normal and shear stresses on the sliding joint surface, respectively. Only the case when the sliding of one part leads to the opening of the other part is considered. During the joint sliding, the incremental stresses on the sliding joint surfaces are assumed to satisfy th Coulomb frictional law;

$$\overline{\sigma}_s^{J'} = \pm \overline{\sigma}_n^{J'} \tan \phi, \quad (11)$$

with the sign depending on joint undulation α.

It is assumed that sliding mechanism and geometry of the joint determine relationship the components of displacement jump in the coordinate system of sliding sub-surface $\Delta[\overline{u}^{J'}]$ and that the joint plane $\Delta[\overline{u}^J]$. Since the sliding joint faces are considered to be smooth and in a close contact, the normal component of the average displacement jump in local coordinates $\Delta[\overline{u}_n^{J'}]$ is assumed to be zero. Thus, the relationship between the other components is given as;

$$\Delta[\overline{u}_n^J] = \Delta[\overline{u}_s^{J'}] \sin \alpha,$$
$$\Delta[\overline{u}_s^J] = \Delta[\overline{u}_s^{J'}] \cos \alpha, \quad (12)$$
$$\Delta[\overline{u}_n^{J'}] = 0.$$

By combining Equations (7), (6), (9), (11) and (12), the average incremental displacement on the joint is obtained as a function of the average incremental stress.

3 FINITE ELEMENT ANALYSIS

3.1 Outline of Kazunogawa power station

Kazunogawa power station cavern constructed by the Tokyo Electric Power Co., Ltd. is a large cavern for a pumped up storage power station having a width of 34.0 m, a height of 54.0 m, a length of 210 m (see; Figure 3). An underground power plant is located at a depth of approximately 500 m below the ground surface to accommodate four 400 MW generators, and its axis is close to north-south direction.

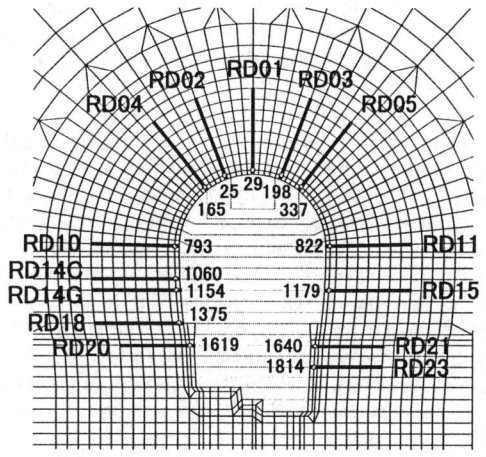

Figure 3. Location of measurement facilities

The rock mass is very cracky and is dominated by one group of joints with high dip angle down to north (see, Figure 4) The cavern axis is selected so that those joints are perpendicular to the cavern axis and their effect on the deformation of rock mass is minimized. There exist two other groups of joints which are nearly parallel to the cavern axis. One group has gentle dip and requires less attention. The other group has south-north strike with a steep dip and intersects the cavern axis at an acute angle. In this analysis, only this group of joints is considered and the dip angles of two dominant joint sets are set to be 80 degrees dipping to west (N00/80W) and 70 degrees dipping to east (N00/70E) and their strike are assumed to be parallel to the direction of the cavern axis.

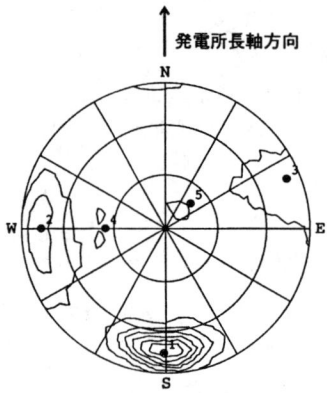

Figure 4. Distribution of joint density

Average distance of these joint sets used for analysis are set, from observation of rock cores, at 1.0 m for the joint set dipping to west and 1.0 m for the joint set dipping to east. Since the effective length of joint is considered to depend on cavern size, it is increased with proceeding the excavation steps. Other input parameters of joints for the MBC analysis, the frictional angle of the distributed joint surface and undulation angle of joint are determined by taking the size of cavern and some past calculation experience into account, and listed in Table 1. Poisson's ratio is set to be $\nu=0.25$. Young's modulus of intact rock mass excluding the dominant joints is determined to be $E=200,000$ kgf/cm^2 by the determination method of unknown parameters of the joints in the MBC analysis proposed by the authors [?]. In the method, the tunnel excavation preceding a large scale cavern is regarded as an in-situ test of stress relaxation type, and measurement data, such convergences are utilized to determine the input parameters for the MBC analysis. This method employs the average strain of the cross section computed from the convergences to represent the behavior of the rock mass around the tunnel.

The support system of the cavern consists of PS anchors, rock bolts and shotcrete. The PS anchors is considered in the analysis. The material properties of PS anchors are shown in Table 2.

For the PS anchor, a truss element is add along the part of the PS anchor to be fixed with concrete grout after each step of excavation is finished. Then, the nodal force is applied at two points (a point on the cavern wall and the internal end of fixed part of the PS anchor), and a truss element is added between those points. The input parameters as well as the installation procedure of arch concrete and the PS anchor follow those of real construction.

The reported initial stress is converted into stress components into the plane perpendicular to the cavern axis for two dimensional analysis. The maximum and minimum values of initial stress used in the analysis are 126 kgf/cm^2 and 112 kgf/cm^2. The direction of maximum stress is 14 degrees clockwise from vertical line.

Table 1. Joints parameters

dip angle	80°(west)	70°(east)
effective length ($\overline{L}^J$)	1.0m~10.0m	1.0~10.0m
average distance (d)	1.0m	0.67m
frictional angle (ϕ)	30~40°	30~40°
undulation angle (α)	10°	10°

Table 2. Parameters of PS anchors

Young's modules	length	nominal area	design load
2,000,000 kgf/cm^2	15 m	8.336 cm^2	24.0 tf

3.2 Numerical results

The numerical results of displacement of rock mass are compared with the measurement data. First, the results of displacement of rock mass along the monitoring lines, RD04, RD05, RD10, RD11, RD14 and RD15 shown in Figure 3, are plotted together with the measurement data in Figure 5 to Figure 10. The results show good agreement with all the measurement data except for RD11.

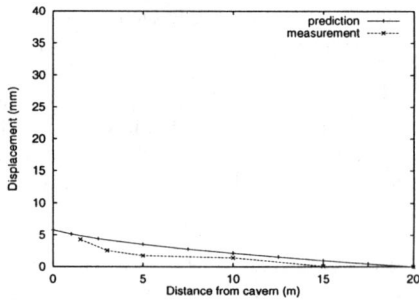

Figure 5. Displacement of rock mass (RD04)

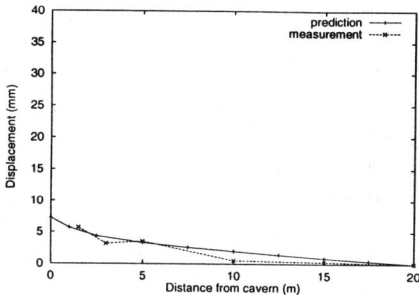

Figure 6. Displacement of rock mass (RD05)

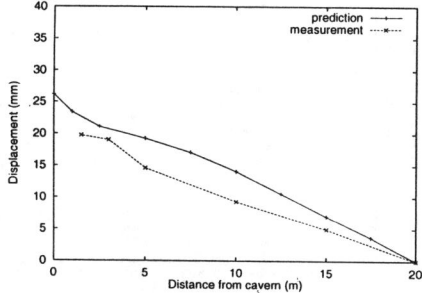

Figure 7. Displacement of rock mass (RD10)

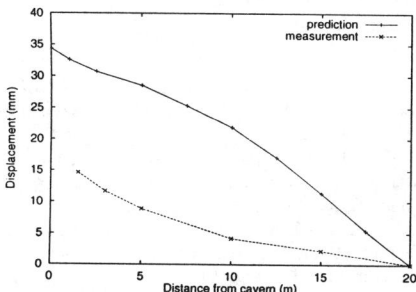

Figure 8. Displacement of rock mass (RD11)

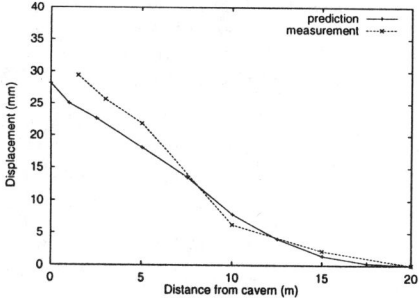

Figure 9. Displacement of rock mass (RD14)

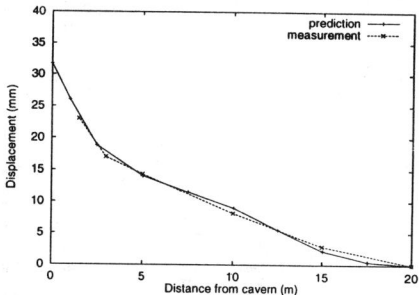

Figure 10. Displacement of rock mass (RD15)

Next, the displacement at a point near the cavern wall along the monitoring lines are plotted against time with measurement data in Figure 11 to Figure 16.

In the case of arch part, the displacement tends to converge after the excavation of arch part. On the contrary, that at the mid-part cavern wall continue to increase till full excavation. The numerical results have good accordance with the measurement data.

4 CONCLUSION

In order to examine the performance of the proposed method, the excavation of Kazunogawa power station cavern having the egg-type cavern shape is analyzed, and results are compared with measurement data. Displacement of rock mass and displacement at cavern wall obtained by the MBC analysis show good accordance with measurement data.

In the analysis, the dip and strike of the dominant joint sets, joint frictional angle and joint average distance are determined consulting with the results of a geological survey and in-situ tests. The other parameters of joints such as joint undulation are assumed on the basis of the past analysis. The effective length of joint increases with proceeding the excavation steps since it is considered to depend on cavern size. Young's modulus of rock mass without dominant joints is determined by a method proposed by the authors (Yoshida et al. 1996). The determination method identifies input parameters from measured data in the tunnel excavation preceding to a power-house cavern excavation. Our future task is to establish a design method for a large-scale cavern using the proposed model.

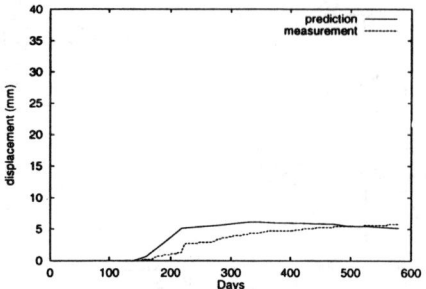

Figure 11. Displacement at cavern wall (RD04)

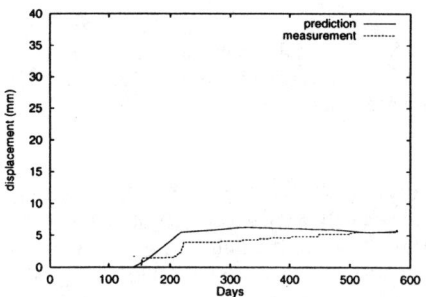

Figure 12. Displacement at cavern wall (RD05)

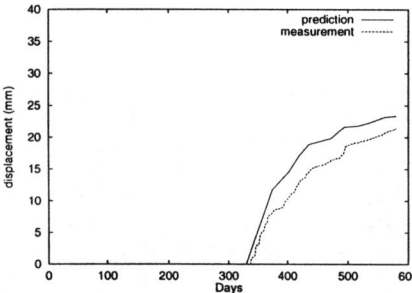

Figure 13. Displacement at cavern wall (RD10)

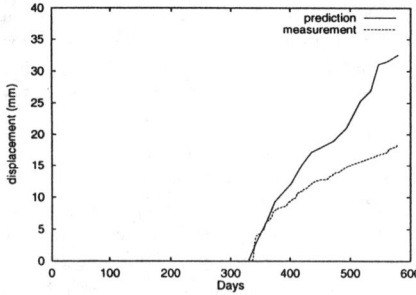

Figure 14. Displacement at cavern wall (RD11)

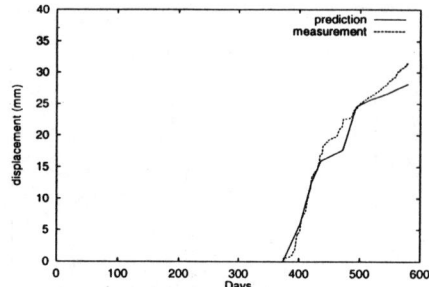

Figure 15. Displacement at cavern wall (RD14)

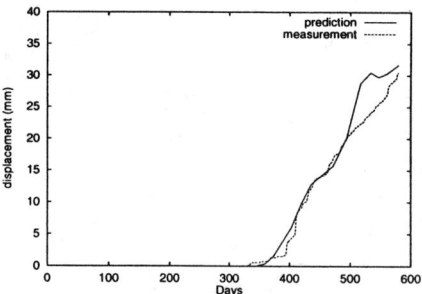

Figure 16. Displacement at cavern wall (RD15)

REFERENCES

Cai, M. and Horii, H.: A constitutive Model of Highly Jointed Rock Masses, *Mechanics of Materials*, 13, pp.217-246, 1992.

Horii, H. and Nemat-Nasser, S.: Over all Moduli of Solids with Microcracks: Load-Induced Anisotropy, *J. Mech. Phys. Solids*, 31, pp.155-171, 1983.

Yoshida, H. and Horii, H.: Micromechanics-based Continuum Model for Rock Masses and Analysis of the Excavation of Underground Power Cavern, *Journal of Geotechnical Engineering of Japan Society of Civil Engineering*, No.535/III-34, pp.23-41, 1996 (in Japanese).

Yoshida, H., Horii, H. and Uchita, Y.: Analysis of the Excavation of Underground Power House at the Okawachi Power Station by Micromechanics-based Continuum Model and Comparison with Measured Data, *Journal of Geotechnical Engineering of Japan Society of Civil Engineering*, No.547/III-36, pp.39-56, 1996 (in Japanese).

Yoshida, H., Horii, H. and Uno, H.: Micromechanics-Based Continuum Theory for Jointed Rock Mass and Analysis of Large-Scale Cavern Excavation, *Proceedings of Eighth International Congress on Rock Mechanics*, pp.689-692, 1995.

Yoshida, H., Hibino, S., Horii, H. and Kudo, K.: A determination Method for Analytical Parameter from Convergences during Tunnel Excavation, *Journal of Geotechnical Engineering of Japan Society of Civil Engineering*, 1996 (submitted, in Japanese).

Development of design criteria for low temperature gas storage

U.E. Lindblom & R. Glamheden
Chalmers University of Technology, Gothenburg, Sweden

ABSTRACT: This paper presents a few simple design criteria that may be used during the first stages of a refrigerated storage project for LPG in site selection and feasibility work. The indications given by the graphs in the paper must be controlled by more careful numerical and other analyses during the detailed design stage. Key factors to be regarded were found to be stability, heat exchange and vapor loss from the cavern. Rock parameters dictating these factors are rock stresses, elastic parameters, thermal expansion coefficient, thermal conductivity, specific heat and depth to the groundwater surface. The difference between temperatures of the rock and the stored product is the driving force for any cavern malfunction.

1 INTRODUCTION

Underground storage of gas is a well known technology with hundreds of installations in operation world wide. The natural capacity of the surrounding rock and groundwater to prevent gas is utilized in these storage installations, and the gas is compressed to a level well below the break-through pressure which is specific for the rock formation at the actual depth. Storage facilities are constructed in porous rock formations, as well as in caverns in salt or in hard rock.

An alternative to conventional gas storage is to store the gas in a state of refrigeration. Thereby, the vapor pressure of the gas reduces and the energy density increases. Except for a number of caverns at Stenungsund, West Sweden, this method has not been in practice elsewhere. The advantage of refrigerated gas storage has however led to a new interest in this storage method.

Non-isothermal design of caverns is a definite departure from conventional rock engineering design. This paper briefly outlines a number of key criteria that have to be observed by the designer: energy loss, stability and containment criteria. Simple guidelines are given as to how to account for these criteria in the cavern design procedure.

2 PERFORMANCE OF A ROCK MASS SUBJECT TO COOLING

A rock cavern is excavated in a rock mass subjected to a natural inherent stress field, primary stresses, see Figure 1 (a). Due to the removal of rock, stress concentrations (positive and negative) will occur in the vicinity of the cavern, secondary stresses, 1 (b). During operation of the cavern with a low temperature gas mass (vapor and liquid), reduction of the near-field stresses occurs by an amount called thermal stress. The development of thermal stresses will release compression in the cavern vicinity, but also enlarge any tensile zones, 1 (c).

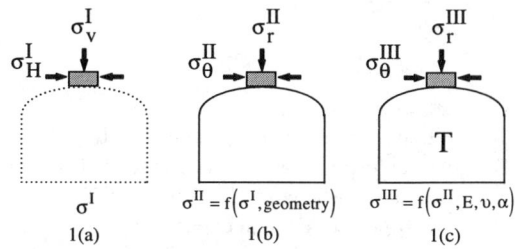

Figure 1. Stress fields to be considered in the design of refrigerated caverns. Denotations I, II and III are used for primary, secondary and tertiary (thermally affected) stresses, respectively.

Whereas the primary stress field is dependent on geologic and tectonic history, the secondary stress field is a function of the primary stresses and the shape of the cavern, independent of rock compressibility under elastic conditions. The influence of temperature is dictated by elastic rock properties, which will vary with temperature, and by the contraction of the rock material due to cooling. The latter effect is described by the coefficient of linear expansion, α.

The secondary stress situation around a cavern may be determined by well established rock mechanical methods. For example, tangential stresses in the cavern roof and walls may be expressed as shown in Hoek & Brown (1980).

roof $\quad \sigma_\theta^{II} = \sigma_v^I (Ak - 1)$ (1)

sidewall $\quad \sigma_\theta^{II} = \sigma_v^I (B - k)$ (2)

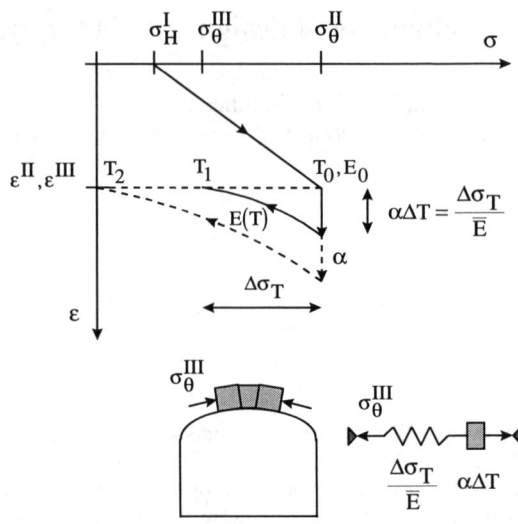

Figure 2. Stress-strain path of compressed rock at constant strain, subject to cooling, incl. analog mechanical model.

where A, B are geometrical shape factors and k the ratio of the horizontal primary stress, acting perpendicular to the cavern, to the vertical primary stress.

Due to the influence of thermal stress, the tangential stresses will reduce. The effect of stress on the thermally induced strain is explained in Figure 2. In this figure, the stress-induced deformation is represented by the spring (stiffening at cooling) and by the box (material shrinkage), respectively. T_0 denotes the normal rock temperature, T the rock temperature at some stage of cooling and T_1 the end temperature. By using the analogy in Figure 2, it follows that:

$$\sigma_\theta^{III} = \sigma_\theta^{II} - \alpha \int_{T_0}^{T_1} E(T)dT \cong \sigma_\theta^{II} - \alpha \overline{E} \Delta T \quad (3)$$

where,

E_0 - elastic modulus at T_0
$E(T)$ - elastic modulus during cooling (at T)
E_1 - elastic modulus at end of cooling (at T_1)
$\overline{E}$ - average elastic modulus from T_0 to T_1
α - coefficient of thermal expansion
ΔT - total temperature decrease ($T_0 - T_1$)

The magnitude of stress reduction is a combined effect of change of modulus and of material shrinkage. The degree of micro-fissuring is also important. By definition, α is a material property to be determined on a sample under zero stress. However, rock mechanical testing of α is commonly performed at either constant strain or constant stress in the sample. Evaluating α from such test results, the change in modulus must be regarded.

Total stress relief is achieved at the temperature T_2, see Figure 2, determined by:

$$T_2 = T_0 - \frac{\sigma_\theta^{II}}{\alpha \cdot \overline{E}} \quad (4)$$

At this temperature, any arching effects vanish in the roof and rock blocks may become unstable if not properly affixed by bolts. In fact, such instability may occur even earlier in the cooling procedure, depending on, among other factors, the fracture geology. Any further cooling will result in opening of rock fractures in the cavern walls. The thermo-mechanical behavior of the cavern near-field will in fact determine the key design factors. In the following, these factors are discussed as criteria for cavern energy loss, containment and stability.

3 ENERGY LOSS CRITERIA

These criteria refer to the liquid phase of the product. The state of equilibrium of the liquid at the cavern pressure requires that the heat exchange occurring in the cavern/product interface is within prescribed magnitude. Any unforeseen increase of this heat exchange will result in costly re-liquefaction processes or flaring of the boil-off gas. Thus, the design of the cavern must aim at avoiding such increases of heat exchange.

A successful design of a refrigerated cavern should safeguard that all heat exchange occurs as heat conduction only. The energy performance of the storage is thereby guaranteed to be reasonable and predictable (Glamheden, 1996). An erratic design of a refrigerated gas cavern would allow the rock wall stresses in the liquid storage region to become totally released and, furthermore, rock fractures to open up, see Figure 3.

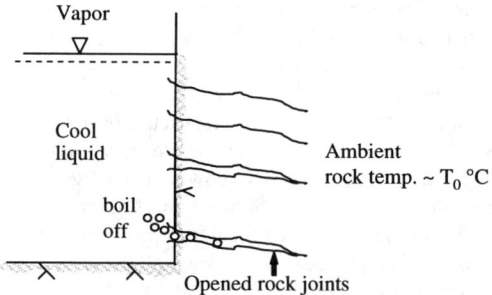

Figure 3. Phenomenological description of the development of rock fracture openings due to thermal effects.

Since the opening of rock fractures allows the stored liquid to penetrate into the warmer rock, boil-off will occur, leading to an undesired increase of heat exchange. Consequently, a design criterion to prevent uncontrolled energy loss in the storage should be to maintain all rock joints closed in the cavern vicinity. The most critical part of the cavern for the development of thermal cracking, in the common case of horizontal rock stresses higher than vertical, is the lower part of the side-wall, cf. Figure 3.

By using eq. (4) one gets:

$$\sigma_\theta^{II} = \alpha \bar{E}(T_0 - T_2)$$

Combining eqs. (2) and (4) yields with $\sigma_v = \rho g z$ the minimum location depth of the cavern with respect to thermal cracking,

$$z_{min} = \frac{\alpha \bar{E}}{(B-k)\rho g}(T_0 - T_2) \quad (5)$$

where,

ρ = rock density
z = location depth
T_2 = lowest allowable temperature

The factor $(B - k)$ in eq. (5) should be replaced by $(Ak - 1)$ in cases when the latter expression is smaller, i.e. when the most likely location of thermal cracking is in the cavern floor. A and B are geometrical factors, typical of the cavern shape, as follows from Table 1 (taken from Hoek & Brown, 1980).

Table 1. Geometrical shape factors A and B for caverns (Hoek & Brown, 1980)

Shape	◯	⌂	◯	⌒	◯	◯	⊂⊃	▭	◯
A	5.0	4.0	3.9	3.2	3.1	3.0	2.0	1.9	1.8
B	2.0	1.5	1.8	2.3	2.7	3.0	5.0	1.9	3.9

For a typical, large-size cavern, A = 4.0 and B = 1.5. In this case, $(Ak - 1) > (B - k)$ for $k > 0.5$. This implies that, for most situations, radial cracks are likely to develop in the cavern side-walls rather than into the cavern floor due to the action of thermal stresses. It shall be noted that conventional, large scale caverns are not suited for rock masses with $k > 1.5$, since they will always develop tensional wall fractures independent of depth. Under such conditions other cavern shapes should be selected, see Table 1.

Figure 4 displays graphical representations of the location depth required to avoid development of thermal cracks and associated boil-off effects in unlined, fully refrigerated caverns for storing (a) propane, (b) propane/butane mix (50/50) and (c) pure iso-butane. The graphs are based on steady-state operation temperatures for the various LPG products. It may be noted from eq.(5) that storage of liquefied natural gas (LNG) will require very large location depths of the caverns to avoid thermal cracking. With $T_0 - T_2$ of the order of 170°C, several hundreds of meters of rock cover will be required to store LNG (Goodall, 1989). This alternative is not treated further in this paper.

4 ROOF STABILITY CRITERIA

The development of de-stressed zones in the cavern vicinity, as discussed above, is also a reason for stability concern in the roof of the cavern. Loss of arching in the crown calls for installation of rock bolts to fix blocks of rock to the stable rock outside.

As a first criterion for roof stability, the diagram in Figure 4 may therefore be used also for roof stability assessment. Numerical analyses should follow in the detailed design stage to provide more adequate information on the size and shape of the zone of stress relief and of the stabilization work required.

5 GAS CONTAINMENT CRITERIA

Earlier in this paper, the negative effects of widening of rock fractures on the heat exchange and the stability of the caverns have been discussed. Despite the fact that fully refrigerated caverns are designed to store the product at its boiling temperature at atmospheric pressure, there will always exist a cushion of vapor of slightly elevated pressure near the cavern roof. The tendency of this gas to migrate upwards in the rock mass is restricted by ice in the rock fractures. The mechanism of gas escape is illustrated in Figure 5. As shown in the figure, the continuing shrinkage of the rock and fracture system may create new passages for gas vapor, should there not be sufficient liquid water outside the ice barrier to form new ice.

Figure 6 is based on calculations of the steady-state location of the ice barrier (0°C-isotherm) for caverns at varying depth in the rock. The depth to the upper face of this barrier is used as a design criterion, together with the predicted, lowest possible location of the groundwater level. Allowing at least 3 m of liquid water to cover the ice, the minimum location depth of the caverns may be estimated from the graphs. Figure 6(a) is for pure propane storage, 6(b) for 50/50 butane/propane mix and 6(c) for pure butane (iso).

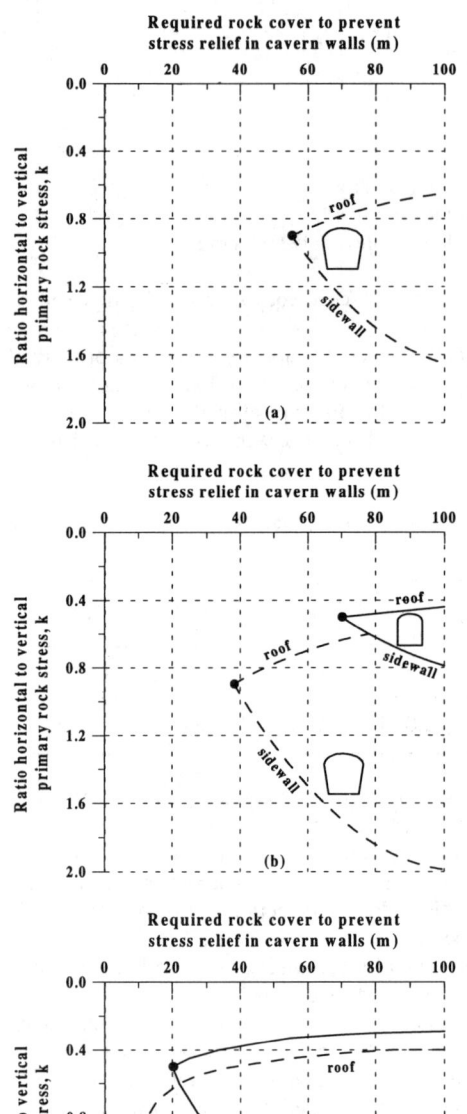

Figure 4. Graphs showing the required location depth for conventional and "bottle-shaped" rock caverns to avoid thermal cracking when storing (a) pure propane, (b) 50/50 butane/propane mix and (c) pure butane (iso). Denotations are found in the text. The graphs is based on $\alpha = 6 \ast 10^{-6}/°C$, $\overline{E} = 10$ GPa, $\rho = 2600$ kg/m^3, $T_0 = 5°C$, Following eq. (5).

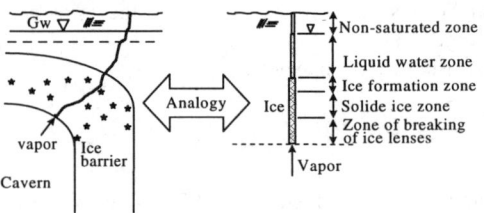

Figure 5. Phenomenological description of the mechanism of gas vapor escape through the frozen zone.

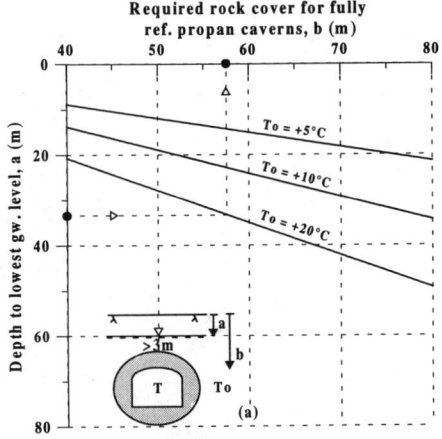
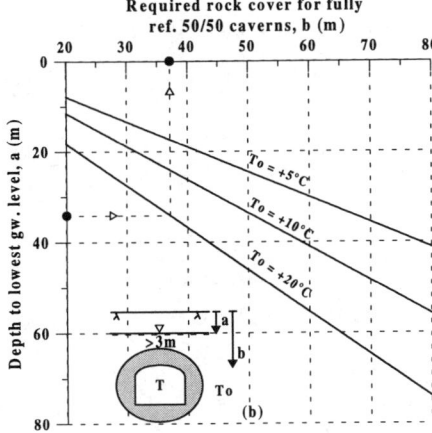
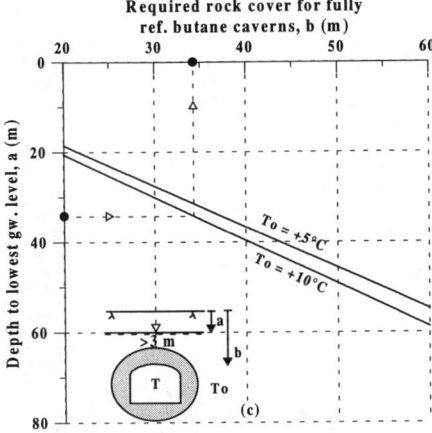

Figure 6. Required rock cover for a fully refrigerated to prevent it from vapor loss from the roof. (a) propane cavern, (b) 50/50 butane/propane and (c) pure butane (iso). The graphs is based on cavern dimension W=20m and H=30m. Thermal conductivity of the rock assumed at λ=3.5 W/m°C

6 SUMMARY AND CONCLUSIONS

This paper presents a few simple design criteria that may be used during the first stages of a refrigerated storage project for LPG in site selection and feasibility work. Key factors to be regarded were found to be stability, heat exchange, and vapor loss from the caverns. Rock parameters dictating these factors are rock stresses, elastic parameters, thermal expansion coefficient, thermal conductivity, specific heat and depth to the groundwater surface. The difference between temperatures of the rock and the stored product is the driving force for any cavern malfunction.

The indications given by the graphs in the paper must be controlled by more careful numerical and other analyses during the detailed design stage.

The rock mechanical consequences of refrigeration has been studied through large-scale field tests in a rock cavern at Chalmers University of Technology (Dahlström, 1992). A comprehensive research reports is presently under preparation.

REFERENCES

Dahlström, L. O., 1992. Rock Mechanical Consequences of Refrigeration - study based on a pilot scale rock cavern. Ph.D. thesis. Chalmers Unv. of Tech., Gothenburg.

Glamheden, R., 1996. Predicted and observed heat transfer around a refrigerated rock cavern. Proc. ISRM Int. Symp. on Prediction and Performance in Rock Mechanics and Rock Engineering, pp. 1395-1401. Eurock'96, Torino. Balkema, Rotterdam.

Goodall, D. C., 1989. Prospects for LNG storage in unlined rock caverns. Int. Conf. on Storage of Gases in Rock Caverns, pp. 237-243, Trondheim. Balkema, Rotterdam.

Hoek, E., Brown, E.T, 1980. Underground Excavations in Rock. The Institution of Mining and Metallurgy, London.

Pre-feasibility study on compressed air energy storage in Korea

Dae-Soo Lee
Korea Electric Power Research Institute (KEPRI), Taejon, Korea

Jung-Youp Kim, Il-Young Han & Soon-Jo Hong
Sunkyong Engineering and Construction Limited (SKEC), Seoul, Korea

ABSTRACT: This paper is to discuss the pre-feasibility of compressed air energy storage (CAES) in Korea. In recent years load leveling becomes very important for electric utilities because the peak load has increased faster than the growth of supply capacity. Although the underground pumped hydro power plant is operating as energy storage plant until now, it was shown that CAES can be a good alternative in Korea.

1. INTRODUCTION

In recent years electric utilities have much interests in facilitating load leveling because peak load has sharply increased in 1990s. The peak load of Korea had increased 11.6% per year, from 17,252 MW in 1990 to 29,878 MW in 1995. But the facility capacity has increased only 8.6% per year, from 21,008 MW in 1990 to 31,793 MW in 1995. As a result of slow increase of facility capacity in comparison to fast increase of peak load, capacity reserve ratio and capability reserve ratio decreased under 10% since 1990. So insufficient power supply became an big socioeconomic problem especially in 1994 when capability reserve ratio was only 2.8%.

From 1993 to 1995 the peak load had increased 16.2% per year that was higher than the average load increasing ratio, 13.1%. For power generation at peak load, three underground pumped hydro power plants - total 1,600 MW - store energy at low load periods and dispatch energy at high load periods. But as it is expected that peak load will be 70,850 MW in 2010 and that nuclear power plant will generate 45.5% of total power which is 10% higher than the role of that in 1996, more energy storage plants should be constructed.

2. ENERGY STORAGE PLANT

Central Research Institute of Electric Power Industry (CRIEPI) in Japan estimated that the optimum capacity of energy storage facility in Japan is 10~16%. As the power generation structure in Korea is very similar to that of Japan, the energy storage facility capacity of

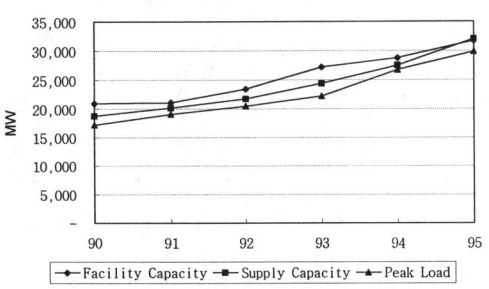

Fig. 1 Power supply and demand in Korea.

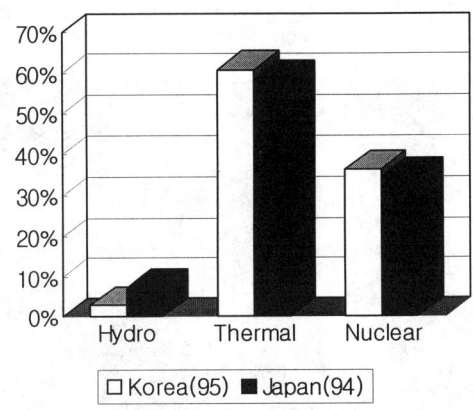

Fig. 2 Power Generation in Korea and Japan

Korea in 2010 can be estimated as 7,085~11,336 MW. As underground pumped hydro power plant needs special geographic condition because there must be 200~300m of height difference, suitable site for underground pumped hydro power plant is limited. Therefore new energy storage plant that does not need any special site requirement should be considered.

Some energy storage systems except underground pumped hydro are emerging nowadays. Those are compressed air energy storage (CAES), superconductor magnetic energy storage (SMES), battery, flywheel and so on. But the only commercialized energy storage system is CAES.

CAES is operated in two mode, storage mode and power generation mode. In storage mode the motor/generator is used as motor to drive the compressor which compresses the air into the underground storage for later use. In the power generation mode, compressed air is mixed with fuel and expanded through a modified gas turbine driving the motor/generator as a generator. The introduction of natural gas in the generating mode of the CAES cycle allows increased 'peak' power production than would otherwise be available from the recovery of compressed air alone.

The first CAES plant in the world is a 290 MW unit, owned and operated by the Nordwestdeutche Kraftwerke. The plant is located in Hontorf, Germany, and has been in commercial operation since 1978. The plant runs on a daily cycle with 8 hours of charging required to fill the cavern. At full load, the plant can then operate for 2 hours of generation.

The second CAES plant in the world is a 110 MW unit owned and operated by the Alabama Electric Cooperative (AEC). The plant is located in McIntosh, USA and has been in commercial operation since 1991. Unlike the Huntorf plant, the AEC plant utilizes an exhaust-gas recuperator to preheat the cold cavern air prior to entering turboexpander train.

Although the storage cavern of mentioned two CAES are constructed in salt dome, it is possible to be constructed in hard rock. In Korea, 60~70% of bedrock is granite and gneiss. As most of bedrock is fresh, there are many underground oil/LPG storage caverns. The hard rock type CAES is almost same as LPG storage cavern except that the cavern depth of CAES is deeper than LPG storage cavern.

3. PRE-CONCEPTUAL DESIGN OF CAES

In this study, pre-conceptual design was done for Seoul LNG steam turbine power plant site. It locates in Dangin-Dong, Seoul, and its capacity is 387.5MW. According to the plan of KEPCO, it will be closed in 2004.

The geology of Dangin-Dong is gneiss (Precambrian Gyeonggi gneiss complex) where many underground storage cavern, for example, K-1, K-1-E, L-1, L-1-E and Pyongtaek LNG storage cavern exist. As the engineering geological condition of Geyonggi gneiss complex is appropriate for large storage cavern, Dangin-Dong can be a candidate for CAES site.

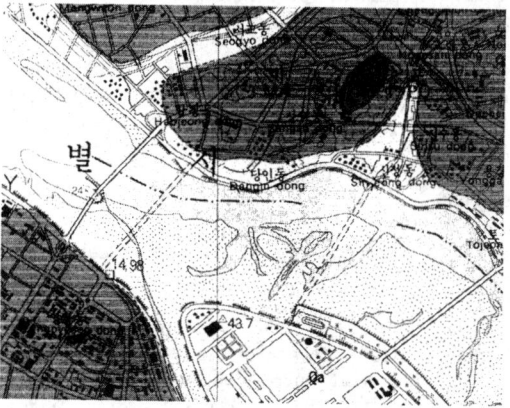

Fig. 4 Geologic map of Dangin-Dong.

The site that CAES requires is only 50~60% of LNG thermal power plant. Consequently it is feasible to construct 700 MW CAES in that site.

Primary factors to design a CAES are capacity, generation time and compression time. Table 1 shows those factors. The capacity factor for this design is set to 20%, so generation time is 4.8 hours per day.

The turbomachinery train used in McIntosh CAES is provided by Dresser-Rand. Its unit capacity is 110 MW. According to the Alabama

Fig 3. McIntosh CAES.

Table 1 Primary factors for conceptual design.

Capacity	700 MW (350 MW × 2 units)
Generation time	4.8 hours
Compression time	less than 8 hours

Electric Cooperative (AEC), the owner of McIntosh CAES, there was a some failure in low pressure expander. Because of low pressure expander failure, McIntosh CAES can not reach its goal of start reliability (95%) until 1996. In 1993, Westinghouse announced turbomachinery train for CAES. Its unit capacity is 350MW. There is some improvement that new turbomachinery train does not need a combustor for high pressure expander. So the heat rate was lowered by 200 Btu/Kwh than Dresser-Rand turbomachinery train used in McIntosh CAES.

In this pre-conceptual design, Westinghouse turbomachinery train was used. As its unit capacity is 350 MW, 2 units are used in this design. Table 2 shows summary of turbomachinery specification.

Table 2. Summary of turbomachinery.

Capacity	350 MW
Hours Comp'n/Gene'n	1.5
Fuel	Gas/Oil/Syngas
Heat Rate	3900 Btu/Kwh
Energy Ratio	0.67
Motive Air Flow	398 kg/sec
Recuperator	Yes

Underground cavern of hard rock type CAES can be constructed by two type, compensated (constant turbine-inlet pressure operation) or uncompensated (sliding turbine-inlet pressure operation). Although compensated type cavern requires reservoir and water shaft, storage efficiency is higher than uncompensated type cavern. The reason is that as higher pressure is needed in uncompensated type, there is much more heat loss during compressing air. For pre-conceptual design, compensated type is used. The reservoir is located near the Han river, which lies adjacent to site, so as to reduce the area for reservoir.

Many factors of underground cavern is determined by turbomachinery train. The determined factors of underground cavern is shown in Table 3.

Table 3 Basic criteria for underground cavern design.

Pressure of Stored Air	50 bar
Depth of Cavern	-550 m
Required Cavern Volume	242,934 m³
Safety Factor	1.2
Final Cavern Volume	291,520 m³

Storage cavern can be constructed by two type, large parallel tunnel method and room and pillar method. Until now large parallel tunnel method is preferred for oil/LPG storage cavern in Korea. As much experience is accumulated from LPG storage cavern design and construction, large parallel tunnel method was chosen for excavation method.

For decision of cross section many parameters, including vertical and horizontal in situ stress, should be known. But in this stage we have no those kind of data. Therefore ovaloid type cross section which is widely used for LPG storage is decided for cross section of storage cavern.

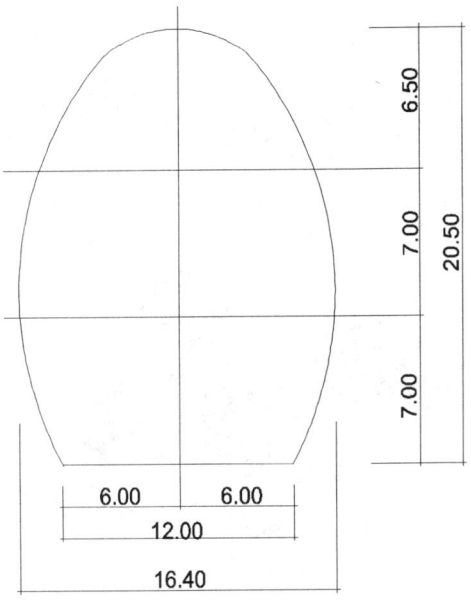

Fig. 5 Cross section of storage cavern (unit:m).

For determination of cavern axis orientation, attitude of major weakness plane should be known. Although joint can be an weakness plane, foliation plane of gneiss is assumed to be major weakness plane. Its attitude is N30°E/45°SE, so the orientation of cavern axis was determined as N60°W~S60°E.

Water shaft will be excavated by blasting and it will be used as main shaft. Air shaft will be excavated by raise boring to lower construction costs.

For air tightness water curtain will be used. If pressurized water is injected to water curtain, storage cavern can be installed at shallow depth. In that case, however, problem such as hydrojacking can be happened. In this design, water curtain with natural water head is considered. If the air conductivity of rock mass is low enough to maintain the tightness of the storage cavern, water curtain will not be required such as many LPG caverns in USA.

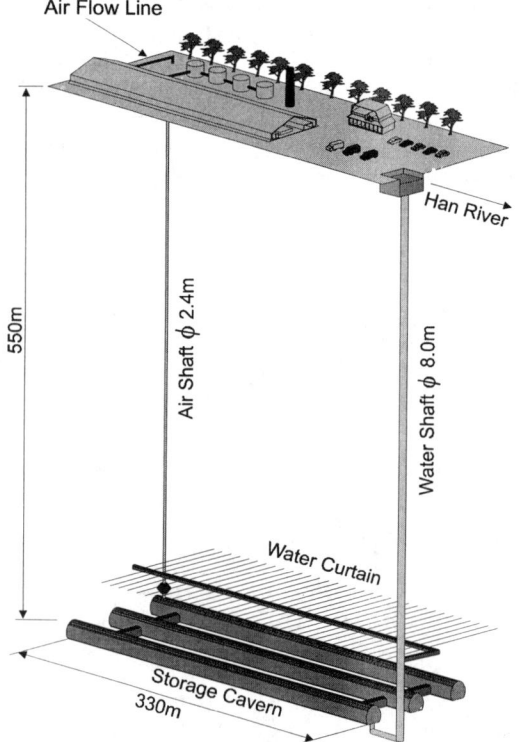

Fig. 6 Pre-conceptual design of CAES.

4. CONCLUSION

Through the pre-feasibility study on compressed air energy storage at the preliminary proposed site in Korea, the followings were concluded.

1. More energy storage plant will be needed for power generation at peak load in Korea.

2. As underground pumped hydro power plant requires special geographic condition for siting, it is necessary to seek a more efficient energy storage plant, which does not need constrained siting condition.

3. Compressed air energy storage is emerging technology for energy saving.

4. As most part of the bedrock in depth in Korea is fresh, hard rock type may be recommended for CAES.

5. In the view of efficient and economical point, CAES of compensated type is preferable to that of uncompensated type.

REFERENCES

Fenix & Scisson, 1988, Evaluation of hard rock cavern construction methods for compressed air energy storage, *EPRI report*

G.W. Gaul, 1993, A case for compressed air energy storage, *Power Magazine: March: 19~23*.

G.W. Gaul, 1993, Compressed air energy storage program status report, *American power conference : 55th annual meeting*

G.W. Gaul, T.M. Cornell, M. Nakhamkin and H. Paprotna, 1993, Compressed air energy storage thermal performance improvements, *Joint ASME/IEEE power generation conference*

I. Kurihara, 1995, Prospect of energy storage in future electric power system in Japan, *CRIEPI report*

Y. Uchiyama and M. Kadoyu, 1990, Technical assessment and economic study of compressed air energy storage in Japan. *CRIEPI report*

Stability evaluation of an underground opening used for special experiment

Y. Mizuta
Department of Civil Engineering, Yamaguchi University, Japan

Y. Kato
Research and Development Department, NOF Corporation, Japan

ABSTRACT: One of the drifts in the Tochibora Mine of Kamioka Mining and Smelting Co., Ltd. was widened and reinforced by a number of cablebolts in order to carry out performance tests of artillery within the opening. The maximum width and length of the opening are 15m and 180m, respectively and height of the opening is about 3m. The original rock stress state around the site was determined by hydraulic fracturing stress measurement. Then, the author carried out numerical calculations by the boundary element method for stability evaluation of the opening and examination of mechanical effect of the cablebolts on roof stability.

1. INTRODUCTION

The site of the performance tests of the heavy guns is underground laboratory for firing experiment by NOF corporation, located 850 meters above the sea in the Tochibora Mine of Kamioka Mining and Smelting Co., Ltd., which is under the steep mountains. The contour map above and around the site is shown in Fig.1.

In-situ measurement of three dimensional initial stress state by hydraulic fracturing method was carried out in the drift near shooting site. Fig.2 briefly shows the plan of the profile of the shooting site and the positions of the mouths of three boreholes used for hydraulic fracturing tests which were carried out 50m distant from the shooting site. The pattern of the cablebolts put in roof of the opening is shown in Fig.3.

In order to examine mechanical effect of the cablebolts put in the roof on rock stability around the opening, The authors carried out numerical simulation by two dimensional boundary element method taking into account the three dimensional initial stress state which was measured in-situ.

2. INITIAL STRESS MEASURED

As shown in Fig.2, the vertical distance from the shooing site to the surface is about 160 meters although the surface plane is steeply inclined. The magnitude of six stress components and three principal stress which were determined by in-situ hydraulic fracturing tests[1] are shown in Table1. The calculated overburden pressure is also shown in Table 1, where horizontal surface is assumed and the depth from the surface is assumed to be 160m.

The directions of the principal stresses are graphically represented in Fig.4.

Table 1 Initial stress state measured and the vertical stress component calculated

Six stress components (MPa)	Magnitude of the principal stresses (MPa)	Vertical stress component (MPa)
$\sigma_N = 3.06 \pm 0.61$	$\sigma_1 = 4.5 \pm 0.9$	$\sigma_V = 3.1 \pm 1.2$
$\sigma_W = 2.24 \pm 0.49$	$\sigma_2 = 2.3 \pm 0.6$	
$\sigma_Z = 3.08 \pm 1.25$	$\sigma_3 = 1.6 \pm 0.8$	
$\sigma_{WZ} = -0.17 \pm 0.46$		
$\sigma_{ZN} = 1.33 \pm 0.66$		
$\sigma_{NW} = -0.55 \pm 0.31$		

N,W and Z indicate North, West and Upward vertical, respectively

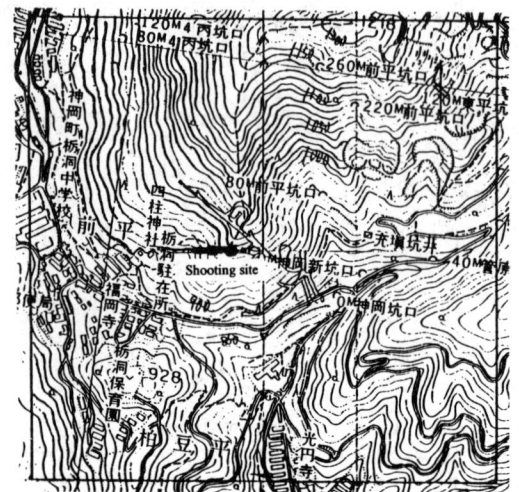

Fig.1 The contour map above and around the shooting site.

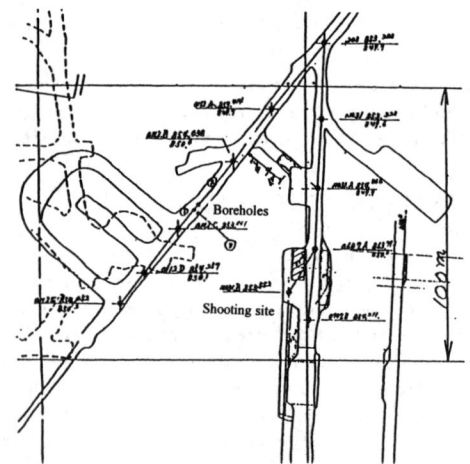

Fig.2 The brief profile of the shooting site and the positions of the boreholes used for hydrofracturing tests.

Fig.3 The pattern of the cablebolts put in the roof rock of the opening(black circle: cablebolt, white circle: rock bolt, double circle: extensometer).

3. INITIAL STRESS CALCULATED

In order to determine the initial stress state at the site of stress measurement, numerical calculation by boundary element method was carried out, assuming that the ratio of the horizontal to vertical stress due to the overburden is $\nu/(1-\nu)$, where ν is the Poisson's ratio. Modelling was carried out by following the procedure of the Fictitious Stress Method. Fig.5 shows the surface boundary divided into the triangular leaf elements. According to the procedure of indirect boundary element method, three components of tractions (uniform distributions of fictitious stresses)over each leaf element are determined first by solving $3N$ linear simultaneous equations, where N is the number of leaf elements on the boundary. The simultaneous equations can be constructed by making the sum of each initial stress component and each corresponding stress component induced by $3N$ tractions be zero for every leaf element [2].

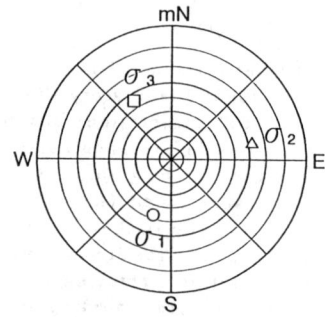

Fig.4 The directions of the measured three principal stress which are represented as the projections into lower sphere.

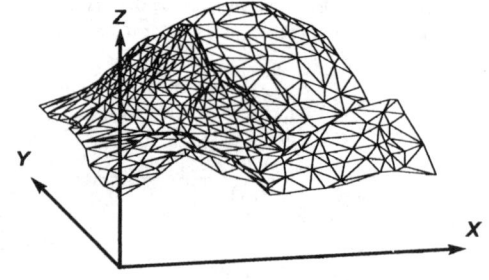

Fig.5 The model of the surface boundary divided into the triangular left elements.

Specific weight of rock, $\gamma = 27.8$ KN/m³, is uniformly distributed over the model, and thus three principal stresses of far field are given as the following expressions.:

$$\sigma_Z = \gamma z, \quad \sigma_N = \sigma_W = \frac{v}{1-v}\gamma z$$

where z(m) is depth from the height of Origin of the coordinate which can be arbitrarily set.

Three dimensional initial stress state at the point corresponding to the stress measurement site can be calculated as the sums of each stress component induced by the corresponding component of traction for every element. Fig.6 shows the stress state on W plane, which was obtained by three dimensional numerical computation.

4. DETERMINATION PROCEDURE OF FAR FIELD STRESS

Fig.7 shows the stress state on W plane, which was measured in-situ. It can be seen Fig.6 and Fig.7 that the calculated stress state is different from the measured stress state. However, if the values of k_N, k_W and k_{NW} in the following formulas are adequately determined, the calculated stress state may agree with the measured stress state.

$$\sigma_N = k_N \gamma z, \quad \sigma_W = k_W \gamma z, \quad \tau_{NW} = k_{NW} \gamma z$$

The coefficients of k_N, etc. which indicate the far field horizontal stresses can be determined by solving the following equation.

$$A^T A k = A^T \sigma$$

where

$$\sigma = \{(\sigma_N)_m, (\sigma_W)_m, (\sigma_Z)_m, (\tau_{NW}), (\tau_{WZ}), (\tau_{ZN})\}^T$$
$$k = \{k_N, k_W, k_{NW}, 1\}$$

$$A = \begin{bmatrix} (\sigma_N)_N, (\sigma_W)_N, (\sigma_Z)_N, (\tau_{NW})_N, (\tau_{WZ})_N, (\tau_{ZN})_N \\ (\sigma_N)_W, (\sigma_W)_W, (\sigma_Z)_W, (\tau_{NW})_W, (\tau_{WZ})_W, (\tau_{ZN})_W \\ (\sigma_N)_{NW}, (\sigma_W)_{NW}, (\sigma_Z)_{NW}, (\tau_{NW})_{NW}, (\tau_{WZ})_{NW}, (\tau_{ZN})_{NW} \\ (\sigma_N)_Z, (\sigma_W)_Z, (\sigma_Z)_Z, (\tau_{NW})_Z, (\tau_{WZ})_Z, (\tau_{ZN})_Z \end{bmatrix}$$

where $(\sigma_N)_m, (\sigma_W)_m, \ldots$ are the measured tress components and $(\sigma_N)_N$, etc., $(\sigma_N)_W$, etc., $(\sigma_N)_Z$ and $(\sigma_N)_{NW}$, etc. are the calculated stress components under far field stress state, $\{\gamma z,0,0,0,\}$, $\{0,\gamma z,0,0\}$, $\{0,0,\gamma z,0\}$ and $\{0,0,0,\gamma z\}$, respectively.

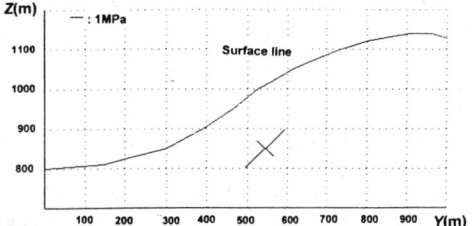

Fig.6 The measured stress state in the plane perpendicular to W axis.

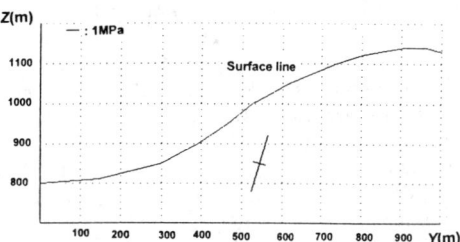

Fig.7 The calculated stress state in the plane perpendicular to W axis.

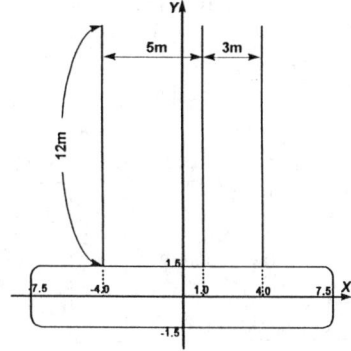

Fig.8 The representative vertical section of the opening.

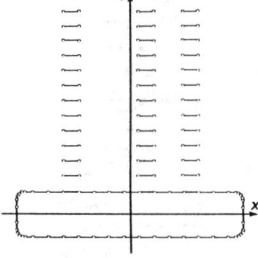

Fig.9 The modelling of the rock surface boundary and the boundaries where the tractions corresponding to bond strength by three line cablebolts of 12m length.

5. NUMERICAL ANALYSIS OF OPENING STABILITY

Fig.8 shows the representative vertical section of the opening of the shooting site. Fig.9 shows the modelling of the rock surface boundary and the boundaries where the tractions corresponding to bond strength of three line cablebolts of 12m length. Input data of the mechanical properties of rock around the opening are shown in Table 2. Fig.10 shows the assumed distribution of the maximum bond strength induced by tension of cablebolt. The value of the maximum adhesion are given by the bond strength of the case that Young's modulus of rock is 50GPa, which is shown in Fig.11. The bond strengths represented in the figure are calculated from Bond Strength Model [3].

Two dimensional Fictitious Stress Method developed by Crouch and Starfield [4] was used for numerical simulation. Fig.12 shows the calculated distribution of the principal stress in rock. The figure (a) is the result without taking the maximum bond strength into account and the figure (b) is the result with taking it into account.

Safety factors were determined by the following formulas from the stresses calculated.

$$F = \min\left(\frac{c + \sigma_s \tan\phi}{\tau_s}, \frac{S_t}{\sigma_n}\right)$$

Table 2 Mechanical properties of rock

Young's Modulus (GPa)	50.5
Poisson's Ratio	0.23
Specific Weight (kN/m³)	27.3
Compressive Strength (MPa)	91.7
Tensile Strength (MPa)	11.4
Frictional Angle (°)	46
Shear Strength (MPa)	18.5

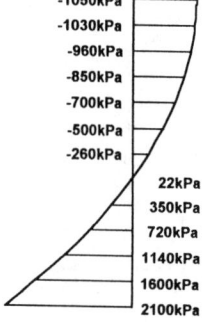

Fig.10 The assumed distribution of the maximum bond strength induced by tension of cablebolt and its integration over each of 12 segments.

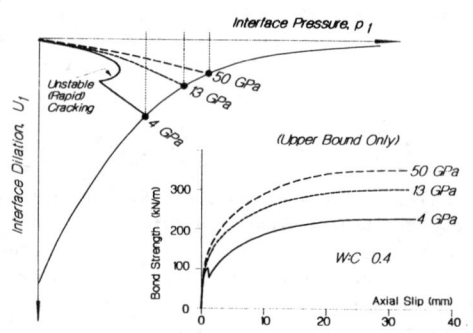

Fig.11 Presumption of bond strength by BSM (Hutchinson & Diederichs, 1996).

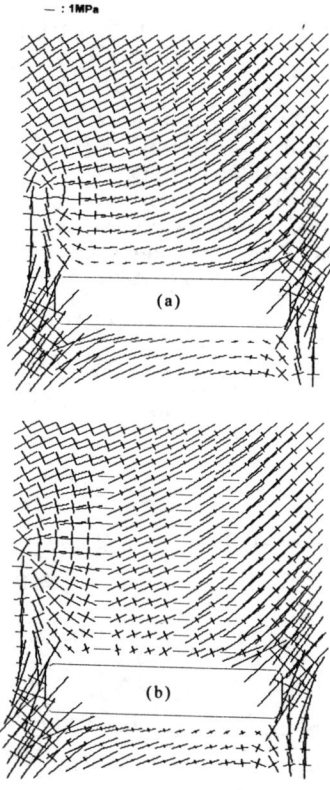

Fig.12 The calculated distribution of the principal stresses in rock (a) without and (b) with taking the maximum bond strength into account.

where σ_s and τ_s are normal stress and shear stress on the plane along which slip is most apt to happen at the specified point, c is shear strength, ϕ is frictional angle, S_t is tensile strength and

σ_n is normal (tensile) stress on the plane along which tensile crack is most apt to be produced at the specified point.

6. EFFECT OF CABLE BOLTING ON SAFETY FACTOR

The distribution of the calculated safety factors is shown in Fig.13. The figure (a) is the results without taking the maximum bond strength into account and the figure (b) is the results with taking the maximum bond strength into account. Fig.14 shows distribution change of the calculated safety factor by taking the maximum bond strength into account. In the area represented by Up, safety factors increase and, in the area represented by Down, safety factors decrease. It is roughly shown from the figure

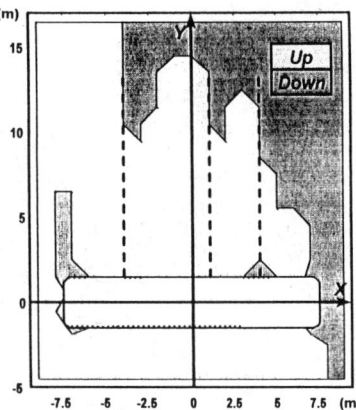

Fig.14 Distribution change of the calculated safety factor by taking the maximum bond strength into account (Up: safety factors increase, Down: safety factors decrease).

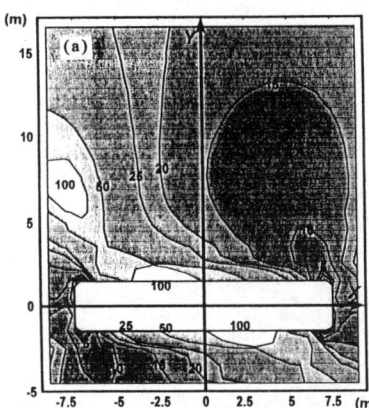

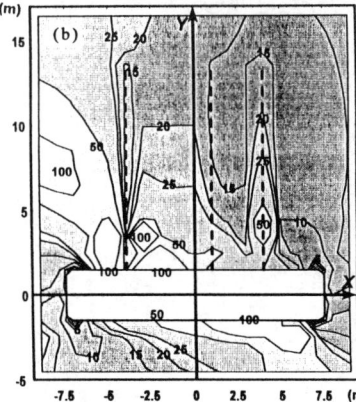

Fig.13 The calculated distribution of the safety factors in rock (a) without and (b) with taking the maximum bond strength into account.

that safety factors increase in the wide area of the roof rock.

7. CONCLUSION

Mechanical effect of cablebolting was shown from the results of numerical simulation by two dimensional Fictitious Stress Method although more precise calculation by three dimensional modelling is desirable. On the other hand, possibility of determination of far field initial stress state from a local initial stress state measured, which is influenced by the earth surface geometry, was presented.

REFERENCES

1. Mizuta,Y., Sano,O., Ogino,S. and Katoh,H., Three Dimensional Stress Determination by Hydraulic Fracturing for Underground Excavation Design, Int. J. Rock Mech. Min. Sci. & Geomech. abstr. Vol.24, No.1, 1987.
2. Kuriyama,k. et al, Three-dimensional Elastic Analysis by the Boundary Element Method with Analytical Integrations over Triangular Leaf Elements, Int. J. Rock Mech. Min. Sci. & Geomech. Abstr. Vol. 32, No.1,1995.
3. Hutchinson,D.J. and Diederichs,M.S., Cablebolting in Underground Mines, BiTech Publishers Ltd., 1996.
4. Crouch,S.L. and Starfield,A.M., Boundary Element Methods in Solid Mechanics, George Allen & Unwin, 1983.

Feasibility study on ACC compressed air energy storage system by water-sealing method

Satoshi Hibino, Eiichi Koda, Kameichiro Nakagawa & Yoji Uchiyama
Central Research Institute of Electric Power Industry, Chiba, Japan

ABSTRACT: The annual load factor of electricity has a tendency of becoming lower year by year in Japan shown in the value of 0.55 in 1994. It is, therefore, very important to develop new effective storage methods of electricity. Among the methods, the compressed air energy storage (CAES) looks very promising.
 In this paper the authors have proposed a new concept of ACC-CAES. The concept includes an economical water sealing method for storage of compressed air and advanced combined turbine system (ACC) for high efficiency.
 At first we performed tests to assure mechanism of water sealing method in a bored hole of 400 m depth. The kind of rocks surrounding the hole was granite. It is revealed from the tests that the water sealing method is effective to store highly compressed air, and the ultimate value of water sealing is equal to or more than the value of pore pressure of ground water. We could store compressed air of 40 kg/cm^2 because of the depth of the hole being 400 m.
 Secondly we estimated cost of electricity on our concept of ACC-CAES, and estimated that 9.3 yen/kWh for a ACC-CAES system of 829 MW.
 From these studies we concluded the ACC-CAES system is fully feasible.

1 INTRODUCTION

Recently, the annual load factor of electric power has a tendency to decrease year by year, dropping to 55.0% in fiscal year 1994. As a result, there is a growing need to develop a new method and technology for storing electricity. At present in Japan, pumping-up power generation is the only method which is commercially available to store electricity. Taking into consideration the limits on the construction sites in the future, efforts to develop new technologies are more than ever indispensable. In response to such needs, intensive research is being advanced to develop new technologies for storing electricity, including superconducting magnetic energy storage, battery electric power storage, compressed air energy storage, flywheel energy storage, etc. Among these technologies, a practical application of compressed air energy storage (CAES) is anticipated in Japan because of its high technical advancement, efficiency and economic advantage. The compressed air energy storage of 290MW was put to practical use in Germany in 1979 and that of 110MW in the United States in 1991, entering into commercial operation in respective countries.

For practical application in Japan, it is important not only to develop technology for storing air energy by water-sealing in hard rocks from the standpoint of economic feasibility, but also to introduce a system that combines the advanced combined cycle (ACC) system with the CAES system from the standpoint of improving efficiency. In this report, the authors discuss the concept of the ACC-CAES system in which both methods mentioned above are applied and at the same time clarify its feasibility.

2 PROPOSAL FOR WATER-SEALING ACC-CAES

In the gas turbine power generation, a large amount of compressed air is required to burn fuel efficiently, and the greater part of fuel which is used for such power generation is spent to produce this compressed air. The objective of CAES is to store compressed air energy at night when the demand for electric power is less and to run gas turbine generators in daytime when there is a high demand for electric power, by the use of compressed air energy stored in the tanks (Figure 1).

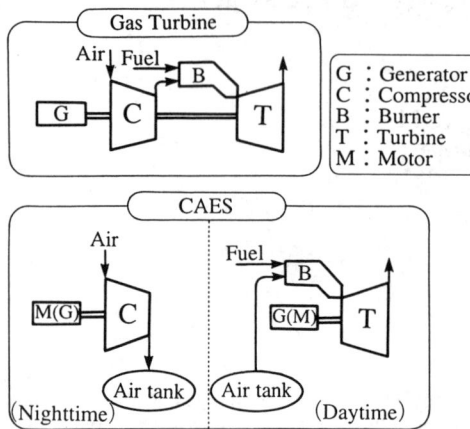

Figure 1 Gas turbine system and CAES system

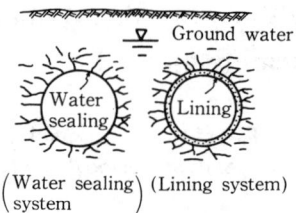

Figure 3 Two types of air tightness

The water-sealing ACC-CAES system proposed herein offers the following two new characteristics in comparison with the existing CAES method used in Europe and America:
(1) A water-sealing method is used to store compressed air energy.

In the CAES method used in Europe and America, salt caverns are employed to store compressed air energy. In Japan, however, there are no salt caverns, so compressed air energy is to be stored in rock caverns (Figure 2). In this case, two methods, i.e., lining method and water-sealing method, may be considered (Figure 3). Comparing both methods, the water-sealing method is more economical. However, the actual storage of compressed air energy by the water-sealing method has been limited to only several atmospheric pressures in the storage of crude oil for example. The pressure required by the CAES system is as high as 30~60 atmospheric pressures. The authors, therefore, conducted on-site tests to identify the water-sealing mechanism in the boreholes, consequently indicating that the water-sealing method is basically feasible even at high pressure, as described below.

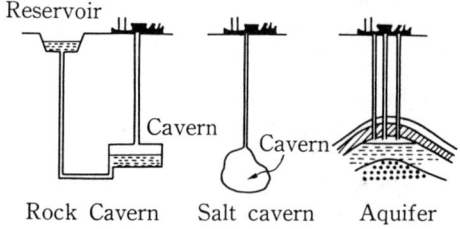

Figure 2 Three types of storing compressed air

(2) Economy and efficiency can be enhanced further by the adoption of ACC.

The conventional CAES systems have been applied only to the heating of compressed air before combustion, by recovering heat wasted from gas turbines through the recuperator. To realize more efficient application, therefore, the method proposed herein is intended to apply the advanced combined cycle (ACC) system, which utilizes heat wasted for steam power generation, to CAES. Furthermore, for application of ACC-CAES, the authors devised a system that permits intermediate-load ACC power generation, and proposed an operational system that can improve economically to a greater degree.

3 VERIFICATION OF WATER-SEALING PERFORMANCE BY IN SITU TEST

When a rock cavern is constructed below the ground water level, underground water oozes out of micro rock cracks toward the cavern walls and the air in the cavern is trapped owing to such oozing water pressure. The water-sealing method is based on this phenomenon. In the rock caverns constructed in Kuji, Kikuma and Kushikino in Japan, petroleum of 5,000,000 kℓ is stored by this method. In these cases, water-sealing pressure is only several atmospheric pressures. In CAES, however, it is necessary to store high-pressure air energy at 30~60 atmospheric pressures. Then, in order to verify the water-sealing mechanism under high-pressure condition in a borehole, testing apparatus was developed and in situ tests were conducted (Nakagawa et al. 1991).

The testing site was located in the Kongo Substation of Kansai Electric Power Co., Inc. in Kawachinagano City, Osaka, where three boring holes were drilled. The deepest hole drilled was 400m below the ground surface. The surrounding rock mass was granodiorite, with rock classification being CH class. The coefficient of permeability was $10^{-4} \sim 10^{-7}$ cm/s. In this borehole, tests were conducted using air-tightness testing apparatus (Figure 4), in accordance with the following procedures:

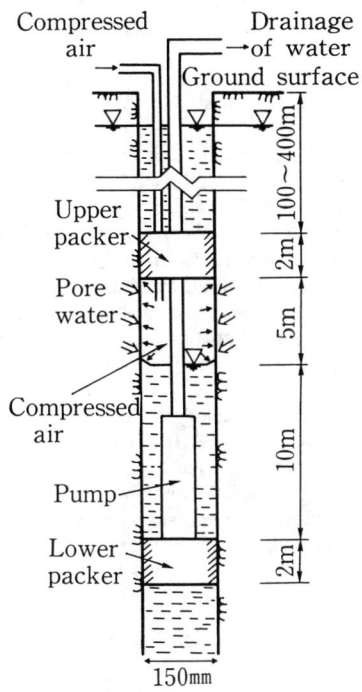

Figure 4 Outline of the air-tightness test apparatus (Nakagawa et al. 1991)

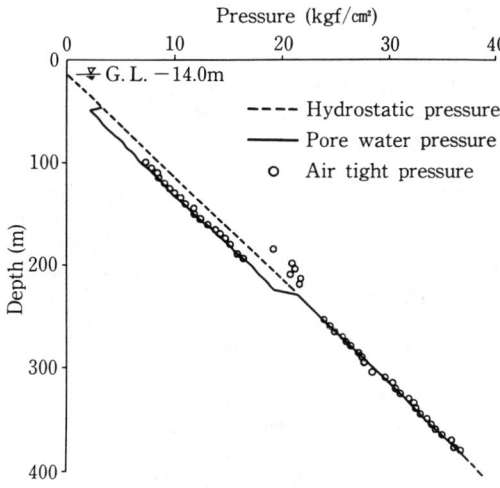

Figure 5 Relationship between air tight pressure and depth (Nakagawa et al. 1991)

(1) Upper and lower packers are installed in the boring hole. (2) Part of groundwater between the two packers is pumped out with a submerged pump and air is introduced. (3) The pressure of the introduced air is raised gradually. (4) The maximum air tight pressure that permits the storage of air energy by water-sealing is examined in accordance with the balance between the pressure of groundwater oozing out of the borehole wall and the pressure of the air. That is to say, the pressure, which causes the air to leak through micro cracks in the borehole wall when the pressure of the air is raised, is the maximum air tight pressure. The test results obtained in this manner revealed that the maximum pressure for the storage of air energy by water-sealing was equal to or greater than the pore pressure in rocks (Figure 5). The depth of the borehole was about 400m and the maximum value of pore pressure was about 40kg/cm^2, thus making it possible to identify the water-sealing mechanism up to about 40 atmospheric pressures.

It is estimated that the actual cavern for the storage of air energy by water-sealing will have a scale of about 8m in diameter, and its length will reach several 100m (Figure 6). Where there are frequent geological composites as in Japan, it is possible that faults, fissured zones or weak-layers

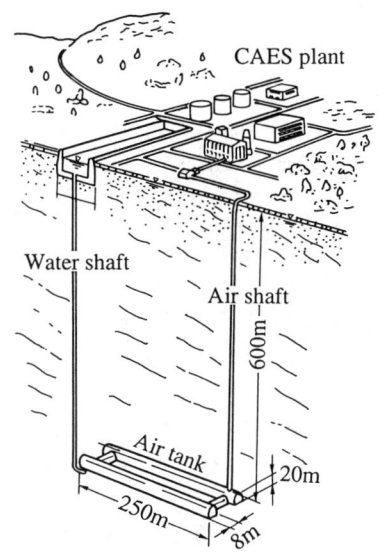

Figure 6 Schematic layout of CAES by water sealing method

exist in the cavern wall surface. As a countermeasure against such faults, fissured zones, etc., replacement or covering by concrete will be necessary. Also, improvements by grouting, etc. will be required for the weak-layers. With regard to the effectiveness of these improvement techniques, there is presently a plan to verify it by on-site test equivalent to full-scale one.

4 ACC-CAES SYSTEM AND ITS OPERATING PATTERN

ACC-CAES is characterized by higher efficiency attained from running the steam turbine with heat recovering from the gas turbine exhaust and by economic advantage attained from operation flexibility so that ACC power generation can be accomplished in the intermediate-load power generation mode as its operating pattern (Moritsuka et al. 1993). In the conventional CAES system, it will also be possible to carry out gas turbine power generation in the intermediate-load power generation mode. However, this method will not be advantageous because fuel efficiency is low. In the case of ACC-CAES, the major attractive point is that large-capacity power generation with output power of 400-800 MW will be possible.

It will be possible to use such gas turbine generator models as G type (made by GE). Items requiring further technical development along with an increase in output capacity include inter-cooler, after-cooler and others, and their technical development will be fully feasible.

5 ECONOMIC FEASIBILITY ASSESSMENT

The economic feasibility of power generation technologies is generally assessed by generating cost. The generating cost is expressed by the sum of fixed cost and variable cost (Uchiyama et al. 1990). The fixed cost is obtained, as a simple method, by multiplying annual fixed rate by construction cost and dividing it by annual generating time. Also, the greater part of variable cost is fuel expense and can be obtained by dividing fuel price by efficiency (equations 1~3).

Studies were conducted in the following cases, i.e., A system (GT 24 by ABB co.) and G system (G type by GE). Plant construction costs in each case are shown in Table 1. Also, for comparison purposes, the pumping-up power generation cost was calculated as follows (equations 4 and 5):

$$\text{Generating cost} = \text{fixed cost (power generating plant + air storage tank)} + \text{fuel cost (nighttime electric power + LNG fuel)} \quad (1)$$

$$\text{Fixed cost} = \frac{[\text{power generating plant construction cost} \times \text{annual fixed rate } \alpha_1 + \text{air storage tank construction cost} \times \text{annual fixed rate } \alpha_2]}{[\text{plant output power} \times \text{annual generating time} \times (1 - \text{plant consumption rate})]} \quad (2)$$

$$\text{Fuel cost} = \frac{[\text{compressor power} \times \text{annual compression time} \times \text{nighttime electric fee} + \text{fuel consumption} \times \text{annual generating time} \times \text{LNG fuel price}]}{[\text{plant output power} \times \text{annual generating time} \times (1 - \text{plant consumption rate})]} \quad (3)$$

$$\text{Fixed cost} = \frac{[\text{plant construction cost} \times \text{annual fixed rate } \alpha_2]}{[\text{plant output power} \times \text{annual generating time}]} = \frac{\text{construction cost (yen/kW)} \times 0.1024}{[8760 \times \gamma]} \quad (4)$$

$$\text{Fuel cost} = \frac{\text{nighttime electric fee (nuclear power)}}{\text{storage efficiency (70\%) in pumping-up power generation}} = \frac{2.56}{0.7} = 3.51 \text{ yen/kWh} \quad (5)$$

where γ is an annual capacity factor.

Table 1 Cost and efficiency of ACC-CAES by water sealing method (in the case of availability being nearly 30%) (Koda et al. 1997)

		System-A	System-G
Capacity	(CAES)	442 MW	829 MW
	(ACC)	245 MW	400 MW
Operating time	(CAES)	7.6 hr	7.6 hr
	(ACC)	6.4 hr	6.4 hr
	(Air compressing)	8 hr	8 hr
Generating efficiency		44 %	41.4 %
Storage efficiency		72.1 %	62.9 %
Cost	(Plant)	35.2 b yen	55.2 b yen
	(Air tank)	13.2 b yen	16.0 b yen
Generating cost		10.1 yen/kWh	9.3 yen/kWh

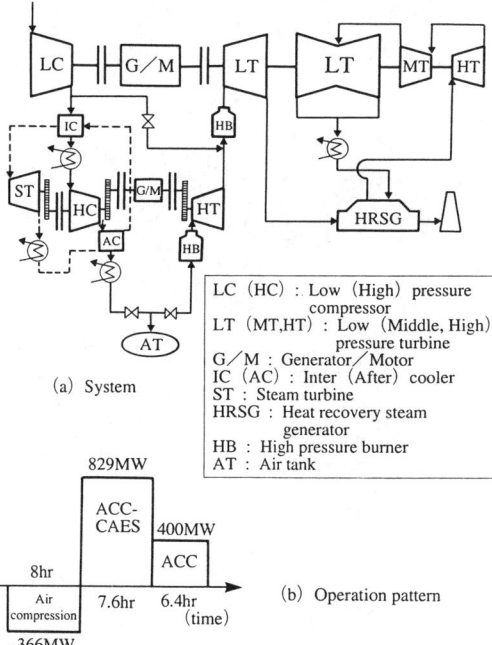

Figure 7 System and operation pattern of ACC-CAES (G system) (Koda et al. 1997)

It is assumed that ACC-CAES is run for 6.4 hours (see Figure 7) in the intermediate-load power by ACC generation mode. The generating cost shows 9~10 yen/kWh (Koda et al. 1997). In pumping-up power generation, also, the generating cost largely varies in the range of 14~28 yen/kWh, depending on the capacity factor. These results indicate that ACC-CAES is very promising as a second-generation electric power storage technology.

6. CONCLUSIONS

The annual load factor of electric power has a tendency to decrease year by year, and it is therefore important to put the compressed air energy storage technology (CAES) to practical use in Japan as a new electric power storage method. CAES has already entered into commercial operation in Germany and the United States. In this paper, the authors proposed a new method in the following aspects as compared with the conventional CAES systems. That is to say,
(1) Compressed air energy is stored by water-sealing in rocks, and
(2) ACC system is combined with CAES to obtain improved efficiency and lower economical cost.

And, with regard to the water-sealing method mentioned in item (1) above, tests were conducted to verify the water-sealing mechanism in a borehole of 400m deep, and it was clarified that "maximum pressure for the storage of air energy by water-sealing is equal to or greater than pore pressure" and at the same time, water-sealing performance up to 40 atmospheric pressure was identified.

Furthermore, assessment was made on the operation of 400-800MW class ACC-CAES, indicating that the generating cost is 9.3~10.1 yen/kWh and its economic advantage is better than that of pumping-up power generation whose cost is 14~28 yen/kWh. Thus, it was clarified that ACC-CAES is feasible in Japan.

This research was carried out as part of Project Research under the cooperation of many researchers, for which the authors would like to express a sincere gratitude.

REFERENCES

Koda E. et al. : Conceptual design of water seal type ACC-CAES power generation system and extraction of development problem, *Central Research Institute of Electric Power Industry (in abbr. CRIEPI) Report*, No.W96503, 1997 (in Japanese)

Moritsuka H. et al. : Study on an integrated compressed air energy storage advanced combined cycle plant, *CRIEPI Report* No.93009, 1993 (in Japanese)

Nakagawa K. et al. : Field tests on air tightness of rock for compressed air storage, *Proc. 7th Cong. ISRM*, pp.135-138. 1991

Uchiyama Y. et al. : Technical assessment and economic study of compressed air energy storage in Japan, *CRIEPI Report* No.Y90002, 1990 (in Japanese)

Stope design in highly fractured area

S.J.Jung & Henry Bogert
University of Idaho, Department of Met. and Min. Engineering, Moscow, Idaho, USA

ABSTRACT: The goal of an efficient mine design is to safely maximize production, grade, and ore recovery while simultaneously minimizing cost, waste, and risk. A good mine design is likely to result when the best available design theory is combined with sufficient geologic information about the ore zone and the surrounding host rock. When the rock data is poor or when the design theory is inappropriate or inadequate, then the success of the underground design becomes less certain.

INTRODUCTION

Over the past century, room-and-pillar mine design theory has been the subject of significant effort toward improvement. The end product has been various design methods based on a diverse assortment of pillar dimensions, rock strengths, or rock mechanics theories. Thus, when several commonly accepted pillar design methods are applied to the same design problem, the solutions can vary significantly. Furthermore, most methods were designed for underground coal mining. Little research has been directed towards the wide spectrum of hardrock pillar mining conditions.

In room-and-pillar mining, a portion of the ore zone is purposely left in place for roof support. The financial value of the ore contained in pillars is lost. Ore recovery can be increased by spacing pillars further apart, making them smaller, or robbing them on retreat. However, the increased recovery is obtained at the expense of safety and cost. As roof spans are increased, the stope becomes more hazardous to miners. If the roof span is too large, the stope will collapse or the expense of support will exceed the cost of another pillar.

In highly fractured or folded hardrock deposits, rock strength is usually very high, but the rock mass is seriously weakened by the presence of folding and the associated fractures. Stope size is often small and limited to a few hundred meter by the sporadic nature of the ore zones. While some coal pillar theories may be applicable in hardrock, most coal pillar design methods are formulated for pillars with width-to-height ratios over one. In hardrock mines, pillar width-to-height ratios less than one are common.

Design in tightly folded metallic hardrock ore zones is further complicated by the fact that drill hole spacing is often not close enough to provide adequate fold and fracture data prior to mining. Therefore, it is not possible to measure individual fracture patterns and include them in geotechnical design.

Determining the load placed on pillars is even a more elusive problem. Several pillar loading theories have been developed. The tributary area theory applies well in extensive horizontal, massive bedded deposits. However, in small stopes, tributary loading theory is an overly conservative approach. The pressure arch theory is a bit more generic, but it is very difficult to quantify the geometry and shape of the arch. Arch characteristics in rock with high fold and fracture intensity are likely to be significantly different than arching in solid, homogenous rock formations.

After a century of advances in rock mechanics and underground mine design, we still lack sufficient technology to confidently maximize efficiency in opening designs in severely folded and fractured rock. Trial and error is often the most accurate design method.

PREVIOUS PILLAR DESIGN WORK

The dimensions of laboratory test specimens are usually significantly less than the spacing of discontinuities in the rock mass. Laboratory tests measure the strength properties of a flawless piece of the rock substance; they do not measure the properties of the rock mass. Pillars on the other hand are large and include discontinuities such as bedding planes, fractures, and changes in the rock structure. Thus pillar design methods based on laboratory strength test data must include some means of adjusting the laboratory test data to accurately reflect pillar strength.

Some pillar design researchers (Obert, et al, 1960; Morrison, 1970) have chosen to apply a deflator to laboratory strength results. The deflator is referred as a safety factor, and is usually based upon comparisons of calculated strength with observed strength. As such, it is a rather subjective final step to an otherwise objective approach.

Other researchers (Bieniawski, 1984) have tested the uniaxial compressive strength of specimens of various sizes and have determined that as progressively larger rock specimens are tested, a "critical size" is reached where the specimen strength is constant for specimens of any size greater than the critical size. Several researchers testing coal have shown that uniaxial compressive strength in coal specimens over 0.9 m in side length is constant, regardless of specimen size (Hustralid, 1976). Other size effect tests have shown that in iron ore (Jahns, 1966) and in quartz diorite (Pratt et al, 1972) strength is also constant in samples over 0.9 m in side length. From these tests (Bieniawski, 1984) determined that the critical size for most rock types is about 0.9 m. Thus, according the critical size theory, pillars of the same rock type and over 0.9 m in size have the same strength regardless of pillar size.

The mechanism of failure in a rock specimen is highly dependent upon the ratio of the width of the specimen divided by the length of specimen measured parallel to its loading axis. Likewise, the failure mechanism in a pillar is also dependent upon the width-to-height ratio. Massive rock pillars with a high width-to-height ratio almost invariably fail by spalling and forming an hour-glass shape when overloaded (Brady and Brown, 1985). Failure begins as small surface spalls and continues until the cross section of the pillar is reduced enough to bring on total failure. Overloaded pillars with a low width to height ratio are prone to fail in inclined shear planes. Furthermore, the presence of regular jointing, high angle fractures, or weak bedding planes can also seriously effect the pillar failure mechanism.

GEOLOGY

The Mineral Hill Mine, near the Jardine Montana, is geologically set in the South Snowy Block of the Bear Tooth Mountains. Regionally, the Jardine area geologic structure consists of over 600 vertical meter of Archean aged metasedimentary deposits intruded by granitic stocks and occasional dikes and sills. With the exception of a single iron formation, the Jardine metasediments are schists characterized by varying proportions of quartz, biotite, and chlorite. Quartz-biotite schist is the most abundant rock type.

Ore at the Mineral Hill is found in quartz lenses and in a banded ironstone formation. The major portion of ore production comes from the ironstone stopes; however, the smaller quartz stopes can produce very high grade ore. The quartz ore zones are hosted in quartz-biotite schist formations and tend to be tabular with dimensions on the order of 30 to 90 m, averaging .3 to 2.4 m in thickness. Ore zones in the banded ironstone formation usually dip steeply and can be up to 270 m long by 240 m high with thickness averaging 4.5 m. Due to a very intense folding, the inclination of stopes in either formation can vary from horizontal to vertical.

The mines uses a room and pillar mining methods in the gently dipping areas and a mechanized cut and fill method in the steeply sipping areas. The mine produces approximately 600 tons of gold ore per day. Room and pillar stopes are usually mined leaving random pillars with a width of 3 m, though some are up to 9 m wide. Pillars are left where ore grade is poor and when roof spans between pillars become excessive. Upon retreat, some of the larger pillars are slabbed and others are removed entirely. During retreat mining, fill is often stowed between the remaining pillars for extra support.

LABORATORY TESTING

In the field of rock mechanics, laboratory testing provides an opportunity to learn more about the properties and behavior of a rock type by testing and observing a representative sample of the rock. The advantages of laboratory testing over field measurements are that the conditions of the test can be controlled, the duration of the test is shorter, and the method of measurement is simpler. Furthermore, the experiments can be tailored to isolate the desired information while holding other variables constant. By use of precise instrumentation and careful observation, valuable information can be gained from lab experiments.

The tensile strength of intact quartz-biotite schist (QBS) is tested and calculated by the Brazilian method. The uniaxial compressive strength, Young's modulus, and Poisson's ratio of intact QBS are experimentally determined by tests on QBS cubes and cylinders. The cohesion and the angle of internal friction of intact QBS are tested in a series of triaxial tests. The effects of varying the specimen width-to-height ratio are examined for the rock substance as well as the rock mass by uniaxial compressive strength tests of 10 square cm QBS blocks of different heights.

During the process of developing the test procedures and testing, over 250 specimens are cut and tested from 450 kg of QBS samples taken from stope of the Mineral Hill Mine.

Stress/Strain Tests on Intact Rock

Axial and diametric strain are measured in quartz-biotite schist NX core cylinders and in quartz-biotite schist cubes while the specimens are gradually loaded uniaxially in compression. The cubes each have horizontally and vertically oriented resistance wire strain gages glued to their sides. Micro Measurements' CEA 350 ohm gages with a 2.08 gage factor are used. Gage resistance measurements are reduced to units of strain with a Shinkoh Digital Strain Indicator. Strain and load are measured and recorded at approximate stress increase intervals of 2.7 MPa for the duration of the tests. Stress/strain curves for the cubes are plotted in Figure 1. The Young's modulus of the linear portion of the curves averages between 62 GPa and 69 GPa, while the Poisson's ratio varies from 0.24 to 0.26.

Uniaxial Compressive Strength Tests with Varied Width-to-Height Ratios

Traditional pillar design begins with laboratory strength tests of nearly perfect rock specimens. Then the laboratory rock strength is deflated by a formula that includes scaling factors for pillar geometry and size (Hustralid, 1976). In homogenous, unfractured rock formations, such testing and calculations might be appropriate. However, the quartz-biotite schist at Mineral Hill is so severely folded and fractured that rock substance strength plays a relatively minor role in geotechnical design. No cases of failure caused by stress exceeding rock substance strength have been observed in the mine. Instead roof and pillar stability are controlled by pre-existing fractures and folds. This series of laboratory tests is designed to include the affects of fractures in the testing program.

Quartz-biotite schist samples from the Mineral Hill Mine are sawed into 10 cm square specimens of various heights ranging from 2.5 to 20 cm. Some specimens contain no visible fractures; others

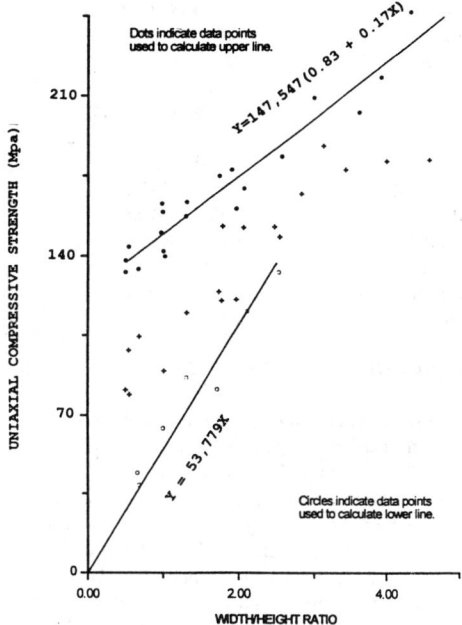

Figure 2. QBS strength vs. width/height ratio

purposely contain as many as four visible fractures oriented at random angles. After surface preparation, the specimens are uniaxially loaded to failure at a rate of 14 MPa per minute. Uniaxial compressive strength is plotted against width-to-height ratio in Figure 2.

As testing progresses, two acceptable failure modes are observed. The specimens with minor pre-existing fractures or no pre-existing fractures fail violently. Specimens with extensive pre-existing fractures fail less violently, with failure following and bridging between the existing fractures. The different width-to-height ratio test results (Fig. 2) fall within a window bounded on the upper edge by the strength of solid, flawless specimens and bounded on the lower edge by the strength of fractured, flawed specimens that are more likely to be representative of actual pillar conditions and the entire rock mass.

The points defining the upper boundary of the data set form a reasonably straight sloping line. This line represents the uniaxial compressive strength of flawless specimens and is in fairly closely agreement with the Obert/Duvall equation. However, the strength of the specimens with pre-existing fractures provides a more useful curve for mine design purposes. The lower boundary of the data set (Fig. 2) appears to be linear for specimens with a width-to-height ratio of less than two and also passes through the origin. Thus, unlike the Obert/Duvall type equations, this curve is more realistic for use in the design of very slender pillars. The lower boundary for specimens with width-to-height ratios greater than two

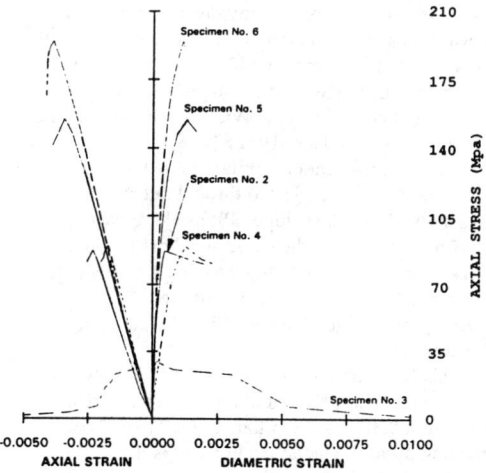

Figure 1. QBS specimens in uniaxial compression

apparently begins tangent to the linear portion of the lower boundary, and the slope decreases until the curve approaches a horizontal line.

Table 1. Rock properties of Quartz-Biotite Schist

Density	2,755	Kg/m^3
Tensile strength	17	MPa
Uniaxial compressive strength	155	MPa
Young's modulus	62	GPa
Poisson's ratio	0.23	
Internal friction angle	23	Degree
Cohesion	51	MPa

UNDERGROUND DATA COLLECTION

In this research, underground data collection is the process of observing and quantifying rock type, fracture conditions, and rock mass behavior, along with the deformation of underground openings and the stress changes that occur during excavation. However, only a tiny fraction of the rock mass is available for observation and measurement at any one time. Drill hole data and instrumentation allow us to "see" further into the rock mass. But still the data available is minuscule compared to the problem of underground mine design.

Several rock mass classification systems exist and provide good methods for quantifying the geologic condition of a rock mass. A RQD contour map of the 750/500 B stope and the "Hockey Stick" area, based on the RQD calculated from exploration drill hole core. The map uses the RQD from the ore zone intercept where the drill hole passed through the ore. Admittedly the RQD at an ore zone penetration does not guarantee the quality of the rock meter away; however, the trend established by many drill holes across a stope can establish a trend. The purpose of the contour map is not identify the RQD at an exact point, but rather to identify the general change in fracture intensity throughout the stope. The ore zone data provides a measure of pillar quality, and the hanging wall data provides a measure of ground support and roof span expectations.

Development to the north out of drift No. 1 in the 750/500 B stope begins in July of 1994. As the drift advances further north, progressively worse ground conditions are encountered. Upon reaching the 49,125 north coordinate, the structure rolls to a 45 degree dip, and the drift has to be turned abruptly left to maintain grade. Attempts are made to drift to the left and initiate room-and-pillar stopes, but poor ground conditions hinder progress. RQD at the face is estimated between 36 and zero at various locations in the vicinity of pillar "H" (Fig. 3) using the following Rule of Thumb for estimating RQD underground:

RQD = 100 - (no. of principle fractures/meter) x 4

Both core and visible estimations of RQD indicate that ground conditions near 49000 north coordination are very weak to maintain openings as well as pillars.

During the early stages of retreat mining, when the ribs in drift No. 1 and drift No. 2 are being slashed, the stress meter in pillar No. 6 shows a 2 MPa increase(Fig. 3). The Young's modulus of pillar No. 6 averages about 5.5 GPa. As pillar recovery continues, the pillar No. 6 meter registers a further 34 MPa stress increase, and the Young's modulus increases to 29 GPa. When mining in the stope ceases, the pillar apparently continues to creep with minimal stress increase. Renewed activity in the stope causes the measured stress change to peak at 39 MPa. Since peaking, the pillar appears to be going into a slow failure mode as it continues deformation while stress decreases. Finite element modeling of the 750/500 B stope indicates the existing vertical stress on pillar No. 6 is about 4.8 MPa when the stress meter was installed. Thus a reasonable estimate of absolute stress at failure is 44 MPa.

During the mining sequence in the 750/500 B stope, pillars No. 1, 2, and 3 begin visible movement along pre-existing fractures. Stress and strain are not measured in these pillars. However, finite element modeling indicates that the stress on pillar No. 3 at failure is less than 23 MPa.

Stress and strain in pillar No. 7 are also being tracked concurrently with pillar No. 6. Pillar No. 7 incurs only a slight load increase during the pillar robbing sequence. After mining in the 750/500 B ceases, pillar No. 7 sheds about 20 percent of the load it acquires during retreat mining. During the entire mining cycle, the Young's modulus for pillar No. 7 is fairly constant and averages 6.9 GPa, and the stress/strain data indicates elastic behavior. The stress profile data gathered from pillar No. 7 and other pillars elsewhere in the mine are very similar. The trend of the measurements strongly indicates that the outer one meter of the pillar provides less long-term load bearing capacity than the inner core. This observation is in agreement with the well established theory in wide pillars where the outer band of the pillar fails, but continues to provide confining pressure for the inner core (Wilson, 1972).

Stress/strain data for pillar No. 6 and pillar No. 7 is successfully collected throughout retreat mining in the 750/500 B stope. Pillar No. 6 bears the brunt of roof support in the stope as pillars to the north and east of it are removed; thus an exceptionally high load is placed on it. Conversely, pillar No. 7 is shielded by pillars No. 4 and No. 6 and incurs minimal load increase.

Underground measurements of Poisson's ratio for pillars at Mineral Hill yield highly variable results ranging from 0.38 in solid pillars to well over 2.5 cm in creeping pillars. The apparent trend is that Poisson's ratio increases as the amount of creep increases.

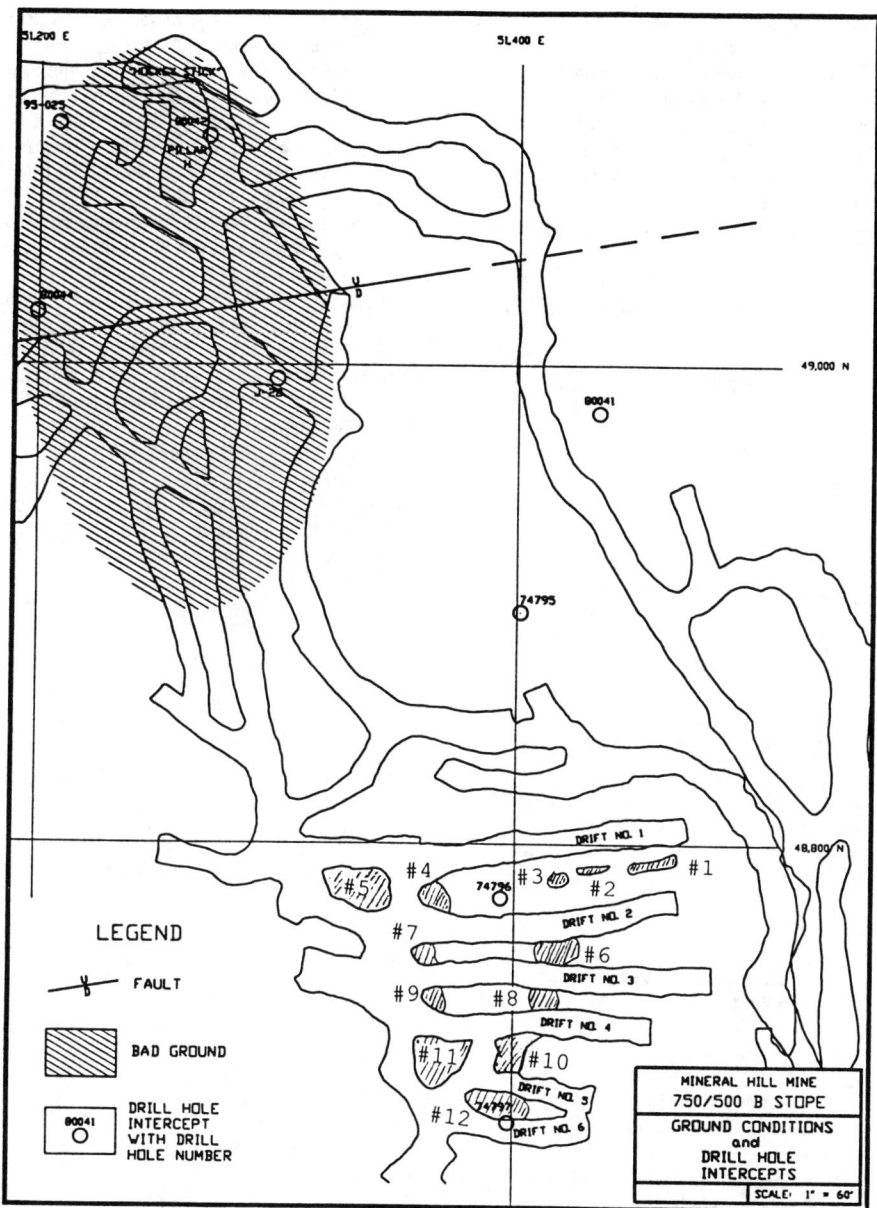

Figure 3. 750/500 B stope map (# pillar number)

Pillar strength and pillar stress are difficult to measure and predict, and even difficult to define. They are dependent upon time, pillar size, and geometry, rock mess strength and many other factors. Further complicating the problem, pillar strength and pillar stress are not constant throughout the volume of a pillar.

The definition of pillar strength could conservatively be defined as the point where a pillar stops behaving elastically, a good definition for long term stability. On the other hand, yield pillars support loads long after departing from elastic behavior. In highly fractured rock at the Mineral Hill Mine, pillar failure is not catastrophic. Movement along fractures begins long before the rock substance fails.

Perhaps the most important conclusion to be

drawn from this study into room-and-pillar mining is that the intensity of fracturing determines pillar strength when the rock substance is strong and the fracture intensity is high. To illustrate correlation between fracture intensity and strength of pillar, RQD and average pillar strength are obtained from underground stress measurement, numerical analysis and laboratory tests. The RQD and average pillar strength for each data point is summarized in the Table 2.

Table 2. Rock Quality Designation and Average Pillar Strength Data

Source	RQD	Average Pillar Strength (MPa)
Pillar No.3	50	13
Pillar No.6	55	19
Pillar No. 8 to 9	60	25
Pillar H	24	2
Lab. Specimen	100	82

The average pillar strength of nearly flawless quartz-biotite schist is estimated from laboratory specimen. Uniaxial compressive strength tests on flawless cubes of various size generate the following equation;

Uniaxial Comp. Strength, MPa
    = (-2800ln (cube, in) + 25,100)x 6894

From the equation, the strength of 0.9 m cubic pillar with RQD equal to 100 is estimated at 82 MPa. The log/log curve fit to the data in Table 2 represent the following equation;

Average Pillar Strength, MPa = $e^{[2.49\ln(RQD)-2.06]}$ x 6894

and can be used to estimate the average pillar strength at the failure in quartz-biotite schist at the Mineral Hill Mine

CONCLUSIONS

The fracture intensity is the most influential variable in rock mass behavior.
RQD is found to be an excellent indicator of rock mass quality in this study.

REFERENCE

Bieniawski, Z.T., 1984, *Rock Mechanics Design in Mining and Tunneling*, Balkema, Rotterdam, 272 p.

Brady, B.H.G and Brown, E.T.,1985, *Rock Mechanics for Underground Mining*, Chapman & Hall, New York, 569 p.

Hustralid, W.A., 1976, "A Review of Coal Pillar Strength Formulas," *Rock Mechanics*, Journal of the International Society for Rock Mechanics, Vol. 8, No. 2, pp. 115-145.

Jahns, H., 1966, "Measuring the Strength of Rock In Situ at an Increasing Scale," *Proc., Ist Int'l Congress on Rock Mech.*, ISRM, Lisbon, Vol. 1, 1966, pp. 477-482.

Morrison, R.G.K., 1970, *A Philosophy on Ground Control*, Ontario Dept. of Mines, Toronto, pp 115-118.

Obert, L., Duvall, W.I. and Merrill, R.H., 1960, *Design of Underground Openings in Competent Rock*, USBM Bul. 587, 36 p.

Pratt, H.R., Black, A.D., Brown, W.S. and Brau, W.F., 1972, "The Effect of Specimen Size on the Mechanical Properties of Unjointed Diorite," *The Intl. Jour. of Rock Mech. and Mining Sci*, pp 513-529.

Wilson, A.H., 1972, "Research into the Determination of Pillar Size, Part 1," *The Mining Engineer*, The Institution of Mining Engineers, Vol. 131, No. 141, pp. 409-417.

A study on the design of the shallow large rock cavern in the Gonjiam underground storage terminal

Eui-Seob Park & Ho-Yeong Kim
Sunkyong Engineering and Construction Ltd, Seoul, Korea

Hi-Keun Lee
Department of Mineral and Petroleum Engineering, Seoul National University, Korea

ABSTRACT: The storage cavern excavated horizontally underneath mountain is generally located at a depth of 30~70m and reqiures for careful site investigation, accurate analysis because of its large width and height. As the GONJIAM underground storage terminal is the first underground cold storage project in Korea, new methods unlike conventional investigations and analyses were introduced. This study discussed emperical, analytical and numerical methods used to analyze the stability of caverns in the base of site investigation results. These results will be verified and revised through the analysis of monitoring and face mapping data collected during the construction of underground storage caverns.

1. INTRODUCTION

For the construction of underground cold storage for foodstuffs, shallow large rock caverns are excavated. For the economic reason, the width of the cavern is to be as large as possible. With minimized rock reinfocement, normally rockbolts and shotcrete, the practical largest size has to be proved in the design stage. Therefore, the careful geological investigation is also required. As the first underground cold storage in Korea, Gonjiam underground storage terminal was planned in 1995. To improve the geological investigation, some new methods were introduced as follows;
- High resolution electrical resistivity (LUND System)
- Seismic tomography
- Borehole ultrasonic televiewer
- Hydraulic fracturing in-situ stress measuremet

With the integrated data from site investgation, in-situ properties of rock mass were carefully evaluated, and finally the possible largest size of rock caverns were studied by numerical experiments.

2. DETERMINATION OF CAVERN WIDTH

Because the dimension of cavern is varied with the characteristics of rockmass, its width is determined as 10, 12, and 15m from the early investigation results.

2.1 Rock Mass Classification

The rock mass of this site is classified as very poor~good by the emperical method such as RMR and Q-System, and most of it is fair and good. But, as the stability of cavern is governed by poorer rock mass, The borehole G-4 was chose out of 6 boreholes, which represents poorer rock condition, and classified according to depth.

Table 1. results of classification (Borehole G-4)

Rock Mass	Depth (m)	Q	RMR	Deformation Modulus (GPa)	Poisson's Ratio	Cohesion (KPa)	Friction Angle (°)
Fair	25	10.3	58.7	17.4	0.23	300	25
Poor	30	1.6	47.2	8.51	0.23	250	30
Very Poor	35	0.11	38.9	5.28	0.23	200	35

The rock mass is classified as very poor, poor, fair by Q-System. The properties of each rockmass such as deformation modulus, cohesion, friction angle were calculated from RMR and Q values.

2.2 Emperical and Analytical Method

To study the stability of cavern, the emperical and analytical methods were used.

2.2.1 Calculation of rock load

The rock load was calculated from various emperical equations as to rock mass conditions.
1) RMR (Unal, 1983)

① $P_s = \dfrac{(100 - RMR)}{100} \gamma B$ (t/m²)

2) Q-System (Barton et al, 1974)

① $P_{roof} = \dfrac{20}{J_r} Q^{-1/3}$ (t/m²)

② $P_{roof} = \dfrac{20}{3} J_n^{1/2} J_r^{-1} Q^{-1/3}$

 : When J_n is less than 3

③ $P_{roof} = \dfrac{4D}{J_r} Q^{-1/3}$ (Bhasin et al, 1995)

 : Considering tunnel diameter when rock mass is poor

where, γ : rock density(t/m³), B: width(m), J_r: joint roughness number, J_n: joint set number, D: tunnel diameter(m)

Table 2 shows the calculation results of rock load. Since each equation requires different types of input data, the range of rock load calculated is widely varied, these results have to be used carefully.

Table 2. Calculation results of rock load
unit : T/M²

Width	Rock mass	RMR ①	Q-System ①	②	③
10 m	Fair	10.7	9.2	7.5	18.4
	Poor	13.7	5.7	6.6	11.4
	Very Poor	15.9	20.9	24.1	41.7
12 m	Fair	12.9	9.2	7.5	22.1
	Poor	16.5	5.7	6.6	13.7
	Very Poor	19.1	20.9	24.1	50.1
15 m	Fair	16.1	9.2	7.5	27.6
	Poor	20.6	5.7	6.6	17.1
	Very Poor	23.8	20.9	24.1	62.6

2.2.2 Calculation of support pressure

Table 3 shows the support pattern used for Gonjiam underground storage cavern. Main supports used are rock bolt and shotcrete. As the rock mass is poor, the spacing of rockbolt is reduced, and the thickness of shotcrete is increased.

Table 3. Support pattern

Rock mass	Rockbolt	Shotcrete
Fair	L=4m, S=2m×2m	100 mm
Poor ~ Very Poor	L=4m, S=2m×2m	150 mm
Very Poor	L=4m, S=1m×1m	200 mm

When rock mass is very poor, two types of support is suggested. The support pressure of each case is calculated, summed up it. Grouted rock arch effect is considered when rockmass is poor.

1) Shotcrete
 ① Semi-emperical Method(B.Singh et al, 1995)
$$P_{s/c} = \dfrac{2q_{s/c} \cdot t}{F_{s/c} \cdot B}$$
 ② Norwegian Method (K.F.Garshol, 1993)
$$P_{s/c} = \dfrac{\sigma_{s/c} \cdot t}{r}$$
 ③ Hoek & Brown (1980)
$$P_{s/c} = \dfrac{1}{2} \sigma_{s/c} \left[1 - \dfrac{r^2}{(r+t)^2}\right]$$
 ④ NATM
$$P_{s/c} = \dfrac{2t \times 0.43 \sigma_{s/c}}{2r \cos(\pi/4 - \phi/2) \sin 30}$$

where, $q_{s/c}$: shear strength of shotcrete, t: thickness of shotcrete, $F_{s/c}$: mobilization factor of shotcrete, $\sigma_{s/c}$: compressive strength of shotcrete, r: tunnel radius, B: tunnel width, ψ : friction angle

2) Rockbolt
 ① Semi-emperical Method(B.Singh et al, 1995)
$$P_{R/B} = \dfrac{2q_{crm} \cdot l' \cdot \sin\theta}{F_s \cdot B}$$
 ② Norwegian Method (K.F.Garshol, 1993)
$$P_{R/B} = \dfrac{F_{r/b}}{a \cdot b}$$

where, q_{crm}: uniaxial compressive strength of reinforced rockmass, l: effective thickness of reinforced arch, F_s: mobilization factor of rockbolt, $\sin\theta \simeq 1.3 B^{-1.6} \leq 1$, $F_{r/b}$: allowable load, a & b: spacing of rockbolt

3) Grouted rock arch
 ① Semi-emperical Method(B.Singh et al, 1995)
$$P_{gt} = \dfrac{2q_{gt} \cdot l_{gt} \cdot \sin\theta}{F_{gt} \cdot B}$$

where, q_{gt}: uniaxial compressive strength of grouted rockmass, l_{gt}: effective thickness of grouted arch, F_{gt}: mobilization factor of grouted arch, $\sin\theta \simeq 1.3 B^{-1.6} \leq 1$,

4) Results

Table 4 shows the results of each support pressure by the emperical, analytical equations. The support pressure of shotcrete is greater than that of rockbolt. The ground improvement by reinforcement and the support pressure by formation of compressive zone also show large values. In the case of shotcrete, the support pressure by Semi-emperical Method (B. Singh et al., 1995) is smaller than that of other equations. Considering that each support material gives different effects as to the characteristics of rock mass, construction method, and so on, it is difficult to rely wholly on the results gained by these equations. But, these equations give not the qualitative estimation of support effect but the basis of quantitative judgement on support.

Table 4. Results of each support pressure
unit : T/M^2

Width	Rock Mass	Shotcrete				Rockbolt		Arch effect
		①	②	③	④	①	②	①
10 m	Fair	16.2	42.0	40.8	40.7	3.1	2.5	23.0
	Poor	23.8	63.0	60.3	64.2	2.0	2.5	22.0
	Very Poor	23.8	63.0	60.3	68.3	1.2	2.5	34.6
	*	31.3	84.0	79.2	91.1	5.2	10.0	34.6
12 m	Fair	13.5	35.0	34.1	33.9	2.5	2.5	18.6
	Poor	19.9	52.5	50.6	53.5	1.7	2.5	17.8
	Very Poor	19.9	52.5	50.6	56.9	1.0	2.5	28.0
	*	26.1	70.0	66.6	75.9	4.2	10.0	28.0
15 m	Fair	10.8	28.0	27.4	27.1	1.9	2.5	14.4
	Poor	15.9	42.0	40.8	42.8	1.3	2.5	13.7
	Very Poor	15.9	42.0	40.8	45.5	0.7	2.5	21.6
	*	20.9	56.0	53.8	60.7	3.2	10.0	21.6

* : Increase support requirement when rockmass is very poor (Rockbolt; length is 4m, spacing is varied from 2m×2m to 1m×1m. Shotcrete; thickness is increased from 150mm to 200mm)

2.2.3 Safety Factor

Table 5 shows the calculation results of safety factor by comparing the required support pressure with the design support pressure. In the case of 10m width, no stability problem occurs for all rockmass. But, in the case of 12m width, the results were different as to support pattern when rockmass is very poor. In the case of 15m width as suggested at the first design stage, it is known that the cavern of 15m width is unstable for all rock mass in this region.

Table 5. Calculation results of safety factor

Width	Rock mass	Required support pressure		Design support pressure		Safety factor		Remarks
		Min.	Max.	Min.	Max.	Min.	Max.	
10 m	Fair	7.5	18.4	41.7	67.5	2.27	9.00	
	Poor	5.7	13.7	47.8	88.7	3.49	15.6	
	Very Poor	15.9	41.7	59.6	105.4	1.43	6.63	
	*	15.9	41.7	71.1	135.7	1.71	8.53	
12 m	Fair	7.5	22.1	34.6	56.1	1.57	7.48	
	Poor	5.7	16.5	39.4	73.8	2.39	12.95	
	Very Poor	19.1	50.1	48.9	87.4	0.98	4.58	Unstable
	*	19.1	50.1	58.3	113.9	1.16	5.96	
15 m	Fair	7.5	27.6	27.1	44.9	0.98	5.99	Unstable
	Poor	5.7	20.6	30.9	59.0	1.50	10.35	
	Very Poor	20.9	62.6	38.2	69.6	0.61	3.33	Unstable
	*	20.9	62.6	45.7	92.3	0.73	4.42	Unstable

* Increse support when rockmass is very poor

2.3 Numerical Analysis

The numerical analysis was carried out to estimate the adequacy of cavern's shape and support design. The program used for analysis is FLAC Ver 3.22, which is adequate for elastoplastic model and support analysis.

2.3.1 Modelling

Table 6 gives the dimensions of cavern for analysis. Since the shape and layout of cavern are not determined yet, this modelling does not agree with the final design. To study the stability of rock pillar, two caverns are excavated in series.

Table 6. Dimensions for Analysis

Case	Width	Height	Spacing
1	10 m	8 m	10 m
2	12 m	8 m	12 m
3	15 m	8 m	15 m

As this analysis was carried out before measuring the in-situ stress of this site, it is assumed that the ratio of the average horizontal to vertical stress is 1.5 because of shallow cavern depth. The support used is rockbolt and shotcrete, the length of rockbolt is 4m, its spacing is 2m, and thickness of shotcrete is 100mm.

2.3.2 Results of analysis

1) Case 1
When rockmass is very poor, the max. displacement is about 4mm, the extent of plastic region is larger than the length of rockbolt, and the resultant stress in shotcrete is near the allowable value. So more support is needed. When rockmass is poor and fair, the results of analysis show that cavern is safe.

Table 7. Analysis results of Case 1

Rock mass	Plastic region (m)		Max. Displacement (mm)		Axial force of rockbolt (ton)		Stress in Shotcrete (kg/cm^2)	
	L	R	L	R	L	R	L	R
Very Poor	A:3-5 W:3-4	A:2-4 W:3-4	4	3.6	2.8	2.6	56 -76	39 -61
Poor	A:2-3 W:2-3	A:1-2 W:2-3	1.7	1.8	1.2	1.0	28 -36	24 -32
Fair	A:1-2 W:1	A:1.5 W:1	0.7	0.8	0.39	0.36	12 -15	13 -16

Note) L, R: Left & right cavern
A, W: Arch & Wall of cavern

2) Case 2
When rockmass is very poor, the extent of plastic region and the resultant stress in shotcrete are larger than the allowable values. It is thought that this unstability can be solved by the additional support, division of excavation region, etc. When rockmass is poor, the total stability is ensured though the extent of plastic region of left cavern exceeds the length of rockbolt. When rockmass is fair, the cavern is stable without additional support.

Table 8. Analysis results of Case 2

Rock mass	Plastic region (m)		Max. Displacement (mm)		Axial force of rockbolt (ton)		Stress in Shotcrete (kg/cm^2)	
	L	R	L	R	L	R	L	R
Very Poor	A:4-7 W:3-4	A:4-5 W:3-4	4.3	3.7	2.9	2.6	62 -86	46 -70
Poor	A:3-4 W:1-2	A:2-3 W:2-3	1.7	1.8	1.3	1.0	32 -40	30 -38
Fair	A:1-2 W:1-2	A:1-2 W:1-2	0.7	0.8	0.39	0.34	14 -17	16 -19

3) Case 3
When rockmass is very poor, the extent of plastic region and the resultant stress of shotcrete exceeds the allowable values. It is thought that the stability of cavern is not ensured. When rockmass is poor, the extent of plastic region at arch is 4~5m, and the stability of pre-excavated cavern is affected by the excavation of new cavern. So the stability of cavern is not ensured. When rockmass is fair, the cavern is stable without the additional support.

Table 9. Analysis results of Case 3

Rock mass	Plastic region (m)		Max. Displacement (mm)		Axial force of rockbolt (ton)		Stress in Shotcrete (kg/cm^2)	
	L	R	L	R	L	R	L	R
Very Poor	A:4-9 W:3-4	A:4-7 W:3-4	4.8	4.4	3.1	3.0	70 -102	53 -87
Poor	A:4-5 W:1-2	A:3-4 W:2-3	2.0	2.1	1.3	1.3	33 -43	34 -46
Fair	A:1-2 W:1-2	A:1-2 W:1-2	0.8	0.9	0.43	0.41	15 -19	17 -21

2.3.3 Disscusions

It is concluded that the stability of cavern was very related to rockmass conditions. In the case of 10m width, cavern is stable for very~fair rockmass with the use of rockbolt and shotcrete. However, It is thought that in the case of 12m width, the cavern is unstable for very poor rockmass. and in the case of 15m width, cavern is unstable for poor~very poor rockmass.

3. STABILITY ANALYSIS OF CAVERN

The layout was determined that the cold room of 12m width, 10m height in two rows and the chilled room of 12m width, 8m height in two rows are excavated parallel by reflecting the intermediate results of geological investigation, condition of loading and automatic facility, etc (Fig 1). Therefore, the detailed study was carried out for 4 caverns.

3.1 Property of rockmass

The representing rockmass of the site is classified as RMR=40, RMR=60. To determine the properties of rockmass for analysis, the results of rockmass classification and lab. test of cores were used. The deformability of in-situ rock was calculated from the various emperical equations suggested by Barton, Bieniawski, Serafin, at al. The characteristics of strength such as cohesion, friction angle and tensile strength were determined from Hoek & Brown failure criterion by using the result of triaxial compressive test, and varies as to the disturbance of rockmass by the effect of blasting. Table 10 shows the estimated rock properties for numerical analyses.

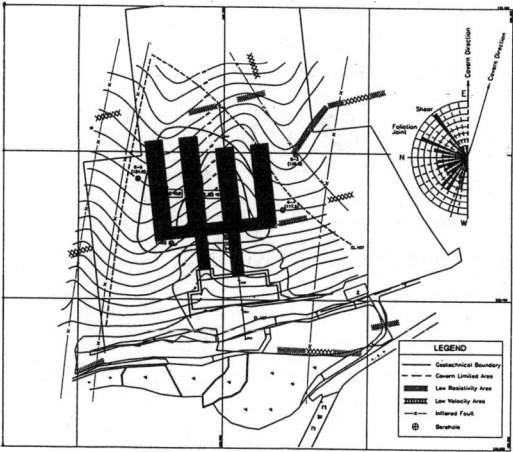

Fig 1. The layout of storage cavern

Table 10. Properities of rockmass

RMR	Distur bance	Elasticity		Strength		
		Deforma bility (GPa)	Poission's Ratio	Cohesion (MPa)	Friction Angle (°)	Tensile Strength (MPa)
40	○	4.5	0.22	0.223	30.5	0.036
	×	4.5	0.22	1.073	43.9	0.12
60	○	9	0.22	0.827	38.9	0.243
	×	9	0.22	2.176	46.8	0.539

3.2 Modelling

In this modelling, the ratio of the average horizontal to vertical stress was determined as 3.0 by the results of hydraulic fracturing method at borehole G-5.

3.3 Conditions of support and load

The support used are rockbolt and shotcrete, and the length of rockbolt is 4m, its spacing is 2m, shotcrete is sprayed twice with thickness 50mm of each layer without considering the disturbance of rockmass. The sequences of excavation are as follows; cavern 2 & 3 are excavated first (Step 1), and then cavern 1 & 4 are excavated (Step 2). The condition of load is applied differently as to the time of excavation and support installation.

Table 11. Specifications of the case study

Case	RMR	Disturbance	Remarks
1	40	○	· Shotcrete 100 mm · Rock Bolt Length : 4 m Spacing : 2 m
2	40	×	
3	60	○	
4	60	×	

3.4 Results

1) Case 1

When RMR is 40 and rockmass is disturbed, The caverns are unstable because the extent of plastic region, displacement and resultant stress in supports are larger than the allowable values(Fig 2). In case of meeting this kinds of rockmass during construction, the controlled blasting method has to be used to prevent the disturbance of rockmass, and the support has to be installed as soon as possible. It is thought that the reduction of stability due to the excavation of adjacent cavern does not occur. As the rock pillar between caverns is apt to be unstable, the sufficient support and monitoring plan have to be established before construction.

Table 12. Results of Case 1

Step	Max. Displacement (mm)				Max. Axial Force of rockbolt (ton)			
	C1	C2	C3	C4	C1	C2	C3	C4
1	-	16.2	14.9	-	-	7.8	7.2	-
2	13	16.7	15.1	11.1	5.68	7.66	7.1	6.1

Step	Stress in Shotcrete (kg/cm²)			
	C1	C2	C3	C4
1	-	195~242	151~193	-
2	121~167	166~214	123~169	111~149

Note) C1, C2, C3, C4 : Number of Cavern

2) CASE 2

When RMR is 40 and rockmass is undisturbed, The extent of plastic region, displacement are greatly reduced unlike case 1(Fig. 3). After the excavation and support of caverns are completed, it is detected that no caverns are unstable. However, the displacement is occured greatly at the wall of cavern 1 & 3, which is related to large horizontal stresses.

Fig 2. The plastic region (RMR=40, Disturbed)

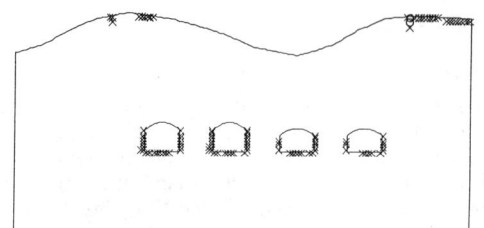

Fig 3. The plastic region (RMR=40, Undisturbed)

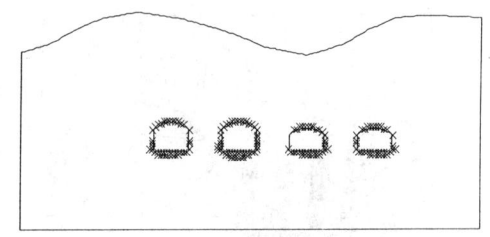

Fig 4. The Plastic region (RMR=60, Disturbed)

Table 13. Results of Case 2

Step	Max. Displacement (mm)				Max. Axial Force of rockbolt (ton)			
	C1	C2	C3	C4	C1	C2	C3	C4
1	-	7.0	5.9	-	-	1.93	1.95	-
2	7.2	4.6	4.3	5.7	1.67	2.2	2.1	1.68

Step	Stress in Shotcrete (kg/cm^2)			
	C1	C2	C3	C4
1	-	57.5~70.9	45.0~57.0	-
2	51.3~64.7	40.6~55.4	45.0~59.0	47.2~58.8

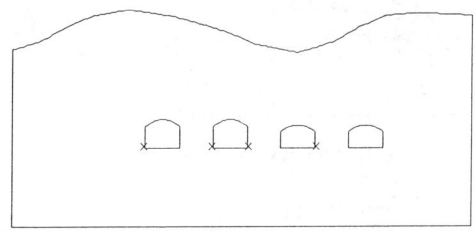

Fig 5. The Plastic region (RMR=60, Unisturbed)

Table 15. Results of Case 4

Step	Max. Displacement (mm)				Max. Axial Force of rockbolt (ton)			
	C1	C2	C3	C4	C1	C2	C3	C4
1	-	3.5	2.9	-	-	0.97	1.0	-
2	3.5	2.3	2.1	2.8	0.86	1.1	1.05	0.9

Step	Stress in Shotcrete (kg/cm^2)			
	C1	C2	C3	C4
1	-	30.1~35.9	23.6~28.8	-
2	27.6~32.4	24.1~29.9	25.5~31.7	24.6~29.8

3) CASE 3, 4

When RMR is 60 and rockmass is good, The caverns are stable for the suggested support pattern. The stability of pre-excavated cavern is not affected by the excavation of adjacent caverns (Fig4, 5). But, when rockmass is disturbed, the extent of plastic region at arch is greater than that at wall, and the tensile failure occurs at the arch of cavern. In the case of the rockmass with high joint density, low friction angle and shear strength, it is required for cautions during construction because of the rock fall at roof. Though it is determined that the caverns are stable during excavation, the continous monitoring and face mapping have to be carried out to maintain the stability of cavern and support.

Table 14. Results of Case 3

Step	Max. Displacement (mm)				Max. Axial Force of rockbolt (ton)			
	C1	C2	C3	C4	C1	C2	C3	C4
1	-	3.6	2.9	-	-	1.33	1.0	-
2	3.6	2.4	2.2	2.9	1.1	1.6	1.3	1.9

Step	Stress in Shotcrete (kg/cm^2)			
	C1	C2	C3	C4
1	-	34.4~42.0	27.3~33.5	-
2	29.8~36.6	24.6~31.8	27.2~34.0	25.9~31.7

4. CONCLUSIONS

1) When rokcmass is poor(RMR = 40) and disturbed by blasting, The caverns are unstable because the extent of plastic region at arch and wall is over 6~8m, the displacement at wall is about 16mm, and the resultant stresses of support are larger than the allowable values. But if rockmass is undisturbed, the results show that the cavern is stable because the extent of plastic region, displacement and stresses of support were reduced greatly.

2) When rockmass is good(RMR = 60), The caverns are stable without having relation to the disturbance. However, when rockmass is disturbed, the tensile failure occured at arch and the possibility of small rock fall at arch can be

greater. So it is required to implement the careful scaling and proper support works.

5. REFERENCES

Bhawani Singh, M. N. Viladkar, N. K. Samadhiya and Sandeep. 1995. *A Semi-empirical Method for the Design of Support Systems in Underground Openings,* Tunnelling and Underground Space Technology, Vol. 10, No. 3, pp. 375-383

Bieniawski, Z.T. 1989. *Engineering Rock Mass Classifications,* John Wiley & Sons, New York

Hoek, E., and E. T. Brown. 1980. *Underground Excavations in Rock,* IMM, London

Hoek, E., and E. T. Brown. 1988. *The Hoek-Brown Failure Criterion - a 1988 Update,* Proc. of 15th Canadian Rock Mechanics Symposium, Toronto

Rural Development Corp. 1996. *A Report on Site investigation and Strucural stability of Gonjiam underground storage terminal*

Sunkyong Engineering Construction Ltd. 1996. A *Report on Test and Stability anaysis of Gonjiam underground storage terminal,* Seoul

Itasca Consulting Group, Inc. 1993. *FLAC Version 3.2,* Minneapolis: ICG

Knut F. Garshol. 1993. *Crossing of the major shear zone in the fjellinjen twin tube highway tunnel in central Oslo. Design of rock support with steel fibre reinforced sprayed concrete,* Int. Symp. on sprayed concrete, Fagernes

Bhasin, R., and Grimstad, E. 1996. *The Use of Stress-Strength Relationships in the Assessment of Tunnel Stability,* Tunnelling and Underground Space Technology, Vol. 11, no. 1, pp. 93-98

Estimation of floor bearing capacity underneath full size pillars in longwall panels

Y.P.Chugh & D.Dutta
Department of Mining Engineering, Southern Illinois University at Carbondale, Ill., USA

ABSTRACT: The design of coal pillars associated with weak floor strata is important in many coal mining basins where actively mined coal seams are associated with weak floor. The current design practice is to consider the ultimate bearing capacity (UBC) of the weak floor strata without considering settlement. A study was conducted by the authors to establish the UBC of full size pillars by instrumenting longwall chain pillars. Earlier studies by the principal author spanning over a decade indicate that a reasonable time-independent weak floor UBC can be obtained from plate load tests and the UBC has a strong statistical correlation with the moisture content (MC) of the floor strata. In this paper, the UBCs obtained from the plate load tests and the MC of the weak floor strata were found to hold the same statistical relationship at two longwall panels. Application of four design techniques to estimate the UBC underneath full size pillars revealed very conservative values of UBC for three design methods. It was found that only one method could predict UBC underneath full size pillars in close agreement with the field observed values.

1 INTRODUCTION

As actively mined coal seams are associated with two to five feet thick weak claystone beds immediately below the coal seams of Illinois Basin coal mines, the conventional design of coal pillars based on the overburden loading of the pillars may not provide safe and satisfactory working conditions in coal mine entries (Chugh et al., 1990). To account for weak floor characteristics in coal pillar design, the ultimate bearing capacity (UBC) of floor strata is compared with the loads transmitted to the floors through the coal pillars. This approach does not consider pillar settlements on weak floor strata. Pillar settlements result in changed mine roadway geometries, floor heave, roof, coal pillar and floor failures; and surface and subsurface movements. It is, therefore, important that pillar design techniques for weak floor strata conditions consider bearing capacity of the floor as well as the pillar settlement.

Vesic's (1973) analysis to estimate the UBC underneath full-size pillars considers floor strata as a two-layer non-homogeneous cohesive soil with a weak layer overlying a stiff layer and negligible angle of internal friction for both the layers. The laboratory determined cohesive strength of the weak floor strata is scaled by a factor of 0.30 to represent rock mass cohesive strength. Speck (1981) has shown that the cohesive strength of the upper layer of the floor can be determined from its natural moisture content. Vesic and Vesic-Speck approaches were the only techniques available to a design engineer before the principal author embarked on systematically studying the pillar design problems *vis a vis* weak floor strata. Studies conducted at the Southern Illinois University at Carbondale (SIUC), spanning over a decade, have led to the development of different design tools for pillar sizing for weak floor conditions. Table 1 summarizes these tools along with their publication sources.

In Illinois longwall mines, tailgate entry failures due to the inadequate bearing capacity of floors have been reported. A study under the sponsorship of Illinois Mine Subsidence Research Program and the then United States Bureau of Mines, was initiated in early 90s to establish the relationships between surface, subsurface and underground movements and stresses on the pillars

Table 1 Design tools for pillar sizing for weak floor conditions developed at SIUC

Tools	Usage
[1] Strength-deformation characteristics of weak floor strata	Predicts strength-deformation characteristics using index properties (natural moisture content and liquid limit) of weak floor strata.
Plate size effects on strength and deformation properties of weak floor strata	Determines reduction factors for calculating UBC underneath a full size pillar from the plate UBC.
[2] Vesic-CHC method	Estimates bearing capacity of single full size pillar using the measured bearing capacity from plate load tests.
[3] Analytical technique	Provides a theoretical solution to determine UBC of a pillar resting on two-layered rock mass system
[4] PANEL.2D and PANEL.3D	Analyzes stresses on pillars using two- and three-dimensional time-dependent roof-pillar-floor interaction models based on the theory of beams on inelastic foundations.
[5] Longwall Ground Mechanics Model (LGM)	Simultaneously analyzes stability of in-mine geometry and predicts surface subsidence using 3-D boundary element model that incorporates displacement discontinuity and Salamon's laminated medium.

Note: Sources: 1. Pula et al. (1990), 2. Chugh et al. (1990), 3. Pytel et al. (1990), 4. Pytel et al. (1988), Pytel and Chugh (1990), 5. Yang et al.(1992).

by instrumenting longwall chain pillars. A part of the study in the overall project included the determination of UBC from the plate loading tests and incremental load-convergence curves of pillars, verification of an earlier established relationship between the moisture content of weak floor strata, and identification of an appropriate design tool for UBC estimation under a full size pillar. This paper discusses some of the results obtained from instrumentation of two longwall panels and their applicability in UBC estimation under a full size pillar. Other findings of the above studies have been published earlier (Yu et al., 1993).

2 INSTRUMENTATION OF LONGWALL CHAIN PILLARS

The mine extracts the Herrin No.6 coal seam in Southern Illinois at an average depth of 198 m from the surface. The thickness of the coal in this area varies from 2.29 m to 3.05 m. The immediate roof consists of 1.37 m to 1.68 m of black shale with a relatively competent 6.4 m thick layer of limestone immediately above the shale. The immediate floor strata consist of light gray underclay, ranging from 0.61 m to 1.52 m in thickness, underlain by 3.05 m to 4.57 of hard calcareous shale. The chain pillars are designed on 36.58-by-18.29 m centers for the headgate of the first panel and 30.48-by-18.29 m for the second panel, using a three-entry system with 4.72 m wide entries as shown in Figure 1. The longwall face is 292.6 m wide and 2134 m long in the East-West direction. Two panels had already been extracted immediately to the north of the study panels. The longwall retreats towards the east approximately 9.14 m per day for the first panel and 12.19 to 13.7 m per day for the second panel.

Instruments were installed at four sites in three panels, which included 182 surface subsidence monuments, 120 convergence stations, 10 MPBX sets, 24 vibration wire stressmeters and 2 inclinometer sets. In addition, plate loading tests were conducted in the field and core samples were taken for laboratory analysis. The surface and underground instrumentation locations are shown in Figure 1. The surface subsidence monitoring networks extend over two longwall panels and are located at three sites. The instrumentation included vibration wire stress meters (VWS), roof-floor convergence points, sag points, multiple point bore-hole extensometers (MPBX) in the pillar and inclinometer in the floor. In addition, core samples were taken from the weak floor for laboratory analysis and plate loading tests were conducted to determine the UBC of the weak floor.

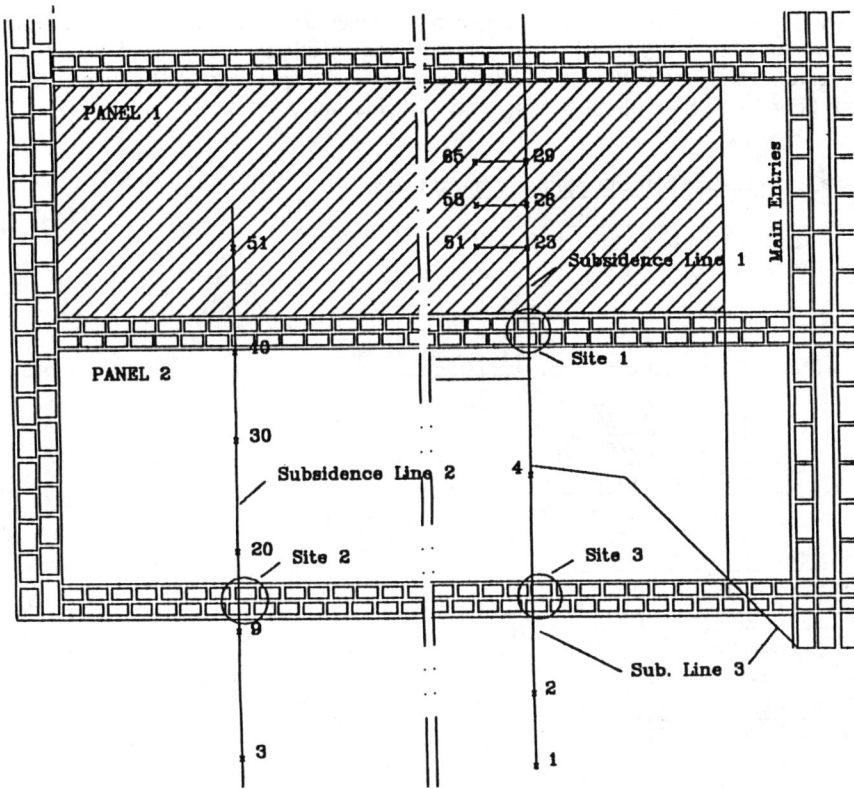

Figure 1 Underground instrumentation locations with subsidence monitoring networks

Pillar stress changes were measured using the vibrating wire stressmeter (VWS). Convergence stations were established to measure roof-floor convergence. Horizontal pillar deformations and vertical floor deformation were monitored using an MPBX probe, and horizontal floor deformation was measured using an inclinometer. All measurements were taken at intervals varying from two (2) to thirty (30) days, depending upon the expected changes.

3 RESULTS AND DISCUSSIONS

Only results pertinent to the ultimate bearing capacity analyses of weak floor are discussed here. Geotechnical data of samples taken at sites 2 and 3 were tested for, *inter alia*, moisture content (MC) and liquid limit (LL). The relationship between MC, LL, and the UBC in MPa (Pula et al., 1990) is:

$$UBC = 2.16 \times 10^{-3} \times e^{(7.38 - 0.145MC)} \quad (1)$$

$$UBC = 2.16 \times 10^{-3} (1405 - 86MC - 1.92LL) \quad (2)$$

Table 2 shows the moisture contents and liquid limits of floor strata at different depth at the two sites along with the UBC calculated using the MC and LL.

Ten plate loading tests were conducted at sites 2 and 3—six under as mined (AM) condition and four under soaked wet (SW) condition. Load-deformation data at each site are plotted in Figures 2 and 3. The UBC was determined using Vesic's (1973) failure criterion which defines the initial point of minimum steady slope as the point of failure and the UBC. The deformation modulus of the immediate floor strata was calculated at 50% (DM_{50}) of the UBC using the following equation:

$$DM = \frac{\Delta P D (1 - \mu^2)}{\Delta W} \quad (3)$$

where, DM is the deformation modulus (MPa), ΔP is the incremental pressure (MPa), ΔW is the incremental plate settlement (mm), D is the equivalent plate diameter in mm, and μ is the Poisson's ratio (an average value of 0.35 was taken).

Table 3 summarizes the UBC and DM_{50} for sites 2 and 3. The weak floor UBC under the SW condition is about 20% of that under the AM condition. The calculated UBC values (based on MC) are close to the field tested UBC from a 305-mm size plate loading test.

Table 2 UBC, liquid limits, and moisture contents of floor strata at two sites.

Site 2				Site 3			
Depth, ft	LL, %	MC, %	UBC, MPa	Depth, ft	LL, %	MC, %	UBC, MPa
2.0	34	2.43		1.75	32	3.34	
2.5	34	1.94		2.3	38	5.43	
3.0	46	3.25		2.7	37	3.12	
3.5	45	4.36		3.0	28	1.91	
4.5	23	0.31		4.2	24	1.37	
Based on MC only			7.22	Based on MC only			7.17
Based on LL and MC			4.92	Based on LL and MC			5.56

Table 3 Plate loading test results

Site	Condition	Plate Size, in.	UBC, psi	DM_{50}, psi
2	AM	152.4	12.30	157.22
2	AM	203.2	11.19	282.06
2	AM	304.8	6.97	219.38
2	SW	152.4	1.79	75.79
2	SW	203.2	1.68	28.21
3	AM	152.4	10.74	308.75
3	AM	203.2	5.03	436.58
3	AM	304.8	4.97	201.49
3	SW	152.4	1.68	31.01
3	SW	203.3	1.79	44.10

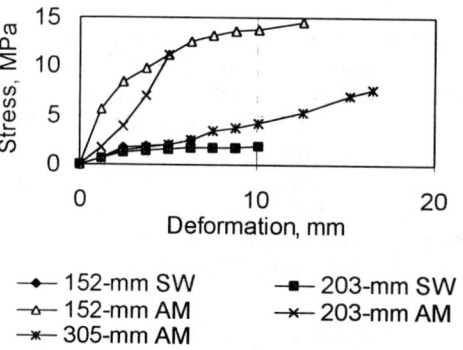

Figure 2 Load-deformation curves for plate loading tests at site 2

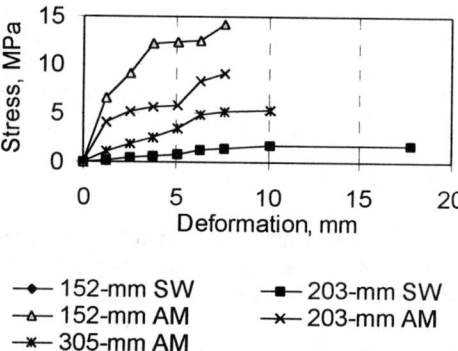

Figure 3 Load-deformation curves for plate loading tests at site 3

In order to estimate the weak floor UBC underneath a full size pillar, the measured pillar loads were plotted against the measured deformations, namely the roof-floor convergence and the pillar settlement. Figures 4 and 5 show the load-convergence curves for the pillars measured sites 2 and 3, respectively. These two curves suggest UBCs of about $5.84 + 1.6\gamma h$ (h is the depth of extraction) for site 2 and $4.08 + 1.6\gamma h$ for site 3. Assuming $\gamma = 0.025$, the UBC at a depth of 198 m (650 ft) for the two sites are 13.79 MPa and 12.02 MPa. The UBCs calculated using equations 1 or 2 or from plate loading tests do not take into account the confining horizontal pressure that is present in the floor underneath full size pillars. The field measurements show that the floors can sustain considerably higher pressure than the UBC obtained from the plate loading tests or calculated using moisture content (MC) of the weak floor because of the confinement of floor underneath full size pillars.

3.1 Design methods for UBC estimation underneath full-size pillars

Detailed analyses of different techniques available for UBC estimation have indicated that four approaches are the most suitable for calculating the UBC of full size pillars on layered strata that have a weak layer overlying a stiff layer (Chugh et al., 1990, Haq, 1990;). These four approaches are shown in Table 4. To verify the applicability of these formulas, UBC was calculated using these available design tools.

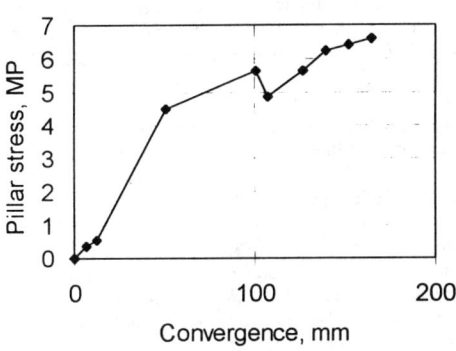

Figure 4 Measured pillar stress versus roof-floor convergence at site 2

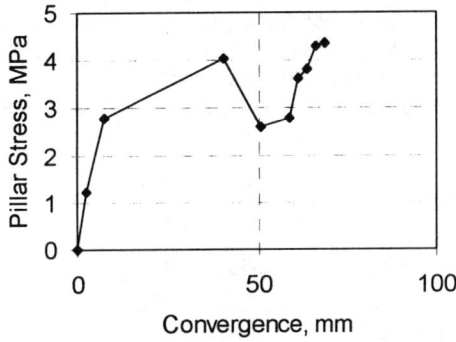

Figure 5 Measured pillar stress versus roof-floor convergence at site 2

Table 4 Different approaches for calculating UBC underneath full size pillars

Design Method	Design Approach
Vesic (1973)	Assumes angle of internal friction to be zero for both layers. Rock mass cohesion of the upper layer stratum is modified using bearing capacity factor that depends on pillar shape, weak layer thickness, and the ratio of cohesion of upper and lower layers.
Vesic-Speck (1981)	Same as Vesic's approach, but rock mass cohesion of the upper layer is calculated by scaling the compressive strength determined from the moisture content of the upper layer stratum.
Pytel-Chugh (1990)	Effects of adjacent pillars and internal angle of friction for both layers of floor strata are used to calculate the bearing capacity factor that modifies plate UBC to obtain UBC of full size pillars.
Vesic-CHC (1990)	Plate UBC (either calculated or field tested) is modified by a reduction factor based on the plate size and moisture content. The reduced UBC is used as the input for weak floor strata cohesion of Vesic's approach.

The parameters used in the calculations are shown in Table 5. Using the parameters of Table 5, the UBCs were calculated and shown in Table 6.

Table 6 suggests that Pytel-Chugh approach of UBC calculation is close to the field observed values. The higher values of UBC in Pytel-Chugh approach is due to the consideration of angle of internal friction of weak and competent floor layers. All the other methods give conservative values of full size pillar UBC.

4 CONCLUSION

The plate UBC can be calculated from the moisture content of weak floor strata. Pytel-Chugh approach of UBC calculation for full size pillars appear to be in close agreement with the field observed values and gives higher UBC compared with the other three methods. This is due to the consideration of internal angle of friction of weak floor layers in Pytel-Chugh approach. Though both Vesic-CHC and Vesic-Speck approaches do not consider the angle of internal friction of weak floor layers, it is found that Vesic-CHC method is more conservative than Vesic-Speck method in estimating UBC underneath a full-size pillar.

REFERENCES

Chugh, Y. P., O. Pula and W. M. Pytel 1990. Ultimate bearing capacity and settlement of coal pillar sub-strata. *International Journal of Mining and Geological Engineering*. Vol. 8. pp. 111-130.

Haq, S. E. 1990. *A Study of The Design Practices For Partial Extraction Room-And-Pillar Mining System In The Herrin (No. 6) Coal Seam*. Master's thesis. Mining Engineering Department. Southern Illinois University at Carbondale. Illinois.

Pula, O., Y. P. Chugh, and W. M. Pytel 1990. Estimation of weak floor strata properties and related safety factors for design of coal mine layouts. In *Proc. 31st U.S. Symp. On Rock Mechanics*. Colorado School of Mines. pp. 93-100.

Pytel, W. M., Y. P. Chugh, O. Pula 1990. An approach for design of coal pillars in partial extraction mining panels with a consideration of roof-pillar-floor interaction. . In *Proc. 31st U.S. Symp. On Rock Mechanics*. Colorado School of Mines. pp. 101-108.

Pytel, W. M. and Y. P. Chugh 1990. Development of a simplified three-dimensional roof-pillar-floor interaction analysis model. In *Proc. of the 8th Annual Workshop of Generic Mineral Technology Center*. Mine Systems Design and Ground Control. Reno. NV.

Pytel, W. M. and Y. P. Chugh, B. Zabel, and R. D. Caudle 1988. A simplified two-dimensional analysis of the roof-pillar-floor interaction problem in coal mines. In Proc. of 7th International Conf. on Ground Control in Mining. Morgantown. WV. pp. 271-281.

Speck, R. C. 1981. The influence of certain geologic and geotechnical factors on coal mine floor stability--a case study. In *Proc. of First Conference on Ground Control in Mining*. Morgantown. WV. pp. 44-49.

Vesic, A. S. 1973. Analysis of ultimate loads of shallow foundations. *J. Soil Mech. And Found. Div.* ASCE. Vol. 99. No. SM1.

Yang, G. Y. P. Chugh, A. Yu, and M. D. G. Salamon 1993. A numerical approach to subsidence prediction and stress analysis in coal mining using a laminated model. *Int. J. Rock Mechanics and Mining Sciences*. Vol. 30. No. 7. pp. 1419-1422.

Yu, Z., Y. P. Chugh, P. E. Miller and G. Yang 1993. A study of ground behavior in longwall mining through field instrumentation. *Int. J. of Rock Mechanics and Mining Science*. Vol. 30. No. 7. pp. 1441-1444.

Table 5 Parameters used for calculation of UBC of full-size pillars.

Parameters	Values
Natural moisture content, %	4
Liquid limits, %	36.5
Plate UBC, MPa	6.94
Plate size (squared shape), mm	304.8
Pillar length, m	25.76
Pillar width, m	13.56
Entry width, m	4.72
Weak layer thickness, m	1.22
Angle of internal friction (upper layer), deg.	20
Angle of internal friction (lower layer), deg.	30
Compressive strength (upper layer), MPa	4.86
Compressive strength (lower layer), MPa	13.89

Table 6 Comparison of calculated and measured UBCs of full-size pillars

Method	UBC, MPa
Vesic	5.10
Vesic-Speck	10.45
Vesic-CHC	8.95
Pytel-Chugh	14.55
Measured	12.02-13.79

Modelling techniques for safety evaluation: Rock characterization

Comparison of Hoek cell and a laboratory made cell in performing triaxial tests on intact and jointed specimens

M. Gharouni-Nik
Department of Civil Engineering, University of Newcastle-upon-Tyne, UK (Presently: Soil and Rock Co., Ltd, Tehran, Iran)

ABSTRACT: Many varieties of triaxial cells are in use in rock mechanics laboratories. In this work two available cells were used and compared. These two cells are the so-called Hoek cell, developed by Hoek and Franklin (1968), and a laboratory made cell (Mk1) that has been designed and manufactured in the Rock Mechanics Laboratories at the university of Newcastle. Both cells have been designed for a maximum confining pressure of 70 MPa and testing of 75 mm diameter by 150 mm long specimens. The comparison showed some differences in the results which are mainly arised from the different configuration of the cells and the type of platens and sleeves which are used. These differences were amplified by increasing the confining pressure. Finally, for intact and jointed specimens, the range of confining pressure in which the Hoek cell and Mk1 cell are suitable to use is concluded.

1 INTRODUCTION

Triaxial compression refers to a test applying axial compression of a rock cylinder with simultaneous application of axisymmetric confining pressure. The specimen is placed inside a pressure cell and is pressurized up to a predetermined confining pressure.

Many varieties of triaxial cells are in use in rock mechanics laboratories and several types are available from commercial suppliers the review of which is beyond the scope of this study. In this work two cells available in the Rock Mechanics Laboratories at the University of Newcastle were used and compared. The first one is the Hoek cell and the second one is Mark One (Mk1) cell which has been designed and manufactured in the Rock Mechanics Laboratories, a detail description of which may be found in the work by Buzdar (1968).

A review of the available literature indicated that previous investigators, that conducted triaxial compression tests with these two type of cells, have not identified any differences between these cells.

One may question whether there are any differences between the results of the tests conducted using these two cells. If indeed there are some differences, are they significant or one may ignore them? Which one is more suitable to conduct triaxial tests on intact and jointed rocks? What criteria one may use in selecting the appropriate cell?

2 EXPERIMENTAL RESULTS

Before embarking in analysing the results obtained from these two tests with the cells, it should be mentioned that an attempt was made to maintain the testing conditions for both of them as follows:
- confining pressure ranged between 2.5 MPa to 15 MPa;
- strain rates of 0.25%/min and 0.02%/min;
- rock types were PrS (Penrith sandstone) and Qu (Quickrock) either intact or with a discontinuity with special emphasis in the latter one, because it is artificial and capable of reproducing the same specimen configuration (GHAROUNI-NIK, 1993);
- both cells were used with the same stiff testing servo-controlled system; and
- confining pressure was applied by means of a pressure intensifier.

2.1 Intact samples

The triaxial tests on intact specimens of PrS and Qu were tested in both the Hoek cell and the Mk1 cell using specimens with identical geometry to permit the comparison of the two cells. The following results from the triaxial tests were obtained:

2.1.1 PrS specimens

The peak strength in both cells increased with increasing confining pressure and the rate of increasing was nearly equal for both cells.

For some reason, perhaps such as slight differences in the cores micro texture, the axial strain at failure (ε_{1f}) for the same confinement, was greater for the Hoek cell than the Mk1 cell.

The results of PrS specimens tested in the cells are listed in table (1). As it is shown, the differences between strength increased with confining pressure and the axial stress at failure obtained from the Hoek cell triaxial tests was always greater than the one for the Mk1 cell. In the highest confining pressure, namely 15 MPa, this difference was approximately 10% (for medium strain rate). This difference for slow strain rate was about 12.3%.

As it can be seen from table (1), an increase in the confining pressure enhanced the effect of strain rate for the Hoek cell, whereas in the Mk1 cell this effect was insignificant even at high confining pressures where the differences were negligible.

Due to the small available internal space in the Hoek cell, the tests were continued after failure up to maximum 3% to 4% to avoid damage to the sleeve and any subsequent difficulties associated with the removal of the failed specimens. As a result, there were noticeable differences in the residual stresses obtained from the Hoek cell and the Mk1 cell, essentially because the latter one has a larger internal space (117.5 mm internal diameter) allowing for higher axial strains. Although this internal space is very important for conducting tests on jointed specimens, its importance is not very significant for intact samples, since after a few percentages of axial strain, the geometry of the specimens would change, thus casting a shadow of a doubt on the validity of the results.

2.1.2 Qu specimens

The comparison between the axial stress at failure obtained from tests with both cells (table 2) showed that an increasing in the confining pressure increases this difference which at the highest confinement, i.e. 15 MPa, exceeded 13%, that is considered a significant difference in engineering design.

The results obtained from both cells have been illustrated in figure (1), where it can be seen for all levels of confinement that the σ_{1f} of the Hoek cell was significantly higher than the MK1 cell and the difference was emphasized at higher confining pressures.

2.2 Specimens containing a discontinuity:

A review of the available published information concerned with the testing of specimens containing a discontinuity, indicated the lack of interest in using the Hoek cell for such tests. The results of specimens

Table 1. Triaxial compression test results on Penrith Sandstone

| | Mk1 Cell | | | | Hoek Cell | | | |
| | Medium Str. Rate | | Slow Str. Rate | | Medium Str. Rate | | Slow Str. Rate | |
σ_3	σ_{1f}(MPa)	ε_{1f}(%)	σ_{1f}(MPa)	ε_{1f}(%)	σ_{1f}(MPa)	ε_{1f}(%)	σ_{1f}(MPa)	ε_{1f}(%)
2.5	102.6	0.97	107.17	1.30	107.01	1.13	108.68	1.20
5	134.17	1.01	136.83	1.28	144.31	1.32	148.09	1.63
7.5	160.94	1.21	158.48	1.37	172.86	1.47	179.31	1.77
10	180.92	1.29	183.39	1.45	188.83	1.61	197.56	1.83
15	203.57	1.35	208.49	1.33	223.56	1.87	234.11	2.00
24	242.40	1.55	-	-	-	-	-	-

Table 2. Triaxial compression test results on Quickrock specimens using various cells

	Mk1 Cell		Hoek Cell	
σ_3(MPa)	σ_{1f}(MPa)	ε_{1f}(%)	σ_{1f}(MPa)	ε_{1f}(%)
2.5	66.04	1.27	70.23	1.31
5	76.86	1.33	77.49	1.31
7.5	78.72	1.53	82.64	1.47
10	85.57	1.70	93.02	1.79
12.5	88.73	1.91	98.34	1.93
15	93.97	3.36	106.34	2.93

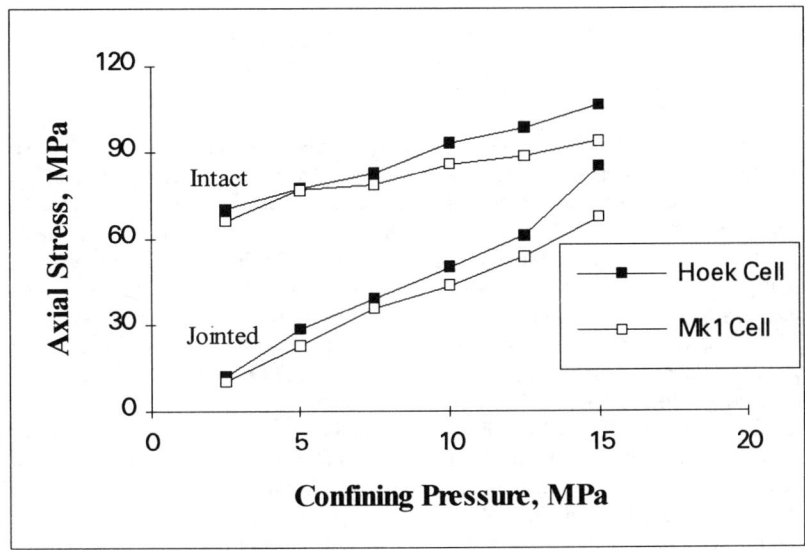

Figure 1 : Comparison of axial stress at failure of intact and jointed Qu specimens in various σ_3 using Hoek cell and Mk1 cell

containing a discontinuity in the two types of cell were compared to assess the suitability of the Hoek cell and to investigate the extent of any shortcomings in using the Hoek cell.

The comparison of the results of the two cells for specimens containing a discontinuity started with the test on PrS specimens at 15 MPa confining pressure and was continued with the Qu specimens for a range of σ_3.

The tests with PrS specimens containing a 45° oriented discontinuity using the two cells showed the existence of the following differences in the results. As one can see from figure (2), two points of deflection in the σ_1-ε_1 curve may be distinguished:

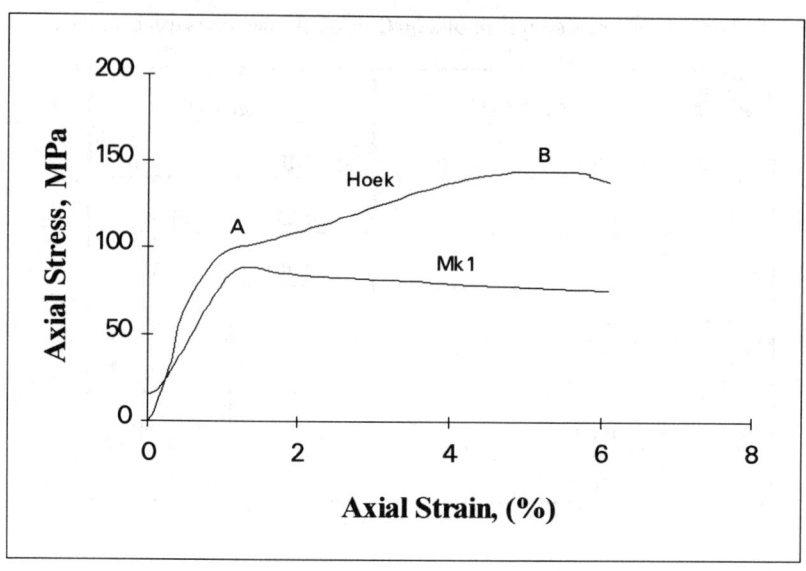

Figure 2 : Stress-Strain relationship of 45° oriented saw-cut PrS using Hoek cell and Mk1 cell (σ_3=15 MPa)

point A corresponding to the initiation of sliding along the discontinuity, and point B defining the failure of the material itself in a surface transcending the discontinuity surface.

The results of the Qu specimens containing a smooth discontinuity at an angle of 30° with respect to the longitudinal axis of the specimen showed the unsuitability of the Hoek cell for testing specimens containing a discontinuities. The σ_{1f} of specimens containing a discontinuity when tested in a Hoek cell was always much more than that for Mk1 cell and this difference was emphasized as the confining pressure increased. At the highest confinement, i.e. 15 MPa, this increase exceeded the level of 26%.

During testing intact specimens with the Hoek cell at higher σ_3, certain difficulties were encountered in removing the specimens from the sleeve after the tests. The same problems were also experienced for the failed specimens containing a discontinuity irrespective of the confinement level.

3 REMARKS

The remainder of this paper is devoted to a discussion of the possible explanation for the aforementioned differences in the results obtained from the two types of cells.

The identification of the differences between the two cells when testing the intact specimens is significant, since as mentioned before, the Hoek cell has been mainly used for the testing of intact specimens. In the case of the specimens containing a discontinuity, however, the reasons for differences are a mixture of various parameters. The reason for conducting the tests on this type of specimens using the two cells was only to detect the extension of variation the results.

Possible reasons for causing the differences between the Hoek cell and the Mk1 cell may be related to:
- The change in geometry of cell-specimen system due to the type of platens;
- The stiffness of the sleeves in the respective cells;
- The type of movement of load piston-platens system inside the cells;
- The available internal space of the cells.

The change in geometry of the cell-specimen system may have some limited effect on the tests carried out on intact specimens. However for specimens containing a discontinuity, the effects are significant.

The possible testing configurations for testing triaxially specimens containing a discontinuity were first introduced by Jaeger (1959). He has identified three systems : a) rigid platens with no spherical seats; b) a single spherical seat; c) two spherical seats. These systems was improved by Rosengren

(1968) who examined end-effects on the cylindrical specimens containing a single discontinuity. His measurements consisted of two pairs of steel discs which were inserted between the specimens and platens at each end. Each pair of discs contained molybdenum grease as lubricant that allowed the two discs to move freely in the lateral direction.

Besides following the system suggested by Rosengren (1968) in substituting the spherical platens with discs when testing specimens containing a discontinuity, Fahimifar (1990) has conducted the same tests on intact rock. He concluded that any modifications of the cell-specimen system are expected to have a significant reduction of the σ_{1f} for both intact specimens as well as specimens containing a discontinuity.

In the case of specimens containing a discontinuity, continuation of sliding in the Hoek cell equipped with spherical seats for 45° orientation at 15 MPa confining pressure is shown in figure (2). As one may see, sliding along the discontinuity eventually resulted in the fracture of the specimen by developing a new shear plane through intact rock, whereas in the Mk1 cell that contains two pairs of steel discs, sliding continued with no fracture through intact material.

The second source of deviations of the results is the type of sleeve used in the cells. As mentioned before, in the Hoek cell the stiff membrane has been used to isolate the specimens from the confining hydraulic oil. Farmer (1983) explained that "... the membrane has the advantage of allowing rapid testing and the disadvantage of restricting deformation after peak stress". As will be shown, beside the influence of the sleeve after peak stress, it affects the stress-strain behaviour in the region between yield and peak points.

The type of sleeve in the Hoek cell introduced another disadvantage that occurred after the test completion and was more pronounced for the Qu specimens. The brittle-ductile transition point for Qu specimens was relatively low in comparison with the PrS specimens. As a result the excessive barrelling of the specimens, that was observed in the range of confining pressure used in this work, caused problems during the removal of the tested specimens from the sleeve. Often damage to the sleeve would occur or structural decomposition of the failed specimen would take place. Consequently, if an inspection of the failed specimen is required, clearly this type of sleeve is unsuitable.

The next identified differences are associated with the mode of movement of the plunger that exercises the axial load on the specimen inside the cell. In the Mk1 cell the plunger is in direct contact with the oil, in contrast with the Hoek cell where the fixed sleeve arrangement is used.

The effect of moving the plunger inside the stiff sleeve of Hoek cell was noticeable in higher confinement, because the friction between the sleeve and plunger became higher. In other words, there was a resistance against the movement of the plunger which caused an overestimation of the applied axial load. Therefore, the load which recorded as axial load was not what was actually applied to the specimen.

The last factor that may considered as being responsible for the differences identified between the two types of cells is the available internal space. The effect of this factor, that is more pronounced for specimens containing a discontinuity, influences the behaviour of intact rock after failure. Essentially, it is the main factor that renders Hoek cell unsuitable for testing specimens containing a discontinuity or investigating the post-failure behaviour of intact specimens. The Hoek cell is suitable for rapid testing on a large number of samples in a short time to obtain a peak strength, while Mk1 cell is a sophisticated cell with a large internal space suitable for tests associated with large axial and volumetric strains of any type of rocks.

It may be concluded that for intact materials, Hoek cell is more suitable at lower confining pressure (i.e. 2 MPa in this work). For high confinement it does not seem to be an appropriate cell. For the tests in which the study of post-failure or examination of the specimens after the test is important, the Mk1 cell must be used. Also for soft rocks whose brittle-ductile transition points are low and the deformation before failure is large, Mk1 cell is more applicable, unless the tests are to be carried out at very low confinement. Furthermore, for the study of jointed specimens either pre- or post-failure, Mk1 cell is to be used.

REFERENCES

BUZDAR, S.A.R.K. 1968. A laboratory investigation into the mechanical properties of some sedimentary rocks with special reference to potash, PhD. Thesis, University of Newcastle upon Tyne

FAHIMIFAR, A. 1990. Experimental investigations on the mechanical properties of rocks containing a single discontinuity, PhD. Thesis, University of Newcastle upon Tyne

FARMER, I.W. 1983. Engineering behaviour of rocks, 2nd. Edition, Chapman and Hall Publications

GHAROUNI-NIK, M. (1993)

Laboratory investigations on the engineering behaviour of rock joints using triaxial and direct shear techniques,
PhD. Thesis, University of Newcastle upon Tyne
HOEK, E. & J.A. FRANKLIN 1968. A simple triaxial cell for field or laboratory testing of rock, Trans., Inst.Min.Metallurgy, 77:22-26
JAEGER, J.C. 1959. The friction properties of joints in rock, Geof. Pur. App. 43:148-158
ROSENGREN, K.J. 1968. Rock mechanics of the Black Star open cut, Mount Isa. PhD. Thesis, Australian National University, Canberra

Influence of porosity on the absorption of moisture of sandstone and siltstone

Herryal Z. Anwar
R&D Centre for Geotechnology, Indonesian Institute of Sciences, Bandung, Indonesia & Kyushu University, Fukuoka, Japan

H. Shimada, M. Ichinose & K. Matsui
Kyushu University, Fukuoka, Japan

ABSTRACT: A common characteristics to all of weak rocks is their high porosity, value may reach 40 % and usually exceed 15 %. Since porosity is an important factor in rock material property, the strength of rock is influenced significantly by the degree of saturation. This paper demonstrates the correlation between porosity and degree of moisture absorption of sedimentary rock of volcanic sandstone, carbonate sandstone and siltstone, all weak rocks are from Indonesia. In this connection the hygroscopicity behaviour of sandstone and siltstone were studied within various humidity conditions, in which saturated solutions for the control of humidity were used, and under immersed water. The degree of moisture absorption of rock tested shows in a higher saturation the variation strongly dependent on porosity. The comparable relation was also shown at the previous test on sandstone from Kyushu area.

1. INTRODUCTION

The porosity is a significant characteristic in assessing the mechanical strength of weak rock such as sandstone and siltstone as reported by Schiller, 1958; Kowalski, Morgenstern and Phukan, 1966 (Hawkes and Mellor, 1970). Therefor porosity is a useful index property in conjunction with mechanical strength. In many type of weak rock, their porosity are relatively high compare with hard rock.

For sandstone and siltstone the porosity affected the mechanical strength is also controlled by the degree of moisture absorption. Hawkins and McConnels (1992) show that for a number of low-porosity sandstone, it was found to be difficult to attain intermediate moisture content using controlled humidity environments. Therefore, for sandstone and siltstone porosity has connection to which rock may absorb water. The percentage of moisture content by weight at saturation for different rock types varied depending on effective porosity. This relationship between porosity and the degree of absorption is important to explain the weakened of rock due increasing of water content.

2. MATERIAL TESTED AND LABORATORY MEASUREMENT

Sandstone and Siltstone which are described in this paper come from Indonesia. There are two type of Sandstone; from Rajamandala area (West Java) called as volcanic Sandstone, and from Kedungombo area (Central Java) called as carbonate Sandstone. Volcanic Sandstone represents the quartenary volcanic deposit of Rajamandala Formation. Otherwise Carbonate Sandstone represents the tertiary deposit of Kerek Formation. In this study was also tested Siltstone, which come from Muaraenim area (South Sumatera), represents the quartenary deposit of Kasai Formation. Petrography and X-ray D and F analysis were carried on both selected Sandstones and Siltstone. and the mineralogy obtained are summarized below :

Volcanic Sandstone : Quartz, Smectite, Plagioclase, Halloysite. The grain size are generally between 0.1 - 0.8 mm. The grains are anhedral and irregular.

Carbonate Sandstone : Calcite, Quartz, Smectite. The grains are anhedral to irregular. The grain size are generally between 0.1 - 0.6 mm.

Siltstone : Quartz, Chlorite, Plagioclase, Kaolinite, Smectite. The grains are irregular. The grain size are generally between 0.1 - 0.2 mm. Silstone shows bedding planes.

Porosity analyses were carried out in porositimeter by using mercury penetration method. In this method three lumps of each specimen were prepared suitable for porosimeter requirement.

In addition to establishing the saturated sample under relative humidity required, three lump specimens for each type material were hygroscopicity tested. Four different relative humidity : 33 %, 48 %, 75 % and 95 %, were established in humidity cabinets with saturated solution present as outlined by Winston and Bates (1960). In this test the saturated solution will maintain a constant relative humidity (RH) and allow the specimens to absorb moisture. To attain the absorption process came along effectively as surrounding humidity, initially the specimens were oven dried under temperature 105° C. At the first stage the specimens were placed inside humidity cabinet of 33 % RH for a period of three weeks under temperature 25° C, and then the experiment would be continued to the following RH in the same period. In this period the rocks is considered may accommodate the relative humidity surrounding. This method is effectively to saturate the specimen under relative humidity required.

This test is also completed by water saturation test, followed the ISSRM suggestion (Brown, 1981).

3. RESULTS

Figure 1 (a, b and c) shows the distribution of pore of the volcanic sandstone, carbonate sandstone and siltstone. The value of pore volume is indicated by $\Delta Vp/\Delta \log R$, in which Vp is pore volume and R is radius of pore. Range of the maximum pore volume of volcanic sandstone is distributed in radius of pore between 10^2 to 10^3 A°, and the range of maximum pore volume of carbonate sandstone is distributed in radius of pore between 10^2 to 10^4 A°. While the range of maximum pore volume of the

Table 1. Porosity of material tested.

Type of Rock	Porosity
Volcanic Sandstone	13 % - 15 %
Carbonate Sandstone	14 % - 16 %
Siltstone	18 % - 19 %

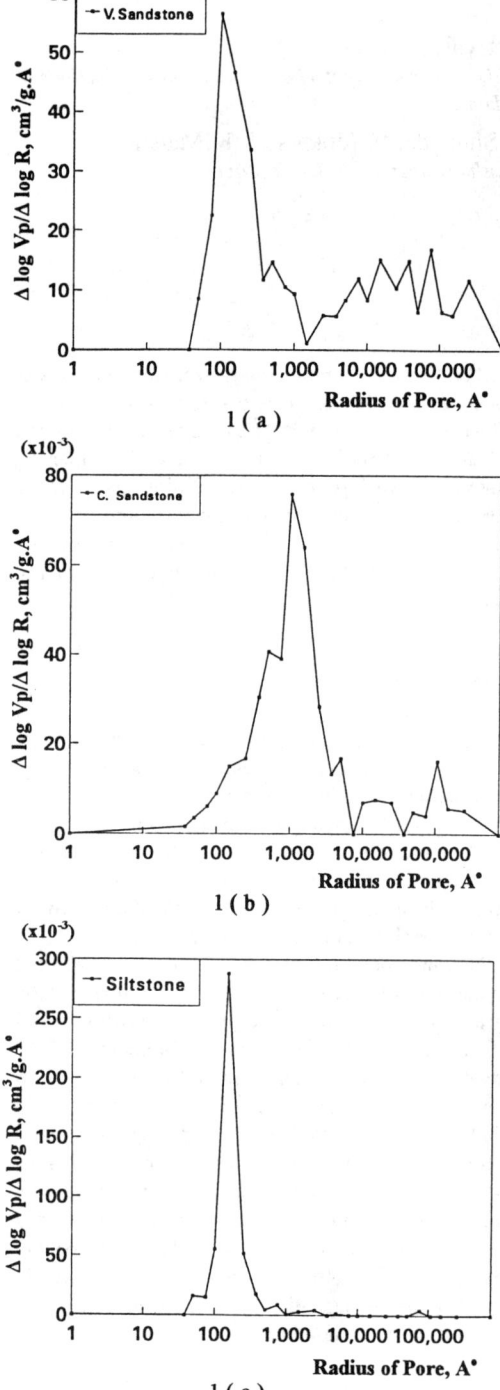

Figure 1. Result of porositimeter analysis for volcanic and carbonate sandstone and siltstone.

siltstone are distributed in radius of pore 10^2 to 10^3 A°. However, the pore volume of siltstone is larger than the two sandstone and the pattern of distribution curve is different compare with the two sandstone.

Compare with the previous test, porosity of volcanic and carbonate sandstone are in the range of Kyushu sandstone (9.4 - 21.6 %), Table 1, as reported by Uchino (1984), and the distribution pattern of radius of pore of is almost identical with the material tested.

Figure 2 and 3 present the increasing of moisture and the absorption rate due to the increasing of relative humidity for each rock.

Hygroscopicity test presents that moisture content increase sharply from dry condition to RH 33 % for every rock. In this stage the moisture absorption rate of rocks is high. From RH 33 % to 75 % the moisture content increase slowly. Nevertheless, the absorption rate decrease sharply, and then increase again from RH 48 % to 75 %. It is surprisingly, in the last stage that the absorption rate tend to increase sharply as the initial stage for all rock.

The result of Hygroscopicity and saturation test demonstrated that the moisture absorption at the highest relative humidity (95 % RH) is only about half of the immersed water condition for Silstone and volcanic sandstone and about 40 % for carbonate sandstone (Table 2). The remarkable difference in water absorption because in water saturation test was used vacuum pump, and hence it will helpful to accelerate the saturation process.

The moisture absorption of carbonate sandstone is

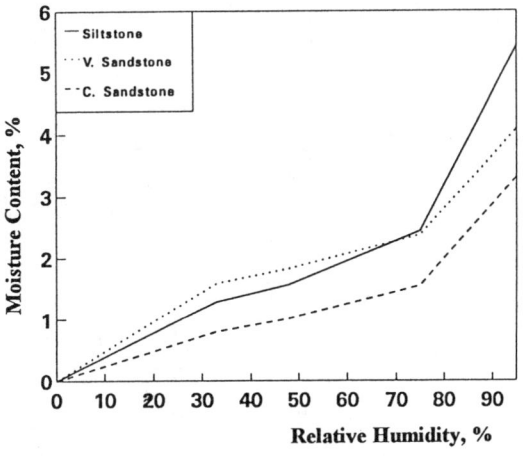

Figure 2. Variation of moisture content in hygroscopicity test.

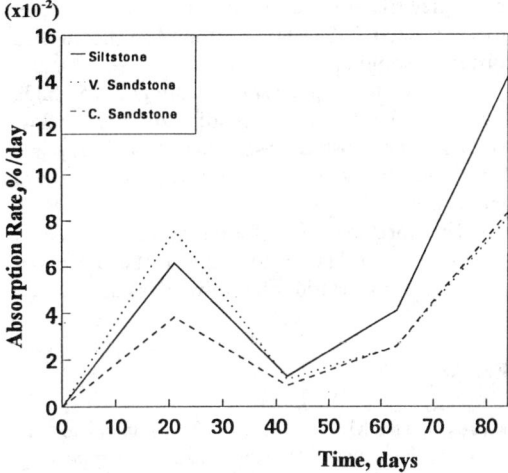

Figure 3. Absorption rate at hygroscopicity test.

Table 2. Moisture content.

Type of Rock	RH 95 %	Water Saturation
Volcanic Sandstone	4.08 %	7.12 %
Carbonate Sandstone	3.31 %	9.17 %
Siltstone	5.41 %	9.64 %

about 60 % of volcanic sandstone below 95 % RH as shown by Figure 2. However, at the highest saturation or in water saturation condition more moisture might be absorbed by carbonate sandstone. Likely, in this stage the porosity dominates the degree of moisture absorption of carbonate sandstone.

The result of porosity and moisture absorption which is shown by Uchino (1984) and Ichinose (1987) demonstrate that in immersed water condition the moisture absorption of some Kyushu's sandstones are porosity dependent. Under low saturation condition the moisture absorption of those sandstones are not affected by the porosity.

In term of mineralogy, volcanic and carbonate sandstone and siltstone contain smectite mineral which is known as mineral that has a tendency to absorb water strongly.

4. CONCLUSIONS

It has been described that in volcanic and carbonate sandstone and siltstone porosity has profound influence the moisture absorption under high degree of saturation or under immersed water. However, since 75 % RH siltstone has shown high moisture absorption.

Although their porosity is categorized high, however, under lower degree of saturation (below RH 75 %), the moisture absorption of sandstone and siltstone did not show a clear relationship toward their porosity.

The variation of moisture absorption has a significance contribution to explain the effect of water content on deterioration of weak rock.

REFERENCES

Hawkes, I. and Mellor, M., (1970), Uniaxial testing in rock mechanics laboratories, *Engineering Geology, vol. 4, 1970*, pp. 177-285.

Hawkins, A.B. and McConnels, B.J., (1992), Sensitivity of sandstone strength and deformability to changes in moisture content. *Quarterly Journal of Engineering Geology*, 25, pp. 115-130.

Ichinose, M., (1984) Study on pore structure of coal measure rocks by mercury penetration. *Research Report*, Kyushu University, Japan.

Ichinose, M., (1987) Hygroscopicity of coal measures rocks. *Journal of the Japan Society of Engineering Geology*, 28-4, pp. 17-26.

Kowalski, W.C., (1966), The interdependence between the strength and voids ratio of limestone and marl in connection with their water saturating and anisotrophy. *Proc. Congr. Intern. Soc. Rock Mech., 1st*, Lisbon, pp. 143-144.

Morgenstern, N.R. and Phukan, A.L.T., (1966), Non-linear deformation of a sandstone, *Proc. Cong. Intern. Soc. Rock. Mech., 1st*, Lisbon, pp. 543-548.

Matthews, M.C. and Clayton, C.R.I., (1993), Influence of intact porosity on the engineering properties of weak rock. In: Anagnostopoulos et.al. (Editor) *Geotechnical Engineering of Hard Soils-Soft Rock*, Balkema, Rotterdam, pp. 693-701.

Schiller, K.K., (1958) Porosity and strength of brittle solids (with particular reference to gypsum). In: W.H. Walton (Editor), *Mechanical Properties of Non-metallic Brittle Materials*, Butterworths, London, pp. 35-49.

Winston, P.W. and Bates, D.H., (1960), Saturated solutions for the control of humidity in biological research. *Ecology*, Vol. 41, No. 1, pp. 232-237.

Relationship between microcrack density and mechanical properties in granite

S.E. Lee & H.M. Park
Department of Mineral and Petroleum Engineering, Chonbuk National University, Korea

ABSTRACT: Micro-defects in granitic rocks can be categorized into microcrack, pore, cleavage, liquid inclusion plane, parting, etc. It has been known that granitic rocks indicate the mechanical anisotropy caused by preferred orientation of microcracks. But most of the studies of anisotropy by the preferred orientation of microcracks have been carried out only qualitatively so far. So this study was performed to examine the relationship between the distribution of microcracks and the mechanical anisotropy quantitatively. For this purpose, crack density and crack length were investigated by using the thin section analysis, and p-wave velocity and tensile strength were also measured. Hamyeol and Nangsan granite in Iksan were chosen as the samples. It appears that there is a close relationship between the crack density and the mechanical properties, and the crack length is related mainly to the particle radius of rock forming minerals.

1 INTRODUCTION

It has been known that granitic rocks indicate the mechanical anisotropy(e.g. uniaxial compressive strength, tensile strength and elastic wave velocity and so on) caused by preferred orientation of microcracks [Birch,1961; Osborne, 1965; McWilliam,1966; Douglass & Voigt,1969; Peng & Johnson,1972; Sano et al.,1987; Park & Lee, 1994]. These preferred orientation planes of microcracks are the three sets of nearly orthogonal planes called as rift plane(R-p), grain plane(G-p) and hardway plane(H-p) in order of ease splitting. These planes have been used in qurrying by quarryman. In this study, the axes vertical to each planes are named as rift axis(R-x), grain axis (G-x) and hardway axis(H-x).

However most of the studies on the anisotropy caused by the preferred orientation of microcracks have been carried out only qualitatively so far. So this study was performed to examine the relationship between the distribution of microcracks and the mechanical anisotropy quantitatively.

For this purpose a distribution of microcracks are studied by using the thin section analysis. Mechanical properties (p-wave velocity, tensile strength) are measured along three orthogonal planes. Lastly, relationships between distribution of microcracks and mechanical properties are considered.

2 DISTRIBUTION OF THE MICROCRACKS IN GRANITIC ROCKS

The origin of formation of microcracks in granite hasn't been validated scientifically. If microcracks were created with closing the pores in magma by the ground pressure or side pressure in the process of solidification when granitic rocks were formed, there may exist a disc crack and it would appear with the line on a cutted plane.

In order to estimate a mechanical behavior of granitic rock mass caused by the preferred orientation of microcracks, it is realized that the distribution of microcracks should be investigated quantitatively.

Geometrical properties such as orientation, density and persistence etc. are the main factors for the purpose of the examination of crack distribution.

Crack density can be calculated from the crack spacing. Average spacing of crack $\bar{x}$ can be estimated from the average of samples as follows

$$\bar{x} = \sum_{i=1}^{n} \frac{x_i}{n} \qquad (1)$$

where n = number of sections which is divided by the intersection between scan line and crack, x_i = length of ith section.

Also, because crack spacing is irregular and distributes randomly, random density function of crack spacing $f(x)$ is expressed as a negative exponential distribution given by equation (2).

$$f(x) = \lambda e^{-\lambda x} \qquad (2)$$

where x=variable of crack spacing, λ=crack density(intersection number per unit length) which is a reciprocal number of $\bar{x}$ and e=natural logarithm.

Crack length can be include as a factor which indicates the persistence of crack. Also, distribution of crack length $f(l)$ can be written as a negative exponential distribution in the same way as random density function of crack spacing.

$$f(l) = \frac{1}{\bar{l}} e^{-\frac{l}{\bar{l}}} \qquad (3)$$

where l=variable of crack length, $\bar{l}$=average length of crack.

3 EXPERIMENT

Distribution of microcracks in granitic rocks was investigated by using thin section analysis and the polarization microscope(Japan, Nicon, Opti photo-66) to observe the rock texture. Thin section was made with a rectangular shape where one side is 1~1.5cm for each R, G and H plane. A scan line was set up in 1 mm spacing and measured the crack spacing and crack length in Figure 1.

Compressive testing system(Shmadzu, UH-A, Japan) was used. The load speed was 2 ton/cm²/min. Tensile strength was measured at an interval of 10 degrees from 0 to 180 degrees for the specimens of each R, G and H axis.

Pundit(England) was used to measure the p-wave velocity. P-wave velocity was measured at an interval of 10 degrees in the same way as Brazilian tensile strength.

4 RESULTS

4.1 Crack density and crack length

The results for the investigation of crack spacing and crack length by thin section analysis are shown in Table 1(relative frequency distribution of crack spacing), Table 2(relative frequency distribution of crack length). In tables, l_s is the total length of scan lines, l_c is the total length of cracks and n is the number of cracks. Also Table 3

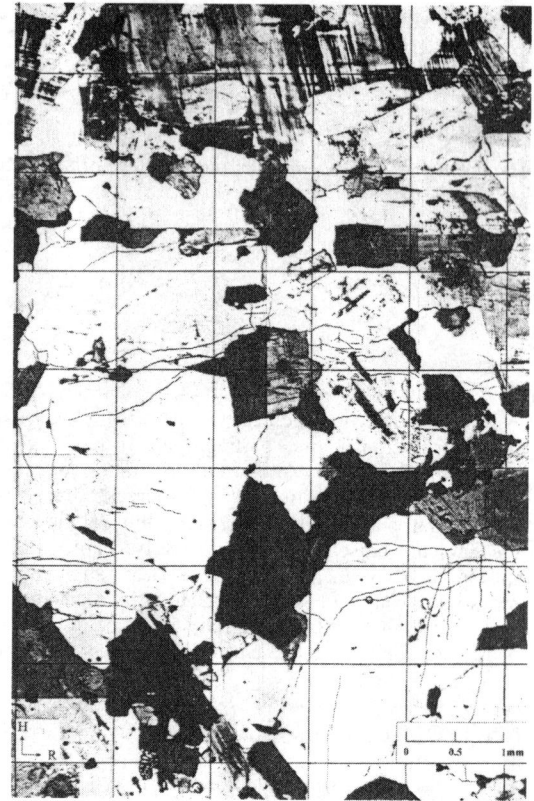

Figure 1. Crack trace and scan lines for the survey of crack distribution.

Table 1. Relative frequency distribution of crack spacing

Crack spacing (mm)	Relative frequency (f_c/n)					
	Nangsan granite			Hamyeol granite		
	R-p	G-p	H-p	R-p	G-p	H-p
n / l_s	99/432	59/287	57/354	139/357	124/411	53/282
0.0~0.49	0.394	0.325	0.298	0.417	0.419	0.366
0.5~0.99	0.192	0.258	0.123	0.255	0.312	0.076
1.0~1.49	0.152	0.125	0.157	0.125	0.123	0.132
1.5~1.99	0.051	0.112	0.133	0.065	0.081	0.113
2.0~2.49	0.071	0.086	0.070	0.065	0.065	0.038
2.5~2.99	0.040	0.034	0.070	-	-	0.094
3.0~3.49	-	0.051	0.035	0.036	-	0.076
3.5~3.99	-	-	0.053	0.022	0.024	-
4.0~4.49	0.040	0.017	0.035	-	-	0.052
4.5~4.99	0.020	0.034	0.018	-	0.008	-
5.0~5.49	-	-	0.018	0.007	-	0.035
5.5~5.99	0.030	0.034	-	-	-	0.025
6.0~6.49	0.010	-	0.035	-	-	0.015
6.5~6.99	-	-	0.053	-	-	-
7.0~7.49	0.010	0.017	0.035	-	0.008	-
7.5~7.99	-	0.017	-	-	-	0.015

Table 2. Relative frequency distribution of crack length

Crack length (mm)	Relative frequency (f_c/n)					
	Nangsan granite			Hamyeol granite		
	R-p	G-p	H-p	R-p	G-p	H-p
l_s / n	58.82/99	32.04/59	30.57/57	82.1/139	80.6/124	25.08/53
~0.29	0.191	0.190	0.193	0.216	0.186	0.243
0.3~0.39	0.111	0.154	0.140	0.151	0.161	0.208
0.4~0.49	0.152	0.170	0.211	0.137	0.113	0.151
0.5~0.59	0.121	0.204	0.123	0.058	0.129	0.113
0.6~0.69	0.081	0.103	0.053	0.108	0.073	0.057
0.7~0.79	0.061	0.086	0.140	0.086	0.081	0.094
0.8~0.89	0.071	0.052	0.035	0.079	0.040	0.019
0.9~0.99	0.040	-	0.053	0.043	0.032	0.038
1.0~1.09	0.040	0.018	0.018	0.029	0.057	0.038
1.1~1.19	0.051	-	0.035	0.036	0.057	0.038
1.2~1.29	-	-	-	0.007	-	-
1.3~1.39	0.020	0.018	-	-	-	-
1.4~1.49	0.020	-	-	0.029	0.032	-
1.5~1.59	-	-	-	0.014	0.024	-
1.6~1.69	0.010	-	-	-	-	-
1.7~1.79	0.020	-	-	0.007	0.016	-
1.8~1.89	0.010	-	-	-	-	-

Table 3. Total length of scan lines, total number of cracks, sum of crack length and density, spacing, length of trace cracks

Samples	Items	l_{total} (mm)	n (ea)	l_{sum} (mm)	$\bar{l}$ (mm)	$\bar{x}$ (mm)	λ (ea/mm)
Nangsan	R-p	432	99	58.82	0.59	4.36	0.23
	G-p	287	59	32.04	0.54	4.86	0.21
	H-p	354	57	30.57	0.54	6.21	0.16
Hamyeol	R-p	357	139	82.08	0.59	2.57	0.39
	G-p	411	124	80.60	0.65	3.32	0.30
	H-p	282	53	25.08	0.47	5.32	0.19

shows the crack density(λ) and numerical means of crack spacing and crack length calculated from total length of scan lines(l_s), number of cracks (n) and total length of crack(l_c) in Table 1 and Table 2.

Figure 2 and Figure 3 are show by a diagram of the relative frequency distribution of crack spacing and crack length. Also Figure 3 illustrates the crack density and numerical means of crack length.

Crack density of a rift plane is bigger and hardway plane's is smaller in Figure 4. As a hole, crack density ranges from 0.16 to 0.39 (ea/mm). Also, crack density in the Hamyeol granite wholly indicates the bigger values for each plane than the Nangsan granite one.

Crack length range from 0.47mm to 0.65mm, but it doesn't indicate a big difference for each plane. That's because crack length is not variable according to the region.

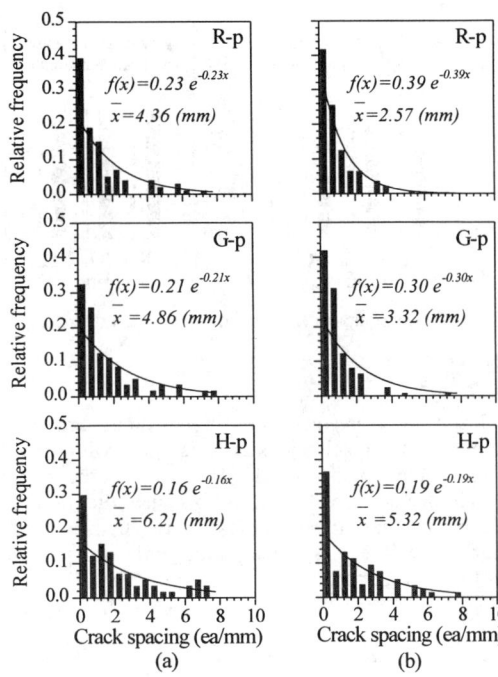

Figure 2. Relative frequency distribution of crack spacing (a) Nangsan granite (b) Hamyeol granite.

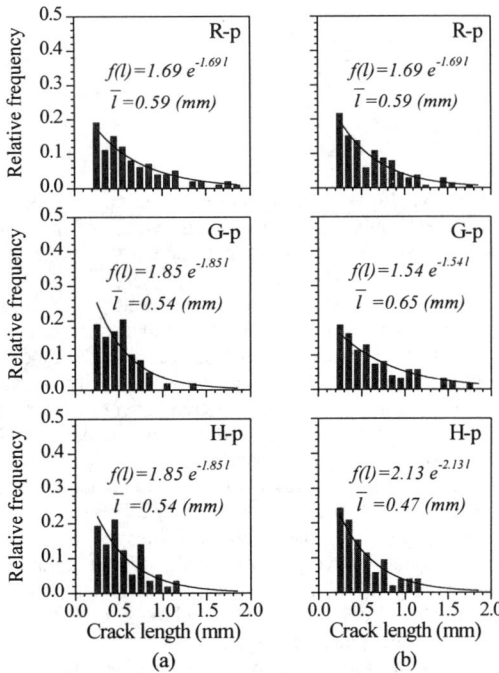

Figure 3. Relative frequency distribution of crack length (a) Nangsan granite (b) Hamyeol granite.

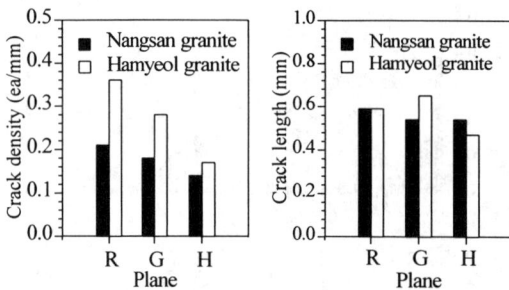

Figure 4. Distribution of crack density and crack length.

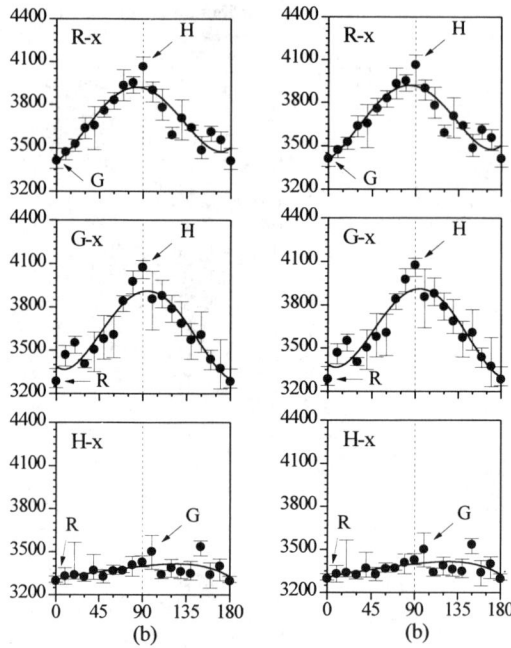

Figure 5. P-wave velocity(y axis, m/sec) versus angle (x axis, degree) (a) Nangsan granite, (b) Hamyeol granite.

From these results, the three planes are concerned with crack density and not with crack length. Additionally, cracks in quartz is the factor which decides crack length, because the number of microcracks in quartz is more than in feldspar. Therefore, crack length depends on the size of the quartz particle.

4.2 Mechanical properties

In this study, the following formulas were used as a criteria for the anisotropic ratio of p-wave velocity and tensile strength for each plane(Kudo et al., 1986).

$$R_{V_p} = \frac{V_{p\,max} - V_{p\,min}}{V_{p\,max}} \times 100\ \% \quad (4)$$

$$R_{\sigma_t} = \frac{\sigma_{t\,max} - \sigma_{t\,min}}{\sigma_{t\,max}} \times 100\ \% \quad (5)$$

where $R_{Vp}/R_{\sigma t}$=anisotropic ratio of p-wave velocity/ tensile strength, $V_{p\,max}$=maximum value of p-wave velocity and $V_{p\,min}$=minimum value of p-wave velocity.

Elastic wave velocity is often used to estimate a mechanical property and the inner structure of rock mass because testing is easy and nondestructive. In general, elastic wave velocity is reduced according to the increasement of crack density because it is sensitive to cracks which exists on the way of transmission.

The measured results of p-wave velocity and tensile strength are shown in Figure 5 and Figure 6. Each points is the average values of five data and fit line is polynomial with degree four. In figure 5, p-wave velocities and tensile strength at the angle of degree 0(=180) and degree 90 nearly indicates the extreme value. These results explain the three orthogonal planes. We realized from the thin section analysis that three planes are not the exact orthogonal planes.

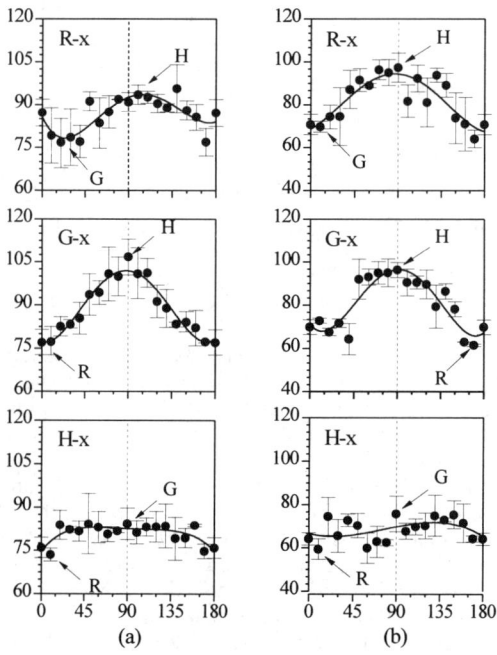

Figure 6. Brazilian tensile strength(y axis, kgf/cm^2) versus angle(x axis, degree) (a) Nangsan granite, (b) Hamyeol granite.

Table 4. P-wave velocities, tensile strength and anisotropic ratio (%)

Specimens	Items	V_p (m/sec)			σ_t (kgf/cm^2)		
		max.	min.	R_{V_p}	max.	min.	$R_{\sigma t}$
Nangsan	R-x	4030	3700	8	94	77	18
	G-x	4300	3600	16	107	77	28
	H-x	3800	3590	6	84	74	12
Hamyeol	R-x	4070	3410	16	97	64	34
	G-x	4080	3290	19	96	62	35
	H-x	3500	3300	6	76	59	22

Table 4 is the anisotropic ratios of p-wave velocities and Brazilian tensile strength for each axis obtained from the extreme values in Figure 5 and Figure 6. In Table 4, anisotropic ratio in the specimen of grain axis is the biggest because rift and hardway plane exist in specimens of grain axis. Also anisotropic ratios is variable in accordance with regions.

4.3 Correlation between crack density and physical properties

Table 5 contains the values of crack density, p-wave velocity and tensile strength. Table 6 is the anisotropic ratios calculated from the values of rift(minimum) and hardway plane(maximum) in table 5. Figure 7 shows the relationship between crack density and mechanical properties. In Table 5, p-wave velocity and tensile strength of the Nangsan granite indicates higher values than of the Hamyeol's one from 2 to 8 percents for each

Table 5. Crack density, p-wave velocity and tensile strength for each planes in korean granites

Samples	Items	λ (ea/mm)	V_p (m/sec)	σ_t (kgf/cm^2)
Nangsan	R-p	0.21	3600	75
	G-p	0.18	3750	81
	H-p	0.14	4160	100
Hamyeol	R-p	0.36	3290	61
	G-p	0.28	3460	70
	H-p	0.17	4070	97

Table 6. Anisotropic ratios

Samples	Items	Anisotropic ratio (%)		
		R_λ	R_{V_p}	$R_{\sigma t}$
Nangsan		32	18	31
Hamyeol		54	21	35

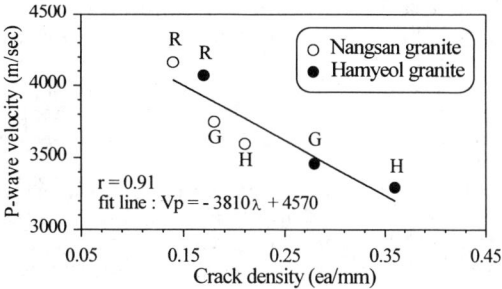

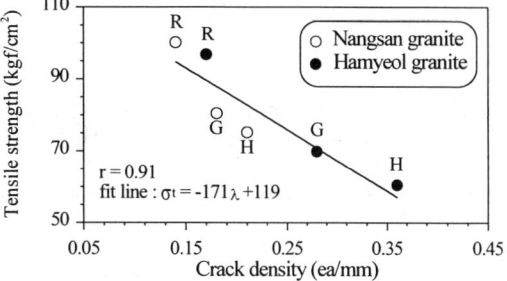

Figure 7. Relationships between crack density and mechanical properties.

plane. Whereas anisotropic ratio of Nangsan granite shows low values between 18 and 42 percents for each plane. Also Figure 7 shows a close relationship between crack density and mechanical properties(coefficient of correlation, $r=0.91$). From these results, it is evident p-wave velocity and tensile strength are reduced according to the increase of crack density. Also, the anisotropic ratio of tensile strength is closer to the anisotropic ratio of crack density than of the p-wave velocity's one. It is shown that tensile strength is a sensitive physical property to the crack density.

5 CONCLUSIONS

The results are summarized as follows ;

1. Crack densities were 0.21, 0.18 and 0.14 (ea/mm) in Nangsan granite, and 0.36, 0.28 and 0.17 (ea/mm) in Hamyeol granite, in order of Rift, Grain and Hardway plane, and also the anisotropic ratio of crack density was 32% in Nangsan granite, and 54% in Hamyeol granite. From these results, crack density is estimated as the factor which determines three orthogonal planes.

2. Since crack length in granite samples showed 0.47~0.65 mm, there is not much variation in Rift, Grain, Hardway planes and the places

of production, so crack length is mainly determined by the particle radius of rock forming minerals.

3. The anisotropic ratios resulted in 18%, 21% (p-wave velocity) and 31%, 35%(tensile strength) for each Nangsan and Hamyeol granite respectively.

4. P-wave velocity and tensile strength tended to decrease according to the increase of crack density, and there was a close relationship between p-wave velocity and crack density and also, between tensile strength and crack density.

6 REFERENCES

Anderson, D. L., B. Minster & D. Cole 1974. The effect of oriented cracks on seismic velocities. *J. Geophys Res.* 79: 4011-4015.

Birch, F. 1961. The velocity of compressional waves in rocks to 10 kilobars. *Part 2, J. Geophys. Res.* 66: 2199-2224.

Brown, E. T. 1981. Rock characterization testing and monitoring. *ISRM suggested methods*: Pergamon Press.

Douglass, P. M. & B. Voigt 1969. Anisotropy of granites: *A reflection of microscopic fabric, Geotechnique* 19: 376-389.

Kudo, Y., K. Hashimoto., O. Sano & K. Nakagawa 1986. Relation between physical anisotropy and microstructure of granite. *Japanese Society of Civil Engineering* 370/III (5): 193.

McWilliams, J. R. 1966. The role of microstructure in the physical properties of rock. In : *Testing techniques for rock mechanics, Am. Soc. Test Mat. STP* 402: 175-189.

Oda, M. 1982. Fabric tensor for discontinuous geological materials. *Soils and Foundations* 22 (4): 96-108.

Osborne, F. F. 1965. Rift, grain and hardway in some Pre-Cambrian granites. *Quebec Econ. Geol.* 30: 540-551.

Park H. M. & Lee, S. E. 1993. Preferred Orientation and Anisotropy of 'Rift' within Granites in the Qurrying Fields. *Journal of the Korean Institute of Mineral & Energy Resources Engineers* 30(6): 463-468.

Peng, S. S. & Johnson, A. M. 1972. Crack growth and faulting in cylindrical specimens of Chelmsford granite. *Int. J. Rock Mech. Min. Sci.* 9: 37-86.

Plumb, R., T. Engelder & D. Yale 1984. Near-surface in situ stress 3. Correlation with microcrack fabric within the New Hampshire Granite. *J. Geophys Res.* 89, No. B11: 9350-9364.

Priest, S. D. & J. A. Hudson 1976. Discontinuity spacing in rock. *Int. J. Rock Mech. Min. Sci. & Geomech. Abstr.* 13: 135-148.

Priest, S. D. & J. A. Hudson 1981. Estimation of discontinuity spacing and trace length using scanline. *Int. J. Rock Mech. Min. Sci. & Geomech. Abstr.* 18: 183-197.

Schroeder, C. 1972. Influence de la lithologique sur le comportement mecanique des roches soumises a essais de compression simple et bresiliens. *Engr. Geol.* 6(1): 31-42.

Soga, N., H. Mizutani, H. Spetzler & R. Martin 1978. The effect of dilatancy on velocity anisotropy in Westerly granite. *J. Geophys Res.* 83: 4451-4458.

Geological engineering research carried out on the Romania's territory concerning the geomechanical properties of the volcano-sedimentary rocks

E. Marchidanu
Technical University of Civil Engineering, Bucharest, Romania

ABSTRACT: The volcano-sedimentary rocks on the Romania's territory are disposed, in the main, on the western board of the Oriental Carpathians on the 10-12% from the surface of country. Detailed research for these rocks has been carried out especially for design and construction some important hydrotechnical storages placed on the such geological formations. The paper submits the mineralogical and petrographycal particularities of volcano-sedimentary rocks, the physical and mechanical properties as well as the behaviour these rocks at the hydrodynamic action of the ground water flows. Finally are presented the conclusions concerning the behaviour volcano-sedimentary rocks in the surface and underground excavations as well as foundation soil for diverse types of constructions.

1 GENERAL DATA

The volcano-sedimentary rocks are distributed on the 10-12% from the Romania's territory (Fig. 1).
From the genetical point of view and of geomechanical properties the volcano-sedimentary rocks can be classified in two main groups:

— Cineritics rocks, deposited and diagenesed into marine environment, constitued by the particles with dimensions to 2 mm, which are meet as the intercalations of volcanic tuff into complexes of sedimentary rocks which are constitued of marl clays, sand-stones, sands, ghips, salt etc.;

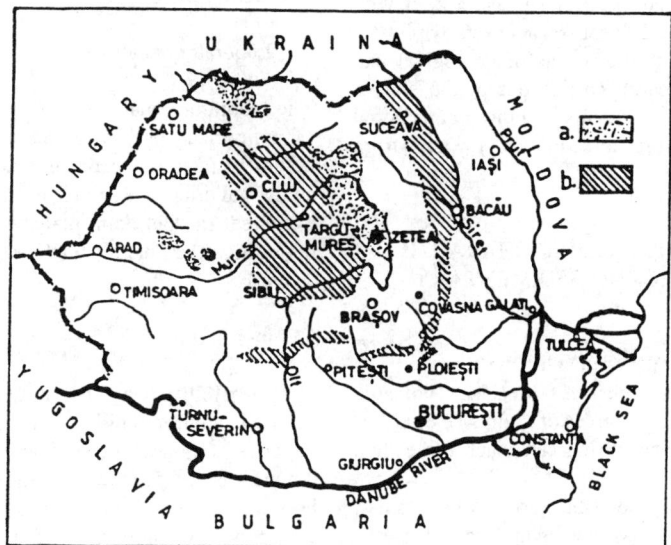

Figure 1. The spreading of volcano-sedimentary rocks on the Romania's territory: a - breccias and volcanic agglomerates; b - volcanic tuff into complexes of sedimentary rocks

– Volcano-sedimentary rocks, constitued of fragments and blocks of hard rocks, included into tuffogene matrix resulted from volcano explosions and deposited in the main, in aerial medium. These rocks are represented of brecciae and volcano-agglomerates.

The great non-homogeneity of volcano-sedimentary rocks, regarding the geomechanical characteristics and particularly vulnerability of tuffogene matrix at the erosion and hydrodynamic antrenment action of ground water flow, have generated a lot of suspicious for promotion the site of dams on the such rocks.

The Romania's economical needs have imposed the carry out a basin storage in a site integral constituted by the brecciae and volcano agglomerates.

Within period 1977-1980 were carried out the geological and hydrogeological studies in the site of Zetea dam (Fig. 1), this site being considered representative for the dams founded on the volcano-sedimentary rocks.

The research carried out have allowed to design and building a dam with 48 m height and a basin storage whose volume is 43 millions cu.m.

At present the basin storage Zetea there is putted into operation at the designed parameters, the dam having a very good behaviour.

The results of research performed in Zetea dam site have led at the promotion other sites of dams placed on the volcano-sedimentary rocks, at present being in the advanced execution phase a dam with the 90 m height and other some sites in study phase.

The paper presents the results of geological and hydrogeological research carried out in the site of Zetea dam, these results being considered representative for volcano-sedimentary rocks from Romania.

2 THE MAIN PETROGRAPHIC CHARACTERISTICS OF VOLCANO-SEDIMENTARY ROCKS

The volcano-sedimentary rocks of the area consists of deposits of agglomerates, brecciae, volcanic cinerites with different degrees of grain size distribution, fissuring and weathering can reach more then one cubic meter.

The predominantly cineritic horizons have small thickness, are gray - whitish, sometimes violet - reddish and have a cryptochrystalline structures and a porous texture. Hornblende crystals and feldspate and pyroxene phenocrystals are frequently encountered within the cinerite mass.

The brecciae and the volcanic agglomerates are predominant, as far as their extension is concerned. They consists of andesite blocks and fragments with angular or rounded shapes caught in a whitish tuffogene matrix.

From the grain size distribution point of view, the rock is non-homogeneous: the predominantly cineritic zones are in alternation with zones consisting of andesite fragments and blocks agglomerations with the interspaces filled with tuffogene material. The andesite fragments and blocks are gray-blackish or reddish: their structure is generally porphyric and their texture massive, and, when in an advanced weathering degree. the texture becomes vacuolar.

From the mineralogical point of view, the andesite fragments contain faldspates, hornblende and subordinately pyroxenes.

The main mass consists of microlites of feldspates, hornblende and volcanic glass.

On the base of chemical composition, determined on 14 representative samples, the volcanogene - sedimentary rocks have an intermediary or neutral character.

3 PHYSICAL AND MECHANICAL CHARACTERISTICS OF VOLCANO-SEDIMENTARY ROCKS DETERMINED IN LABORATORI AND IN-SITU

3.1 Laboratory analyses

The laboratory analyses were made with test pieces of cylindrical form with diameter of 50 mm and height of 50 mm, made of rock blocks taken from study galleries and cores extracted from drillings executed in the dam placement. The main data regarding the made analyses are presented in Table 1.

3.2 In-situ tests

<u>Compressibility tests</u>. The determinations regarding the rock compressibility were made in three study galleries, G_2, G_3, G_4, in underground rooms with dimensions of 5.0x5.0x2.5 m with plate press of Freyssinet type, fixed inside of steel concrete cylindres with diameter of 1,500 mm.

The rock settling was measures in the gallery hearth on the loading surface outline.

Table 1. Main physical and mechanical characteristics of the volcano-sedimentary rocks determined in laboratory

Characteristic		Symbol	Unit measure	Number of the tested samples n^0	Quadratic mean value deviation S	Mean value A^a
PHYSICAL	Apparent density	ρ_a	g/cm³	192	0.263	2.01
PHYSICAL	Porosity	n	%	192	8.69	24.04
PHYSICAL	Compressive dry strength	σ_{cd}	MPa	91	27.6	18.81
PHYSICAL	Compressive saturation strength	σ_{cs}	MPa	113	33.05	16.90
MECHANICAL	Modulus of elasticity	E	MPa	3	-	4870
MECHANICAL	Modulus of deformation	D	MPa	3	-	3385
MECHANICAL	Angle of internal friction — Top	Φ	Degrees	3	-	35°
MECHANICAL	Angle of internal friction — Residual	Φ_r	Degrees	3	-	30°
MECHANICAL	Cohesion — Top	c	MPa	3	-	0.9
MECHANICAL	Cohesion — Residual	c_r	MPa	3	-	0.3

The applying of the loading pressure was made in increasing-decreasing cycles, every pressure stage being mentained until the W_z settling decreased under 0.01 mm within 60 minutes.

The deformation modulus D and the elastic modulus E are calculated with the relations:

$$D = \frac{1.275\,(1-\nu^2)}{W_z} \cdot qa \quad (3.1)$$

$$D = \frac{1.275\,(1-\nu^2)}{\Delta W} \cdot qa \quad (3.2)$$

where ν is the Poisson's coefficient, a - the radius of the loading surface, W_z - the displacement in the direction of the applied pressure.

The compression-settling diagrams (q-W_z) for the compressibility tests made in G_2, G_3, G_4 galleries are presented in Figure 2.

<u>The rock-rock shearing tests.</u> In order to determine the shearing strength on rock blocks with dimensions of 80x80x40 cm, there were executed three tests, in study galleries G_2, G_3, G_4, every test consisting of shearing four rock blocks.

Due to the great rock unhomogenity from the structural and textural point of view, from a parameters of shearing strength are included in a relatively extended dispersion field. For obtaining

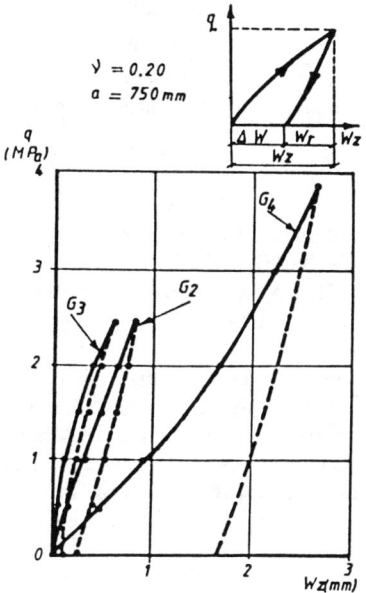

Figure 2. The compression-settling diagrams for the compressibility tests made in G_2, G_3, G_4 galleries

the average values of parameters of shearing strength, by statistic calculation using the method of the smallest squares, they processed the point values of the parameters σ - τ obtained from the tests made inside the three galleries.

The mean values of the parameters of shearing strength are presented in Figure 3.

The tested rocks was practically saturated.

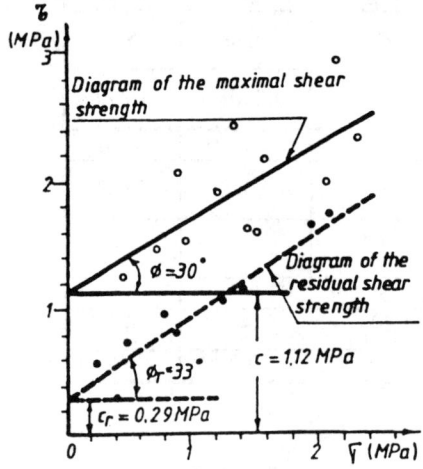

Figure 3. The medium Coulomb's line of the shearing strength determined by the tests made inside G_2, G_3, G_4 galleries

4 EXPERIMENTAL INVESTIGATIONS REGARDING THE CINERITIC MATRIX BEHAVIOUR AT WATER HYDRODINAMIC ACTION

The investigations carried out on the cineritic material in which andesite fragments are embedded had the purpose of establishing the conditions which make the mentioned material be unstable under the hydrodynamic action of water with this purpose, laboratory tests and in-situ determinations were performed.

4.1 Laboratory tests

In laboratory were made following tests:
- Evaluating the permeability coefficient and the critical gradient of hydrodynamic entrainment of the disintegrated cineritic material;

The investigation of the hydrodynamic erosion on fissures on a hydraulic model consisting of a rock sample which undisturbed material.

The first test consisted of introducing the cineritic material, with the dry volumic weight γ_d = 13.7-14 kN/cu.m, into a sufosimeter between two sand and gravel filters. The experiment was performed on the material sample subjected to vertical, decreasing loading, as follows: σ = 20; 14; 10; 5 and 0 kPa.

For every loading step the permeability coefficient was determined at increasing hydraulic gradients which ranged between i = 0.33 and i = 5.35 (Fig. 4).

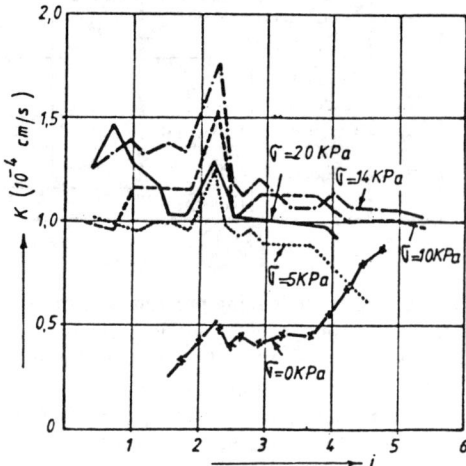

Figure 4. Relation between permeability coefficient k, hydraulic gradient i and overload σ

A transport of fine particles was observed from the base of the sample towards the upper part in flow sense, which led to the sample silting towards the upper part.

By reducing the overload σ to zero and increasing the gradient gradually, the hydraulic rupture was produced at i = 4.75; the upper part of sample, silted with fine particles separates from the looser lower part by means of a rupture surface.

Other test on disintegrated cineritic material was performed modeling a fissure which is represented circular plate situated at the upper part of the sample, on which the overload σ = 20 kPa acts, and the inner wall of the sufosimeter.

The permeability coefficient decreases gradually, due to the self-silting till reaching the value of

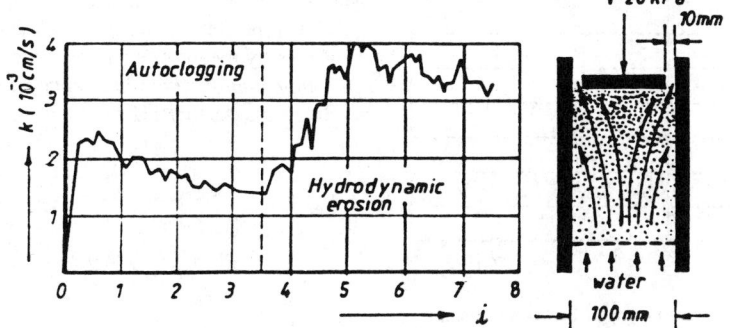

Figure 5. Illustration of self silting process and hydro-dynamic erosion on disintegrated cineritic material

approximated i = 3.5 of the hydraulic gradient; after that the fine particles begin to be expelled from the sample through the annular space which models the fissure and the permeability increases.

The evaluation of the phenomenon is illustrated in figure 5.

The recent test was performed on undisturbed cineritic material, in which small andesite fragments were embedded, which porosity n = 26-29% and the permeability k = 3-6x10^{-3} cm/s.

The test was achieved on hydraulic models consisting of rock samples with 10 cm length, in which a coaxial orifice of Φ = 6 mm was made. Every sample was introduced into a special case, with the watertighening between the sample and cell insured by means of a special paste.

Through the orifice into the sample, a water jet was passed at an increasing speed applied by steps. The time variation of the discharge flowing through the orifice was recorded for each step of the applied hydraulic gradient (Fig. 6).

The analysis of the diagrams reveals the fact that the internal erosion began only at flow gradients bigger than 20.

4.2 In-situ experiments

A group of drillings (Fig. 7) were carried out in order to determine the critical gradient of hydrodynamic entrainment of the cineritic material. The tested section was located on the depth interval of 10-15 m depth under the ground surface.

The experiment consisted of the injection water in central drilling and surveillance of the time variation of the water absorption q and the permeability coefficient k respectively (Fig. 8).

In the beginning, larger absorptions were

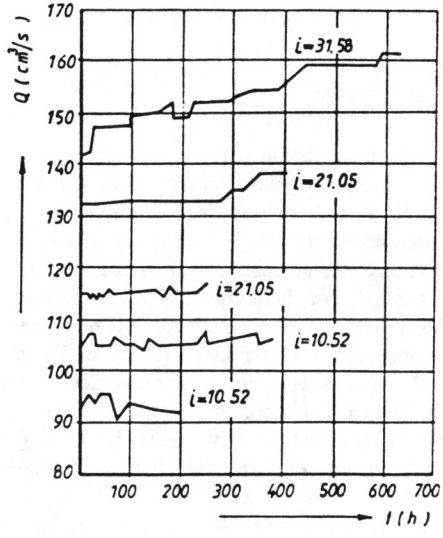

Figure 6. Influence of hydraulic gradients on cemented cineritic material stability

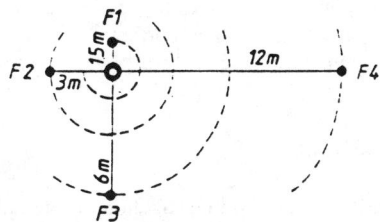

Figure 7. Drilling location in the experimental group

recorded with water refusal in drillings F_1 and F_3, followed by a self-silting process. After 31 hours, a new refusal of short duration took place in drilling

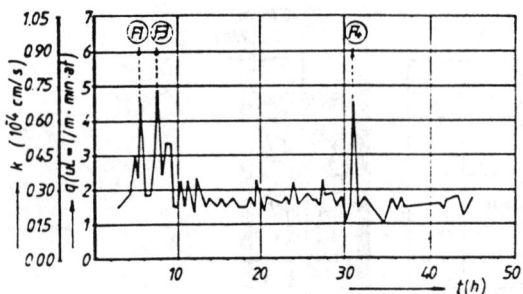

Figure 8. Time variation of permeability at a constant gradient

Marchidanu, E. & N. Sima 1995. Improvement of The slope stability on the volcanic-sedimentary Rocks by means of anchors. *Proc. of the Int. Symposium Anchors in Theory and Practice.* Salzburg, Austria : 221-228.

F_4, followed by the self-silting of the preferential water circulation ways towards drilling F_4; afterwards, the flow got established.

5 CONCLUSIONS

The volcano-sedimentary rocks on the Romania's territory have got a great unhomogeneity concerning the mineral and petrographic composition, the structure and texture being reflected in a large field of variation of the geomechanical characteristics.

In general, for buildings, inclusive for heavy buildings, volcano-sedimentary rocks can be used as foundation soil don't being necessary to be taken especial measures for improvement their geomechanical characteristics.

The volcano-sedimentary rocks have a good behaviour in the excavations. In the site Zetea dam were achieved the excavations with depth of 15-20 m and slopes of 60-80°. The galleries for dam, with length bigger than 350 m and 7 m diameter were carried out without propping up.

These rocks are considered as difficult concerning their using as foundation soil for dams due especially to the susceptibility of being instable to the action of infiltration currents by erosion and hydrodynamic carrying of the tuffogene matrix.

REFERENCES

Marchidanu, E. 1994: Investigation of internal hydraulic erosion of volcanogene sedimentary rocks in Zetea dam site and technical solutions adopted at the performance of the grout curtain of the foundation ground to prevent erosion. *Proc. of the Int. Symposium Geological Engineering and Environmental Protection.* Constantza, Romania :64-74.

The dynamic deformation moduli of some metamorphic rocks from Sri Lanka

U.de S.Jayawardena
Department of Civil Engineering, University of Peradeniya, Sri Lanka

ABSTRACT: The dynamic deformation moduli of different rock samples from various localities were determined using the ULTRASONIC MATERIALS TESTER model UCT2/1822.
Rectangular block specimens were used for the test. The travel times for compression (p) and shear (s) waves through the specimens were obtained by operating the particular equipment. After measuring the bulk density, the pulse velocities, Young's Modulus, Poisson's Ratio, Modulus of Rigidity and Bulk Modulus were calculated.
Charnockitic gneiss has higher values for Young's Modulus (90GPa) and thinly banded garnet biotite gneiss shows the lowest values (16GPa). Modulus of rigidity (G) and bulk modulus (K) also vary in the same way as Young's module (E). Poisson's Ratio varies from 0.21 to 0.38, decreasing with higher E values. The results which do not show the actual ranges of values at this preliminary stage, are tabulated for future references.

1 INTRODUCTION

1.1 General

The determination of the engineering properties of rocks in relation to geotechnical problems is required for each construction site. The various ways of determining these properties of rocks are:(a) by referring to values given in the appropriate literature; (b) by in situ experiments; or (c) by a programme of laboratory testing. As far as Sri Lankan rocks are concerned there are no freely available data to use as literature reviews. A classification of the various engineering properties of Sri Lankan rocks is, therefore, a necessity, and the author expects to carry out a programme of research to determine these properties. This paper is a part of the research and highlights various dynamic deformation moduli of some metamorphic rocks from Sri Lanka.

1.2 Summary of geology

About nine-tenths of Sri Lanka are underlain by Precambrian crystalline rocks. These are mainly high grade metamorphic rocks, which have been subdivided into three groups, namely, the Highland Series, the Vijayan Complex and the Southwestern Group (Cooray, 1967). The remaining rocks are sedimentary rocks of predominantly Miocene age in the north west, with some Jurassic sediments preserved in small faulted basins. The various geological formations in Sri Lanka can be summarized as given below:
Quaternary: Alluvium, beach sands, dune sands, clays, beach rocks, Red Earth, mottled gravels.
Miocene: Jaffna limestone
Jurassic: Sandstones, shales, grits and arkoses.
Precambrian:
(a) Vijayan Complex-Migmatites, biotite gneiss, hornblende biotite gneiss, granite and granitic gneiss.

(b) Highland Series-Quartzites, granulites, garnet sillimanite graphite schists, sillimanite gneiss, charnockitic gneiss, marble and dolomite.

(c) Southwestern Group: Wollastonote scapolite gneisses, cordierite gneiss, charnockitic gneiss and granitic gneiss.

Highland Series rocks are present mainly in the hill country of the Island and Vijayan rocks are found in the low-lying areas on either side of the Highland Series. The Southwestern group rocks occupy the southwestern region of the Island (Fig.1).

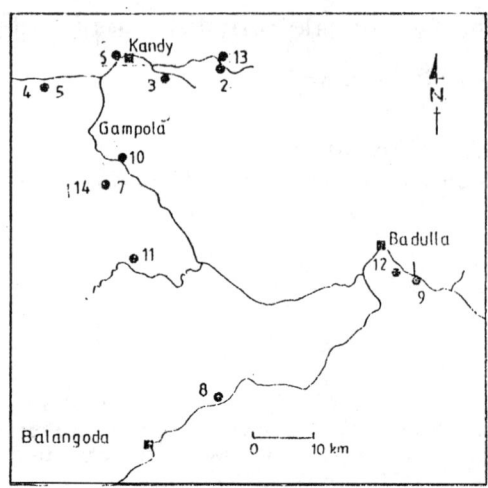

Fig.2 Locations of rock samples used in the tests (refer Table 2 also).

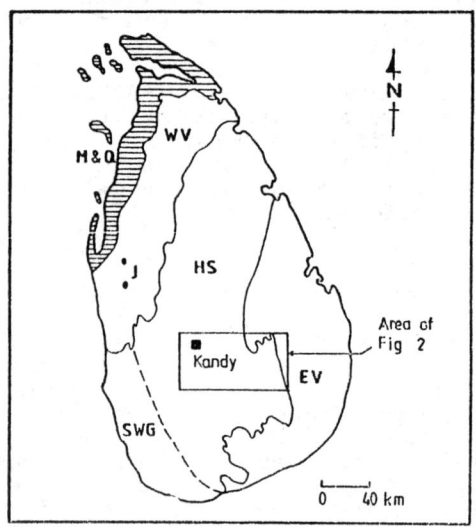

Fig.1 Main geological subdivisions of Sri Lanka, HS=Highland Series, EV=Eastern Vijayan Complex, WV=Western Vijayan Complex, SWG=South Western Group, J=Jurassic, M=Miocene.

2. EXPERIMENTAL METHOD

2.1 Materials

The investigation was limited to some major rock types from Hill Country, all of which belong to the Highland Series. Irregular rock samples were collected from quarries at different localities, the samples were being selected randomly to represent the major rock type in the region.

Rectangular block specimens were prepared according to the ASTM, No.19 (1980) for the tests. The length of the specimens vary from 120mm to 150 mm; the width and thicknesses were between 100mm and 80mm. The end planes of the specimens were smoothned for better contact with the transmitter and the receiver of the testing equipment.

2.2 Laboratory Tests

The following tests were made in the laboratory.

(i) The mineralogy, texture and planes of weakness of each sample were observed. Ten samples of each rock were used to find the bulk density of the rock.

(ii) The ULTRASONIC MATERIALS TESTER type UCT2/1822 (DAWE Instruments Limited, London) was employed to measure the signal transmission time through the rock materials. Figure 3 shows a layout of electronic components for sonic velocity testing of rock using P and S waves (the elastic waves that travel through the interior of a rock mass). These are called body waves. The P type of body waves induces longitudinal oscillatory particle motions in the direction of transmission; and the S type body waves vibrates in a vertical plane normal to the axis of transmission. The velocity of S waves is less than that of P waves (Atwell and Farmer, 1976). The transmitter converts electrical pulses (which were generated by pulse generator) into mechanical pulses and the receiver converts mechanical pulses into electrical pulses. The travel time for compression (P) and shear(S) waves between transmitter and receiver were measured from the cathode-ray oscilloscope. Two specimens of each rock type were tested.

(iii) The dimensions of each specimen were measured and the natural water content and the bulk density according to ASTM, No14 (1980) were determined.

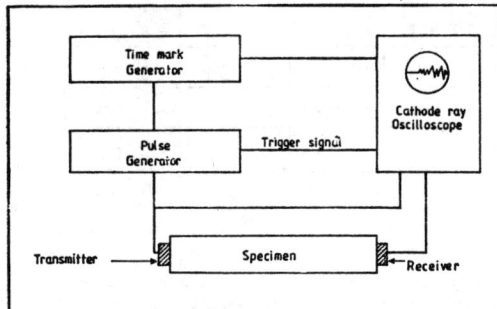

Fig.3 Layout of electronic components for sonic velocity testing of rock using P and S waves.

3 CALCULATIONS

According to the theory, the rock is assumed to be a continuous, homogeneous, linear-elastic, isotropic material. This material obeys Hook's law of proportionality between stress and strain; that is, the strains are linear functions of stresses. Various relationships between stresses and strains are given by the following constants. These basic elastic constants are referred to as dynamic deformation moduli of materials. They are:

Young's Modulus of Elasticity (E)
= normal stress / normal strain

Poisson's Ratio (P)
= lateral strain/axial strain

Modulus of Rigidity (G)
= shear stress / shear strain

Bulk Modulus (K)
= hydrostatic stress field/dilation
(Obert and Duvall, 1967)

Wave velocities and the deformation moduli were calculated using the following equations (Attwell and Farmer, 1976).

$$V_p = d/t_p \quad \ldots \ldots \ldots \ldots \ldots (1)$$
$$V_s = d/t_s \quad \ldots \ldots \ldots \ldots \ldots (2)$$

where
V_p = velocity of P waves,
V_s = velocity of S waves,
d = distance between the two transducers, in meters.
t_p and t_s = time taken by P and S waves to travel the distance d in seconds.

$$E = \frac{\{DV_s^2 [3-4(V_s/V_p)^2]\}}{\{1-(V_s/V_p)^2\}} \ldots \ldots (3)$$

$$P = \frac{[1-2(V_s/V_p)^2]}{2[1-(V_s/V_p)^2]} \ldots \ldots \ldots (4)$$

$$G = DV_s^2 \ldots \ldots \ldots \ldots \ldots \ldots (5)$$

$$K = \frac{\{DV_p^2[3-4(V_s/V_p)^2]\}}{3} \ldots \ldots (6)$$

where D = bulk density of rock in kgm^{-3}. E, G and K are in GPa.

4 RESULTS AND DISCUSSION

Table 1 shows the values that were obtained in this preliminary investigation and Figure 4 shows bar charts of the dynamic deformation moduli and porosity for the various rock types.

Except for quartzite, all tests were carried out on fresh rock samples which showed no fractures. According to the theory of elasticity the rock is assumed to be homogeneous and isotropic, but gneisses and impure marbles are not always homogeneous due to the different thicknesses of mineral bands and randomly distributed mineral concentrations.

Young's modulus (E) is generally high in charnockitic gneiss, and lowest in garnet-biotite gneiss. In the later, the biotite flakes, garnet grains and quartz and feldspar are finer grained than in the other rocks. E values for other rock types fall within these limits.

Table 1. Values of dynamic deformation moduli and related physical properties of some metamorphic rocks from Sri Lanka.

Sample No	Vp m/s	Vs m/s	D, 10^2 kgm^{-3}	n%	E, GPa	P	G, GPa	K, GPa
1	5233	2616	25.95	0.04	47.38	0.33	17.75	47.37
2	5133	2204	28.34	0.1	38.23	0.38	13.76	56.35
3	5243	2527	28.27	0.03	48.76	0.34	18.05	53.70
4	3579	1879	27.84	0.18	25.76	0.31	9.82	22.58
5	4725	2304	28.65	0.18	40.91	0.34	15.20	43.73
6	3450	1493	26.82	0.05	16.56	0.38	5.97	23.97
7	5343	3017	28.72	0.03	66.22	0.26	26.14	47.22
8	6270	3274	30.81	0.02	88.06	0.33	33.02	77.20
9	5261	2751	27.14	0.03	53.89	0.31	20.54	47.8
10	4780	2677	26.52	0.04	48.35	0.27	19.00	35.31
11	5571	2851	29.12	0.02	62.69	0.32	23.66	58.89
12	6352	3514	28.70	0.05	90.93	0.27	35.44	68.66
13	6260	2975	29.81	0.06	71.46	0.35	26.38	81.72
14	6131	3687	27.18	0.04	89.91	0.21	36.95	53.02

Table 2. Rock types of samples in Table 1.

Sample No	Rock Type	Locality
1	Quartzite	Passara
2	Impure marble	Digana
3	Pure Marble	Ampitiya
4	Hornblende biotite gneiss	Kadugannawa
5	Garnet hornblende biotite gneiss	Kadugannawa
6	Garnet biotite gneiss	Dodanwela
7	Garnet biotite sillimanite gneiss	Kotmale
8	Garnet graphite sillimanite gneiss	BelihulOya
9	Garnet biotite granulite	Passara
10	Pink granite	Atabage
11	Garnet granulite	Talawakele
12	Charnockitic gneiss-intermediate	Badulla
13	Mafic pyroxene granulite	Digana
14	Garnetiferous charnockitic gneiss-acidic	Kotmale

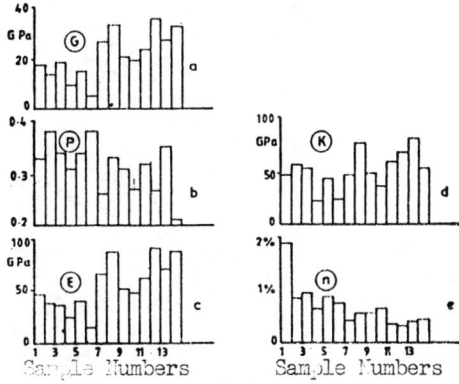

Fig.4. Variation of dynamic deformation moduli and porosity of rocks from Sri Lanka a=Modulus of Rigidity, b=Poisson's Ratio, c= Young's Modulus, d= Bulk Modulus, e=porosity

Quartzite does not show real values owing to the irregularity and intensity of fractures in the rock samples. G and K also vary in the same way as E in the various rock types. It is also seen that less porous rocks have higher E. Poisson's Ratio (P) decreases with higher E values.

These results are preliminary and do not show the actual range of the values for each rock type as tests were done on a limited number of specimens in this investigation.

Acknowledgment: The author thanks to the staff of the materials and soils laboratories of the Department of Civil Engineering, University of Peradeniya for their assistance to carry out this investigation.

REFERENCES

American Society for Testing and Materials (ASTM), Part 19, 1980, USA.
American Society for Testing and Materials (ASTM), Part 14, 1980, USA.
ATTEWELL, P.B., and FARMER, I.W., 1976: Principles of Engineering Geology, Chapman & Hall Ltd., London.
COORAY, P.G., 1967, An Introduction to the Geology of Ceylon, Ceylon National Museums Dept, Colombo.
OBERT, L. and DUVALL, W.I., 1967< Rock Mechanics and the design of Structures in Rock, John Wiley & Sons, Inc. New York.

Mechanical properties of shales for estimation of damage zone dimensions around waste repositories

P.A. Nawrocki, M.B. Dusseault, B. Davidson & M. Kim
University of Waterloo, Ont., Canada

ABSTRACT: Laboratory results of the extensive testing program conducted on shales are summarized in this paper. The program consisted of hydrostatic and triaxial compression tests, and has been designed to collect constitutive parameters needed for realistic, non-linear modelling of stresses around openings. Results obtained are interpreted in the context of a general, semi-analytical model proposed for that purpose. The model introduces stress-dependent elasticity functions $C(\sigma)$ (compressibility modulus) and $D(\sigma,\tau)$ (inverted shear modulus). A special method of interpreting triaxial test results to obtain constitutive parameters needed for the model is proposed, and use of obtained stress predictions for estimation of damage zone dimensions around waste repositories or other openings is outlined.

1 DISCUSSION AND SCOPE

The problem of opening stability assessment and damage zone evaluation is of importance for the petroleum industry and for deep waste repositories, because of environmental and economical factors. The development of a damaged zone dominated by microfissures implies a change in permeability, with clear consequences on rock sloughing around boreholes, or impairment of flow properties in the case of waste containment structures. Damage can be of different types (grain damage, cohesion loss, microfissuring), and can be expected to result in a reduced stiffness modulus E, what, in turn, will result in a redistribution of stresses around an opening. Thus, from a stress point of view, damage can be estimated as a Young's modulus variation with distance from the opening wall.

A zone of significantly reduced stiffness is probably generated for at least several opening radii away from the opening wall. Different factors can contribute to damage zone growth. For example, damage around a borehole may be expected during drilling, production of fluids, high pressure injection, or thermal effects (Dusseault and Gray, 1993). In the case of shafts and tunnels, boring leads to damage, and blasting cycles may lead to cumulative damage. Heating a borehole will result in an increase in the mean stress σ near the borehole, and this may increase E if the rock is stress-sensitive. However, if the heating degrades the cohesion, the stiffness may drop considerably, c.f. Hojka et al. 1993.

For full analysis of repositories, the transport properties should be functionally linked to the degree of damage and to the state of stress. This article stops short of this step, but we emphasize its importance and note that the amount of experimental data relating damage to transport properties is extremely small. We believe that this area is worthy of extensive effort to permit rational evaluation of waste repositories.

Because the amount of damage that a material experiences is related to the stress level (or strain level, as there is a direct coupling), a realistic solution for borehole stresses should address issues of damage and non-linear (NL) behaviour. In general, however, formal damage mechanics approaches (Krajcinovic, 1989) are difficult to incorporate directly into analytical treatments of simple geometries such as generally axisymmetric or cylindrical plane strain, thus alternatives will be sought. Both stress distributions and mechanical properties (e.g. triaxial test data) are required to estimate damage zone extent (Young's modulus variation) away from an opening. Different predictive models have been developed for that purpose, linking rock stresses to rock deformation through experimentally determined elasticity constants of the rock material. For realistic simulation of rock stresses, such "constants" have to be considered as elasticity functions rather then constants. There is now general agreement that linear elasticity (LE) analyses invariably underpredict opening stability, and that models which are more realistic (and less conservative) in their predictions should be utilized. Examples are NL and elasto-plastic models.

NL rock properties have not yet attracted adequate attention. One limited but widely used function for simulation of stress-strain curves in finite element (FE) analysis was formalized by Duncan and Chang (1970). Except for this hyperbolic model, incorporation of NL and inelastic material behaviour has been avoided except in complex models suitable only for FE analysis (e.g. Vaziri 1986). On the other hand, NL finite element models are seldom used for waste repositories design, in part because the constitutive parameters are seldom measured with a precision that would justify use of a complex analysis.

Different semi-analytical, NL models postulate different mathematical representations for the functions mentioned above. The Pressure-Dependent Modulus model (PDM model; Santarelli et al. 1986) and the Radius-Dependent Modulus model (RDM model; Nawrocki and Dusseault 1995) provide examples of models where E is not constant. Santarelli in his model introduced a confining stress dependent Young's modulus, i.e. $E=f(\sigma_3)$, and Nawrocki and Dusseault used an assumption of stiffness related to damage or radial distance measured from the opening wall, $E=f(r)$. An alternative approach to introducing material non-linearities into the modelling of borehole stresses has been proposed by Nawrocki et al. (1996), in an attempt to fill the gap between simple, LE solutions and FE approaches. It is based on the assumption that mechanical behaviour of geomaterials at any stress level can be presented as a superposition of hydrostatic and deviatoric states of deformation which are governed by stress dependent elasticity functions $C(\sigma)$ and $D(\sigma_3,\tau)$.

In this paper we summarize the experimental results of the extensive testing program conducted on shales. We do that in the context of a more general version of the NL model mentioned above, assuming that hydrostatic deformation is governed by a mean stress-dependent compressibility modulus, $C(\sigma)$, whereas deviatoric deformation is governed by both mean stress- and shear stress-dependent inverted shear modulus, $D=D(\sigma,\tau)$. Such an assumption seems to be a natural one and it may represent a better method of borehole stress analysis, as current models are based almost exclusively on linear elasticity or σ_3-dependent elasticity models. It is also more general than PDM or RDM models because both σ and τ depend on σ_3. Also, we propose a special method to interpreting triaxial test results to obtain constitutive parameters needed for the model proposed. The functions obtained can then be used in modelling of opening stresses based on NL elastic relationships similar to those proposed by Nawrocki et al. (1996). They, in turn, can be used in the experimentally obtained $E(\sigma_3)$ function, thus assuming that the radial stress in the field plays the role of confining stress σ_3. In this way, the Young's modulus variation with distance from an opening wall can be defined, providing an estimate of damage zone dimensions.

2 EXPERIMENTAL PROCEDURES

The main goal of the testing program discussed below is to determine constitutive parameters for our two proposed elasticity functions $C(\sigma)$ and $D(\sigma,\tau)$. The hydrostatic test is intended to measure the volumetric deformation of rock specimens and to provide data necessary to establish the $C(\sigma)$ dependency, and triaxial tests provide information on shale behaviour in shear and D-modulus variation with both σ and τ.

2.1 *Sample preparation*

The testing program has been conducted on shale samples collected from the Queenston Formation near Windsor, Ontario. Queenston shale is a non-fissile compaction shale with harder and more durable shale bands parallel with the bedding and occasionally at right angles. It's colour is brick red with occasional green bands (calcium or durable material); it is homogeneous and isotropic with approximately 60% clay minerals.

Upon collection, the cores were wrapped, marked, and stored in a refrigerator at a temperature of about -10 to 0 degrees Celsius. The cores were cut into right circular cylinders with a height to diameter ratio of 2.0 to 3.0 and a diameter of about 45 mm. The ends of the specimen are cut flat and perpendicular to the longitudinal axis, and the sides are smooth. Specimens are taped, wrapped in Saran Wrap, marked and then stored in a cool place. Once the specimen is properly prepared, four strain gauges are glued onto the specimen in the middle of the sample, and 90 degrees apart. The specimen is sealed using a rubber membrane, and then attached to the lid of the hydrostatic cell. The cell is completely filled with hydraulic fluid and is connected to a hydraulic line and pressure intensifier.

2.2 *Hydrostatic compression test*

To accurately determine the $C(\sigma)$ function, hydrostatic compression tests have been performed using unloading-reloading cycles, with bulk modulus K being determined at unloading, Cristescu (1989). The compressibility modulus can then be calculated as $C=1/K$. Both in hydrostatic and triaxial tests load on samples was increased in steps. In the case of hydrostatic test, load on the sample has been increased in small increments $\Delta\sigma=5$MPa, and with each load increment an instantaneous volumetric strain $\Delta\varepsilon_v$ takes place. In the first-order approximation K can be obtained from a relationship $\Delta\sigma = K(\sigma)\Delta\varepsilon_v$, where $K(\sigma)$ depends on the mean stress. After each load increase sample was kept under constant load for 20 minutes before partial unloading. Keeping load constant allows deformation

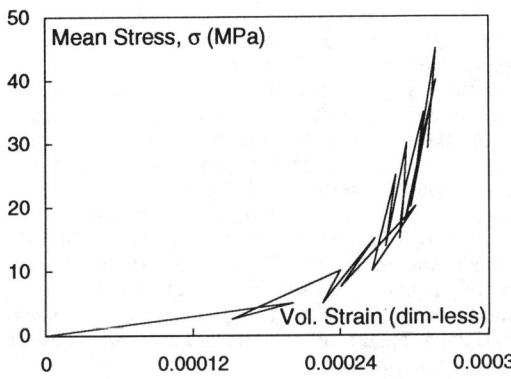

Fig.1 Hydrostatic test results.

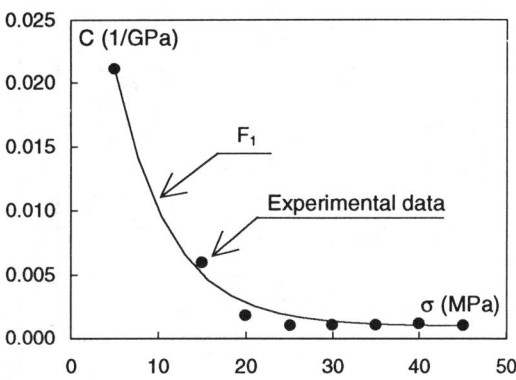

Fig. 2: Hydrostatic test results

to stabilize and reduces the effect of creep on the measured values of elastic moduli. During unloading the pressure has been reduced by 5% of its current value each time at one minute intervals. Unloading continued until 75% of the initial pressure has been reached, and then the reloading sequence begun until the initial pressure increased by 5 MPa (the second loading step). This loading-unloading sequence has been repeated until high stresses are achieved, where virtually no deformation increase has been observed in 20 minutes periods of constant load. Elastic moduli have been measured on unloading parts of the compression curve. We believe this to be more representative of unloading moduli which clearly govern the deformation response when a borehole or a tunnel is advanced. Experimental data show that the end of the unloading process can be somewhat non-linear. However, for simplicity we usually approximate the entire unloading portion with a linear relationship, choosing for K a 'secant' value.

2.3 Triaxial compression test

In order to obtain useful information on $D(\sigma,\tau)$ behaviour, triaxial tests should be performed for many different values of σ_3, so as to delineate the deviatoric response over the range of stresses expected in situ. For this project we have used confining stresses ranging from 2.5 to 35 MPa, differing by 2.5 MPa from test to test. About 30 triaxial tests were performed using ISRM Suggested Methods for Rock Testing, and the best results were used in the analysis. Triaxial tests have been carried out using successive loadings similar to the procedure for the isotropic tests. Each triaxial test consisted of two parts; the hydrostatic loading and deviatoric loading. During the deviatoric stage, σ_1 has been gradually increased to the desired deviatoric stress level and then is held constant for 20 minutes to allow creep deformation to stop and not influence the measured elastic moduli. Unloading began at deviatoric stresses σ_{dev}=3.75, 11.25, 18.75 MPa, and so on, with 7.5 MPa increments. These values of σ_{dev} were selected because of the way we intended to read test results (see below). When creep deformation was negligible, the specimen was partially unloaded to allow the measurement of elastic constants corresponding to the current level of σ_{dev}. The rock specimen is reloaded, deviatoric stress increased to its next unloading value, and the same unloading-reloading sequence repeated to obtain elastic constants corresponding to next value of σ_{dev}, or τ. Elastic moduli (E and ν) have been determined by taking the slope of the unloading curve. Then, the shear modulus is calculated using the relationship $G=E/2(1+\nu)$, and $D=1/G$.

3 TEST RESULTS AND INTERPRETATION

In the case of isotropic tests, laboratory results are presented in a σ vs ε_v format, and the bulk modulus, K, corresponding to the current stress level σ, is determined by finding the slope of the curve during the unloading sequence of the hydrostatic test, and C=1/K. For $C(\sigma)$, we should in principal obtain a strictly decreasing function of σ with a horizontal asymptote. Thus, to obtain stable values of compressibility modulus, it is necessary to conduct hydrostatic compression tests at high stresses. However, for porous materials such as shales, it is not possible to "close" all the pores, and a pseudo-asymptote can be chosen, beyond which the non-linearity becomes unimportant for the application considered.

Results of one of the hydrostatic tests are shown in Fig. 1. The maximum mean stress attained during that test was 45.08 MPa. The compressibility modulus C calculated for the data shown in Fig. 1 is presented in Fig. 2 as a function of σ. The asymptotic trend of $C(\sigma)$ dependency is clearly visible in Fig. 2. Data in Fig. 2 are approximated by a function:

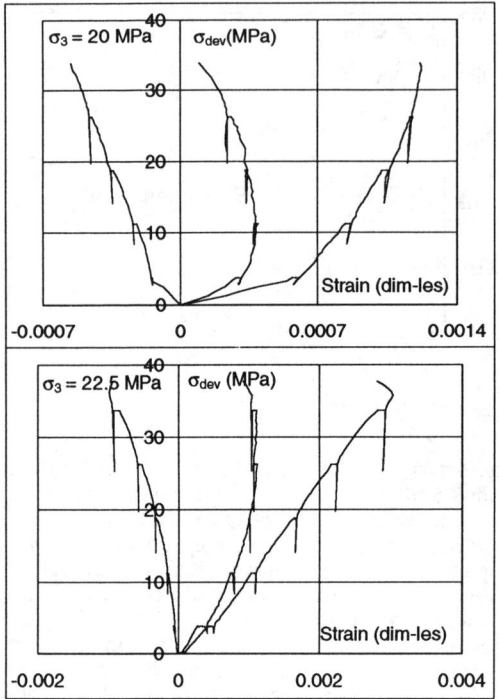

Fig.3: Triaxial test results

Table 1: Elastic constants for the tests above.

σ_3=20 MPa					
σ_{dev}@unload (MPa)	E (GPa)	v	K (GPa)	G (GPa)	D (1/GPa)
3.75	31	0.25	20	12	0.0802
11.25	127	0.28	97	49	0.0203
18.75	201	0.34	211	75	0.0134
26.25	265	0.40	435	95	0.0105
σ_3=22.5 MPa					
σ_{dev}@unload (MPa)	E (GPa)	v	K (GPa)	G (GPa)	D (1/GPa)
3.75	99	0.13	44	44	0.0227
11.25	247	0.16	119	107	0.0094
18.75	290	0.20	161	121	0.0083
26.25	263	0.26	184	104	0.0096
33.75	223	0.32	208	85	0.0118

$$F_1: \quad C(\sigma_n) = C_1 - \gamma [1 - e^{-\delta(\sigma_n - 1)}] \quad (1)$$

Note, that F_1 is an exponential function which has a horizontal asymptote $C_1 - \gamma$. In Eqn.(1) σ_n is a normalized mean stress, $\sigma_n = \sigma/\sigma_{(1)}$; γ and δ are constants; C_1 is the compressibility modulus corresponding to the first data point, and normalization is with respect to the mean stress $\sigma_{(1)}$ of that point. Thus, $C_1 = C(\sigma_{n(1)})$, and γ can be determined from the boundary condition at the last data point ($\sigma_{(2)}$; C_2) giving $\gamma = (C_1 - C_2)/[1 - 1/\exp(\delta(\sigma_{n(2)} - 1))]$, where $\sigma_{n(2)} =$ $\sigma_{(2)}/\sigma_{(1)}$. By changing δ it is possible to obtain a good approximation of data with F_1. In the case of the curve shown in Fig. 2, $\delta = 0.8$.

Note, that compressibility modulus decreases with σ and tends towards a constant value C_{ult} which is about 0.001 1/GPa, when all pores and microcracks are closed. Thus, $C(\sigma)$ is a monotonic function of σ in the interval $0 < \sigma < \sigma_o$ only. It is expected that for $\sigma = \sigma_o$ all the cracks are already closed, and that for $\sigma > \sigma_o$ the behaviour of the rock is nearly linearly elastic and reversible. In our case, σ_o is approximately equal to 40 MPa. This value can be a reference value for $C(\sigma)$ used to generate dimensionless variables. Therefore, F_1 decreases and tends towards a horizontal plateau at high stresses.

3.1 Triaxial test results

Typical results of two of the triaxial tests (σ_3=20 MPa and σ_3=22.5 MPa), following the procedure discussed previously, are shown in Fig. 3. The elastic moduli E and v were calculated directly from unloading slopes of the compression curve. With E and v known, G, K, and D can be calculated. Thus, the set of five elastic moduli is available for each point of unloading. Elastic moduli corresponding to the compression curves shown in Fig. 3 are given in Table 1.

To read the triaxial test results we use the method recently recommended by Nawrocki et al. (1997), who also presented all the details of their recommended approach. Fig. 4 provides a schematic explanation of the method used to read test results. It is plotted in a σ_{dev} - σ format, and each inclined line in Fig. 4 corresponds to one triaxial test. During the isotropic part of testing, σ_{dev} is is approximately zero. Thus the hydrostatic loading coincides with the horizontal axis, whereas inclined lines correspond to deviatoric loading. Moreover, crossing points of test lines and vertical lines where we read test results define the deviatoric stress of unloading; that is, the elastic moduli E, v, K, G, and D are determined for stress combinations corresponding to each crossing

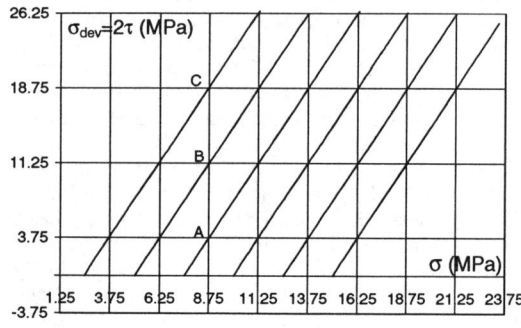

Fig. 4: Triaxial tests design chart.

Table 2: Elastic parameters from several tests.

σ=18.75 MPa						
σ₃ (MPa)	τ (MPa)	E (GPa)	ν	K (GPa)	G (GPa)	D (1/GPa)
17.5	1.88	17	0.24	11	7	0.1495
15.0	5.63	163	0.40	283	58	0.0172
12.5	9.38	270	0.39	402	97	0.0103
10.0	13.13	243	0.91	x	x	x
σ=26.25 MPa						
σ₃ (MPa)	τ (MPa)	E (GPa)	ν	K (GPa)	G (GPa)	D (1/GPa)
25.0	1.88	-53	0.46	x	x	
22.5	5.63	247	0.16	119	107	0.0094
20.0	9.38	237	0.21	137	98	0.0102
17.5	13.13	339	0.31	296	129	0.0077
σ=38.75 MPa						
σ₃ (MPa)	τ (MPa)	E (GPa)	ν	K (GPa)	G (GPa)	D (1/GPa)
35.0	1.88	-153	0.29	x	x	x
32.5	5.63	31	0.35	35	12	0.0866
30.0	9.38	354	0.28	270	138	0.0072
27.5	13.13	165	0.29	128	64	0.0156
25.0	16.88	314	0.89	x	x	x

point. In other words, if we use the data along the inclined lines of equation σ-σ'=const or σ-τ_{oct}/√2=const (here σ' is the equivalent stress and τ_{oct} the octahedral shear stress) we will determine the dependency of the elastic parameters on mean stress and on the combination of the invariants, say σ-τ_{oct}/√2. In order to directly determine the dependency on the two invariants σ and τ_{oct}, one reads the data along the vertical line first to find the moduli dependency on τ (as if the tests would be done in "true" triaxial conditions), and afterwards one reads the data along the horizontal line to find the dependency of the coefficients on mean stress. Once elastic moduli have been determined for all the triaxial tests, results can be interpreted by graphing the elastic modulus D as a function of τ. Thus, to obtain the D(σ,τ) function, we look at D dependency on τ first (for given σ), reading triaxial test results on vertical lines, that is, for σ=const. Then D dependency on σ can be introduced by comparing coefficients of functions approximating the D(τ)-dependency obtained for different mean stresses.

Thus, for given value of σ, D modulus can be plotted as a function of shear stress, τ, by picking moduli values from the unloading points of triaxial tests crossed by the vertical line being used (points A, B, and C in Fig. 4). In our case, 13 such graphs can be plotted (the case of σ=3.75 MPa is disregarded because only one inclined line is crossed by the vertical line plotted at σ=3.75). Table 2 presents the results of such analysis for several values of mean stress (σ=18.75, 26.25, and 38.75 MPa). Then D modulus can be plotted as a function of τ, and data can be approximated with functions. Such a D(τ) function can be found for any value of σ for which valid triaxial test results are available. It has been found that the experimental data obtained can be best modelled by logarithmic functions. Some of the functions found are listed below:

$$D(\tau) = -0.0919\ln(\tau) + 0.1997 \quad (for\,\sigma = 18.75)$$
$$D(\tau) = -0.0016\ln(\tau) + 0.0127 \quad (for\,\sigma = 26.25)$$
$$D(\tau) = -0.0334\ln(\tau) + 0.1083 \quad (for\,\sigma = 28.75)$$
$$D(\tau) = -0.0008\ln(\tau) + 0.0099 \quad (for\,\sigma = 31.25)$$
$$D(\tau) = -0.0897\ln(\tau) + 0.2319 \quad (for\,\sigma = 38.75)$$

(2)

The final step involved is to examine the graphs (functions) corresponding to different σ's to find the coefficients of a function approximating how the D-dependency on τ varies as a function of σ. If the general form of the logarithmic function is D(τ)=Aln(τ)+B, then we plot A and B against σ to find their dependency on mean stress. The following dependencies have been found for our data:

$$A(\sigma) = -0.001\sigma^2 + 0.047\sigma - 0.681$$
$$B(\sigma) = 0.002\sigma^2 - 0.099\sigma + 1.447$$

(3)

Knowing A(σ) and B(σ), one can finally write the full D(σ,τ) function:

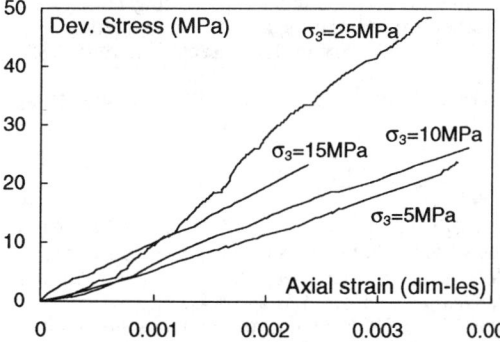

Fig. 5: Triaxial test results

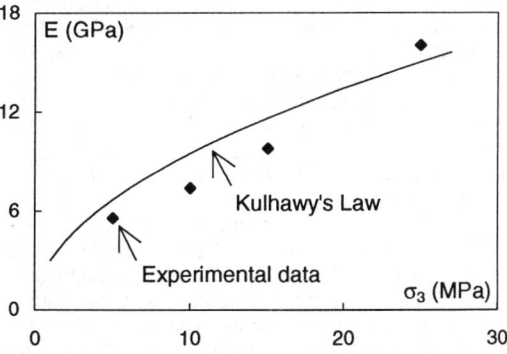

Fig. 6: Young's modulus as a function of σ_3.

$$D(\sigma,\tau) = (-0.001\sigma^2 + 0.047\sigma - 0.681)\ln(\tau) \\ + 0.002\sigma^2 - 0.099\sigma + 1.446 \quad (4)$$

Together with the previously determined $C(\sigma)$ functions (1), such a function can be used in NL analysis of borehole stresses using a method presented by Nawrocki et al. (1996). The radial stress obtained can then be used in the experimentally determined $E(\sigma_3)$ function, thus assuming that $\sigma_3 = \sigma_r$.

Several compression curves corresponding to triaxial tests run at different σ_3's are presented in Fig. 5. Young's moduli determined for each curve shown in Fig. 5 are shown in Fig. 6, where the Kulhawy's law was used to approximate data with a single function. It has the following form:

$$E(\sigma_3) = E_o \sigma_3^m \quad (5)$$

$E_o = 3.0$ and $m = 0.5$ for the Kulhawy's function shown in Fig. 6. This function can now be used together with the NL radial stresses to obtain information how E varies as a function of distance from the opening wall. The extent of damage zone can thus be estimated.

4 SUMMARY AND CONCLUSIONS

In this paper we propose methodology for estimation of damage zone dimensions around boreholes or waste repositories. It consist of realistic modelling of stresses using NL models, a dedicated interpretation of compression test results, and using opening stresses obtained in such a way in experimentally determined $E(\sigma_3)$ function to obtain information on Young's modulus variation away from the opening wall. Although we have not yet done so in our experiments, one may also simultaneously measure acoustic emissions and permeability changes associated with damage, which be of value in evaluating the consequences of a microfissured and damaged zone around a repository structure in shales, or in carrying out coupled stress-flow-damage transient analyses.

Our data show several important aspects of shale compressive behaviour. First, shale deformation is markedly NL. Second, there is a significant effect of the mean stress on bulk modulus, as well as shear modulus dependency on σ and τ. However, the effects of both hydrostatic and the deviatoric components can be measured, analyzed, and included in modelling. Furthermore, results allow us to study the Young's modulus dependency on σ_3, provide failure envelopes, and provide information on shale behaviour in unloading.

It is believed that the experimental procedures and methods of analysis presented in this paper should be effective in better characterizing mechanical properties of rocks to aid NL analyses. They provide means for estimating constitutive parameters both for linear and NL models. Generally, such models are used first before refering to more complex viscoelastic or viscoplastic models, which may nonetheless be warranted by experimental and field performance data. They can play a vital role in petroleum engineering (borehole stability, reservoir engineering, natural gas storage projects), allowing parametric analyses varying rock strength and stiffness, in situ stresses, geometry, and so on. As such parametric analysis tools seem to be lacking, development of these into a comprehensive analysis package for such applications seems warranted. Perhaps these approaches can aid in this task.

ACKNOWLEDGEMENTS

Special thanks go to Dr.N.Cristescu for his help in designing experimental procedures reported in this paper. We would also like to thank Mr.D.Hirst and Ms.J.Fooks who helped in testing. Finally, financial assistance of NSERC, Saga Petroleum, and several other companies involved in the Waterloo Shale Project is gratefully acknowledged.

REFERENCES

Duncan, J.M. & C.Y. Chang 1970. Non-linear analysis of stress and strain in soils. *J. Soil Mech. Found. Div. ASCE.* 96, No. SM5 Sept. 1970: 1629-1653.

Dusseault, M. & K.Gray 1992.Mechanisms of stress-induced wellbore damage. *Proc. Conf. on Wellbore Damage*, Lafayette, SPE #23825: 511-521

Hojka, K., Dusseault, M.B., Bogobowicz, A.D. 1993. Analytical solution for transient thermoelastic fields around a borehole during fluid injection into permeable media. *J. Can. Pet. Techn.* 34 (4): 49-57

Krajcinovic D., 1989. Damage mechanics. *Mech. Mater.* 8: 117-197.

Nawrocki,P.A. & M. Dusseault 1995. Modelling of damaged zones around openings using radius-dependent Young's modulus. *Rock Mech. and Rock Engng.* 28 (4): 227-239.

Nawrocki, P.A. Dusseault M.B. & R.K. Bratli 1996. Semi-analytical models for predicting borehole stresses around openings in NL geomaterials. EUROCK'96; Torino, Italy, Sept. 2-5, 1995: 785-792.

Nawrocki P.A., Cristescu, N.D., Dusseault M.B. & R.K. Bratli 1997. Experimental methods for deter mining constitutive parameters for non-linear rock modelling. IACMAG'97, Wuhan, China, Nov.2-7, 1997.

Santarelli F., Brown E.T. & V.Maury 1986. Ana lysis of borehole stresses using pressure-depen dent linear elasticity, *Int. J. Rock Mech. Min. Sci & Geomech. Abstr.* 23: 445-449.

Vaziri, H.H. 1986. Non-linear temperature and con solidation analysis of gassy soils, PhD Thesis, Uni versity of British Columbia.

Thermal conductivity of saturated quartz-illitic and smectitic shales as a function of stress and temperature

David A. MacGillivray, Brett Davidson & Maurice B. Dusseault
University of Waterloo, Ont., Canada

ABSTRACT: The thermal conductivity of quartz-illite and smectitic shales as a function of temperature, stress, pore pressure, and bedding anisotropy is being investigated. Experiments are performed using a one-dimensional, steady state, heat flow device. The use of optical quartz as a standard ensures a reliable calibration factor for the cell. Preliminary results for the Queenston shale indicate substantially higher values for thermal conductivity (1.74 - 1.82 $Wm^{-1}K^{-1}$) than the currently accepted range for shales (1.05 - 1.45 $Wm^{-1}K^{-1}$). This is attributed to the saturated conditions and the high muscovite and quartz content of the Queenston shale. The thermal conductivity of the Pierre shale ranges from 1.30 - 1.70 $Wm^{-1}K^{-1}$ and was found to increase as a logarithmic function with stress. The thermal conductivity of the Mancos shale also increases with applied stress and values ranged from 1.80 - 2.25 $Wm^{-1}K^{-1}$.

1 INTRODUCTION

The thermal history of sedimentary basins and how this affects petroleum generation and organic matter maturation is of particular interest to petroleum geologists. Thermal conductivity data are used in basin modelling techniques applied to the thermal history and organic maturation of these basins. It has been suggested (Blackwell and Steele, 1989) that laboratory measurements of shale thermal conductivities are unreliable and 25-50% lower than literature values, and that thermal conductivity of shale may change unpredictably during deformation. It is therefore of interest to determine the thermal conductivity of individual shales under a range of stress and temperature conditions.

2 BACKGROUND

The one-dimensional expression for heat flux through a solid of cross-sectional area, A, is

$$q_x = -kA\frac{\delta T}{\delta x} \qquad (2.1)$$

and it is called the Fourier law of heat conduction, or the Fourier law of thermal diffusion (Bejan, 1993). This expression also serves as a definition for thermal conductivity, k, a value that must be measured experimentally. Units of thermal conductivity are expressed as $Wm^{-1}K^{-1}$ (SI) and mcal/cms°C (CGS). Units of heat flow are expressed in mWm^{-2} (SI) and $\mu cal/cm^2 s$ (CGS).

Equation (2.1) is analogous to the Darcy law of fluid flow through a porous medium. Heat flows in the direction of lower temperatures as predicted by the second law of thermodynamics.

2.1 Thermal conductivity

Thermal conductivity can be determined by using equation (2.1) and a simple one-dimensional experiment set up to measure the temperature gradient between two points with a known heat flux at the entry point. The value of k measured in such an experiment will depend on the thermodynamic state (temperature and pressure) of the material being tested, the internal structure of the sample, and the point at which the temperature is measured in the sample. According to the theory of dielectric solids, thermal transport is associated with acoustical waves, or phonons, whose velocity depends on the elastic moduli of the solid (Davis, 1984). The thermal conductivity is then partially determined by the degree of phonon scattering and as a result such things as crystal lattice imperfections, impurities, grain boundaries, and crystal size will be important in the determination of thermal conductivity (Davis, 1984). The simplest case of internal structure is a homogeneous and isotropic sample, and the most complicated would be a sample that is inhomogeneous and generally anisotropic. If the sample were inhomogeneous but isotropic, it would not matter in which direction the sample was oriented for the heat flow measurement, but data would depend on where the measurement was made. For a homogeneous but anisotropic sample, the measurement point would not be as important as the orientation, provided that the layering is much smaller than the sample. The value of the thermal conductivity coefficient is dependent on temperature, and this function is different for various materials.

3 THERMAL CONDUCTIVITY OF POROUS MEDIA

The thermal conductivity of porous media (Figure 1) depends on composition, geometric microstructure, and interphase relationships (ie. air, water, solid, oil). Heat transfer in porous media is conductive when the pores of the material are smaller than 10^{-4} m in diameter and when the temperature is in the range -50 to 100°C (Kovalenko and Flanders, 1991). Convective and radiative heat transfer are negligible except when there are very large pores such as those found in thermal insulation or large celled concrete mixtures (Kovalenko and Flanders, 1991).

Because of heterogeneity, averaged values of the medium will result in averaged values of the process. The volume content of components, temperature, heat flow, and thermal conductivity will all be statistically derived, as there is no absolute value that can be measured. A representative sample of the appropriate scale is of the utmost importance.

Microstructure determines specific heat capacity, c [J/kg•K]; density, D [kg/m^3]; and thermal conductivity, k [W/m•K]. A detailed knowledge of microstructure is not feasible due to its great variability and random structure but the average properties of the media over a representative elementary volume can be determined.

Several experiments have been performed (Sugawara and Yoshizawa, 1961, Woodside and Messmer, 1961, Brigaud and Vasseur, 1989) to investigate how thermal conductivity changes with respect to various fundamental properties of porous materials. Qualitative predictions have been made of the effect of temperature, porosity and pore fluids, composition, anisotropy, and pressure.

3.1 Temperature

The relationship between thermal conductivity and temperature is linear: thermal conductivity increases with an increase in temperature (Sugawara and Yoshizawa, 1961). This has been found true for various porosities of multicomponent systems such as air and glass or saturated sandstone (Sugawara and Yoshizawa, 1961).

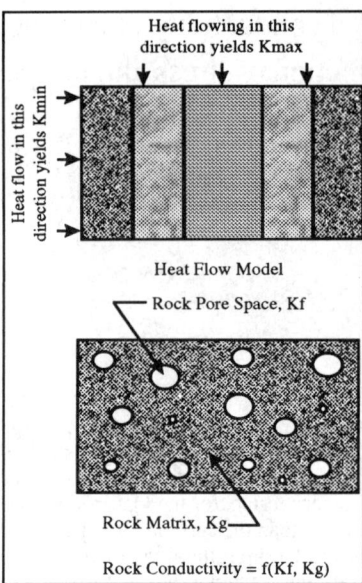

Figure 1: Heat Flow in Porous Media

3.2 Porosity and pore fluids

The porosity and the fluid contained within the pores greatly affects the thermal conductivity of the porous medium, especially when there is a large difference in the thermal conductivity of the pore fluid and the solid component. If the thermal conductivity of the solid component of a porous material is greater than the thermal conductivity of the pore fluid, then the overall thermal conductivity is decreased with a porosity increment (Sugawara and Yoshizawa, 1961; Woodside and Messmer, 1961). If the thermal conductivity of the pore fluid is greater than the thermal conductivity of the solid component, then the thermal conductivity of the porous material is increased with an increase in porosity (Sugawara and Yoshizawa, 1961; Woodside and Messmer, 1961). These functions are non-linear. The grain-to-grain contacts in the pore structure were shown to promote heat conduction by noting that when air is evacuated from a grain pack test chamber, the conductivity approached specific values (Zierfuss, 1969). Pore fluid convection and radiation have little effect on the conduction of heat when the grains are smaller than 2 mm (Zierfuss, 1969).

3.3 Pressure

An increase in pressure increases the thermal conductivity of an evacuated porous medium in a non-linear fashion. Woodside and Messmer (1961) showed that with an application of 4000 psi overburden pressure the thermal conductivity of an evacuated Berea Sandstone sample is almost doubled (37% increase in thermal conductivity when subject to a simulated overburden pressure of 3500 psi). The conductivity tends to level off at a pressure between 3000 - 4000 psi, and will return to its initial value when this pressure is released. The application of an overburden pressure increases the grain contact area within the rock, thus increasing the overall conductivity. The higher the conductivity of pore fluid, the smaller the increase in effective conductivity of the porous medium (Woodside and Messmer, 1961). Overburden pressure will have a small effect on the thermal conductivity of a fully saturated rock when the pore fluid conductivity is greater than solid component conductivity (Woodside and Messmer, 1961). A 0.5 - 3% rise in thermal conductivity per 1000 atm has been reported for various materials (Zierfuss, 1969).

3.4 Mineralogy

To show the mineralogy effect on thermal conductivity in porous rocks, the relative proportions of two minerals or mineral groups (ie. quartz and clays) are changed while the other components remain the same. Brigaud and Vasseur (1989) compared sandstones with varying clay content at various porosities and found that there was a decrease in thermal conductivity with an increase in clay content. They also compared kaolinitic materials with various porosities and found that there was an increase in thermal conductivity with increasing quartz fraction. Both of these relationships are non-linear. It appears that quartz has a conductive effect on bulk conductivity and clays have more of an insulating effect.

3.5 Anisotropy

Stratification of sedimentary rocks, produced by changes in mineralogy, granulometry, or compaction, leads to anisotropy. The value of thermal conductivity parallel to bedding is almost always greater than perpendicular. Brigaud and Vasseur (1989) measured thermal conductivity parallel and perpendicular to bedding in water-saturated and air-saturated sedimentary well cores with low clay fractions. It was found that the thermal conductivity anisotropy factor (ratio of $q_{\parallel}$ to $q_{\perp}$) was 1.06. For the air-saturated samples the anisotropy factor was determined to be 1.1. This can be explained by realizing that the structural origin of the anisotropy is due to bedding of the minerals in the stratification and to porosity layering. When the pores are filled with a poor conducting fluid (ie. air) the anisotropy due to porosity layering is enhanced, increasing the anisotropy factor slightly.

4 LABORATORY ANALYSIS

4.1 Waterloo Thermal Conductivity Cell

There are two well established methods for determining the thermal conductivity of porous media and rocks: steady state and transient. To quantify the thermal properties of shales, a one-dimensional, steady-state, heat flow compaction device, the Waterloo Thermal Conductivity Cell (WTCC) was fabricated (Figure 2). The WTCC is designed to withstand axial stresses of 100 MPa which, in a sedimentary basin, is equivalent to burial depths of 3-4 km. The lateral stress is not controlled, but by virtue of the lateral confinement afforded by the cell, probably has a value of 25-50% of the axial stress. Temperatures can be accurately controlled from 23°C to 80°C.

Thermal conductivity is determined by creating and maintaining a constant steady-state heat gradient (flow) from upper and lower heat exchangers. Ethylene glycol is circulated through the exchangers with the upper platen warmer and the bottom platen cooler, which creates a thermal gradient. For the water-saturated sample, the gradient is recorded by monitoring temperature at eight locations along the centerline of the compaction cell using embedded thermistors. Radial heat loss through the cell walls is controlled by a lucite liner inside an outer steel

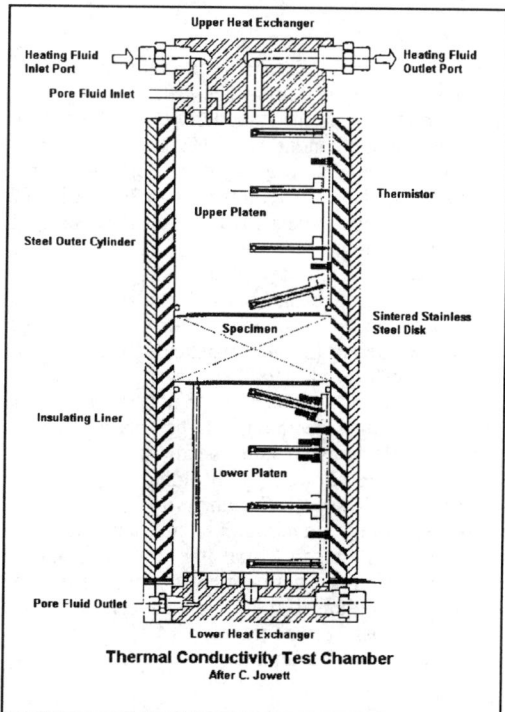

Figure 2: Waterloo Thermal Conductivity Cell

body and by enclosing the cell in a temperature controlled environment.

4.2 Test Measurements

Thermal conductivity for each test specimen is determined with reference to a calibration obtained using an 2.54 cm thick quartz optical quartz disc. The thermal conductivity of a geomaterial is easily obtained by establishing a steady, linear flow through the heat exchangers and measuring the temperature gradient through a sample of equal thickness to that of the calibration standard. From the known thermal conductivity of the optical quartz disc, thermal conductivity of the sample is calculated.

Great care must be taken in the preparation of the samples to ensure accurate results. The thickness should be constant at every point in the sample, flat parallel surfaces are necessary, and the surfaces should be as smooth as possible. The edge and the side of the sample needs to be perfectly insulated.

Introducing a sample into the test system disturbs the natural flow of heat, creating a thermal discontinuity. A thermal contact resistance is created which can be made negligible and can be accounted for during calibration of the thermal cell. The apparent conductivity (k_{app}) in the presence of a contact resistance (Beck, 1988) is defined as:

$$\frac{D+2l}{k_{app}} = \frac{2l}{k_f} + \frac{D}{k} \qquad (4.1)$$

where D is the thickness of the sample; l is the thickness of the contact film between the sample and the cell platen; k_f is the conductivity of the contact film, and k is the conductivity of the sample (Figure 3). If the contact resistance is negligible then $\Delta T = T_2 - T_1$. T_2 and T_1 are known, $D >> 2l$, and if $2l/k_f$ is negligible with respect to D/k, then there is small error in the system.

A pressure of a 1.0 MPa is applied to stabilize the system and to make the thermal contact resistance from specimen to specimen uniform. It has been shown that contact thermal resistances will stabilize after a few kPa (Beck, 1988).

Errors due to diameter mismatch between the platen and the specimen, slight asymmetry of specimen placement, and small chips in the specimen are minimized by the long bar design of the Waterloo Thermal Conductivity Cell, and by careful specimen preparation. The mean ambient temperature is set to equal the mean specimen temperature and the temperature change across the system is less than 10°C to reduce errors due to lateral heat loss. From equation (4.1) the thermal conductivity can be calculated by:

$$\frac{D}{k} = \frac{\Delta T}{k_p \Gamma_p} - \frac{2l}{k_f} \qquad (4.2)$$

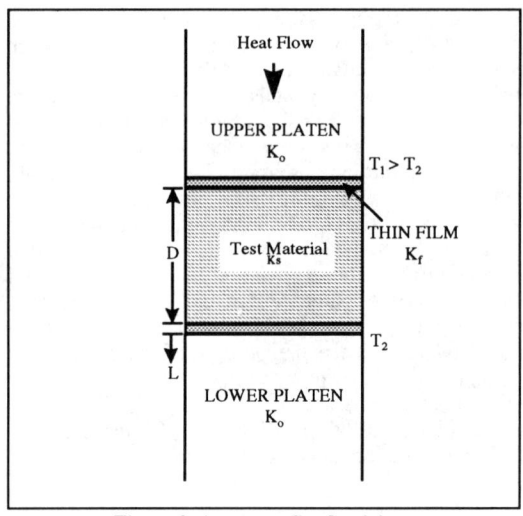

Figure 3: Apparent Conductivity

where k_p is the thermal conductivity of the platens; Γ_p is the temperature gradient in the platens; and $2l/k_f$ represents the thermal contact resistance.

Equation (4.2) can also be written:

$$\frac{\Delta T}{\Gamma_p} = k_{eff}\left(\frac{D}{k}\right) \qquad (4.3)$$

where k_{eff} is the effective thermal conductivity and incorporates the thermal contact resistance created in the system.

The effective thermal conductivity is calculated by substituting a disk of fused quartz of known thermal conductivity (Beck, 1988):

$$k = \frac{1323 + 1.93T - 0.0067T^2}{1000}$$

The effective thermal conductivity is thereby determined for the temperature range using Eq. 4.2.

5 RESULTS

A comparison of the thermal conductivities of the Queenston, Pierre, and Mancos shales is shown in Figure 4.

5.1 Queenston Shale

Samples of Queenston shale were tested and the results for two samples are shown in Figure 5. Tests were performed at saturation (deionized water) with negligible pore pressure and various axial stresses (1, 3.5, 13.8, and 24.1 MPa). The thermal conductivity ranges from 1.74 to 1.95 Wm^{-1}K^{-1}.

The thermal conductivity increases slightly with

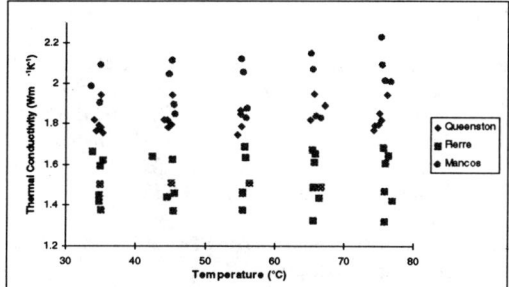

Figure 4: Thermal Conductivity Comparison

temperature and stress, but more samples need to be analyzed to completely characterize these functions. Sample 2 has a greater response to temperature at 1 MPa but also a greater error associated with the individual measurements.

An error analysis was performed on the initial results of samples one and two resulting in some rejected measurements primarily due to the samples not reaching thermal equilibrium. The total error in the measurements due to temperature fluctuations of the system are limited to ± 0.01 $Wm^{-1}K^{-1}$ with error for sample one at 1 MPa slightly higher (± 0.02 $Wm^{-1}K^{-1}$).

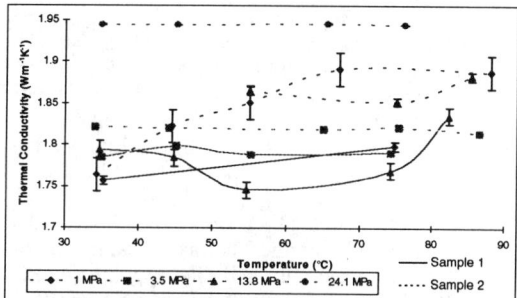

Figure 5: Thermal Conductivity of Queenston Shale

An analysis of the clay size fraction of a crushed sample of the Queenston shale was performed by standard x-ray diffraction techniques using the seven standard pretreatments on oriented specimens. These tests indicated the presence of muscovite, chlorite, calcite, and quartz in the coarse-grained fraction (>0.2 µm), and the presence of muscovite and chlorite in the fine-grained fraction (<0.2 µm). There was a high amount of iron substitution in the chlorite, indicated by the decrease in intensity of odd numbered peaks and increase in intensity of even numbered peaks in the x-ray diffraction patterns. This explains the red coloured bands commonly observed in the Queenston shale. A moisture content of 2.6% before testing and 4.6% after testing indicates that samples were being slightly resaturated during testing.

The thermal conductivity of dry shale is suggested to range from 1.05 to 1.45 $Wm^{-1}K^{-1}$. The values calculated for the Queenston shale are slightly higher. This can be attributed to the saturated test conditions, the stress, and the high quartz and muscovite content of the material.

5.2 Pierre Shale

Results for two samples of the Pierre shale are shown in Figure 6. Samples were kept saturated by using a fluid with a chemistry similar to the shale's natural pore water. Similar conditions of negligible pore pressure and uniaxial confining pressure were used. The thermal conductivity ranges from 1.30 - 1.70 $Wm^{-1}K^{-1}$.

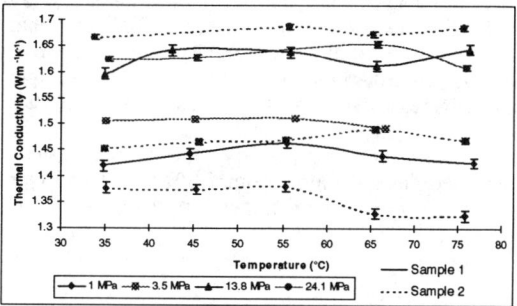

Figure 6: Thermal Conductivity of Pierre Shale

...The thermal conductivity of the Pierre shale trends as a logarithmic function of axial stress (Figure 7). Values of thermal conductivity are much lower than the Queenston shale values and this reduction is probably due to the insulating effect of the smectitic mineralogy of the Pierre shale. The moisture content of samples before testing was determined to be 11% and increased sligtly during testing to 12%. The error associated with these measurements was limited to ± 0.01 $Wm^{-1}K^{-1}$

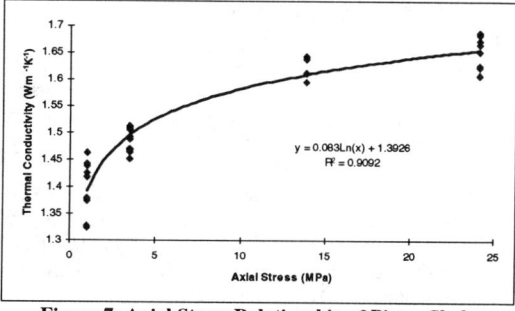

Figure 7: Axial Stress Relationship of Pierre Shale

5.3 Mancos Shale

Results for one sample of the Mancos shale with similar test conditions are shown in Figure 8. Samples were kept saturated with deionized water.

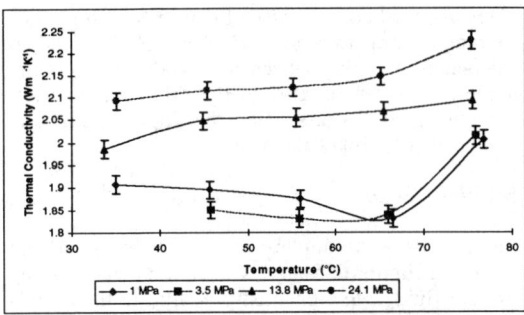

Figure 8: Thermal Conductivity of Mancos Shale

The thermal conductivity ranges from 1.80 - 2.25 $Wm^{-1}K^{-1}$, slightly higher values than those calculated for the Queenston shale. A powdered x-ray diffraction analysis of these samples indicates that greater than 50% of these samples are composed of quartz. The thermal conductivity of the sample tested increases with axial stress as shown in Figure 9.

The moisture content of the Mancos shale increased during testing from 2.8% to 3.1%. Error associated with random temperature fluctuations was limited to ± 0.02 $Wm^{-1}K^{-1}$.

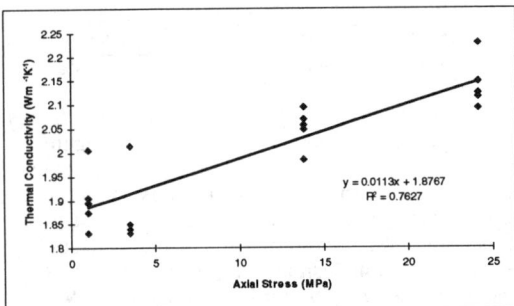

Figure 9: Axial Stress Relationship of Mancos Shale

6 DISCUSSION AND APPLICATIONS

Our data obtained under the careful controls used, with high-quality specimens and realistic values of stresses and temperatures imply that thermal conductivity values reported in the literature are substantially underestimated, by as much as a factor of 1.3-2.

A brief note of potential applications of the data is warranted. The data have value in basin analysis, in borehole stability assessments, and in estimating heat losses from thermal stimulation methods used in the petroleum industry. Heat flux from radioactive waste repositories in shale or sedimentary rocks, from heated tanks or subsurface structures, and other similar cases can use these data as well. Some of these cases also require thermal expansivity and constitutive stress-strain laws for shale deformation and yield calculations. We are generating such data for shales and will publish it in future articles.

REFERENCES

Beck, A.E. (1988). Methods for Determining Thermal Conductivity and Thermal Diffusivity. *Handbook of Terrestrial Heat-Flow Density Determination.* R. Haenel, L. Rybach, and L. Stegna (eds.). Kluwer Academic Publishers, pp. 87-124.

Bejan, A. (1993). *Heat Transfer.* John Wiley & Sons, Inc. 675 pp.

Blackwell, D.D., and J.L. Steele (1989). Thermal Conductivity of Sedimentary Rocks: Measurement and Significance. *Thermal History of Sedimentary Basins*, N.D. Naeser and T.H. McCulloh (eds.). Springer-Verlag New York Inc. 24 pp.

Brigaud, F., and G. Vasseur (1989). Mineralogy, Porosity, and Fluid Control on Thermal Conductivity of Sedimentary Rocks. *Geophysical Journal.* Vol. 98, pp. 525-542.

Davis, B.W. (1984). Thermal Conductivities of Reservoir Rocks. Technical Memorandum - Chevron Oil Field Research Company, La Habra, CA #TM84000293. 32 pp.

Kovalenko, Y.A., and S.N. Flanders (1991). Thermal Conductivity of Porous Media and Soils - A Review of Soviet Investigations. U.S. Army Corps of Engineers, Cold Regions Research & Engineering Laboratory, Special Report 91-6. 12 pp.

Sugawara, A., and Y. Yoshizawa (1961). An Investigation on the Thermal Conductivity of Porous Materials and its Application to Porous Rock. *Australian J. Phys.* Vol. 33, pp. 3135-3138.

Woodside, W., and J.H. Messmer (1961). Thermal Conductivity of Porous Media. I. Unconsolidated Sands. II. Consolidated Rocks. *Journal of Applied Physics.* Vol. 32, No. 9, pp. 1688-1706.

Zierfuss, H. (1969). Heat Conductivity of Some Carbonate Rocks and Clayey Sandstones. The American Association of Petroleum Geologists Bulletin. Vol. 53, No. 2, pp. 251-260.

Triaxial extension tests of cemented soils and soft rocks

Kazuo Tani
Central Research Institute of Electric Power Industry, Chiba, Japan

ABSTRACT : The strength characteristics of such geomaterials as possess appreciable amounts of tensile strengths, i.e. cemented soils and soft rocks, are usually evaluated by triaxial compression tests for shear failure and by either Brazilian tests or unconfined extension tests for tensile failure. In this study, a triaxial extension test equipment was developed, in which a cylindrical specimen can be directly extended in an axial direction under confining pressures. Shear failures were observed in the specimens if confining pressures were higher than the uniaxial compression strength, while tensile failures were observed under lower pressures. The tensile strength, i.e. the minimum principal stress at tensile failures, was found to be independent of confining pressures; thereby this strength value is considered to be unique for the material.

1 INTRODUCTION

Cemented soils and/or sedimentary soft rocks may possess appreciable amounts of tensile strengths as a result of cementation and/or bonding of composed particles provided by some kinds of cementing agents or through diagenetic processes. In general, this sensitive quantity can be conveniently evaluated by Brazilian tests (ISRM, 1978 : Hiramatsu et al., 1964). In this test, although the tensile minimum principal stresses σ_3 (<0) may become ideally uniform along the relevant failure plane which is in alignment with the axial load, the compressional maximum principal stresses σ_1 (>0) inevitably become non-uniform. Consequently, this test method is not regarded as a kind of element tests, in which stresses and strains are principally uniform throughout the specimen. Furthermore, since stress concentration is significant around the vicinity of the line of load application, the results cannot escape from the skepticism that the failure may involve unwanted shear failures around the loading point. Since the failure plane cannot develop freely as its potential location is preset depending upon the orientation of the specimen's axis and the loading direction, special attention should be paid if the material is of anisotropic nature. With all these reasons, the Brazilian tests may not be considered as of rigorous nature in evaluation of element characteristics of material properties (Malan et al., 1994).

An alternative technique to evaluate the tensile strength is a less popular method of unconfined extension tests where a cylindrical specimen with metal caps glued to both ends is axially extended under the no confining pressure condition, σ_c =0 and σ_3 <0. On the other hand, variety of compression type of laboratory element tests using triaxial cells are available in which the failure stress condition can be rigorously specified. However, since application of tensile stresses can not be achieved, only the shear failures under compressional stresses are attainable, thus $\sigma_1 > \sigma_3 \geqq 0$. Few research work has been attempted to extend rather than compress the specimens under confining pressures. As a result, by these conventional types of compression tests, it has been impossible to study the failure under specific stress condition where the maximum principal stress stays compressional, σ_1 >0, while the minimum principal stress become extensional, σ_3 <0; thereby the Mohr's stress circle at failure includes the origin of $\sigma \sim \tau$ diagram. In other words, neither tensile failure under some confining pressures (σ_1 >0), nor shear failure under the tensile stress conditions (σ_3 <0) has been investigated by conventional test methods.

In this study, triaxial extension test technique has been developed to cover the gap between triaxial compression tests and unconfined extension tests. This is to investigate the failure criterion for the stress condition where the Mohr's stress circle includes the origin (σ_1 >0>σ_3). Geomaterials of interest are such as experienced diagenesis to create a cemented structure, i.e. treated soils and sedimentary soft rocks. This element tests allow us to investigate the tensile strength under

Table 1 : Mixing proportions of the cast mortar and consolidation pressure

M.P.	sand[1] S (%)	clay[2] $C\ell$ (%)	cement[3] C (%)	water[4] W (%)	consolidation pressure pc (MPa)
S3	71.4	7.4	3.4	17.8	12.2
S4	65.0	6.7	12.1	16.2	12.2

1) Tone river sand, 2) Kibushi clay, 3) High-early-strength portland cement, 4) Tap water

compressional maximum principal stresses, $\sigma_1 > 0$, as well as the shear strength under extensional minimum principal stresses, $\sigma_3 < 0$.

2 TEST APPARATUS AND TEST METHOD

Figures 1 illustrate the triaxial extension test apparatus developed in this study. A cylindrical specimen can be directly extended in an axial direction under a confining pressure. Particular attention was paid to the design of the apparatus so that the axial load of extension can be exerted to the specimen without applying even a slightest amount of moment load.

First, a pair of metal cap attachments are glued onto both ends of the cylindrical specimen. These glued surface should be carefully prepared to be smooth and desiccated by a hand dryer to help the adhesive, two liquid mixing type of epoxy resin, work properly (Nagayama et al., 1990). Then these attachments are rigidly connected to the cap and pedestal so that it can be pulled in an axial direction in a triaxial cell.

Since the non-homogeneous nature of natural geomaterials such as sedimentary soft rocks may affect the test results with gross scattering, artificial soft rocks of consolidated mortar were used in this study (Tani et al., 1992 & 1995). These materials have proved to demonstrate similar mechanical behavior to natural sedimentary soft rocks of uniform nature. The mixing proportions of the mortar are shown in Table 1. Since the main composition is sand, the resultant materials are close to weak sandstones.

The consolidation pressure of 12.2MPa was applied for a period of a fortnight immediately after the mortar was cast in a large mold, 500mm in diameter and 1200mm in height. Specimens, 50mm in diameter and 110mm in height, were drilled from the cylindrical blocks which were subsequently cured under water for approximately one month to ensure that the material reached the stable state.

Axial strains are measured locally in the middle of the specimens. A pair of rings were fixed on the side of the specimen by 4 set-screws respectively at different heights of 65mm separation. The bottom one is denoted as the transducer holder carrying two proximeters at diagonally-opposite positions; and the top one is denoted as the target holder. The

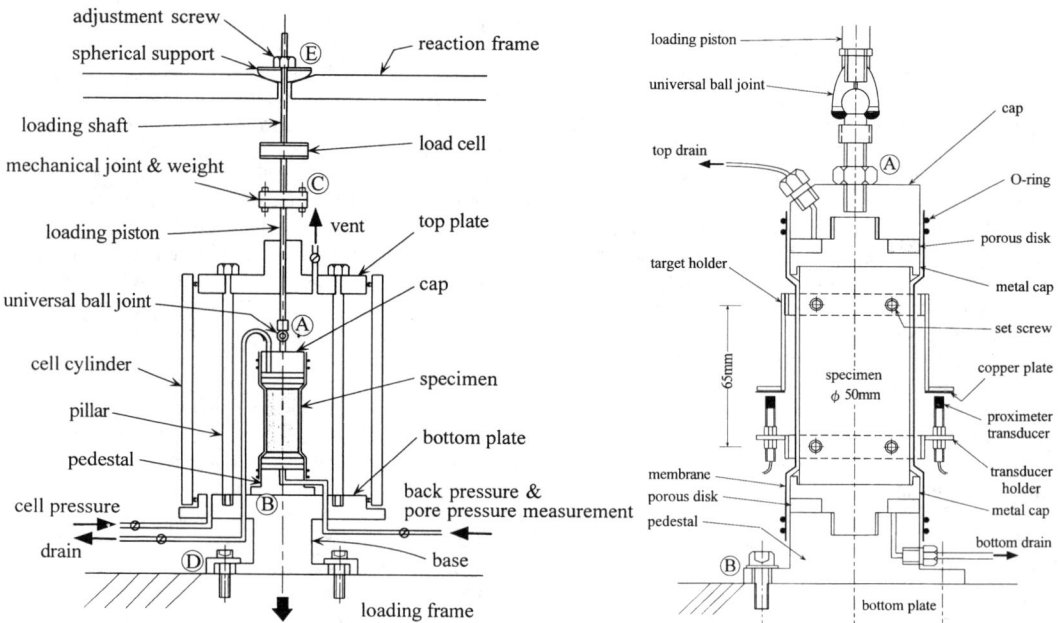

Figure 1 : Triaxial extension test apparatus (Left : general view, Right : around the specimen)

saturated specimen was first consolidated isotropically. Then the loading piston was pulled axially by displacement control at a rate of 0.05mm per minute until either shear or tensile failure occurred under the undrained condition.

The utmost attention should be paid for extension tests is undoubtedly that the specimen should be extended completely parallel to the specimen's axis without applying any moment force to it. Therefore, all the parts were precisely manufactured and assembled carefully so that the loading piston should be held by the top platen vertically and the base platen should be fixed horizontally. The specimen is to be placed in the triaxial cell in complete alignment with the loading piston. The following procedure is taken to achieve this requirement. The locations of the connecting points are shown in Figures 1.

(1) The specimen with the cap and the pedestal on both ends was hung under the loading piston by way of the universal ball joint (Connection A);
(2) The suspended specimen was carefully lowered until the pedestal reached the bottom plate, where they were connected rigidly (Connection B);
(3) Deaired water was circulated; supplied through the pedestal by way of the side porous disk and then discharged through the cap. The proximeter transducers were placed around the side of the specimen. The outer cell was placed followed by the application of the initial confining pressure of 0.01MPa;
(4) Adjust the position of the triaxial cell exactly under the loading shaft suspended by way of a roller joint from the reaction frame. Then connect the loading shaft and the loading piston, while releasing the bolt of the roller joint in paying attention not to apply any tensile load to the specimen (Connection C & Release E);
(5) Circulate the deaired water in order to saturate the specimen, obtaining the Skempton's B-Value greater than 0.9. Then consolidate it up to the specified pressure;
(6) Fix the base of the cell to the reaction frame in the right position (Connection D);
(7) Apply a marginal amount of tension in the piston rod and loading shaft by the adjustment screw. Note that no tensile stress develops inside the specimen because of overwhelming confining pressure plus the weight of the pieces above the specimen (Connection E).

3 DEVELOPMENT OF SPECIAL EQUIPMENT TO GLUE METAL ATTACHMENTS TO THE SPECIMEN

In all uniaxial and triaxial element tests, it is ideal, or supreme requirement to apply axial load without applying any moment on the specimen. The utmost difficulty in conducting extension tests compared to the counterpart, compression tests, is expected because even marginal amount of the moment force may exert an immense effect on the resultant tensile strength. In case of compression type of tests, since the stress and strain relationship is non-linear hardening type, the extent of non-uniformity in stress field induced by unwanted moment force tends to be eased gradually. As a result, the stresses and strains within the shear zone developed across the specimen may become fairly uniform at the moment of failure. It is needless to say that development of shear band itself implies the localization of deformation, i.e. non-uniform deformation. Consequently, shear strengths can scarcely be underestimated by a marginal amount of moment force in compression tests. On the contrary, in case of extension type of tests, the onset

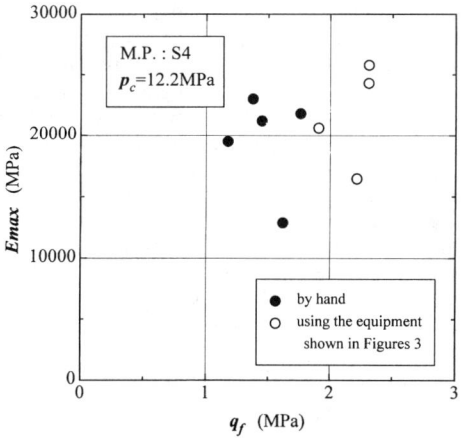
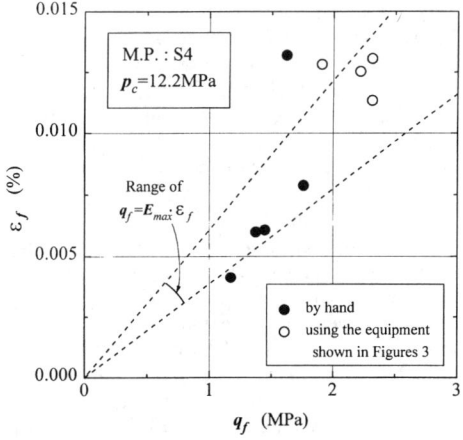

Figure 2 : Test results of M.P. : S4 at σ_c'=3.92MPa (Left : E_{max}~q_f relationship, Right : ε_f~q_f relationship)

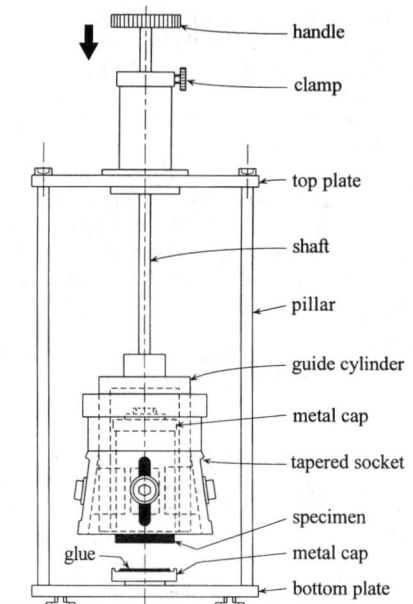
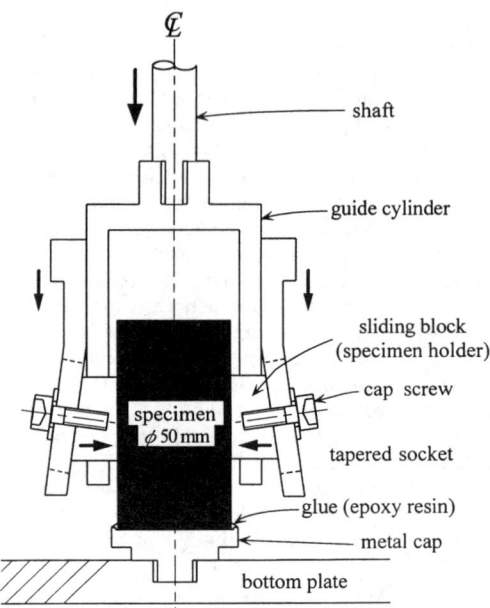

Figure 3 : Equipment for gluing the metal attachments (Left : general view, Right : around the specimen)

of even a tiny localized failure directly indicates the reduction of the resisting area against tension. Thus such a preceding local failure immediately results in the overall failure of the specimen leading to a seriously underestimated tensile strengths.

Since the influence of this marginal moment force is considered as reflected on the scattering of the experimental data, the stability and credibility of this test were examined by using the artificial rock denoted as M.P. : S4 which is of relatively high strengths and hence of ease to prepare the ends of the specimen. A total of 9 tests were carried out under the effective confining pressure σ_c' of 0.392MPa and the back pressure u_{BP} of 0.098MPa.

The experimental results are show in Figures 2. q_f and ε_f are the maximum deviatric stresses (=σ_1'−σ_3') and the axial strains at failure, and E_{max} is the secant modulus at axial strain level of $\varepsilon_a=1\times10^{-5}$. The solid circles indicate that the gluing of the metal attachments to the specimen's ends was done by hand and the open circles indicate that this procedure was done carefully by using a specifically designed apparatus illustrated in Figures 3. The objective to develop this gluing equipment is to glue the metal attachments to the ends of the specimen correctly in as accurately coaxial as possible.

It should be noted that the deviatric strengths were increased from q_f=1.1∼1.8MPa to q_f=1.9∼2.3MPa by using this equipment. When this operation is done by hand, the attachment is glued depending upon the surface of the specimen's ends. This immediately reflects the fact that if the specimen's ends are not prepared in such a way as perpendicular to the specimen's axis, the attachment and specimen can not be aligned practically. Although the extent of misalignment may neither be measured exactly nor be appreciably identified, it should be noted that a marginal moment could result in drastic underestimation of the tensile strengths.

Interestingly enough, however, whether the gluing equipment was used or not, or whether a marginal moment might be exerted or not, appeared not to influence the initial stiffness E_{max} significantly. Moreover, since the tensile failure took place before the non-linear nature of the stress and strain relationship becomes serious, the axial stain at failure, ε_f, are found to be smaller for the cases without using the equipment shown in Figures 2, thus possibly by the moment. Thus, it is concluded that even a tiny amount of misalignment, which is inevitably associated with even a most careful handling, could lead to underestimation of tensile strengths and failure strains; therefore some special equipment is definitely needed.

When the gluing equipment was used, the resultant minimum principal stresses at failure, i.e. tensile strengths, seem to be σ_{3f}=−1.9∼−1.5MPa, and the excess pore water pressures at failure were found to be rather small, u_f=−0.03∼−0.02MPa.

4 INFLUENCE OF CONFINING PRESSURE ON TENSILE STRENGTH

Using the specimens of mixing proportion S3, a

series of triaxial extension tests, a total of 27 tests, were carried out varying the effective confining pressures $\sigma_c'=0.005 \sim 4.91$MPa. The gluing equipment shown in Figures 3 was used to adhere the metal attachments onto the specimen's ends.

Figures 4 demonstrate the relationships of the effective and the total minimum principal stresses, σ_{3f}' & σ_{3f}, and the axial strains, ε_f, at failure, and the effective confining pressures, σ_c'. If the tests were conducted under lower confining pressures, $\sigma_c' \leqq$ 2.94MPa, failure was observed by tension with failure zones observed perpendicular to the specimen's axis, which are shown by solid circles in the figures. Whereas if the tests were conducted under higher pressures, $\sigma_c' \geqq 3.92$MPa, failure was observed by shear in a triaxial extension mode with failure zones observed to bear around 60° to the specimen's axis, which are shown by open circles in the figures. The axial strains at failure, ε_f, were found to be much smaller for tensile failure compared to shear failure, and tend to become larger for greater confining pressures, thus greater deviatric stresses at failure. It is of interest to note that the particular confining pressure which separates the resultant failure mode to either shear or tensile failures is comparable to the uniaxial compression strength, $qu=2.7 \sim 3.2$MPa.

It may be justified to conclude that the effective minimum principal stresses at failure, $\sigma_{3f}' \fallingdotseq -0.13 \sim -0.07$MPa, appear to be constant irrespective of the effective confining pressures, σ_c'. Whereas, the trend of the total minimum principal stresses at failure, σ_{3f}, may not be clear due to significant scattering. As a consequence, the criterion for tensile failure may be depicted as $\sigma_{3f}'=-\sigma_t'$, where σ_t' is the effective tensile strength. This fact favorably suggests that the Brazilian test may deserve as an effective materiel test in evaluating the tensile strengths despite that the maximum principal stresses distribute non-uniformly along the failure plane. However, it should be borne in mind that the Brazilian test is not an element test but a kind of boundary value problem. Hence, further research is needed to examine whether the failure involves local shear-mode failure and whether the minimum principal stresses distribute uniformly along the failure plane. Additionally, it should also be noted that the pore water pressure measurement may not be reliable since the measurement could only be made by way of the side drainage due to the sealed specimen's ends.

Furthermore, in a special test case of $\sigma_c'=3.92$MPa, failure of specimens by shear was observed under the particular condition that the minimum principal stresses were tensile, $\sigma_{3f}'<0$. This has never been achieved in conventional kinds of compression tests. As shown in Figure 5, the Mohr's circle a failure is located in the immediate extrapolation of the triaxial

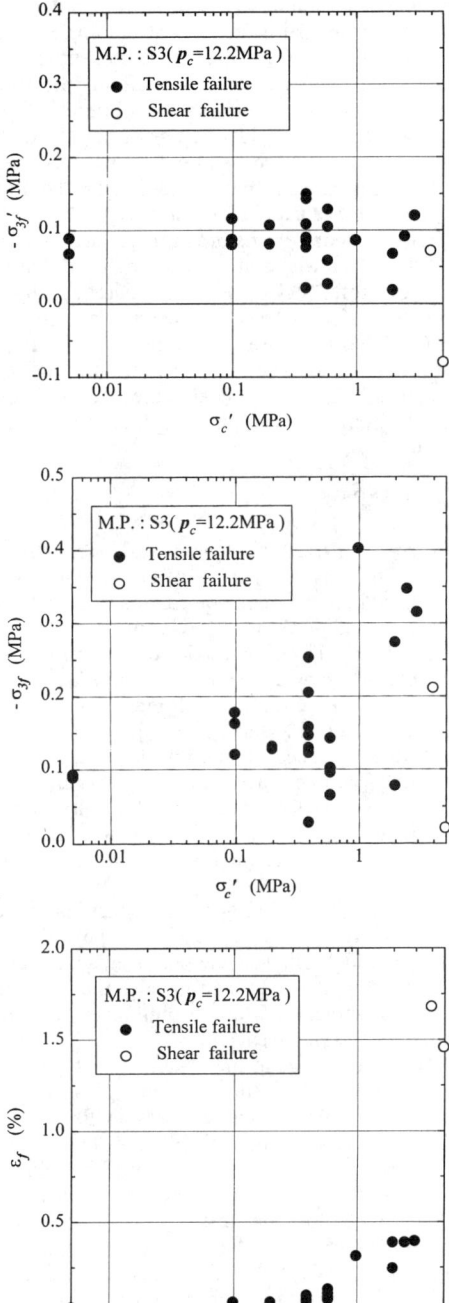

Figure 4 : Test results of M.P. : S3
(Top : $-\sigma_{3f}' \sim \sigma_c'$ relationship,
Middle : $-\sigma_{3f} \sim \sigma_c'$ relationship,
Bottom : $\varepsilon_f \sim \sigma_c'$ relationship)

extension shear failure of σ_c'=4.91MPa where the minimum principal stress at failure was compressional, σ_{3f}'>0. This fact demonstrates that the failure mechanism under low confining pressures, which covers between shear failure under compressional stress field and tensile failure under unconfined uniaxial stress field, thus has been unable to investigate by conventional laboratory testing, can be evaluated by extrapolating the failure criterion of triaxial extension shear failure (Tani et al., 1995). As a whole, it can be concluded that the failure of soft rocks can be specified by a combination of the shear failure criterion, such as Mohr-Coulomb's failure criterion, and a simple tensile failure, σ_{3f}'=-σ_t'. The failure of the specimen is to be seen under either failure mode upon which the stress point reaches the relevant criterion.

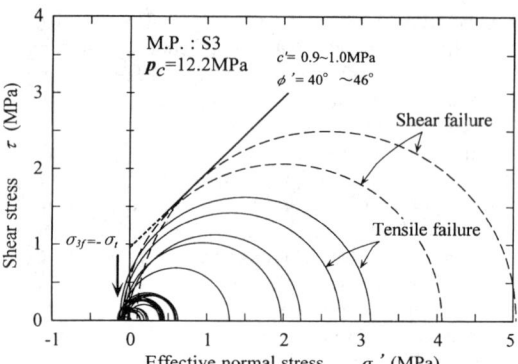

Figure 5 : Mohr's stress circles for failure conditions (M.P. : S3A)

5 SUMMARY

Two series of triaxial extension tests were carried out of uniform nature using artificial soft rocks of consolidated mortar. The objective is to investigate the failure condition under low confining stresses, between shear failure by triaxial compression tests and tensile failure by uniaxial extension tests.

In the first series of the program, 9 tests were conducted under the same confining pressure to evaluate the influence of possible misalignment of the loading (pulling) direction with respect to the specimen's axis. The experimental results clearly demonstrated that even a small amount of moment load applied to the specimen may reduce the tensile strengths to a significant extent. In order to overcome the difficulty to achieve perfect loading of absolutely coaxial, a special equipment was developed, which allowed the metal attachments to be glued on to the specimen's ends; thus ensured complete alignment of the loading axis with the longitudinal axis of the specimen.

In the second series of the program, 27 tests were conducted under different confining pressures σ_c' ranging from 0.005MPa to 4.905MPa. The specimens' failure can be defined either by shearing or by tension. Shear failures, with the oblique shear plane developed across the specimen's axis, were observed for the test cases of confining pressures higher than the uniaxial compression strength qu=2.7~3.2MPa. Whereas tensile failures, with the failure plane developed perpendicular to the specimen's axis, were observed under the lower confining pressures. The effective tensile strength σ_t', the minimum effective principal stress at tensile failure -σ_{3f}', was found to be independent of the confining pressures σ_t'=-σ_{3f}' =0.07 ~ 0.13MPa. Another finding is that, in terms of the effective stresses, the failure criterion for shear failure can be extended towards the tensile stress region up to that for tensile failure. On the other hand, significant scattering was observed for the tensile strength in terms of the total stresses. Further research is needed to ensure accurate measurement of the pore water pressures.

ACKNOWLEDGMENTS

Thanks are due to Mr. T.Yokokura, Mr. K.Seo and Ms. A.Ebihara of C.R.S. Co. Ltd. for the experimental work and for preparation of the figures.

REFERENCES

Hiramatsu,Y., Nishihara,M. and Oka,Y. (1964) "Investigation on extension test on rock", *J. of Soc. of Japanese Mining Industry*, Vol.70, No.793, pp.285~188 (in Japanese).

ISRM (1978) "Suggested method for determining tensile strength of rock materials", *Int. J. of Rock Mech. and Min. Sci. & Geomech. Abst.*, Vol.15, No.3, pp.99~103.

Malan,D.F., Napier,J.A.L. and Watson,B.P. (1994) "Propagation of fractures from an interface in a Brazilian test specimen", *J. Rock Mech. and Min. Sci. & Geomech. Abst.*, Vol.31, No.6, pp.581~596.

Nagayama,I., Watanabe,K. and Obata,N. (1990) "Experimental study on direct tensile strength of dam concrete", PWRI report, No.2914 (in Japanese).

Tani,K. and Yoshida,Y. (1992) "Preparation of artificial soft rock", *Proc. 24th Nat. Sym. on Rock Mechanics*, JSCE, pp.251~285 (in Japanese).

Tani,K., Nishi,K. and Yoshida.Y. (1995) "Mechanical properties of artificial soft rock material developed for laboratory model tests", *Proc. 26th Nat. Sym. on Rock Mechanics*, JSCE, pp.529~533 (in Japanese).

The acoustic emission properties of faulted rock under compression

Dongyan Liu & Keshan Zhu
Faculty of Civil Engineering, Chongqing Jianzhu University, People's Republic of China

ABSTRACT: To simulate faulted rocks, cement mortar specimens with slanting central slits are tested under uniaxial compression and the acoustic emission techniques are used to locate the fracture process zone and to study the various damage mechanisms of the mixed mode fracture in different loading stages. Test results have shown that the strength of faulted rock under compression might be differentiated into initial tensile strength, peak strength and residual strength.

1. INTRODUCTION

It is well-known that the failure mode and strength of jointed rockmass is mainly dependent on the structural features of the rock mass. From engineering practice and experimentation (Zhu et al 1990), we all acknowledge that crack growth in brittle rock is governed by the stress concentration near tips of the fissure and is always associated with a fracture process zone. The high stress concentration leads to the brittle failure of the rock bridges and the whole collapse of the rock is the result of the interconnection of some dominant fissures and cracks. In order to estimate the compressive strength and predict the failure mode of faulted rock, it is necessary to study the micro-fracture mechanisms of the brittle rock and develop a rational constitutive model to describe the behavior of the jointed rock mass.

The brittle fracture of rock has been attracted wide attention for a very long time, but most of the earlier studies were mainly on the tensile-shear mode of fracture and the associated criteria (Lawn et al 1974, Maji et al 1987).

For the important compression-shear mode of fracture in rocks, there appeared very few researches having been reported. To rationally and accurately describing the fracture processes in rock under compression, it is essential to be able to determine them in rock by suitable experiments, a task hardly to be fulfilled.

Thus, cement mortar specimens were used to determine the location of the fracture process zone by acoustic-emission measurements. Acoustic emissions (AE) are transient elastic waves caused by sudden localized changes in the micro-scale stress state, such as a crack is initiated or it extends suddenly with a sudden release of the strain energy. In rock-like materials as cement mortar, acoustic emission can be detected due to microcracking, intergranular friction, debonding and other damage phenomena.

AE signals generated during loading are detected by piezoelectric transducers and analyzed in several ways: (1) The rate of occurrence of AE events of is used to predict the extent at internal damage caused by stress concentration. (2) Frequency content of AE signals are used to distinguish various damage mechanisms occurring at different loading stages. (3) Several transducers can be used to detect an AE activity. Based on the time differences between detection of the event at different transducers, its source can be located. These methods of analyzing AE signals can provide a powerful tool to study the nature of fracture processes occurring in rock.

2. TEST SPECIMENS

A total of 50 cement mortar specimens of 200 x 200 x 50 mm with a slanting central slit were tested under uniaxial compression, as shown in Fig. 1.

Specimens were cast in metal molds and cured in atmospheric steam at the temperature of $100°C$ for 6 hours following the standard procedure. The specimens were removed from the molds after the autoclave curing. The slits in the specimens were formed by presetting thin metal shims, 0.07 to 0.2 mm thick, and then withdrawn from the mortar carefully after the mortar set. The inclinations of the

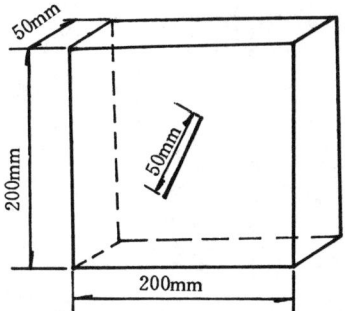

Fig. 1　Sample geometry

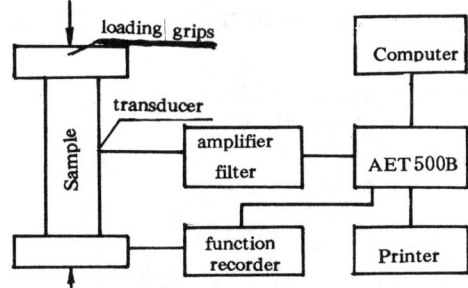

Fig. 2　Acoustic-Emission monitoring system

slits to the loading axis were divided into six groups: $0°, 15°, 30°, 45°, 60°$, and $90°$.

The mix proportion of the mortar was designed to simulate the brittle nature of rock and the acoustic wave properties through rock as cement : fine sand : water = 1 : 2.5 : 0.52 by weight. The index properties of the mortar were as follows:

Table 1. the index properties of the mortar

Property	Value
Uniaxial compressive strength	σ_c =10.5 MPa
Cleavage tensile strength	σ_t =1.46 MPa
Young's modulus	E = 11.4 GPa
Poisson's ratio	μ = 0.168

3. TEST PROCEDURE

The samples were loaded in a 2 MN capacity testing machine of an electric-servo-hydraulic Instron-1346. Both loading and displacement control were used. The loading rate was 1.525 kN/min. , and the displacement rate was 0.1 mm/min. All specimens were monotonically loaded until the residual load in the post-peak region dropped below 50 percent of the peak load. The load-displacement curves were recorded by the function recorder.

In order to eliminate the noise at the ends of the specimen and those from the load system, specially designed attenuators were inserted between the ends of the specimen and the load platens, so the true AE signals of crack growth could be detected. The PTFE foils were also inserted between the attenuators and the specimen to reduce the end restraint effect. The acoustic emission monitoring system is schematically shown in Fig. 2, which is comprised of (a) the eight-channel acoustic emission apparatus, Model AET-5000B, (b) the North Star personal computer, (c) transducers, amplifiers and filters, (d) Epson printer, and (e) software for signal analysis.

The sensor resonant frequency is 175 kHz, the band width of the filters is 125 to 250 kHz, the preamplifier gain is 60 dB, the main amplifier gain is 23 dB, and the threshold voltage is 0.1 V. The AE events and the load and displacement signals can be detected and stored for detailed analysis with the help of the computer.

4. ANALYSIS OF TEST RESULTS

The main results obtained from the tests are summarized in the following sections.

4.1 *Properties of AE events from a transducer*

In order to display the behavior of crack tips under compression-shear modes, an AE transducer was mounted near the crack tip, as shown in Fig. 3. The load-AE events distribution and accumulation curves are shown in Fig. 4.

As can be seen when $\beta = 15°$, the rate of AE events increases gradually with the increase of the loading , and the accumulation curve is rather smooth. The curve steepens up gradually with the increase of the load, which means that the crack propagates steadily and the AE events become stronger and stronger. Thus we may infer that the fracture mechanism should be the propagation of the shear fracture under the influence of the compressive stress concentration.

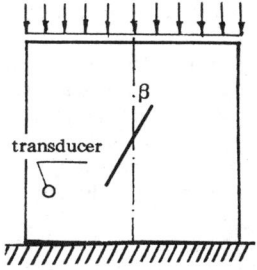

Fig. 3　Test setup to detect AE from a transducer

With the increase of the inclination, shear effect at the crack tip decreases, while the tensile stress concentration increases until new crack initiates. This is evidenced by the sudden outburst of AE events and a peak appears in the AE-versus-load curves, when $\beta = 30°$ or $\beta = 45°$ as shown in Fig. 4. With further increase of the load, a second peak appears and is much higher than the first one, which implies that a brittle fracture mechanism of the rock-like material really exhibits.

When $\beta = 60°$, two distinct portions of peaks can be seen in the AE-versus-load curve. The first peak corresponds to the initiation of the crack somewhat like that corresponding to the initial fracture strength of the specimen. The second one corresponds to an unstable propagation of the crack, which takes place when first a sudden drop and then a rapid rise to several higher peaks of the AE-versus-load curve can be obviously observed and thus corresponds to the peak strength of the specimen.

4.2 *Location of the AE source*

Based on the time delay between the arrival of the AE events detected by different transducers, the AE source can be located by signal-analysis software installed in AET-5000B.

There are four piezoelectric transducers mounted

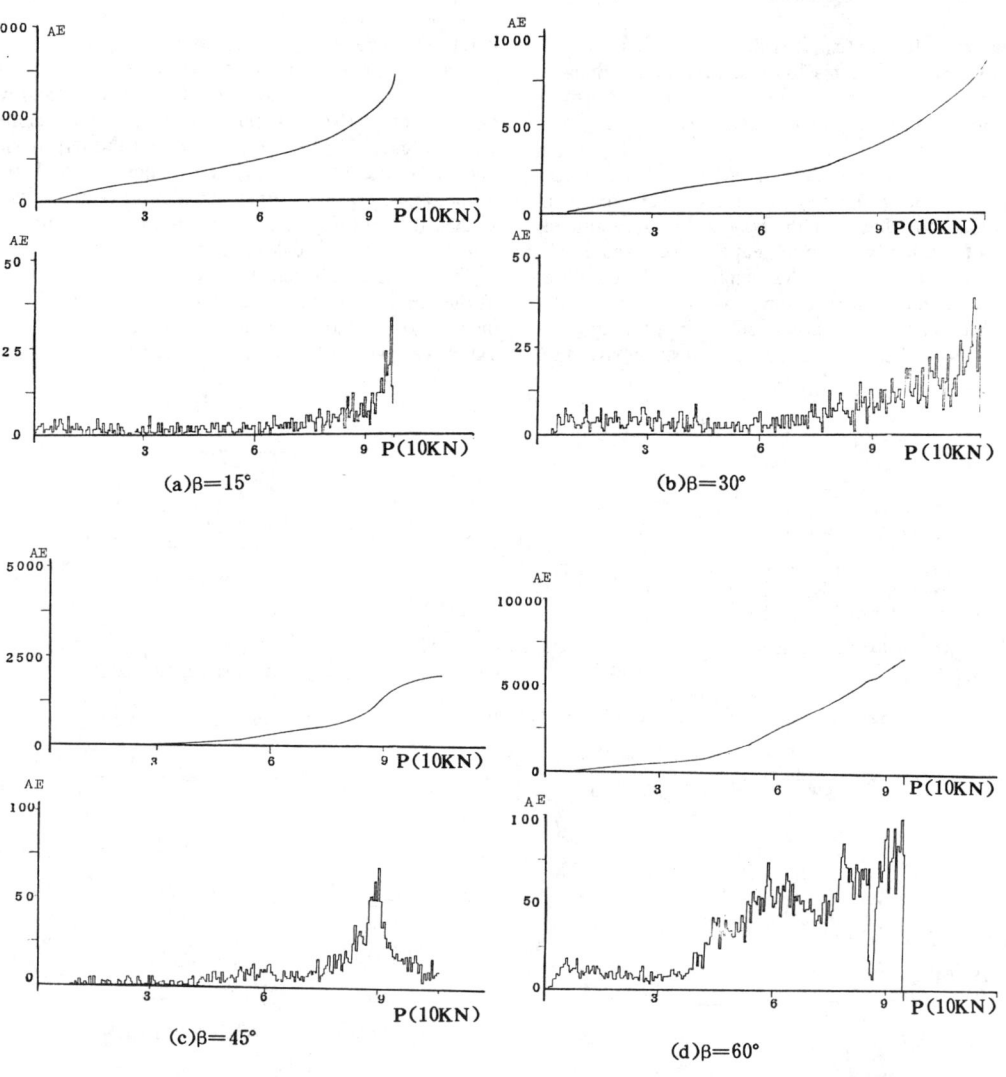

Fig. 4 The load - AE events distribution and accumulation curves

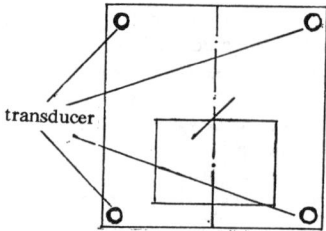

Fig. 5 Test setup to locate AE source

at the four corner points on the specimen, as shown in Fig. 5. By measuring the sonic velocities before,
during, and after the test, it is discovered that the difference in the transmission velocities is less than five percent. So, for all practical purposes, uniform and isotropic velocities can be accepted for the source location.

The crack trajectories and extent of the crack region can be predicted by theoretical analysis. According to these estimates, a predesiganated region of 100 mm x 100 mm near the predicted crack tip is setup on the face of the specimen. The region is further divided into 10 mm x 10mm elements and 80 internal nodes are used to input a known impulse to calibrate the source location by the installed software. The test results of the AE source location for mortar specimens at different compressive loading stages and the corresponding load-displacement curves are shown in Fig. 6. The AE events in the given region against the load are shown in Fig. 7.

In order to elucidate the fracture mechanism, the load-displacement curve is further divided into several smaller stages before and after the peak load. In the initial stage of the loading, the load-displacement curve is essentially nonlinear and the corresponding AE events are distributed over a rather large area and the counts of the events are small. Most of them are generated owing to the compaction of the mortar. By further increasing the load, the AE events start to increase gradually until initiation of crack appears accompanying with a rapid increase of the AE events. On the other hand, the AE sources seem to gather together in the fracture process zone. After reaching the peak load, the stable crack extension comes to the end and the AE events decrease gradually. In the subsequent second and third stages, the AE sources are denser than all other stages. It is in these two stages, the initial strength is transformed to the peak strength.

These AE measurements display clearly that most of the AE events occur near and after the peak load in a narrow band along the direction of crack extension, which may be explained by the fact that

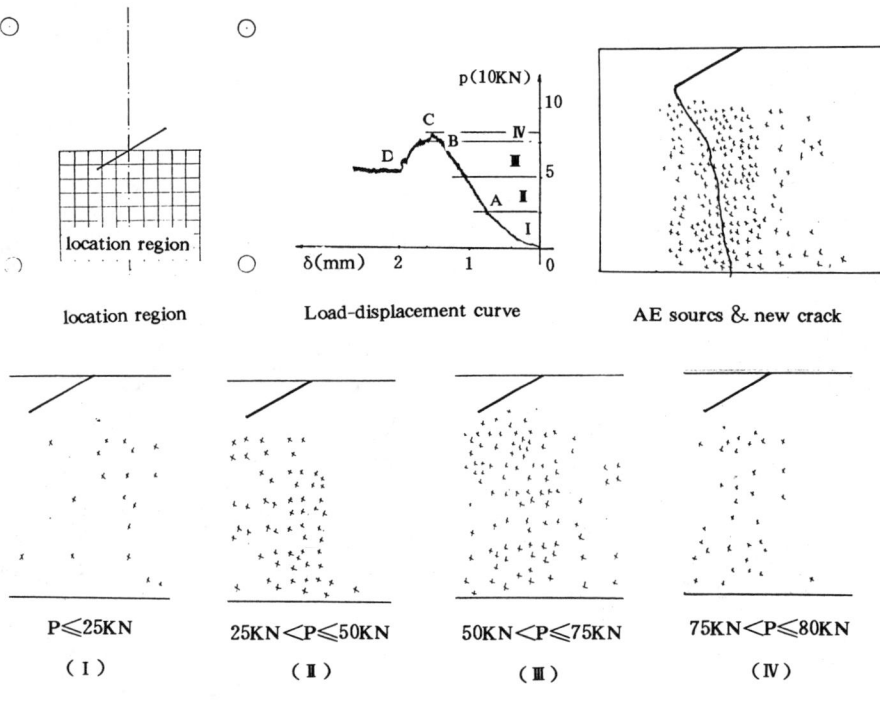

Fig. 6 AE source location for mortar specimens at different loading stages

410

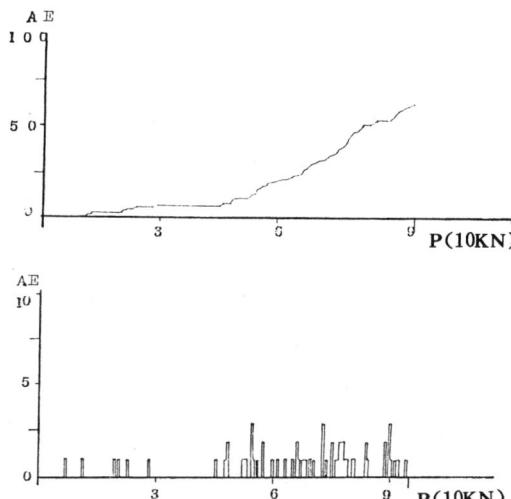

Fig. 7 AE events in the given region (LD60C)

AE events are generated mainly owing to the release of strain energy during crack widening and extension. Since rather a large number of fractures may be cleavage cracks, the dominate cracks are accompanying with microcracks nearly parallel to the main cracks. This is why AE events continue to occur near and behind the extended crack tip and some of them occur along the surface cracking.

Surface cracks are examined with an optic microscope (40 x) when the specimen is removed from the loading assemblage. The post-mortem crack lengths are plotted as shown in Fig. 6. They confirm that crack initiates near the crack tips and extends toward the direction of the major principal stresses. Despite the fairly large scale of crack propagation, the specimen does not lose load carrying capacity until the peak strength is reached.

5. CONCLUSIONS

By dint of the AE monitoring system, AE events generated and the AE sources around and near the crack tips in a total of 50 cement mortar specimens with a slanting central slits are carefully examined under uniaxial compression. The following conclusions can be obtained:

1. The failure fracture of rock with slits may be differentiated into three stages. The first stage corresponds to the tensile initiation strength where the AE events are infrequent and the source of AE events is diffused in a rather large area. The second stage corresponds to the strength between the initiation strength and the peak strength, in which the AE events are frequent and the AE source zone is clustered in a fracture process zone, reflecting the stable development of the crack under the combined compressive and shear stresses. The last stage corresponds to the post peak strength where the AE events are few and the strain energy has been released to compensate the fracturing process.

2. The failure of the rock-like cement mortar manifests itself as splitting when the inclination angle β is small (such as $15°$ tested). When the inclination angle β is larger, a wing-shaped shear band zone is developed. Thus, inclination of the slits dictates the type of failure mode either shear banding in a ductile manner or cleavage splitting in the brittle manner.

REFERENCES

Lawn, B. R. & T. R.Wilshaw 1974. Fracure of brittle solids. London: Cambridge

Maji, A. & S. P. Shah 1988. Process zone and acoustic-emission measurements in concrete. *Experimental Mechanics*. 27:27-33.

Zhu, Keshan , Dongyan Liu & Jinwei Pan 1990. Strength properties of rock mass with bi-directional intermittent cross joints. *Proceedings of the international symposium on rock joints.Leon, 4-6 June 1990*:791-796. Rotterdam: Balkema.

A comparison of the Barton-Bandis joint constitutive model with the Mohr-Coulomb model using UDEC

Rajinder Bhasin & Nick Barton
Norwegian Geotechnical Institute (NGI), Oslo, Norway

ABSTRACT: A numerical study is performed to investigate the influence of a joint constitutive model on the stress-strain behaviour of a rock mass. Distinct element simulations are carried out on 3 different block size models of a rock mass using the Barton-Bandis (BB) and the Mohr-Coulomb (MC) joint constitutive models. The results show that the peak shear strength of a rock mass depends on the constitutive law used. The BB model, which allows the modelling of the dilation accompanying shear, predicts results similar to those from reported physical model tests on jointed slabs of a rock model material. A closely jointed rock mass in which block rotations occur exhibits a lower stiffness but a higher strength than a rock mass with widely spaced joints. The MC model, in which the dilation angle is constant, is relatively insensitive to the effects of different block sizes on the stress-strain behaviour of a rock mass.

INTRODUCTION

Numerical models serve as useful tools in simulating the response of discontinuous media subjected to loading. The Discrete Element Method, UDEC [Universal Distinct Element Code, (Cundall (1980), Cundall and Hart (1993)] is a powerful discontinuum modelling approach for simulating the behaviour of jointed rock masses subjected to quasi-static or dynamic loading conditions. In this method, the deformations and volumetric changes of the intact rock material (blocks) as well as the shear and normal displacements along the joints are included.

Due to the high degree of non-linearity of the systems being modelled, explicit (as opposed to implicit) numerical solution techniques are favoured for codes like UDEC (Universal Distinct Element Code). In this technique no matrices are formed as the procedure marches forward in small steps ensuring final equilibrium at each material integration point in the model.

The mechanical behaviour of a jointed rock mass is strongly affected by the behaviour of discontinuities. Therefore, an inevitable component of many numerical techniques is the constitutive model of discontinuities. During the excavation of an underground opening, the jointed rock may slip or separate along the discontinuities and the movement of the rock blocks may occur through translational or rotational shear. A clear understanding of the mechanical behaviour of rock joints is important for analysing and predicting underground structures in jointed rock masses. Several joint constitutive models have been developed in the past two decades for providing a realistic simulation of the mechanical behaviour of rock discontinuities. (*e.g.*, Barton (1982) and Barton and Bandis (1990), Cundall and Hart (1984), Saeb and Amadei (1992), and Souley and Homand (1995). However, it is still customary among many numerical modellers to use the non-realistic linear-elastic Mohr-Coulomb joint constitutive model. This may be attributed to its computational efficiency in numerical codes and the assumed availability of the Mohr-Coulomb parameters for the cohesion intercept and friction angle in the literature.

This paper compares the results from numerical modelling of the stress-strain behaviour of a rock mass using the non-linear Barton-Bandis (BB) joint constitutive model with those from the Mohr-Coulomb (MC) model. The numerical modelling of block size effects and the influence of joint properties in multiply jointed rock using UDEC-BB has been investigated by Bhasin and Høeg (1997). A brief description of the BB model and some UDEC-BB results are given below:

A BRIEF DESCRIPTION OF THE BB MODEL

The BB non-linear rock joint model describes the shear strength of a joint using the following equation of Barton and Choubey (1977):

$$\tau = \sigma_n \tan\left[JRC \log\left(\frac{JCS}{\sigma_n}\right) + \Phi_r\right] \quad (1)$$

where
- σ_n = effective normal stress across the joint
- JRC = joint roughness coefficient (0-20) (In Equation 1 the dimension of JRC is in degrees)
- JCS = joint wall compressive strength
- Φ_r = residual friction angle

The normal stress (σ_n) versus closure (V_z) relation of a joint in terms of initial normal stiffness K_{ni}, and maximum closure (V_{zm}) is described by Bandis et al. (1981). Their relation for normal stiffness, K_n is as follows:

$$K_n = \left(\frac{K_{ni}}{(1 - V_z/V_{zm})^2}\right) \quad (2)$$

In the BB model, the secant shear stiffness up to peak (usually called peak shear stiffness) is calculated using the following relationship:

$$K_s = \frac{\tau_p}{\delta_p} \quad (3)$$

where the peak shear displacement δ_p is given by the following empirical relationship Barton (1990):

$$\delta_p = \frac{L_n}{500}\left(\frac{JRC_n}{L_n}\right)^{0.33} \quad (4)$$

where L_n = length of joint in meters and JRC is a coefficient between 0-20 depending on the joint roughness.

The size dependent joint roughness and strength are derived using the following equations proposed by Barton and Bandis (1982):

$$JRC_n = JRC_0 \left[\frac{L_n}{L_0}\right]^{-0.02 JRC_0} \quad (5)$$

$$JCS_n = JCS_0 \left[\frac{L_n}{L_0}\right]^{-0.03 JRC_0} \quad (6)$$

where the subscripts (0) and (n) refer to the reference joint test length (100 mm) and *in situ* joint length (L_n), respectively.

An appealing feature of the BB-model is that the basic input parameters for rock discontinuities JRC, JCS and ϕ_r are easily measurable in the field or can be obtained through index testing in the laboratory.

NUMERICAL EXPERIMENTS USING UDEC-BB

Table 1 shows the mechanical properties of the intact rock and the BB joint shear strength parameters (JRC, JCS and ϕ_r) adopted for the UDEC-BB modelling studies. The intact rock consists of fully deformable blocks subdivided into finite difference zones to calculate internal stress and strain. These blocks were modelled as linear-elastic and isotropic materials.

Table 1. BB joint parameters and intact rock parameters assumed for the UDEC simulations.

Parameters	Values
Joint roughness coefficient JRC_0	10
Joint compressive strength JCS_0 (MPa)	35
Laboratory scale length L_0 (m)	0.1
Residual friction angle $\phi_r(°)$	25
Uniaxial compression strength σ_c (MPa)	50
Density (ρ) kN/m^3	27.5
Poisson's ratio (υ)	0.25
Deformation modulus E_d (GPa)	20

Distinct element simulations were performed on models of a rock mass of given dimensions but with three different block sizes. The three 1m×1m rock mass models (vertical slabs assumed), shown in Fig. 1, contain two sets of equally spaced persistent joints dipping at an angle of 45° from the horizontal. Joint spacing ranges from 0.3m in model 1 to 0.1m in model 3. For investigating the influence of joint spacing on the stress-strain behaviour of a rock mass, stress boundaries ($\sigma_v=\sigma_1$, $\sigma_h=\sigma_3$) are applied as shown in Fig. 2. As a first step the models were consolidated under the stress $\sigma_1 = \sigma_3 = 2.5$ MPa and from then on σ_1 was increased keeping σ_3 constant.

The strains caused by the deviatoric stresses for the three rock mass models using the BB input data are shown in Fig. 3. For this comparison, the three models had identical input parameters except for the joint spacing. No scaling of the properties was introduced as would be usual in view of the different joint lengths in the three models.

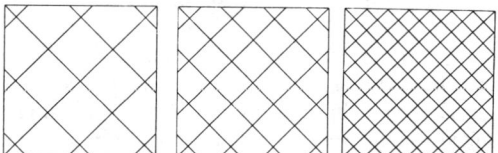

Figure 1. Three discretised assemblies of blocks used for the numerical experiments

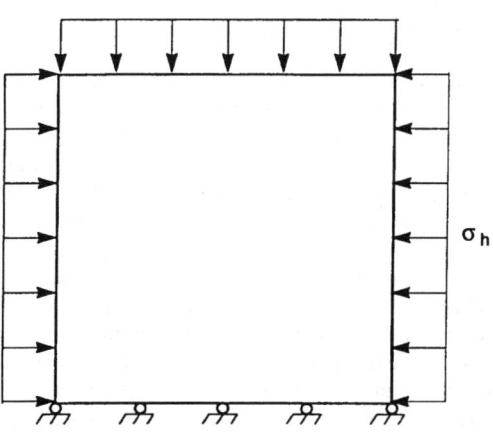

Figure 2. Stress boundary conditions.

The large lateral expansion (mass "Poisson's ratio") is apparent in all the three cases. As block size was reduced, the secant modulus up to a deviatoric stress of 10 MPa reduced from 1.2 to 0.7 GPa (see Fig. 4). At the same level of $\sigma_1 - \sigma_3$, the axial strain in model 1 was smaller than in models 2 and 3. However, a higher deviatoric stress is required to fail model 3 than the two others. The results indicate that the shear strength of a closely jointed rock mass is significantly higher than for a rock mass with wider spacing.

Models 1 and 2 failed as shown in Fig. 5, which presents a displacement vector plot visualising the shear failure for model 1. The contrasting stress-strain behaviour exhibited by the three different models indicates some block size effects on the mass "Poisson's ratio". The ratio of lateral strain/axial strain increases to well beyond = 0.5 as shear strength is increasingly mobilised (see Fig. 3). The

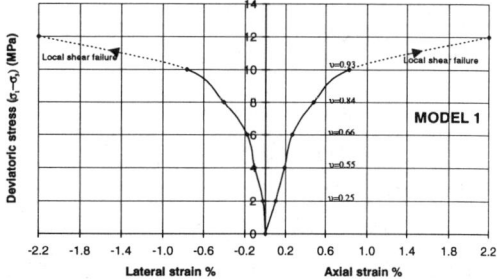

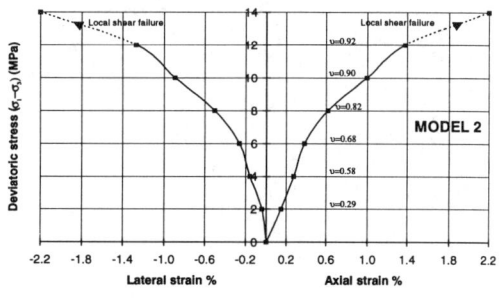

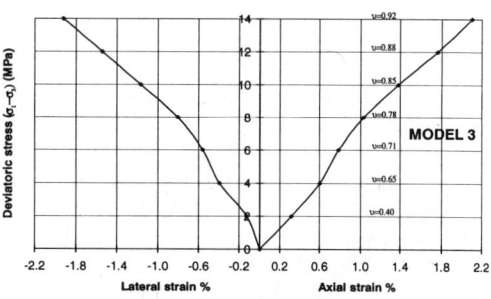

Figure 3. Deformation behaviour exhibited by the three models of different sized blocks.

larger size block assemblies (models 1 and 2) deform mainly through translational shear resulting in a rapid increase in mass "Poisson's ratio" to about 0.92 before the onset of shear failure. At a deviatoric stress of about 8 MPa the mass "Poisson's ratio" in model 3 falls below that of models 1 and 2. At a deviatoric stress of about 14 MPa the mass Poisson's ratio for model 3 is 0.92. Note that at this stress level models 1 and 2 would probably have had a mass "Poisson's ratio" beyond 1.0, but they failed before that level was reached.

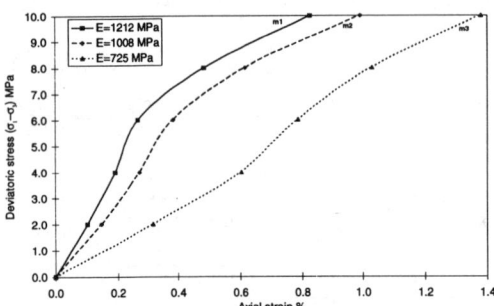

Figure 4 Equivalent secant modulus upto 10 MPa for the three different block size models

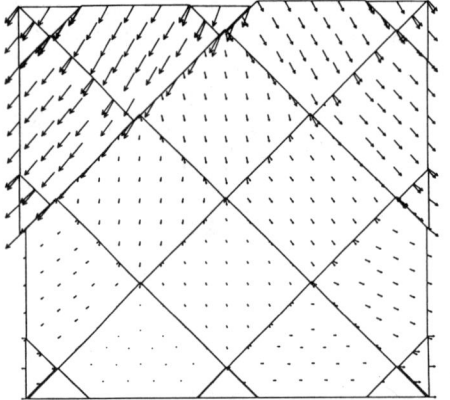

Figure 5. Displacement vector plot showing the shear failure for model 1.

These numerical results are very similar to the physical model tests reported by Barton and Bandis, 1982 (see Fig. 6). In these physical model tests, jointed slabs simulating different cross joint spacing were stressed to failure in a biaxial frame. A double bladed "guillotine" was used for generating model tension fractures (joint sets) through slabs of the weak brittle material, which was formed from an oven-cured combination of red lead-sand-ballotini-plaster-water. The peak strength (τ) of the joints were described closely by Equation (1). The large increase in mass "Poisson's ratio" due to joint shear in the numerical experiments (see Fig. 3) have also been observed in the physical model studies reported by Barton and Hansteen (1979) and Barton and Bandis (1982), and will depend on the orientation of the joints.

The distinctly convex load-deformation behaviour shown by models 1 and 2 in Fig. 4 are typical of a rock mass with diagonal joints. As illustrated schematically in Fig. 7, the shear components are largely responsible for the deformation of a rock mass with diagonal joints. Horizontal joints such as those parallel to the bedding are mainly subjected to normal closure and show a concave load-deformation behaviour. In this case the lateral expansion is limited, and the shear components are largely absent. It is interesting to note that model 3 shows a linear type of load-deformation behaviour due to the combined effect of shearing and closure on an increased number of joints (see Fig. 3). This resembles the axial strain results from the physical model test on 4000 blocks (see Fig. 6).

The present numerical simulations are clearly illustrative of the block size effects on the deformation behaviour of jointed rock masses. Small block sizes which allow rotation to occur, exhibited a higher strength than the larger blocks which failed by translational shear along a number of continuous joints.

Tunnels in jointed rock masses are likely to fail by translational shear if block sizes are large compared to the tunnel dimensions. Block rotations occur only when partial failure through translational shear has occurred creating the space for rotations. Figure 8 shows a special case of failure for model 3 in which a tunnel was excavated at a deviatoric stress of 14 MPa. The increased degree of freedom of the individual joint blocks is apparent from this figure. However kink band development as observed in a 4000 block model [see Fig. 6 and Barton and Bandis (1982)] did not develop in the present 144 block numerical model.

ESTIMATION OF EQUIVALENT c AND ϕ VALUES FROM THE BB JOINT PARAMETERS

In the following, the Mohr-Coulomb parameters c and ϕ are derived from the BB parameters to have consistent comparison. These are then introduced in the 3 models so that the effect of linear joint properties on the predicted behaviour of the rock mass can be investigated. In the BB-model the dilation along the joint develops gradually with shear

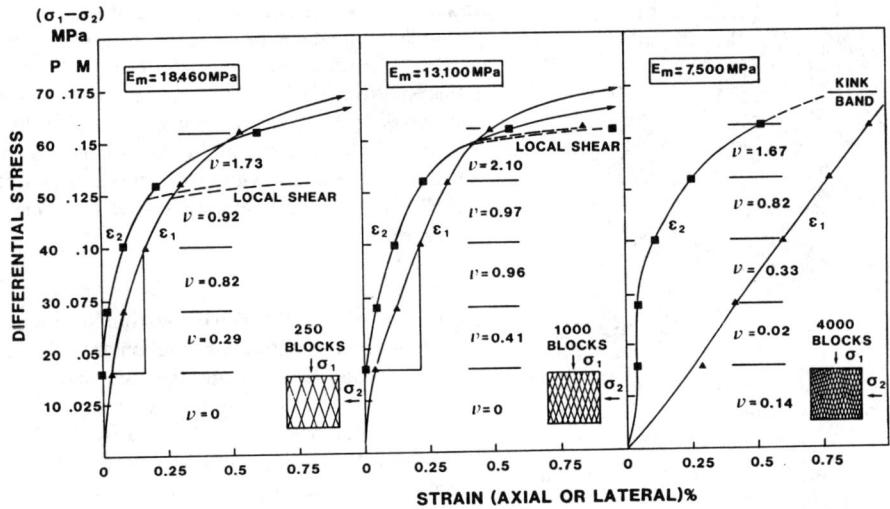

Figure 6 Stress-strain behaviour exhibited by assemblies of different sized blocks, Barton and Bandis, 1982.

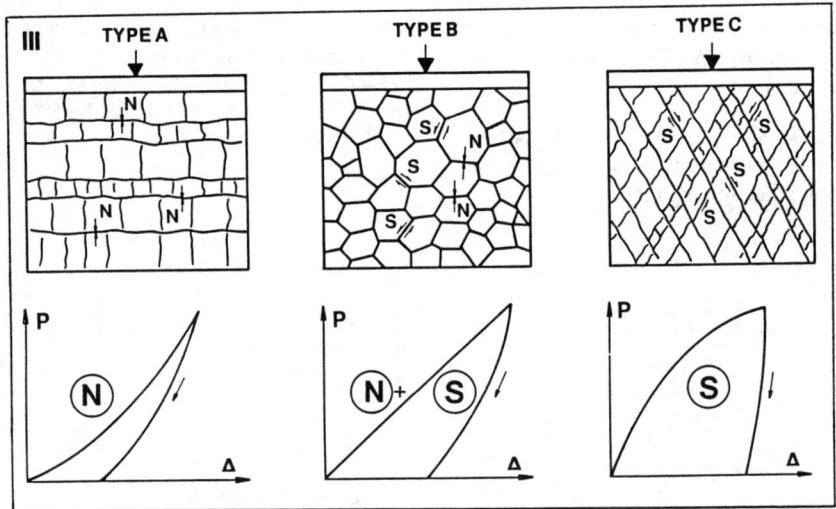

Figure 7 Contrasting load-deformation behaviour for rock masses with different magnitudes of joint shear (S) and normal deformation (N) components, Barton, 1990.

displacement (using the JRC mobilised concept of Barton (1982), and the peak dilation angle is estimated from the following equation:

$$\phi_d = 0.5 JRC_n \log\left(\frac{JCS}{\sigma_n}\right) \quad (7)$$

In the Mohr-Coulomb model, the dilation angle is constant and represents an average value. Therefore a reduction factor of ½ is applied to the above expression to calculate ϕ_d for the Mohr-Coulomb model.

Inside the range of normal stresses relevant (5 and 10 MPa,), we can find the shear strength (τ) based on the BB formulae presented in Equation (1). To be equivalent to the BB-model in the same stress range,

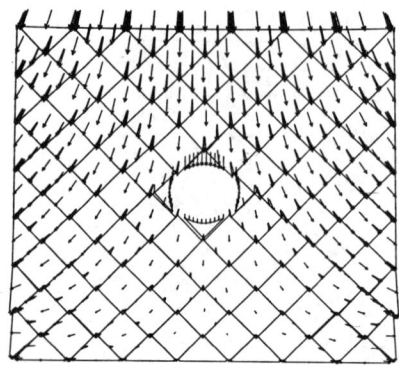

Figure 8 Magnified (×3) visualisation of block movements in model 3 after tunnelling

the Mohr-Coulomb model should give the same shear strength as the BB-model, i.e.,

$$\tau_5 = \sigma_n \tan\phi + c \quad (\sigma_n = 5\text{MPa}) \quad (8)$$

$$\tau_{10} = \sigma_n \tan\phi + c \quad (\sigma_n = 10\text{MPa}) \quad (9)$$

By solving the above system of equations, the friction angle and cohesion intercept for the equivalent Mohr-Coulomb criterion are obtained. They turned out to be $\phi=24.8°$ and $c=0.45$ MPa for the BB-parameters shown in **Table 1**.

Fig. 9 shows the strains caused by the deviatoric stresses for the three rock mass models. It may be seen from the figure that although model 3 exhibits a lower stiffness than models 1 and 2, the three models fail at the same deviatoric stress. The results indicate a relative insensitiveness to the effect of different block sizes on the peak shear strength of a jointed rock mass. This insensitivity is attributed to the fact that in the MC model, the dilation angle is constant and does not increase with the shear displacement. In the BB model the dilation angle rises to a peak and slowly reduces thereafter.

Figures 10 a and 10b show the rotation of blocks (shown in the figure through rotation arcs) for model 3 after consolidation using the BB and the MC model respectively. It may be seen that more block rotations occur when using the BB model than the MC model. Block rotation occurs only when translational shear has already occurred. Translational shear in fact causes dilation of the joints creating the space for rotations. It is widely acknowledged that the degree to which a rock joint dilates when sheared has important consequences in rock mechanics. Thus, for non-planar joints the BB-model may render itself more appropriate for numerical analyses of underground excavations if dilation effects are expected to be significant. However, the computational effort is larger when using the BB-model than the MC model.

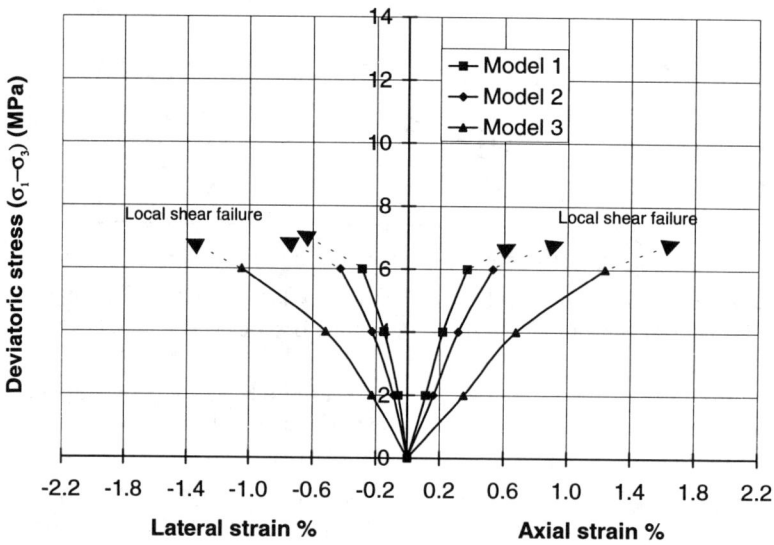

Figure 9 Deformation behaviour exhibited by the three models when using the MC model

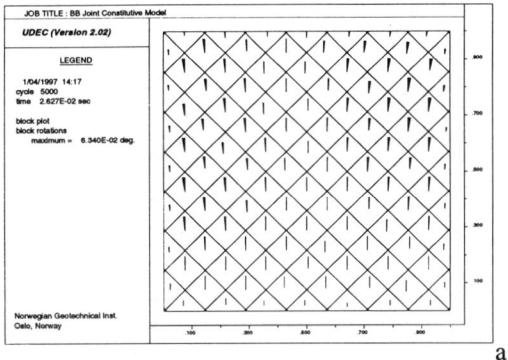 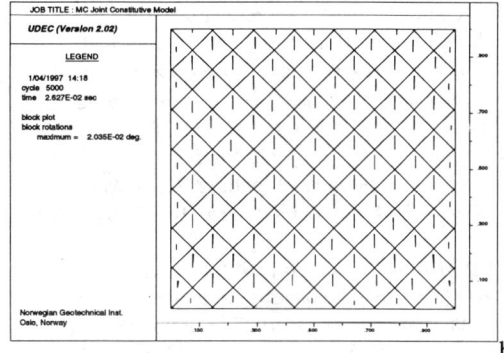

Figure 10 A comparison of block rotations (shown through rotation arcs) when using the BB model (a) and the MC model (b)

CONCLUSION

In this paper the effect of joint constitutive laws on the stress-strain behaviour of a rock mass was examined using the numerical code UDEC. The results have shown that the peak shear strength of a rock mass depends on the constitutive law used. In this respect it is revealed that the Mohr-Coulomb model, in which the dilation angle is constant, is relatively insensitive to the effect of different block sizes on the peak shear strength of a rock mass.

The BB model, which allows the modelling of variable dilation accompanying shear, predicts results similar to those from earlier physical model tests on a rock model material. These results indicate that the size of the individual blocks controls both the shear strength of the rock mass and its deformational characteristics. A closely jointed rock mass in which block rotations occur exhibits a lower stiffness but a higher peak shear strength than a rock mass with widely spaced joints.

REFERENCES

Barton, N., and Choubey, V., 1977. The Shear Strength of Rock Joints in Theory and Practice. *Rock Mechanics,* No. 1 / 2, pp. 1-54, Springer, Vienna, 1977.

Barton, N. and Hansteen, H., 1979. Very Large Span Openings at Shallow Depth: Deformation Magnitudes from Jointed Models and F.E. Analysis, *Proc. 4th Rapid Excavation and tunnelling Conference,* pp. 1331-1353, Atlanta, USA

Bandis, S.C., Lumsden, A.C., and Barton, N., 1981. Fundamentals of Rock Joint Behaviour. *Int. J. Rock Mech. Min. Sci. & Geomech. Abstr.* Vol. 20, No. 6, pp. 249-268 (1981)

Barton, N., 1982 Modelling Rock Joint Behaviour from Insitu Block Tests: Implications for Nuclear Waste Repository Design, ONWI-308, prepared by Terra Tek, Inc. *for Office of Nuclear Waste Isolation, Columbus, OH, USA.*

Barton, N. and Bandis, S., 1982. Effects of Block Size on the Shear Behaviour of Jointed Rock, Keynote Lecture, *Proc. 23rd U.S. Symposium on Rock Mechanics,* pp. 739-760, Berkeley, California (1982)

Barton, N., 1990. Scale Effects or Sampling Bias? *Proc. Int. Workshop on Scale Effects in Rock Masses,* pp. 31-55, Loen, Norway, 1990

Barton, N., and Bandis, S., 1990. Review of Predictive Capabilities of JRC-JCS Model in Engineering Practice. *Proc. Int. Symp. on Rock Joints, Loen, Norway, 1990,* pp. 603-610

Bhasin, R. and Høeg, K., 1997. Numerical modelling of block size effects and influence of joint properties in multiply jointed rock. *Submitted to Tunnelling and Underground Space Technology*

Cundall P. A., 1980 A generalised distinct element program for modelling jointed rock. Report PCAR-1-80, Contract DAJA37-79-C-0548, European Research Office, U.S. Army, Peter Cundall Associates

Cundall P.A. and Hart R.D., 1984 Analysis of block test No. 1 in elastic rock mass behavior: phase 2 - a generalization of joint behaviour. *Itasca Consulting Group,* Rockwell Hanford Operations,

Subcontract SA-957, Final Report (1984)

Cundall P. A. and Hart R.D., 1993 Numerical modeling of discontinua. *Comprehensive Rock Engineering, Principles, Practice and Projects, Vol. 2 (Ed. J.A. Hudson),* pp. 231-243 (1993)

Saeb S. and Amadei B., 1992 Modelling rock joints under shear and normal loading. *Int. J. Rock Mech. & Min. Sci. & Geomech. Abstr.* 29, 267-278

Souley M. and Homand F., 1995 An extension of the Saeb and Amadei constitutive model for rock joints to include cyclic loading paths. *Int. J. Rock Mech. & Min. Sci. & Geomech. Abstr.* 32, 101-109

Shear behaviour of soft joints using large-scale shear apparatus

B. Indraratna, A. Haque & N. Aziz
Department of Civil and Mining Engineering, University of Wollongong, N.S.W., Australia

ABSTRACT: Shear tests have been conducted in the laboratory on simulated rock joints by using a large-scale shear apparatus, under constant normal load (CNL) and constant normal stiffness (CNS) conditions. Test results show that the CNL condition underestimates the peak shear stress which is attained at a lower horizontal displacement in comparison to the CNS conditions. The strength envelopes for three triangular interfaces (inclinations of 9.5°, 18.5° and 26.5°) show that a linear envelope for a low asperity angle, and a bilinear envelope for a higher asperity angle are appropriate for the CNS condition. In contrast, a bilinear envelope is obtained even for a low asperity angle under CNL shearing. The shear strength of infill joints is observed to diminish with the increase in infill thickness. At an infill thickness to asperity height ratio of 1.60, the shear strength approaches that of the pure infill.

1 INTRODUCTION

Usually, the shear strength of discontinuities have been investigated in the laboratory by direct shear tests, where the shearing takes place under a constant normal load (CNL) condition. In this situation, the dilation of joint occurs against a constant stress, that may not represent the confinement from the surrounding rock. However, in reality, some of the deformation is inhibited by the surrounding rock mass during shearing, and inevitably increases the normal stress applied over the interface. Especially, the shearing of non-planer interfaces no longer remain under constant normal load conditions but rather their behaviour is governed by changing normal load conditions. Johnstone & Lam (1989) has demonstrated that the normal stiffness remains constant during shearing of rock socketed piles, which is a representation of vertical non-planar interfaces. It is understood that the shear behaviour of non-planar joints need to be investigated under constant normal stiffness (CNS) condition. In order to understand the shear behaviour of soft model joints, a large-scale shear apparatus has been designed by the authors which can be employed for testing joints under both CNS and CNL conditions. Shear tests have been conducted in the laboratory by Seidel & Haberfield (1995), Ohnishi & Dharmaratne (1990) and Archambault et al. (1990) using the CNS technique for concrete-rock and artificial hard rock joints. In comparison, only a limited number of studies have been reported on the CNS testing of soft rock joints. In one such study, Cheng et al. (1996) mentioned that the CNS shear strength envelope for infilled joints is purely frictional. This paper extends further the aspects of shear behaviour of soft, synthetic rock joints (triangular asperities) under CNS conditions, with particular reference to the effect of infill material.

2 LARGE-SCALE SHEAR APPARATUS

The shear apparatus consist of two top and bottom boxes of size 250x75x150mm and 250x75x100mm, respectively. The top box can only move vertically on ball bearings as shown in Figure 1. The bottom box which is rested on bearings can move only in the direction of shearing. Figure 1 illustrates how the whole assembly is firmly fixed using two rigid frames. The normal stiffness of the rock mass is modelled in the laboratory by a set of four springs placed over the top specimen. The normal and shear loads are applied through hydraulic jacks. The piston of the shear loading jack is connected to a strain controlled device which operates under a prefixed strain rate, within the range of 0.30 to 1.70 mm/min. The apparatus has normal and shear load capacities of 120 kN and 180 kN, respectively. Load cells are connected with digital strain meters to monitor the change in normal and shear loads during shearing. In order to measure the dilation and horizontal displacements, strain gauges were located on the center of the top specimen, and also along the shear direction.

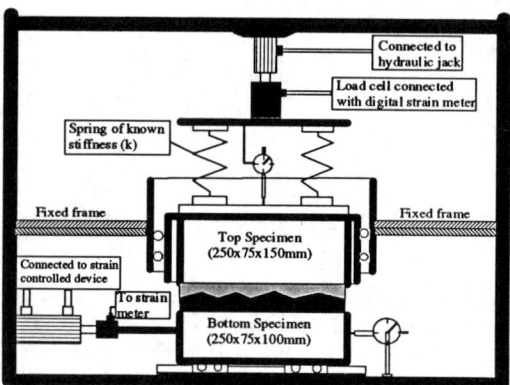

Figure 1. Details of the large-scale CNS shear apparatus.

3 SPECIMEN PREPARATION

Specimens of mating joints were cast inside the shear boxes by mixing gypsum plaster with water in the ratio of 5:3. Special moulds were built to cast triangular interface profiles having inclinations of 9.5° (Type I), 18.5° (Type II) and 26.5° (Type III). The moulds could be stripped after an hour and left inside an oven at a temperature of 50°C for two weeks for curing to be completed. The following physical properties of the cured specimens were obtained by conducting compression tests on 50mm diameter specimens: uniaxial compressive strength (σ_c)=11-13 MPa; Young's modulus (E)=1.9-2.3 GPa; and Poissons's ratio (ν)=0.25. These material properties typically represent the behaviour of soft rock joints based on dimensionless strength parameters (Indraratna, 1990).

4 LABORATORY INVESTIGATION

Shear tests were conducted on Type I, II and III interfaces under both CNL and CNS conditions, for various initial normal stresses (σ_{no}). The shear displacement rate was maintained constant at 0.50 mm/min for all the tests. The values of σ_{no} varied from 0.05 to 2.43 MPa while a constant normal stiffness (k) of 8.5 kN/mm was used for all the CNS tests. Changes in normal and shear stresses together with the corresponding dilations were recorded for each test as a function of the horizontal displacement.

5 RESULTS AND DISCUSSIONS

5.1 Shear stresses under CNS and CNL conditions

The change in shear stress corresponding to the initial normal stress (σ_{no}) for Type I profile under both CNL and CNS conditions are plotted in Figure 2. The shear stress of the joint is observed to increase with the shear displacement until the "peak-to-peak" contact is reached. At lower normal stress, the shear

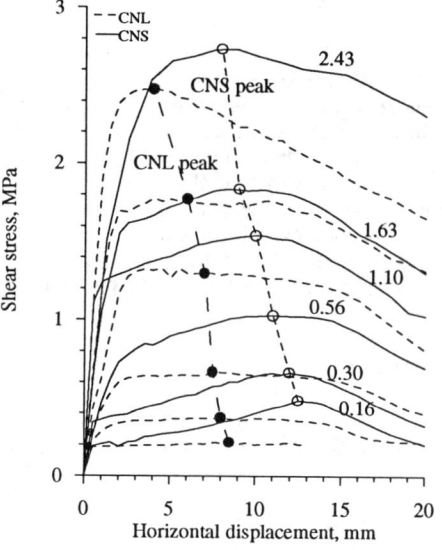

Figure 2. Variation of shear stress with horizontal displacement under CNL and CNS conditions for Type I profile.

stress changes gradually, as little shearing occurs across the asperities. In contrast, the asperities are easily sheared at higher normal stresses, thus resulting in a flatter variation of the stress-displacement curves. As expected, the peak shear stress under the CNS condition is always higher than that of the CNL condition for a given σ_{no}. It is also interesting to note that the dilation measured for CNL conditions is greater than that recorded under CNS conditions for the same horizontal displacement. The peak shear stress is attained at a lower horizontal displacement with the increased σ_{no} for both CNS and CNL conditions. However, the peak shear stress under CNL condition is always occurs at a smaller horizontal displacement, and also the CNL peak is always below the CNS peak, as shown in Figure 2.

5.2 Effect of asperity angle on shear behaviour of joints under CNS conditions

Types I, II and III surface profiles were tested under the same σ_{no} of 1.63 MPa and normal stiffness (k) of 8.5 kN/mm, respectively. The variations of shear stress, normal stress and dilation with the horizontal displacement are plotted and compared in Figure 3. It is evident that the shear stress increases with the horizontal displacement as the asperity angle is increased from 9.5° to 26.5°. As shearing progressed, the peak shear stress was attained at a lower horizontal displacement with increased asperity angle. This may be attributed to the high stress concentration surrounding the asperity. The normal stress and dilation for Type I and II joints are observed to increase gradually with the horizontal displacement until the peak contact is reached. In contrast, the normal stress and dilation curves for Type III profiles indicate a more rapid increase followed by a gradual decrease. For Type III joints, the drop in shear stress after the peak is considerable in comparison with Type I and II joints. The higher value of peak shear stress obtained for the higher asperity angle was to be expected due to enhanced friction. Nevertheless, the degradation of prominent asperities (Type III) of smaller horizontal displacements is subsequently associated with a rapid post-peak drop in shear strength (Fig. 3).

5.3 Peak shear strength envelopes

The shear stress against the normal stress relationship (CNS) measured for Type III (i=26.5°) profiles is plotted in Figure 4, for various initial normal stresses. The individual shear stress vs normal stress curves represent the stress paths of each test. The peaks of the stress paths define a bilinear strength envelope under CNS conditions, for these soft joints. In separate studies, Ohnishi & Dharmaratne (1990) and Seidel & Haberfield (1995) have discussed the shear strength envelopes for harder rock joints under CNS testing. A similar, bilinear envelope is also observed for Type II profile (Figure 5) having a lower asperity angle (i=18.5°). In contrast, a linear strength envelope is sufficient for Type I joints (i=9.5°) under CNS conditions. The apparent friction angle for Type I joints was measured to be about 47° (approx. ϕ_b+i), where the basic friction angle (ϕ_b) of 37.5° for plaster joints was obtained by conducting shear tests on planar surfaces. In comparison, the CNL strength envelope for Type I joints is bilinear. The slope of the initial part of the strength envelopes for Type II and III joints are considerably higher due to the enhanced shear resistance offered by the higher asperity angles. At a normal stress lower than 1.50 MPa, the apparent friction angles are approximately equal to 56° and 64° for Type II and III joints, respectively. As the normal stress is increased further, the slope of the second part of the envelope tends to coincide with the basic friction angle, confirming the asperity behaviour described initially by Patton (1966).

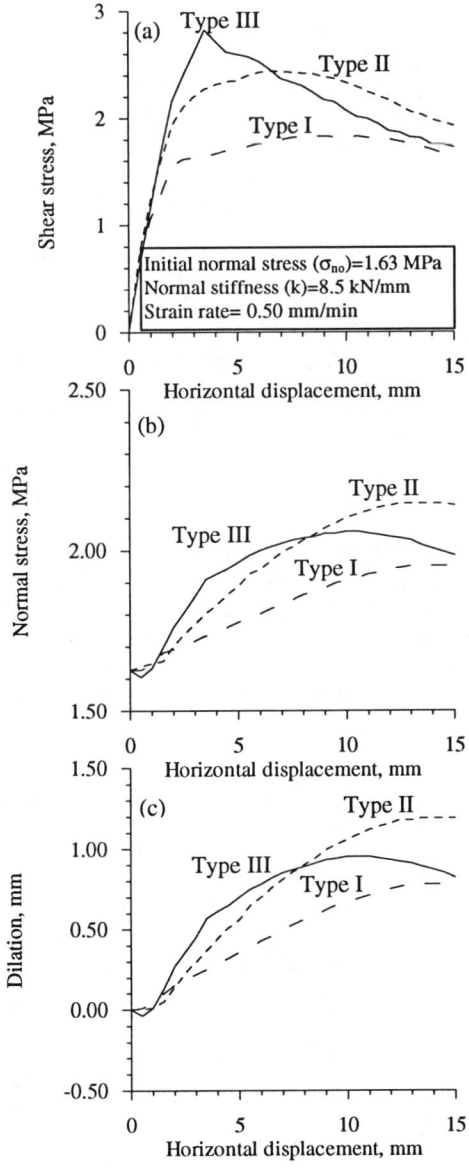

Figure 3. Variations of shear, normal stress and dilation with horizontal displacement for Type I, II, III profiles.

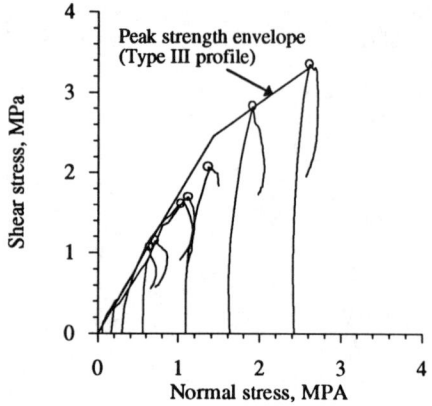

Figure 4. Strength envelope for Type III profile under CNS condition.

Based on this study, the following equation (Eqn.1) is proposed by the authors to describe the peak strength envelope for soft rock joints under CNS condition:

$$\tau_p = (\sigma_{no} + k \cdot d_v/A) \tan(\phi_b + i) \qquad (1)$$

where, τ_p=peak shear stress, σ_{no}=initial normal stress, k=normal stiffness, d_v=dilation corresponding to peak shear stress, ϕ_b=basic friction angle and i=initial angle of asperity. It is understood that the value of i in Eqn.1 may be neglected for normal stresses exceeding 1.50 MPa due to asperity degradation.

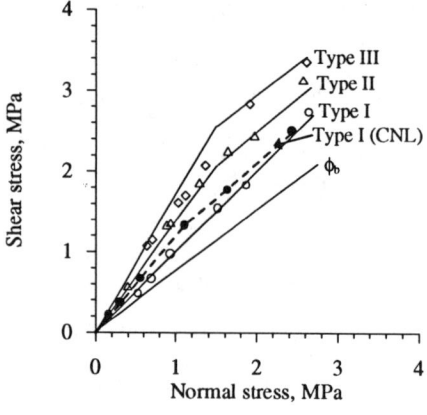

Figure 5. Peak shear strength envelopes for Type I, II, III interfaces.

The magnitude of d_v in Eqn.1 can be obtained from the following empirical equation (Eqn.2):

$$d_v/a = \alpha \, e^{(-\beta \cdot \sigma_{no})} \qquad (2)$$

where, α and β are surface profile coefficients whose magnitudes for the three type of joints are summarised in Table 1.

Table 1. Values of α and β for different interfaces.

Constant	Type I	Type II	Type III
α	0.67	0.63	0.33
β	0.78	0.97	0.82

5.4 *Effect of infill thickness on shear behaviour*

In order to study the effect of infill on the shear behaviour of joints, bentonite was placed between the mated joints. The variation of shear stress with the horizontal displacement for Type I infill joint was measured and plotted in Figure 6. The thickness of infill was varied from 1 to 4mm for the same initial normal stress of 0.16 MPa. The peak shear stress was found to decrease considerably even at very small infill thickness (t/a=0.40). At a t/a ratio of 1.60, the shear strength of joint was no different to that of the infill material.

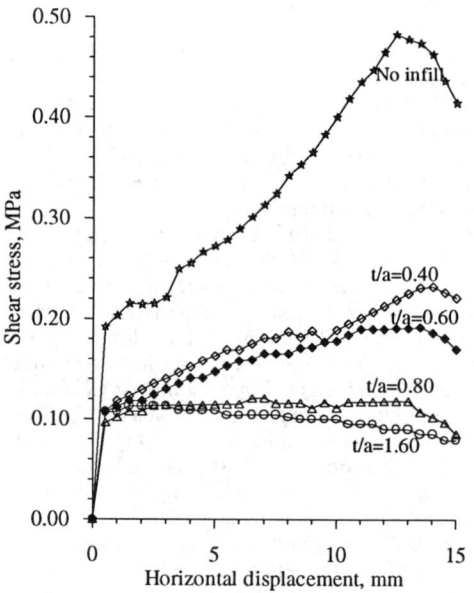

Figure 6. Effect of infill thickness on shear stress of Type I joints under CNS condition.

6 APPLICATION OF THE RELATIONSHIPS

The proposed bilinear peak strength relationship (Eqn.1) can be used to predict the peak shear stress of soft joints under CNS condition, provided that the values of σ_{no}, stiffness (k) and asperity height (a) are known. The dilation of joints corresponding to the peak strength can be calculated using Eqn.2. The predicted dilation is then substituted in Eqn.1 to determine the peak shear strength of joints. The predicted and experimental values of the peak shear strength for Type I, II and III profiles are summarised in Table 2 for comparison. It can be seen that the predicted results are very close to the measured results, especially in the high normal stress region. However, the predicted results generally overestimate the measured stresses. The proposed equations consider the sliding behaviour along the plane of weakness, but can not incorporate accurately the exact shearing modes and the extent of degradation of the asperities at higher normal loads. Ladanyi & Archambault (1970) have pointed out that the loss of interlocking and non-uniform stress distribution on the surface of the asperities that cause their premature breakage can be responsible for a reduction in peak strength.

7 CONCLUSIONS

The shear behaviour of soft rock joints is different to the conventional direct shear tests conducted on non-planar discontinuities. The peak shear strength measured under CNS condition is always higher than that of CNL condition. At increased initial normal stress, both the CNS and CNL peaks are attained at lower horizontal displacements. The asperity angle has considerable effect on the peak shear strength of joints. As expected, the peak strength increases with the increase in asperity angle. In CNS condition, the shear strength envelope is bilinear for higher asperity angles (Type II & III) and linear for smaller asperity angle (Type I). However, the shear strength envelope is still bilinear for Type I joints under CNL conditions. The bentonite infill has a significant influence on the shear strength of joints. An infill thickness to asperity height of 1.60 is sufficient to decrease the strength of joints to almost that of the pure infill.

The proposed empirical relationships describe the shear strength and dilation characteristics of soft model joints. It is important to note that this study has described the behaviour of soft model joints only. Therefore, while the findings of this study are appropriate to interpret the behaviour of soft rock joints, they should not be extended to describe the behaviour of much harder rock joints.

Table 2. Summary of the experimental and predicted results.

Profile Type	σ_{no} (MPa)	Experimental results			Predicted results		
		τ_{peak} (MPa)	σ_n (MPa)	d_v (mm)	$\sigma_n=\sigma_{no}+(k.d_v/A)$ (MPa)	$\tau_p=\sigma_n\tan(\phi_b+i)$ (MPa)	$\tau_p=C+\sigma_n\tan(\phi_b)$ (MPa)
Type I	0.16	0.48	0.53	1.39*	0.46	0.49	-
	0.30	0.66	0.69	1.32*	0.54	0.58	-
	0.56	1.01	0.94	1.04	1.04	1.12	-
	1.10	1.54	1.50	0.90	1.43	1.53	-
	1.63	1.80	1.83	0.50	1.84	1.98	-
	2.43	2.72	2.63	0.24	2.54	2.72	-
Type II	0.16	1.30	0.85	2.60*	1.01	1.49	-
	0.30	1.36	0.92	2.28*	1.02	1.50	-
	0.56	1.86	1.29	1.88	1.39	2.06	-
	1.10	2.25	1.65	1.29	1.59	-	2.12
	1.63	2.44	1.97	0.76	1.92	-	2.38
	2.43	3.12	2.57	0.28	2.56	-	2.86
Type III	0.16	1.14	0.73	2.47*	0.78	1.60	-
	0.30	1.61	1.08	2.25*	0.81	1.67	-
	0.56	1.68	1.13	1.51	1.26	2.58	-
	1.10	2.05	1.36	0.73	1.55	-	2.60
	1.63	2.82	1.88	0.57	1.92	-	2.87
	2.43	3.35	2.58	0.42	2.58	-	3.38

(*) Measured dilation is corrected by the initial compliance of the test apparatus $(d_o)=0.80$ mm for $\sigma_{no}<k/A$.

8 ACKNOWLEDGMENTS

The financial support from the Key Centre for Mines (KCM) is gratefully acknowledged. The authors wish to thank Mr. Alan Grant for his help during testing and sincere work to build the large-scale CNS shear apparatus at the Workshop of University of Wollongong.

REFERENCES

Archambault, G., Fortin, M., Gill, D.E., Aubertin, M. and Ladanyi, B. (1990). Experimental investigations for an algorithm simulating the effect of variable normal stiffness on discontinuities shear strength. *Proc. Int. Symp. on Rock Joints*, (Barton & Stephansson eds.), Leon, Norway, Balkema Publishers, Rotterdam, pp.141-148.

Cheng, F., Haberfield, C.M. and Seidel, J.P. (1996). Laboratory study of bonding and wall smear in rock socketed piles. *Proc. 7th ANZ Conf. on Geomechanics*, Adelaide, Inst. of Engineers, Australia, pp.69-74.

Indraratna, B. (1990). Development and applications of synthetic material to simulate soft sedimentary rocks. *Geotechnique*, 40:2, pp.189-200.

Johnstone, I.W. and Lam, T.S.K. (1989). Shear behaviour of regular triangular concrete/rock joints - analysis. *J. Geotech. Eng.*, ASCE, Vol.115, No.5, pp.711-727.

Ladanyi, B. and Archambault, G. (1970). Simulation of the shear behaviour of a jointed rock mass. *Proc. 11th Symp. Rock Mech.*, (Somerton, W.H. eds.), AIME, Berkeley, Calif., pp.105-125.

Ohnishi, Y. and Dharmaratne, P.G.R. (1990). Shear behaviour of physical models of rock joints under constant normal stiffness conditions. *Proc. Int. Symp. on Rock Joints*, (Barton & Stephansson eds.), Leon, Norway, Balkema Publishers, Rotterdam, pp.267-273.

Patton, F.D. (1966). Multiple modes of shear failure in rock and related materials. *Ph.D. thesis*. Univ. of Illinois, Urbana, 282p.

Seidel, J.P. and Haberfield, C.M. (1995). The application of energy principles to the determination of the sliding resistance of rock joints. *Rock Mechanics & Rock Eng.* Vol.28, No.4, pp.211-226.

Studies on contact mechanism and closure behaviour of rock joints

Caichu Xia
Department of Geotechnical Engineering, Tongji University, Shanghai, People's Republic of China

Zongqi Sun
Central South University of Technology, Changsha, People's Republic of China

ABSTRACT: The experimental technique of joint closure and the measurement method of initial aperture of a joint are introduced in this paper, The mechanism of non-linear unrecoverable deformation during joint closure is inquired into. The characteristics of composite topography and closure behaviour of joints under different contact states are studied. The relation equation of normal load and closure deformation containing the topographic parameters of composite profiles and dimension less initial aperture is obtained.

1 INTRODUCTION

The closure test curves of joints are basically similar. They possess high non-linearity and as normal stress increases their slopes tend gradually to a vertical asymptote which corresponds to the maximal closure of joints. Loading and unloading cycles exhibit hysteresis and pavement set which diminish rapidly with successive cycles. The un-mated joint have lower stiffness and bigger hysteresis than the mated one. Goodman(1980) attributed most of the non-linear deformation of closure-stress curves to inelastic crushing and splitting of contacting asperities and thought that the closure-stress curves during unloading would follow the corresponding curves of intact rock. Sun(1983) thought that the most of non-linearity of closure-stress curves was still elastic which could be explained by Hertz contact theory. This conclusion has been confirmed by the closure tests of joints on marble, slate et al..

For the same a joint, the initial aperture and closure behaviour under different contact states are not identical either. Thus, the closure behaviour of a joint is related with its composite topography and the empirical equations reflecting the law of closure deformation of a joint should also reflect the influence of its initial aperture and the characteristic parameters of composite topography on closure behaviours. The relationships between the initial aperture under different contact states, the characteristic parameters of composite topography and normal deformation of a joint are analysed in this paper.

2 CLOSURE TESTS OF JOINTS

The samples are taken from marble. The two loaded faces of a sample are formed by cutting with parallel double blade cutter and polishing with parallel double faces grinder. The sample size is 100 × 100 × 100mm. At first the basic elastic parameters E, μ of rock samples are determined, then artificial rock joints are produced by splitting the rock sample at middle into two parts with Brazil splitting method. The topography of the two surfaces of every joint is measured by RSP-I type intellectualization profilometer of rock surface manufactured by ourselves. At least three profiles at the same interval in two perpendicular direction are measured. The sampling interval is 0.5 mm. The accuracy of the profilometer is estimated at 0.01 mm.

The tests are carried out on Instron-1345 servo stiffness material testing machine. The average deformation of distance d between two pre-set signs on joint sample is measured with two extensometers. The normal displacement-load curves are simultaneously recorded with microcomputer and function recorder during test. After the closure test of a joint under mated state is finished the two surfaces of the joint are moved apart 1, 2, 3 and 4 mm respectively in order to attain the closure curves of un-mated joints. Because both composite topography and initial aperture of a joint sample under different contact states have been changed after two surfaces of a joint are mismatched different amount of displacement, the closure tests results

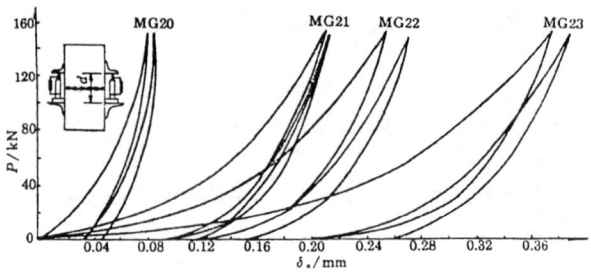

Figure 1. The Typical Closure Test Curves of Joints

would be regarded as the closure test curves of un-mated joints with different characteristics of topography and different mated degree. The closure tests of two or three loading and unloading cycles are made for each of contact states.

In order to measure the initial aperture of a joint under different contact states, actuator situation of testing machine is recorded during every cycle of every test when actuator just starts to contact sample(contacting load is 0.3% of the maximal normal load for each circle) which is regarded as datum point determining initial aperture of un-mated joints. The initial aperture of an un-mated joint is the difference between actuator situation and the datum point.

The normal deformation of mated joints is calculated by the normal deformation recorded during testing subtracting the deformation of intact rock from the two pre-set signs. The normal deformation of an un-mated joint is calculated by subtracting the value of average normal deformation at the central line of loading and unloading curve at the last cycle of the joint at mated state from the recorded value . The typical closure test curves of joints are shown in figure 1. The meaning of number in this paper is that MG is rock kinds which stands for marble sample, the first numeral is sample number, the second numeral is the amount of mismatched displacement between two faces of a joint, the third numeral is cycle sequence. The test results show that the closure curves of the same joint under different contact states at different amount of mismatched displacement are quite different. This indicates that the closure behaviour of a joint depends more on its characteristics of composite topography under specific contact state than on the characteristics of respective topography of its two faces.

3 CONTACT STATES AND COMPOSITE TOPOGRAPHY OF JOINTS

In order to study the initial aperture and closure behaviour of joints under different contact states, it is necessary to study composite topography of joint surfaces which is the sum of topography heights of corresponding points on two surfaces of a joint under specific contact states. The method to form composite topography of the same joint under different contact states from topography measurements is adding the measurement data of joint topography of upper surface to those of lower surface on corresponding profile at several sampling interval. Several artificial tensile joints of marble are mismatched 1, 2, 3, 4 mm respectively. The composite topography of a joint under these kinds of contact states is illustrated in figure 2. The parameters of composite topography are calculated from the data of composite topography (Tab.1). With the help of the analysis of profile characteristics of composite topography and the parameters of composite topography it is found that different amount of mismatch placement between two faces of a joint will result in big change of its composite topography and the difference of parameters of composite topography. Because the parameters of height characteristics of surface topography (the

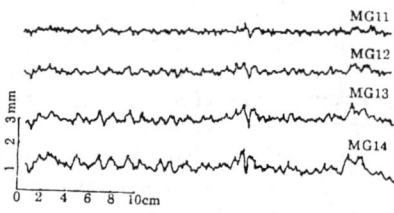

Figure 2. Variations of Composite Topography for a Joint under Different Contact States

Table 1. Parameters of Composite Topography and Regressive Coefficient of Closure Curve

Test No.	z_0 (mm)	z_1 (mm)	E_0	D_0 (mm)	R_z (mm)	z_2	η (mm^{-2})	β (mm)	n	m	P_0 (KPN)	σ_b
MG11	.057	.0709	6.435	.40	.456	.130	.191	4.48	.763	4.15	6.98	.0616
MG12	.097	.1301	5.402	.65	.703	.133	.181	7.33	.734	4.53	7.84	.0999
MG13	.132	.1819	4.813	.85	.876	.138	.144	7.45	.636	5.23	6.09	.0957
MG14	.157	.2163	5.190	1.05	1.123	.150	.140	5.81	.719	13.25	3.70	.1045
MG21	.062	.0795	5.568	.40	.443	.137	.206	8.18	.635	10.39	3.29	.0544
MG22	.077	.1037	5.874	.55	.610	.112	.217	7.75	.597	15.11	2.91	.0640
MG23	.107	.1424	5.724	.80	.815	.125	.224	8.82	.570	25.81	2.39	.0912
MG24	.176	.2265	4.962	1.00	1.124	.149	.194	6.01	.507	35.68	2.10	.0893
MG31	.054	.0683	5.612	.45	.383	.119	.224	9.75	.528	12.26	2.52	.0368
MG32	.089	.1151	5.286	.60	.609	.118	.235	7.47	.643	11.98	3.76	.0683
MG33	.127	.1683	5.091	.80	.834	.142	.202	7.88	.673	12.18	4.22	.0814
MG34	.158	.2047	4.927	1.20	1.009	.152	.224	7.86	.638	20.86	2.65	.0972
MG41	.053	.0673	5.501	.35	.370	.129	.209	7.92	.550	20.63	1.68	.0547
MG42	.088	.1085	5.040	.70	.547	.128	.188	9.02	.596	26.64	1.78	.0862
MG43	.112	.1426	4.932	.80	.698	.131	.171	7.04	.681	15.54	3.48	.0970

centre line average roughness z_0, the root mean square roughness height z_1 and ten points height of irregularities R_z) are mainly dominated by waviness components which are low frequency and large amplitude, while the parameters of texture characteristics (the root mean square roughness gradient z_2, the density of asperity peaks η and the radius of a hemispherical asperity β) are mainly dominated by components of irregularities with high frequency small amplitude. Thus the variation of the former change with the amount of mismatching displacement is analogous to the waveform of waviness, that of the latter is random and of smaller amplitude.

Among the parameters of composite topography of joints, ten points height of irregularities R_z is fairly in accord with the measuring value of initial apertures of joints, and there is a relationship as follows:

$D_0 = R_z + 3s$; $s = 0.02$ (1)

It is reasonable that the initial aperture of a joint is controlled by the several height peaks at the first contact. Considering the dimensionless initial aperture E_0 that is the ratio of initial aperture D_0 to the root mean square roughness height z_1, it is found that E_0 of a joint with different degree of mating formed by different amount of mismatching displacement between two surfaces of a joint is basically constant (Tab. 1). As a result it could be inferred that fresh joints with the same rock material and the same cause of formation as well as without loading history and damage has the same E_0.

4 MECHANISM OF CONTACT DEFORMATION OF JOINTS

In order to study the mechanism of contact deformation of joints, the closure tests with two or three loading and unloading cycles for a joint sample of marble are made with the maximal loads of 5 Mpa, 20Mpa, 25Mpa respectively. For fresh joints without damage the recoverable ratio of normal deformation of every loading and unloading cycle is listed in table 2. According to test data, the recoverable ratio of normal deformation decreases with the cycle sequences and slightly varies as normal stresses too. Because of this, most of the non-linear closure deformation of joints is recoverable and non-linear deformation still belongs to elasticity. A small part of unrecoverable non-linear deformation is due to the reason that the highest several peaks at the first contact on joint surfaces are crushed and some disturbances happen between two surfaces of a joint resulting in engaging more tightly. The loading curve of next cycle will follow the closure curves of a joint the initial aperture of which has diminished so that the stiffness closure deformation becomes bigger. Since the highest peaks have been crushed during first cycle, the numbers of peaks crushed during the second and the third cycles reduced and the recoverable ratio of normal deformation increases as the cycle sequences.

Table 2. The Proportion of Recoverable of Normal Deformation under Different Maximal Normal Stresses

Maximal stresses (MPa)	Recoverable ratio(%) 1st	2nd	3rd
15	55.8	82.6	90.6
20	60.0	86.8	90.5
25	60.0	87.0	93.3

5 CLOSURE AND CONSTITUTIVE RELATION OF JOINTS

For the closure test data of unmated joints in marble sample during loading cycles, the regressive analysis is made with different types of function. It is found that the following exponential relation and power relation give the highest fit correlation coefficient to the test data:

$$\delta_n = mP^n$$
$$\delta_n = A + \sigma_n \ln P = -\sigma_b \ln(P_0 + P) \quad (2)$$

where: $m, n, A, \sigma_n, \sigma_b$ are regressive coefficient. P_0 is initial closure load of joints.

The regressive coefficient and some parameters of composite topography of joints are listed in table 3. The power n of power function equation fluctuates in the range of 0.53- 0.76. The average value is 0.63. This agrees with the result (2/3) of Greenwood's contact model on two rough surfaces, whose probability distribution density function is uniform distribution. Therefore, the power n could be regarded as a constant which is not relative to the initial aperture and topographic characteristics. Table 3 listed the results according to one element regressive analysis between the main topographical parameters and other regressive coefficients. The influence of the root mean square of composite topography height and dimensionless initial aperture $E_0(D_0/z_1)$ of joints on their closure behaviour are the most notable. The two element regressive analysis of E_0 and z_1 using different functions is made further. Following regressive equation are attained($N=15$)

$$n = 0.631 + 0.075$$
$$m = -0.176 + 0.355z_1 + 0.005E_0^2$$
$$\sigma_b = -1.221 + 0.433z_1 + 1.051E_0^{0.1} \quad (3)$$
$$A = -246.5 + 424.2z_1 + 6.0E_0^2$$

Substituting each of the above formula for two formula of equation(2), the constitutive equation of normal force-normal deformation of joints containing initial aperture and the parameters of composite topography of joints is derived as follows:

$$\delta_n = (-0.176 + 0.355z_1 + 0.005E_0^2)P^{2/3}$$
or $\delta_n = -246.5 + 424.2z_1 + 6.0E_0^2 +$
$$(-1.221 + 0.433z_1 + 1.051E_0^{0.1})\ln P \quad (4)$$

Sun(1987) noted: the compliance of joints reflects the force-deformation behaviour of joints at constant stress direct shear test better than the stiffness of joints and is more agreeable with field practice situation such as slope sliding. By differentiating each formula of equation sets (2) about P and putting corresponding regressive results in the differentiated equation. The relationship between stiffness coefficients of joints and normal force is given as following:

$$C_{11} = mnP^{n-1} = 0.0032(E_0^2 + 70.96z_1 - 35.32)P^{-1/3} \quad (5a)$$
or
$$C_{11} = \sigma_b/P = (-1.221 + 0.433z_1 + 1.051E_0^{0.1})/P \quad (5b)$$

Referring to each formula of regressive equation (5), the following conclusions are obtained:

1. When E_0 is a constant, the regressive coefficients m, A, σ_n are all in linear relationship with z_1 for fresh joints in the same rock with random composite topography. E_0 of fresh joints with random composite topography and without loading history is approximately a constant. The closure curves of joints are only relative to z_1, and this was verified by predecessors.

2. When z_1 is a constant, the regressive coefficients of closure curves of joints decrease with E_0 so that the compliance of joints diminishes. In the second and third cycle E_0 gets smaller and the compliance of joints get smaller too. Because the variation of initial aperture of a joint causing by crushing the highest several peaks decreases with cycle sequences under the same maximal normal force, E_0 and closure curves of joints tend gradually to a constant when cycle sequences increase.

According to this, the corresponding parameters of normal force-normal deformation of joints in the second and the third loading cycle could be predicted with formula (4) and (5) after dimensionless initial aperture E_0 is figured out through of practice measurement initial aperture D_0 of a joint in corresponding loading cycles. Taking exponential relationship as an example, regressive coefficient σ_b is very close to predicted value σ_b^*, and relative error E_0 does not exceed 6%. The compliance curves calculated through by experimental data are very close to predicted compliance curves.

Table 3. Joint Compliance Predicted and Calculated

Test No.	z_1 (mm)	E_0	D_0 (mm)	σ_b^*	σ_b	E_r
MG111	.0709	5.642	.400	.0580	.0616	6.0
MG112	.0709	4.654	.330	.0342	.0342	0.0
MG113	.0709	4.616	.327	.0331	.0319	3.8
MG221	.1037	5.304	.550	.0645	.0640	0.8
MG222	.1037	4.388	.455	.0412	.0404	2.0
MG331	.1638	5.092	.834	.0804	.0814	1.3
MG332	.1638	4.103	.680	.0591	.0558	6.0
MG333	.1638	4.017	.658	.0566	.0548	3.5
MG431	.1416	5.643	.800	.0888	.0970	8.5
MG432	.1416	4.336	.614	.0562	.0537	4.9
MG433	.1416	4.195	.594	.0522	.0451	15.0

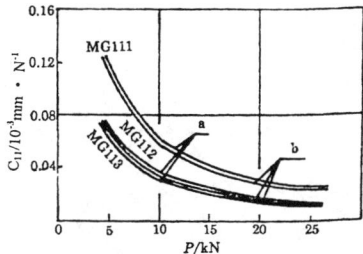

Figure 3. Compliance Curve
a. Calculated from Test Data
b. Predicted from formula (5b)

6 CONCLUSION

1. The composite topography and normal force-normal displacement test curves of a joint under different contact states are obtained. The conclusion is derived from tests that ten points height of irregularities R_z of composite topography of a joint can be used to predict more accurately its initial aperture under the contact state.

2. The closure behaviour of a joint depends on primarily its characteristics of composite topography under specific contact state more than the characteristics of respective topography of its two faces singularly. z_1 and E_0 are primary elements dominating closure behaviour of a joint. The constitutive equation of normal force-normal deformation of a joint containing these two parameters is obtained, which provides more accurate input information for analysis and calculation.

3. The path of cyclical loading and unloading only causes to change E_0 and not to change z_1. The various cyclical constitutive equations of normal force-normal deformation of joints can be predicted with that of last cycle and the initial aperture of the cycle, This provides theoretical basis for making full and reasonable use of test information.

Acknowledgement-The authors gratefully acknowledge support for this work obtained from National Natural Science Foundation of China (No:49402037) and Engineering Geology Open Research Laboratory Foundation of Institute of Geology, The Chinese Academy of Sciences.

REFERENCE

Cook, N. G. W. 1988. Natural Joints in Rock Mechanical, Hydraulic and Seismic Behaviour and Properties under Normal Stress. Fist Jaeger Memorial Lecture, 29th US Symp. on Rock Mech., Berkeley: Chapman and Hall.

Goodman, R. E. 1976. *Methods of Geological Engineering in Discontinuous Rock*. West Publishing, New York.

Sun Zongqi. 1983. Fracture Mechanics and Tribology of Rocks and Rock Joints. Lulea: Lulea University of Technology.

Xia Caichu, Sun Zongqi. 1995. RSP-I Type Intelligent Profilometer of Rock Surface. *J. Hydraulic Engineering*. (6):62-66.

Xia Caichu. 1993. The waviness Characteritics and Mechanical Effects of Topography of Rock Structure Faces. *J. Tongji University*. 21(3):371-377.

Greenwood, J. A. & Tripp, J. H. 1971. The Contact of Two Nominally Flat Rough Surfaces. *Pro. Ins. Mech. Engrs.*, 185:625-633.

Effect of structural anisotropy on deformation properties of granite under cyclic loading

H. Kusumi & K. Nishida
Department of Civil Engineering, Kansai University, Japan

Y. Mine
Central Japan Railway Co., Ltd, Japan

ABSTRACT: The effects of the structural anisotropy on the deformation properties of granite under cyclic loading have not been clarified. In this paper, four types of granite specimens are prepared with an angle θ of the structural anisotropic plane to the loading axis of 0°, 30°, 60° and 90° respectively and cyclic loading tests are performed under uniaxial compression. The stress amplitude levels used in this experiment are 0-30%, 20-50% and 40-70% of the uniaxial compressive strength for this granite. As the results, for the stress amplitude level of 0-30%: the elastic modulus under the cyclic loading is approximately constant. However it is recognized that the increase of axial strain during the cyclic loading is largest for the specimen with θ of 30°, and it's value under ten thousands cyclic numbers corresponds to 1.27 times as much as the axial strain given by 20% stress for the uniaxial compressive strength, and that it's uniaxial compressive strength after cyclic loading is lowest for the all stress amplitude conditions. Further, It is clarified that it's volumetric strain only indicates the negative value under 50% and 70% stress levels under cyclic loading.

1. INTRODUCTION

The Dynamic properties of hard and soft rocks are becoming to be important because compressed Air Energy Storage System and Superconducting Magnetic Energy Storage System are planning or developing in various country. These systems will be constructed in rock masses, and the cyclic stresses apply on the wall of the rock mass, therefore it is necessary that the mechanical behaviors of rock masses under cyclic loading will be investigated. Especially, the Superconducting Magnetic Energy Storage is planning and constructing in hard rock mass because the cyclic large loads apply on it. Granite is often treated as homogeneity rock, but in general, when the mechanical tests using granite are done, it is recognized that its physical properties (strength, strain, AE behavior, elastic modulus, etc.) indicate the anisotropy. The former studies in relation with the mechanical behavior of hard rocks under cyclic loading ,which can be given about the effect of stress amplitude, the dependence for the frequency of cyclic loading, etc., however, these do not consider the anisotropy characterization of granite. The problems of the structural anisotropy in granite are treated by the static conditions, and it is clarified that the strength, the elastic modulus, the elastic wave velocity etc. indicate the anisotropy under the these conditions. It is well known that the three planes exist in the granite, and those are called at hardway, grain and rift plane, these perpendicularly intersecting each other. The granite specimens using the experiments are four types which the angle θ between the rift plane and the loading axis is 0°, 30°, 60° and 90° respectively. These specimens are tested under uniaxial compressive cyclic loading. And the relationship between the axial, volumetric strain or the elastic modulus and the structural anisotropy of granite under cyclic loading conditions are discussed. Therefore. the relationship between the uniaxial compressive strength after cyclic loading and the structural anisotropy of it is investigated.

2. GRANITE SPECIMEN

The granite using this experiment is occurred from Ohshima island at Ehime prefecture in Japan. Table 1 indicates the physical properties of Ohshima granite. The height and the diameter of this granite specimen is 10cm and 5cm respectively. Granite has three plane, and these perpendicularly intersecting each other. The plane of the most easily crack is called as rift plane, the second being grain plane and another plane is called on hardway plane. The each normal line of rift, grain and hardway plane is defined as R,G and H axis respectively.

Figure 1 shows the example of one specimen, the specimens using this experiment are distributed by the angle θ between the loading axis and rift plane, the angle θ being 0°, 30°, 60° and 90° and each specimen is named as G0, G30, G60 and G90 respectively.

Table 2 represents the P-wave velocity and S-wave velocity of each specimen. The direction which these waves propagate is the same to the loading axis. The P-wave velocity of G0 and G90 are 3810m/sec and 3010m/sec respectively, the specimen which the angle θ is becoming to large indicates the smaller velocity of the elastic wave. It is recognized that these specimens have the anisotropy of elastic wave velocity.

Table-1 Physical properties of Ohshima granites.

Dry density (gf/cm^3)	Specific gravity	Water absorption	Porosity (%)
2.64	2.50	0.34	0.86

Table-2 Elastic velocities of axial direction with each specimen.

	V_p(m/s)	V_s(m/s)
G0	3810	1900
G30	3640	1720
G60	3200	1450
G90	3010	1430

3.TESTING PROCEDURE

The triaxial compressive testing machine is used to the apparatus of cyclic loading machine taking the electric oil pressure servo-controlled system. The cyclic loading is only given by the axis direction for a specimen. In based on the stress- strain curves given by the uniaxial compressive test for each specimen, the stress amplitude level using to the cyclic loading test is decided on 0-30%, 20-50% and 40-70% of the uniaxial compressive strength for each specimen. Figure 2 shows the stress-strain curve and the stress amplitude levels applied on the cyclic loading tests for G60 specimen, and in this figure,σc20,σc30,σc40, σc50 and σc70 correspond to 20%, 30%, 40%, 50% and 70% of the uniaxial compressive strength of G60 respectively. ①, ②

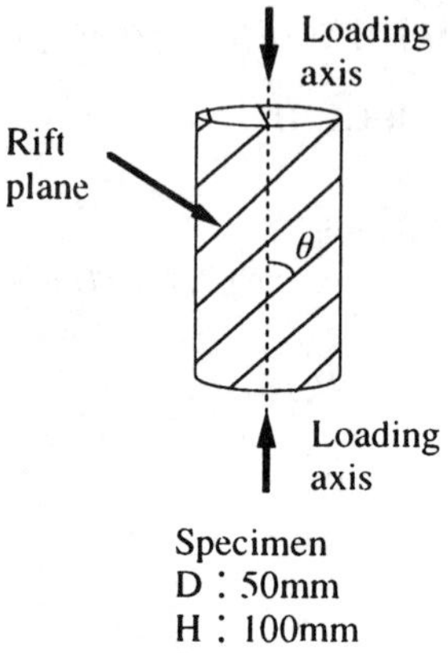

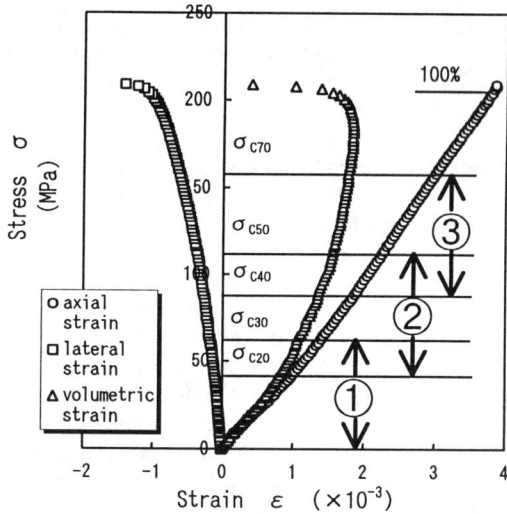

Fig-1 Relationship between the rift plane and the loading axis for the Ohshima granite specimen. θ means the angle between rift plane and loading axis.

Specimen
D：50mm
H：100mm

Fig-2 Stress-strain curve and the stress amplitude levels applied on cyclic loading tests for G60 specimen.

and ③ indicate the stress amplitude levels applied on the cyclic loading test of G60 specimen, and the testing conditions of the other specimens are the same, and therefore, the magnitude of the all stress amplitudes for these testing conditions are 30%. The loading used to these tests are 0.1Hz and maximum number of cyclic loading are 10,000,and the frequency and wave adopts the stress controlled method. The specimens for the cyclic sine wave. This cyclic loading test do which not break during

Table-3 Uniaxal compressive strength before and after cyclic loading for each specimen.

Specimen	Uniaxial compressive strength before cyclic loading.(kgf/cm²)	Uniaxial compressive strength after cyclic loading.(kgf/cm²)		
		Stress amplitude level		
		0-30%	20-50%	40-70%
G0	1857	2202	1871	2188
G30	2046	1592	1366	--*
G60	2156	2367	1661	--*
G90	2270	2267	2167	--*

* : breakage during cyclic loading

the ten thousands cyclic loadings are applied on the uniaxial compressive test.

4. UMAXIAL COMPRESSIVE STRENGTH BEFORE AND AFTER CYCLIC LOADING

Table 3 indicates the uniaxial compressive strength of each specimen before and after cyclic loading test. In this table, the uniaxial compressive strength after cyclic loading test of all the specimens except G30 under 0-30% stress amplitude level condition don't decrease compared with that of before cyclic loading test, and the uniaxial compressive strength of G30 specimen after cyclic loading test takes the dropping of 22% for the uniaxial strength before cyclic loading test. The degree of the dropping of uniaxial strength for G30 and G60 under 20-50% stress amplitude level condition is 35% and 23% respectively. In case of 40-70% stress amplitude level testing condition, all the specimens except G0 break before the number of cyclic loading reaches ten thousands. These causes are discussed the following.

5. ELASTIC MODULUS DURING CYCLIC LOADING

Figure 3 indicates the stress-strain hysteresis loop of each cyclic number for G60 under 0-30% stress amplitude level. From this figure, increasing the number of cyclic loading, the maximum and minimum strains of each hysteresis loop are taking the large value, therefore it is recognized that the hysteresis loop of each cyclic number moves parallel toward the increase of axial strain. Figure 4 shows the schematic diagram of the elastic modulus for each cyclic number. In this figure, the elastic modulus for each cyclic number is defined as the inclination of the straight line connecting the maximum axial stain to the minimum it.

Figure 5,6 and 7 show the relationship between elastic modulus and cyclic number under each

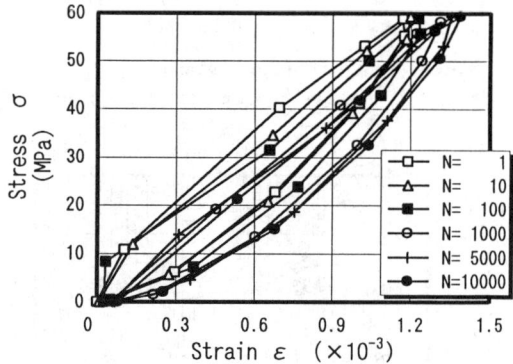

Fig-3 Hysteresis loop of each number for G60 specimen.

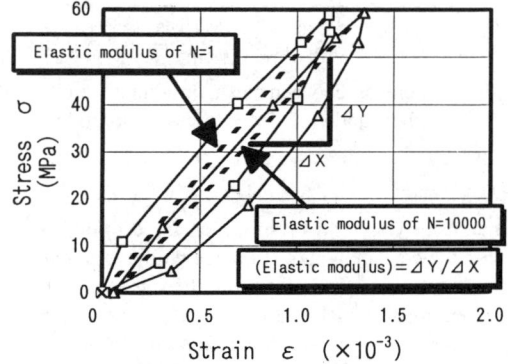

Fig-4 Schemastic modulus for each cyclic number.

stress amplitude level condition. In case of 0-30% stress each specimen indicates almost constant value. The elastic modulus of G0 specimen is taking the maximum compared with other specimens though ten thousands cyclic loading, and under this amplitude level testing condition, the specimen which θ is becoming to bigger takes the smaller elastic modulus. In relation with these amplitude level condition, increasing the number of cyclic loading, the elastic modulus of phenomena,

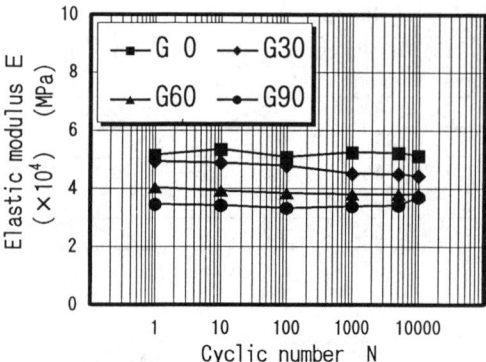

Fig-5 Relationship between elastic modulus and cyclic number under 0-30% stress amplitude level.

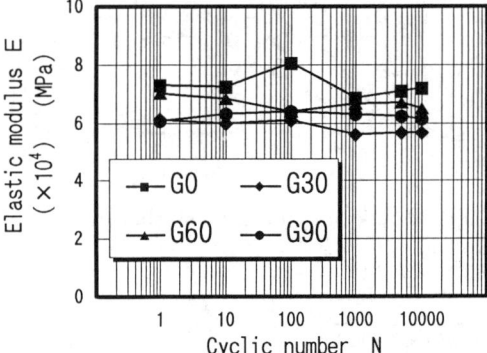

Fig-6 Relationship between elastic modulus and cyclic number under 2~50% stress amplitude level.

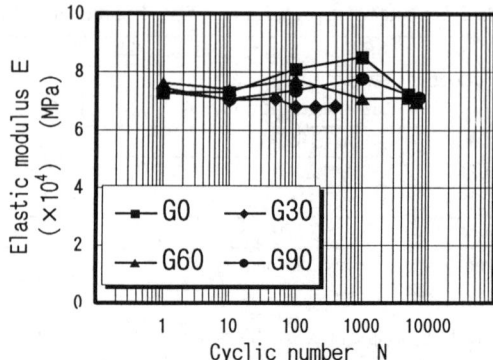

Fig-7 Relationship between elastic modulus and cyclic number under 4~70% stress amplitude level.

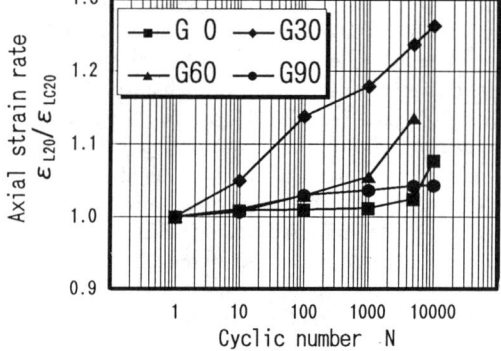

Fig-8 Relationship between axial strain rate and cyclic number under 20-50% stress amplitude level.

it is considered that θ of specimen is becoming to bigger, the deformation of rift plane within granite is taking the large value: therefore these is similar to the phenomena which the pore or the pre-existing cracks with the rock specimen close on the initial static loading condition. Representing Figure 6, in case of the stress amplitude level condition of 20-50%. the elastic modulus of G0 and G90 are not remarkable changing with cyclic condition, however those loading of G30 and C60 can be observed in a little dropping as the number of cyclic loading proceeds. Figure 7 shows the relationship between the elastic modulus and cyclic number under the stress amplitude level condition of 40-70% range. As this figure, the elastic modulus of G30 and G60 are observed in much dropping. and then be getting to the breakage at a certain number of cyclic loading.

6. AXIAL & VOLUMETRIC STRAIN DURING CYCLIC LOADING

Figure 8 shows the relationship between the axial strain rate($\varepsilon_{L20} / \varepsilon_{LC20}$) and the number of cyclic loading under the stress condition of 20% level against the uniaxial compressive stress before the cyclic loading test, where ε_{L20} is the accumulative axial strain occurred by cyclic test of the stress condition of 20% level against the uniaxial compressive stress before the cyclic loading test, and ε_{LC20} represents the axial strain which gives 20% of the uniaxial compressive strength for each specimen. In this figure, the axial strain rate($\varepsilon_{L20} / \varepsilon_{LC20}$) for C30. G60 and G90 are increasing as the cyclic number N proceeds. Especially, ($\varepsilon_{L20} / \varepsilon_{LC20}$) of G30 takes the maximum increased tendency in comparison with G60 and G90, and ($\varepsilon_{L20} / \varepsilon_{LC20}$) of

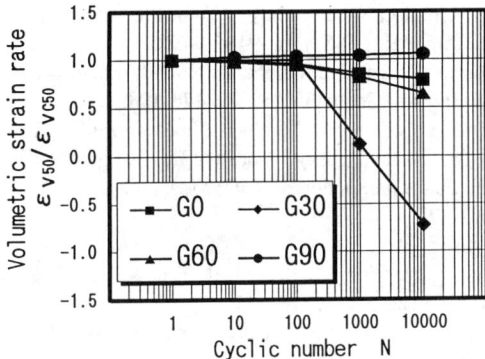

Fig-9 Relationship between volumetric strain rate and cyclic number under 20-50% stress amplitude level.

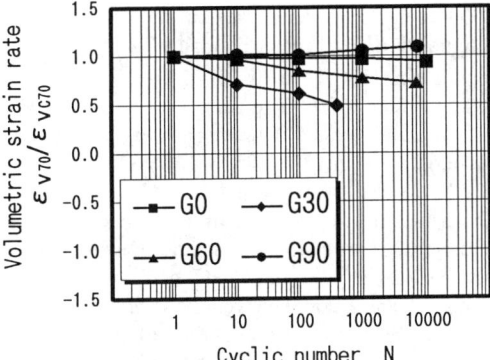

Fig-10 Relationship between volumetric strain rate and cyclic number under 40-70% stress amplitude level.

that under N=10,000 is getting to 1.27. It is recognized that the accumulative axial strain under cyclic loading test reached 1.27 times as much as the axial strain corresponding to 20% stress against the uniaxial compressive strength, and this strain is equivalent to the axial strain which gives 27.9% stress for the uniaxial compressive strength of G30. ($\varepsilon_{L20}/\varepsilon_{LC20}$) of G60 takes 1.15, and this strain is equivalent the axial strain which gives 24.9% stress for the uniaxial compressive strength of G60. ($\varepsilon_{L20}/\varepsilon_{LC20}$) of G0 and G90 indicate 1.08 and 1.04 respectively, and these accumulative strain is small compared with G30 and G60.

Figure 9 shows the relationship between the volumetric strain rate ($\varepsilon_{v50}/\varepsilon_{vc50}$) and the number of cyclic loading under the stress condition of 50% level against the uniaxial compressive stress before the cyclic loading test, where ε_{v50} is the accumulative volumetric strain occurred by cyclic test of the stress condition of 50% level against the uniaxial compressive stress before the cyclic loading test, and ε_{vc50} represents the volumetric strain which gives 50% of the uniaxial compressive strength for each specimen. In this figure, the volumetric strain rate($\varepsilon_{v50}/\varepsilon_{vc50}$) of each specimen except G90 indicates the same decreasing tendency until about N=100. This phenomenon is owing to the volumetric expansion of these specimens. After N=100, ($\varepsilon_{v50}/\varepsilon_{vc50}$) of G30 is rapidly dropping, especially after N=1500, ($\varepsilon_{v50}/\varepsilon_{vc50}$) becomes to the negative value.

This indicates which the volume of G30 specimen after N=1500 is dilating, and it is considered that this behavior is due to which the increasing degree of the lateral strain is large in comparison with that of the axial strain under 20-50% stress amplitude level condition. In detail, it can be supposed that this behavior which the volumetric strain of the specimen during cyclic test becomes to be negative value generates the cracks conducted toward the breakage within the specimen, and that it depends on the relationship between the direction of cyclic load and the strike of rift plane. On the volumetric strain rate ($\varepsilon_{v50}/\varepsilon_{vc50}$) of G0 and G60 specimens, these decreasing trends after N=100 in figure 9 are different, ($\varepsilon_{v50}/\varepsilon_{vc50}$) of G0 and G60 on N=10,000 are 0.78 and 0.64 respectively. ($\varepsilon_{v50}/\varepsilon_{vc50}$) of G90 specimen takes about 1.0 independent of cyclic number N. It is considered that the volumetric strain of the granite specimen included θ=90° is nearly equal to the volumetric strain corresponded to 20% stress as much as the uniaxial compressive strength before cyclic loading test, and that the accumulation of that which is caused by the cyclic loading is not observed.

Figure 10 shows the relationship between the volumetric strain rate ($\varepsilon_{v70}/\varepsilon_{vc70}$) and the number of cyclic loading under the stress condition of 70% level against the uniaxial compressive stress before the cyclic loading test, where ε_{v70} is the accumulative volumetric strain occurred by cyclic test of the stress condition of 70% level against the uniaxial compressive stress before the cyclic loading test, and ε_{vc70} represents the volumetric strain which gives 70% of the uniaxial compressive strength for each specimen. ($\varepsilon_{v70}/\varepsilon_{vc70}$) of G30 can not be only observed until N=500 because it breaks at this cyclic number. It is clear that ($\varepsilon_{v70}/\varepsilon_{vc70}$) of G30 decreases during cyclic loading in the same manner as the behavior of G30 represented by Figure 9. ($\varepsilon_{v70}/\varepsilon_{vc70}$) of G0 and G60 gradually decrease due to the proceeding of cyclic number, and these at N=10,000 take a little values compared with those of Figure 9. It is considered that the reason of this phenomenon is owing to which ε_{v70} of G0 and G60 already shifted from the compressive side to the dilation it. ($\varepsilon_{v50}/\varepsilon_{vc50}$) of G90 specimen takes about 1.0 independent of cyclic number N in the same manner as the behavior of G90 represented by Figure 9.

7. CONCLUSIONS

In this paper it has been shown that the mechanical and the deformation behavior of granite specimen under cyclic loading depend on the relation between the strike of rift plane and the cyclic loading axis and the magnitude of the stress amplitude level range. The elastic modulus of each cyclic number is defined as the inclination of the straight line connecting the maximum axial stain to the minimum it, and although on the specimens using this experiment ,these angles between cyclic loading axis and rift plane within granite specimens are variable, the changing characterizations of the elastic modulus during cyclic loading at each stress amplitude level was not recognized the remarkable difference. However the uniaxial compressive strength after cyclic loading test of the specimen which includes the rift plane at 30 deg, with the ache loading axis decreases in comparison with that of before cyclic loading test under all the stress amplitude level conditions. In this reason, it is recognized that the proceed of the axial strain of this granite specimen under cyclic loading is largest, and the volumetric strain of this specimen changes from the compressive site to the dilatancy site under almost stress amplitude level conditions. It is clarified that the axial strain of the specimen which the rift plane meets at right angle with cyclic loading axis is not proceed under each stress amplitude level, and that the value of this specimen's volumetric strain under cyclic loading is nearly equal to that given by uniaxial compressive test. It is recognized that the effect of the decreasing strength caused by cyclic loading on this specimen is not remarkable.

REFERENCES

Cho. T.F. and Haimson B.C., 1987, Effect of cyclic loading on circular openings; results of a laboratory simulation", Proc. of 28th U.S.Symp. on Rock Mechanics, 805-812.

Costin, L.S., 1983. A microcrack model for the deformation and failure of brittle rock, J. Geophys. Res., 88, B11, 9485-9492.

Douglass, P. M. and Voight, B., 1969, Anisotropy of granites, Geotechnique, 19, 376-398.

Haimson, B.C., 1974, Mechanical behavior of rock under cyclic loading, Proc. of 3rd Congress of Intemational Society of Rock Mechanics, 3, 2-A, 373-378.

Haimson, B.C., 1978, Effect of cyclic loading on lock, Dynamic Geotechnical Testing (ASTM STP-654), ASTM, 228-245.

Heroesewojo. R, Nishimatsu. Y and Suzuki. K., 1989, The effects of repeated compressive or tensile load on the mechanical properties of rock, Journal of the mining and metallurgical institute of Japan, 87, 1001, 515-520,(in Japanse).

Ishizuka. Y, Abe. T, Koyama. H and Komura. S, 1993, Effects of strain rate and frequency on fatigue strength of rocks, Journal of geotechnical engineering, JSCE 469/III-23, 15-24

Kranz, R.K., 1980, The effects of confining pressure and stress difference on static fatigue of granite, J. Geophy. Res., 85, B4, 1854-1866.

Masuda, K., Mizutani, H and Yamada, I., 1987, Experimental study of strain-rate dependence and pressure dependence of failure properties of granite, J. of Phys. Earth, 35, 37~66.

Matsuki. K and Kudo. H, Cyclic fatigue process of rocks under compression, Journal of the mining and materials processing institute of Japan, 106, 13, 781-786, (in Japanese).

Scholz, C.H. and Koczynski, T.A., 1979, Dilatancy anisotropy and response of rock to large cyclic loads, J. Geophys. Res, 84, 110,5525-5534.

Taniguchi. K, Kusumi. H, Matsui. Y and Teraoka. K, 1989, Longitudinal wave velocity of anisotropic rocks under biaxial stress loading, Butsuri-tansa(Geophysical exploration), 42, 4, 271-278, (in Japanese).

Taniguchi. K, Kusumi. H, Morimoto. K and Teraoka. K 1987, Relations between deformation characteristic and P-wave velocities of rock specimen under biaxial stress loading, Butsuri-tansa (Geophysical exploration), 40, 1, 11-21

Yanagitani, T, Nishiyama. T and Terada. M, 1987, Anisotropic development of dilatancy in uniauially compressed granites, Journal of geotechnical engineering, JSCE 382/3-7,63-72

Postfailure lateral deformation of rock specimen under the triaxial compression test

Toshiaki Saito, Sumihiko Murata & Hideaki Takehara
Kyoto University, Department of Earth Resources Engineering, Japan

ABSTRACT : To estimate the stability of rock caverns is one of the most important problem on mining and civil engineering. The stability is not estimated correctly on the stage of design, because the deformation behavior of rock in the loosed region is not sufficiently clear. Therefore, investigation of the postfailure lateral deformation of rock is necessary by conducting triaxial compression tests.

In this study, we used the measuring system composed of eight displacement transducers of cantilever type and a data logger with very fast channel scanner and large capacity of data storage for measuring the lateral deformation of rock specimen under the triaxial compression test. Consequently, we could measure the lateral deformation in every 45 degree of radial direction even under the postfailure condition. Moreover, the behavior of the lateral deformation could be understood quantitatively by the analysis with classifying the deformation into the four modes of deformation, (a) isotropic deformation, (b) anisotropic deformation caused by the anisotropy of rock specimen, (c) horizontal movement of the specimen caused by the testing conditions and (d) anisotropic deformation caused by the shear displacement along the fracture surface. In addition, the point where the apparent fracture was generated could be detected by the starting point of mode (d) even though the apparent stress drop was not seen on the stress-strain curve.

1. INTRODUCTION

To estimate the stability of rock caverns is one of the most important problem on mining and civil engineering. During the excavation, the stability is usually estimated from the measurement of rock wall displacement such as convergence, but on the stage of design, the stability is not always estimated correctly. This is because the deformation behavior of rock in the loosed region induced by the excavation around the cavern is not sufficiently clear. In the loosed region, rock should be in the failed condition and it will deform plastically or slide along the fracture surfaces. On the other hand, the tangential and the normal stresses on the wall of the cavern are the maximum and the minimum principle stresses respectively. This stress condition corresponds to the condition of triaxial compression test and the rock wall displacement also corresponds to the lateral displacement of a specimen under the test. Therefore, the investigation of the postfailure lateral deformation of rock specimen under a triaxial compression test will be supposed to present much useful information to understand the behavior of deformation of rock in the loosed region.

Under a triaxial compression test, the lateral displacement of a specimen is usually measured by using some strain gauges or a circumferential displacement meter. Strain gauges can measure the lateral deformation of the specimen precisely when the deformation is small, but they cannot measure it when the deformation becomes large under the postfailure condition. On the other hand, a circumferential displacement meter can measure the large deformation even after the specimen has failed, but it cannot detect the anisotropy of the deformation.

In this study, the eight displacement transducers of cantilever type mounted around the rock specimen in every 45 degree of radial direction were used. The measured lateral deformation of the rock specimen was classified into four modes and the variation of the magnitude of each mode was calculated. Here, we would present the procedures and the results of the measurement and the analysis of the lateral deformation of rock specimens under the triaxial compression test.

2. EXPERIMENTAL PROCEDURE

All specimens were cored perpendicular to the bedding plane from the large block sample of Shirahama sandstone. The cylindrical specimens had a diameter of 3cm and a length of 6cm. None of the specimens had apparent heterogeneity. Typical uniaxial compression strength, Young's modulus and Poisson's ratio of the specimen are shown in Table 1.

Table 1: The typical material properties of Shirahama sandstone specimens.

Compressive Strength (Mpa)	50
Young's Modulus (Gpa)	20.2
Poisson's Ratio	0.23

Under a triaxial compression test, a displacement meter should be as small as it can be set in a narrow pressure vessel and work precisely under a very high hydrostatic pressure. Moreover, the measuring points were needed as many as possible to detect the anisotropy of deformation correctly. For this reason, eight displacement transducers of cantilever type manufactured by Tokyo Sokki Kenkyujo Co., Ltd. were adapted in this study. Each transducer has a sensitivity of 1750×10^{-6}/mm and a capacity of 3mm. The specimen was set according to the following procedures.

First, the specimen was glued on the lower iron end block to reduce the relative displacement between the specimen and the displacement transducers. Second, the specimen and both of upper and lower end blocks were covered with double heat shrinkage tube to separate the confining pressure from the pore pressure of the specimen. Third, eight displacement transducers were fixed around the jig in every 45 degree of radial direction, and the jig was fixed to the lower end block with screws. At this time, jig's position was adjusted so as to measure the lateral displacement at the center of specimen. Last, the Teflon film was attached at the measuring points to reduce the friction of specimen surface. The feature of a specimen set with the transducers is shown in Figure 1.

Every triaxial compression test was carried out with the constant axial displacement rate of 0.036mm/min until the axial displacement reached 3mm. Four kinds of confining pressure, 15Mpa, 25Mpa, 35Mpa and 45Mpa, were selected in this study and they were kept constant during each test. The eight lateral displacements, axial load and axial displacement were measured by the data logger, UCAM70A, manufactured by Kyowa Electronic Instruments Co., Ltd. The apparatus has very fast channel scanner and large capacity of data storage. It took only 0.5 seconds for scanning the ten measuring points. The scanning was conducted every 5 second during the test.

3. EXPERIMENTAL RESULTS

The stress-strain curves for each confining pressure are shown in Figure 2. In this figure, the curves in the right hand side of the stress axis are for

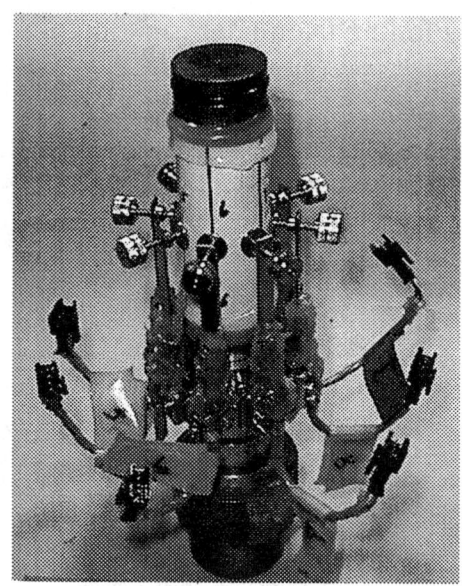

Figure 1: Photograph of a specimen set with the displacement transducers of cantilever type.

the radial strain and the curve in the left hand side of it is for the axial strain. The radial strain is the ratio of the measured lateral displacement to the original radius of the specimen and this value should be apparent because it may contain the displacement caused by the horizontal movement of the specimen. The diametric stress-strain curves of each direction for the confining pressure of 15Mpa are also shown in Figure 3 as an example. From Figure 3, it can be seen that the diametric stress-strain curves of every direction correspond each other before the axial stress reaches the peak strength. This explains that the specimen deforms with isotropic manner in the lateral direction under the elastic condition. On the other hand, a little difference can be seen among the radial stress-strain curves on all figures in Figure 2 even though the specimen is in the elastic condition. This difference can be explained by the horizontal movement of the specimen caused by the testing conditions such as parallelism between the ends of the specimen or the loading plates of the testing machine.

After the peak strength, that is, postfailure condition, the discrepancy of the lateral strain among each direction becomes large. This shows that the anisotropy of lateral deformation appears just before or after the axial stress has reached peak strength and the anisotropy increases with increasing the deformation. In this anisotropy deformation, two components of deformation may be included. One is caused by the anisotropy of the rock specimen and the other is caused by the shear displacement along

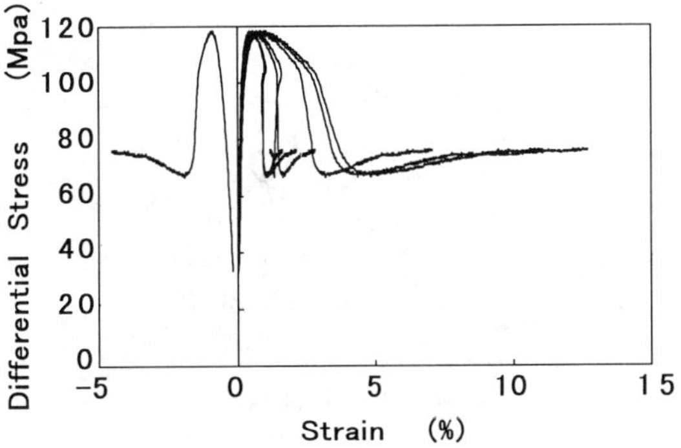

(a) Confining pressure is 15Mpa

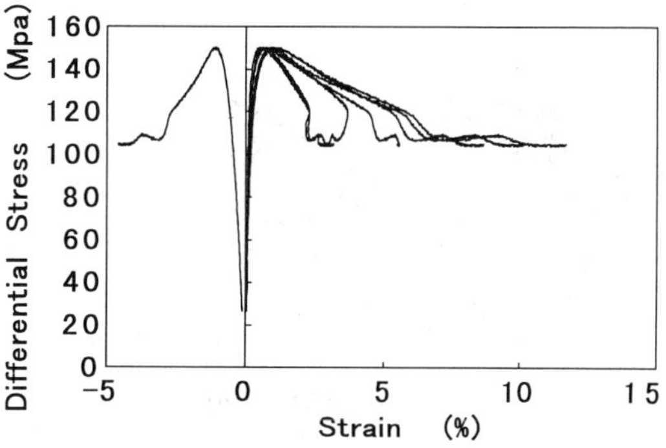

(b) Confining pressure is 25Mpa

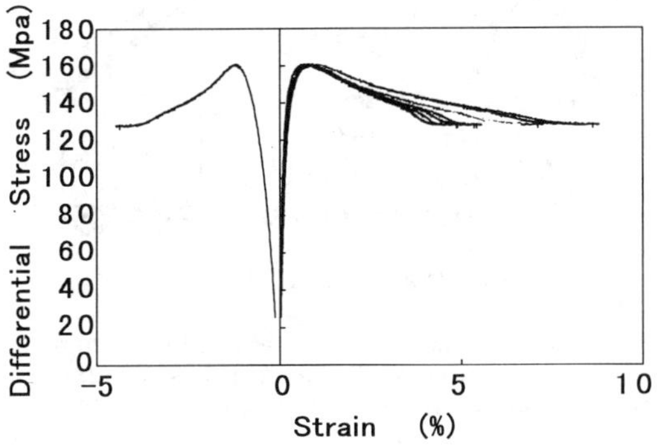

(c) Confining pressure is 35Mpa

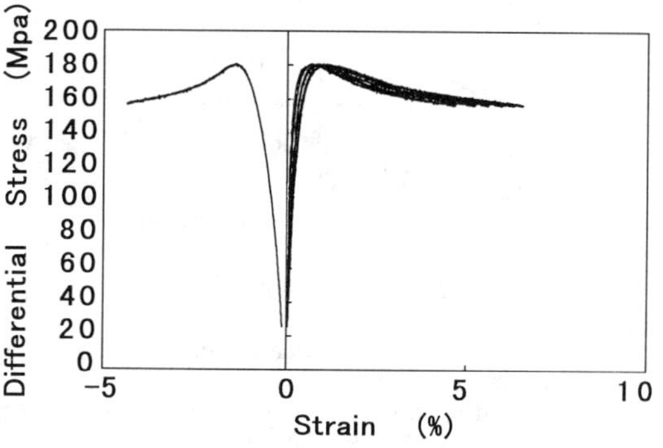

(d) Confining pressure is 45Mpa

Figure 2 : The stress-strain curves for each confining pressure. The lateral stress-strain cureves is based on the radial strain.

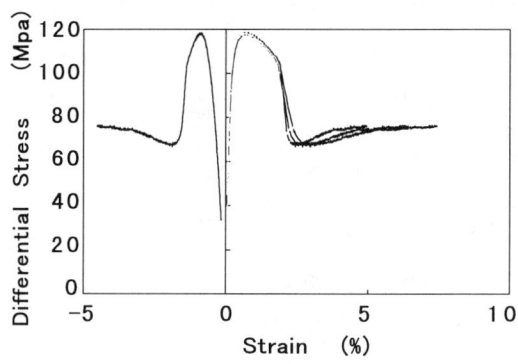

Figure 3 : The diametric stress-strain curves of each direction for the confining pressure of 15Mpa

the generated fracture surface. It can be assumed that the former mainly appears before the apparent fracture is generated and the latter should appear after that.

Moreover, the remarkable stress drop can be observed when the confining stress is 15Mpa and 25Mpa. In these cases, where the apparent fracture was observed on the tested specimen, the lateral displacement of one or two directions stops increasing and that of opposite directions keeps increasing after the stress drop. Therefore, it can be assumed that the stress drop indicates the generation of apparent fracture. However, the stress drop cannot be seen in spite of observing the fracture on the tested specimen when the confining pressure is 35Mpa. In this case, we cannot know when the apparent fracture was generated from the stress-strain curves. In addition, when the confining pressure is 45Mpa, the stress drop is not seen on the stress-strain curve and

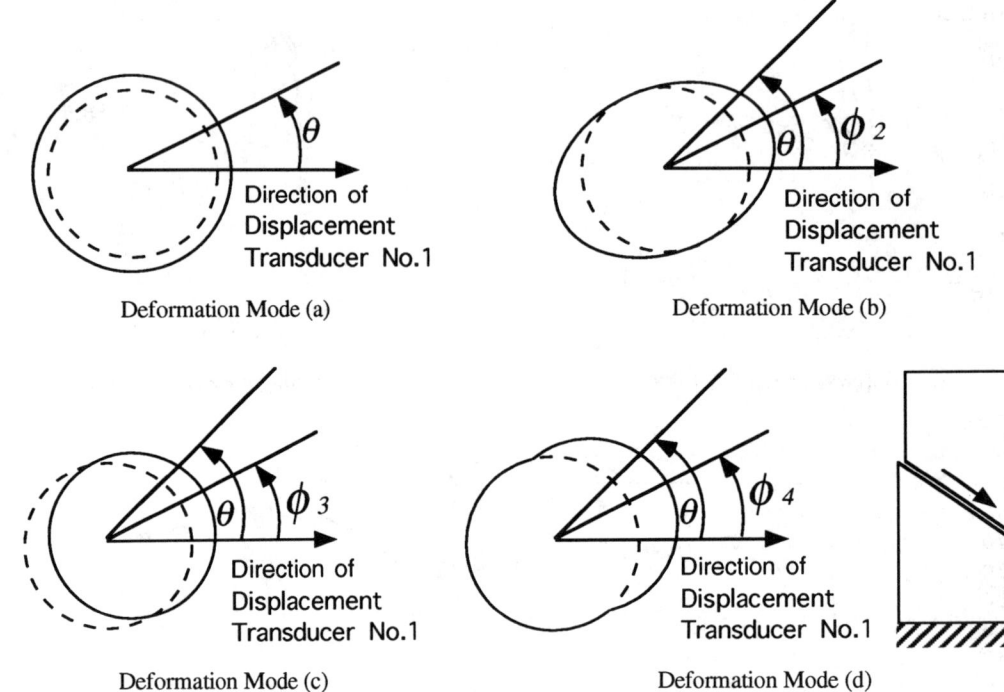

Figure 4 : Deformation mode of the rock specimen unter the compression test. They were (a) isotropic deformation, (b) anisotropic deformation caused by the anisotropy of rock specimen, (c) horizontal movement of the specimen caused by the testing conditions and (d) anisotropic deformation caused by the shear displacement along the fracture surface.

the fracture was not observed on the tested specimen. As shown in these results, the fracture of the specimen varies from the apparent localized fracture to the entire fracture with increasing the confining pressure.

From these experimental results, the behavior of lateral deformation can be explained qualitatively. In the next section, a quantitative approach will be presented to describe the lateral deformation of rock specimen.

4. DESCRIPTION OF LATERAL DEFORMATION

In this study, the lateral deformation of the rock specimen was classified into four modes as shown in Figure 4. They were (a) isotropic deformation, (b) anisotropic deformation caused by the anisotropy of rock specimen, (c) horizontal movement of the specimen caused by the testing conditions and (d) anisotropic deformation caused by the shear displacement along the fracture surface. The mode (a), (b) and (c) can be assumed to appear from the beginning of the test and mode (d) should appear after the apparent fracture has generated. Each mode of lateral displacement can be described as the function of the radial direction, θ, as follows.

Mode (a) :

$$C_1 \text{ (constant)} \tag{1}$$

Mode (b) :

$$C_2 \cos^2(\theta - \phi_2) \tag{2}$$

Mode (c) :

$$C_3 \cos(\theta - \phi_3) \tag{3}$$

Mode (d) :

$$C_4 \{0.3183 + 0.5 \cos(\theta - \phi_4) \\ + 0.2122 \cos 2(\theta - \phi_4) \\ - 0.04244 \cos 4(\theta - \phi_4)\} \tag{4}$$

where ϕ_2, ϕ_3, and ϕ_4 are the phase angle of each mode measured from the direction of the displacement transducer No. 1, and the coefficients, C_1, C_2, C_3 and C_4, are the constants that indicate the magnitude of each mode. Assuming the fracture is passing through the just center of the measuring cross area of the specimen, the equation (4) can be derived from the Fourier series of following equation.

$$\begin{cases} \cos\theta & (|\theta| \leq \pi/2) \\ 0 & (|\theta| > \pi/2) \end{cases} \quad (5)$$

Equation (5) shows the displacement of only the upper part of the specimen separated by the fracture, because the lower part of the specimen is glued to the lower end block where the displacement transducers were fixed. Consequently, the lateral displacement of the specimen can be described as the summation of these equations from (1) to (4). This yields

$$u(r) = C_1 + C_2 \cos^2(\theta - \phi_2) + C_3 \cos(\theta - \phi_3) + C_4 f(\theta - \phi_4) \quad (6)$$

where

$$f(\theta - \phi_4) = \{0.3183 + 0.5 \cos(\theta - \phi_4) + 0.2122 \cos 2(\theta - \phi_4) \quad (7) - 0.04244 \cos 4(\theta - \phi_4)\}$$

First, we measured the phase angle, ϕ_4, assuming that it should correspond to the direction of the maximum lateral displacement detected from the radial stress-strain curves, and then calculated the coefficients and another phase angles of equation (6) from measured lateral displacements of every 45 degree by using the least square method.

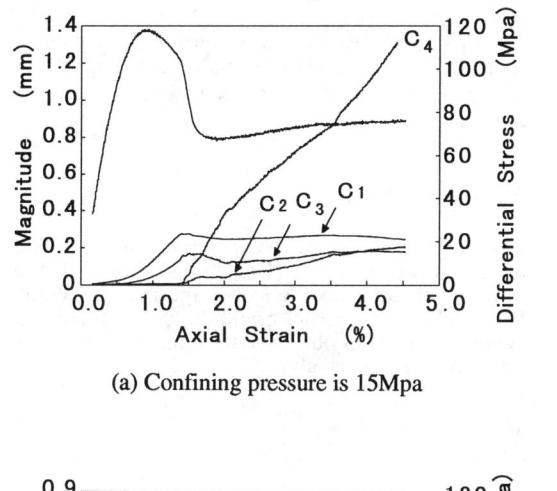

(a) Confining pressure is 15Mpa

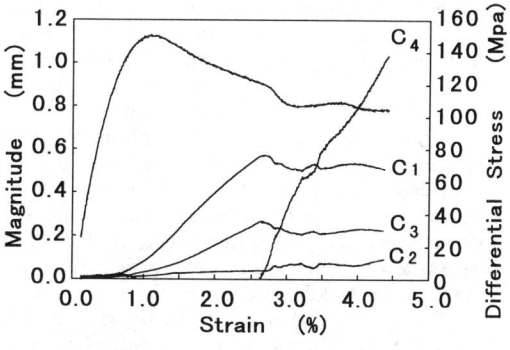

(b) Confining pressure is 25Mpa

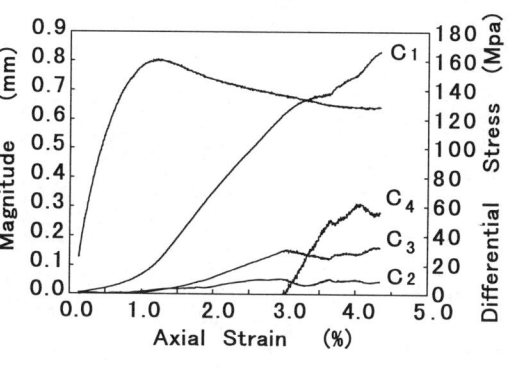

(c) Confining pressure is 35Mpa

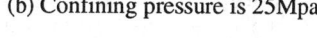

(d) Confining pressure is 45Mpa

Figure 5 : The variations of the magnitude of each deformation mode according with the axial strain

5. ANALYTICAL RESULTS

The variations of the magnitude of each deformation mode according to the axial strain are shown in Figure 5 for each confining pressure. In this figure, axial stress-strain curve is also drawn to indicate the variation of stress for each confining pressure. In this section, we would use the labels, (a), (b), (c) and (d), defined in the previous section to indicate each deformation mode.

It can be seen that the every mode except mode (d) appears from the beginning of the test. The magnitudes of mode (a) and (c) increase with increasing the axial strain until the mode (d) appears and after that they are almost constant in the case of confining pressure of 15Mpa and 25Mpa. On the other hand, mode (a) continues increasing even after the mode (d) appears in the case of confining pressure of 35Mpa. In addition, mode (a) increases rapidly from just before the axial stress reaches the peak value. This tendency becomes clearer with increasing the confining pressure. These results show that the isotropic deformation and horizontal movement of the specimen are dominant until the apparent fracture is generated in the case of low confining pressure and the isotropic deformation is dominant all through the test in the case of high confining pressure.

The reason mode (c) varies with mode (a) can be assumed that mode (a) caused the mode (c) when the loading axis does not correspond to the center axis of the specimen. Moreover, in this study, mode (b) is small in every case of confining pressure. This shows that Shirahama sandstone used for a specimen is isotropy in the bedding plane.

Since mode (d) is caused by the sliding along the generated fracture surface, the point where mode (d) appears corresponds to the point where the apparent fracture is generated. In the case of low confining pressure such as 15Mpa and 25Mpa, this point is detectable from the apparent stress drop on the stress-strain curves, but in the case of high confining pressure such as 35Mpa and 45Mpa, it is impossible because the apparent stress drop is not clear. However, the point where the apparent fracture is generated can be detected by the start point of mode (d) even thought the apparent stress drop cannot be seen on the stress-strain curves. For example, when the confining pressure is 35Mpa, where the apparent fracture was generated, mode (d) appears in spite of never seeing the apparent stress drop on the stress-strain curves Therefore, in this case, the point where the apparent fracture is generated can be known by the start point of mode (d).

In the case of confining pressure of 45Mpa, not only mode (d) but also apparent stress drop cannot be seen, because the specimen deformed as a barrel, and the apparent fracture was not generated.

6. CONCLUSION

The measuring system composed of eight displacement transducers of cantilever type and a data logger with very fast channel scanner and large capacity of data storage was used for measuring the lateral deformation of rock specimen under the triaxial compression test. It could measure the lateral deformation in every 45 degree of radial direction even under the postfailure condition. Moreover, the behavior of the lateral deformation could be understood quantitatively by the analysis with classifying the deformation into the four modes of deformation, (a) isotropic deformation, (b) anisotropic deformation caused by the anisotropy of rock specimen, (c) horizontal movement of the specimen caused by the testing conditions and (d) anisotropic deformation caused by the shear displacement along the fracture surface. In addition, the point where the apparent fracture was generated could be detected by the starting point of mode (d) even though the apparent stress drop could not be seen on the stress-strain curves.

We will try to estimate the quantity of maximum lateral deformation and its direction and to apply them to the analysis of the stability estimation of rock caverns as a next step.

ACKNOWLEDGMENT

This study was supported by the Grant-in-Aid for Scientific Research (B), No.08455485, of The Ministry of Education, Science, Sports and Culture. We would show special gratitude to Mr. Imamura, graduate student of our laboratory, for helping on the experiment and the analysis.

Anisotropic behaviour of schistose rocks and effect of confining pressure on them

M. H. Nasseri
Department of Mining Engineering, University of Tarbiat Modaress, Tehran, Iran

K. S. Rao & T. Ramamurthy
Department of Civil Engineering, Indian Institute of Technology, New Delhi, India

ABSTRACT: Tensile strength, compressive strength and modulus anisotropic behaviour of four schistose rocks has been analysed critically. An attempt is made to find out the effect of confining pressure of up to 100 MPa on detorioration of the strength and modulus anisotropy of four schistose rocks obtained from two power house sites at the foothills of the Himalayas.

1 INTRODUCTION

Rational and realistic approach to the analysis and design of the on ground or underground engineering structures on anisotropic rocks necessiates indepth evaluation of anisotropy in the measurement of strength (tensile and compressive) and deformation. Inherent anisotropy acquired by metamorphic rocks make them weaker and dictate its overall behaviour. Anisotropic behaviour of schists is due to the process of metamorphic differentiation (segregation of constituents). In response to high pressure and temperature gradients associated with mountain building activities, rocks flow and recrystallize under new tectonic conditions and develop layers of contrasting mineralogical facies. Such weak foliation planes or schistosity will effect the strength and deformational responses of anisotropic rocks when orientation of stresses with respect to these weak planes changes. Prediction of the anisotropic respenses of strength and deformation of anisotropic rocks involves preparation of specimens at different orientation angles, β (the angle between major principal stress direction and the foliation planes).

The measurement of the strength anisotropy for various anisotropic rock types have been carried out by many investigators e. g.(Donath 1964), (Chenevert and Gatline 1965), (McLamore and Gray 1967), (Hoek 1968), (Attewell and Sandford 1974), and (Brown et al. 1977), on shales and slates, (Deklotz et al. 1966), (McCabe and Koerner 1975), (Nasseri 1992), (Nasseri et al. 1996), on gneisses and schists, phyllite by (Ramamurthy et al. 1988), (Horino and Ellicksone 1970), and (Rao et al. 1986) on sandstones, (Pomeroy et al. 1971), on coal and (Allirote and Boehler 1979) on diatomite. An overall analysis and review of their works exhibit that maximum failure strength is either at $\beta = 0°$ or $90°$ and it is minimum usually arround $\beta = 30°$, The Shape of the curve between the uniaxial compressive strength (σ_c) and the orientation angle β designated as the "type of anisotropy" and is found to be generally of three types such as "U - shape", "shoulder" and "wavey" (Fig. 1).

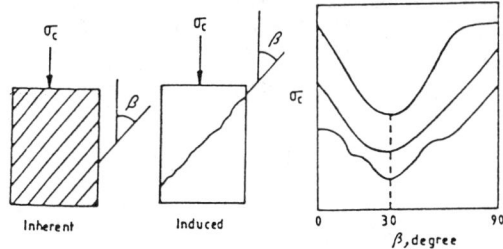

Fig. 1: Possible variation of σ_c versus β for inherent and induced anisotropies (after Ramamurthy 1993)

Keeping the above observations in view and inadequate knowledge on anisotropy of deformational and strength responses of schists for entire range of β, an analytical approach has been attempted through a comprehensive investigation into four varieties of schists obtained from two Hydroelectrical power project sites in Himalayas.

2 ROCKS TESTED

Four anisotropic schistose rocks have been collected from the foothills of Himalayas for laboratory investigation. Quartzitic and Chlorite schists were obtained from Uri Hydroelectrical power project whereas Quartz mica and Biotite schists were obtained from Nathpa - Jhakri Hydroelectrical power scheme.

3 EXPERIMENTAL PROGRAMME AND TEST PROCEDURE

Tests on schists have been carried out under the following three major catagories, (i) petrography and petrofabric, (ii) physical properties and (iii) geotechnical properties. To evaluate petrography and petrofabric of the rocks, thin sections across and along the schistosity were prepared and observed under polarized transmitted light. For detailed quantitative and semi-quantitative mineralogical composition, X-ray diffraction powder method has been adopted. Scanning Electron Micrographs were used to find out the surfacial grain to grain contact.

Determination of physical properties of the schists such as density, specific gravity, water absorption, porosity have been carried out based on ISRM (1972). Out of ten tests for each property an average value has been evaluated and presented in Table 1.

Table 1. Geotechnical properties of the schists

Properties	Quartzitic schist	Chlorite schist	Quartz mica schist	Biotite schist
Specific gravity	2.66	2.90	2.83	2.85
Dry density g/cm^3	2.63	2.88	2.72	2.74
Porosity (%)	0.81	0.26	1.7	0.76
σ_c MPa $\beta = 90°$	190	110	50	50
$E_t \times 10^4$ MPa	2.0	1.2	0.4	0.5
Poission's ratio	0.25	0.2	0.15	0.1

To determine the compressive strength, and deformational anisotropies of these rocks, uniaxial compressive strengths at different orientations ($\beta = 0$, 15, 30, 45, 60, 75 and 90°) have been determined. Table 1 also shows the results of compressive strengths at "representative orientation" i. e ($\beta = 90°$) for the four schists. Triaxial compressive tests at different β and confining pressure ($\sigma_3 = 5$, 15, 35, 50 and 100 MPa) have been carried out to understand the shear strength characteristics and effect of confining pressure on compressive strength and deformational anisotropy of schistose rocks. A high pressure triaxial cell with a maximum lateral pressure capacity of 140 MPa and ram load capacity of 5 MN was used. A 10 channel autoscanning digital strain indicator has been employed to measure the axial and diametral strains in rock specimens, through electrical resistance strain gauges connected to a printer to record the strains and axial load at different stress intervals. The specimens were first oven dried at 105 ±1°C for 24 h and cooled in desiccators over calcium chloride.

4 RESULTS AND DISCUSSION

4.1 Anisotropy

Schistose rocks exhibiting strong anisotropy due to the varied distribution of minerals as well as their orientation and packing. More over all the geotechnical characteristics of these rocks show varied results at different orientations. In the following paragraphs, anisotropic behaviour under compression, tension as well as variation in modulus for the schists tested are discussed.

4.1.1 Compressive strength anisotropy

It may be discerned from Fig 2 that all schists exhibit "U - shape" anisotropy as a result of presence of the one set of cleavage or foliation planes. Similar anisotropic curves were obtained for Texas slate (McLamore and Gray 1967). Martinsburg slate (Donath 1964), Penrhyn slate (Attewell and Sandford 1974) Chlorite and Graphite schists (Akai, 1971) and Phyllite (Singh et al., 1989). The values of "anisotropy ratio" which is defined as "the ratio of the

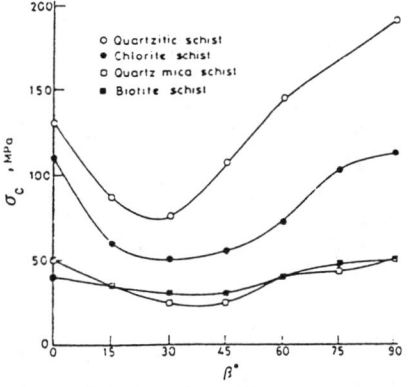

Fig. 2: Variation of σ_c with β

representative compressive strength to the minimum compressive strength observed over a range of β from 0 to 90 degrees $(\sigma_{c90}/\sigma_{c\,min})$" for number of anisotropic rocks are tabulated in Table 2.

Table 2. Strength anisotropy of different rocks

Sl. No.	Rock type	σ_{cmax} ($\beta°$)	Anisotropy ratio	Source
1.	Martinsburg slate	90	13.46	Donath, 1964
2.	Fractured sandstone	90	6.37	Horino & Ellickson, 1970
3.	Barnsley hard coal	90	5.18	Pomeroy et al., 1971
4.	Penrhyn slate	90	4.85	Attewell & Stadford, 1974
5.	South African slate	0	3.68	Hoek, 1964
6.	Texas slate	90	3.00	McLamore & Gray, 1967
7.	Permian shale	90	2.33	Chenevert & Gatlin, 1967
8.	Green river shale-1	0	1.62	McLamore & Gray, 1967
9.	Green river shale-2	0	1.41	McLamore & Gray, 1967
10.	Green river shale	0,90	1.37	Chenevert & Gatlin, 1965
11.	Kota sandstone	0	1.12	Seshagiri Rao, 1986
12.	Arkansas sandstone	0	1.10	Chenevert & Gatlin, 1965
13.	Quartzitic phyllite	90	2.19	Singh, 1989
14.	Carbonaceous phyllite	90	2.19	Singh, 1989
15.	Micaceous phyllite	90	6.00	Singh, 1989
16.	Quartzitic schist	90	2.70	Present work
17.	Chlorite schist	90	2.24	Present work
18.	Quartz mica schist	90	2	Present work
19.	Biotite schist	90	1.6	Present work

Analysis of "anisotropy ratio" for the present schistose rocks is also included in Table 2. As per this, the ratio is 2.7 for Quartzitic schist, 2.24 for Chlorite schist, 2 for Quartz mica schist and 1.6 for Biotite schist. It is noticed that the schist with high compressive strength shows high anisotropic ratio and vice versa.

4.1.2 Tensile strength anisotropy

From Fig. 3 it may be observed that all the schists reveal maximum Brazilian tensile strength (σ_{tb}) at $\beta=90°$ (with respect to loading direction) and the minimum at $\beta=0°$. Same behaviour is observed in terms of variation of axial point load strength with β (Fig. 4). The tensile strength anisotropy ratio $(\sigma_{tb90}/\sigma_{tb0})$ has been found to be maximum in Quartzitic schist, followed by Quartz mica, Biotite and Chlorite schists with the magnitudes of 5.1, 3.6, 2.8 and 2.6 respectively. The respective values obtained

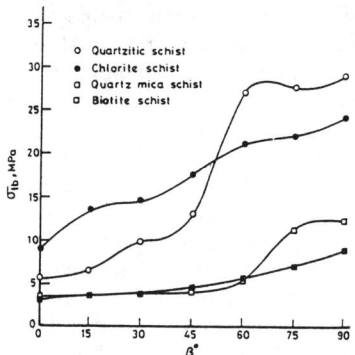

Fig. 3: Variation of Brazilian strength (σ_{tb}) with β

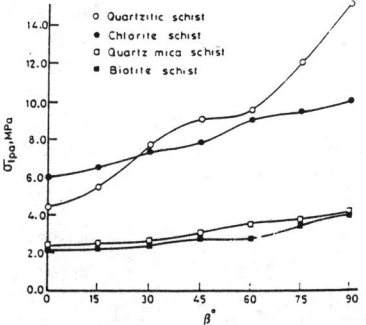

Fig. 4: Variation of axial point load strength (σ_{tpa}) with β

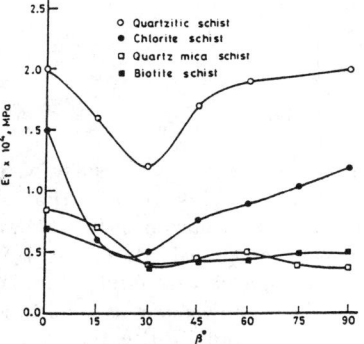

Fig. 5: Variation of modulus (E_t) with β

for tensile strength anisotropy obtained from axial point load test $(\sigma_{tpa90}/\sigma_{tpa0})$ exhibit similar order of variation as observed in Brazilian strength anisotropy, with a comparatively lower magnitude of 3.4, 1.9, 1.7 and 1.5 for Quartzitic, Biotite, Quartz mica, and Chlorite schists respectively.

4.1.3 Modulus anisotropy

From figuer 5. it has been found that the modulus anisotropy curves of rocks under investigation may fall under two categories i.e. "U-shape" and "Irregular shape". Chlorite schist shows a maximum modulus anisotropy followed by Quartz mica, Biotite and Quartzitic schists with respective magnitudes of 3, 2.1, 1.73 and 1.6.

4.1.4 Strain ratio anisotropy

Strain anisotropy ratio (ratio of axial strain at $\beta=90°$ to that at $\beta=0°$) for all the schists has been calculated from their stress - strain curves. It is observed that the strain anisotropy ratio ($\varepsilon_{90}/\varepsilon_0$) for Quartzitic schist is the maximum value of 1.5, followed by 1.45 for Chlorite schist, 1.2 for Quartz mica schist and 1.14 for Biotite schist.

In Table 3 all the anisotropy ratios in terms of compressive and tensile strengths along with modulus and strain are tabulated for the schistose rocks.

Table 3. Anisotropy ratio from mechanical properties of schists

Rock Type	Anisotropy				
	$\dfrac{\sigma_{c90}}{\sigma_{c30}}$	$\dfrac{\sigma_{tb90}}{\sigma_{tb0}}$	$\dfrac{\sigma_{tpa90}}{\sigma_{tpa0}}$	$\dfrac{E_{to}}{E_{tmin}}$	$\dfrac{\varepsilon_{90}}{\varepsilon_0}$
Quartzitic schist	2.70	5.1	3.4	1.60	1.50
Chlorite schist	2.24	2.6	1.5	3.00	1.45
Quartz mica schist	2.00	3.6	1.7	2.10	1.20
Biotite schist	1.60	2.8	1.9	1.75	1.14

4.1.5 Permeability anisotrpy

The variation of the coefficient of permeability, K, at different water heads ranging upto 7 MPa for schistose rocks at different orientation angles is shown in Fig. 6 for Quartz mica and Biotite schists. It is discerned from the figures that coefficient of permeability for all the rocks decreases as water head increases. Quartz mica and Biotite schists are characterized with higher coefficient of permeability at $\beta=0°$ with k value of 4×10^{-8} and 1.9×10^{-8} cm/sec at 7 MPa of water head, than that at $\beta=90°$ where k values for the two rocks are 4.5×10^{-9} and 3.5×10^{-9} cm/sec at 7 MPa of water head respectively. As is observed in Fig. 6a Quartz mica schist shows more reduction of coefficient of permeability as water head pressure increases than Biotite schist. There is a systematic reduction of k for both the aforementioned rocks as β varies from 0° to 90°. Quartzitic schist being more porous than Chlorite schist exhibits higher k values at all orientations. For both the rocks there appears to be no influence of orientation of foliation planes on the variation of coefficient of permeability due to their very fine grained nature.

5 EFFECT OF CONFINEMENT ON ANISOTROPY

The reduction of strength anisotropy of schistose rocks as a function of confining pressure can be demonstrated through a plot of stress ratio (σ_1/σ_3) versus β for a range of confining pressure (σ_3).

Figure 7a and 7b demonstrate the variation of stress ratio (σ_1/σ_3) with β for Quartzitic, and Biotite schists respectively. It is clear from the figures that the values of stress ratio decreases and the effect of

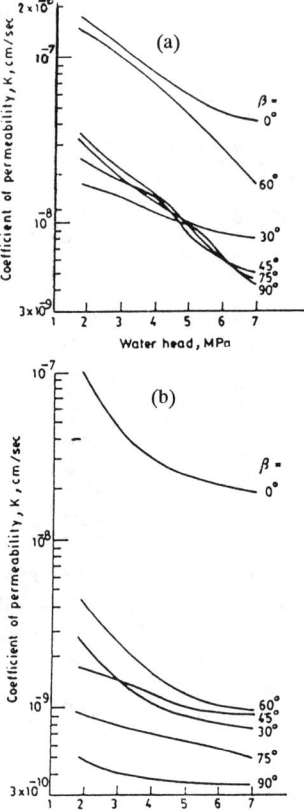

Fig. 6: Variation of K with water head at different β for (a) Quartz mica and (b) Biotite schists

anisotropy diminishes as a consequence of rise in confining pressure. For example Quartzitic schist shows a maximum stress ratio of 65 and 47 for $\beta=90°$ and 0° orientations respectively and minimum ratio of 37 for $\beta=30°$ at 5 MPa of confining pressure. Similarly Biotite schist show minimum stress ratio (Fig. 7b) among all these rocks with minimum stress ratios of 24 and 25 for $\beta=90°$ and 0° orientation respectively and a stress ratio of 15 for $\beta=30°$ at 5 MPa of confining pressure. The respective values of stress ratios for Chlorite and Quartz mica schists fall in between these two extremes.

Another approach to demonstrate the effect of confining pressure on reduction of anisotropy is through a plot of $(\sigma_1-\sigma_{min})/\sigma_3$ versus σ_c/σ_3, as suggested by Ramamurthy (1993). In this plot σ_1 is the failure strength at $\beta=90°$ orientation and σ_{min} is the minimum failure strength observed at that particular confining pressure (usually at $\beta=30°$) and σ_c is the uniaxial compressive strength at $\beta=90°$ orientation. Reduction of anisotropy of schistose rocks as a function of

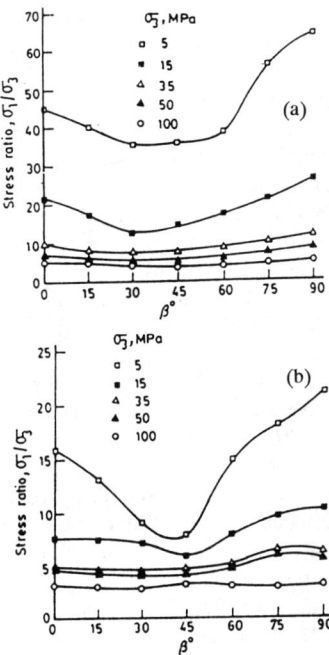

Fig. 7: Variation of stress ratio with β at different σ_3 for (a) Quartzitic and (b) Biotite schists

confining pressure from the literature is shown in Fig. 8. Schists prove to behave similar to that of other anisotropic rocks and it is observed that strength anisotropy detoriates faster below σ_c/σ_3 ratio of 5 and anisotrpy almost vanishes in most of the rocks at $\sigma_c/\sigma_3=1$, where the effect of anisotropy is as low as 10% which does not play significant role for practical purposes. It is also observed that strength reduction at all stages of σ_c/σ_3 is the maximum in Biotite schist followed by Chlorite schist, Quartz mica and the least in Quartzitic schist.

5.1 Effect of confinement on deformation Modulus

Degree of anisotropy in modulus gets suppressed at higher confining pressure for different orientation angles. Variation of the ratio of modulus to confining pressure (E_t/σ_3) versus orientation angle β, at different confining pressure is shown in Fig. 9a and 9b for Quartzitic, and Quartz mica schists respectively. It is observed that this ratio (E_t/σ_3) is highest at $\beta=0°$ for schists at lower confining pressure and being highest for Quartzitic schist, i. e. 0.6×10^4, 0.3×10^4 and 0.5×10^4 MPa at $\beta=0°$, 30° and 90° respectively for 5 MPa of confining pressure (σ_3). These values reduce to almost 0.03×10^4 MPa at all the orientation angles for 100

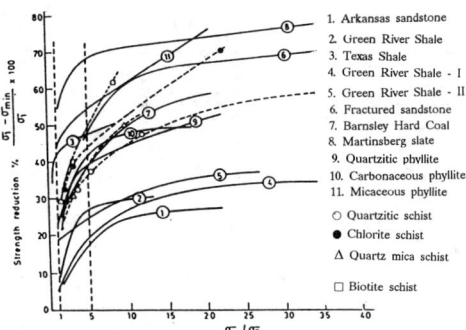

Fig. 8: Effect of confining pressure on anisotropy

MPa of confining pressure. This reduction being minimum for Quartz mica schist as it shows least values of modulus, varying from 0.2×10^4, 0.15×10^4 and 0.1×10^4 MPa at $\beta=0°$, 30° and 90° for 5MPa of confining pressure to a value of 0.03×10^4MPa at almost all the orientation under $\sigma_3=100$ MPa. Quartz mica schist undergoes maximum deformation due to its thinly laminated foliation planes and higher degree of porosity.

Another way of expressing the effect of confining pressure on anisotropy in modulus adopted in the present study is by plotting the ratio of deformation modulus at $\beta=0°$ to that at $\beta=90°$, i.e. (E_{t0}/E_{t90}) verses

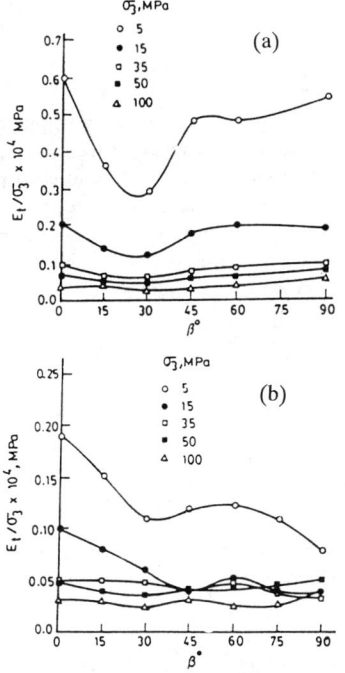

Fig. 9: Variation between E_t/σ_3 versus β at different σ_3 (a) Quartzitic and (b) Quartz mica schists

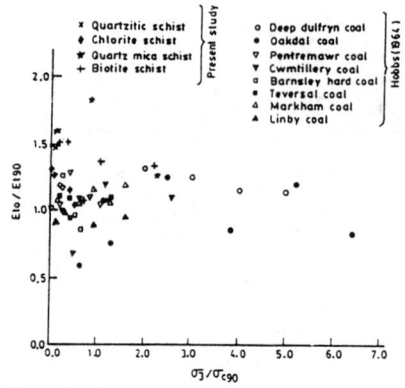

Fig. 10: Variation between E_{t0}/E_{t90} with σ_3/σ_{c90} at different σ_3

the ratio of σ_3/σ_{c90} (Fig. 10). On the basis of the analysis of the available data on modulus at different confining pressures for orientations parallel and perpendicular to foliation planes of 8 different coals studied by Hobbs (1964) for confining pressure ranging upto 35 MPa, and the modulus obtained from the four schists, it can be observed that when the ratio σ_3/σ_{c90} approaches to one, the ratio E_{t0}/E_{t90} varies from a value of 1.6 to almost 1. This behaviour indicates that anisotropy in modulus almost vanishes when the confining pressure applied is of the same magnitude as the uniaxial compressive strength (σ_{c90}).

6 CONCLUSIONS

Schistose rocks studied here exhibiting anisotropy due to the varied distribution of minerals as well as their orientation and their genetic history. This anisotropy has a great influence on their geomechanical properties such as tensile, compressive strength and modulus behaviour. It has been observed that when the confining pressure rises to a value almost equal to σ_c of an anisotropic rock, anisotropy in terms of strength and modulus detoriates.

REFERENCES

Akai, K. 1971. The failure surface of isotropic and anisotropic rocks under maltiaxial stresses. *J. Soc. Mater. Sci.* Japan, 20: 122-128.

Attewell, P. B. & Sandford, M. R. 1974. Intrinsic Shear strength of brittle anisotropic rock - I; Experimental and mechanical interpretation *Int. J. Rock Mech. & Min. Sci.*, II: 423-430.

Brown, E. T., Richards L. R. & Barr, M. V. 1977. Shear strength characteristies of Delabole slate. *Proc. Conf. Rock Engy.*, New Castle upon Tyne 31-51.

Chenevert, M. E. & Gatlin, C. 1965. Mechanical anisotropies of laminated sedimentary rocks. *Soc. Pet. Engrs. J.*, 2: 67-77.

Deklotz, E. J., Brown, J. W. & Stemler, O. A. 1966. Anisotropy of schistose gneiss. *Proc. 1st Cong. Int. Soc. Rock Mech.* Lisbon I: 465-470.

Donath, F. A. 1964. Strength variation and deformational behaviour of anisotropic rocks. *State of stress in the earth's crust. W. R. Judd, ed.*, 281-298. Elsevier publ. comp. New York.

Hobbs, D. W. 1964. The strength and stress - strain characteristics of coal in triaxial compression. *J. Geol.* 72(2): 214-231.

Horino, F. G & Ellickson, M. L. 1970. A method of estimating the strength of rock containing planes of weakness. *U. S. Bureau Mines, Report Investigation*, 7449.

ISRM Committee on laboratory Tests 1972. Suggested methods for determining water content, porosity, density, absorption and related propereies and swelling and slake durability index properties, document 2:36.

McCabe, W. M. & Koerner, R. M. 1972. High Pressure shear strength of an anisotropic mica schist rock. *Int. J. Rock Mech. Min. Sci.*, 12: 1015-1021.

McLamore, R. & Gray, K. E. 1967. The mechanical behaviour of anisotropic sedimentary rocks, *Trans. Am. Soc. Mech. Engrs., Series B.* 89:62-76.

Nasseri, M. H. 1992, Strength and deformational responses of schistose rocks, *A Ph. D. thesis submitted to the Indian Institute of Technology New Delhi, India.*

Nasseri M. H., Rao, K. S. & Ramamurthy, T. 1996. Prediction of Strength and deformational responses of schists. *ISRM Int. Sym. Eurock 96 Torino* 2-5 sep. 1996, 1.

Pomeroy, C. D., Hobbs, D. W. & Mahmoud, A. 1971. The effect of weakness plane orientation on the fracture of Barnsly hard coal by triaxial compression, *Int. J. Rock Mech. Min. Sci.*, 8(3): 227-238.

Ramamurthy, T. Rao, G. V. & Singh, J. 1988. A strength criterion for anisotropic rocks. *Proc. Int. Symp. on rocks at great depth* I:37-44.

Ramamurthy T. 1993. Strength and modulus respenses of anisotropic rocks. *Comprehensive rock engineering. J. A. Hudson (ed).* 1: chapter 13.

Seshagiri, Rao, K., Rao, G. V. & Ramamurthy, T, 1986. A strength criterion for anisotropic rocks. *Ind. Geotech.J.* 16(4): 317-333.

Singh, J. Ramamurthy, T. & Rao, G. V. 1989. Strength anisotropies in rocks. *Ind. Geotech. J.*, 19(2): 147-166.

… _Environmental and Safety Concerns in Underground Construction_, Lee, Yang & Chung (eds)
© 1997 Balkema, Rotterdam, ISBN 90 5410 910 6

Development of a micromechanical crack model based on crack information

Seokwon Jeon
Department of Mining and Geological Engineering, The University of Arizona, Tucson, Ariz., USA

ABSTRACT: Rock contains discontinuities at all scales. These discontinuities make rock behave in a complex way. The mechanism of deformation and failure of coal obtained from the McKinley Mine, New Mexico, and the Twenty Mile Coal Mine, Colorado, were studied by observing the distributions of length, orientation, and spacing of the pre-existing as well as stress-induced cracks. Different types of laboratory tests were employed to observe the different scales of cracks and to obtain different types of crack information. The crack information is dependent on the scale used. The cracks propagate along the intersections of the pre-existing cracks, and both extensile and shear crack growth occur depending on the direction of the load relative to the bedding planes. An analytical model that takes into account both shear and extensile crack growth was developed to predict the nonlinear stress-strain behavior of coal including strain-hardening and strain-softening.

1 INTRODUCTION

In coal, the fracture and the forces required to break it greatly depend on the innate weakness (Szwilski, 1985; Singh, 1986; Friesen, 1987; Zipf, 1990). The bedding planes are one obvious source, where bonding between successive layers of deposition has not fully taken place. Other planes of weakness in coal beds, particularly in bituminous coals, are generally referred to as 'cleats'. They are represented by sets of parallel fractures, usually oriented perpendicular to the bedding of the seam. The characteristics of these features of weakness are their planar geometry and their persistence. There are two engineering properties of fractures which are significant in an underground mining context. The first is the low tensile strength in the direction perpendicular to them. The second is the relatively low shear strength of the surfaces. Both of these properties should be considered in the excavation design procedure. Usually, working faces tend to be established parallel to the face cleat direction, and advance takes place mainly perpendicular to it to obtain maximum efficiencies of production (Khair, 1989; Prucz, 1989).

In this study, a new approach to modeling the deformation and the failure of coal is developed and applied to coal pillar design. The approach is to model explicitly the process involved in coal deformation and failure. The progressive evolution of the crack network under compressive loading also results in nonlinear stress-strain behavior. There are many advantages to this approach. First of all, size effects are built in. If a statistical distribution of initial cracks is assumed, this distribution will be scale-dependent and scale effects will be predicted. Secondly, coupled effects are also built into this approach. The changes in the thermal and hydraulic properties due to the evolution of the crack network can be calculated in a straight forward manner using the principles of rock fracture mechanics. Finally, the crack relationships from this approach can be easily implemented into a finite element method to model complex pillar geometries. This approach has been used successfully in modeling tuff (Wang & Kemeny, 1992), sandstone (Myer et al., 1992), granite (Kemeny & Cook, 1991), and other rocks.

2 OBSERVATIONS FROM LABORATORY TESTS

2.1 *Uniaxial compression tests*

Coal has a characteristic systematic fracture network consisting of bedding planes and two mutually perpendicular sets of cleats. These discontinuities strongly affect the deformation and failure of coal. In most cases, the bedding planes and major (face) cleats lie horizontally and vertically, respectively. However, depending on the excavation geometry and the in-situ stress state, the maximum principal stresses may be at various angles relative to the discontinuities. Therefore, uniaxial compression tests were conducted with the direction of the load at various angles relative to the bedding planes. The primary purpose of the tests was to understand the mechanisms involved in coal deformation and failure when the direction of the principal stresses varied with the orientation of discontinuities.

Several coal samples of 2 inch (5 cm) diameter were cored out of lumps from the McKinley Coal Mine, New Mexico, with three different angles of 0°, 45°, and 90° relative to the bedding planes. From the tests, details of the deformation and failure were carefully examined. In all samples, the stress-strain behavior was almost linear up until the point of failure. Distinct differences in the way the samples failed were found depending on the orientation of the bedding planes relative to the load.

The stress was applied parallel to the axes of the cylindrical samples. All the samples broke along the intersections of discontinuities. The sample with horizontal bedding planes had a relatively higher strength than others resulting in crushed zone with fine powder in the diagonal plane of the sample which was caused by shear localization. Cracks propagated along the intersection of the bedding planes and the cleats where extensile crack propagation was predominant. The sample with the bedding planes angled with 45° showed shear failure along the bedding planes. At the top and the bottom of the sample, local crushed zones developed. There were some vertical planes of failure due to the extension of the existing cleats. The sample with vertical bedding planes showed splitting failure due to extensile crack growth. In summary, it was observed throughout the tests that the cracks propagated along the intersections of pre-existing discontinuities, and both shear and extensile crack growth occurred depending on the direction of the stress relative to the bedding planes.

2.2 *Uniaxial compression tests with resin injection*

In order to examine the damage in a coal sample due to uniaxial stress, a new test scheme was developed. In the new test, a 2 inch (5 cm) diameter coal sample subjected to uniaxial stress was simultaneously injected with epoxy resin mixed with fluorescent dye. After drying, the sample was cut into several slices to observe the three dimensional crack distribution. The idea was that the resin with dye would penetrate through all the connected discontinuities, and when cut and illuminated with a fluorescent light, the pattern of pre-existing cracks and their growth could then be observed. The injection technique has been successfully used by Cavanaugh & Knutson (1960), Pittman (1970), Gardner (1980), and Wang & Kemeny (1992).

For the test, a special resin injection cell was made of transparent plastic as shown in Figure 1. The cell consists of a plastic hollow cylinder, two o-rings, and ports. The o-rings at the top and the bottom of the cell hold the sample inside the cell and more importantly prevent the leakage of vacuum pressure and resin. The ports are designed to connect the vacuum pump and resin reservoir to the cell.

The test procedure was as follows. A coal sample which is slightly longer than the cell was inserted in the cell applying silicon grease around the o-rings. The cell was put in the loading frame, and a small amount of load (0.44 MPa) was applied. Under the load, the cell was connected to a vacuum pump, and a pressure of 200 millitorr (= 2.632×10^{-4} atm) was applied for two days. Then epoxy resin mixed with fluorescent dye and a few drops of methyl ethyl ketone was injected into the cell, while the vacuum pump was still on. When air bubbles stopped coming out of the sample, the cell was pressurized at about 200 psi with nitrogen gas to keep the resin inside the small cracks.

The resin was allowed to set completely for a few days. Then, the cell was carefully cut into 3 slices with a thin diamond saw, from which six digital images of cracks were taken with a high resolution digital video camera. The images were processed to binary images by thresholding the 256 gray scale into white background and black fractures. From the binary images, the computer program called *Image* calculated the position and the orientation of each crack. Figure 2 presents the relationship between the length and the orientation of cracks collected from the six different images. In was observed that there were many short cracks shorter than 25 mm and a few longer cracks. In terms of the orientation, there were distinct clusters of cracks subparallel to 0° and 90°, which are directions of face and butt cleats. After a series of tests, histograms of crack length and orientation were obtained as shown in Figures 3 and 4. The distribution of crack length appears to follow a Poisson's distribution. But, if a large interval is employed on the horizontal axis of the histogram, it would follow an exponential distribution. In addition, it should be noted that the limit of the resolution of the digital video camera prevented the small cracks from being detected. It is, therefore, reasonable to assume that there exists many small cracks that so unidentified, and that the distribution of crack length would follow an exponential distribution. In Figure 4, there is a peak near 90° which is one set of fractures, and it seems to have a normal distribution around the peak. The other peak near 180° is not as clear as the peak near 90°, since it is known that one set of cracks (face cleats) is dominant over the other set of cracks (butt cleats). Another measure of crack distribution is the crack density as defined by

$$\chi = N \cdot a^2 / A \qquad (1)$$

where, χ is the crack density (dimensionless), N is

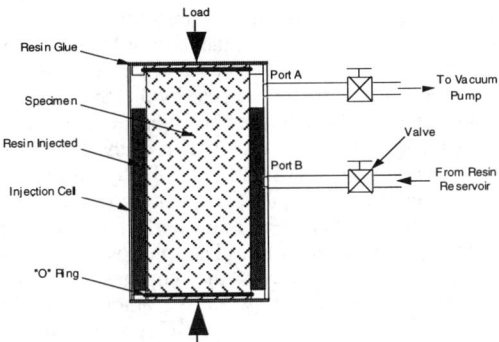

Figure 1. Special cell devised to inject resin during applying uniaxial stress.

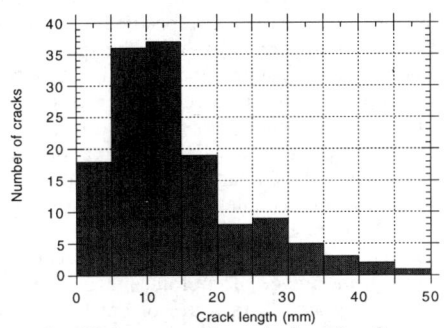

Figure 3. Histogram of crack lengths from resin injection tests.

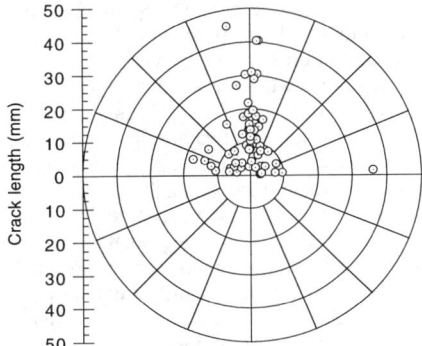

Figure 2. Relationship between crack length and orientation

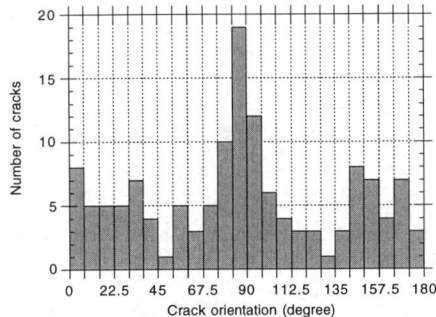

Figure 4. Histogram of crack orientations from resin injection tests.

the number of cracks in a specific area A, and a is the average crack length. The average crack density obtained in the resin injection tests is 1.17.

2.3 *Creep test*

One of the difficulties in monitoring crack growth in brittle materials in laboratory compression tests is the short period of time over which the crack growth takes place. In order to slow down the progression of the crack growth under a uniaxial stress, a creep test has been conducted. A rectangular coal sample of 5 cm x 3 cm x 3.58 cm was prepared so that one of the faces of the sample could be imaged using a digital video camera. The bedding planes were placed horizontally, and the stress was applied perpendicular to the bedding planes. Based on the strength of the coal obtained from the preliminary laboratory tests, the initial creep stress of 13.8 MPa (= 2000 psi) was determined. In the difficulty in maintaining the stress constant, incremental stress was used.

The strain increased monotonously until the sample failed except at 480,000 seconds, where a sharp increase of strain occurred due to a local spalling out along the left side of the sample. The increase of the strain as a function of time is presented in Figure 5, where the primary, secondary, and tertiary creep behaviors are distinct. In Figure 5, Burgers model was fitted to the observed strain as well. It should be noted that the model only traces strain up to the onset of the tertiary creep.

The deformation of the sample during the creep test was monitored using a digital video camera. From the record of the deformation, six images were selected for detailed image processing analyses. These images are referred to as Stage 1 through 6, 1 being the image in its initial state, and 6 immediately prior to the failure. The images were corrected to fit the same area and to have the same overall gray level distribution, which was necessary to take into account changes in lighting conditions and changes in zoom setting of the camera. After the same technique of image processing as used in the resin injection tests was used, the crack length and orientation were calculated in each stage. The crack distribution for

Stages 2 and 5 represent the initial state and the state near failure of the sample, respectively. In Stage 2, most cracks were parallel to either the face cleats or the bedding planes. On the other hand, in Stage 5, the cracks subparallel to the direction of the load have grown and new shorter cracks have also formed with random orientations. It was also noted that the overall gray scale value of the image correlated with the level of cracking in the sample. Since the cracks appear dark in the image compared with the light background, as cracks grow, dark gray levels become prevalent in the image. Therefore, the average gray level provide a simple measure of crack growth in coal samples subjected to load.

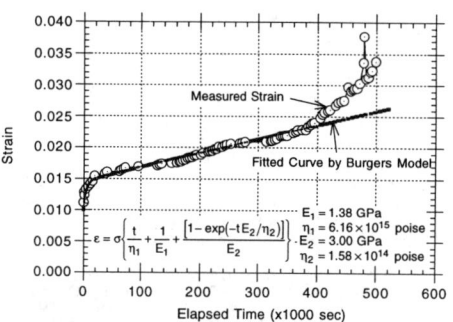

Figure 5. Increase of strain as a function of time in the creep test.

2.4 Scanline survey

Statistical data on the distribution of microcracks in coal is required for numerical modeling. The distributions of crack length and orientation have been obtained from a series of tests including the creep test and compression tests with resin injection. The other parameter is crack spacing, which determines the number of cracks in a given area or volume.

The approach used, called a scanline survey, is a procedure that involves counting the number of intersections which an array of parallel lines make with microcracks in a plane section (Underwood, 1970). The scanline surveys were conducted on coal samples from the McKinley Coal Mine, New Mexico.

An unloaded rectangular coal sample of 140 mm x 120 mm x 120 mm as shown in Figure 6(a) was used in the scanline analysis. The sample was cut to have horizontal bedding planes which are parallel to the xy plane in the figure. Three scanlines were surveyed along two perpendicular directions on 2 sides of the rectangular sample. A total of twelve scanlines were surveyed in Plane A and Plane B, as shown in Figure 6(a). The results are presented in Table 1. Since the scanlines in the z-direction intersect mostly bedding planes as shown in Figure 6(a), the crack frequencies in the z-direction which are perpendicular to them give the best information of bedding planes. On the while, the scanlines in the x- and y-directions mostly intersect with face and butt cleats, respectively. The average crack frequencies of the face and butt cleats are about half of the frequency of the bedding planes. This shows that cleats are not as frequent as bedding planes in the sample analyzed.

Scanline surveys were also performed on two uniaxially loaded samples. The purpose of these surveys was to examine the changes of crack frequencies and spacings when the amount and the direction of the applied load vary. Figure 6(b) shows one of the samples. A standard cylindrical 2 inch (5 cm) diameter coal sample was prepared with vertical bedding planes. The uniaxial load which is parallel to the direction of the bedding planes was applied up to 15.4 MPa which caused cracks to grow but did not break the sample. The sample was then cut in half across the central cross section. As shown in Figure 6(b), four scanlines were surveyed across the bedding planes (z-direction) and face cleats (x-direction), where the z-direction is always across the bedding planes for convention. The applied uniaxial load induced many cracks parallel to the direction of the load which in this case is parallel to the bedding planes. This resulted in a large increase of crack frequency in the z-direction as shown in Table 1. There is also an increase in crack frequency of the cleats. Therefore, the crack spacing decreased significantly in both z- and x-directions.

The other standard cylindrical coal sample was prepared with horizontal bedding planes. The sample was uniaxially loaded up to 31 MPa without failure. The sample was cut in the middle as shown in Figure 6(c). Four and three scanlines were surveyed across the bedding planes and face cleats, respectively. In this case, the applied uniaxial load caused many stress-induced cracks in the direction of the face cleats as presented in Table 1. The crack frequency of the bedding planes also increased. In the same way, the crack spacings decreased as evident in Table 1.

The results in Table 1 show that the average crack frequencies of the bedding planes are about 0.575 and 0.18 at 10x and 1x magnifications, respectively. The crack frequencies of face cleats (0.23) and butt cleats (0.26) are almost the same at 10x magnification. The data at 1x magnification are less precise as evident in Table 1, due to the poor resolution. When loaded, the crack frequencies increase by approximately a factor of three in the direction perpendicular to the direction of the load: for bedding planes (1.55/0.575) and cleats (0.73/0.23). In the direction parallel to the direction of the load, crack frequencies doubled as shown by the numbers of 0.59/0.23 and 1.03/0.575. Crack spacings have almost the same variation as crack frequencies.

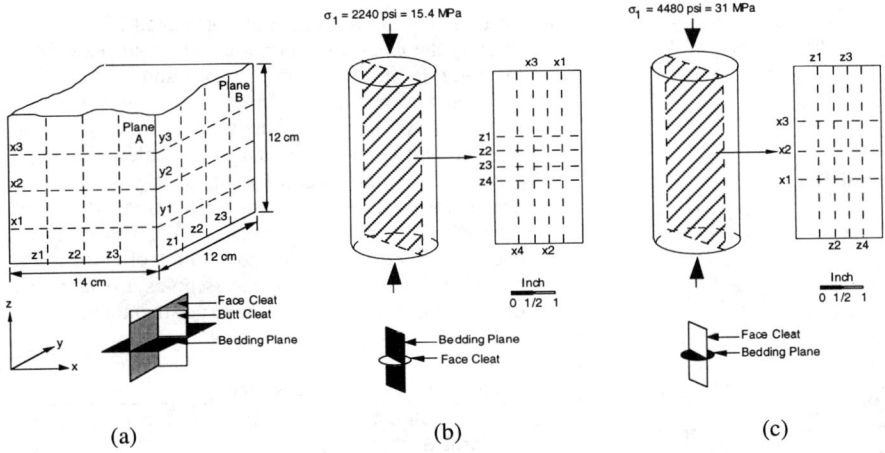

Figure 6. Scanline survey scheme of (a) an unloaded coal sample, (b) a loaded coal sample when the applied uniaxial stress is 15.4 MPa (=2240 psi) and the bedding planes are parallel to the direction of the stress, and (c) a loaded coal sample when the applied uniaxial stress is 31 MPa (=4480 psi) and the bedding planes are perpendicular to the direction of the stress.

Table 1. Average crack frequencies and spacings from the scanline surveys.

Sample	Plane	Line	Crack frequency[1,*]	Crack spacing[2,*]
Unloaded	A	z	0.56(0.23)	1.73(4.33)
	A	x	0.23(0.16)	4.11(6.03)
	B	z	0.59(0.13)	1.69(7.07)
	B	y	0.26(0.093)	3.74(10.2)
Loaded Load ∥ BP		z	1.55	0.0962
		x	0.59	0.131
Loaded Load ⊥ BP		z	1.03	0.0628
		x	0.73	0.164

[1] (Number of cracks/mm)
[2] (mm)
*Numbers in parentheses are in 1x magnification, otherwise in 10x magnification.

3 DEVELOPMENT OF A MECHANICAL MODEL

From the laboratory tests, the mechanism of deformation and failure was observed. Initially, the coal consists of an orthogonal network of cracks that represents the face cleats, butt cleats and bedding planes. Under load before the peak stress, long cracks tend to grow in the direction subparallel to the maximum principal stress, and small cracks are formed in relatively random orientations. As the load approaches the peak stress, spalling starts to occur in the areas of high crack densities. Finally, at the peak stress, major fractures are created subparallel to the maximum principal stress, along with significant failure of the sample. The microscopic examinations indicated that two primary mechanisms of fracturing occur in coal; sliding crack growth from pre-existing cracks, and shear cracking either from preferentially aligned discontinuities or through the connection of en-echelon arrays of extensile or pre-existing cracks.

To take into account crack interaction, the sliding crack model of Kemeny and Cook (1987) is considered. The model consists of an axially aligned column of sliding cracks. A sliding crack has an initial length of $2l_o$, angle θ, and coefficient of friction μ. Crack coalescence into a single splitting crack occurs when $l/b=1$, where $2l$ is crack length in the column and $2b$ is the separation of cracks. The stress intensity solution is given by (2).

$$K_I = \frac{2l_o \tau^* \cos\theta}{\sqrt{b\sin\left(\frac{\pi l}{b}\right)}} - \sigma_2 \sqrt{2b \tan\left(\frac{\pi l}{2b}\right)} \quad (2)$$

where

$$\tau^* = \sigma_1(\sin\theta \cdot \cos\theta - \mu\cos^2\theta) \\ -\sigma_2(\sin\theta \cdot \cos\theta + \mu\sin^2\theta) \quad (3)$$

where μ is the coefficient of friction, σ_1 and σ_2 are the applied axial and confining stresses at the boundary of the body. The cracks start to grow simultaneously when the crack growth criterion $K_I = K_{IC}$ is met. The axial stress σ_c and the axial strain ε_a^{total} at that moment are calculated as given below.

$$\sigma_c = \left\{ \frac{\left[K_{IC} + \sigma_2 \sqrt{2b \tan\left(\frac{\pi l}{2b}\right)}\right] \sqrt{b \sin\left(\frac{\pi l}{b}\right)}}{2l_o \cos\theta} \right.$$

$$\left. + \sigma_2 (\sin\theta\cos\theta + \mu\sin^2\theta) \right\} \Big/ (\sin\theta\cos\theta - \mu\cos^2\theta) \quad (4)$$

$$\varepsilon_a^{total} = \frac{1-v^2}{E}\left[\sigma_1 - \frac{v}{1-v}\sigma_2 + \frac{16\chi(sc-\mu c^2)c}{\pi}\right.$$

$$\left.\left\{\tau * c \ln\frac{\tan\left(\frac{\pi l}{2b}\right)}{\tan\left(\frac{\pi l_o}{2b}\right)} - \sigma_2 \frac{1}{l_o} \ln\frac{\tan\left\{\frac{\pi}{4}\left(1+\frac{l}{b}\right)\right\}}{\tan\left\{\frac{\pi}{4}\left(1+\frac{l_o}{b}\right)\right\}}\right\}\right] \quad (5)$$

where K_{IC} is mode I fracture toughness, χ is crack density, v is Poisson's ratio, c is $\cos\theta$, and s is $\sin\theta$.

For shear cracks, the stress intensity solution is given by Equation (6).

$$K_I = K_{III} = 0$$
$$K_{II} = \tau * \sqrt{\pi l} \quad (6)$$

The stress for crack growth and the associated strain for a body containing N shear cracks are given by

$$\sigma_c = \frac{\sqrt{\frac{G_c E}{(1-v^2)\pi l}} + \sigma_2(sc + \mu s^2)}{sc - \mu c^2} \quad (7)$$

$$\varepsilon_c^{total} = \frac{1-v^2}{E}\left[\sigma_1 - \frac{v}{1-v}\sigma_2\right.$$

$$\left. + 2\chi\pi\tau *(sc - \mu c^2)\left\{\left(\frac{l}{l_o}\right)^2 - 1\right\}\right] \quad (8)$$

where $s = \sin\theta$, $c = \cos\theta$, G_c is shear fracture energy.

4 NONLINEAR STRESS-STRAIN CURVES FOR COAL

From the micromechanical crack model, nonlinear stress-strain curves for coal are calculated and compared with laboratory test results. The laboratory tests provided the average crack length and spacing of the pre-existing cracks in coal of about 10 mm and 5 mm, respectively. Then, two hundred pre-existing cracks are considered in a 5 cm x 10 cm standard coal specimen with an exponential distribution of crack length. Table 2 presents the input parameter values.

The nonlinear stress-strain curves of coal at three different confining stresses are presented in Figure 7. Initially the crack growth results in strain hardening behavior. In the strain hardening portion of the curve, more and more cracks start to grow as the stress increases. As the cracks become longer, they start to interact with each other and shear cracks begin to grow. Both crack interaction and shear crack growth contribute to the start of strain softening where the sample progressively weakens. The peak stress occurs immediately prior to strain softening. The model subjected to uniaxial stress gives a peak stress of 10 MPa, which matches the experimental results.

Table 2. Parameter values for coal.

Volume	0.005 m^2 (5 cm x 10 cm)
Number of cracks	200
Initial crack length	0.5-5 cm (Exponential)
Initial crack orient.	45°-225° (Bimodal normal)
Cohesion	0.7 - 1.5 MPa
Coefficient of friction	0.2
Young's modulus	1 GPa
Poisson's ratio	0.15
Fracture toughness	0.06 MPa√m
Shear fracture energy	800 Joules/m^2

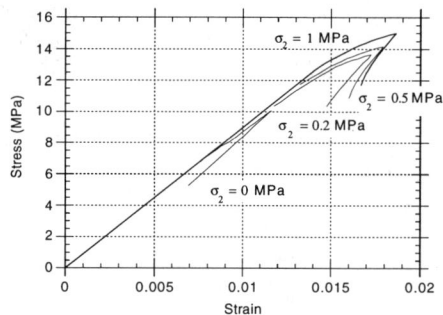

Figure 7. Nonlinear stress-strain curves for coal.

5 CONCLUSIONS

A mechanical model for the deformation and failure of coal has been developed based on the observations made from laboratory tests. A few different types of laboratory tests were employed to examine different scale of cracks and different types of information of cracks. It was successful to model the complex process of crack growth in coal. The model was used to predict the strain hardening, strain softening, and nonlinear stress-strain behavior of coal given crack information in a specific area or volume. An example was taken to simulate the nonlinear stress-strain behavior of coal in the laboratory scale, which validated the usefulness of the model.

REFERENCES

Cavanaugh, R.J. & Knutson, C.F. 1960. Laboratory technique for plastic saturation of porous rocks (Geological Notes), *Bulletin of the American Association of Petroleum Geologists*, Vol. 44, No. 5: 628-640.

Friesen, W.I. & Mikula, R.J. 1987. Fractal dimensions of coal particles, *Journal of Colloid and Interface Science*, Vol. 120, No. 1: 263- 271.

Gardner, K.L. 1980. Impregnation technique using colored epoxy to define porosity in petrographic thin sections, *Canadian Journal of Earth Science*, Vol. 17: 1104-1107.

Hadley, K. 1976. Comparison of calculated and observed crack densities and seismic velocities in Westerly Granite, *Journal of Geophysical Research*, Vol. 81, No. 20: 3484-3494.

Kemeny, J.M. & Cook, N.G.W. 1987, Crack models for the failure of rocks in compression, *Constitutive Laws for Engineering Materials, Theory and Application*, Vol. 2: 879-887.

Kemeny, J.M. & Cook, N.G.W. 1991. Micromechanics of deformation in rocks, *Toughening Mechanisms in Quasi-Brittle Meterials*: 155-188.

Kemeny, J.M. & Wang, R. 1991. Damage progression in rocks under compressive stress, *Proceedings of the Ninth Annual Workshop: Generic Mineral Technology Center - Mine Systems Design and Ground Control*, Lexington, KY: 13-20.

Khair, A.W. & Reddy, N.P. 1989. Mechanisms of coal fragmentation by a continuous miner, *Mining Science and Technology*, Vol. 8: 189-214.

Myer, L.R. & Kemeny, J.M., 1992. Extensile cracking in porous rock under differential compressive stress, *Appl. Mech. Review*, Aug.

Pittman, E.D. & Duschatko, R.W. 1970, Use of pore casts and scanning electron microscope to study pore geometry, *Journal of Sedimentary Petrology*, Vol. 40, No. 4: 1153-1157.

Prucz, J.C. & Fu, S.H. 1989. Prediction of dynamic fracture modes in coal mining (Technical Note), *Int. J. Rock Mech. Min. Sci. & Geomech. Abstr.*, Vol. 26, No. 2: 161-167.

Singh, S.P. 1986. Brittleness and the mechanical winning of coal, *Mining Science and Technology*, Vol. 3: 173-180.

Szwilski, A.B. 1985. Relation between the structural and physical properties of coal, *Mining Science and Technology*, Vol. 2: 181-189.

Underwood, E.E. 1970. *Quantitative stereology*, Addison Wesley.

Wang, R. & Kemeny, J.M. 1992. Fracturing mechanicsms in Apache Leap Tuff under compressive stress, *Fractured and Jointed Rock Masses*: 381-388, Lake Tahoe, CA.

Wang, R. & Kemeny, J.M. 1993. Micromechanical modeling of tuffaceous rock for application in nuclear waste storage, *Int. J. Rock Mech. Min. Sci & Geomech. Abstr.*, Vol. 30, No. 7: 1351-1357.

Wong, T-f. 1982. Micromechanics of faulting in Westerly Granite, *Int. J. Rock Mech. Min. Sci & Geomech. Abstr.*, Vol. 19: 49-64.

Wong, T-f. 1985. Geometric probability approach to the characterization and analysis of microcracking in rocks, *Mechanics of Materials*, 4: 261-276.

Zipf, R.K. & Bieniawski, Z.T. 1990. Mixed-mode fracture toughness testing of coal, *Int. J. Rock Mech. Min. Sci & Geomech. Abstr.*, Vol. 27, No. 6: 479-493.

Development of a numerical tool for the treatment of the data supplied by compression tests

J.P.Tshibangu Katshidikaya
Service Mécanique des Roches et Exploitation des Mines, Faculté Polytechnique de Mons, Belgium

ABSTRACT

A software for microcomputer of PC type has been developed to set an automatic treatment of the stress-strain curves drawn from rock testing systems. The software implements procedures for smoothing and detecting the characteristic points of stress-strain curves. The smoothing is made by three points sliding mean, while the detection of the characteristic points is made by an original treatment using numerical derivatives. This treatment allows the setting of the onset of the linear portion of the curve, the elastic limit and the failure strength. The software implements also an original algorithm to build Mohr's envelopes. It has been tested with satisfactory results on many rocks ranging from fragile to ductile materials.

1. INTRODUCTION

Testing in Rock Mechanics is a field of research in which the standard deviation of measurements is globally higher than in other more classical experimental systems. This dispersion is due to the heterogeneous nature of rocks and to the presence of preexisting defaults like cracks in the materials. The presence of cracks induces important noises on the stress-strain curves registered during testing, especially when the experimental compression system is strain controlled.

Beside the dispersion due to the heterogeneous nature of rocks, we have to notice the manual interpretation of data. In spite of the development of automated systems to do mechanical tests, the quality of characteristics derived from those tests depends in general on the operator's ability. That introduces more errors in the results and contributes to increase the dispersion in the results.

We have developed a software for micro computer of PC type to set an automatic treatment of stress-strain curves for drawing directly some mechanical properties of rocks. This software presents three advantages:
- It enables a more precise estimation of mechanical properties by setting some strict criterions for the curves interpretation;
- It makes the interpretation more rapid, and lowers the cost of tests,
- It makes the interpretation more simple because one needs just a few commands to treat a mechanical test measurements.

2. THE GENERAL PROCEDURE APPLIED ON A STRESS-STRAIN CURVES

The software has been designed to treat data supplied by computer acquisition systems from a general true triaxial compression test (polyaxial). These data concern three stresses and three strains.

Figure 1 shows the method developed to treat a typical stress-strain curve ($\sigma_1 - \varepsilon_1$). It presents a such idealized curve with three characteristic points A (onset of the linear part), B (the elastic limit), and C (the failure point); and the first and second derivatives of σ_1 (σ' and σ'').

For making calculations according to the procedure described we need some steps as shown bellow.
- Smoothing the curves obtained from experimental benches: that is made by the three points sliding mean in three steps:
  1. Sorting the numerical values of the X vector (abscissa) to get increasing values;
  2. Making the abscissa discrete with a defined number of constant intervals;
  3. Computing the sliding mean .

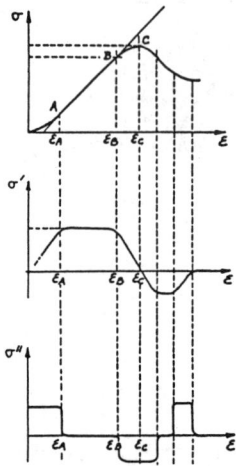

Figure 1: First and second derivative of the stress-strain curve, and characteristic points

- Computing the first and second derivatives of the stress-strain curve: we use for that the central differences technique of derivation based on the 5 points Lagrangian formula.
- Smoothing the derivative curves.
- Setting the beginning and the end of the linear part of the stress-strain curve: points A and B.
- Setting the yielding point C.
- Calculating the mechanical properties: these are given by the following formulas:

$$E = \frac{\sigma_{1B} - \sigma_{1A}}{\varepsilon_{1B} - \varepsilon_{1A}}$$

$$\nu = \frac{(\varepsilon_{2B} - \varepsilon_{2A}) + (\varepsilon_{3B} - \varepsilon_{3A})}{2(\varepsilon_{1B} - \varepsilon_{1A})} \quad (1)$$

$$\sigma_E = \sigma_{1B}$$

$$\sigma_R = \sigma_{1C}$$

In these formulas E is the Young's modulus, ν the Poisson's ratio, σ_E the elastic limit, and σ_R the failure strength. Subscripts 1, 2 or 3 refer to the stresses directions (major, intermediate and minor).

In the working method set to find the points A, B and C and to calculate the mechanical properties, first we make 10 iterations to smooth the curves $\sigma_1 - \varepsilon_1$ (Sig1 - Eps1), $\varepsilon_2 - \varepsilon_1$ (Eps2 - Eps1), and $\varepsilon_3 - \varepsilon_1$ (Eps3 - Eps1); and after that the $\sigma_1 - \varepsilon_1$ curve is plotted before looking for the characteristic points.

The point C is searched first and then the points A and B. Some tests are set in the algorithm to detect the occurrence of some problems during the treatment. If that is so, a warning message is displayed and the user can then use a semi manual procedure implemented in the software to determine the points A, B and C. In general the treatment gives satisfactory results and problems like a brutal variation in the measurements are solved during smoothing.

3. SETTING OF CHARACTERISTIC POINTS OF THE STRESS-STRAIN CURVE

3.1. The maximum of the curve: point C

It corresponds to the first zero of σ' or the minimum of σ'', and the algorithm developed to find the point C is presented on figure 2a. We begin by computing the first derivative of the $\sigma_1 - \varepsilon_1$ curve and search the occurrence of the first zero of that curve. If the zero exists (case of curves presenting a maximum), the corresponding value of the strain ε_C (found by interpolating the strain vector) is the abscissa of the point C. To check the validity of the point C, the second derivative is computed and its minimum searched. If the difference between the two values is less than a tolerable error R, the position of C is validated, otherwise, a warning message is displayed.

If the zero of the first derivative does not exist, as is the case of a strictly hardening material, we search the minima of the first and the second derivatives. By interpolating the values of the strains ε_{C1} and ε_{C2} corresponding to the two minima and by averaging them we found the position of point C. The position of the so calculated point C is generally situated at the onset of the linear part of the hardening line.

3.2. The elastic zone: points A and B

In an ideal situation, the point A is situated at the beginning of the first flat part of σ' and/or the first zero of σ''. The value of σ' corresponding to the flat part between A and B is the Young's modulus of the material. The point B is characterized by the decreasing of σ' from the flat portion or by the negative sign taken by the second derivative. On the real curves obtained from tests, the variations of the derivatives are not so simple, and we have to use some special techniques to set the position of points A and B.

The research of the points A and B is done in the strain interval ranging from zero to the abscissa of C. The procedure elaborated for that purpose is presented on figure 2b. The numerical value of the Young's modulus is given by the maximum of the first derivative in the defined range between 0 and the abscissa of C. But, due to the non perfect nature of the derivative calculations, we will rarely observe a flat zone between A and B as stated in the idealized case of the figure 1. This is why we set a margin below the maximum of the first derivative (80% of the maximum E_{max}).

That margin has been determined according to the most usual values obtained by manual treatment of stress-strain curves in our laboratory. The plotting of an horizontal line corresponding to the margin allows the finding of points A and B at the intersections with the σ' curve.

If the σ' curve begins with a value greater than the usual $0.8 \times E_{max}$ ($\sigma'[1] > E$), the value of the margin is increased in order to intersect the σ' curve in two points. If the adjusted value of Margin becomes greater than 1, the point A is chosen at the onset of the curve ($\varepsilon_A = X[1]$).

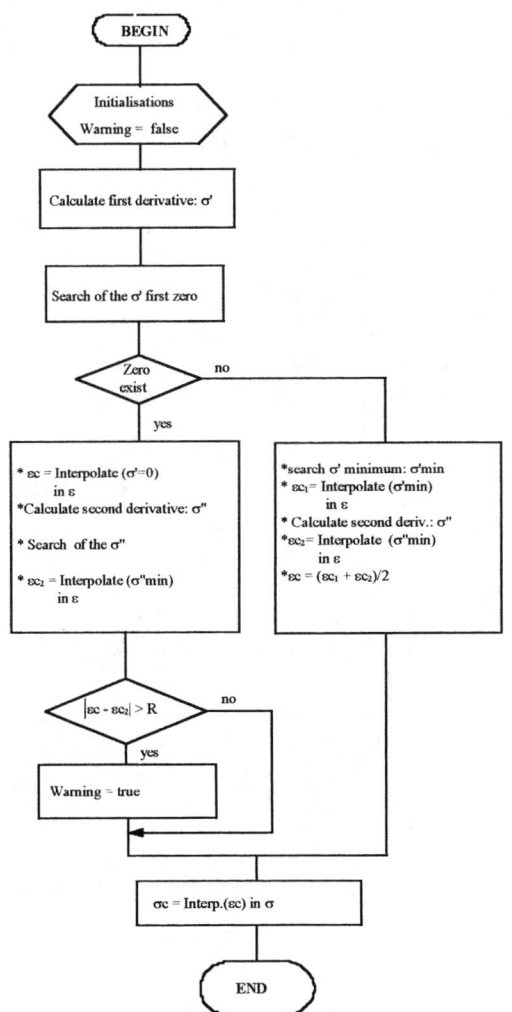

Figure 2a: searching of point C

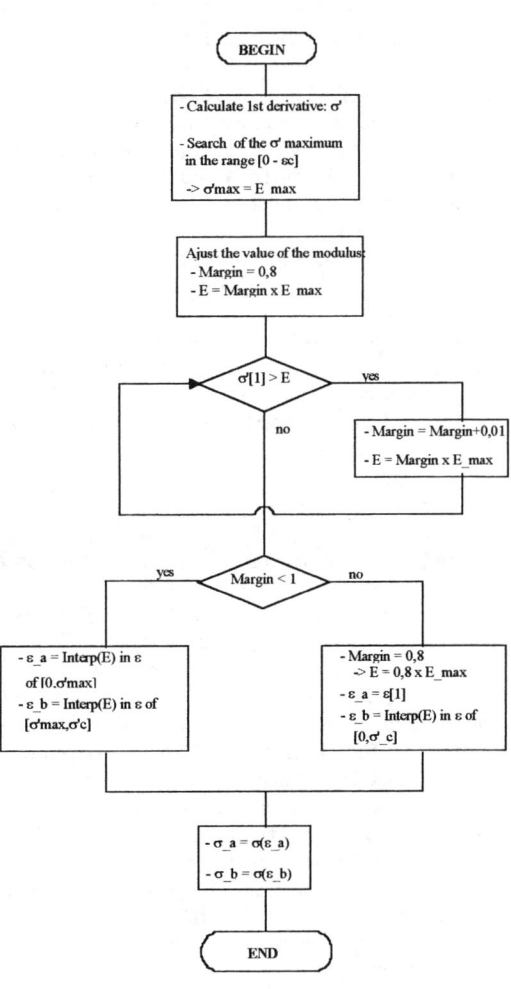

Figure 2b: Searching of points A and B

4. A TYPICAL RESULT OBTAINED AFTER THE TREATMENT OF A STRESS-STRAIN CURVE

In general the results supplied by the automatic treatment are very satisfactory and in accordance with the results coming from well manually interpreted data. The figure 3 presents the steps of a such automated treatment on a soft limestone named MOKA marble from Portugal.

- The figure 3.a shows a raw curve coming from a polyaxial test automated system.
- The figure 3.b gives the smoothed curve with the characteristic points calculated automatically.
- The figures 3.c and 3.d give the shape of the first and second derivatives.

5. THE MOHR'S ENVELOPE OF A ROCK

The building of Mohr's envelopes of rocks is the second mode of treatment implemented in the software. Results obtained from the stress-strain curve treatment for different confining conditions on a specific rock can be stored in a rock file. These results can then be retrieved to build the Intrinsic curves (elastic limit and failure) of the material being studied.

The set of mechanical characteristics (which constitute a record) from the stress-strain treatment can be supplied in any order to the Mohr's envelope tool. When the file of a specified rock is loaded, the sets of mechanical properties are sorted according to an increasing order of the minor principal stress σ_3. To keep that order during the treatment, a chained list is created by pointing the first field of a record on the last field of the preceding record. The use of a chained list allows the insertion and the deletion of records, and then makes possible the manual supplying, if not directly available, of results from uniaxial compressive and tensile tests. From the chained list, the Mohr's circles corresponding to different confining conditions can be displayed, and the building of the polygonal line enveloping those circles initiated. Figure 4 shows the principle of developed method to make a such treatment.

The treatment starts with the point A situated on the first circle of the list, from which a tangent line is drawn to the second circle. This gives the point T which becomes the new starting point to draw a next tangent line to the following circle.

The starting point A is chosen such that its abscissa is given by the following Eqn 2.

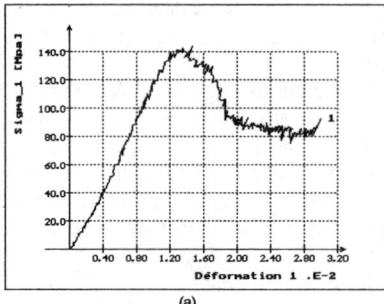

(a)

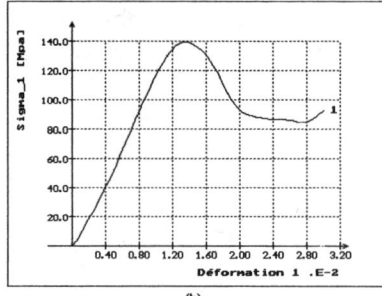

(b)

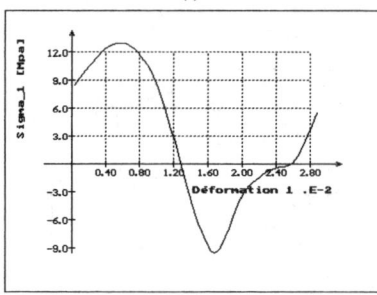

(c)

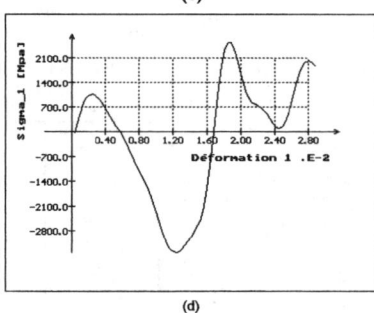

(d)

Figure 3: Steps of the treatment of a stress-strain curve

$$X_A = \sigma_3 + 0.3 \times (\sigma_1 - \sigma_3) \quad (2)$$

We can then determine its corresponding Y value by

$$Y_A = \sqrt{\tau_{max}^2 - (X_A - X_{CE})^2} \quad (3)$$

with

σ_1 and σ_3: the major and minor principal stresses of the Mohr's circle being treated;
X_{CE} and τ_{max}: the abscissa of the center and the radius of the circle which are given by

$$X_{CE} = \frac{\sigma_1 + \sigma_3}{2}$$
$$\tau_{max} = \frac{\sigma_1 - \sigma_3}{2} \quad (4)$$

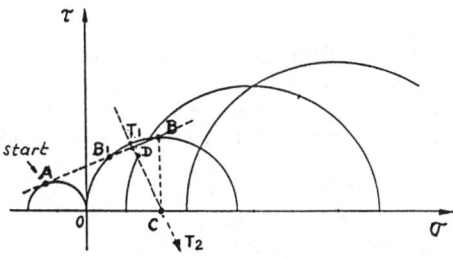

Figure 4: Method developed to build the Mohr's envelope

The general working method set to build the Mohr's envelope is presented on the figure 5. When the position of A is determined, we take on the next circle which will be marked B, a point also noted B corresponding to the maximum Y value (τ_{max}). If A is outside the area of the circle B, the line AB must intersect that circle in two points, B and another marked B_1.

If C represents the center of the circle B, its orthogonal projection on the line AB will give the point D, and the intersection of line CD with the circle B will give two points T_1 and T_2. Of these two points we choose the one situated between B1 and B2 (B) as the tangent point T. But this point T is not really the true tangent on the circle B; so the position of point B will be adjusted and set on T, and the described procedure repeated. Iterations will stop when the distance CD will be approximately equal to the radius of the circle B.

The treatment starts with the point A situated on the first circle of the list, from which a tangent line is drawn to the second circle. This gives the point T which becomes the new starting point to draw a next tangent line to the following circle. We proceed in that manner until all circles are treated.

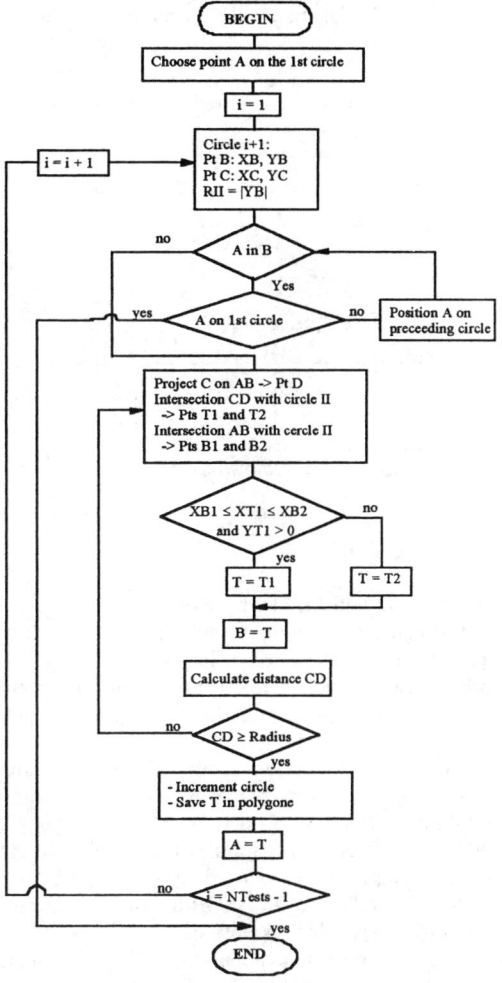

Figure 5: general algorithm for building the Mohr's envelopes

The algorithm allows an automatic detection of an A point situated inside the area of the circle B and makes an appropriate treatment to build a true envelope of a lot of circles having different radii not necessarily increasing with the value of minor principal stress. It means that the software can treat the situation in which tests give for a given confining state, a deviatoric stress ($\sigma_1 - \sigma_3$) which is less than that obtained with a smaller value of σ_3.

After this treatment, the result is supplied as a table of numbers which can be used to fit some mathematical models (regression analysis). Figure 6 presents an example of a failure intrinsic curve built on the Soignies limestone (Belgium). The user has the ability of displaying the elastic curve or the failure one.

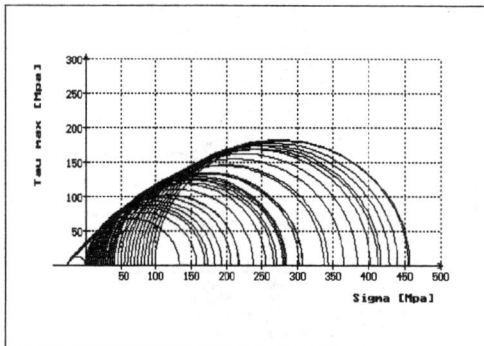

Figure 6: Intrinsic curve of the Soignies limestone

6. CONCLUSIONS

The treatment software that we developed to run on PC type computers allows an automatic treatment of data from compressive tests in rock mechanics. It works in two complementary modes: the treatment of stress-strain curves, and the building of Mohr's envelopes.

In the first working mode of the software, stress-strain curves are treated in order to compute the usual mechanical properties like the Young's modulus, the Poisson's ratio, the elastic limit, and the failure strength of a rock. Those properties are evaluated directly from an ASCII file of a mechanical compressive test which is now currently supplied by automated compressive apparatus. The algorithms implemented to complete the defined tasks are very original and use procedures based on smoothing by three points sliding mean, and numerical derivations to detect characteristic points of a stress-strain curve. The results obtained are very satisfactory and seem to be more reproducible than those from manual interpretation.

The second mode of the software allows the elaboration of two Mohr's envelopes for each rock being investigated: one corresponding to the elastic limit, and another to the failure strength. This software module is designed so that it can sort the results supplied in any order, and the size of different experimental circles can be detected to build a true Mohr's envelope including all circles.

The use of this treatment software in our laboratory has enabled a complete characterization of rocks in a very short time compared to the same job done manually a few years ago. Possibilities of developing the software exist, and we are working notably on the automatic construction of tridimensional envelopes of rocks and the creation of interfaces with numerical codes to facilitate simulations about the behavior of underground openings. The software is very easy to use because it is menu driven with mouse pointing.

REFERENCES

[1] J. BRYCH: "Contribution à l'étude des paramètres physico-mécaniques fondamentaux des roches"; internal publication FPMs, Nr 35, October 1979.

[2] J. BRYCH, TSHIBANGU K., B.FROMENT:"Une presse polyaxiale pour échantillons cubiques de petite dimension"; 7th International Congress of Rock Mechanics, Aachen 1991.

[3] J. BRYCH, TSHIBANGU K.:"Rapport du service de Génie sur recherche FRFC concernant la rhéologie des roches"; October 1991.

[4] N. CRISTESCU: "Elastic/viscoplastic constitutive equations for rock"; Int. J. Rock Mech. Min. Sci. & Geomech. Abstr., vol.24 N°5, pp. 271-282, Great Britain 1987.

[5]. J.C. JAEGER, N.G.W. COOK: "Fundamentals of rock mechanics"; Chapman and Hall, 3rd edition, 1976.

[6] J.P. TSHIBANGU K.:"Etude des effets du confinement sur le comportement mécanique des roches pour application au calcul des ouvrages souterrains; Doctoral Thesis, November 1993.

The effect of a polyaxial confining state on the behavior of two limestones

J.P.Tshibangu Katshidikaya
Service Mécanique des Roches et Exploitation des Mines, Faculté Polytechnique de Mons, Belgium

ABSTRACT: The work presented in this paper was guided by the need of getting reliable mechanical informations on rocks and their potential use by numerical codes for modeling. For this purpose, a polyaxial compressive test apparatus completely driven by a micro-computer has been built; and the data supplied by this system are treated numerically to find the desired characteristics. A lot of tests have been achieved on two carbonates. We have investigated the changing in the behavior of the rocks when the confining state increases: change of the rheological model and evolution of the mechanical properties. We have also built tridimensional envelopes at failure and tried to set some constitutive laws using a modified Mohr-Coulomb criterion.

1. INTRODUCTION

Sophisticated numerical tools are developed for modeling the behavior of underground openings and related problems. This development has been accelerated by the lowing cost of the computer treatment.

But drawing good mechanical properties from rocks is a difficult problem. This can be solved partly by developing computer driven loading frames. If this is so, some deep studies can be conducted on specified rock materials to reach a better understanding of their behavior when submitted to general stress systems.

We conducted tests on two ornamental limestones: the "Soignies limestone", a hard and brittle material from Belgium; and the "Moka marble", a more soft material from Portugal. The general physico-mechanical properties of those rocks are as follows:

Rock	γ	Rt	Rc	ν	E
Soignies limestone	2,7	7,5	102	0,34	13200
MOKA Marble	2,4	4,5	57	0,33	11500

with γ the density of the rock (gf/cm^3), Rt the tensile uniaxial strength (MPa), Rc the compressive uniaxial strength (MPa), ν the POISSON's ratio, and E the YOUNG's modulus (MPa).

2. THE EXPERIMENTAL SYSTEM

It is a polyaxial test system working on 3 cm cubes built in our laboratory of the Technical University of Mons (J.P. Tshibangu K., 1993). The cell is of three rigid pistons type and can develop stresses up to 500 MPa in each direction, this allows confining minor stress σ_3 to take values greater than 100 MPa.

The overall experimental system comprises the polyaxial cell, an hydraulic bench, a box containing electronic components, and a micro-computer of PC type equipped with acquisition/command boards to drive the hydraulic bench. The PC is supplied with a program that we developed to make the following tests in polyaxial conditions: strain or stress controlled compressive tests, creep and relaxation, biaxial tests.

3. EFFECTS OF THE CONFINEMENT ON THE MECHANICAL PROPERTIES OF THE ROCKS

3.1. The classical triaxial conditions

The confining conditions are such that $\sigma_2 = \sigma_3$. The mechanical properties studied are: the Young's modulus E (M. Young), the Poisson's ratio ν (C. Poisson), the elastic limit σ_E (Sigma_El), the failure strength σ_R (Sigma_Ru), the maximum elastic strain ε_E (Eps_El), and the failure strain ε_R (Eps_Ru).

Figure 1 presents, in deviatoric mode, smoothed curves drawn from polyaxial tests for confining stresses ranging from 0 to 100 MPa. It shows that the Soignies limestone has a failure type from fragile to perfectly plastic, while the Moka marble behaves like a strain hardening material at high values of σ_3.

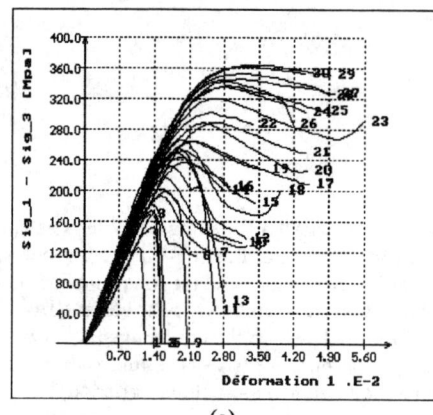

(a)

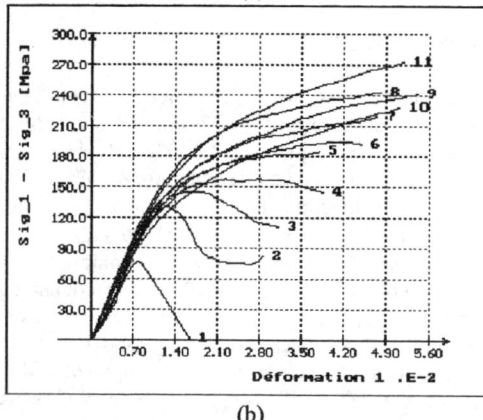

(b)

Figure 1: Stress-strain curves get on the two rocks for confining levels ranging from 0 to 100 MPa: (a) Soignies limestone, (b) Moka marble.

The treatment of such results gives the charts of the figure 2 presenting the way in which the mechanical properties previously enumerated evolve. To put all parameters on the same chart, values are reduced with respect to the uniaxial compressive test.

It can be seen on figure 2 that the Young's modulus E is the parameter which is the least affected by the confining stress σ_3, but it shows an increasing tendency (about 20% for the two rocks). Different fitting models have been tested and the linear one chosen for that parameter.

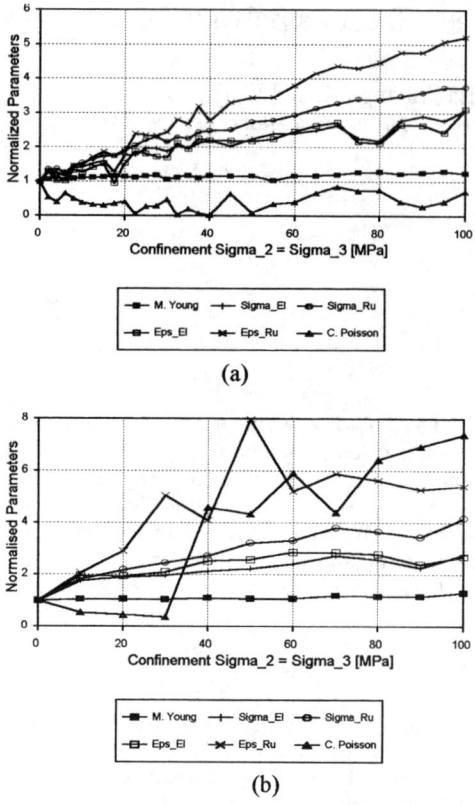

Figure 2: Graphical evolution of the mechanical properties versus the confining stress ($\sigma_2 = \sigma_3$) for the Soignies limestone (a) and the Moka marble (b).

The result obtained on the Soignies limestone with a correlation coefficient of 0.80 is given by

$$E = 14299 + 25{,}25\, \sigma_3 \qquad (1)$$

For the Moka marble, the regression line with a correlation coefficient of 0.89 is given by

$$E = 11527 + 26{,}62\, \sigma_3 \qquad (2)$$

The evolution of the Poisson's ratio seems to be random for the Soignies limestone, and we think that this is due to the quality of the measurements done with the polyaxial cell. For the Moka marble, the important variations shown by this parameter is probably due to the increasing viscoelastic behavior.

The other parameters show a clear increase with the confining stress. They characterize the yielding of the material and will be discussed further.

3.2. Importance of the intermediate principal stress

To study the effect of that stress component, tests in true triaxial conditions have been achieved by varying σ_2 for some chosen values of σ_3.

Figures 3 shows the results obtained on the failure strength. For the Soignies limestone (figure 3a), it seems that the failure strength increases with the intermediate stress especially for the low confining levels ($\sigma_3 = 0$ and $\sigma_3 = 20$ MPa). At higher values of σ_3, this effect becomes less significant.

The Moka marble (figure 3b) presents a pronounced effect of σ_2 at high confining levels while this is less visible at low values of σ_3.

The same observations can be derived from the tendency shown by the variation of the elastic limit versus the intermediate confining stress σ_2.

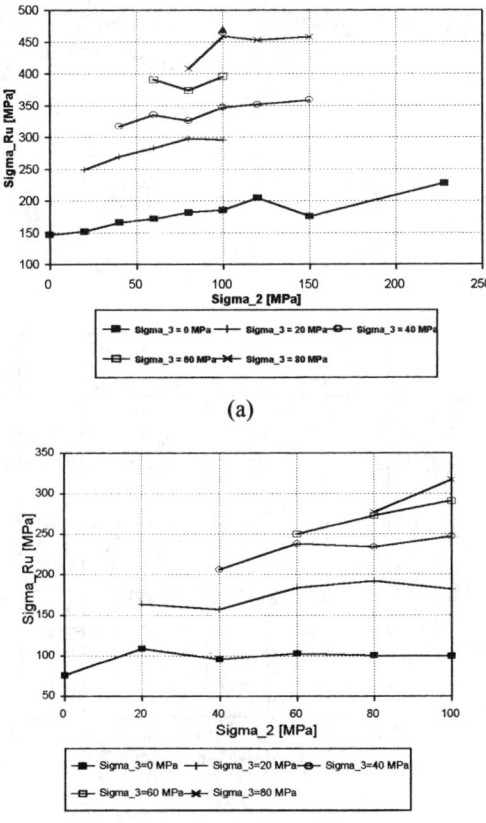

(a)

(b)

Figure 3: Evolution of the failure strength versus the intermediate stress for the Soignies limestone (a) and for the Moka marble (b).

4. BUILDING TRIDIMENSIONAL ENVELOPES

If a good strategy is applied in the choice of the confining conditions, it is possible to build true tridimensional envelopes at failure or at the elastic limit and to set hence some elaborated constitutive laws of the materials. When working in the principal stresses space with isotropic material, tridimensional envelopes are expressed analytically by the general formula:

$$\sigma_1 = f(\sigma_2, \sigma_3) \quad (3)$$

A more convenient representation of the limiting envelope can be done in the octahedral (or deviatoric) plane as shown on figure 4. In this representation each state of stress is decomposed in an hydrostatic part $h = \sigma_{oct}$ and a deviatoric one $d = \tau_{oct}$ on the octahedral plane. Another coordinate parameter has to be introduced to describe completely the state of stress, this is the so called "Lode angle ψ". This is the angle made in the octahedral plane by the directions of the vectors τ_{oct} corresponding to the triaxial state ($\sigma_2 = \sigma_3$) and to the state of stress being represented. The σ_{oct} and τ_{oct} are numerically given by following formulas (Jaeger & Cook, 1979) in which I_1 and I_2 are stress invariants.

$$\sigma_{oct} = \frac{1}{3}(\sigma_1 + \sigma_2 + \sigma_3) = \frac{1}{3} I_1$$

$$\tau_{oct} = \frac{1}{3}\sqrt{(\sigma_1-\sigma_2)^2 + (\sigma_2-\sigma_3)^2 + (\sigma_3-\sigma_1)^2} \quad (4)$$

$$= \frac{\sqrt{2}}{3}[I_1^2 + 3I_2]^{1/2}$$

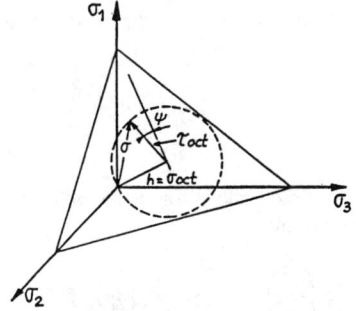

Figure 4: Representation of a state of stress in the octahedral plane.

To derive the value of the Lode angle we have to make some definitions. In a rectangular axes system Oxyz, a line passing through the origin and a determined stress point is characterized by its direction cosines l, m, and n. These direction cosines are related by the following relation

$$l^2 + m^2 + n^2 = 1 \qquad (5)$$

Numbers a, b, c known to be proportional to the direction cosines of a line, so that $a = kl$, $b = km$, $c = kn$, will be called direction ratios, and it follows from Eqn. 5 that the constant k must be

$$k = \sqrt{a^2 + b^2 + c^2} \qquad (6)$$

It can be demonstrated (Jaeger & Cook, 1979) that the angle ψ between two lines whose direction cosines are l, m, n and l', m', n' is given by

$$\cos\psi = l'l + m'm + n'n \qquad (7)$$

Further developments show also that the direction ratios of the octahedral shear stress are given by

$$a = \sigma_2 + \sigma_3 - 2\sigma_1$$

$$b = \sigma_3 + \sigma_1 - 2\sigma_2 \qquad (8)$$

$$c = \sigma_1 + \sigma_2 - 2\sigma_3$$

To solve the problem, we need just calculate the direction cosines of the octahedral shear stress corresponding to the triaxial state ($\sigma_2 = \sigma_3$), and do the same for an any polyaxial stress state. The Lode angle ψ can then finally be calculated by the Eqn. 7. The searching of the direction cosines of the octahedral shear stress which corresponds to the triaxial state gives

$$l' = -\frac{\sqrt{2}}{3}, \quad m' = \frac{\sqrt{6}}{6}, \quad n' = \frac{\sqrt{6}}{6} \qquad (9)$$

So, the cosine of the angle ψ for an any polyaxial stress state will be given by

$$\cos\psi = \frac{1}{k}(al' + bm' + cn') \qquad (10)$$

where a, b and c are calculated according to Eqn. 8 and k from Eqn. 6.

A very careful planning of the experimental conditions (i.e. the values of σ_2 and σ_3) is necessary to get a complete envelope for the chosen octahedral planes. Figure 5 shows two examples of such envelopes obtained on the two rocks: figure 5a is the failure envelope of the Soignies limestone on the 110 MPa octahedral plane, and figure 5b deals with the 80 MPa octahedral plane of the Moka marble.

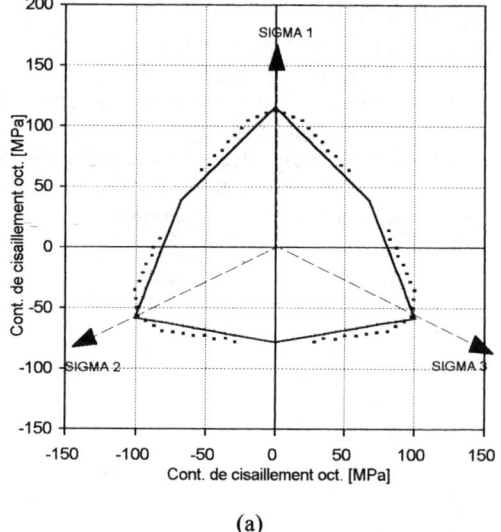

(a)

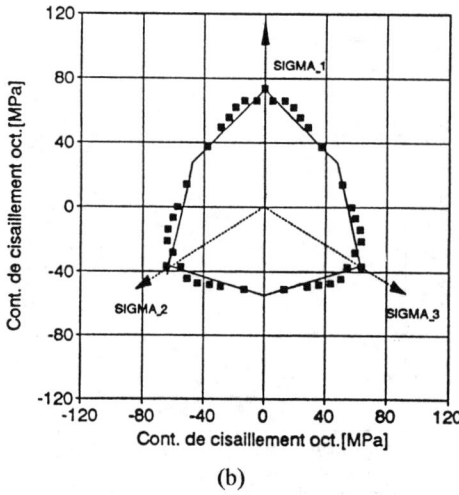

(b)

Figure 5: Octahedral failure envelopes of the Soignies limestone on the 110 MPa plane (a) and of the Moka marble on the 80 MPa plane (b).

The shape of the envelopes obtained on the rocks may be fitted to some theoretical criterions. We have tried to make such fitting with a Mohr-Coulomb criterion where the values of the friction angle and the cohesion are adjusted with the confining stress according to the shape of the intrinsic curves. The result of this fitting is shown on figure 5 where this modified Mohr-Coulomb criterion is represented by the polygonal line.

5. ELASTO-PLASTIC MODELING

5.1. The hardening law

The hardening law for a rock material is a mathematical relation linking the state of stress to the corresponding state of strain. This law is the generalization of the model drawn from the classical stress-strain curve. To model the stress-strain curve, different mathematical laws can be used: linear, power, exponential, etc. The general hardening law retained for our rocks may be represented by the following equations

$$\varepsilon_e = \frac{\sigma}{E(\sigma_3)} \qquad (11)$$

$$\varepsilon_p = g(\sigma) = \left\langle \frac{\sigma - \sigma_E(\sigma_3)}{K_E(\sigma_3)} \right\rangle^{1/M_E(\sigma_3)}$$

ε_e, ε_p: elastic and plastic strains; K_E and M_E are the parameters of the power law; σ_E is the elastic limit.

To set the evolution of the parameters K_E and M_E of the power law with σ_3, we have two steps:
- we fit the power law to stress-strain curves for points lying above the elastic limit and draw the numerical values of the parameters;
- we fit the variation of the obtained parameters versus the confining stress to different mathematical models.

The application of this method to the results supplied by mechanical tests gave the following results for the Soignies limestone:

$$\varepsilon_e = \frac{\sigma}{14299 + 25.3\,\sigma_3}$$

$$\varepsilon_p = \left\langle \frac{\sigma - (151 + 1.98\,\sigma_3)}{1156} \right\rangle^{1.96} \qquad (12)$$

As can be observed, the influence of the confining stress σ_3 on the power law parameters was not significant and had been neglected.

For the Moka marble the hardening law is represented by the two equations

$$\varepsilon_e = \frac{\sigma}{11527 + 26.7\,\sigma_3}$$

$$\varepsilon_p = \left\langle \frac{\sigma - 100\,\sigma_3^{0.103}}{164 + 17\,\sigma_3} \right\rangle^{1.88} \qquad (13)$$

For this material we have observed a clear increase of the parameter K_E with the confining stress σ_3. The value of the second parameter of the power law M_E is assumed to be not dependent on σ_3.

5.2. The Plastic behavior

We showed on the figures 5 how the Mohr-Coulomb criterion in which we vary the characteristic parameters φ (friction angle) and C (cohesion) fitted to the octahedral envelopes obtained on the limestones.

A general plastic behavior law can be presented as follows

$$\Delta \varepsilon_p = \Delta \lambda \frac{\partial Q}{\partial \sigma} \qquad (14)$$

with λ the plastic multiplier and Q the plastic potential.

The hardening laws presented above will be used to identify the plastic multiplier. If we assume the simple hypothesis of associated plasticity, the plastic potential can be considered identical to the failure envelope which is represented by the Mohr-Coulomb criterion in the actual situation. That assumption leads to

$$Q = \sigma_1 - K_p\,\sigma_3 - \sigma_c$$

$$\frac{\partial Q}{\partial \sigma_1} = 1 \qquad (15)$$

$$\frac{\partial Q}{\partial \sigma_1} = -K_p$$

σ_c is the uniaxial compressive strength and K_p depends on the friction angle and is given by

$$K_p = tg^2\left(\frac{\pi}{4} - \frac{\phi}{2}\right) \quad (16)$$

By replacing in the Eqn. 14 the plastic deformation and the plastic potential by their corresponding expressions, we can draw the value of the plastic multiplier λ. If we use the general equation of the power law we get

$$\Delta \varepsilon_p = \left(\frac{\Delta \sigma}{K_E}\right)^{\frac{1}{M_E}} = \Delta \lambda \quad (17)$$

The plastic behavior law can then be represented by

$$\begin{Bmatrix} \Delta\varepsilon_1^p \\ \Delta\varepsilon_3^p \end{Bmatrix} = \left(\frac{\Delta \sigma}{K_E}\right)^{\frac{1}{M_E}} \begin{Bmatrix} 1 \\ -K_p \end{Bmatrix} \quad (18)$$

In this law the parameters K_E and M_E can be expressed as functions of the confining stress and their expressions for the two limestones can be drawn from Eqn. 12 and 13.

6. CONCLUSIONS

This paper gives an idea of the quantity of mechanical informations which can be drawn from the polyaxial test system developed in the rock mechanics laboratory of the Technical University of Mons. The great advantages of this experimental system concern the low dimensions of cubes required to make polyaxial tests (3 centimeters), the completely automated mechanical tests and the numerical treatment of the stress-strain curves.

Due to those possibilities, we have been able to make a deep study of two carbonate rocks: the Soignies limestone and the Moka marble. We have investigated the influence of the confining stress on the mechanical properties of the rocks and especially on the elastic parameters. This analysis has shown for example that the Young's modulus increases of about 20% when the confining stress varies from 0 to 100 MPa.

The importance of the intermediate stress on the failure strength has also been emphasized. This leads to building true triaxial failure envelopes for which we have presented some octahedral sections. Many such sections will lead to the complete knowledge of the limiting envelope.

We have also shown how a modified Mohr-Coulomb criterion fits the to the shape of the octahedral envelopes. This has led to the formulation of elasto-plastic behavior laws in which the parameters vary with the confining stress.

It is obvious that the obtaining of this so great quantity of mechanical informations was possible because the experimental system used allows a complete automated treatment. This lowers the cost of laboratory characterization and gives the opportunity of supplying numerical codes with good parameters.

7. REFERENCES

[1] S. AMPADU, F. TATSUOKA:"An automated Stress-Path Control system Triaxial system"; American Society for Testing and Materials, 1989.

[2] H. ARMEN, A.B. PIFKO, H.S. LEVINE, G. ISAKSON: "Plasticity"; Finite Element technics in structural mechanics, Tottenham and Brebbia, 1970.

[3] R. AZEVEDO, M.M. DE FARIAS: "Elasto-Plastic modeling of compressibility and strength characteristics of sand"; International Conference on Computationnal Plasticity, pp. 1527 -1540, Barcelona, Spain 1987.

[4] E.T. BROWN: "Rock Characterization, Testing and Monitoring, ISRM Suggested Methods"; Pergamon Press, Oxford 1981.

[5] J. BRYCH, TSHIBANGU K., B.FROMENT:"Une presse polyaxiale pour échantillons cubiques de petite dimension"; 7ième Congrès International de Mécanique des Roches, Aix-la-Chapelle 1991.

[6] C.R. CALLADINE: "Engineering plasticity", Pergamon Press, 1969.

[7] M.J. COCKRAM, W. KAMP: "True triaxial compression experiments on Felser sandstone"; 7ème Congrès International de Mécanique des Roches, tome 1 pp. 447 - 450, Aix-La-Chapelle, Allemagne 1991.

[8]. J. GUSTIEKIEWICZ: "Synoptic view of mechanical behaviour of rocks under triaxial compression"; Rock at great depth pp. 3-10, Balkema Rotterdam, 1989.

[9]. J.C. JAEGER, N.G.W. COOK: "Fundamentals of rock mechanics"; Chapman and Hall, 3rd edition, 1979.

[10] TSHIBANGU K.:"Etude des effets du confinement sur le comportement mécanique des roches pour application au calcul des ouvrages souterrains; Doctoral Thesis, November 1993.

A laboratory evaluation of grout jointed specimens composed of different rocks

R.K.Srivastava, K.K.Sharma & D.S.Soni
M.N.R. Engineering College, Allahabad, India

ABSTRACT : The geological features of rock mass, particularly rock joints, effect the failure mode of rockmass significantly under the type of loads and stresses, the rock mass is usally subjected to. For example in a dam foundation. The situation is more complicated when the joint is an interface between two types of rocks or same type of rock with different intact rock strength. To improve the engineering behaviour of rock mass, especially in case of dam foundations, grouting is carried out, but the data available on grout performance is very little, especially when different types of rocks are grouted togather. And since the strength and deformation behaviour of rock mass is greatly influenced by joint orientation, shape of joint and number of joints, it becomes very difficult to evaluate the grout performance in fissured rock mass. The present study has been carried out under controlled laboratory conditions with specific shape of joints and grout thickness to develop an understanding and characterise the strength and deformation behaviour of grout jointed rock specimens composed rocks of same formation but different strengths. Planar joints at an angle of 0°, 30°, 45° and 90° have been created and filled with cement fine sand grout (w/c = 0.7) of thicknesses 3 and 4 mm. Triaxial tests have been carried out at confining pressures of 0, 5, 7.5 and 10 MPa. Geotechnical characterization and evaluation of strength and deformation behaviour of intact and grout jointed rocks are reported.

1 INTRODUCTION

The number, magnitude and size of civil engineering projects are increasing day-by-day to fulfill the demands of human society. Higher dams and deeper underground excavations works under comparitively more difficult geological conditions have posed unique challenges to geotechnical engineers. The rock which one feels, is very strong, actually behaves in such an unpredictable and fragile manner, under the kind of stresses it is subjected to that one has to be very cautious in characterising its behaviour and design structures in or over it. Under a comparatively large extent, which is usually the case of the river valley projects, under investigation or construction stages in India, it has been observed that the rock mass that is encountered is invariably discontinuous and contains planes of weakness like bedding planes, joints, fissures and faults etc. These geological features, particularly rock joints, effect the failure mode of rock mass. Foundations located on jointed rocks may settle excessively due to tendancy of closing of joints under the imposed load. Dam underlain by discontinuous rock may under go rotation and slip as a result of sliding of rock blocks along one or more planes of weakness. Stability problems may occur in case of rock slopes and underground openings. Thus, a need is generally felt to improve the properties of the rokc and rock mass. This is usually carried out mainly by rock reinforcement (by rock bolting, steel supports, shotcreting or guiniting etc.) or by pressure grouting.

Particularly in case of dam foundation, grouting has been more commonly used for the last many years. But the data available on the actual performance of such grouted foundations is rather scarce. In fact large scale field tests before and after grouting are required to actually evaluate the performance of a grout. There are still many aspects that are yet to be deciphered, such as whether grouting is really necessary at a particular site, if so how much of grouting is needed and above all an inadequate understanding of the extent of reduction in factor of safety due to excessive grouting etc. when certain limits are exceeded (grouting pressure), some deterimental effects may result in the rock mass. Prominent among them are the destruction of overall structure of rock mass, leading to irreversible situations and consequent destruction of contact between joint surfaces.

In addition, increasing attention is paid by dam design engineering to the concept of bulk safety of structure and rock mass. Keeping in view the complexity of rock mass and the limitations of insitu tests, many investigators reconcile to study various phenomena connected with rock mass either by refined experimental study or computer based method. For a rational understanding, it is very desirable to perform tests and characterise the shear behaviour of grouted rock under controlled conditions.

The present study has been carried out in controlled

laboratory conditions and with specific shape of joints and grout thickness to develop an understanding and characterize the strength and deformation behaviour of grout jointed rocks.

Keeping the above facts in view an experimental programme has been carried out on grout jointed sand stones (with different intact rock strengths) from Vindhyachal-Mirzapur region of Uttar Pradesh, India, belonging to Bhander series of the upper Vindhyans. The strength behaviour of planar jointed rocks with variable grout thickness have been studied under triaxial conditions. Cement-fine sand grout (w/c = 0.7) of thickness 3 and 4. mm. has been used with joint plane orientation varying from 0° to 90° and confining pressure varying from 0 to 10 MPa.

2 LITERATURE REVIEW

The literature available on rock grouting is very meagre. The present practice in India is based on the practices in U.K. and U.S.A. On occassions, grouting trials have been carried out on major projects to decide various parameters of the grouting process. But emergence of a general code of practice has been difficult.

The important feature of a grouting programme is a decision on the type of grout itself which would be successful at that particular site. Vaughan (1963) has concluded that rocks with an effective permeability coefficient as low as 10^{-7} ms^{-1} can be successfully treated with cement grouts. This indicates that cement particles will penetrate fissures efficiently.

According to Ewert (1985), w/c ratio in design of cement grout should be determined by weight to avoid any difficulty caused by changing density of the cement. A w/c ratio > 6.0 (by volume) has been preferred in American practice while in Europe w/c ratio < 3.0 (by weight) has been preferred.

Kunert (1976) and Kutzner (1982 a) suggest that in case of wide fissures (e.g. Haune dam), thickest possible mixture should be used. Wittke (1967) have reported a study on injections using very thick grout slurries (paste grouting). Recommendation of Houlsby (1982 a) on w/c ratio is < 3.0 (by volume) and > 2.0 (by weight). The influence of grouting pressure was analysed by Ewert (1979) in case of Tavera dam. His analysis showed that fracturing occurrence has led to a misestimation of the natural permeability due to greater absorption of grout caused by the fracturing.

In view of the rock fracturing behaviour, selection of grouting pressure becomes very important. Jaeger (1969) has summarised various practices and opinions and trends in American practice. Two somewhat older recommendations are by Grundy (1955) and Zaruba and Mencl (1961) Grundy suggested grouting pressure p=44. D (D being depth in meters) and Zaruba suggested that for the case of rocks with steeply dipping joints, $p=30 D + 2D^2$ and for rocks with horizontal joints, $P=24 + 0.5D^2$ Where P is the pressure in Kpa and D is depth in mm.

Kunert (1976) has also discussed the prevailing opinion on grout pressures to be used and has concluded that grouting pressure must ultimately be different for every grouting programme and best determined by insitu test. Recently Albritton (1982), Bruce (1982), Deere (1982) and Houlsby (1982 b) have given recommendation on this aspect. Model test have also been carried out to determine various aspects of grouting by Hassler et. al. (1987). They concluded that determination of the properties of grouts is important in order to compare different materials and for fruther development in the field of grouting science. Studies on grouted rock specimen are very few. Coulson (1970) examined artificial joints in coarse and fine grained granite with grout filling of 0.8 mm to 6.4 mm. He concluded that there is a detrimental influence in peak and ultimate strengths at higher values of normal stress. At lower values (<4 kg/cm^2) of normal stress, the peak strengths of grouted joints was observed to be higher than that of the natural joints. Borroso (1970) observed that on planar surface, there is an improvement in the value of friction angle from 25° to 30° after grouting. But in case of Shales no improvement was observed. He also observed the importance of relative strengths of grout and rock because beneficial or negative effect of grout depended upon w/c ratio and grout strength and strength gain could be expected only in case of weak rocks. The shear behaviour of grout-rock system depended/dominated the component having higher shear strength.

Studies on rheological properties of grout has been reported by Kutzner (1982 b) and Hassler et. al. (1987).

Barla et. al. (1987) have also recognised the need of increasing the present state of knowledge in shear behaviour of filled joints and reported about an ongoing research project.

Vishawanathan (1988) has carried out a laboratory study on cement grouted sandstone. Some of his important conclusions are (i) the variation of friction angle is not much irrespective of joint inclination or thickness of grout. (ii) cohesion intercept is very sensitivie to orientation and thickness of grout and the value increases with increasing grout thickness. (iii) There is an increase in modulus value with increase in grout thicness upto 2 mm. thereafter it decreases. (iv) The initial peak strength observed is maximum for 4 mm. thick grout. (v) Failure in most cases is by tensile splitting and shear across the grouted joints.

Recently Srivastava et. al. (1989) have reproted a laboratory study on strength behaviour of grout jointed ($\beta = 45°$) Shakteshgarh sandstone using two types of grouts (cment-sand-water and cement-flyash-water) of 4mm thickness. Performance of the two types of grout material and overall shear behaviour of the grout jointed rocks has been discussed.

Sharma (1989) has opined that grouting is an art which cannot be quantitatively analysied and mathematically modelled to the same extent as some other branches of dam engineering (where it has its maximum use). Cement based grout remains to be main grouting material for rock treatment. Flyash and pozzolanic materials are added in a number of cases. w/c ratio, cement granulometry, additives, grouting pressure constitutes the main parametres which define the properties of the mix, the outcome of grouting and its interaction with the fissured rock. there are also the areas where difference of opinion exists and research work is needed.

Greco et. al (1990) carried out study by experimental measurements and numerical modelling. It has the aim to analyze the behaviour of cylindrical specimens composed by different alternate rock disks, subjected to uniaxial compression. Various combinations of material types are utilized in the laboratory investigations with different ratios between strength and stiffness, with the purpose to determine overall mechanical parameters. Particularly the tests involved intact rock specimens and disk specimens, both of the same rock and of combined rocks, with different contact between the disks (free contanct, glued contact and dumped contact). The conclusions refer on the comparision between experimental results and the numerical simulation results. The laboratory tests have brought out a loss of uniaxial compressive strength due to mainly two factors : the different shape of the test specimens made up of disks compared with the intact specimens and the contacts between materials having different mechanical characteristics.

3 EXPERIMENTAL PROGRAMME

The sandstone rock specimens used are from Vidhyachal region of Uttar Pradesh, India and belong to Bhander Series of Upper Vindhyans. The rock is isotropic, light yellowish in colour due to presence of more silica and feldsper. Physical properties (e.g. density, sp. gr., water absorption, porosity and slake durability) and strength indices (e.g. Brazilian and Point load strength, UCS of intact rock) have been determined first to characterise the rock. Subsequently a series of tests have been carried out under triaxial stress conditions on various types of specimen as shown schematically in figure 1.

4 RESULTS AND DISCUSSIONS

4.1 *Intact rock characterization*

The physical, index and engineering properties of the two rocks are presented in Table 1. Figures 2 and 3

Table 1 : Properties of sandstone rocks

Property	Values	
	Rock (S1)	Rock (S2)
Water Absorption (%)	2.81	2.84
Specific gravity	2.663	2.645
Density (KN/m^3)		
Dry	23.20	23.25
Saturated	24.53	23.95
Prosity (%)		
Apparent	8.691	8.813
Total	12.692	12.880
Slack durability (%)	96.37	99.38
UCS (σc) MPa	99.607	63.404
Brazilian strength (σtb) MPa		
Oven dried	6.986	5.4579
Saturated	3.1656	4.1481
Air dried	6.1129	4.803
Point load strength (σtp) MPa		
Oven dried	9.9206	5.3155
Saturated	6.172	5.0069
Air dried	9.43	5.144
Cohesion c (MPa)	20.86	17.71
Angle of internal friction	48.31	32.12

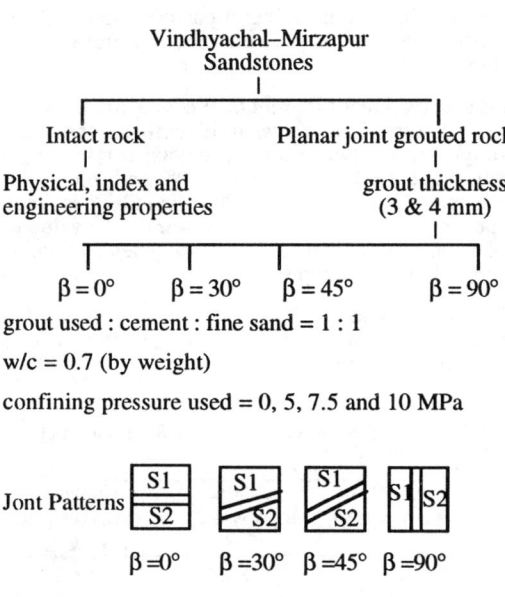

S1—relatively softer rock
S2—relatively harder rock

Fig. 1 : Schematic diagram-specimen tested

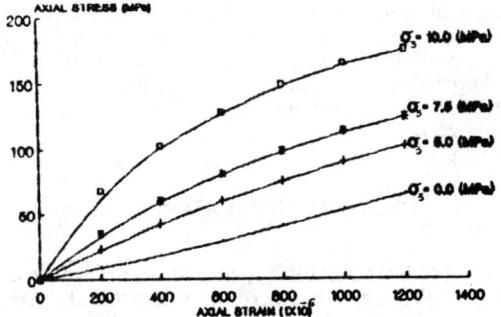

Fig. 2 Stress-strain variation for Rock type S1

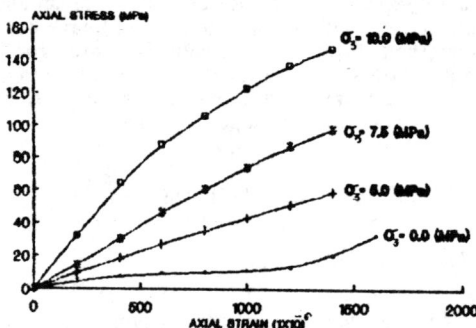

Fig. 3 Stress-strain variation for Rock type S2

shows the stress-strain variation for the two rocks under different confining pressures. It is observed that

(a) The ratio of UCS/(σtp) a around 10.56 for S1 and 16.38 for S2 sandstones specimens.

(b) The ratio of UCS/(σtb) a round 16.30 for S1 and 17.54 for S2 sandstones specimens.

(c) The ratio of (σtb) a/(σtp) a round 0.654 for S1 and .933 for S2 sandstones specimens i.e. observed to be varying between 0.5 to 1.0.

Based on Deer and Miller's (1968) compressive strength modulus ratio classification, Vindhyachal sandstones falls in CM (S1) and CL (S2) category ued in present study.

4.2 *Grout jointed rock*

Tables 2 and 3 present variation of UCS value for different joint orientations for grout thicknesses 3 mm and 4 mm respectively.

It is observed that the UCS value varies, obviously, with joint orientation and is least around $\beta=30°$. In case of specimens composed of two rock types, the

Table 2 : Variation of UCS with β (grout thickness=3mm)

β	UCS (MPa) Rock type S1	UCS (MPa) Rock type S2	UCS (MPa) specimen composed of S1 & S2
0°	75.8	69.4	73.4
30°	64.2	47.2	55.8
45°	90.2	35.4	54.1
90°	94.9	42.8	41.5

Table 3 : Variation of UCS with β (grout thicness = 4 mm)

β	UCS (MPa) Rock type S1	UCS (MPa) Rock type S2	UCS (MPa) specimen composed of S1 & S2
0°	85.2	58.9	49.7
30°	54.8	44.5	31.8
45°	59.9	33.8	47.2
90°	74.1	32.7	21.8

UCS value is close to average UCS value of individually grouted rock. For grout thickness 4 mm, the UCS values in general are lower than that for 3 mm grout thickness and in case of specimens composed of different rocks the UCS value is in general less that of individually grouted rocks.

The variations in the value of cohesion and angle of internal friction for grouted rock are presented in tables 4 to 6.

It is observed that for grout thickness 3 mm the value of cohesion is in between in case of specimen composed of two rock types as compared to individual rocks, where as 30° and 45° joint orientation it is more than that for individual rock type. In case of grout thickness 4 mm, the value of cohesion reduces drastically for specimen compound with different rock types.

The angle of internal fiction is marginally effected in case of grout thickness 3 mm but significantly effected for grout thickness 4 mm.

Table 4 : Variation in cohesion with β (grout thickness = 3 mm)

β	c (MPa) Rock type S1	c (MPa) Rock type S2	c (MPa) specimen composed of S1 & S2
0°	16.7	8.7	12.6
30°	8.9	4.1	10.7
45°	8.5	6.3	18.2
90°	12.2	5.2	7.1

Table 5 : Variation in cohesion with β (grout thickness = 4 mm)

β	c (MPa) Rock type S1	c (MPa) Rock type S2	c (MPa) specimen composed of S1 & S2
0°	16.3	21.7	0.7
30°	10.8	8.4	2.7
45°	17.2	6.3	4.3
90°	16.5	7.5	5.9

Table 6 : Variation in angle of internal β

β	φ Rock type S1 grout thickness		φ Rock type S2 grout thickness		φ specimen composed of S1 & S2 grout thickness	
	3mm	4mm	3mm	4mm	3mm	4mm
0°	47.7	44.4	46.5	26.4	45.5	57.1
30°	54.1	56.1	50.1	44.4	39.6	35.3
45°	54.7	34.4	44.4	44.4	51.3	38.2
90°	42.4	43.4	47.7	41.4	37.8	18.9

Variation of stress ratio (σ_1/σ_3) with joint orientation β presented in Table 7 for grout thickness 3 mm.

Table 7 : Variation of σ_1/σ_3 with β, grout thickness 3 mm

β	σ_3 (MPa)	σ_1/σ_3 Rock type S1	σ_1/σ_3 Rock type S2	σ_1/σ_3 Specimen composed of S1 & S2
0°	5	25.4	15.9	15.5
	7.5	18.1	12.2	13.2
	10	15.9	11.9	12.4
30°	5	23.4	12.1	15.1
	7.5	19.1	12.1	12.4
	10	16.3	10.8	10.5
45°	5	18.3	12.7	11.3
	7.5	18.8	9.8	9.1
	10	16.6	10.3	10.4
90°	5	19.2	11.0	10.4
	7.5	13.2	10.2	9.2
	10	11.2	10.6	7.5

It is observed that in planar grout jointed rocks in case of rock type S1, anisotropic behaviour is higher as compared to rock type S2. In case of specimen composed of rock types S1 & S2, anisotropy is slightly, reduced.

5 CONCLUSIONS

An attempt has been made in the present study to develop an understanding of strength behaviour in case of planar jointed grouted rocks (with joint orientations at 0°, 30°, 45° and 90° from vertical and grout thicknesses of 3 and 4 mm.)

Following observations have been made from the present study.

1. The physical and strength properties indicate that the sand stones used in the present study are a medium strength rock (S1) that may be classified as CM and low medium strength rock (S2) that may be classified as CL from Deere and Miller (1966) classification.

2. The specimens fail in general by sliding for β=45°, otherwise in general, the failure is by a combination of tensile splitting and shear across the joint at all confining pressure.

3. Specimen composed of two rock types have UCS value in between that of individual rock type, it varies with joint orientation and reduces when grout thickness is increased from 3 mm to 4 mm.

4. The effect on cohesion value is very significant when grout thickness is higher (4 mm) in case of specimen composed of different rock types. The effect on angle of internal friction is marginal.

5. The grout jointed rock has behaviour similar to that of anisotropic rocks.

REFERENCES

Albritton, J.A. 1982. Cement grouting practics. U.S. Army of Engineers. Proc. Conf. Grouting. Geotech. Engg., New Orleans, ASEC, New York, pp. 264-278.

Barla, G., Forlati, F., Bertacchi, P. and Zaninetti, A. 1978. Shear behaviour of natural filled joints. 4th International Congress on Rock Mechanics, Montreal, Canada, pp. 291-299.

Borroso, M. 1970. Cements grouts and their influence on shear strength of fissured rock mass. Proc. Second International Congress on Rock Mchanics, Belgrade Vol. 3, pp 189-195.

Bruce, D.A. 1982. Aspects of rock grouting practice on British dams. Proc. Conf. Grouting. Getotech. Engg., New Orleans, ASCE, New York pp. 301-316.

Coulson, J.H. 1970. The effect of surface roughness on the shear strength of joints in rocks. Ph.D. Thesis, Illinots, Urbans, U.S.A.

Deere, D.U. 1982. Cement-Bentonite grouting for dams. Proc. Conf. grouting. Geotech. Engg., New Orleans, ASCE, New York, pp. 270-300.

Deere, D.U. and Miller, R.P. (1966) Engineering classification and Index Properties for Intact Rocks. Tech. Report No. AFWL-TR-65-116, Air Force Weapons Lab., New Mexico.

Ewert, F.K. 1979. Die untergrundabdichtung am Tavera damm and die Entwicklung des Durchalassigkeitsver haltens nach dem Einstan. Munster Forsch Geol Palaontol 49 : 1-79.

Ewert, F.K. 1985. Rock grouting with emphasis on dam sites. Springer Verlag, Berlin.

Greeco, O.D., A, Ferrero and Peila, D. (1990) Behaviour of laboratory specimens composed by different rocks. Proc. 7th ISRM, p. 241-245.

Grundy, C.F. 1955. The treatment by grouting of permeable foundations for dams. 5th ICOLD Congress, Paris, 1955, O 16, R. 66.

Hassler, L., Stille, H. and Hakansson, U. 1987. Simulation of grouting in jointed rock. Sixth International Congress on Rock Mechanics, Montreal, Canada, pp. 943-946.

Houlsby, A.C. 1982a. Optimum water : Cement ratios for rock grouting. Proc. Conf. Grouting Geotech. Engg. New Orleans, ASCE, New York, pp. 317-331.

Houlsby, A.C. 1982b. Background Talk : Cement grouting for dams. Proc. Conf. Grouting Geotech. Engg. New Orleans, ASCE, New York, pp. 1-34.

Jaeger, J.C. 1969. Rock mechanics and engineering. Univ. Press. Cambridge.

Kunert, N. 1976. Injektionsmittel and Technlogoie bel Verpressungen im Festgestein. Ber Techn Akad Wuppertal 12 pp. 36-44.

Kutzner C. 1982a. Injektionsmittel. Wasser Boden 10 : 444-449.

Kutzner, C. 1982b. Grout mixes and grouting work. Symposium on recent developments in ground improvement Techniques. Bangkok, 29 Nov. 3 Dec. 1982, pp. 288-298.

Sharma, V.M. 1989. Foundation problems, seepage and grouting. Int. Workshop on Research needs in dam safety, 7-14 Feb. 1989, CBIP, New Delhi, pp. 21-25.

Srivastava, R.K., Jalota, A.V., Amir, A.A.A. and Rao, K.S. 1989. Laboratory studies on strength behaviour of grout jointed Shakteshgarh sandstone : National sympossium on rok mechanics, Dec. 4-6, 1989, Roorkee, India.

Vaughan, P.R., 1963. Contribution of discussion, Grouts and drilling Muds in engineering practice. British Geotechnical Society, London, p. 54.

Vishwanathan, B.N. 1988. Strength and deformation behaviour of cement grouted sandstone, M.E. Thesis, IIT, Delhi.

Wittke, W. 1967. Zur Reichweite Von Injektionen im Kliiftigen fels. Felsmech Ing. Geol. Suppl. IV : 79-89.

Zaruba, O. and Mencl V. 1961. Ingeniurgeologic. Akademic-Verlag, Berlin 606 pp.

Analysis of nonlinear stress-strain behavior of intact rock

Moon-Kyum Kim & Phil-Kyu Lee
Department of Civil Engineering, Yonsei University, Seoul, Korea

ABSTRACT : The constitutive model with a single yield surface has been developed for predicting the stress-strain relationship of frictional materials such as clay, sand, and concrete. In this paper, an application of this model is extended to predict the behavior of intact rock subjected to three-dimensional loads. In this model, elasticity theory is employed together with incremental plasticity theory which uses an isotropic hardening and softening rule. These components are all expressed in terms of stress invariants and material parameters. The concept of translated stress space considering the effective cohesion is used for calculating the stress invariants. The material parameters can be determined by the curve-fit with the results of the simple tests such as isotropic compression test and triaxial compression test. Examples of prediction of the intact rock under three-dimensional loading conditions are presented.

1. INTRODUCTION

A constitutive model has been developed on the basis of a through review and evaluation of data from experiments on frictional materials such as sand, clay, concrete (Kim and Lade, 1988; Lade and Kim, 1988a, 1988b). The framework for the evaluation and subsequent development consisted of concepts contained in elasticity and work-hardening plasticity theories.

The model employs a single, isotropic yield surface shaped as an asymmetric tear-drop with the pointed apex at the origin of the principal stress space. This yield surface, expressed in terms of stress invariants, describes the locus at which the total plastic work is constant. The total plastic work (due to shear strains as well as volumetric strains) serves as the hardening of the yield surfaces. The use of constant plastic work contours(or any other measure of hardening) as yield surfaces' results in mathematical consistency in the model, because the measure of yielding and the measure of hardening are uniquely related through one monotonic function. The nonassociated flow rule is derived from a potential surface shaped as a cigar with an asymmetric cross-section.

The model is devised such that the transition from hardening to softening occurs abruptly at the peak points at which the hardening modulus is zero, but the pointed peak is hardly noticeable in actual comparison with experimental data.

The main principles of the model and the governing equations for each component are reviewed below. Values of material parameters are then determined from the tests on intact rock performed by Bieniawski(1969), and predictions of three-dimensional behavior are compared with experimental data for intact rock.

2. SINGLE HARENING STRESS-STRAIN MODEL

The total strain increments observed in a material under the loads are divided into elastic and plastic components such that

$$d\varepsilon_{ij} = d\varepsilon_{ij}^e + d\varepsilon_{ij}^p . \qquad (1)$$

These strains are calculated separately. Both are expressed in terms of effective stresses.

Below is a brief review of the framework and the components of the constitutive model. In order that the presentation follows a logic developmental sequence, the components are presented in the following order : Elastic behavior, failure criterion, flow rule, yielding criterion and work-hardening/softening law.

2.1 Elastic behavior

The elastic strain increments, which are recoverable upon unloading, are calculated from Hooke's law, using a model for the unloading-reloading modulus defined as:

$$E_{ur} = K_{ur} \cdot p_a \cdot \left(\frac{\sigma_3}{p_a}\right)^n \qquad (2)$$

The dimensionless constant values of the modulus number, K_{ur}, and the exponent, n, are determined from triaxial compression tests performed with various values of the confining pressure, σ_3. In Eq. (2), p_a is the atmospheric pressure expressed in the same units as that of E_{ur} and σ_3.

In order to include the effective cohesion and the tension which can be sustained by rock, a translation of the principal stress space along the hydrostatic axis is performed. Thus, a constant stress, $a \cdot p_a$ is added to the normal stresses before substitution

$$\sigma_{ij} = \widetilde{\sigma}_{ij} + \delta_{ij} \cdot a \cdot p_a \qquad (3)$$

in which a is a dimensionless parameter, δ_{ij} is Kronecker's delta, $\widetilde{\sigma}_{ij}$ are stress tensors, and σ_{ij} are translated stress tensors. The value of $a \cdot p_a$ reflects the effect of the tensile strength of the material. In the followings, the translated stress tensors are used throughout.

The value of Poisson's ratio has often been found to be close to 0.2 for the elastic parts of unloading-reloading stress-paths. This value is therefore used in the following calculation.

2.2 Failure criterion

A general, three-dimensional failure criterion has been developed for soils, and concrete. The criterion is expressed in terms of the first and third stress invariants of the stress tensor:

$$\left(\frac{I_1^3}{I_3} - 27\right) \cdot \left(\frac{I_1}{p_a}\right)^m = \eta_1 \qquad (4)$$

in which
$$I_1 = \sigma_x + \sigma_y + \sigma_z \qquad (5)$$
$$I_3 = \det|\sigma_{ij}| \qquad (6)$$

The parameters η_1 and m are constant dimensionless numbers.

In principal stress space, the failure criterion is shaped like an asymmetric bullet with the pointed apex at the origin of the stress axes, and the cross-sectional shape in the octahedral plane is triangular with smoothly rounded edges in a fashion that conforms to experimental evidence.

The three material parameters, η_1, m, and a, may be determined from the results of simple tests such as triaxial compression tests.

2.3 Flow rule

The plastic strain increments are calculated from the flow rule:

$$d\varepsilon_{ij}^p = d\lambda_p \cdot \frac{\partial g_p}{\partial \sigma_{ij}} \qquad (7)$$

in which g_p is a plastic potential function and $d\lambda_p$ is a scalar factor of proportionality. A suitable plastic potential function for frictional materials was developed and presented by Kim and Lade(1988).

This function is different from the yield function and the nonassociated flow is consequently obtained. The plastic potential function is written in terms of the three invariants of the stress tensor:

$$g_p = \left(\psi_1 \frac{I_1^3}{I_3} - \frac{I_1^2}{I_2} + \psi_1\right) \cdot \left(\frac{I_1}{p_a}\right)^\mu \qquad (8)$$

in which I_1 and I_3 are given by Eq.(5) and (6) and the second stress invariant is defined as:

$$I_2 = \tau_{xy} \cdot \tau_{yx} + \tau_{yz} \cdot \tau_{zy} + \tau_{zx} \cdot \tau_{xz} \\ -\left(\sigma_x \cdot \sigma_y + \sigma_y \cdot \sigma_z + \sigma_z \cdot \sigma_x\right) \qquad (9)$$

The material parameters ψ_2 and μ are dimensionless constants that may be determined from triaxial compression tests. The relationship between two parameters, ψ_1 and m is as follows:

$$\psi_1 = 0.00155 \cdot m^{-1.27} \qquad (10)$$

The parameter ψ_1 acts as a weighting factor between the triangular shape (from the I_3 term) and the circular shape (from the I_2 term). The parameter

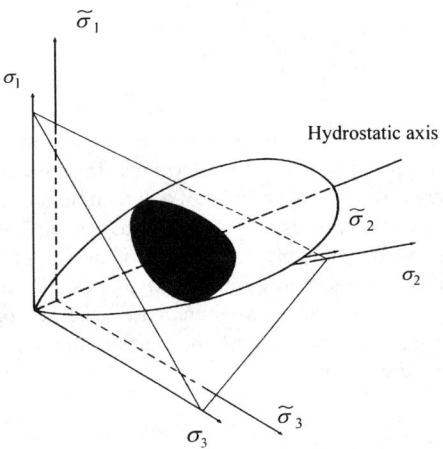

Fig.1. Plastic potential surface in principal

ψ_2 controls the intersection with the hydrostatic axis, and the exponent μ determines the curvature of meridians. The corresponding plastic potential surfaces are shown in Fig. 1. They are shaped as asymmetric cigars with smoothly rounded triangular cross-sections similar but not identical to those for the failure surfaces.

The derivatives of g_p with regard to the stresses are

$$\frac{\partial g_p}{\partial \sigma_{ij}} = \left(\frac{I_1}{p_a}\right)^{\mu} \begin{cases} G - (\sigma_y + \sigma_z) \cdot \frac{I_1^2}{I_2^2} - \psi_1 \cdot (\sigma_y \cdot \sigma_z - \tau_{yz}^2) \cdot \frac{I_1^3}{I_3^2} \\ G - (\sigma_z + \sigma_x) \cdot \frac{I_1^2}{I_2^2} - \psi_1 \cdot (\sigma_z \cdot \sigma_x - \tau_{zx}^2) \cdot \frac{I_1^3}{I_3^2} \\ G - (\sigma_x + \sigma_y) \cdot \frac{I_1^2}{I_2^2} - \psi_1 \cdot (\sigma_x \cdot \sigma_y - \tau_{xy}^2) \cdot \frac{I_1^3}{I_3^2} \\ 2 \cdot \frac{I_1^2}{I_2^2} \cdot \tau_{yz} - 2 \cdot \psi_1 \cdot (\tau_{xy} \cdot \tau_{zx} - \sigma_x \cdot \tau_{yz}) \cdot \frac{I_1^3}{I_3^2} \\ 2 \cdot \frac{I_1^2}{I_2^2} \cdot \tau_{zx} - 2 \cdot \psi_1 \cdot (\tau_{xy} \cdot \tau_{yz} - \sigma_y \cdot \tau_{zx}) \cdot \frac{I_1^3}{I_3^2} \\ 2 \cdot \frac{I_1^2}{I_2^2} \cdot \tau_{xy} - 2 \cdot \psi_1 \cdot (\tau_{yz} \cdot \tau_{zx} - \sigma_z \cdot \tau_{xy}) \cdot \frac{I_1^3}{I_3^2} \end{cases}$$

(11a-f)

where

$$G = \psi_1 (\mu + 3) \frac{I_1^2}{I_3} - (\mu + 2) \frac{I_1}{I_2} + \frac{\mu}{I_1} \psi_2 \quad (12)$$

These derivatives are used to obtain the plastic strain increment from Eq.(7).

Once the parameter ψ_1 is evaluated, the other parameters, ψ_2 and μ can be determined using triaxial compression test data. To do this, the incremental plastic strain ratio is first defined as

$$v_p = -\frac{d\varepsilon_3^p}{d\varepsilon_1^p} \quad (13)$$

Substitution of Eqs.(7) and (11) for the plastic strain increment under triaxial compression conditions $(\sigma_2 = \sigma_3)$ into Eq.(13) produces the following equation:

$$\xi_y = \frac{1}{\mu} \cdot \xi_x - \psi_2 \quad (14)$$

where

$$\xi_x = \frac{1}{1+v_p} \left\{ \frac{I_1^3}{I_2^2}(\sigma_1 + \sigma_3 + 2v_p\sigma_3) + \psi_1 \frac{I_1^4}{I_3^2}(\sigma_1\sigma_3 + v_p\sigma_3^2) - 3\psi_1 \frac{I_1^3}{I_3} + 2\frac{I_1^2}{I_2} \right\} \quad (15)$$

and

$$\xi_y = \psi_1 \frac{I_1^3}{I_3} - \frac{I_1^2}{I_2} \quad (16)$$

Thus, $1/\mu$ and $-\psi_2$ in Eq.(14) can be determined by linear regression between ξ_x and ξ_y determined from several data points.

2.4 *Yield criterion and work hardening/softening law*

The yield surfaces are intimately associated with and derived from surfaces of constant plastic work, as explained by Lade and Kim(1988a). The isotropic yield function is expressed as follows:

$$f_0 = \left(\psi_1 \frac{I_1^3}{I_3} - \frac{I_1^2}{I_2}\right) \cdot \left(\frac{I_1}{p_a}\right)^h \quad (17)$$

where ψ_1 is given by Eq. (10). The exponent h is introduced to simulate the meridional curvature of yield surface.

In order to include work softening behavior, the yield surface should contract after the peak stresses. This can be achieved by using a modified yield function.

$$f_p = f_0 \cdot e^q \quad (18)$$

where the exponential function e^q is incorporated to simulate hardening-softening behavior and q is devised as function of stress level such that:

q = 0   during isotropic compression
q = 1   at failure stresses
0 < q < 1 during hardening
q > 1   during softening

For isotropic compression state Eq. (18) has a value of:

$$f_p = (27\psi_1 + 3) \cdot \left(\frac{I_1}{p_a}\right)^h \qquad (19)$$

The plastic work equation is expressed by

$$W_p = D \cdot p_a \cdot f_p^\rho \qquad (20)$$

where

$$D = \frac{C}{(27\psi_1 + 3)^\rho} \qquad (21)$$

$$\rho = p/h \qquad (22)$$

in which C and p are used to model the plastic work during isotropic compressions:

$$W_p = C \cdot p_a \cdot \left(\frac{I_1}{p_a}\right)^p \qquad (23)$$

The constant parameter h is determined on the basis that the plastic work is constant along a yield surface. Thus, for two stress points, A on the hydrostatic axis and B on the failure surface, the following expression is obtained for h:

$$h = \ln \frac{\left(\psi_1 \cdot \frac{I_{1B}^3}{I_{3B}} - \frac{I_{1B}}{I_{2B}}\right) \cdot e}{27\psi_1 + 3} \cdot \ln \frac{I_{1A}}{I_{1B}} \qquad (24)$$

in which e is the base of natural logarithms.

During work hardening and softening, the level of a stress state may be defined through a normalized function f_n/η_1 nothing that f_n/η_1 varies form 0 at the isotropic compression state to 1 at the failure state.

In order to the variation of q against the stress level for intact rock the following two explicit functions of f_n/η_1 are proposed for q:

$$q = 1 - \left(1 - \frac{f_n}{\eta_1}\right)^1 \quad \text{for hardening} \qquad (25)$$

$$q = \left\{1 + \left(1 - \frac{f_n}{\eta_1}\right)^1\right\} / \left(\frac{f_n}{\eta_1}\right)^1 \quad \text{for softening} \qquad (26)$$

where parameter l is required to define the rate of increase in q and it can be determined from the data at a stress point in the middle of a hardening process.

The yield surfaces are shaped as asymmetric tear drops with smoothly rounded triangular cross-sections and traces in the triaxial plane as shown in Fig.2. As the plastic work increases, the isotropic yield surface inflates until the current stress point reaches the failure surface.

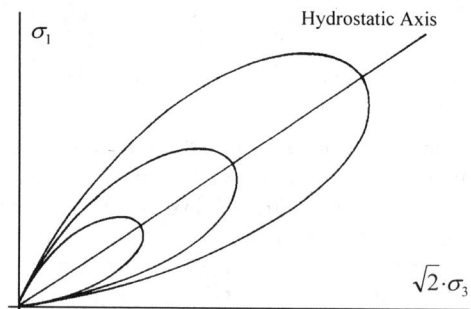

Fig. 2. Yield surface shown in triaxial plane.

The relation between f_p and W_p is described by a monotonically increasing function whose slope decreases with increasing plastic work, as shown in Fig.3.

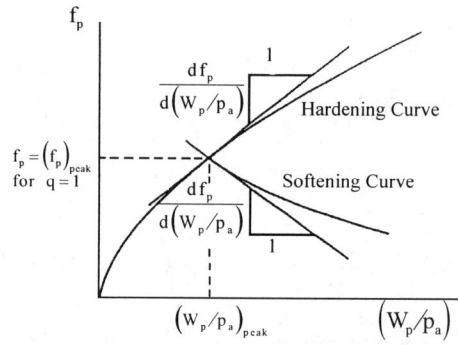

Fig. 3. Modeling of work hardening and softening

Using the expression for the plastic potential in Eq.(8), the relation between plastic work increment and the scalar factor of proportionality $d\lambda_p$ in Eq.(7) may be expressed as:

$$d\lambda = \frac{dW_p}{\mu \cdot g_p} \tag{27}$$

in which the increment of plastic work can be determined by differentiation of the hardening and softening equations.

3. PREDICTIONS FOR INTACT ROCK

The experimental data for intact rock tested by Bieniawski(1969) were used to demonstrate the capacities of the single hardening model.

The required number of parameters for a material with effective cohesion is twelve, i.e., elastic moduli (K_{ur}, n, ν), failure criterion(a, m, η_1), plastic potential(ψ_2, μ), hardening function (C, p) and, yield criterion(h, l). Table 1. summarizes the values of the parameters used in the predictions.

Table 1. Material parameters for intact rock

	Material parameter	Value
Elastic Moduli	K_{ur}	194792
	n	0.0083
	ν	0.2
Failure Criterion	a	36.304
	m	1.258
	η_1	2.009×10^6
Plastic Potential	ψ_2	-4.035
	μ	9.745
Hardening Equation	C	1.373×10^{-6}
	p	1.473
Yield Function	h	1.188
	l	0.37

Fig. 4,5,6,7, and 8 show examples of comparisons between measured and predicted behavior for the triaxial compression tests on intact rock. An extensive application to the intact rock shows that the criterion is effective in modeling the failure(Fig.9). Plastic yielding of the intact rock can be related to the plastic work contours(Fig10). The method used in this study provides straight-forward means for describing yield functions and relating them to the hardening variables. The proposed yield function can handle hardening and softening behaviors of the intact rock with single surface. The behavior under isotropic compressions is also embedded in the consistent yield function. Plastic potential surfaces form similar surfaces in stress space as yield surfaces.

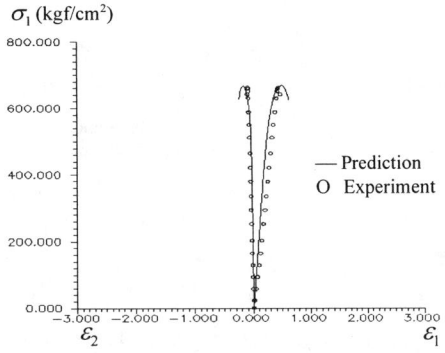

Fig. 4. Comparison of predicted and measured stress-strain curves on intact rock($\sigma_2 = \sigma_3 = 0$) [tested by Bieniawski(1969)]

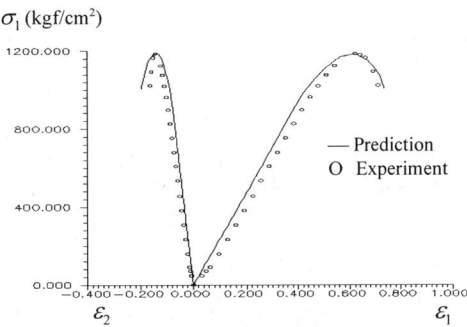

Fig. 5. Comparison of predicted and measured stress-strain curves on intact rock($\sigma_2 = \sigma_3 = 70$ kgf/cm^2) [tested by Bieniawski(1969)]

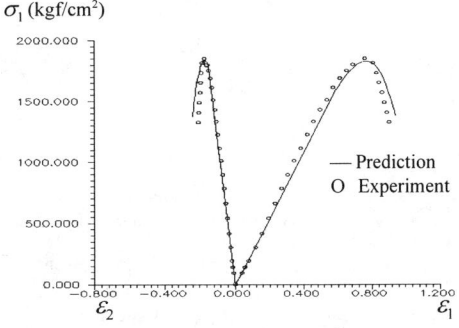

Fig. 6. Comparison of predicted and measured stress-strain curves on intact rock($\sigma_2 = \sigma_3 = 140$ kgf/cm^2) [tested by Bieniawski(1969)]

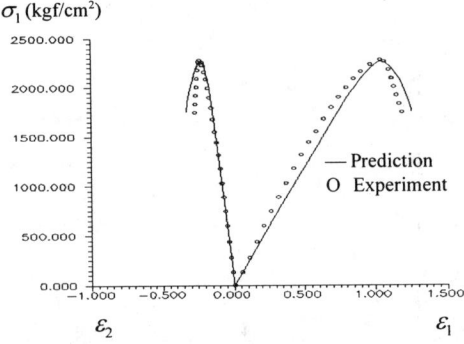

Fig. 7. Comparison of predicted and measured stress-strain curves on intact rock ($\sigma_2 = \sigma_3 = 280$ kgf/cm^2) [tested by Bieniawski(1969)]

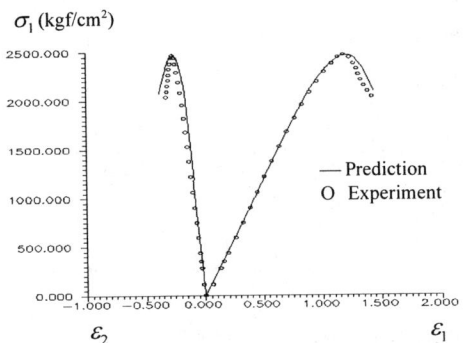

Fig. 8. Comparison of predicted and measured stress-strain curves on intact rock ($\sigma_2 = \sigma_3 = 350$ kgf/cm^2) [tested by Bieniawski(1969)]

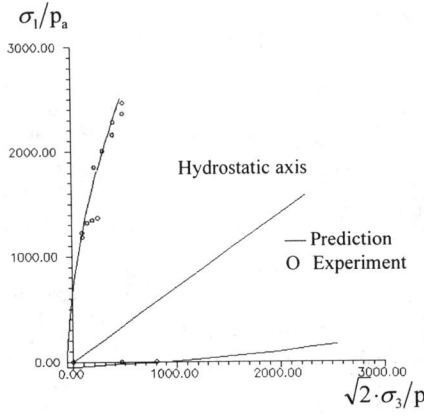

Fig. 9. Comparison of failure criterion in triaxial plane

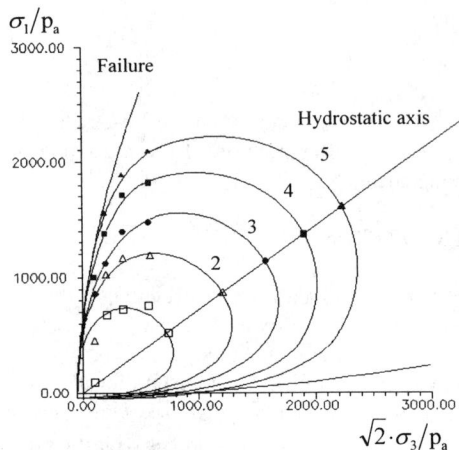

Fig. 10. Analytic contours of plastic work on triaxial plane

4. CONCLUSUSION

Compared with the experimental results, this model predicts accurately the behavior of intact rock subjected to three-dimensional loads. The proposed model, which combines all the governing functions studied, is valid for predicting stress-strain curves for the intact rock.

REFERENCES

Bieniawski, Z. T. 1969. Deformation Behavior of Rock under Multiaxial Compression. *Solid Mechanics and Engineering Design. Vol.1* : 589-598.

Bienawski, Z. T. 1972. Propagation of brittle fracture in rock. *Proceedings 10th Symposium on Rock Mechanics (AIME)*. : 409-427.

Kim, M. K. 1984. *A study of constitutive models for frictional materials* : 1-380.

Kim, M. K. and Lade, P. V. 1988. Single hardening constitutive model for frictional materials, I. Plastic potential function. *Computers & Geotech.* 5 : 307-324.

Lade, P. V. and Kim, M. K. 1988a. Single hardening constitutive model for frictional materials, II. Yield criterion and plastic work contours. *Computers & Geotech.* 6 : 13-29.

Lade, P. V. and Kim, M. K. 1988b. Single hardening constitutive model for frictional materials, III. Comparisons with experimental data. *Computers & Geotech.* 6 : 30-47.

Lee, P. K., 1996. *Analysis of non-linear stress-strain behavior of intact rock using single hardening model* : 1-64

The influence of stress ratio and confining pressure on the weakly cemented sandstone

H.R. Nikraz & M. Press
School of Civil Engineering, Curtin University of Technology, Perth, W.A., Australia

ABSTRACT: This paper reports the finding of a investigation carried out to evaluate the deformation characteristics of saturated sandstone aquifers in the Collie Coal Basin under conditions anticipated during dewatering operations.

A drained triaxial compression test technique for various combinations of confining pressure and back pressure was designed to examine the applicability of the effective stress theory suggested to Terzaghi (1), with particular reference to the Collie sandstone. Further, the results obtained were utilised to characterise the strength and deformability properties of the Collie sandstone.

1 INTRODUCTION

The Collie Coal Basin is located approximately 200 km south-east of Perth in Western Australia. It contains extensive reserves of good steaming coal which is currently being mined by both open cut and underground methods.

The Collie Coalfield has a long history of strata control problems. They manifest themselves in the form of localised poor roof control, surface subsidence, slope instability and mine abandonment (due to a sand-slurry inrush). Major sources of these problems include the very extensive, weak, saturated, sandstone aquifers. As a result, underground operations have been limited to room and pillar extraction, presently carried out by continuous miners and road-heading machines. Approximately 30-40% recovery by volume is being achieved by this method.

In order to increase the recovery to approximately 70%, the Wongawilli method of short-wall mining has been introduced. Caving of the immediate roof is integral with this method. Extensive aquifer de-watering was carried out to enable this mining method to be applied. The porous and weak nature of the aquifers provides a potential source of subsidence (due to pore closure), and strata failure (due to increasing the effective stress), as a result of pore pressure reduction upon de-watering. The proposed development of multiple seam extraction below areas sensitive to surface subsidence has increased the need to establish the strata mechanical properties. This will assist in confident application of rock mechanics principles for predictive modelling of strata behaviour. The importance of a knowledge of the effects of induced stress changes as a result of de-watering a weakly-cemented sandstone aquifer has been previously referred to by Kawecki et al. (2). The two aspects that are of primary interest are (1) deformations produced by changes in the state of stresses of the aquifer as a result of de-watering, and (2) an erosional effect due to the flow of water induced by de-watering, which would tend to exacerbate any detrimental effect of pore pressure reduction.

Although seepage and drainage of both soils and hypothetical media have been investigated theoretically and experimentally by geotechnical engineers, drainage of poorly-consolidated rock has been neglected. There is little reference in the literature to the basic properties of rock deformation caused by migration of rock particles. Furthermore, most failure theories and experimental work are related to non-porous and low-porosity rock types, while major problems during aquifer de-watering relate to relatively weak formations such as loose sand and/or poorly-consolidated sandstone, as is the case in the Collie Basin.

The role of fluid and fluid pressure on the deformation behaviour of rock is often an important consideration in studies of rock mass stability. The influence of pore pressure, particularly that of pore pressure on the strength of rocks has been reported by Handin, et al (3), Robinson (4), Heard (5), Serdengecti, et al (6), Handin, et al (7), Jaeger (8), Vutukuri (9), and others. These investigations have often produced conflicting results and have tended, particularly in the region of less porous rocks, to cause confusion as the validity of the effective stress principle.

A drained triaxial compression test technique for various combinations of combining pressure and back pressure has been designed to examine the applicability of the effective Stress Theory suggested by Terzaghi (1), with particular reference to the Collie Sandstone. Further, the results obtained have been utilised to characterise the strength and deformability properties of Collie Sandstone.

The influence of stress ratio and confining pressure on deformability modulus has been examined and a mathematical function for non-linear stress-dependent characteristics of the Collie sandstone is proposed.

The paper contains a description of the equipment commissioned, test techniques, results, analysis and interpretation of the data obtained.

The testing evaluation techniques are general in nature and can be applied to field situations in locations where similar weak sandstones occur.

2 GEOLOGY

The geological formation, in the Collie Basin is the sedimentation of an impervious igneous basin with the water trapped within the sediments leaving the sediments heavily saturated from basement to water table. The water table lies in some proximity to the river drainage system, and so is reasonably close to the surface.

The groundwater is contained within the Collie Coal Measures, which constitute a multi-layer of sandstone beds separated by confining beds of shale, mudstone and coal beds. Sandstone and grits in the Collie Basin constitute approximately 65 to 75 percent of the coal measures. The water influence has prevented the normal consolidation of the sediments and the sandstone and siltstone are very weak, with little adhesive characteristics. The resultant hydraulic pressure with an absence of consolidation presents a formidable mining problem, which greatly increases with depth.

3 TRIAXIAL EQUIPMENT

The scope of the research required unique triaxial testing of rock. As no commercial system was available, a system with the appropriate capabilities was designed.

An automated data capture system utilising transducers and dynamic recording were designed and commissioned. The overall system was designed to withstand a maximum predicted hydraulic pressure of 14 MPa.

While the equipment developed for use in this and in other associated programmes of research was similar in principle to that described by Bishop and Hankel (10), a number of refinements had to be introduced because of the markedly higher strength and stiffness characteristics of soft rock which required the use of significantly higher testing pressures.

The system developed for this study consisted of six integrated units :

1. A triaxial cell
2. A confining pressure system
3. A pore pressure system
4. A stiff load frame and servo-controlled loading ram
5. Monitoring equipment for applied loads, water flows, and specimen deformation, and
6. Data capture and display system

Full details of the equipment design may be found in Nikraz (11).

4 DRAINED TRIAXIAL COMPRESSION TEST

Thirty two specimens of Collie sandstone were prepared and tested at various combinations of cell pressure and pore pressure to evaluate the applicability of the effective stress theory and also to characterise the strength and deformability properties of the sandstone.

The drained triaxial compression tests consisted of four stages, namely :-

1. specimen mounting in triaxial cell
2. specimen saturation
3. back pressure adjustment
4. load application

Only a brief description of test procedures will be presented herein but full details may be found in Nikraz (11).

After the specimen was mounted in the triaxial cell and saturated with water, a predetermined confining pressure was applied and the back pressure was adjusted to the desired level. Vertical load was only applied after the confining pressure had been adjusted. During the test the confining pressure and the back pressure were maintained at a constant level. The specimen then tested to failure by application of a constant vertical loading rate (3×10^{-4} strain/min).

The back-pressure connected to the drainage outlet, the function of which can be described as to equate levels of the test's pore water pressure to in situ pore water pressure. However, in absence of detailed information with regard to in situ pore water pressure, different levels of pore water pressure and thus, back pressure were adopted.

5 RESULTS ANALYSIS

The stress/strain relationships typical of sandstones tested in drained triaxial compression at various combinations of cell pressure and pore pressure are shown in Figures 1. These curves emphasise the significant increase in triaxial strain at failure as the effective confining stress (σ_3') increases. The variation of axial strain at failure $\varepsilon_{1(F)}$ is shown in Figure 2. This indicates that the maximum differential stress is essentially the same for tests with equal effective confining pressures and increases with an increase in effective confining pressure. Thus, it may be concluded that the effective stress theory is valid for the specimens tested.

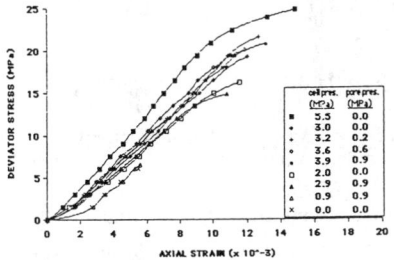

Figure 1 - Typical average stress-strain curves for drained test specimens with the same effective confining pressure.

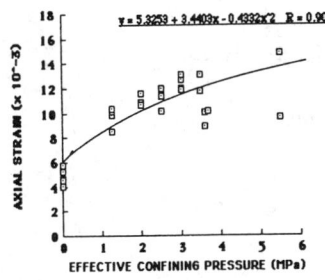

Figure 2 - Variation of axial strain at failure with effective confining pressure.

This theory states that the effective stress is the difference between the total stress and the pore pressure and is the controlling factor influencing frictional strength of rock, other parameters remaining constant.

Comparing the stress/strain behaviour, as shown in Figure 1, with that of strong rocks, as is documented by Jaeger and Cook (12), and Brady and Brown (13), it can be noted that for the sandstone tested no clear transition is apparent due to their increased plasticity. In contrast in competent strata, the transition from elastic to plastic behaviour is readily recognisable.

The source of this plasticity is the non-recoverable strain arising from the closure of pores and fissures and the onset of microfractures with increased stress. The magnitude of plastic deformation depends on rock type, stress level and rock environment (14, 15, 16, 17, 18, 19).

Classification of the rocks according to the method of Deere and Miller (20) showed them to have a lower modulus ratio than that defined for weak rocks (see Figure 3). The modulus ratio is relatively constant (between about 160 and 360) irrespective of the strength or the modulus. This, however, means that the modular ratio is less suitable for characterising and classifying the sandstone in the Collie Basin.

The influence of effective confining stress and deformability modulus is illustrated in Figure 4. The deformability modulus varies considerably with the increase in effective confining pressure (σ_c'). The tangent modulus was obtained from triaxial stress/strain curves at 50% of the peak strength. Volumetric strain was not monitored and therefore it is not known if these values are above the onset of dilatancy.

Furthermore, typical variations of strains with stress ratio, $K = \sigma_3'/\sigma_1'$ under different confining pressures are shown in Figure 5. This indicates a parabolic trend with strain becoming excessive as the stress ratio approaches the failure ratio, K_f. To analyse the data mathematically a function of type :-

$$\varepsilon_{1\,(K)} = a \ln \frac{1}{(K - K_f)} \quad (1.1)$$

was found to be reasonably adequate (Figure 6).

where $\varepsilon_{1(K)}$ = Axial strain at any stress ratio

    K = Stress ratio, $K = \sigma_3'/\sigma_1'$
    K_f = Stress ratio at failure
    σ_1' = Effective axial pressure
    σ_3' = Effective confining pressure
    a = Constant depends on the observed stress/strain characteristics of the rock.

Typical values for 'a' are presented in Table 1.

$$x = \frac{1}{(K - K_f)} \quad (1.2)$$

reduces to :

$$\varepsilon_{1\,(K)} = a \ln x \quad (1.3)$$

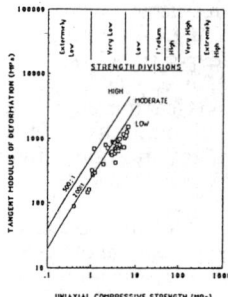

Figure 3 - Summary of strength and modulus data for Collie Sandstone (after Deere and Miller, 1966).

Substitution of the type :

The stress and strain at a point are dependent upon the elastic constants, deformability modulus E and Poisson's Ratio ν, (21) and the symmetry attainable with triaxial apparatus, yields the relationship :

$$E \varepsilon_1 = \sigma_1' - \nu\, 2\, \sigma_3' \quad (1.4)$$

and

$$E \varepsilon_3 = \sigma_3 - \nu\, (\sigma_1' + \sigma_3') \quad (1.5)$$

in which ε_1 and ε_3 = the axial and lateral strains respectively.

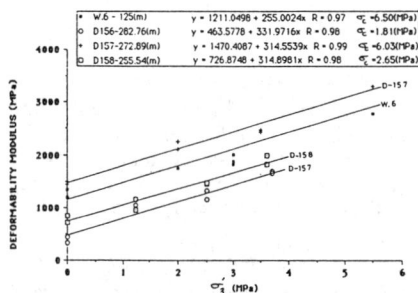

Figure 4 - Variation of deformability modulus with effective confining pressure

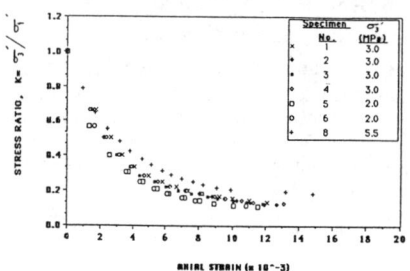

Figure 5 - Typical variation of axial strain with stress ratio (bore hole No. - WD.6 - 125.00)

Rearranging equations (4) and (5) provides :

$$E = \frac{(\sigma_1' + \sigma_3')\, \sigma_1' - 2\, (\sigma_3')^2}{(\sigma_1' + \sigma_3')\, \varepsilon_1 - 2\, \sigma_2'\, \varepsilon_3} \quad (1.6)$$

Equation (1.6) is valid when deformability modulus, E, is material constant. However, as shown in Figure (1) the non-linear relationship between stress and strain for the sandstone tested indicates that deformability modulus is stress-dependent. To account for the non-linear and stress dependent behaviour of the rock, it is assumed that deformability modulus is constant in infinitesimal stress and strain changes and Equation (1.6) can be written as :-

$$E = \frac{(\Delta\sigma_1' + \Delta\sigma_3')\, \Delta\sigma_1' - 2\, (\Delta\sigma_3')^2}{(\Delta\sigma_1' + \Delta\sigma_3')\, \Delta\varepsilon_1 - 2\, \Delta\sigma_3'\, \Delta\varepsilon_3} \quad (1.7)$$

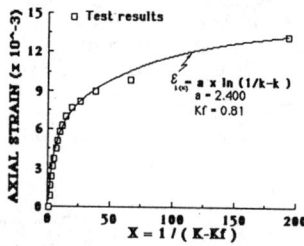

Figure 6 - Example of transformed stress-strain curve to determine parameter "a" (specimen No. 8)

Value σ_3' = constant = σ_0; and failure of the sample with σ_1' increasing, in which σ_0 = initial confining pressure.

Under constant effective confining pressure the term $\Delta\sigma_3'$ is zero and :

$$\Delta\sigma_1' = \sigma_1\, (K + \Delta K) - \sigma_1\, (K) = \frac{\sigma_0}{K + \Delta K} - \frac{\sigma_0}{K} \quad (1.8)$$

Rearranging equation (1.8) yields :

$$= - \frac{\sigma_0\, \Delta K}{K\, (K + \Delta K)} \quad (1.9)$$

Also the increment in axial strain, $\Delta\varepsilon_1\,_{(K)}$ can be written as :

$$\Delta\varepsilon_1\,_{(K)} = + \varepsilon_1(K + \Delta K) - \varepsilon_1\,(K) \quad (1.10)$$

From Equations (1.7), (1.9) and (1.10), and using differential difference equation technique :

$$E = - \frac{\sigma_0}{K^2}\, \frac{1}{\varepsilon'_{1(K)}} \quad (1.11)$$

Table 1 - Parameters 'a' and 'K_f' for W6 - 125 specimens

Specimen No.	Constant Effective Confining Pressure, $\sigma_3 - \sigma_o$ (MPa)	a	K_f
1	3.0	2.6689	0.131
2	3.0	2.6188	0.122
3	3.0	2.5462	0.130
4	3.0	2.6091	0.126
5	2.0	2.3100	0.109
6	2.0	2.2390	0.188
8	5.5	2.4500	0.181

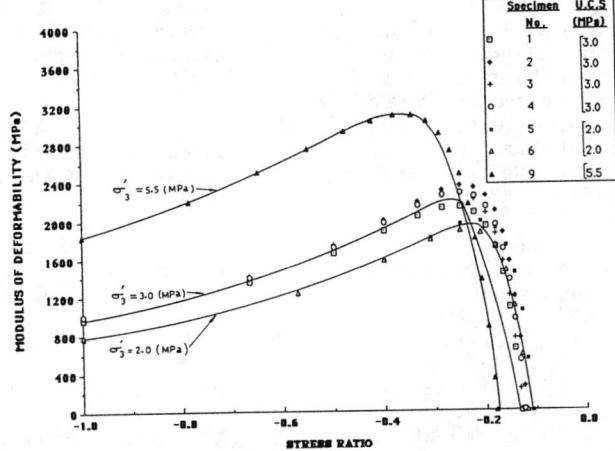

Figure 7 - Variation of modulus of deformability with stress ratio, $k = \sigma_3'/\sigma_1'$.

in which $\varepsilon'_{1(K)}$ is the first derivative of $\varepsilon_{1(K)}$.

From Equation (1):

$$\varepsilon'_{1(K)} = -\frac{a}{\bar{x}} \qquad (1.12)$$

in which $\bar{x} = (K - K_f)$.

Values ε_1 expressed as percentage and σ_0:

$$E = \frac{\sigma_0 \bar{x}}{K^2} \frac{1000}{a} \qquad (1.13)$$

The variation of deformability modulus of W6 - 125 specimens with stress ratios for various confining pressures are shown in Figure (7). The deformability modulus, E, increases at or near, K = 1, to its peak value and reduces to zero at the stress ratio, $K = K_f$. Figure (7) indicates that the peak deformability values are stress ratio dependent.

More studies are necessary to confirm these results for other types of weak sandstones. However, deformability modulus obtained from a uniaxial or triaxial compression tests must be regarded with reservation.

6 CONCLUSIONS

The results, analysis and discussions derived in the course of this study allow the following conclusions to be drawn.

(a) A drained triaxial compression test at various combinations of confining pressure and back pressure has been developed to attempt to characterise the deformability properties of the Collie sandstone.

(b) Test results indicated that the effective stress theory as given by the original Tarzaghi Equation (1.1) is applicable to the Collie sandstone.

(c) The influence of stress ratio and confining pressure on deformability modulus has been examined and a mathematical function for the non-linear stress-dependent characteristics of the Collie Sandstone is proposed (Equation 1.13):

$$E = \frac{\sigma_c \bar{x}}{K^2} \frac{1000}{a} \qquad (1.13)$$

REFERENCES

1. Terzaghi, K. "The shearing resistance of saturated soils", Proc. 1st Inst. Conf. Soil Mechanics, Vol. 1 (1936) 54-56

2. Kawecki, M.A., Evans, A.W. & Nikraz, H.R.: "The Influence of Pore Pressure Variation on Sandstone Behaviour, and its Contribution to Subsidence and Strata Behaviour Interpretation in the Collie Basin", Western Australian Mining and Petroleum Research Institute - Project 74, Final Report, June 1988

3. Handin, J. and Hager, R.V. : "Experimental Deformation of Sedimentary Rocks under Confining Pressure : Tests at Room Temperature on Dry Samples", Bull. Am. Ass. Petrol. Geol., 41 (1957) 1-50

4. Robinson, L.H. : "Effect of Pore and Confining Pressure on the Failure Process in Sedimentary Rocks", Colo. Sch. Mines, 54, (1959), 177-199

5. Heard, H.C. : "Transition from Brittle to Ductile Flow in Solenhoften Limestone as a Function of Temperature, Confining Pressure and Interstitial Fluid Pressure in Rock Deformation", Geol. Soc. Amer. Mem. 79, (1960), 193-226

6. Serdengecti, S., Boozer, G., and Hiker, K.H. : "Effects of Pore Fluids on the Deformation Behaviour of Rocks Subjected to Triaxial Compression", Proc. fifth symposium on Rock Mechanics, Univ. of Minnesota, Pergamon, (1962), 579-625

7. Haudian, J. Hager, R.V., Frideman, M. and Feather, J.N. : "Experimental Deformation of Sedimentary Rock under Confining Pressure : Pore Pressure Tests", Bull. American Assoc. Petrol. Geol., 47, (1963), 717-755

8. Jaeger, J.C. : "Brittle Fracture of Rocks", 8th symp. Rock Mech. on Failure and Breakage of Rock, Univ. Minnesote, (1966) 3-57

9. Vutukuri, V.S., Lama, R.D., and Saluja, S.S. : "Handbook of Mechanical Properties of Rocks", Vol. I-IV, Trans, Tech. publication (1974).

10. Bishop, A.W. & Henkel, D.O. : "The Measurement of Soil Properties in the Triaxial Test", 2nd Ed., Edward Arnold, London (1962) 227

11. Nikraz, H.R. : "Laboratory Evaluation of the Geotechnical Design Characteristics of the Sandstone Aquifers in the Collie Basin", PhD Thesis, Curtin University of Technology, (1991) 317

12. Jaeger, J.C. & Cook, N.G.W. : "Fundamentals of Rock Mechanics", publ. London, Champman & Hall, 2nd ed., (1976), 585

13. Brady, B.H.G., & Brown, E.T., "Rock Mechanics for Underground Mining", George Allen & Unwin, London, (1985), 527

14. Brace, W.F. : "Some New Measurements of Linear Compressibility of Rock", J. Geophys. Res., Vol. 70, (1965) 391-398

15. Walsh, J.B. "The Effects of Cracks on the Compressibility of Rock", J. Geophys. Res. Vol. 70, 2, (1965) 381-399

16. Simmons, G., Siegfried, R.W., & Feves, M. : "Differential Strain Analysis : A New Method for Examining Cracks in Rocks", J. Geophys. Res., Vol. 79, (1974) 4383-4385.

17. Podneiks, E.R., Chamberlain, P.G., & Thill, R.E. : "Experimental Effects on Rock Properties", Basic & Applied Rock Mechanics, (1968) 1275-1314.

18. Van Eeckhout, E.M. and Pong, S.S. : "The Effect of Humidity on Compliances of Coal Mine Shales", Int. J. Rock Mech. Min. Sci. Vol. 12, No. 11, (1975) 335-340

19. Priest, S.D. & Selvakumar, S. : "The Failure Characteristics of Selected British Rocks", Transport and Road Research Lab. Report, (1982) 189

20. Deere, D.V. & Miller, R.P. : "Engineering Classificiation and Index Properties of Intack Rock", Technical Rerport NAFNL-TR-65-116, US Air Force Weapons Lab, Kirkland Air Force Base, New Mexico (1966)

21. Timoshenko, S & Goodier, J.N. : "Theory of Elasticity" Mc-Graw Hill Book Co. Inc., New York, (1951)

Strength and geophysical behaviour of metagraywacke rock

C.S.Gokhale & J.M.Kate
Civil Engineering Department, Indian Institute of Technology, New Delhi, India

A.M.Deshmukh
Civil Engineering Department, Government Engineering College, Farmagudi, Goa, India

ABSTRACT : Metagraywacke which occurs in many parts of India is derived from metamorphism of graywackes which has complex matrix resulting from different processes of formation. These rocks are encountered during various civil engineering activities, tunnelling, mining and quarrying operations and are also used as construction material. This paper briefly presents experimental data on geotechnical and geophysical characteristics of metagraywacke rock collected from the state of Goa in India. The geophysical behaviour is assessed through elastic wave velocity and electrical resistivity and geotechnical behaviour through uniaxial compressive strength, Brazilian tensile strength and punch shear strength. The relationships between geophysical and geotechnical properties are evaluated for field core samples. The results presented in the paper may be employed in the rational interpretation of geophysical sounding data for extracting certain engineering parameters required for analysis and design of underground construction in similar rock formations.

1 INTRODUCTION

Graywacke rocks occur prominently in India. Such rocks form one-fifth to one-fourth of all sandstones and are most common in the Paleozoic and earlier orogenic belts (Pettijohn, 1984). The matrix of graywacke gives evidence of more than one process of origin. It is basically sedimentary rock. The graywacke gone under process of metamorphism is commonly called metagraywacke. These rocks are widely encountered in mining, tunnelling and civil engineering activities and also used as construction material.

The proper assessment of nature and engineering behaviour of such rocks is essential for rational analysis, design, construction and stability to understand workings in such rocks for structures, excavation and quarrying materials. The successful implementation of numerous construction projects in metagraywacke deposits requires their characterization under in-situ stress and environmental conditions.

The geophysical methods namely seismic wave velocity and electrical resistivity are increasingly being employed in geotechnical investigations. These geophysical techniques are non-destructive, rapid and highly cost effective when used specially for subsurface exploration for large area coverage. The applicability for variety of subsurface conditions and encouraging results of these investigations have attracted the investigators to study together strength and geophysical parameters of rocks and rock formations, so as to understand interrelationships between them.

To understand strength and geophysical behaviour together, normally laboratory investigations are conducted on intact rock specimens obtained from rock blocks through coring in laboratory. The specimens obtained through field drilling of boreholes are more representative of in-situ conditions, then the laboratory drilled intact specimens.

In view of the above in the present study an attempt has been made to understand strength and geophysical behaviour of metagraywacke rock specimens obtained from the field.

2 EXPERIMENTAL INVESTIGATIONS

2.1 *Sample collection and specimen preparation*

The rock samples for present study were obtained from the vicinity of Panaji, Goa, India, in the form of cylindrical cores of diameter of about 54 mm and length larger than 110 mm from various boreholes with depth ranging from 22 to 30 m. These boreholes were located in an area of about 3500 m² near Rua-De-Orem creek. The study of the cross-sections through various boreholes and continuous profiles revealed that rock was dipping towards South-East. The predominant type of rock encountered here is metagraywacke underlain by meta-basalt.

The rock cores obtained from various boreholes were cut and finished as per ISRM tolerance limits to achieve length to diameter ratio of 2.0, 0.50 and 0.25 for uniaxial compression tests, Brazilian tensile strength tests and punch shear tests respectively.

2.2 Tests conducted

The macro and micro features of the rock specimens were studied carefully. The physical properties such as dry and saturated unit weights, water absorption, and porosity were determined for all specimens. The specific gravity and Schmidt rebund number for rock were also determined. The compressional and shear wave velocities and uniaxial compressive strength were obtained both for dry and saturated specimens. The electrical resistivity for all specimens was determined in saturated condition. The Brazilian tensile strength and puch shear strength test were conducted on dry specimens.

The punch shear strength tests were conducted using specially designed and fabricated device as used by Kate (1993). The arrangement adopted for measurement of electrical resistivity was similar to that by Kate and Rao (1989). All other tests were conducted as per ISRM suggested methods (Brown, 1981).

3 RESULTS AND DISCUSSIONS

3.1 Rock quality designation

On the basis of borelog data from 12 boreholes the value of rock quality designation, RQD (Deere, 1968) was estimated which varies from 10 to 76%. The corresponding value of core recovery, CR was observed to range from 29 to 95%. The Figure 1 shows the plot of RQD with CR. As can be seen from this figure the points can be represented by two lines A and B. The regression analysis yields the following equations for line A and B with

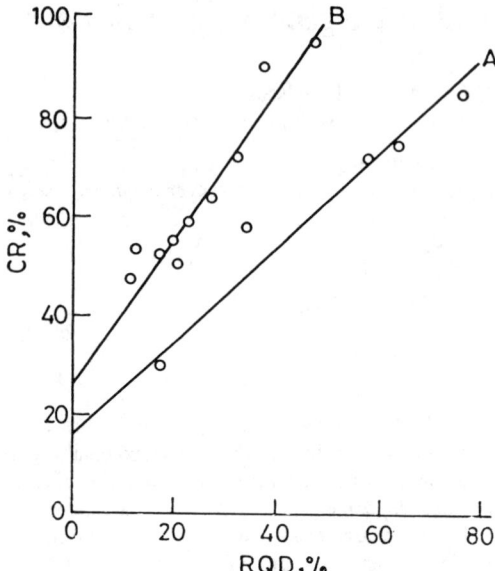

Figure 1. Variation of rock quality designation with core recovery.

coefficient of correlation, r, 0.968 and 0.960 respectively.

For line A   RQD = 1.08CR - 18.9   (1)

For line B   RQD = 0.65CR - 17.1   (2)

This moderate variation in RQD-CR relationship for line A and B may be due to the fact that the data of line A pertains to Boreholes on North-East part of the site where in a creek runs parallel to NW-SE direction. The deposits on this part might have been strengthened due to deposition of sedimentary cementing materials having access through creek. This is supported by the close examination of the samples from the boreholes located on remaing part (line B) of the site were observed to have more micro fractures compared to those from North-East part.

3.2 Physical characteristics

The ash-grey hard compact fine to medium grained Goa metagraywacke has specific gravity of 2.82. As reported by Katti et al. (1996) this rock type contains around 50% of SiO_2, 17% of Al_2O_3, 14% of Fe_2O_3 and rest other oxides. Visual observations revealed that some of the specimens used for uniaxial compression test were having micro fractures, where as all the specimens used to determine Brazilian tensile strength and punch shear strength were almost intact. The dry unit weight varied from 25.473 to 27.027 KN/m³ and that saturated unit weight from 25.481 to 27.078 KN/m³. The porosity varied in a narrow range of 0.796 to 1.092%. The corresponding range of water absorption obtained was 0.288 to 0.486%. The average value of Schmidt rebound number was around 44 for dry specimens.

3.3 *Geophysical behaviour*

3.3.1 *Elastic wave velocities*

The compressional wave velocity, Vp, for dry and saturated specimens varied from 2.91 to 3.96 km/s and 2.93 to 4.02 km/s respectively. The shear wave velocity, V_s, for dry specimens varied from 1.63 to 2.12 km/s and that for saturated specimens from 1.64 to 2.14 km/s. The plot of V_p with V_s presented in Figure 2, shows linear relationship which can be expressed by the following best fit equation with r equal to 0.948.

$$V_s = 0.51 \ V_p + 0.17 \quad (3)$$

The dynamic modulus of elasticity E_D, calculated on the basis of V_p, V_s and rock density varies from about 18.62 to 32.92 GPa, while dynamic Poisson's ratio ranged from 0.246 to 0.302.

3.3.2 *Electrical resistivity*

The electrical resistivity, ρ_s, for saturated specimens ranges from 5.4 x 10³ to 10.8 x 10³ Ohm.m. The resistivity of saturating water was about 10.3 Ohm.m. The Figure 3 and 4 presents variation of formation factor, F, (ratio of rock resistivity to the resistivity of pore fluid) with V_p and V_s respectively. Although the points are scattered a trend of increase in both V_p and V_s with increase in F can be discerned.

The variation of ρ_s, with acoustic impedance, Z, (product of compressional wave velocity and density) is illustrated in Figure 5 for saturated specimens. This figure shows the linear relationship between ρ_s and Z with moderate value of coefficient of correlation of 0.74. The equation of relationship is

$$\rho_s = 1.05Z + 1.22 \quad (4)$$

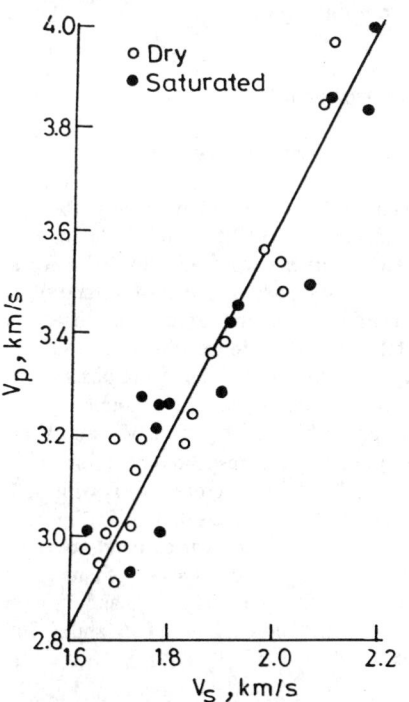

Figure 2. Interdependence of compressional wave velocity and shear wave velocity.

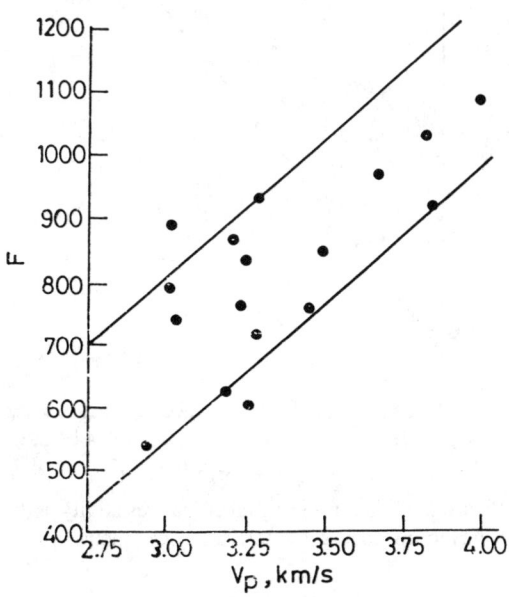

Figure 3. Variation of formation factor with compressional wave velocity.

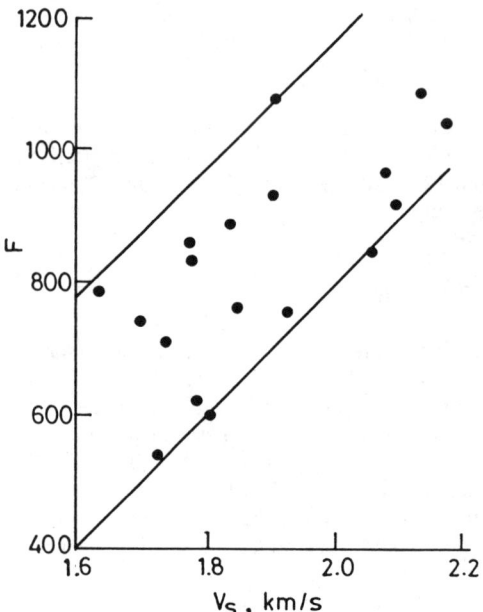

Figure 4. Variation of formation factor with shear wave velocity.

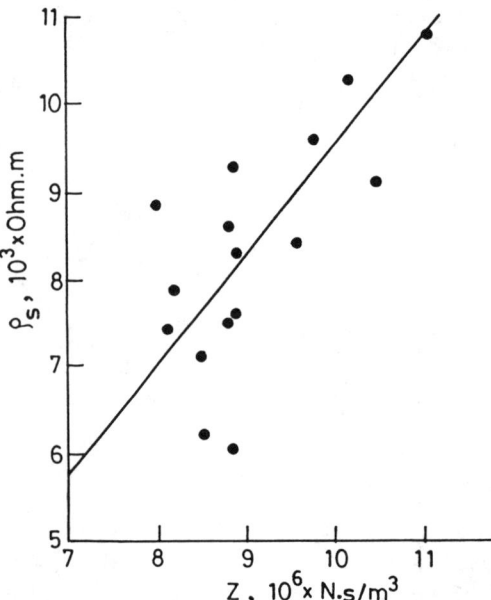

Figure 5. Relationship of electrical resistivity with acoustic impedence for saturated specimens.

The above observation indicate that Z which takes into account rock density may be a better parameter to relate two geophysical properties.

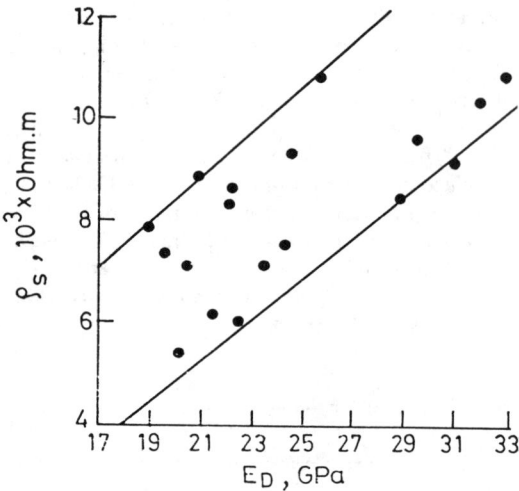

Figure 6. Dependence of dynamic modulus of elasticity on electrical resistivity for saturated rock.

The dependence of dynamic modulus of elasticity with resistivity for saturated specimens is presented in Figure 6. A linear increase in dynamic modulus of elesticity with increase in resistivity can be discerned from this figure.

3.4 Strength behaviour

3.4.1 Uniaxial compressive strength

The uniaxial compressive strength, σ_c, for dry specimens ranges from 30.2 to 120.8 MPa and that for saturated specimens from 45.2 to 101.6 MPa. This wide range of σ_c and also that of V_p and ρ_s, is due to the fact that the specimens used in experimental work were those obtained from the field through borehole drilling and were observed to contain micro fractures in varying amount.

The correlationships of σ_c with V_p and V_s is shown in Figure 7 and 8 respectively. From these figures a general trend of increase in σ_c with both V_p and V_s can be observed. Although in Figure 7 and 8 the relationship is shown as band covering most of points, the regression analysis of data gives the following best fit equations for V_p and V_s with coefficient of correlation being 0.830 and 0.869 respectively.

$$\sigma_c = 63.87\ V_p - 140.13 \qquad (5)$$

$$\sigma_c = 127.99\ V_s - 167.07 \qquad (6)$$

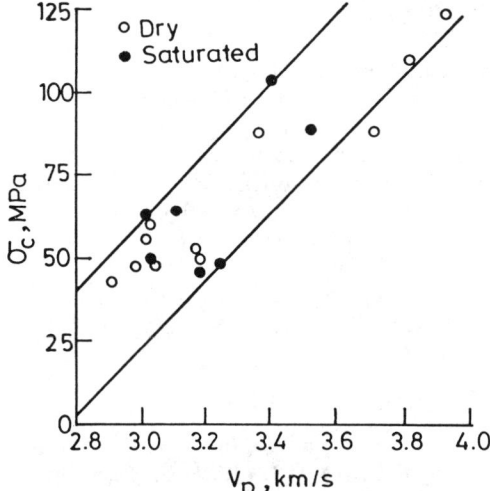

Figure 7. Correlationship of uniaxial compressive strength with compressional wave velocity.

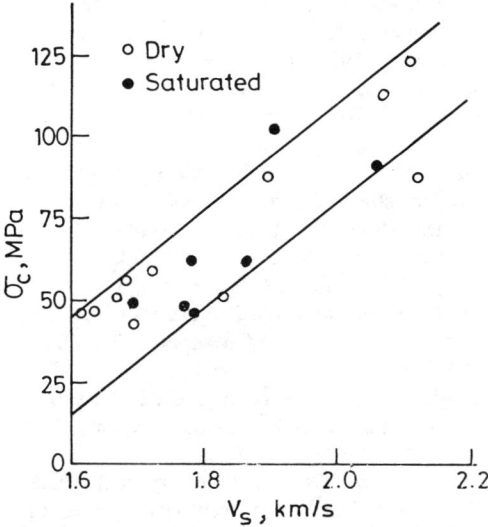

Figure 8. Correlationship between uniaxial compressive strength and shear wave velocity.

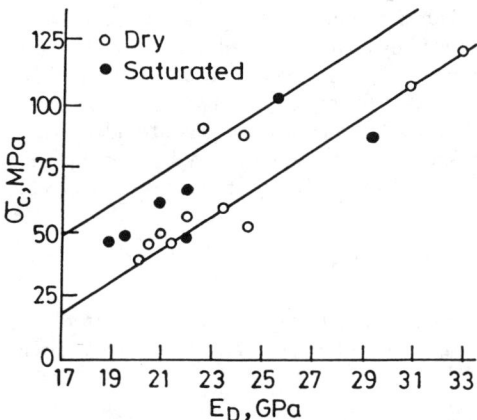

Figure 9. Inter-relation between dynamic modulus of elasticity and uniaxial compressive strength.

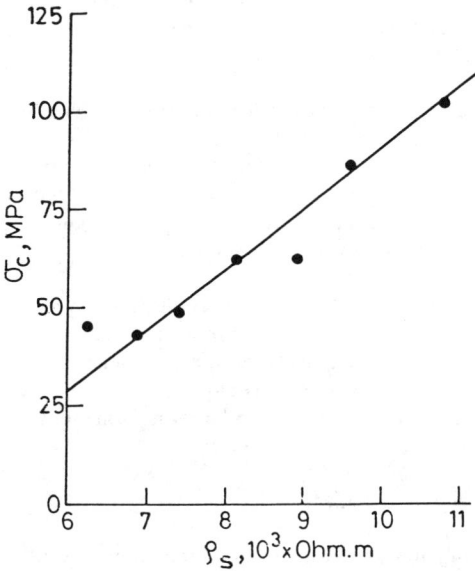

Figure 10. Correlation of uniaxial compressive strength with electrical resistivity for saturated metagraywacke.

The correlationship of E_D with σ_c as shown in Figure 9 again indicate linearly increasing band. The following best fit equation with high coefficient of correlation, r, of the order of 0.84

$$\sigma_c = 6.67\ E_D - 81.63 \qquad (7)$$

The Figure 10 which presents the variation of σ_c with ρ_s for saturated specimens indicate linear increase in σ_c with increase in ρ_s. Mutgi (1989) reported similar trend for Magod graywacke. The relationship between σ_c and ρ_s has following expression with very high valve of coefficient of correlation of order of 0.939.

$$\sigma_c = 13.21\ \rho_s - 46.26 \qquad (8)$$

3.4.2 *Tensile and shear strength*

The tensile strength which is of relevance for

analysis and design of underground workings was determined by commonly used indirect method i.e. Brazilian tensile strength test. It varies from 16.38 to 22.26 MPa for dry specimens. The punch shear which is used in stability analysis of underground mine support pillers was observed to range from 22.35 to 30.56 MPa for dry specimens of this rock.

4 CONCLUSIONS

Based on present study on cores obtained from the field drilling, following conclusions have been arrived at.

1. The study shows that the ash-gray compact, fine to medium grained metagraywacke which has low, narrow range of porosity and water absorption exhibits moderate range of elastic wave velocity and comparatively wide range of electrical resistivity and uniaxial compressive strength.

2. The two geophysical characteristics viz. elastic wave velocity and electricl resistivity which are observed to be interdependent exhibit good correlationship with uniaxial compressive strength, even for these cores from the field containing varying amount of micro fractures.

The knowledge of such relationships may prove to be very useful not only for rock characterization but also for extracting the design parameters from field geophysical soundings alone, minimising the expensive in-situ tests in similar rock formations.

ACKNOWLEDGEMENTS

The authors express their sincere gratitude towards Mr. A.V. Karindikar, Senior Engineer, DESCON Company, Panaji, Goa, India and Mr. P.R. Mutgi, Assistant Professor, Department of Civil Engineering, Government Engineering College, Farmagudi, Goa, India for providing rock samples, field data, encouragements and valuable suggestions. The help rendered by the staff of Rock mechanics laboratory Civil Engineering Departments of Government Engineering College, Goa and Indian Institute of Technology, New Delhi is gratefully acknowledged.

NOTATIONS

CR   Core Recovery, %
E_D Dynamic Modulus of Elasticity, GPa
F   Formation Factor
RQD   Rock Quality Designation, %
r   Coefficient of Correlation
V_p Compressional wave velocity, km/s
V_s Shear Wave Velocity, km/s
Z   Acoustic Impedence, Ns/m^2
ρ_s Saturated Electrical Resistivity, ohm.m
σ_c Uniaxial Compressive Strength, MPa

REFERENCES

Brown, E.T.(ed.) 1981. Rock characterization testing and monitoring : ISRM suggested methods. Pergamon press, Oxford, U.K.

Deere, D.U. 1968. Geological considerations. In : K.G. Stagg and O.C. Zienkiewiez (eds.), Rock Mechanics in engineering practice. John Wiley & Sons, London, U.K., Ch.2, pp. 1-20.

Kate, J.M. 1993. Geotechnical behaviour of intact rock specimens. *Proc. Int. Symp. on Hard Soils and Soft rocks*. Athens, Greece, Vol.I, pp. 569-573.

Kate, J.M. and Rao, K.S. 1989. Effect of large overburden stress on geophysical behaviour of sandstones. *Proc. Int. Symp. on Rocks at great depth.*, Pau, France, Vol. 1, pp. 171-178.

Katti, R.K., Katti, A.R. and Khatri, V.S. 1996. Properties of tropical residual soil series with depth. *Proc. Indian Geotech. Conf.* Madras, India, Vol. 1, pp. 154-157.

Mutgi, P.R. 1989. Geotechnical Characteristics of graywacke. M.E. dissertation, Goa University, Goa, India, pp. 124-154.

Pettijohn, F.J. 1984. Sedimentary rocks. CBS Publishers and distributors, New Delhi, India, pp. 195-259.

Direct observation of progressive microcrack development in relaxation tests on granites and their main component minerals

Y.S.Seo, N.Fujii & Y.Ichikawa
Department of Geotechnical and Environmental Engineering, Nagoya University, Japan

ABSTRACT: The micro-structure of rock plays an important role for its mechanical properties. Recently, we developed a new relaxation testing equipment. By using water-saturated granite specimens, we observed microcrack development. In this paper, the relationship between relaxation behavior and microcrack development is discussed.

1 INTRODUCTION

The long-term behavior of granitic rocks at shallow crustal depth is a crucial property for various problems such as radioactive waste isolation and earthquake hazards reduction. The creep under constant stress and the relaxation under constant strain are typical time dependent phenomena of rock, and through these we can guess the true stress-strain behavior of rock or rock mass.

The relaxation test has ever been performed scarcely in contrast to the creep test. Peng et al.(1972) carried out the relaxation tests by using a servo-controlled hydraulic testing machine and obtained stress relaxation curves. Haupt(1991)used a microprocessor-supported closed loop control system with a digital strain gauge for relaxation test.

The change of microcracks plays an essential role to predict the long-term behavior of rock. In order to observe the state of micro-structures, we developed a new relaxation testing equipment which enables us to take photographs of the specimen during the test. Series of relaxation tests have been performed with microscopic observation by using the above mentioned equipment in a constant-temperature room.

2 EXPERIMENTAL PROCEDURES

2.1 Specimen preparation

The specimens used are two kinds of granite, that is, coarse- and fine-grained types, produced in Inada, Japan. Specimens are made from these two types of granite. The composed minerals of the coarse-grained granite are quartz; 39.7%, feldspar; 51.5%, and biotite; 8.7%, and the minerals of the fine-grained are similarly quartz; 33.7%, feldspar; 62.5% and biotite; 4.7%. They involve three mutually perpendicular cleavages, that is, rift-, grain- and hardway-planes. As shown in Figure 1, the specimens called as the No.1, No.2 and No.3 types are prepared parallel to the grain-, hardway- and rift-planes, respectively. Furthermore, as mensioned above, we provided two kinds of granite, that is, the coarse- and fine-grained granites, so we call the coarse-grained No.1 specimen as C-1 type, the fine-grained No.1 specimen as F-1 type, and so on.

The size of the specimens are $40 \times 20 \times 5mm$. All of the samples are prepared to ensure a surface tolerance of 4/1000mm concerning parallelism and perpendicularity of the faces. The observing surface is polished by $1000gr$. emery powder.

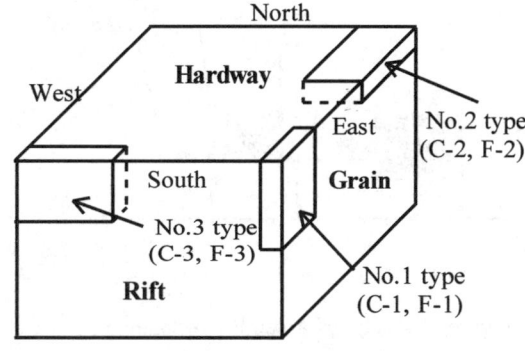

Figure 1. Specimen preparation for laboratory test.

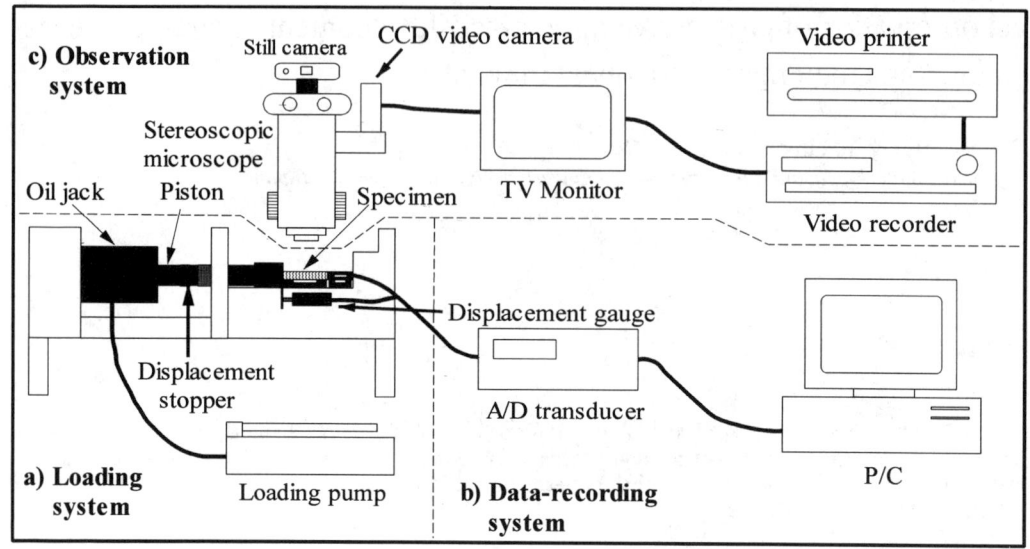

Figure 2. Schematic diagram of experimental system.

2.2 Experimental system

To observe the time-dependent propagation of microcracks and its affect to the macroscopic response of the rock material under stress relaxation state, we developed a new testing system which consists of three subsystems a)the loading system, b)the data-recording system and c)the observation system (Figure 2).

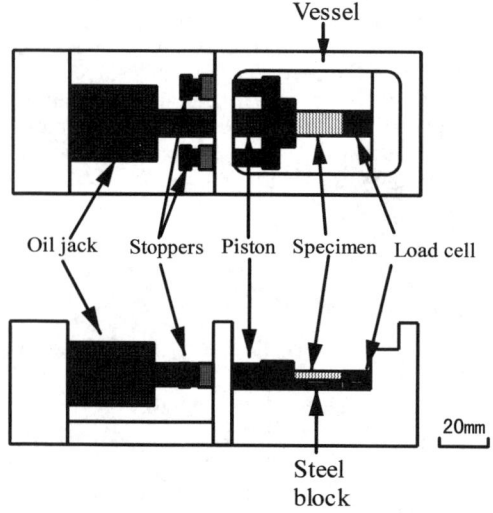

Figure 3. Each part assembly of loading system.

The loading system consists of a vessel, an oil jack, a piston, a loading platen, a load cell and two stoppers (Figure 3). The main difficulty in developing the loading unit arises from keeping constant axial deformation. For overcoming this, two screw-stoppers are introduced for fixing the loading platen. We provide a load cell on which two parallel strain gauges are affixed to make sure balance of the loadings.

The data-recording system consists of an A/D transducer and a personal computer.

The observation system consists of a stereoscopic microscope(Nikon, SMZ-U) with magnifying power of 110 times, a still camera, a CCD video camera, TV monitor, a video recorder and a video printer.

2.3 Relaxation test procedure

The temperature change seriously affects the relaxation behavior. To eliminate this effect, the experiments are performed in a constant temperature room under 20 ± 0.5 °C. The specimen used for the test is saturated for a week in a vacuum container.

The testing procedure is as follows: First, a specimen is placed on a concave-shaped steel block which is put on the bottom surface of the vessel. Next, a load is applied to the specimen at the velocity of 0.2MPa/sec by using a manually controlled pump. When a specified stress level of the specimen is reached, the position is fixed by the two stoppers. Then, the relaxation test is started. The time-dependent propagation of microcracks is observed by the observation system.

Six kinds of tests were performed at several different stress levels and eleven relaxation test results were obtained. One test required about one week.

3 EXPERIMENTAL RESULTS

3.1 Uniaxial Strength Test

Before the relaxation test started, we checked the uniaxial strength of 5 samples for each type of specimens by using the same apparatus under the same conditions as the relaxation test. Figure 4 shows the distribution and the mean value of uniaxial test results of each specimen type. C-2 and F-2 samples show about 15MPa higher uniaxial strength than other ones. The reason may be that C-2 and F-2 samples are made parallel to the hardway plane.

Based on the mean values of the uniaxial strength, the stress levels of relaxation tests are determined. That is, the relaxation test with the stress level of 50% implies that we apply the 50% load of the mean uniaxial strength at t=0, then fix the displacement.

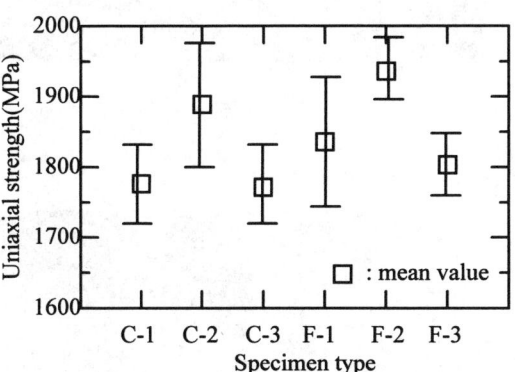

Figure 4. Distribution of uniaxial strength(saturated Inada granite).

3.2 Relaxation Test

Stress relaxation tests have been performed at 50 - 55% stress level (Figure 5) and at 70 - 75% stress levels (Figure 6) by using each type of samples. We can observe that all of the specimens show rapid decreasing of the stress just after the start of experiments, and gradually the stress converges to a constant level.

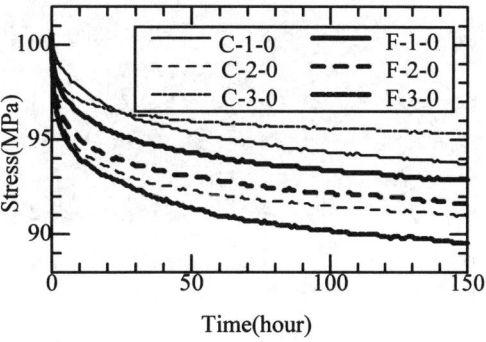

Figure 5. Relaxation curves at 50% - 55% stress level.

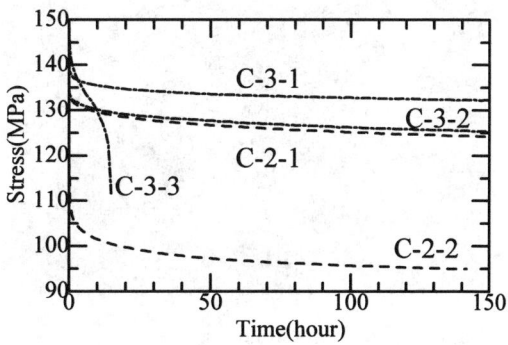

Figure 6. Relaxation curves at several stress level.

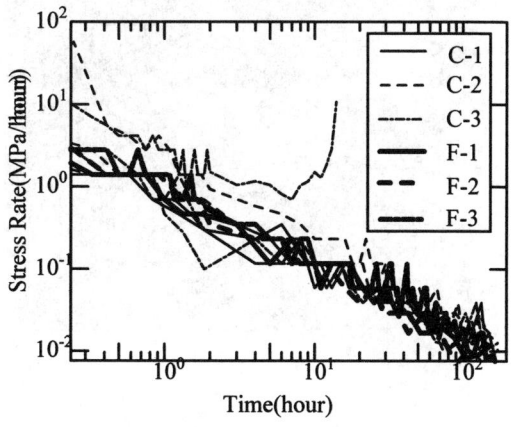

Figure 7. The rate of relaxation stress.

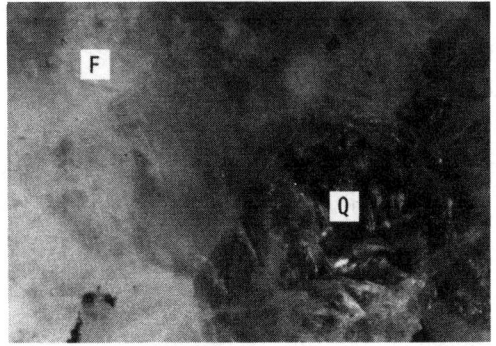

Figure 8(1). Photo micrograph of C-2-2 at t=0 hour in relaxation test.

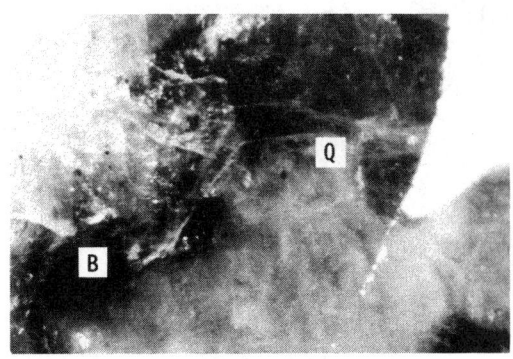

Figure 9(1). Photo micrograph of C-3-2 at t=0 hour in relaxation test.

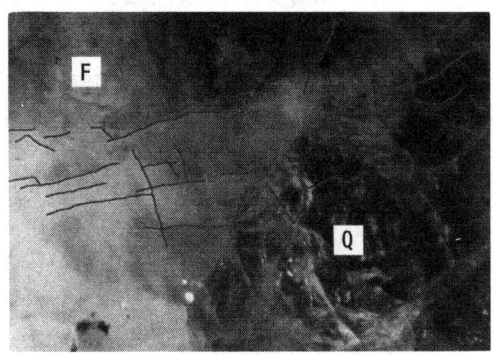

Figure 8(2). At t=25 hours.

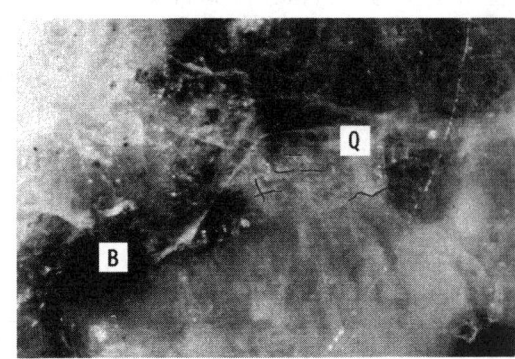

Figure 9(2). At t=25 hours.

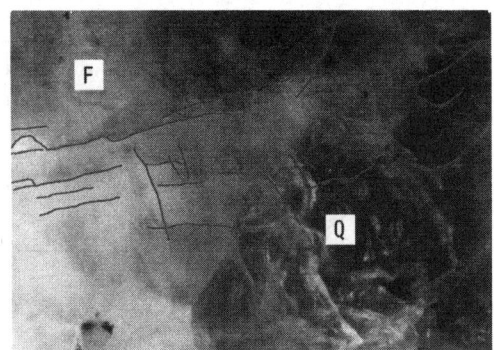

Figure 8(3). At t=150 hours.

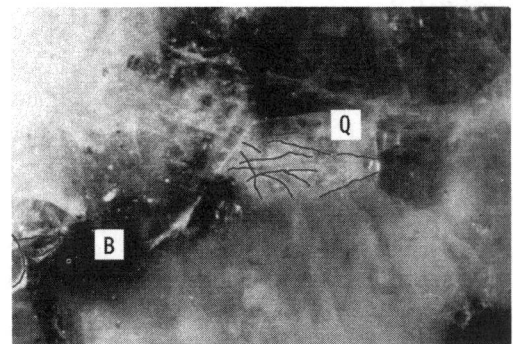

Figure 9(3). At t=150 hours.

That is, the totally relaxed stress is 5 - 9MPa, and among it about 20% appeared within 10 minutes. Note that the C-3-3 specimen which implies the 3rd specimen of C-3 type exhibits failure within 15 hours (Figure 6) because the test was started at a higher value of the relaxation stress level.

The rate of stress relaxation was calculated by using the results. As shown in Figure 7, the relaxation rate is of almost same tendency for all specimens except for the case of C-3-3, while the relaxation stress itself is quite different in each specimen. It should be noted that the difference of relaxation behavior was not found between coarse and fine-grained granites.

We took several sequential photographs of the microcrack development. Figure 8(1)-(4) are for the C-2-2 specimen in which we applied the load of the relaxation stress level of 60%, and the total stress relaxation was 20MPa. Note that among it, about 50% occurred quickly within 10 minutes. It was found that most microcracks were developed in feldspar in the direction of loading within 25 hours, and after this period we observed little development of microcracks. On the other hand, Figure 9(1)-(4) are for C-3-2 in which we applied the 70% stress, and the stress relaxation was 9MPa. In this case, we could find that microcracks are developed in the quartz and grain boundaries, and the development is lasted until the last stage.

According to these observation, the initial microcracks developed in the loading direction at the first stage and are developed gradually in other directions.

From these photographs, it was found that there is strong relationship between the relaxation behavior and microcrack development.

4 CONCLUSIONS

The uniaxial strength of granite is related to microcracks which are mainly distributed in rift-, grain- and hardway-plane. The rate of stress relaxation is not related to the applied initial stress level. According to the results of the relaxation test and direct observation of micro-cracking, it is found that the relaxation behavior is strongly affected by the time-dependent propagation of microcracks.

REFERENCES

Haupt, M. 1991. A constitutive law for rock salt based on creep and relaxation tests. Rock Mechanics and Rock Engineering. 24:179-206.

Peng, S. et al. 1972. Relaxation and the behavior of failed rock. Int.J.Rock Mech.Min.Sci.9:699-712.

Fatigue behavior of cyclically compressed rock under the confining pressure

Chung-In Lee, Jai-Oong Im & Jae-Joon Song
Division of Civil, Urban and Geosystem Engineering, Seoul National University, Korea

ABSTRACT: This paper presents the effects of applied maximum stress and confining pressure on fatigue life of rock. Cyclic compression tests were conducted on granite and marble under uniaxial and triaxial compression states. Stress ratios(S) of the applied maximum stress to the dynamic compressive strength were plotted against the number of cycles(N) at failure, and the relationship was expressed by the equation: $\log(S-S_{re})=A-B \cdot \log N$. Under the uniaxial compression, the fatigue limit(S_{re}) was 0.60 in granite and 0.75 in marble. As confining pressure increases, however, the fatigue limit is increased in granite, but decreased in marble. In this paper, strain change according to the increase of loading cycles and the relationship between stress-strain curve and total strain at various stress ratios are also discussed.

1 INTRODUCTION

When underground structures such as tunnels for mining or traffic, storage caverns for crude oil or liquified petroleum gas are constructed, rock masses around the openings are under the influence of dynamic loading induced by drilling and blasting. Rock foundations of atomic power plants, bridges and dams experience earthquakes, machinary vibrations and repeated stress from traffic. The rock, as it is with other materials, possesses a characteristic of failing by cyclic loading. Therefore, investigations on the fatigue feature of rocks can give designers important information in estimating a long-term stability and determining necessary strength and safety factors of rock structures.

The first fatigue test of rocks was conducted with a servo-controlled testing system by Hardy and Chugh(1970), who concluded that rocks evidently show fatigue behavior. After this, many investigators cleared the influence of a loading frequency on the failure of rocks, the fatigue life of rocks revealed from the tension and the compression-tension fatigue tests etc(M. Borsetto et al. 1981, E.S. Harry et al. 1986 & Y.M. Tien et al. 1990). What need to be examined more closely, however, are the influences of the amplitude of the cyclic load, the average stress, the rate and the order of loading, and the confining pressure under the cyclic triaxial compression the fatigue feature.

In this paper, the influences of the confining pressure and the maximum stress on the fatigue behavior, especially the fatigue limit were studied. Here, Chechon granite and Guatemalan marble which are in contrast with each other for their genesis condition, the structure of mineral grains and mechanical behavior under the static compression, are used as specimens. Sinusoidal cyclic compression of 1Hz with various amplitudes was loaded on to these specimens until they fail under the uniaxial and the triaxial condition.

2 TEST APPARATUS AND PROCEDURE

2.1 Rock Specimen and Equipments

In this study, Chechon granite in Korea and Guatemalan marble were used for fatigue tests. For each rock type, two hundred specimens with the diameter of 30mm and the length of 60mm were prepared. They were dried more than 24 hours at 105℃, left as they were for more than a week and were used in experiments. The tests of mechanical properties such as the specific gravity, porosity, uniaxial compressive strength, elastic modulus, Poisson's ratio, seismic wave velocity, internal frictional angle, and tensile strength revealed that the deviation from the mean value of each item did not exceed ±5%. Therefore, all of the specimens were assumed to be homogeneous.

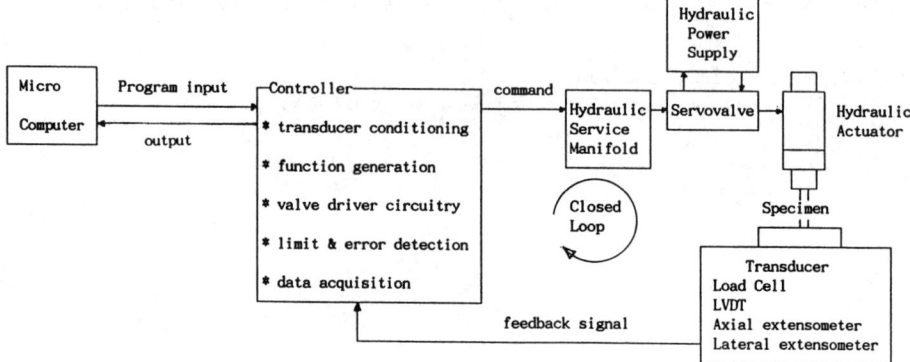

Figure 1. Block diagram of the testing system

The equipment used in this study is the servo-controlled testing system of Interlaken Co. which is composed of three parts: the main body, the controller and the air pump. Figure 1 is a block diagram of the testing system. The response time of this system is about 0.005sec and the minimum sampling time is 0.001sec. Therefore, it can be said that the system has a comparatively short response time and data aquisition capability.

2.2 Procedure

To find out the stress ratio for a certain load level in the fatigue test, static and dynamic compressive strength tests for granite and marble were carried out under uniaxial and triaxial states. Static compressive strengths were determined by mean values from 10 specimens for both rock types at the loading rate of 0.7MPa/sec. Dynamic compressive strengths were measured with each 10 specimens at the rate of dynamic compressive strength × 2/sec, that is, corresponding to loading frequency of this fatigue test(1Hz) under the assumption that the dynamic compressive strength is 125 percent of static compressive strength.

The maximum stress of the first fatigue test for each rock was set to be 95% of dynamic compressive strength, signifying that the stress ratio is 0.95. Cyclic load was applied to a specimen continuously until the specimen failed, and this procedure was repeated while decreasing the maximum stress by 5% at each trial. When there was no failure till the 200,000th stress cycle, the maximum stress at that time was considered to be a fatigue limit. For each stress level, 5 specimens were used for the test.

Fatigue tests of marble under triaxial compressive condition were performed with confining pressures, 5MPa and 10MPa, and in case of granite, 10MPa was applied to the specimen.

3 RESULTS AND DISCUSSIONS

3.1 Static and Dynamic Compressive Strength

In marble, an uniaxial dynamic compressive strength(125.4MPa) is higher than static compressive strength(100.8MPa) by 25% and the same results can be observed in the test of 5MPa and 10MPa confining pressures. In granite, dynamic uniaxial compressive strength is revealed to be higher than the static strength by 17%. The ratio, however, decreases by 14% under 10MPa confining pressure.

In marble under 5MPa confining pressure, which is 4% of dynamic compressive strength, there is a 32% increase of dynamic compressive strength over uniaxial dynamic strength. In granite, however, applying 10MPa confining pressure, which is 4.6% of uniaxial dynamic compressive strength, results into 54% increase of dynamic compressive strength. From the above results, it is obvious that granite is more dependent on confining pressure than marble.

The stress-axial strain and stress-lateral strain curves after applying the load with the rate corresponding to 1Hz-frequency under uniaxial and triaxial conditions are drawn in Figures 2 and 3. These figures convey that granite deforms in a more brittle manner than marble.

3.2 Relationship between Stress Ratio(S) and the Number(N) of Cycles at Failure

Figures 4 through 8 show the S-N curve obtained by cyclic loading with 1Hz frequency. The maximum stress decreases from 95% to the fatigue limit by 5%. The minimum stress is equal to 2% of the dynamic compressive strength under uniaxial state and 0MPa under triaxial condition.

For cyclic compression applied to rock specimen, to express the relation between the stress ratio and the logarithmic number of cycles to failure, several equations suggested untill present are as follows;

① $S = A - B \log N$
② $S = S_{re} + \dfrac{C}{\log N + D}$
③ $\log(S - S_{re}) = E - F \log N$

where, S : stress ratio of the maximum stress
S_{re} : stress ratio of fatigue limit
A, B, C, D, E, F : constants

Coefficients of determination for each equation were determined from the method of least squares and the result is illustrated in Table 1. In this table, since the coefficients of determination of equation ③ are the highest, it can be said that equation ③ expresses best the test results. Solid lines in Figures 4 through 8 represent these equations with different constants for each condition.

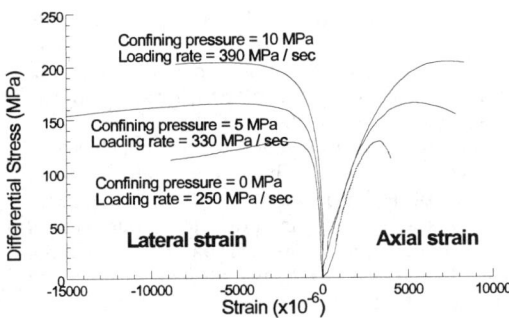

Figure 2. Stress-strain curves for marble at dynamic loading rate under uniaxial and triaxial conditions

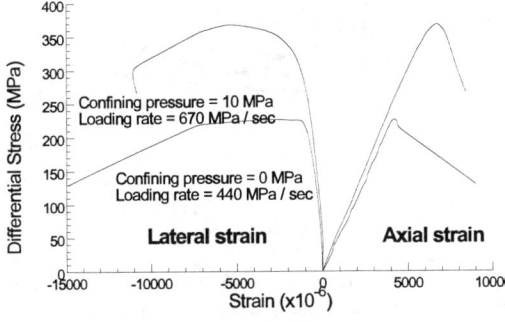

Figure 3. Stress-strain curves for granite at dynamic loading rate under uniaxial and triaxial conditions

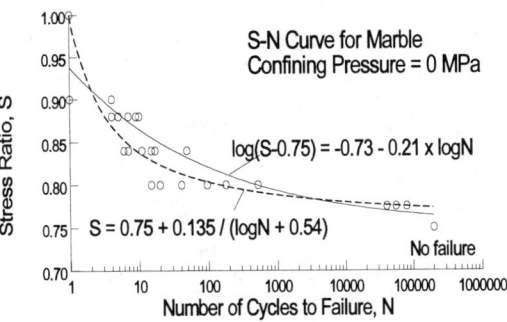

Figure 4. S-N curve for marble under uniaxial compressive condition

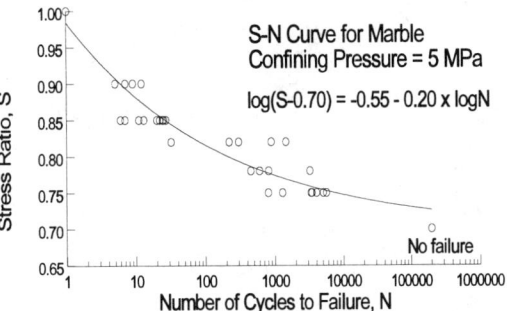

Figure 5. S-N curve for marble under triaxial condition (confining pressure; 5MPa)

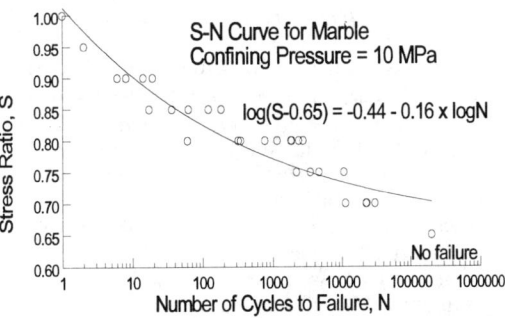

Figure 6. S-N curve for marble under triaxial condition (confining pressure; 10MPa)

While the S-N curve equation for marble under uniaxial condition has the low coefficient of determination, R^2 of 0.83, the coefficient of determination from equation ② is 0.96. Equation ② is shown in a dashed line in Figure 4.

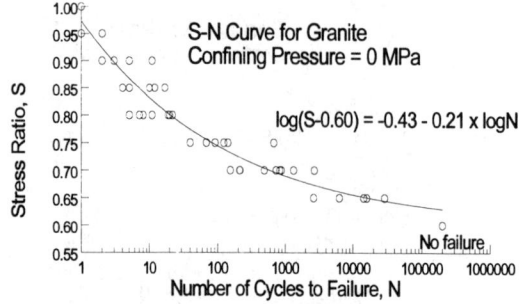

Figure 7. S-N curve for granite under uniaxial compressive condition

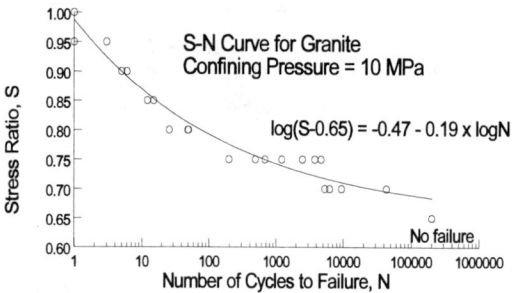

Figure 8. S-N curve for granite under triaxial condition (confining pressure; 10MPa)

3.3 Change of Fatigue Limit According to the Increase of Confining Pressure

S_{re} decreases from 0.75 to 0.70 and 0.65 at the confinging pressure of 0 to 5 and 10MPa for marble; but, increases from 0.6 to 0.65 at the confining pressure of 0 and 10MPa for granite.

When stress ratios are converted to differential stresses, the stress at fatigue limit increases from 94.1MPa to 115MPa and 126.6MPa for marble with increasing confining pressure. For granite, it also increases from 131.6MPa to 220MPa.

For granite, the differential stress at fatigue limit is 67% higher under the confining pressure of 10MPa than that under the uniaxial stress, which shows a higher increment of stress than for marble. This seems to be due to the fact that the porosity of granite(0.54%) is greater than that of the marble(0.32%); the granite comprehends relativley more small cracks and pores, and the increment of confining pressure prevents such small openings from expanding. One of those reasons, the cohesive force between particles in granite is relatively weaker than the strength of its grains, whereas it is relatively stronger in case of marble.

3.4 Strain Change According to Increase of Loading Cycles

Axial strain at the maximum stress behaves in three stages as the number of loading cycles increase. In the first stage, axial strain increases rapidly and in the second, the strain goes up linearly at a low speed. In the final stage, it increases abruptly to the failure of a specimen. In triaxial compressive test, the first stage becomes clear and the second tends to expand.

Lateral strain behaves differently from axial strain with the increase of the number of loading cycles. The second and third stages can only be observed in the lateral strain. Figures 9 and 10 show the axial and lateral strain behavior for granite under 10MPa confining pressure with various stress ratios.

3.5 Relationship between Stress-strain Curve and Total Strain at Various Stress Ratios

For marble, a complete uniaxial stress-strain curve at a loading rate of 250MPa/sec and the permanent strain lines by cyclic compression with various stress ratios are shown in Figure 11. In this figure, the length of a horizontal line indicates the magnitude of total permanent strain just before failure. It is observed that all the axial and lateral permanent strain lines are included in the stress-strain curve. This agrees with the result of study of Haimson et al.(1978) that the accumulated axial strain just prior to failure cannot exceed a complete stress-strain curve without regard to a stress level, and also means that the same rule is applicable to a lateral strain.

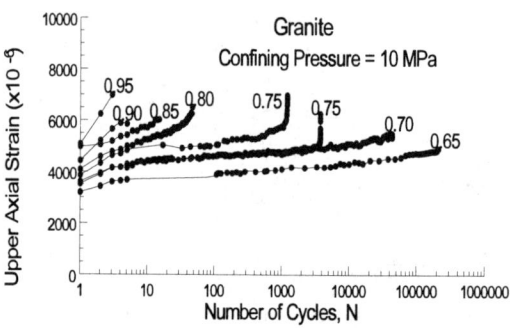

Figure 9. Relationship between upper axial strain and number of cycles for granite under triaxial state; confining pressure; 10MPa
(Numerals on the curves are stress ratios)

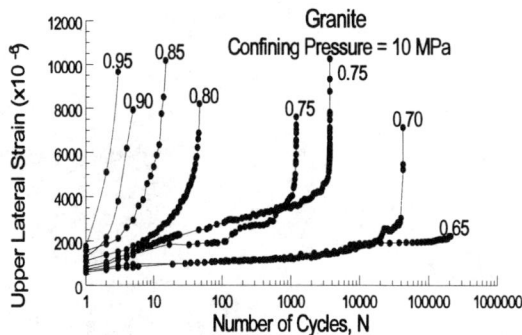

Figure 10. Relationship between upper lateral strain and number of cycles for granite under triaxial state; confining pressure; 10MPa
(Numerals on the curves are stress ratios)

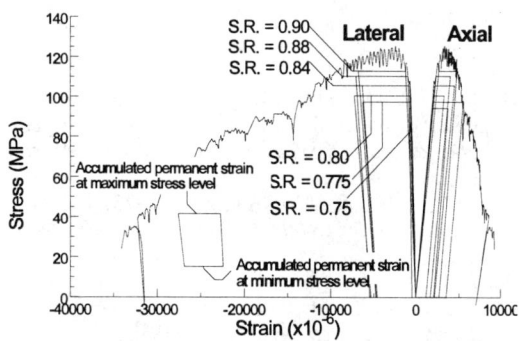

Figure 11. Relationship between complete stress-strain curve and accumulated strains for marble under uniaxial state (S.R. indicates stress ratio)

4 CONCLUSIONS

In this study, fatigue tests were done using Ch-echon granite and Guatemalan marble to examine the fatigue limit and fatigue life with different stress amplitudes and confining pressures. These are needed as the basic input parameter in estimating the long-term stability and in determining the safety factors. In particular, to consider the effect of confining pressure on the fatigue behavior of rocks which is not well known yet, experiments were conducted in the triaxial state of 5MPa and 10MPa confining pressure.

The main conclusions are as follows;

1. Dynamic compressive strength is 25% higher than static compressive strength for marble and 17% for granite.
2. Constants in S-N regresssion equations are calculated for marble and granite under uniaxial and triaxial compressive states.
3. The stress ratio of fatigue limit decreases with the increase of conifing pressure for marble but increases for granite, caused by the difference in porosity of each rock, the bonding force between rock forming minerals, and the strength of rock particles.
4. Under cyclic compression, axial strain at the maximum stress behaves in three stages and lateral strain in two stages.
5. It is observed that all of the axial and lateral permanent strains just before failure by cyclic compression are included in a complete uniaxial stress-strain curve.

Table 1. Equations of S-N curves under various confining pressure for marble and granite

Confining Pressure (Rock Type)	Equation of S-N curve ; Coefficient of Determination, R^2	
0MPa (Marble)	① $S = 0.888 - 0.0119 \log N$;	0.62
	② $S = 0.75 + \dfrac{0.135}{\log N + 0.54}$;	0.96
	③ $\log(S - 0.75) = -0.73 - 0.21 \log N$;	0.83
5MPa (Marble)	① $S = 0.920 - 0.02 \log N$;	0.84
	② $S = 0.70 + \dfrac{0.27}{\log N + 0.90}$;	0.89
	③ $\log(S - 0.70) = -0.55 - 0.20 \log N$;	0.90
10Mpa (Marble)	① $S = 0.960 - 0.0246 \log N$;	0.92
	② $S = 0.65 + \dfrac{0.43}{\log N + 1.23}$;	0.78
	③ $\log(S - 0.65) = -0.44 - 0.16 \log N$;	0.89
0MPa (Granite)	① $S = 0.903 - 0.0289 \log N$;	0.85
	② $S = 0.60 + \dfrac{0.31}{\log N + 0.77}$;	0.90
	③ $\log(S - 0.60) = -0.43 - 0.21 \log N$;	0.94
10MPa (Granite)	① $S = 0.932 - 0.0251 \log N$;	0.90
	② $S = 0.65 + \dfrac{0.32}{\log N + 0.91}$;	0.90
	③ $\log(S - 0.65) = -0.47 - 0.19 \log N$;	0.95

REFERENCES

Hardy, H.R. & Y.P. Chugh, 1970, Fatigue of geologic materials under low cycle fatigue, *Proc. of the 6th Symp. on Rock Mech.*, Montreal, pp. 33-47.

Attewell, P.H. & I.W. Farmer, 1973, Fatigue behavior of rock, *Int. J. Rock Mech. Min. Sci.* 10:1-9.

Peng, S.S., E.R. Podnieks & P.J. Cain, 1974, The behavior of Salem limestore in cyclic loading, *Soc. Petr. Eng. J.* 14:19-24.

Haimson, B.C. & K.S. Kim, 1977, Accoustic emission and fatigue mechanism in rock, *TRANS. TECH. S. A.*, pp. 35-55.

Haimson, B.C. et al., 1978, Effect of cyclic loading on rock, Dynamic Geotechnical Testing, *ASTM STP 654*, pp. 228-245.

Matsuk, K. & H. Kudo, 1988, Effect of pore pressure on the cyclic fatigue characteristics of water saturated rocks under confining pressure - A study on mechanical bahavior of rocks under cyclic loading in confining pressure(2nd report), *J. Mining and Metallurgical Institute of Japan* 104:157-164.

Lee, W.K., C.I. Lee, K.W. Lee, M.K. Kim & S.C. Kim, 1987, On the mechanical behavior of granite in cyclic loading, *J. Korean Institute of Mineral and Mining Engineers.* 24:1-7.

Vutukuri, V.S., R.D. Lama & S.S. Saluja, 1978, *Rock fatigue*, Handbook on Mechanical Properties of Rocks, Tech. Publications 1:216-219.

Tien, Y.M., D.H. Lee & C.H. Juang, 1990, Strain, pore pressure and fatigue characteristics of sandstone under various load conditions, *Int. J. Rock Mech. Min. Sci. & Geomech. Abstr.* 27:283-289.

Harry, E.S. & M. Asce, 1986, Permanent strains from cyclic variable amplitude loadings, *J. Geotech Eng. Div. ASCE.* 112:646-660

Borsetto, M., P.P. Rossi & R. Ribacchi, 1981, Long-term and cyclic plate loading tests in weak rocks, *Proc. Int. Symp. on Weak Rocks*, Tokyo, pp. 143-148.

Detection of mode of failure of sandstone through image processing

D.Chakravarty & S.K.Pal
Department of Mining Engineering, I.I.T. Kharagpur, India

ABSTRACT: With the advent of new generation computers the area of image processing has experienced a vigorous growth. The probable use of this technique finds avenues to study the rock characters at the micro-level, where distinction is otherwise impossible with naked eye. In this paper an attempt has been made to examine the SEM images of broken surfaces of sandstone failed under tension and shear modes with the help of fractal geometry. The results are in agreement with the general concept that rock surfaces failed under tension should have more number of sharp asperities and thus higher fractal dimension compared to those failed under direct shear.

1 INTRODUCTION

The failure in rock excavations is a common phenomenon in almost all underground and surface mines. Sudden failures of rock surfaces often lead to huge disruption of normal activities as well as cause possible damage to valuable properties and human lives. The nature and description of fractured surface are a store house of important information on post failure analysis where the pre-failure state of stress is sought for. In order to explore the micro-level characteristics of the broken rock surfaces the present study utilises the techniques of image processing and artificial machine vision. The distinction that is impossible to be made with naked eye can be done with the use of digital images taking the aid of modern mathematical techniques of micro-textural feature selection and classification. These images represent mathematically a two-dimensional light intensity function, $I = f(x,y)$, where x and y denotes the spatial co-ordinates. The values of $f(x,y)$ are proportional to the intensity of the reflected light from scanned surfaces, which in turn depend on the nature of surface roughness. $f(x,y)$ can have values ranging from 0 to 255 for 8 bit grey scale images. Out of the various micro-level information that can be derived from acquired images, the fractal dimension has been used to correlate with mode of failure in naturally occurring rocks because of its inherent complexity, which is not possible to be represented with the help of conventional concepts of Euclidean geometry.

2 DIRECT IMAGE ACQUISITION

Images can be acquired or sampled by sensors working in infra-red, visible light or x-ray range of electromagnetic spectrum. For industrial applications visible range of light is found to be most useful. Images are acquired using CCD cameras and the standard resolution of these devices is typically 640 x 512 non-square pixels, but the trend is towards 1024 x 1024 square pixels. In sampling of images one important factor is the illumination technique used. The different lighting techniques are - diffused lighting, back lighting, structured lighting and directional lighting (Figure 1). Out of all these methods, the last one is used for detection of surface defects (Fu et. al. 1987). In this method a beam of light is made incident on the surface at an angle and the intensity of upward light depends mainly on pits and scratches on the surface when other factors remain unaltered. Rocks are composed of various minerals that have different light attenuation properties. Thus a smooth and polished rock surface having differently textured particles may produce widely varying grey-level values of the acquired images thereby giving a false representation of the actual elevations. In order to verify this a test study was carried out with two types of smoothened rock samples - first one is a natural uncoloured rock surface and the other one is an artificially blackened rock surface. In each type two steps of 1.21 mm and 3.39 mm elevation differences were imaged from the same horizon using a CCD camera and the difference in average grey-level values between the

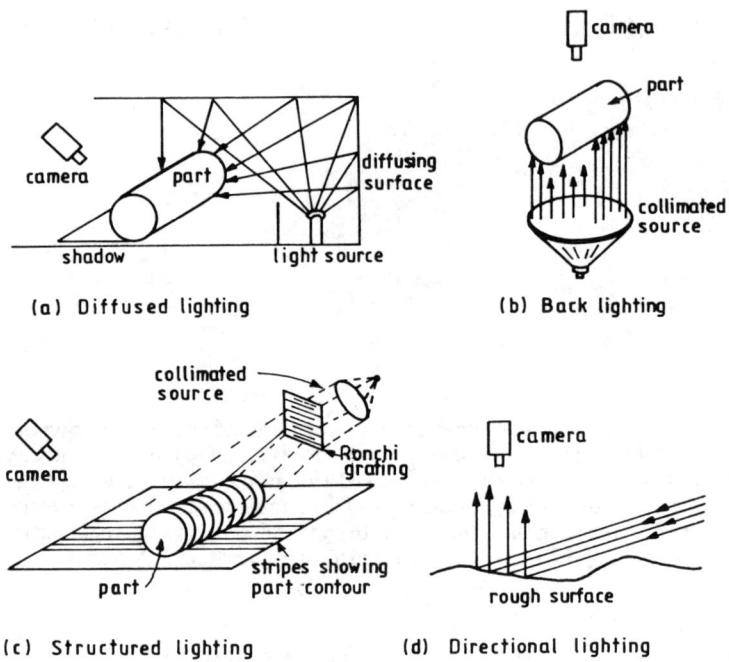

Figure 1. Four basic illumination schemes.

steps were found to be 16 & 29 for natural rock surface and 7 & 19 for the blackened rock surface. From this the range of variation of the number of pixels per mm elevation difference is computed as 8.62 to 13.2 for uncoloured natural rock surface and 5.6 to 5.7 for blackened rock surface. This indicates that the pixel values per mm elevation difference for uncoloured sample are inconsistent whereas, for blackened samples the accuracy of detection of elevation differences (0.18 mm per pixel) is not sufficient enough to bring out minute differences in surface relief between broken surfaces under different modes of failure. From this study it was found that direct image acquisition of broken surfaces was not suitable for further analysis.

3 SEM PHOTOGRAPHY AND IMAGE DIGITISATION

In order to bring out the minute differences in surfaces features of broken rock surfaces the digitised image analysis of photographs taken through a Scanning Electron Microscope was attempted. In this method nine sandstone cores of 50 mm diameter broken under tension by Brazilian method, six cores of 22 mm diameter broken under shear using Direct Shear Box Apparatus and four broken cores of 22 mm diameter failed under unknown mode were used. The representative broken rock surfaces were first sputtered with gold to improve their reflectance and then held at 30 degrees inclination to the horizontal so as to achieve directional lighting effect. These gold coated rock surfaces were photographed in the Scanning Electron Microscope (Model No. 2, DV Cam Scan, UK) at a magnification of 900X. Since the influencing surface parameters like nature of surface, its reflectance, absorption, etc. are kept unchanged for all the samples the photographs represent their relative elevation differences only. However, the choice of magnification as 900 was made based on an earlier study by the authors (Chakravarty 1994) where it was found that SEM photographs at this magnification provide consistent fractal measures for distinguishing micro-level textures. All the photographs were then digitised from the same horizontal elevation with identical lighting conditions with the help of a Benchmark Vision System which uses a Punix CCD camera and a high resolution monitor of 512 x 512 square pixels. Finally the photograbbing process was completed with the help of IMAGER-AT software. In this way the 8-bit images were digitised, where each pixel can assume grey-level values ranging from 0 to 255. Figure 2 shows eight SEM photographs and their corresponding digitised images, four samples under each mode of failure.

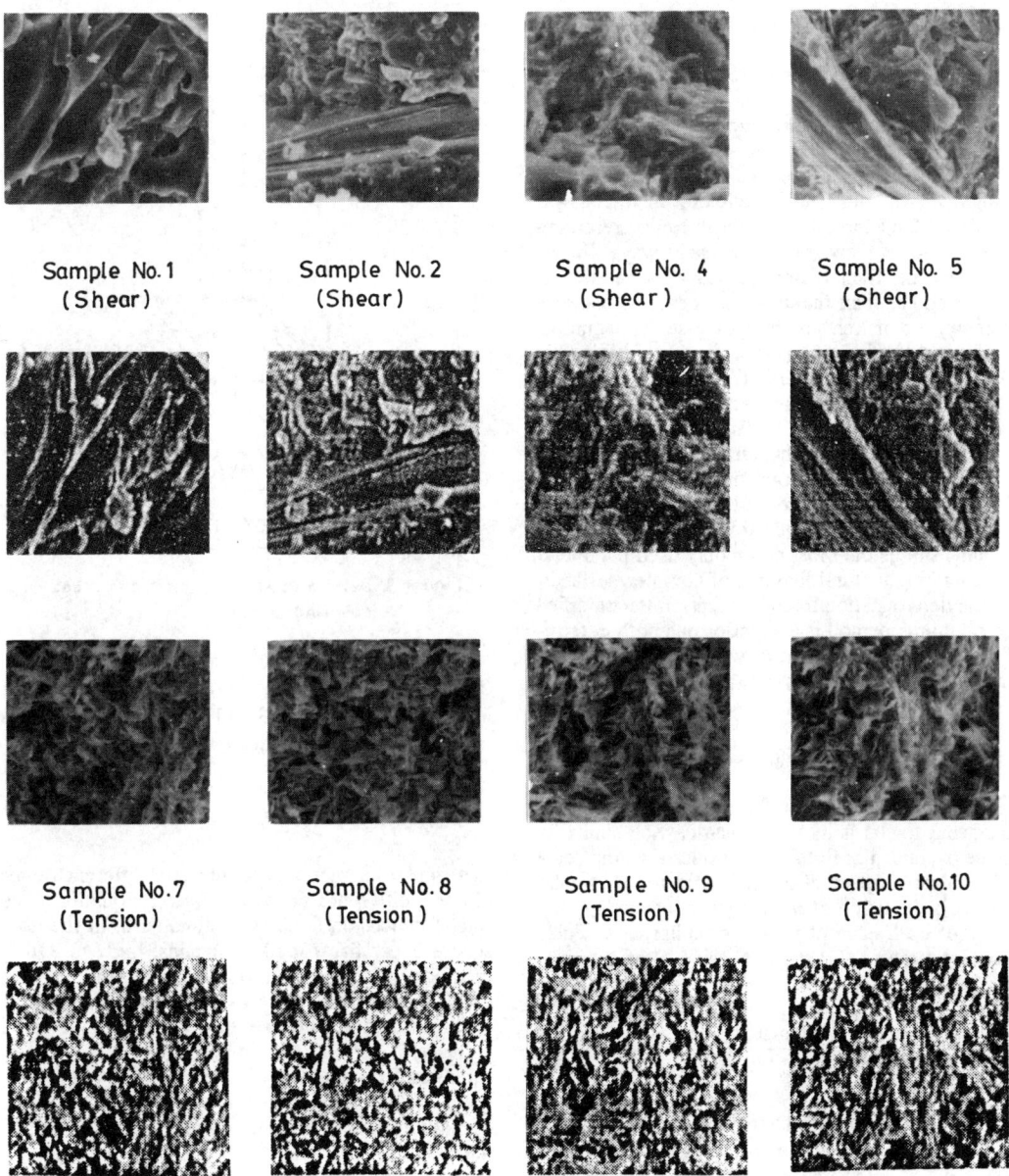

Figure 2. SEM photographs and their corresponding digitised images

4 FRACTAL ANALYSIS OF THE IMAGES

Traditional geometry fails to model naturally occurring objects because of their inherent complexity that does not reduce at any level of magnification. In order to study such complex surface character the applicability of fractal geometric concepts was investigated. Although initial definition of fractal was given by Mandelbrot (Mandelbrot 1967), researchers now-a-days view fractals in many ways. In one such concept, fractals are defined as a set that is more irregular than the sets considered in classical geometry, no matter how much the set is magnified, smaller and smaller irregularities become visible (Xie 1993). It is further established that natural surfaces are self similar and fractals can be used to describe textural features of these surfaces at any magnification (Pentland 1984). Thus fractal analysis of the digitised data obtained from magnified photographs of rock surfaces can be utilised to reveal its textural features. Out of the various fractal parameters fractal dimension is the most commonly used parameter for describing textural features of complex surfaces. For the determination fractal dimension the modified box counting method is a superior one both in terms of saving in computer time as well as accuracy in the analysis of complex surfaces (Sarkar 1993).

4.1 *Modified Box Counting Method*

According to Mandelbrot the criterion for the surface being fractal is its self similarity. Self similarity can be explained as follows. Consider a bound set A in Euclidean n-space. The set is said to self similar when A is the union of n_r distinct (non-overlapping) copies of itself each of which is similar to A scaled down by a ratio r. Fractal dimension D of A can be derived from (Mandelbrot 1982)

$$1 = n_r r^D \quad or \quad D = \frac{\log(n_r)}{\log(1/r)} \tag{1}$$

In this method, for computation of n_r the image of $M \times M$ pixels is scaled down to a size $s \times s$ where $M/2 > s > 1$ and s is an integer. Then r is estimated as $r = s/M$. Considering image as a 3D space with (x,y) denoting 2D position and the third co-ordinate z denoting grey-level, the (x,y) space is partitioned into grids of size $s \times s$. On each grid there is a column of boxes of size $s \times s \times s'$. If the total number of grey-levels is G then $[G/s'] = [M/s]$. Referring to Figure 2, where $s = s' = 3$, the boxes were assigned numbers 1,2, ... Let the minimum and maximum grey-level of the image in $(i,j)^{th}$ grid fall in box number K and L respectively. Then $n_r(i,j) = L - K + 1$ is

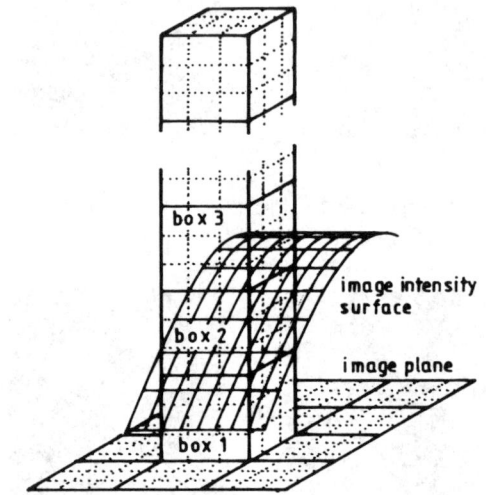

Figure 3. Determination of n_r by modified box counting method.

the contribution of n_r in the $(i,j)^{th}$ grid. For example in Figure 3, $n_r(i,j)$ = 3-1+1. Taking contributions from all grids n_r can be determined as

$$n_r = \sum_{i,j} n_r(i,j) \tag{2}$$

In a similar manner n_r is counted for different values of r (i.e, different values of s). Using Equation 1 the fractal dimension D can be estimated from the least square linear fit of $\log(n_r)$ against $\log(1/r)$. Other methods of computing fractal dimension do not cover the image surface so well and hence cannot capture the fractal dimension for rough textured surfaces.

Based on the above method a computer program was developed for the computation of the fractal dimensions of different images. The input to the program was the name of the image file mentioning its size. The output of the program was the fractal dimension values of the broken surfaces, obtained from the negative slope of the plot between $\log(1/r)$ and $\log(n_r)$.

5 EFFECT OF FAILURE MODE ON FRACTAL DIMENSION

Fractal dimension (FD) values of all the images as computed by modified box counting method are given in Table 1. For samples failed under shear mode the mean value of fractal dimension is 2.424 with a standard deviation of 0.016. Similarly the mean fractal dimension value of samples failed under tension is 2.463 with a standard deviation 0.026. Thus, it can be said that the samples failed under shear have, in general, lower values of fractal dimension than those failed under tension. This is in agreement with the general concept that samples failed under tension will maintain the sharpness of grains at the failure surfaces, thus introducing a very rough broken surface with a large number of sharp asperities. This will definitely lead to higher fractal dimension compared to the samples failed under direct shear, where the tips of sharp asperities are expected to have sheared off creating a flatter broken surface. Assuming the FD values to be normally

Table 1. FD values of broken surfaces.

Sample No	Failure mode	FD values	Mean (SD)
1	Shear	2.438	
2	Shear	2.419	
3	Shear	2.402	2.424
4	Shear	2.434	(0.016)
5	Shear	2.441	
6	Shear	2.439	
7	Tension	2.454	
8	Tension	2.453	
9	Tension	2.482	
10	Tension	2.518	
11	Tension	2.479	2.463
12	Tension	2.453	(0.026)
13	Tension	2.439	
14	Tension	2.455	
15	Tension	2.437	
16	Unknown	2.457	
17	Unknown	2.456	2.423
18	Unknown	2.340	(0.056)
19	Unknown	2.439	

distributed, about 70% of them for sheared surface is expected to be in the range of 2.424 + 0.016. Similarly, the expected range of FD values for surfaces failed under tension is 2.463 + 0.026. From this the overlapping range of the FD values is found to be (2.463 - 0.026) to (2.424 + 0.016), i.e., 2.437 to 2.440.

Using the above findings, the FD values for the broken sandstone samples of unknown mode of failure are interpreted as follows:
I. Sample Nos. 16 and 17 have failed under tension as their FD values are higher than 2.440.
II. Sample No. 18 must have failed under shear as its FD value is much lower than 2.437.
III. Sample No. 19 with an FD value of 2.439 lies in the overlapping range. Therefore, no definite inference can be made regarding its mode of failure.

6 CONCLUSION

From the present study the following conclusions are drawn:
1. Magnified SEM images of broken surfaces of sandstone has been successfully used to estimate their 2D fractal dimension values.
2. The mean value of fractal dimension of broken sandstone surfaces failed under shear is lower than that of surfaces failed under tension.
3. Broken sandstone surfaces having FD values lower than 2.437 are expected to have failed under shear, whereas those having FD values higher than 2.440 are expected to have failed under tension.

The present study however, suffer from a few drawbacks. Firstly, the method of image digitisation from SEM photographs might have introduced some errors in spire of our best efforts to minimise the same. Secondly, owing to the availability of limited resources a large number of SEM photographs could not be taken and hence the statistical interpretation of the data does not have a strong basis. In spite of these drawbacks the present study has indicated a bright scope for future application of 2D fractal analysis in detecting the mode of failure during post failure investigation.

REFERENCES

Chakravarty, D. 1994. *Fractal image analysis for rock surface characterisation.* M.Tech. Disser-tation. I.I.T. Kharagpur, India.

Fu, L. & Gonzalez, R. 1987. *Robotics: control sensing, vision and intelligence.* New York : McGraw-Hill.

Mandelbrot, B. B. 1967. How long is the coast of Britain? statistical self-similarity and fractal dimension. *Science.* 155: 636-638.

Mandelbrot, B. B. 1982. *The fractal geometry of Nature.* New York: W. H. Freeman.

Pentland, A. P. 1984. Fractal based description of natural scenes. *IEEE Trans. Pattern Analysis Mach. Intell.* PAMI-(6): 661-674.

Sarkar, N. 1992. An efficient approach to estimate fractal dimension of textural images. *Pattern Recognition.* 25(9): 1035-1041.

Xie, H. 1993. *Fractals in Rock Mechanics.* Rotter-dam: A. A. Balkema.

Sub-critical damage in brittle rock around underground storage caverns

J.F.Shao, G.Duveau, M.Sibai & M.Bart
LML-URA CNRS, EUDIL, Cité Scientifique, Villeneuve d'Ascq, France

N.Hoteit
ANDRA, Parc de la Croix Blanche, Chatenay Malabry, France

ABSTRACT: Laboratory tests on a brittle rock have been performed to characterise damage initiation and growth, dilatation, induced anisotropy, and sub-critical propagation of microcracks. A continuous damage model has been proposed to describe both stress induced 'instantaneous' damage and time dependent sub-critical damage. An example is given in the field of underground nuclear waste storage.

1 INTRODUCTION

Damage by microcracking is the main cause of non-linear deformation and failure of brittle rocks. Many laboratory investigations have shown various mechanisms by which cracks can initiate and grow under compressive stresses (we do not give an exhaustive list here). In addition, crack growth can also occur in time dependent way even if the applied stresses are smaller than the short term elastic limit. Physical interpretations of the subcritical crack growth have been discussed by Atkinson (1984) and Atkinson and Meredith (1987). Various approaches have been developed for damage modelling of brittle rocks, mathematical crack growth models, phenomenological models and micromechanical models (we do not give an exhaustive list of damage models because of the limited length of paper).

In this paper, the emphasis is put on the study of time dependent damage due to sub-critical microcrack growth. This feature represents a great interest in the field of underground nuclear waste deposit. Indeed, galleries for deposit purpose are often excavated in hard rocks with low permeability like granite and hard clay. Due to the high mechanical strength of these rocks, the damage induced by applied stresses during the excavation is generally very small. However, instability problems can occur in long term. For this purpose, laboratory tests will be performed on a representative granite. Material behaviour related to sub-critical crack growth will also be analysed. A continuous damage model, able to describe both the 'instantaneous' mechanical damage and time dependent sub-critical damage will be proposed. The calibration of the model from laboratory tests and an application example will be presented.

2 SUMMARY OF THE EXPERIMENTAL STUDY

The studied rock is a granite with small size grains, mainly composed of feldspaths and automorphes of 2 to 5 millimetres. There are also angular quartz and other secondary minerals. For a first order of approximation, an average grains size of 3 millimetres can be suggested.

Laboratory experiments composed of various tests have mainly focused on the experimental characterisation of damage in the material. The tests performed are composed of triaxial compression with unloading cycles, and during each test, acoustic emission hits are registered and ultrasonic velocities measured. In some tests, the evolution of permeability with crack growth was studied. In this paper, it is not possible to give a detailed presentation of the results obtained from all the tests conducted. Only some representative results will be summarised. In figure 1, typical stress-strain curves in a triaxial test are shown for a given confining pressure, and in figure 2, the corresponding cumulative number of acoustic emission hits is presented. Some interesting remarks can be made. (1) The lateral strain (ε_3) presents a earlier and more significant non linearity than the axial strain (ε_1), and as a consequence a large volumetric dilatancy is obtained. This is due to a oriented cracks growth in the axial direction. (2) Unloading-reloading cycles allow to determine the effective elastic moduli of the rock at different damage level. By making a detailed analysis of the obtained results, one can notice that the effective elastic moduli decrease as the damage grows and an anisotropic behaviour (transversely isotropic under triaxial loading) is obtained due to the oriented microcracking. (3) There is a good agreement between the strain non linearity (or dilatancy) and the increase of acoustic emission. However, in the present case, it seems that the acoustic emission is very sensitive to the microcrack growth. Indeed, significant acoustic emission hits are registered in the first stage of microcracking where the lateral strain non linearity is not yet seen. During unloading, microcracks are partially closed and no further growth occurs. Accordingly, no further acoustic

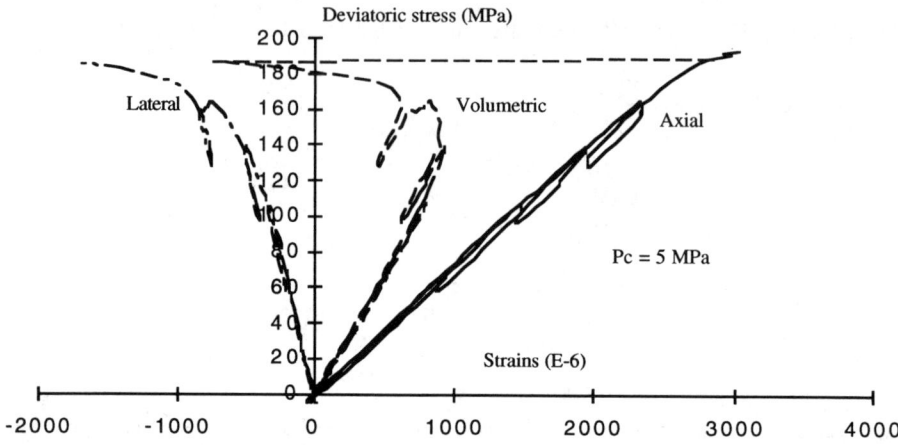

Figure 1 : Stress strain curves of the granite in a triaxial test with unloading cycles
(5 MPa confining pressure)

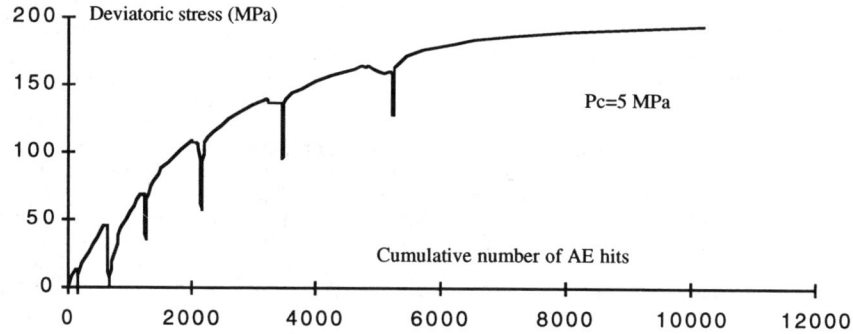

Figure 2 : Cumulative number of acoustic emission hits in a triaxial tests unloading cycles
(5 MPa confining pressure)

emissions should be registered. This is clearly shown in figure 2. A good agreement is also obtained during reloading phase, further AE increase occurs when the applied stress exceeds its previous value, inducing further microcrack growth.

3 ANISOTROPIC CONTINUOUS DAMAGE MODEL

The present study is limited to open penny-shaped cracks. Based on the works of Hill (1963) and, Kachanov (1992), the following second order damage tensor is proposed:

$$\overline{\overline{D}} = \sum_j \hat{d}_j(s)(\vec{n} \otimes \vec{n})_j \quad (1)$$

with $\hat{d}_j(s) = \left[\dfrac{\hat{a}^3 - a_0^3}{a_0^3}\right]_j$ for the 3-D case

and $\hat{d}_j(s) = \left[\dfrac{\hat{a}^2 - a_0^2}{a_0^2}\right]_j$ for the 2-D case

In these equations, a_0 is the initial radius of cracks which is supposed to be uniform in undamaged rock. $\hat{a}$ is the statistical average radius of the jth set of cracks. $\hat{d}_j(s)$ is considered as the relative change of crack density induced by the applied stresses and sub-critical growth. The use of such a relative density allows to conserve the macroscopic character to the damage tensor $\overline{\overline{D}}$ which is an internal state variable of the material damage. In addition, only the sets of open microcracks are accounted for in the evaluation of the damage tensor $\overline{\overline{D}}$.

Two kinds of crack growth criteria, mechanical growth due to applied stresses and subcritical growth due to stress corrosion, are considered. The first type of growth is based on the principles of the linear fracture mechanics. By supposing that the crack growth mainly occurs in the tensile mode, the stress intensity factor K_1 for a small penny-shaped crack with the unit normal vector $\bar{n}$ in a local region of tension may be approximated by:

$$K_1 = \frac{2}{\pi}\sqrt{\pi \hat{a}}\left[\frac{\sigma_{kk}}{3} + t\left(1 - \frac{(\hat{a}-b)^2}{a_0(a_0-b)}\right)\bar{n}\bar{\bar{S}}\bar{n}\right] \quad (2)$$

where $\bar{\bar{S}}$ is the deviatoric stress tensor, t is a proportionality constant between the applied deviatoric stress and the local tensile stress concentration, and b is a limit crack size at which the coalescence of cracks occurs. The above equation is similar to that proposed by Costin (1985). Eqn. 2 is proposed for a single crack in a given orientation. In order to obtain a continuous description of damage, it is assumed that an ensemble of cracks contained in some region of the solid behaves in a manner similar to an individual crack. Therefore, Eqn. 2 is extended to describe the response of an ensemble of cracks with $\hat{a}$ being the statistical average size of cracks with normal $\bar{n}$.

The presence of the mean stress term $\sigma_{kk}/3$ in the expression of K_1 is primordial and allows to account for the influence of the mean stress on the growth of microcracks. In fact, a positive (tensile) value of the mean stress increases K_1 and thus crack growth. On the other hand, a negative (compressive) value decreases K_1 and thus prevents crack growth. Therefore, the mean stress is included in the expression of K_1 to be able to describe one of the main properties of brittle rocks, the pressure sensitivity.

Mechanical growth of crack occurs when the following condition is verified:

$$K_1 = K_{1c} \quad (3)$$

where K_{1c} is the critical stress intensity factor of the rock. From Eqn.2 and Eqn.3, the actual size of cracks $\hat{a}$ is determined for any stress state.

For the time dependent damage due to sub-critical growth of microcracks, the previous experimental studies (Atkinson 1987) have shown that the crack growth rate is proportional to the stress intensity factor. According to the DT tests conducted on the studied granite, the Atkinson theory was quite well verified. Therefore, the following well known equation is retained:

$$\dot{a} = AK_1^N \quad (4)$$

where A and N are two material constants generally depending on the environment (water, air, etc..).

To obtain the constitutive equations of the damaged material, the free energy function proposed by Dragon (1993) is used in this work:

$$w(\bar{\bar{\varepsilon}},\bar{\bar{D}}) = g\,\text{tr}(\bar{\bar{\varepsilon}}.\bar{\bar{D}}) + \frac{1}{2}\lambda(\text{tr}\bar{\bar{\varepsilon}})^2 + \mu\,\text{tr}(\bar{\bar{\varepsilon}}.\bar{\bar{\varepsilon}})$$
$$+ \alpha(\text{tr}\bar{\bar{\varepsilon}})\text{tr}(\bar{\bar{\varepsilon}}.\bar{\bar{D}}) + 2\beta\,\text{tr}(\bar{\bar{\varepsilon}}.\bar{\bar{\varepsilon}}.\bar{\bar{D}}) \quad (5)$$

where λ and μ are Lame's constants, g, α and β are three constants defining the damage induced change of the strain energy. The term $g\,\text{tr}(\bar{\bar{\varepsilon}}.\bar{\bar{D}})$ in Eqn.5 allows to take into account the residual stress effects without explicit reference to plasticity concept. By differentiating Eqn.5 with respect to $\bar{\bar{\varepsilon}}$, we obtain the constitutive equations as follows:

$$\bar{\bar{\sigma}} = \frac{\partial w}{\partial \bar{\bar{\varepsilon}}} = g\bar{\bar{D}} + \lambda(\text{tr}\bar{\bar{\varepsilon}})I + 2\mu\bar{\bar{\varepsilon}}$$
$$+ \alpha\left[\text{tr}(\bar{\bar{\varepsilon}}.\bar{\bar{D}})I + (\text{tr}\bar{\bar{\varepsilon}})\bar{\bar{D}}\right] + 2\beta(\bar{\bar{\varepsilon}}.\bar{\bar{D}} + \bar{\bar{D}}.\bar{\bar{\varepsilon}}) \quad (6)$$

By applying Eqn.6 to the case of conventional triaxial loading, the following stress strain relationships are obtained:

$$\begin{cases} \sigma_1 = (\lambda + 2\mu)\varepsilon_1 + (2\lambda + 2\alpha D_3)\varepsilon_3 \\ \sigma_3 = gD_3 + (\lambda + \alpha D_3)\varepsilon_1 + (2\lambda + 2\mu \\ + 4\alpha D_3 + 4\beta D_3)\varepsilon_3 \end{cases} \quad (7)$$

where $D_3 = D_2$ is the lateral component of the damage tensor.

The proposed damage model contains 5 parameters for the constitutive equations (Eqn.6), 4 for the mechanical crack growth criterion (Eqn.2) and 2 for the sub-critical growth criterion (Eqn.4). All the parameters can be determined from classic laboratory tests. For instance, μ and λ are determined from the linear part of stress-strain curves in a triaxial test, while g, α and β from the non-linear (damaged) part with unloading slopes and using Eqn.7. Note that in a short term test, the damage evolution is given by Eqn.2, the damage tensor $\bar{\bar{D}}$ is known at each stress level, the values of (g,α,β) are obtained directly by solving a linear set of 3 equations. However, for the granite studied in this work, it is was found that the damage induced residual stress (strain) was negligible. Therefore, it is possible to take $g = 0$. The initial crack size a_0 is determined from a microscopic observation of the rock and usually taken to be equal to the grain size. The constant b representing the critical size for crack coalescence and the proportionality constant t, can be determined together from damage threshold and peak stress in a triaxial test. Two constants (A, N), determining the sub-critical growth of cracks are usually determined from a Double Torsion (DT) test. Finally, the critical stress intensity factor K_{1c} is obtained from a Three Point Bending (TPB) test.

For the granite studied, TPB tests under different confining pressures have been conducted. It is found that the value of K_{1c} varies with the confining pressure. The following simple correlation is obtained:

$$K_{1c}(MPa\sqrt{m}) = 1,41 + 0,09 < \sigma_3 > \qquad (8)$$

Accordingly, the values of t and b are determined from some triaxial tests with different confining pressures. Due to the variation of K_{1c}, the value of t is found depending also on the confining pressure while the value of b remains quasi constant:

$$t = 1,37 + 0,014 < \sigma_3 > \qquad (9)$$

The values of the other parameters are given in Table 1. By using these values, some triaxial tests are simulated. In figure 3, comparisons between numerical results and experimental data are presented. A good agreement is obtained.

4 EXAMPLE

In this section, an application example of the proposed model is presented in the filed of underground nuclear waste storage. For this purpose, the damage model has been introduced into a FEM code. A circular gallery excavated at the 500 meters depth is considered. The geometric form of the gallery is illustrated in figure 4.

In this example, the emphasis is put on the analysis of time dependent damage evolution around the gallery. Therefore, a suitable value of the in situ horizontal stress is chosen so that the stress induced instantaneous damage is negligible during excavation phase (the vertical stress is taken to be equal to the earth weight). Calculations are made for the investigation of time dependent damage induced by sub-critical growth of microcracks after the excavation of gallery. The calculations are made under plane strain conditions.

In this paper, only one of the calculations is presented and the boundary conditions used are

Table 1
Values of material parameters for the granite studied

E (MPa)	v	g (MPa)	α (MPa)	β (MPa)	a_0 (mm)	b / a_0	A (m/s)	N
85000	0.25	0	-116	-6642	3	1.69	1.7×10^{-11}	55

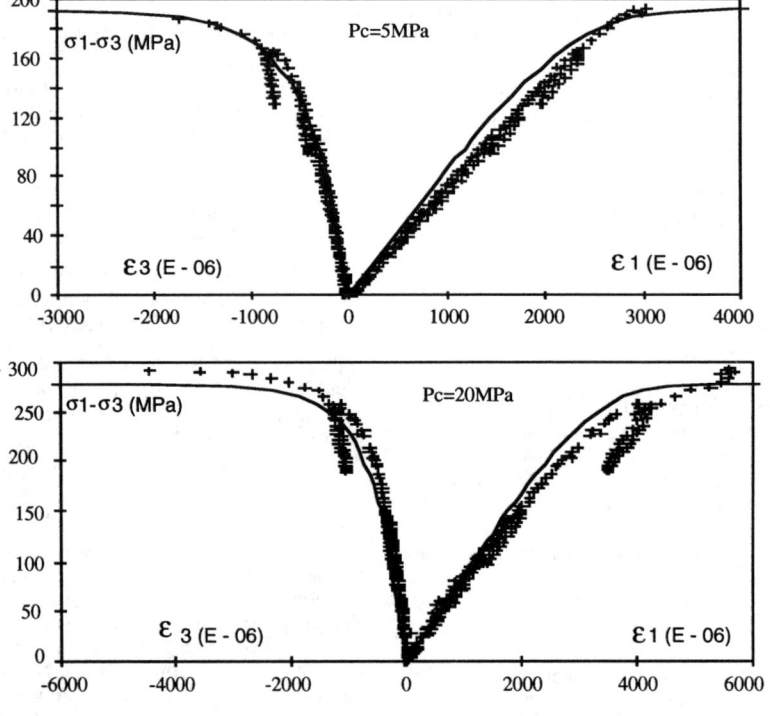

Figure 3: Simulation of triaxial tests (continuous lines are numerical predictions) for two confining pressures (5 and 20 MPa)

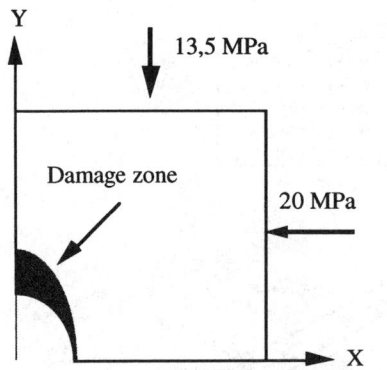

Figure 4: Schematic presentation of the gallery form and boundary conditions

shown in figure 4. In Figure 5 and 6, the evolution of damage tensor and the maximum crack length at the most damaged point around the gallery are presented. One can notice that during a short period after excavation, there is a rapid growth of microcracks. During a later period, the damage growth rate is smaller but remains significant. After a period of three year (the end of the calculation), the cumulated damage becomes very significant and the microcrack size in the damaged zone approaches slowly (but continuously) to the critical value (coalescence and failure). In figure 7, damage distributions around the gallery at various time steps are presented. One can notice that both the damage level and the damage zone increase with time.

5 CONCLUSION

Laboratory tests have been used to characterise the deformation and damage in a brittle granite. The oriented microcrack growth results in a strong dilatancy and a transversely isotropic behaviour. There is a good agreement between the acoustic emission measurement and the macroscopic stress strain behaviour. A detailed analysis of the damage evolution in this material, characterised by different methods (microscopic observation, AE measuring, ultrasonic velocity, permeability evolution) will be presented in a forthcoming paper. A continuous damage model is used to describe both the stress induced 'instantaneous' damage and time dependent sub-critical damage. TPB and DT tests are conducted to determine the material parameters related to the sub-critical growth of microcracks. The proposed model is calibrated from laboratory tests and a good agreement with the experiment data is obtained. A simple application example of the model in the field of underground waste storage is presented. The results obtained clearly show that even under small

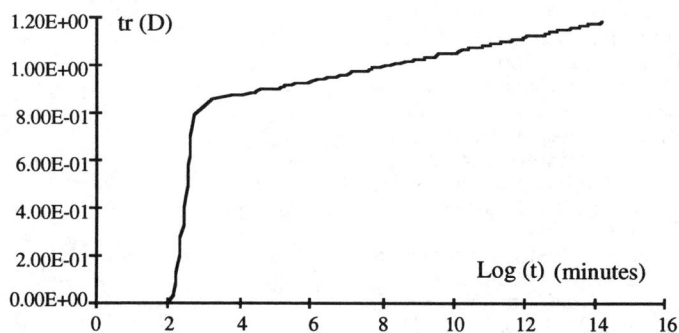

Figure 5: Evolution with time of the maximum damage around the gallery

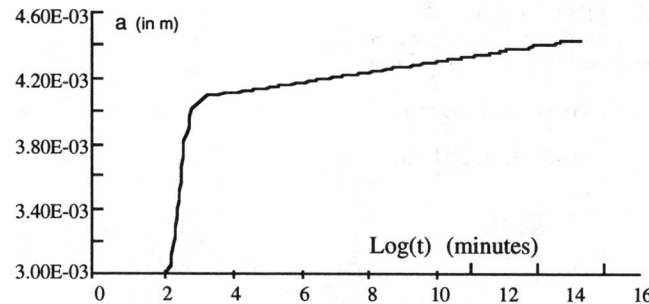

Figure 6: Evolution with time of the maximum microcrack length around the gallery

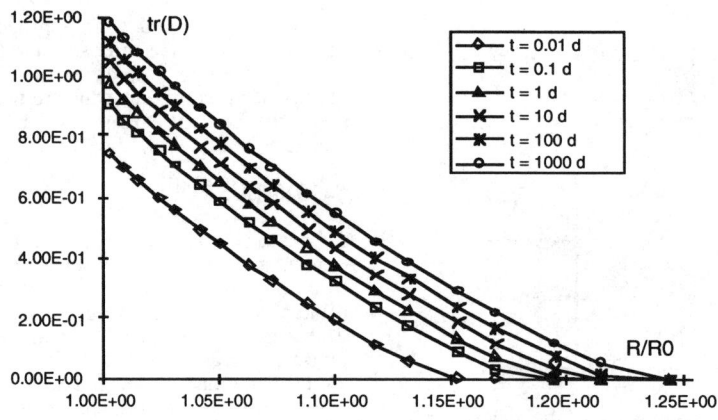

Figure 7 : Damage distribution around the gallery at various time steps

stress changes, time dependent damage due to sub-critical growth of microcracks can develop very significantly and cause structure failure.

REFERENCES

- Atkinson B.K. (1984), Subcritical crack growth in geological materials, J. of Geophysical Research, Vol.89, N°B6, 4077-4114
- Atkinson B.K. & Meredith P.G. (1987), The theory of subcritical crack growth with applications to minerals and rocks, in Fracture Mechanics of Rock, Academic Press Inc. (London), 113-166
- Costin L.S. (1985), Damage mechanics in the post failure regime, Mechanics of Materials 4, 149-160
- Dragon A., Cormery F., Désoyer T., Halm D. (1994), Localized failure analysis using damage models, in Localisation and bifurcation theory for soils and rocks, Chambon R., Desrues J. & Vardoulakis I. (eds), Balkema, 127-140
- Ewy R.T. & Cook N.G.W. (1990), Deformation and fracture around cylindrical opening in rock-I; Observations and analysis of deformations, Int. J. Rock Mech. & Min. Sci., vol.27, No.5, 387-407
- Ewy R.T. & Cook N.G.W. (1990), Deformation and fracture around cylindrical opening in rock-II; Initiation, growth and interaction of fractures, Int. J. Rock Mech. & Min. Sci., vol.27, No.5, 409-427
- Hill R. (1963), Elastic properties of reinforced solids: some theoretical principles, J. Mech. Phys. Solids, 11, 357-372
- Kachanov M. (1992), Effective elastic properties of cracked solids: critical review of some basic concepts, Appl Mech Rev, Vol.45, No.8, 304-335

A study on the measurement of the shear strength of the Seoul granite by the multiple direct shear test

Du-Young Kim, Ji-Seon Yoon, Hee-Seong Lee, Hyun-Ick Yoon
Department of Mineral and Energy Resources Engineering, In-ha University, Korea

ABSTRACT: This paper describes multiple failure state direct shear test for the purpose of determining failure envelope. The major concern in this paper is the behavior of artificial joint specimen under the shear and normal stress. In a multiple failure state direct shear test, the specimen is preserved in a state of perpetual sliding while the shear and normal stress are being stepwise changed. In this test the normal stress was forced respectively 1.0, 2.0, 3.0 and 4.0MPa. In a individual direct shear test c(cohesion force) was 2.7MPa and ϕ (internal friction angle) was 58.0° meanwhile c, ϕ was 1.9MPa and 59.1° in multiple direct shear test. This paper intends to confirm the use of multiple direct shear test deciding strength constants c and ϕ with one specimen.

1. INTRODUCTION

In general, rock mass deformation is basically controlled by the discontinuities exists inside. It is necessary that the deformation of a rock mass be measured on the object of naturally existing rock mass including discontinuities. However, it is very difficult to collect undisturbed specimens including enough number of discontinuities and to be equipped with the same quality of specimens.

Triaxial compressive test and direct shear test are being done to determine the failure criteria. In general, the maximum limit of shear strength is given by the strength of the intact rock specimen and the lowest limit is given by that of joint surface specimen in failure criteria. So it is important procedure to determine the strength of joint surface affects on the basis of the stability analysis and the design of the rock mass structure.

In this study various kinds of physical property test, Brazilian test, uniaxial compressive test, triaxial compressive test and direct shear test were made in a laboratory setting in order to determine the failure criteria of a rock mass in Seoul granite. In particular, we introduced the multiple failure state test for direct shear test, which has been applied only to triaxial compressive test in domestic, and compared it with the result of individual direct shear test.

The results of triaxial compressive test for intact rock specimen, direct shear test and multiple direct shear test for artificially fractured joints were compared. A shear strength constants(c, ϕ) between direct shear test and multiple direct shear test were very approximate considering the specimen inhomogeneous and the roughness degree of joint. Direct shear test for saw-cut joint specimen was also made to investigate the strength variation with the condition of joint.

Hoek & Brown's empirical equation-(1980) was used to determine the shear strength of the rock mass according to rock mass quality and compared with the upper results.

2. THE PRINCIPLE OF MULTIPLE DIRECT SHEAR TEST

As the result of direct shear test on artificially fractured rock specimen, Fig 1-(a) shows the property of shear deformation of joint well. After applying

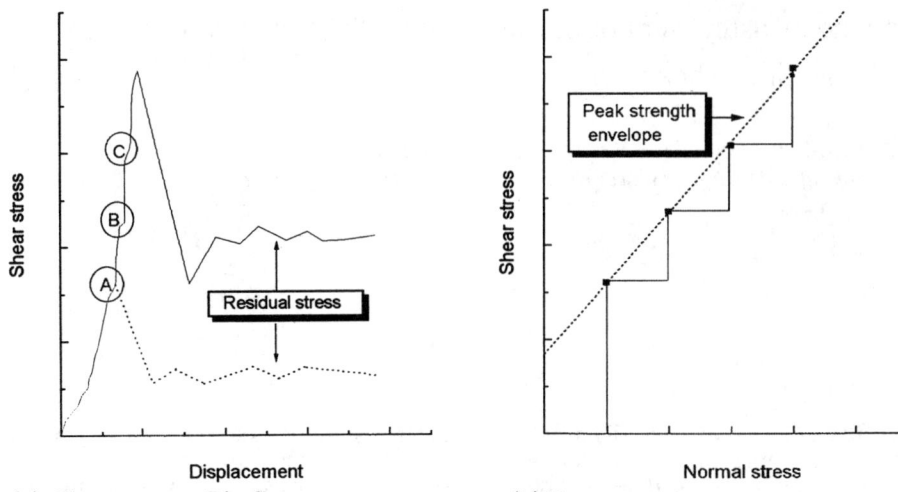

(a) Shear stress-Displacement curve (b) Shear stress-Normal stess curve
Fig 1. Schematic diagram showing the principle of multiple direct shear test

the normal force, the shear force is applied horizontally and it reaches at the peak shear strength, the shear force can continued in order to obtain the residual strength values.

The existing direct shear test is the method that we determine the failure envelope changing each normal load on the different joint specimens. But the problems of this procedure are to have restriction on the homogeneousness of specimen and the similarity of joint roughness. Even though we can determine the general failure criteria through the test at least four times, it is not the efficient method from time and economy point of view.

It is the multiple direct shear test method was introduced to compliment these disadvantages. Although the test method is similar with the direct shear test, it has the characteristic that we increase normal load at the earlier stage of nonelastic curve (A) under the first normal, continue to shear load until examine closely shear deformation of (B, C) position. These stepwise procedure continued to a chosen normal load and we can obtain failure envelope like Fig 1-(b) using one specimen. In general, multiple direct shear test is not the method that measures the unique rupture point under normal load but the method that obtains stress path suitable to failure envelope. This test method was introduced by 'A Tisa and K. Kovari' ("*Continuous Failure State Direct Shear Tests*, 1983)

3. TEST PROCEDURE

3.1 The properties of rock core sample and preparation of core specimen

The specimens of uniaxial compressive test, Brazilian test, triaxial compressive test, direct shear test and multiple direct shear test are granite collected from Seoul Bulkawngdong site which final excavation depth was 10~50m. Seoul granite is coarse grain granite. It is consisted mainly of Quartz, Feldspar, and Biotite. Augite, Microcline, Hornblend and Opaque minerals are included as accessories. Through the test for physical properties, the results were obtained as Table 1.

In this study, for the sake of performancing direct shear test and multiple direct shear test for joint, rock specimen was made in NX core, artificially separated by the splitter.

3.2 Triaxial compressive test

Triaxial compressive test instrument used in this test is the triaxial confinement chamber made by Sinco Co. Considering the excavation depth, confining pressure was loaded each 1.5, 3.0 and 4.5MPa.

The result of triaxial compressive test is shown in Table 2. Plotting the relation of the compressive strength and confining pressure, the linear regression analysis has been done to determine the c, ϕ by the

Table 1. Physical properties of Seoul granite

Properties	Range	Avg
Unit weight (g/cm^3)	2.48~2.72	2.57
Apparent porosity (%)	0.68~1.23	1.03
Uniaxial compressive strength(MPa)	91.5~265.9	127.5
Tensile strength (MPa)	6.8~9.5	8.1
Vp(m/s)	5670~5920	5770
Vs(m/s)	3110~3480	3220
Dynamic poisson's ratio	0.21~0.30	0.26

following equation,

$$\sin\phi = \frac{m-1}{m+1}$$

$$c = b\frac{m(1-\sin\phi)}{2\cos\phi}$$

where, m is the tangent of the inclination, b is its Y intercept.

3.3 Direct shear test and multiple direct shear test

Direct shear test instrument used in this test is potable shear box (model PHI-10) made by ROCTEST CO. The structure of direct shear test instrument is shown in Fig 2. The main constituents are one set of normal ram, two set of horizontal ram, shear box and roller carriage, and so on. Also mould set for setting the rock specimen in shear box is included. Normal ram loads vertical force and horizontal ram loads shear force, these rams have a 10 tons capacity and a working load limited to 5tons. In particular, pressure maintainer of normal load system is essential for successful shear tests, ensuring a constant normal load throughout the test.

The procedure of multiple direct shear test using the same instrument is as follows Fig 3.

According to this procedure, the relation of the maximum shear stress and each normal stress is plotted. Through the linear regression analysis, we calculate the cohesion force(c) and internal friction angle(ϕ) by Mohr-Coulomb failure criterion.

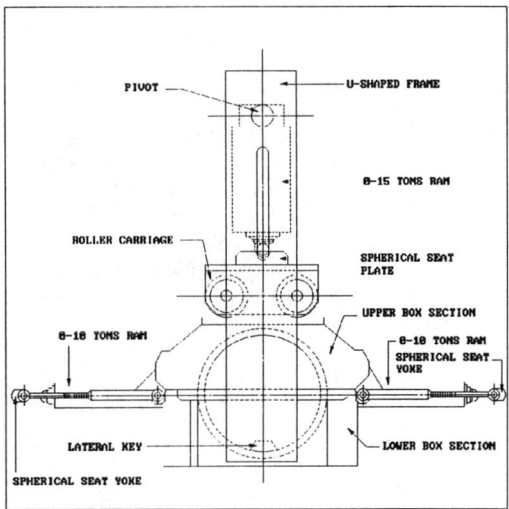

Fig 2. Portable shear box assembly model PHI-10

```
Cut off and polish the Seoul granite rock.
              ↓
Separate the specimen by the splitter.
              ↓
Adhere the specimen to the mould using
the epoxy resin, set it in the shear box.
              ↓
Establish the applications of normal force
and pressure maintainer setting.
              ↓
Load the first step of normal load
using the hydraulic pump.
              ↓
Load the shear force, report the shear
displacement(velocity of shear loading was
100.0~150.0N/min).
              ↓
Load the next step normal force when
shear stress reaches at near the yield point
in stress-displacement curve.
              ↓
Continue the stepwise procedure until the
chosen maxim normal load. When the shear
stress reaches at residual stress, the
multiple direct shear test is completed
```

Fig 3. Flow of the multiple direct shear test

4. RESULTS

4.1 Triaxial compressive test

The results of triaxial compressive test are followed as Table 2. Triaxial compressive strength is 141.8MPa, 184.3MPa and 225.7MPa at each confining pressure-1.5, 3.0 and 4.5MPa. It is obtained that the cohesion force is 9.5MPa, internal friction angle is 68.5°. It is thought that the reason of large internal friction angle is due to the narrow scope of the confining pressure, which do not reduce the triaxial compressive strength.

Table 2. Results of triaxial compressive test.

Confining pressure (MPa)	Triaxial compressive strength(MPa)
1.5	141.8
3.0	184.3
4.5	225.7
Shear strength (MPa)	9.5
Internal friction angle(°)	68.5

4.2 Direct shear test

The direct shear test results for the saw-cut flat joint of rock specimen are shown in Fig 4 and Fig 5. The normal stress was applied to three stage. As the normal stress increased, the increase of shear strength was decreased and the cohesion force was 0.6MPa, the basic friction angle of granite was 22.8°. After the maximum strength, the residual shear strength was calculated by the average value at the scope of few difference. The cohesion force for residual shear stress was decreased to 0.4MPa, but internal friction angle was increased to 28.8°. This result is due to continuous increase of residual shear stress after the maximum shear strength.

Direct shear test for artificially jointed specimen was performed under the normal stress of four stage 1.0, 2.0, 3.0 and 4.0MPa. The five specimens were tested at each normal stress, results are shown as Fig 4 and Fig 5. The cohesion force of Seoul granite is 2.7MPa, internal friction angle is 58.0°. The cohesion force for residual shear stress is 0.7MPa, internal friction angle is 32.0°. Though the result of this test was unlike the one for flat joint, it shows the rapid increase of shear strength because of the surface roughness. It is observed that the residual shear stress is 34~36% level of the maximum shear strength.

4.3 Multiple direct shear test

The multiple direct shear test for artificially jointed specimen was performed, the results are shown as Fig 4 and Fig 5. The results of multiple direct shear test for

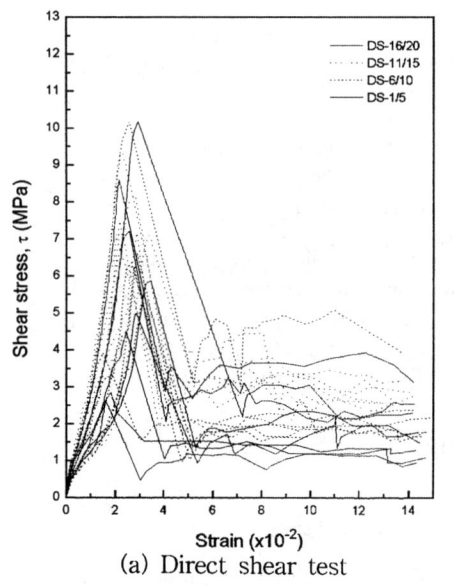
(a) Direct shear test

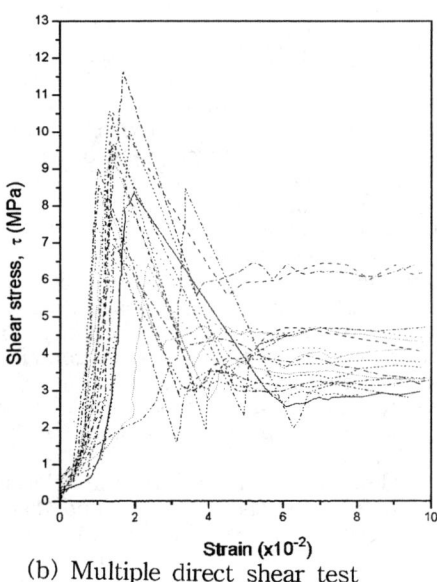
(b) Multiple direct shear test

Fig 4. Stress-strain curve of individual and multiple direct shear test

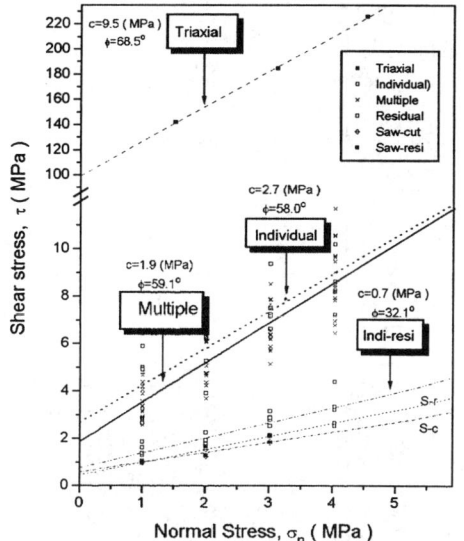

Fig 5. Relationship between normal stress and peak shear strength for joint specimens

17 specimens, cohesion force is 1.9MPa, internal friction angle is 59.1°

In direct shear test, the shear stain at peak strength was 1.48% ~ 3.47%. Though the shear strain at peak strength was 0.67% ~ 3.16% under the first normal stress, 0.13% increase on an average under the second normal stress(2.0MPa), 0.07% increase under the third normal stress(3.0MPa) and 0.16% increase of the shear strain to reach the failure point under the fourth normal stress(4.0MPa).

4.4 Determination of the shear strength using the Hoek & Brown's empirical equation

Hoek & Brown's empirical equation determines the shear strength of intact rock specimens and heavily jointed rock masses is

$$\tau_n = A(\sigma_n - \sigma_{tn})^B$$

where, τ_n, σ_n are normalized shear and normal stresses, τ/σ_c and σ/σ_c and σ_{tn} is the normalized uniaxial tensile strength which is defined by

$$\sigma_{tn} = \sigma_t/\sigma_c = \frac{1}{2}(m - \sqrt{m^2 + 4s})$$

m and s are material constants which depend upon the properties of the rock. A and B are empirical constants from the regression analysis. The results of application compared with the values of triaxial compressive test, direct shear test and multiple direct shear test are shown as Table 3.

5. CONCLUSION

From the results of the shear tests on Seoul granite specimens with fractured joint samples, following(Table 4.) can be obtained.

The shear strength of intact rock specimen was differ from artificially fractured joint specimen's due to the effect

Table 3. Comparison of the shear strength by the tests with that of Hoek & Brown's empirical equation on Seoul granite

Classification	Empirical equation	Result (MPa)	Triaxial compressive strength (MPa)	Direct shear strength (MPa)	Multiple direct shear trength (MPa)	Residual shear strength (MPa)	SC shear strength (MPa)
Intact rock samples	$\tau_n = 1.220(\sigma_n + 0.040)^{0.705}$	$\tau = 18.9$					
Very good quality rock mass	$\tau_n = 0.998(\sigma_n + 0.008)^{0.712}$	$\tau = 7.4$	↑ 9.5				
Good quality rock mass	$\tau_n = 0.603(\sigma_n + 0.002)^{0.707}$	$\tau = 3.4$					
Fair quality rock mass	$\tau_n = 0.346(\sigma_n + 0.0002)^{0.700}$	$\tau = 1.8$		↑2.7	≒1.9		
Poor quality rock mass	$\tau_n = 0.203(\sigma_n + 0.0001)^{0.686}$	$\tau = 1.1$					
Very Poor quality rock mass	$\tau_n = 0.078(\sigma_n)^{0.556}$	$\tau = 0.8$				≒0.7	↓0.6

Table 4. Comparison of shear strength for Seoul granite

	Triaxial compressive test	Direct shear test		Multiple direct shear test	Saw-Cut direct shear test	
		peak	residual		peak	residual
Shear strength (MPa)	9.5	2.7	0.7	1.9	0.6	0.4
Internal friction angle (°)	68.5	58.0	32.0	59.1	22.8	28.8
Value,%	100	28.4	7.4.	20.0	6.3	4.2

of joint. Though there was few difference between the results of individual direct shear test and multiple direct shear test. Individual direct shear test result was 28.4% degree of triaxial compressive test's on intact rock and multiple direct shear test result was 20.0% degree.

There was few difference between the shear strength (6.3%) of saw cut flat joint and the residual shear strength (7.4%) of artificially fractured joint in direct shear test. It explains that the shear behavior of joint which asperity ruptured approaches to the contact behavior of the rock surface itself.

In direct shear test, the shear stain at peak strength was 1.48% ~ 3.47%. In multiple direct shear test, the shear strain at peak strength under the first normal stress was 0.67% ~ 3.16%, 0.13% increase on an average under the second normal stress(2.0 MPa), 0.07% increase under the third normal stress(3.0MPa) and 0.16% increase of shear strain to reach the failure point under the fourth normal stress(4.0Mpa).

The comparison of the shear strength through the test result with that result by the Hoek & Brown's empirical equation is Table 3. The shear strength by the triaxial compressive test is near to the 'Very good quality' condition, that of individual direct shear test is more close to the 'Fair quality' condition than 'Good quality' condition and that of multiple direct shear test is approximate at the 'Fair quality' condition. Consequently, the difference of test result between direct shear test and multiple direct shear test is so little as not to affect on the rock mass classification.

6. REFERENCES

1) A. Tisa & K. Kovari 1984. *Continuous Failure state Direct shear Tests* :83-85. Rock Mechanics and Rock Engineering 17.
2) E. Hoek & Brown. 1980. *Underground Excavation in Rock* : 95-177. The institution of Mining & Metallurgy, London.
3) E. T. Brown. 1981. *Rock characterization Test & Monitoring* : 123-127. Pergamon press Ltd.
4) Lee, S.D. , Kang, J.H. & C.I., 1994. S*hear strength and deformation behavior of rock joint roughness*, J. of Korean Rock Mechanics Society 4:261-273.
5) Yoon, J. S. 1991. *Rock mass Investigation and Test.* 559-573. Kumiseokwan

Fundamental fracture modes of granite using new testing devices

M.P. Luong
CNRS-LMS, Ecole Polytechnique, Palaiseau, France

N. Hoteit
ANDRA, Châtenay Malabry, France

ABSTRACT: The paper presents an investigation on fundamental fracture modes of granite, using new testing devices in order to highlight the salient features of fracture processes in granite such as path dependence (mode I tension fracture, mode II shear fracture and mode III torsion fracture), and shear transfers due to aggregate interlock (crack dilatancy), cross effect (frictional contact slip). Experimental results provided useful mechanical characteristics such as cracking strength, stiffness, and fracture energy according to the three fundamental fracture modes, and particularly the Hillerborg characteristic length of granite in tension readily derived from direct tension testing. Computerised X-ray tomography analysis evidences localisation of cracking in specimens subject to different fracture modes, and finally a non destructive ultrasonic technique has been used to detect the occurrence of damaging non linearities.

1 INTRODUCTION

The literature on rock excavation technology often includes such terms as grinding, wearing, cutting, breaking, shearing, scraping, melting, chipping, scabbing, slabbing, or spalling, but these terms are seldom defined. A realistic mechanical understanding of fundamental rock fracture modes is thus necessary in the search for greater efficiency and effectiveness of excavation techniques, particularly relevant in cases of rock fragmentation by blasting (Fourney 1983). The majority of rock drilling and boring techniques employ principles of mechanical attack by means of which stresses are induced in the rock by impact, abrasion, wedge penetration or cutting, erosion, and sometimes by a combination of these methods.

Fracture processes in granite are substantially influenced by the stiffness and strength of their constituent particles. An important aspect of crack propagation in granite is to examine the effects of dilatancy during the evolution of failure and to detect the early occurrence of micro cracking.

Experimental observations show that the shear displacement or shear slip is affected not only by the shear stress, but also, by the normal displacement owing to rough asperities of aggregate or surface roughness and variable granite strength. The normal stresses across the interface in mode II shear tests and mode III torsion tests are measured and applied in different ways, which result in the different numbers of Belleville springs, ensuring different shear stiffnesses. These factors must be taken into account when measuring toughness properties of brittle disordered materials such as granite.

2 DIRECT TENSION STRENGTH

Tensile strength of quasi brittle materials is one of their most fundamental and important properties, particularly because these materials are very much weaker in tension than in compression. In excavation technology, fracture processes in blasting mainly involve tensile strength.

Most rock materials are more or less brittle. When unconfined, the test samples cannot yield plastically to relieve the stress concentrations that are produced at the localised points around the specimens, where these are gripped to the pulled apart by the testing machine. Consequently premature failure is generated from these points. Difficulties in ensuring truly axial loadings also exist, so that the specimen is liable to be twisted or bent when gripped and pulled from either end. It has been reported that different fracture energy values were obtained if different specimen geometries were used (Jenq & Shah 1989). Does this difference result from different testing set-ups involved: different degree of eccentricity, alignment and support conditions, etc.?

The proposed test specimen (Luong 1989) is a cylindrical tube (Fig. 1) easily prepared, with two parallel flat ends and two inverse tubular coaxial borings. The external surface requires no particular preparation. The proposed configuration converts the applied load on the specimen into a tension, so that the usual compressive test machine can be used. The cylindrical symmetry permits self alignment of loads parallel to the axis of the specimen. The test requires no special device for specimen gripping and compressive loads for example can be applied without any precaution. There will be no tendency to cause bending so that abnormal stress concentrations are avoided.

The main advantage of the proposed direct tensile test is evidenced by the uniformity of uniaxial tension stress throughout the test volume. Thanks to its cylindrical symmetry, there is no bending or torsion stresses, no stress concentrations arising from geometrical irregularities of the specimen and no end restraint effects perturbing the stress field: most of observed failures occur in the central part of the specimen.

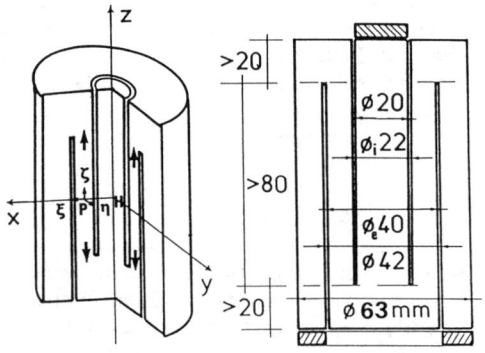

Fig. 1 - Direct tension (mode I) test specimen.

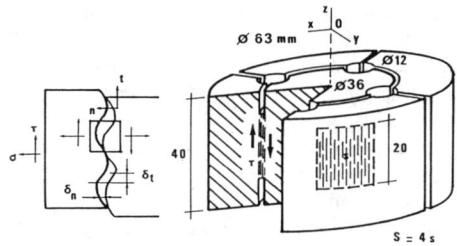

Fig. 2 - Direct shear (mode II) test specimen.

3 DIRECT SHEAR STRENGTH

Shear strength is often used to cover several different concepts (Everling 1964) such as (i) strength against pure shear, (ii) shear stress required for failure without normal stress, (iii) shear diagram on solid interface depending upon normal stress and (iv) Mohr's stress envelope. Several methods have been devised for measuring the shear strength of rock: torsion test, single shear test, double shear test, punch test, shear loaded beam with starter notches according to mode II of fracture mechanics.

The main requirements of a mode II test specimen must be: simple compact geometry, ease of preparation, simple loading system and stress conditions little affected by extremely small geometric alterations. Attempts to realise mode II crack propagation on a laboratory scale usually fail because mode I growth becomes predominant. On the other hand, traces of mode II growth are often detected at or after earthquake slipping. The present paper uses a direct shear test (Luong 1988) that maximises mode II conditions of crack propagation.

The axial load F is applied on the central cylinder of the specimen that is retained by the four concentric tubular parts (Fig. 2). The notches define the sheared zone $S = 4s$. The specimen geometry has been optimised by a numerical analysis of stress intensity factors in mixed modes' problems by path independent integrals J_I and J_{II} (Bui 1982). A criterion for deciding whether the crack will grow in mode I by formation and extension of a kink or in mode II in the original crack direction, was established similarly as done by Melin (1986), by comparing the stress intensity factors K_{Imax} for a small kink with K_{IImax} for the main crack. If K_{Ic} and K_{IIc} denote the critical stress intensity factors for mode I and II respectively, mode II growth will be preferred to mode I if

$$\chi = K_{IImax} / K_{Imax} > K_{IIc} / K_{Ic} = \chi_c$$

Several series of tests have been performed on specimens of granite subjected to diverse values of normal stress and different types of loading on the sheared interface in order to analyse the dilatancy effect. The nominal mode II shear strength is given by $R_{II} = F/4s = \tau$ in connection with the normal stress $\sigma = F_n/s$ where F_n denotes the lateral force directly applied on the failure surface using a variable number of coned discs or Belleville springs for the control of normal loading stiffness.

It is commonly considered that the strength of rock will become constant if sufficient relative movement is developed (Barton 1976). Some authors suggested that the displacement necessary to develop the residual state increases with rock strength (Krsmanovic et al 1966). Various rocks behave differently with regard to relations of the peak and the residual strengths. The main parameter is the stiffness of the normal loading, able to partially prevent the aggregate interlock breakdown, which softens the mechanical response of the materials.

Fig. 3 - Measuring dilatancy effects in direct shear testing.

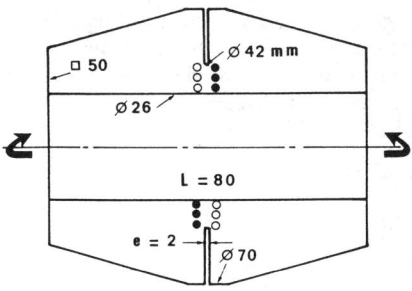

Fig. 4 - Direct torsion (mode III) test specimen.

When the tangential displacement or shear slip δ_t occurs along the crack interface, the shear stress τ working parallel to the crack is induced and is accompanied by the compressive stress σ and the normal displacement δ_n or crack width normal to the crack plane. These four parameters define the deformational characteristics of the cracked interface (Yoshikawa et al 1989). The specimen subjected to direct shearing, defined as mode II of fracture mechanics, is loaded as shown in Fig. 3 in order to monitor the dilatancy effects.

4 DIRECT TORSION STRENGTH

When studying the torsion responses of members, it is convenient to identify two different modes of torsion resistance: (i) circulatory torsion and (ii) warping torsion. In the former case, the torsion shear stress remains constant in magnitude of the longitudinal stresses. In the latter case, torsion shear stresses are not uniform and account must be taken of the variable effects of dilatancy in regard to the difference of normal stiffnesses.

Conventional techniques of determining shear strength of rock materials used indifferently tests that promote predominantly mode II or mode III of fracture mechanics (Lama & Vutukuri 1978). In this study, a tubular cylindrical specimen for direct torsion testing is proposed for the determination of mode III fracture strength of quasi brittle materials. The specimen resists the applied torque by circulatory torsion. The shear flow lies some small distance inside the outside surface of the notch tip. The two ends are prepared as square cross section prisms. A thin cut is turned in the central part of the specimen.

It is assumed that, in spite of the low tensile strength of quasi brittle materials, fracture will occur from the shearing stresses in mode III, since the screw-shaped surface of tensile fracture would be much larger than the thin round neck under torsion. When it is subjected to torsion T by holding its ends in chucks, maximum mode III shear stress develops at the outermost fibre. With an external radius r and an internal radius mr of the same order of magnitude, the nominal mode III strength R_{III} (Fig. 4) can be given by

$$R_{III} = T / \pi r^3 (1 - m^4)$$

The fracture surfaces appear complex, consisting of inclined facets, known as *factory roof* fracture surfaces by metallurgist workers, formed by the intersection of a large number of independently initiated mode I cracks. Thanks to the thin round neck, mode III or torsion fracture has been preferred as shown by the fracture surfaces, mode I fracture being controlled by the stiffness of normal stress loading or dilatancy effect.

The normal stress $\sigma = F_n / \pi r^2 (1 - m^2)$ is applied on the specimen by means of a long screw. The stiffness of normal stress loading on the failure surface is controlled by one or several Belleville springs just as described by Fig. 5.

Fig. 5 - Measuring dilatancy effects in direct torsion testing.

5 FRACTURE ENERGY

The majority of rock drilling and boring techniques employ principles of mechanical attack by means of which stresses are induced in the rock by impact, abrasion, wedge penetration or cutting, erosion, and sometimes by a combination of these methods. Thus there is a need to evaluate energy spent for fragmentation of rock. The performed tests in this work readily provide energy characteristics to bring the rock materials to fracture in different modes (Fig. 6).

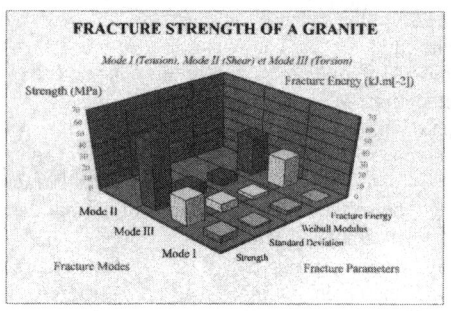

Fig. 6 - Fracture characteristics of the tested granite.

6 DETECTION OF MICRO CRACKING

In most rocks, both the acoustic velocity and the attenuation vary greatly. According to Koltonski and Malecki (1958), granite, for instance, depending on its grain structure can reach values from 1.7 to 5.0 km.s^{-1}. The non destructive testing (NDT), most extensively used at present, is probably the ultrasonic technique. Several researchers (Jones 1952 for instance) used sonic and ultrasonic method to study the fracture of concrete and reported that crackling noises occurred at about 25% to 75% of ultimate load. It was recognised that these significant changes in the property of concrete and rock materials are probably caused by micro cracking, resulting in the increase of volume under compression and the increase of Poisson ratio.

An input-output non parametric procedure (Liu & Vinh 1991), based on ultrasonic pulse propagation and using a non linear analyser for data reduction, has been chosen to portray the non linear behaviour of granite (Luong 1995).

The pulse transmission method, using video scan piezoelectric transducer V150 (250 kHz), has been applied in conjunction with a pulser-receiver Panametrics 5052PR that provides high energy broadband performance. The pulse travels through the specimen that is subjected to a given static shearing, to a second transducer acting as a receiver. The transducer converts the pulse into an electric signal that is then amplified and conditioned by the receiver section and made available for non linear analysis.

Fig. 7 - Arrangement for microcracking detection.

Signals records indicate that at 42% of mode II shear strength, the high order kernels are still negligible (Fig. 8a). They become very significant at 70% (Fig 8b).

7 X-RAY TOMODENSITOMETRY OF GRANITE

Originally used for medical analysis (Hounsfield 1973), the X-ray computerised tomography (CT) technique has been recently used for engineering materials (Vinegar, 1986, Latière & Mazerolle 1987, Withjack 1987, Colleta et al 1991).

As for conventional radiography, the basic principle of the CT technique is the attenuation of X radiations through materials. This attenuation is mainly dependent on material density, effective atomic number, and thickness. In computerised tomography, the object is penetrated by a planar fan of rays, and attenuation is measured by 1024 detectors. Source and detector are attached to a rotor and move on a circular track around the object.

Attenuation is determined at a rate of 150 measurements per second on each of the 1024 detectors during a 360° rotation, which lasts for 3.4, 6.8 or 13.6s. During the rotation the X-ray fan investigates a slice of the object that may be 1.5mm, 3mm or 8 mm thick. The large number of measurements (about 10^6 for a 6.8s time of acquisition) enables attenuation values to be computed for a 512×512 matrix of elements of volume (voxels) in the investigated slice. The computed values produce a 512×512 pixel image, which represents a cross section of the object. The scanned specimen is translated horizontally through the investigation plane by a motor-driven bed with an accuracy of 1mm. Thus it is possible to obtain serial cross sections of the model. Attenuation values, which are given in Hounsfield units (HU) for 130kV energy, are compatible with the range of the medical scanner. One of the most significant advantages of computed tomography is its sensitivity to very small differences in density. Voids and inclusions can be identified and localised. The internal cracks are visualised (Fig. 9 - 10 -11).

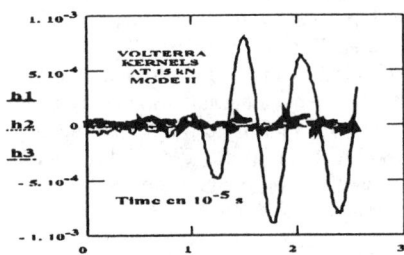

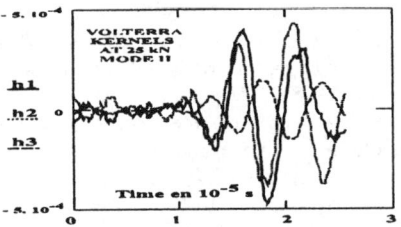

Fig. 8 - Non linearities corresponding to micro cracking occurrence.

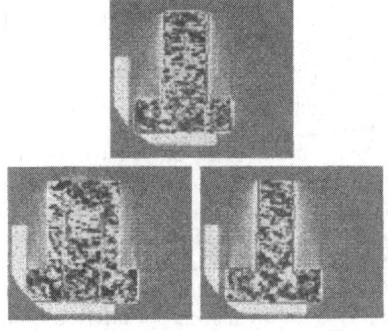

Fig. 9 - Tomography of a mode I specimen.

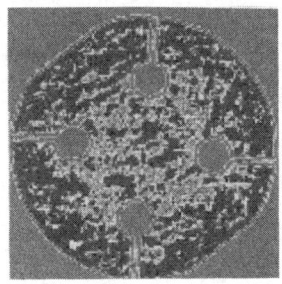

Fig. 10 - Tomography of a mode II specimen.

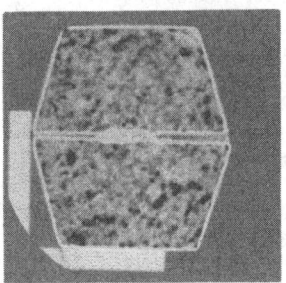

Fig. 11 - Tomography of a mode III specimen.

8 CONCLUDING REMARKS

The versatility of these new testing devices allows a ready evaluation of the behaviour of granite during mode I direct tension, mode II direct shear and mode III direct torsion loadings.

The proposed testing arrangements for both direct shear and torsion shear strength measurements simplify the loading equipment by the use of a uniaxial test machine with the combined compression, shear and torsion loading frames described in the paper. The test set-up allows the normal stresses to be monitored and measured during the experiments under varying shear stresses. The resulting responses are shear stress versus slip and shear stress versus normal stress, which are of course very useful in the interpretation of conventional pull-out tests. They are practical and reliable. They also facilitate the testing procedure when the materials are subjected to severe and hostile environmental conditions. Finally they can be used as routine tests more readily achieved than on the conventional triaxial tests in the cases of low mean stress level.

In addition they demonstrate the significance of fracture modes on the mechanical behaviour of granite.

9 REFERENCES

Bui, H.D. 1982. Découplage du mode mixte de rupture par utilisation de deux nouvelles intégrales indépendantes du contour. *C.R. Acad. Sci. Paris,* 235:521-525

Barton, N. 1976. Rock mechanics review: the shear strength of rock and rock joints. *Int. J. Rock Mech. Min. Sci. & Geomech. Abstr.* 13:255-279.

Colletta, B., J.Letouzey, R.Pinedo, J.F.Ballard & P. Balé 1991. Computerized X-ray tomography analysis of sandbox models: Examples of thin-skinned thrust systems. *Geology:* 19:1063-1067.

Everling, G. 1964. Comments upon the definition of shear strength. *Int. J. Rock Mech. Mining Sc.* 1:145-154.

Fourney, W.L. 1983. Mechanisms of rock fragmentation by blasting. *Comprehensive Rock Engineering* 4(2):39-69.

Hillerborg, A. 1989. Fracture mechanics and the concrete code. *Fracture Mechanics: application to concrete* ACI, SP-118:157-169.

Hounsfield, G.N. 1973. Computerized transverse axial scanning (tomography): Part I. Description of system. *British J. Radiology* 46:1016-1022.

Jenq, Y.S. & S.P.Shah 1989. *Geometrical effects on mode I fracture parameters.* Report to RILEM Committee 89-FMT, January 1989.

Jones, R. 1952. A method of studying the formation of cracks in a material subjected to stresses. *British J. Applied Physics,* V 3(7):229-232.

Koltonski, W & I.Malecki 1958. Ultrasonic method for the exploration of the properties and structure of mineral layers. *Acoustica,* 8:307-314.

Krsmanovic, D., M.Tufo & Z.Langof 1966. Shear strength of rock masses and possibilities of its reproduction on models. *Proc. 1st Cong. Int. Soc. Rock Mech.* Lisbon, 1:537-542.

Lama, R.D. & V.S.Vutukuri 1978. *Handbook on mechanical properties of rock.* Trans Tech Pub. 1.

Latière, H.J. & F.Mazerolle 1987. The X-ray scanner. A tool for the examination of the intravoluminal crystalline state of aluminium. *Engng Fract. Mech.* 27(4):413-463.

Liu, H. & T.Vinh 1991. Multidimensional signal processing for nonlinear structural dynamics. *Mechanical Systems and Signals Processing* 5(1):61-80.

Luong, M.P. 1990. Dilatancy effects on mode II and mode III shear strength of quasi-brittle materials. *Micromechanics of failure of quasi-brittle materials,* S.P.Shah, S.E.Swartz & M.L.Wang (eds) Balkema:157-162.

Luong, M.P. 1992. Fracture testing of concrete and rock materials. *J. Nuclear Enginering and Design* 133:83-95.

Luong, M.P. 1995. Non destructive detection of micro cracking in rock. *Rock Foundation,* Yoshinaka & Kikuchi (eds), Balkema:157-162.

Melin, S. 1986. When does a crack grow under mode II conditions? *Int. J. Fracture* 30:103-114.

Thomas, J.J. 1995. *Contrôle non destructif des matériaux et des structures par analyse dynamique non linéaire.* Doctorat Ecole Polytechnique, Palaiseau France.

Vinegar, H.J. 1986. X-ray CT and NMR imaging of rocks. *J. Petroleum Technology* March:257-259.

Withjack, E.M. 1987. Computed tomography for rock-property determination and fluid-flow visualisation. *Society of Petroleum Engineers* SPE 16951:183-196.

A new empirical failure criterion using data from triaxial tests for intact rocks

Chulwhan Park & Chan Park
Korea Institute of Geology, Mining and Materials, Taejon, Korea

Yoenjun Park
Department of Civil Engineering, University of Suwon, Korea

ABSTRACT : Many failure criteria have been developed theoretically and empirically based on Mohr's postulation which is still valid for rock failure. But none of them can satisfactorily be applicable to all types of rocks. A new empirical failure criterion is proposed in this paper to predict rock strength under $\sigma_2 = \sigma_3$ condition. The equation of failure envelope is expressed in the σ-τ domain in order to estimate the physical properties easily and to be applied to numerical analysis program directly. The equation is represented with 2 constants as follows,

$$\tau = B \cdot (\sigma - T)^a$$

and expresses the parabolic curve.($a = 0.5 \sim 1.0$) Two empirical constants can be determined using data from triaxial compressive tests while T may be tensile strength. To verify the validity of the new equation, triaxial data from other criteria, selected papers and authors' tests are used. All data mentioned in the analysis are those for intact rocks. It will be the next problem that this new criterion can predict the failure strength if in-situ rock mass.

1 INTRODUCTION

The maximum principal stress is proportional to the minimum principal stress under triaxial compression test. The problem is that the relationship is not linear for rocks, and furthermore does not show the same tendency for every type of rock. Many criteria have been developed to describe the failure envelopes theoretically and empirically.

Mohr's postulation is still valid for rock failure. Coulomb-Navier expresses the linear Mohr's envelope which is well satisfied for most igneous and hard crystalline rocks. Griffith's criterion modified by Murrell (1963) represents the parabolic Mohr's envelope, which can be effective in some sedimentary rocks.

The empirical criteria proposed by von Karman, Bieniawski and Hoek-Brown are well known. Von Karman described that its relation is linear for brittle rocks, and USBM in 1953 reported to show its validity with empirical constants $k = 5 \sim 20$ for many types of intact rocks through triaxial compression tests (Obert 1967). This relationship can be expressed with normalized stress terms as in eq.(1) where C_0 is unconfined compressive strength.

$$\frac{\sigma_1}{C_0} = 1 + k \frac{\sigma_3}{C_0} \quad (1)$$

Bieniawski proposed eq.(2), in which the constant B varies 3 to 5 and the exponent a is 0.75 for all types of rocks (Bieniawski 1984).

$$\frac{\sigma_1}{C_0} = 1 + B \left(\frac{\sigma_3}{C_0}\right)^a \quad (2)$$

Empirical criterion proposed by Hoek-Brown is expressed in eq.(3) as the exponent a is 0.5 and the constant m varies 5 to 30 or more. This equation is more popular because it can be used for in-situ rocks by adjusting the constant s, whereas the above two are verified from the tests on intact rocks (Hoek 1980 and Vutukuri 1994).

$$\frac{\sigma_1}{C_0} = \frac{\sigma_3}{C_0} + \left(s + m \frac{\sigma_3}{C_0}\right)^a \quad (3)$$

Fig.1 shows the σ_1-σ_3 relationships for sandstone for which $k = 7.0$, $B = 4.0$ and $m = 9.6$ respectively in three different equations. Failure envelopes derived from those data are different for each case, which means none can be satisfactorily applicable to all types of rocks. Furthermore as equations are expressed in terms of principal stresses, it is inconvenient to identify the shear strength(c) and the angle of internal friction(ϕ) to apply directly to numerical programs for the design and safety analysis of the structures in rock.

2 SUGGESTION

Authors are going to suggest a new empirical criterion for rock failure applicable to all types of intact rocks. It is based on Mohr's shear failure criterion. The equation representing the failure envelope is expressed in the σ-τ domain in order to estimate the physical properties easily and to be applied to numerical analysis program directly.

It is presented with 3 constants basically as follows,

$$\tau = B \cdot (\sigma - T)^a \quad (4)$$

This equation is defining that the exponential constant a varies in the range of 0.5~1.0, therefore failure envelope is represented by a parabolic curve. No other criteria express the constant out of this range. If it is 1, this equation becomes linear envelope and practically identical to the Coulomb-Navier's theory and eq. (1), and if 0.5, exact parabola, Griffith-Murrell's and eq.(3). The triaxial test data which is outside of the given range of exponent in analyzing will be rejected. In this criterion, σ_2 is neglected and triaxial test may be carried out under $\sigma_2 = \sigma_3$ condition.

If T is the tensile strength (T_0), ϕ can be 90° and tensile stress is limited at T_0 which coincides with Mohr's theory. If tensile test is not carried out, T_0 can be estimated from the unconfined and confined compression data. T_0 can be assumed as C_0/k or C_0/m because constants k and m are very close to the brittleness index for most intact rocks (Park 1996). Its more discussion will be entered into details with actual data in next section.

After identifying the T value, we can compute the other 2 constants by non linear regression analysis of data from triaxial compression tests including C_0. Constants are determined as an envelope is found, which adjoins the Mohr's circles tangentially with the least error. This suggested criterion is followed that the functional relationship must be determined experimentally as Mohr explained. The mechanical values of c and ϕ can be easily recognized not only from the functional equation but from the figure plotted in the σ-τ domain.

With the data sets for sandstone predicted by eq.(1) and eq.(3), two failure envelopes can be drawn respectively by the new equation as in Fig.2. Five values of minimum principal stresses as high as $0.25C_0$ are used for every envelope in this analysis. Two of five Mohr's circles are selected to plot in Fig.2. The envelope of Hoek-Brown's sandstone is expressed with $a = 0.70$ while $a = 0.94$ for von Karman's sandstone, almost linear as the results of analysis of the new equation.

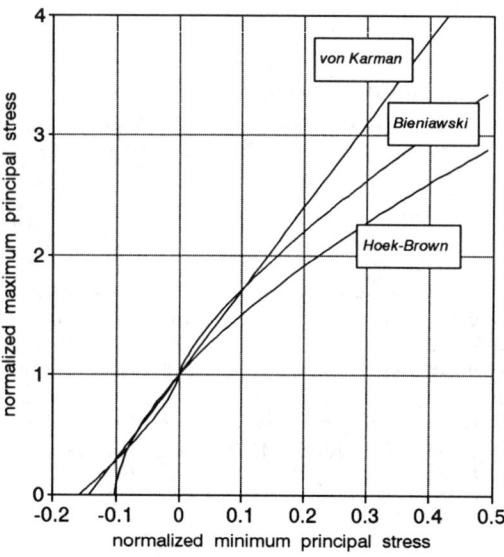

Fig. 1. Comparison of σ_1-σ_3 relationships for three different criteria

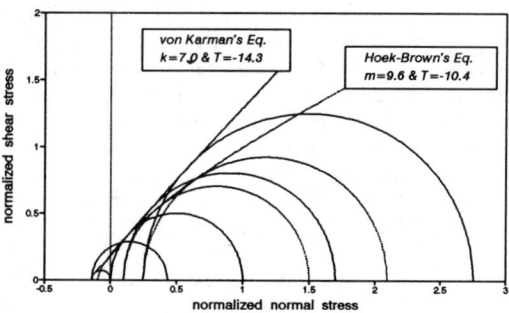

Fig. 2. The failure envelopes for sandstones from the analysis of the suggested criterion

3 DATA ANALYSIS

In order to verify its application, the suggested failure criterion will be discussed with the experimental data in selected reports. 12 types of rocks are mentioned to analyze in this section. Their unconfined compressive strengths are various, as from 24 MPa to 350 MPa. Tensile strengths of 2 rocks which were not announced in reports are assumed from m value. All the triaxial compression data were taken under $\sigma_2 = \sigma_3$ condition. The analyzed results are summarized in Table 1.

The error which is the percentage gap between the actual failure strength and the predicted one from analyzed criterion, has the range of 0.1% to 5.5% in average and less than 4% in most cases. Its individual maximum value is 13.3%, which comes from potash ore rock - No.1 triaxial test. The predicted one is analyzed as 51.2 MPa while the actual one is 44.4 MPa for $\sigma_3 = 3$ MPa.

The exponents are determined in the range of 0.62~0.97. It is perfectly independent from C_0 or T_0, but dependent on constant m with R=0.95 shown as filled square mark and relation line in Fig.3. Blank square mark represents the exponent - brittleness index relationship which is not good, R=0.76 so much as expected.

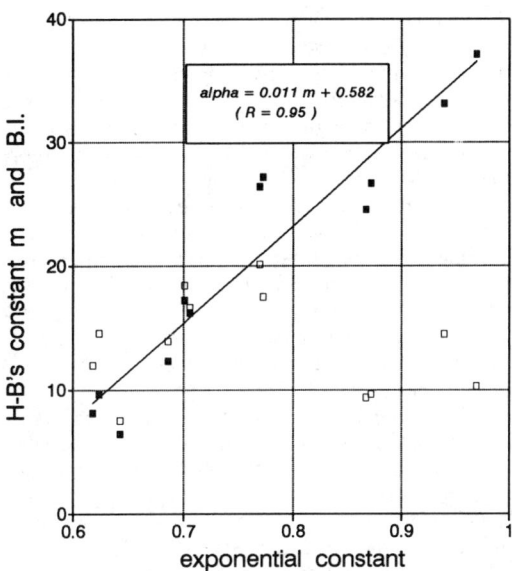

Fig. 3 Hoek-Brown's constant and brittleness index vs exponential constant

For the third constant T, more detailed data for Ainoura sandstone is listed in Table 2. T_0 may be estimated as 5 MPa from the m values, which can cause the least error in analysis

Table 1. The exponential constants and errors for various types of rocks

rock type	C_0 (MPa)	T_0 (MPa)	constant m	constant a	error (%) mean & max.	sampling site and experimenter
halite	24	2.5	26.7	0.872	3.7 & 10.7	Spain, Campos De Orellana 1996
potash ore	28	3.0	24.6	0.866	5.5 & 13.3	Spain, Campos De Orellana 1996
granite (w)	39	3.8	37.1	0.969	3.2 & 7.3	Taejon, Korea, Authors 1995
sandstone	56	7.5	6.4	0.642	4.0 & 7.4	Samtan mine, Korea, Park 1987
marble	78.2	5.6	12.3	0.686	0.3 & 0.4	Jeonju, Korea, Lee 1984
sandstone	83	5*	16.3	0.653	1.3 & 2.7	Ainoura, Japan, Kimura 1988
granite (f)	116	8.0	33.1	0.939	1.7 & 5.0	Taejon, Korea, Authors 1996
gneiss	130	10.9	8.1	0.618	3.9 & 4.8	Kyungki, Korea, Authors 1995
tuff	147	8	17.2	0.701	1.9 & 5.8	Nevada, USA, Wang 1995
granite	167	8.3	26.4	0.769	2.7 & 4.2	Hwangdung, Korea, Authors 1995
limestone	197	13.5	9.6	0.624	0.1 & 0.1	Jangsung mine, Korea, Park 1987
quartzite	350	20*	27.2	0.772	2.5 & 4.7	South Africa, Sellers 1996

remark : * tensile strength was not reported.

among the other values, 3~9 MPa. With this range of T_0, nevertheless a varies from 0.65 to 0.80 in Table 2, the envelope is similar to each other as shown in Fig. 4. The predicted triaxial compressive strengths will change greatly only for normal stresses less than 4 MPa or more than 80 MPa on various values of T_0. From the analyzed result in Table 2, a becomes larger as T_0 is selected with a higher value. Also envelope will be closer to linear and larger error will be computed for data with smaller σ_3. In addition, if choosing $T = -0.166C_0$ by Mohr-Coulomb's linear envelope for von Karman's sandstone in Fig.2, analysis yields $a = 0.998$ with 0.1% error.

Fig.5 which is another example, shows 3 envelopes for Apache Leap tuff for maximum, minimum and average of a analyzed. The tensile strength was not mentioned in Wang's report (1995), but Ghosh (1995) tested it on scale effects and reported 5.1~8.1 MPa. Even if the average was estimated as 6.5 MPa, the higher value 8 MPa is used here by the reasons mentioned above. As a result of this analysis, the failure envelope for welded Apache Leaf tuff is expressed in eq. (5) and plotted as intermediate curve in Fig.5. The linear envelope which adjoins the the Mohr's circles tangentially is also plotted in the lower level of normal stresses.

$$\tau = 4.954 \cdot (\sigma + 8)^{0.701} \quad (5)$$

Values of the shear stresses and the angles

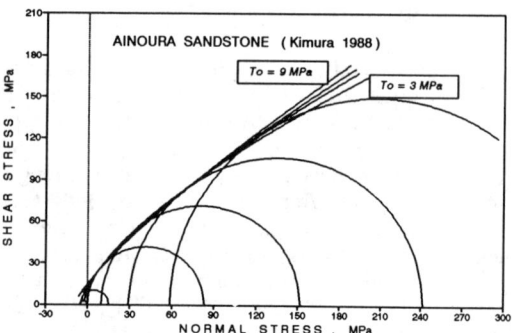

Fig. 4 Comparison of failure envelopes with the various tensile strengths for Ainoura sandstone

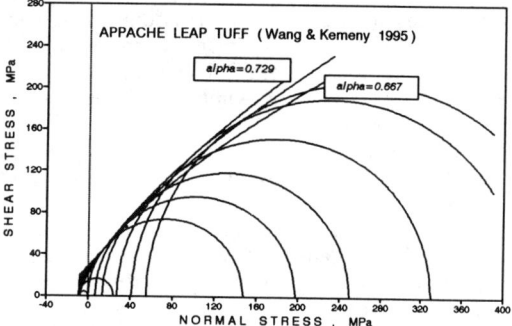

Fig. 5 Maximum and minimum failure envelopes for Apache Leap tuff

Table 2. Comparison of constants for different tensile strengths for Ainoura sandstone

variables	$T_0 = 3$ MPa	$T_0 = 5$ MPa	$T_0 = 7$ MPa	$T_0 = 9$ MPa
σ_3 and m	a and error	a and error	a and error	a and error
9.8MPa, 16.3	0.644, 1.7%	0.714, 1.6%	0.781, 5.4%	0.847, 7.3%
19.6MPa, 17.0	0.657, 1.3%	0.717, 2.7%	0.773, 5.2%	0.826, 6.2%
29.4MPa, 15.6	0.649, 1.0%	0.700, 1.0%	0.748, 0.2%	0.793, 0.2%
39.2MPa, 15.3	0.648, 1.2%	0.696, 2.4%	0.739, 2.1%	0.779, 2.9%
49.0MPa, 16.2	0.656, 1.0%	0.701, 1.0%	0.742, 1.5%	0.781, 3.1%
58.8MPa, 16.8	0.661, 2.7%	0.705, 0.0%	0.744, 1.2%	0.781, 3.3%
average	0.652, 1.35%	0.706, 1.34%	0.755, 2.59%	0.801, 3.63%
constant B, avr.	5.13	3.99	3.14	2.52
shear intercept	10.5 MPa	12.4 MPa	13.6 MPa	14.7 MPa

of internal friction will be easily obtained with various values of normal stresses from the equation. Shear intercept is 21.3 MPa and the angle of internal friction is 61.8° on $\sigma = 0$ whereas 28.4 MPa and 47.8° for Mohr - Coulomb's linear envelope which is also plotted with tension cut-off. This is another big differences between the suggested envelope and the linear one under the condition of tensile stress. The stability of every element subjected by tensile or low compressive stresses (black painted zone in Fig.5) is estimated poorer for this criterion than for Mohr-Coulomb's criterion. This phenomenon is also seen under the high compressive normal stresses.

4 CONCLUSIONAL DISCUSSION

A new empirical failure criterion is suggested based on Mohr's postulation for intact rock. It is expressed by a parabolic equation with 2 constants, and is verified within 4% of error for several experimental data in this report. The equation is defined in σ-τ domain, and can be easily understood than those expressed in terms of principal stresses.

The critical shear stress which decide the failure of every interesting element is determined lower for this criterion than for Mohr-Coulomb's criterion. It has the tendency that the safety factor is estimated low. This difference which is able to be proved after application is not mentioned here. It will be the next study also with applicability to in-situ rock masses.

REFERENCES

Bieniawski Z. T. 1984; *Rock mechanics design in mining and tunneling.* Chap. 5 & 8. A.A. Balkema.

Campos De Orellana A.J. 1996. Pressure solution and non-associated plasticity in the mechanical behavior of potash mine openings. *Int. J. Rock Mech. Min. Sci. & Geomech. Abstr.* 33-4: 347 -370. Pergamon Press.

Ghosh A., K.Fuenkajorn and J.J.K.Daeman 1995. Tensile strength of welded Apache Leap tuff : Investigation of scale effects. *Proc. 35th U.S. Symposium on Rock Mechanics*: 459-464. A.A. Balkema.

Hoek E. and E.T. Brown 1980. *Underground excavation in rock.* Chap. 6. London. IMM.

Kimura T. et al 1988. Experimental and theoretical studies on strain softening behavior of rocks. *J. mining and Metallurgical Institute of Japan.* 104-1199: 11-16.

Lee H.W. 1984. On the strength and deformation behavior of rocks with artificial joints under triaxial compression. *M.S. Thesis for Seoul Nat'l Univ.*

Murrell S.A.F. 1963. A criterion for brittle rock and concrete. *Proc. 5th Symposium on Rock Mechanics.*

Obert, L. and W.I.Duvall 1967. *Rock mechanics and the design of structures in rock.* Chap. 10. John Wiley & Sons.

Park Chulwhan 1996. Analysis of empirical failure criteria and suggestion of new equation for intact rocks. *J. Korean Society for Rock Mechanics.* 6-3: 234-238

Park Chulwhan 1987. A study on drilling speed and rock fracure for percussive rock drill. *Ph.D. Thesis for Seoul Nat'l Univ.*

Sellers E. and F.Scheele 1996. Prediction of anisotropic damage in experiments simulating mining in Witwatersrand quartzite blocks. *Int. J. Rock Mech. Min. Sci. & Geomech. Abstr.* 33-7: 659-670. Pergamon Press.

Vutukuri V.S. and K. Katsuyama 1994. *Introduction to rock mechanics.* Chap. 3. IPC Inc.

Wang R. and J.M.Kemeny 1995. A new empirical failure criterion for rock under polyaxial compressive stresses. *Proc. 35th U.S. Symposium on Rock Mechanics*: 453-458. A.A. Balkema.

Strength properties and their relations with abrasiveness of some Indian rocks

A.K.Giri, C.Sawmliana, T.N.Singh & D.P.Singh
Centre of Advance Study, Department of Mining Engineering, Institute of Technology, Banaras Hindu University, Varanasi, India

ABSTRACT: Drillability of rock, which can be defined as the rate of penetration of drill bit into that rock by a particular drill depend on a number of parameter like hardness, abrasiveness, specific gravity of rock, drilling environment such as thrust, rotational speed, flushing condition etc. Among them abrasiveness has a great influence in drilling performance.

Abrasiveness affects adversely to the penetration rate of drill machine by wearing of a bit during relative movement of the cutting edge at the contact surface of rock. It also increases the magnitude of the thrust to maintain the magnitude of the stress concentration at the cutting edge. If we calculate the abrasiveness index properly it will be helpful for designing of bit as well as material selection for manufacturing bits for better drilling performance and provide longer life of drill bits.

Strength properties of rock are intimately related to the rockmass. The present paper deals with the effect of strength properties and their relation to the abrasiveness.

Experiments were carried out for different types of rocks. The results indicates a close relationship between the strength properties and abrasiveness.

1. INTRODUCTION

In depth, knowledge of physico mechanical properties is an integral part of any mining and civil engineering constructions. Any project needs detail study of rock behaviour under different environment. The rock friendly machine can be only design after understanding the strength properties. Drilling and blasting process are the main component of any excavation process. Drillability of a rockmass is directly related to the strength properties of rockmass. During the drilling process rock fails either due to compression, shear and tensile stress environment. So, drilling machine should be of such type that it can handle the rock smoothly and properly. Abrasivness is one of the important properties of rock that is measured in order to assess their suitability for mechanical and economical excavation (West,1989). The abrasiveness factor determines the rate of wear. The rate by which the thrust force have to be increased to maintain the stress concentration at the cutting point necessary for penetration. Generally, it was reported that increase in silicon content rapidly increase the abrasiveness and retarding the drillability index.

Abrasivity is considered to be a major impediment and main factor which control the rate of wear and ultimately affect the penetration rate. Singh (1968, 1973) carried out a detail work on abrasivity of rock. He reported that bit wear increased with abrasivity and consequently penetration of the drag bit decreased. The rate of wear of the drill bit increase with increase in abrassiveness of the rock and ultimately penetration of the bit decreased (Shiromura & Takata, 1958 ; Farmer et al, 1979; Miranda and Mendes, 1983).

2. METHODS OF ABRASIVITY MEASUREMENT

There are number of method available for determination of abrasiveness of rock that is measured in order to asses their suitability for mechanical and economical excavation (West, 1989). The abrassiveness factor determines the rate of wear. The rate by which the thrust force have to be increased to maintain the stress concentration at the cutting point necessary for penetration. Generally it was reported that increased in silica content rapidly increase the abrasiveness and g the drillability index.

There are number of methods available for determination of abrasiveness of rock, i.e.
  Burbank test,
  Microbit drilling test, and
  Cerchar test.

2.1 Burbank Test:

This test is designed to determine the relative abrasivity of rocks interacting with metal part of crushing equipment(Burbak, 1955). The apparatus consisted of a steel paddle, rotating at 627 revolution/ minute, within a steel drum rotating at 74 revolution /minute in the same direction. The drum is lined with a wire-mesh to lift the crushed charge upwards so that it cascades into the path of the steel paddle.

2.2 Microbit Drilling:

Goodrich (1957) designed and fabricated a microbit drilling machine. The lower part of the machine consists of an assembly of two mild steel plates fixed with four bolts and a motor which rotate the drill bit at a fixed rpm of 120 through a shaft. The middle part is a moveable steel plate which act as a dead weight to provide a constant thrust. The upper surface of the plate contains a bush to connect it with a vice and tightening screw to hold the specimens. Afterward the drilling was done for fixed time. After that, the bit was removed and width of the wear flate was again measured up to 3 decimal places. The wear depends upon the abrasivity and strength of rock, cutting edge and bit geometry (Singh, 1973; Pandey, 1991).

2.3 Cerchar Test:

The cerchar apparatus was fabricated according to the information available from Valantin & Guillon, 1971 and Bougard, 1974 (Fig.1). This test gives the abrasivity of rock expressed as it ability to cause wear to a tool scratching on the rock surface. The cylindrical or cubical specimens fixed in the jaw of the vice and clamped tightly. The upper surface of the rock is arranged to be leveled. The abrasivity determination correspond to the wear diameter in 1/100 mm increments of flate plane produced by scratching the pin over a length of 100 mm on the surface of the sample under a static load of 7.00 Kg. The static load is provided by a compressed spring which is fixed inside a hollow tube and spring holds the testing pins. The mean result of 5 scratches is considered to give good result (Brown & Phillip, 1977).

3. EXPERIMENTAL WORK

Different types of rock specimens were collected from different localities. The rock samples were prepared as per recommendation of International Society of Rock Mechanics (ISRM) for determination of physico-mechanical properties. The uniaxial compressive strength was determined on Universal Testing Machine (UTM). The specimen was placed between the two platens. The specimen was then subjected to a uniform rate of loading up-to-failure. The load at which the specimen fails was recorded directly from the calibrated scale attach with the machine. Compressive strength was determined from the formula

$$\sigma_c = P/A, \text{ Mpa},$$

where, σ_c = compressive strength, MPa
  P = Failure load, Kg and
  A = cross-sectional area of the specimen, cm^2

Similarly, tensile strength and single shear strength values were also determined with the help of UTM and standard formulae. The mean values of compressive, tensile and shear strength of different rock types is given in Table 1. The abrasivity of different rock types were determined with the help of cerchar abrasiveness test apparatus. The abrasiveness of each samples was measured using the apparatus(Fig. 1) and method of testing described earlier. Five tests being made on each sample and the mean value determined (Table 1).

Table 1. Strength properties and relative abrasivity of different rock types.

Sample	Type of rock	Compressive Strength (MPa)	Tensile Strength (MPa)	Shear Strength (MPa)	Abrasivity
1	Biotite gniees	63.50	5.90	34.07	0.79
2	Sandstone	78.00	5.79	31.58	0.69
3	Pure quartzite	78.00	5.25	30.90	0.69
4	Sandstone	55.42	5.75	22.55	0.63
5	Quartzite	69.00	8.82	29.15	0.69
6	Impure quartzite	66.00	9.59	33.14	0.63
7	Phyllite	65.00	6.62	31.72	0.69
8	Coarse grain sandstone	31.16	3.70	15.15	0.70
9	Impure marble	188.12	20.82	98.15	0.90
10	Coarse grain sandstone	62.10	6.50	31.23	0.61
11	Fine grain sandstone	100.14	11.72	54.18	1.11
12	Impure quartzite with calcite silicate	132.13	16.82	61.20	0.90

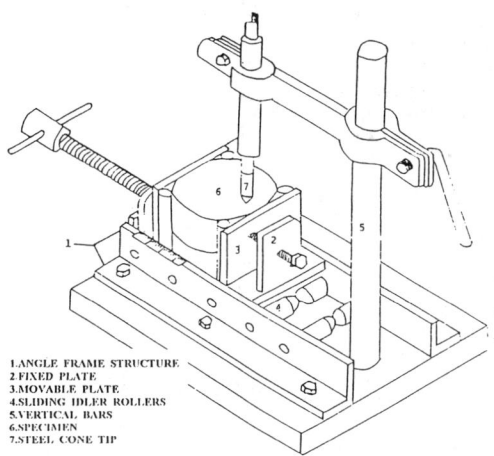

Fig.1 Cercher Test Apparatus

1.ANGLE FRAME STRUCTURE
2.FIXED PLATE
3.MOVABLE PLATE
4.SLIDING IDLER ROLLERS
5.VERTICAL BARS
6.SPECIMEN
7.STEEL CONE TIP

4. RESULTS AND DISCUSSION

The abrasiveness of some typical rocks that were encountered during the mining as well as tunnelling activity in various part of northern India have been determined. The tested rocks were from different localities and varieties. Mostly rock samples are from sedimentary and metamorphic origin.

A graph is plotted between uniaxial compressive strength and abrasiveness (Fig. 2). Normally, it is observed that higher the uniaxial compressive strength(UCS) higher abrasiveness except in case of impure marble having content of iron leaching. The UCS of impured marble is 188.12 Mpa but lower abrasivness 0.90 as compared to fine grained sandstone (Table. 1). The regression analysis was done and an equation was obtained for the best fit line in linear way. Similarly in case of tensile strength and shear strength plote equations were derived. The plot (Figs. 3 & 4) shoes that as the strength increases, the abrasiveness increases. But in few cases, we got the reverse values and it was probably due to the effect of geological factors and it depends upon the silica content and the orientation of grains with respect to scratching and loading. West (1989) found that the abrasiveness of sandstone of different localities gives wide variation from 1.30 to 6.24. The calcareous rock shows low abrasiveness as compared to arenaceous rock.

The obtained equations for all the three conditions i.e., compressive, tensile and shear are given below respectively.

$$A = 0.248812 \times C^{0.252861} \quad (1)$$
$$A = 0.267253 \times T^{0.220902} \quad (2)$$
$$A = 0.290794 \times S^{0.262779} \quad (3)$$

The results show good correlation between strength parameter and abrasiveness.

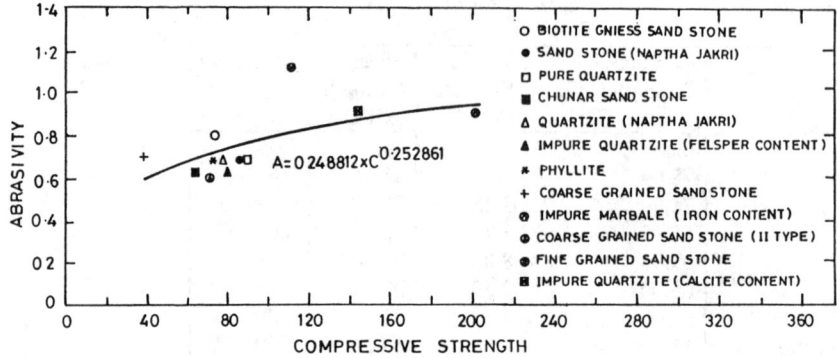

FIG. 2 ABRASIVITY AND COMPRESSIVE STRENGTH

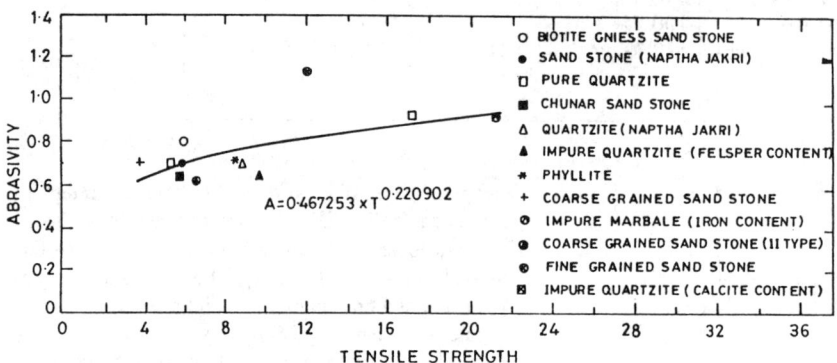

FIG. 3 ABRASIVITY AND TENSILE STRENGTH

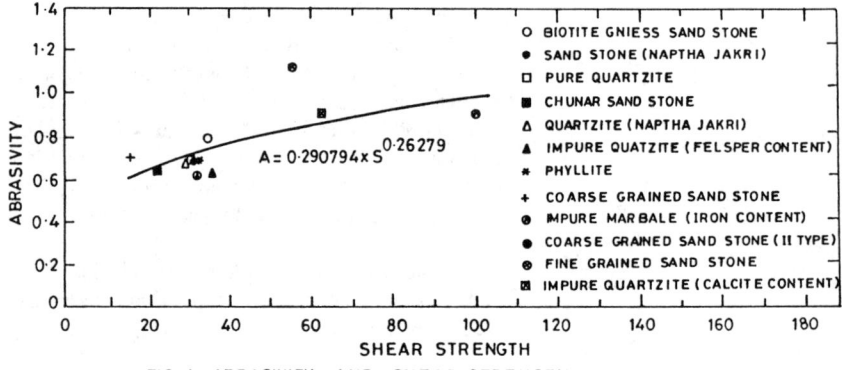

FIG. 4 ABRASIVITY AND SHEAR STRENGTH

5. CONCLUSIONS

The present study gives some of the important conclusions.
- It is clear from this study that abrasiveness is dependent upon a number of rock parameters.
- It is found that, generally, strength increases abrasiveness increases linearly.
- It is not necessary to apply high thrust for getting maximum penetration rate. Penetration rate increases with an increase in the value of strength properties of rock.
- Higher abrasiveness values may give high rates of wear and drillability of rock decreases subsequently.

REFERENCES

Bougard, J.F., 1974. Propositions relatives aux measures et essais a effectuer dans le cadre d'um chantier de-cresement mecanique Tunnels et Ouvrages Souterrains 5:215.

Brown, E.T. and Phillips, H.P., 1977. Recording drilling performance for tunnelling site investigation, CIRI, A Rep. No. 8:120.

Burbank, B.B., 1955. Measuring the relative abrasiveness of rock minerals and ores, Pit and Quarry, :114-118.

Farmer, I.W, Hignett, H.I. and Hudson, J.A., 1979. The role of geotechnical factors in the cutting performance of tunnelling mechanics in rrocks, Proc. IV th. Congress of Int. Soc. Rock Mech. , Montreux, 1: 371-377.

Goodrich, R.H., 1957. High pressure rotary drilling mechanics , Bull. Sch. Min. Metal. University of Missoury, USA, 94: 25-45.

Miranda, M.A.,and Mendes, F., 1983. Drillability and drilling methods, Proc. 5th. Cong. Int. Soci. Rock Mech. , Melbourne, 5: E195-200.

Pandey, A.K., 1981. Mechanical rock performance in percussive drilling, Ph. D Thesis, BHU, Varanasi, India.

Shiromura, Y. and Takata, A., 1958. Research on relation of physical properties of rocks and drilling rate of diamond bit, Min. Met. Inst. Japan, 74 (844) : 854-860.

Singh, D.P., 1968. Drillability and physical properties of rocks, M. Sc Thesis, University of Melbourne , Australia.

Singh, D.P., 1973. A study of bit wear in rotary drag drilling, Australian Mining : 17-20.

Valantin, A. and Guillon, P., 1971. Determination de. lo nocivite des roches vis-a vis des pics test de. durete. et. d'abrasivite Ref. AMO-PGu/ ES-71-73-37/1.

West, G., 1989. Rock abrasiveness testing for tunnelling, Int. J. Rock Mech. Min. Sci. and Geomech. Abst., 26(2) : 151-160.

Nikraz, H.R. 483
Nilsen, B. 227
Nishida, K. 433
Noma, T. 119

Oh, D.Y. 267
Ohnishi, Y. 3
Okamoto, A. 85, 93
Otsuka, Y. 19

Pal, S.K. 507
Park, C. 147, 531
Park, E.S. 345
Park, H.M. 371
Park, N.S. 267
Park, Y. 531
Press, M. 483
Pronin, K.K. 193

Ramamurthy, T. 445
Rao, K.S. 445
Ruan, L. 255
Rychlicki, S. 183
Ryu, C.H. 147
Ryu, D.W. 79

Saito, T. 439
Sakurai, S. 261
Sawmliana, C. 153, 537
Scott, C. 171
Seifabad, M.C. 131
Sellberg, B.T. 55

Seo, Y.S. 495
Shabarov, A.N. 137
Shao, J.F. 513
Sharma, K.K. 471
Shimada, H. 221, 367
Shimogochi, T. 61
Shmouratko, V.I. 193, 233
Sibai, M. 513
Singh, D.P. 537
Singh, T.N. 153, 537
Soliman, M.A. 3
Song, J.J. 501
Song, M.K. 13
Song, W.K. 141
Soni, D.S. 471
Srivastava, R.K. 471
Stopa, J. 183
Stryczek, S. 183
Suenaga, H. 299
Sun, Z. 427
Sundaram, M. 113
Swarup, A. 245
Synn, J.H. 147

Tajika, H. 3
Takehara, H. 439
Takemura, T. 125
Tanaka, M. 3
Tani, K. 401
Tezuka, M. 101
Thiel, K. 239
Tosaka, H. 19, 93, 299

Tshibangu K., J.P. 459, 465
Tsuchiya, T. 119
Tsurumi, K. 125
Tsutsui, M. 261
Tyler, D. 171

Uchiyama, Y. 333

Villaescusa, E. 171

Wang, Y. 287

Xia, C. 427
Xie, M. 217
Xu, X. 287
Xu, Z. 287

Yamatomi, J. 125
Yan, S. 217
Yokozawa, K. 107
Yoon, H.I. 519
Yoon, J.S. 519
Yoshida, H. 311
Yu, J. 217

Zabuski, L. 239
Zapryagaev, A.P. 137
Zemisev, V.N. 197
Zeng, X. 251
Zhang, Q. 251
Zhou, G. 177
Zhu, K. 407

Author index

Ando, K. 3
Anwar, H.Z. 367
Asche, H. 165
Aziz, N. 421

Baek, D. 67
Bart, M. 513
Barton, N. 413
Bhasin, R. 413
Bilak, R.A. 25
Bogert, H. 339
Broch, E. 227
Bruno, M.S. 33
Bulychev, N.S. 211

Chakravarty, D. 507
Cherkez, E.A. 193, 233
Chikahisa, H. 261
Choi, O. 93
Choi, S.I. 13
Chugh, Y.P. 353
Ciccu, R. 41
Constantinescu, A. 199

Davidson, B. 33, 389, 395
Deshmukh, A.M. 489
Dimova, V.I. 161
Dube, A.K. 245
Dusseault, M.B. 25, 33, 389, 395
Dutta, D. 353
Duveau, G. 513

Esaki, T. 177

Fotieva, N.N. 211
Fowell, R.J. 293
Fujii, N. 495

Garagash, I.A. 205
Gharouni-Nik, M. 361
Giri, A.K. 537
Glamheden, R. 317
Goel, R.K. 245
Gokhale, C.S. 489
Gonet, A. 183

Ha, E.R. 267

Hada, M. 119
Han, I.Y. 323
Haque, A. 421
Hasegawa, M. 85
Hasui, A. 101
Hibino, S. 333
Hong, S.J. 323
Horii, H. 311
Hoteit, N. 513, 525
Huh, D.H. 187

Ichikawa, Y. 495
Ichinose, M. 367
Im, J.O. 501
Inada, Y. 61
Indraratna, B. 421
Ismail, Z. 3
Itoh, K. 19

Jayawardena, U.de S. 383
Jee, W.R. 275
Jeon, S. 451
Jiang, Y. 177
Jo, Y.D. 49
Jung, S.J. 339

Kameda, N. 177
Kate, J.M. 305, 489
Kato, Y. 327
Kazikaev, J.M. 9
Kim, B.C. 187
Kim, B.Y. 49
Kim, D.Y. 519
Kim, H.Y. 345
Kim, J.Y. 323
Kim, M. 389
Kim, M.K. 227, 477
Kim, T.K. 79
Kinoshita, N. 61
Koda, E. 333
Kohmura, Y. 61
Kojima, K. 85, 93, 299
Komoo, I. 113
Koo, J.H. 275
Kozlova, T.V. 233
Krotov, N.V. 137
Kudo, K. 311

Kudo, Y. 101
Kusumi, H. 433
Kwon, Y.D. 267

Lee, C. 67
Lee, C.I. 501
Lee, D.H. 79
Lee, D.S. 323
Lee, H.K. 73, 79, 345
Lee, H.S. 519
Lee, K.J. 73
Lee, K.W. 147
Lee, P.K. 477
Lee, S. 67
Lee, S.E. 371
Li, H. 287
Li, Y. 217, 255
Lindblom, U.E. 317
Liu, D. 407
Luong, M.P. 525

Ma, S.J. 293
MacGillivray, D.A. 395
Manca, P.P. 41
Marchidanu, E. 377
Matsuda, H. 101
Matsui, K. 221, 367
Matsumoto, K. 261
Matsuo, T. 61
Mine, Y. 433
Mizuta, Y. 327
Mogi, G. 125
Moon, H.K. 13, 187
Moon, S. 67
Mori, N. 177
Mori, S. 19
Murata, M. 107
Murata, S. 439

Nakagawa, K. 101, 333
Nakagawa, T. 125
Nakahara, H. 261
Nakayama, S. 49
Nakazawa, Y. 85
Nasseri, M.H. 445
Nawrocki, P.A. 389
Nguyen Minh, D. 199